P9-CKS-225

APPLIED ELECTROMAGNETISM

Third Edition

Liang Chi Shen

University of Houston

Jin Au Kong

Massachusetts Institute of Technology

PWS Publishing Company

I(T)P An International Thomson Publishing Company

Boston · Albany · Bonn · Cincinnati · Detroit · London · Madrid · Melbourne · Mexico City · New York · Paris · San Francisco · Singapore · Tokyo · Toronto · Washington

PWS PUBLISHING COMPANY
20 Park Plaza, Boston, MA 02116-4324

Copyright © 1987 by PWS Publishers. Copyright © 1995 by PWS Publishing Company, a division of International Thomson Publishing Inc.
All rights reserved. No part of this book may be reproduced, stored in a retrieval system, or transcribed in any form or by any means—electronic, mechanical, photocopying, recording, or otherwise—without the prior written permission of PWS Publishing Company.

I(T)P™
International Thomson Publishing
The trademark ITP is used under license

For more information, contact

PWS Publishing Co.
20 Park Plaza
Boston, MA 02116

International Thomson Publishing Europe
Berkshire House 168-173
High Holborn
London WC1V 7AA
England

Thomas Nelson Australia
102 Dodds Street
South Melbourne, 3205
Victoria, Australia

Nelson Canada
1120 Birchmount Road
Scarborough, Ontario
Canada M1K 5G4

Sponsoring Editor: Tom Robbins
Assistant Editor: Ken Morton
Editorial Assistant: Lai Wong
Production and Interior Design: Pamela Rockwell
Marketing Manager: Nathan Wilbur
Manufacturing Coordinator: Lisa Flanagan
Compositor: Santype International Ltd
Cover Printer: Henry Sawyer Company
Text Printer and Binder: Quebecor, Martinsburg

International Thomson Editores
Campos Eliseos 385, Piso 7
Col. Polanco
11560 Mexico C.F., Mexico

International Thomson Publishing GmbH
Königswinterer Strasse 418
53227 Bonn, Germany

International Thomson Publishing Asia
221 Henderson Road
#05-10 Henderson Building
Singapore 0315

International Thomson Publishing Japan
Hirakawacho Kyyowa Building, 31
2-2-1 Hirakawacho
Chiyoda-ku, Tokyo 102
Japan

Library of Congress Cataloging-in-Publication Data

Shen, Liang Chi.
Applied electromagnetism/Liang Chi Shen, Jin Au Kong.—3rd ed.
p. cm.
Includes bibliographical references (p.) and index.
ISBN 0-534-94722-0 (alk. paper)
1. Electric engineering. 2. Electromagnetic theory. I. Kong, Jin Au, 1942- . II. Title.
TK 153.S475 1995
621.3—dc20 95–47
CIP

This book is printed on recycled, acid-free paper.

Printed and bound in the United States of America.
95 96 97 98 99 — 10 9 8 7 6 5 4 3 2

To our parents:

How could a blade of grass
Repay the warmth from the spring sun
–Meng Chio (751–814)

COURSE OPTIONS/*APPLIED ELECTROMAGNETISM**

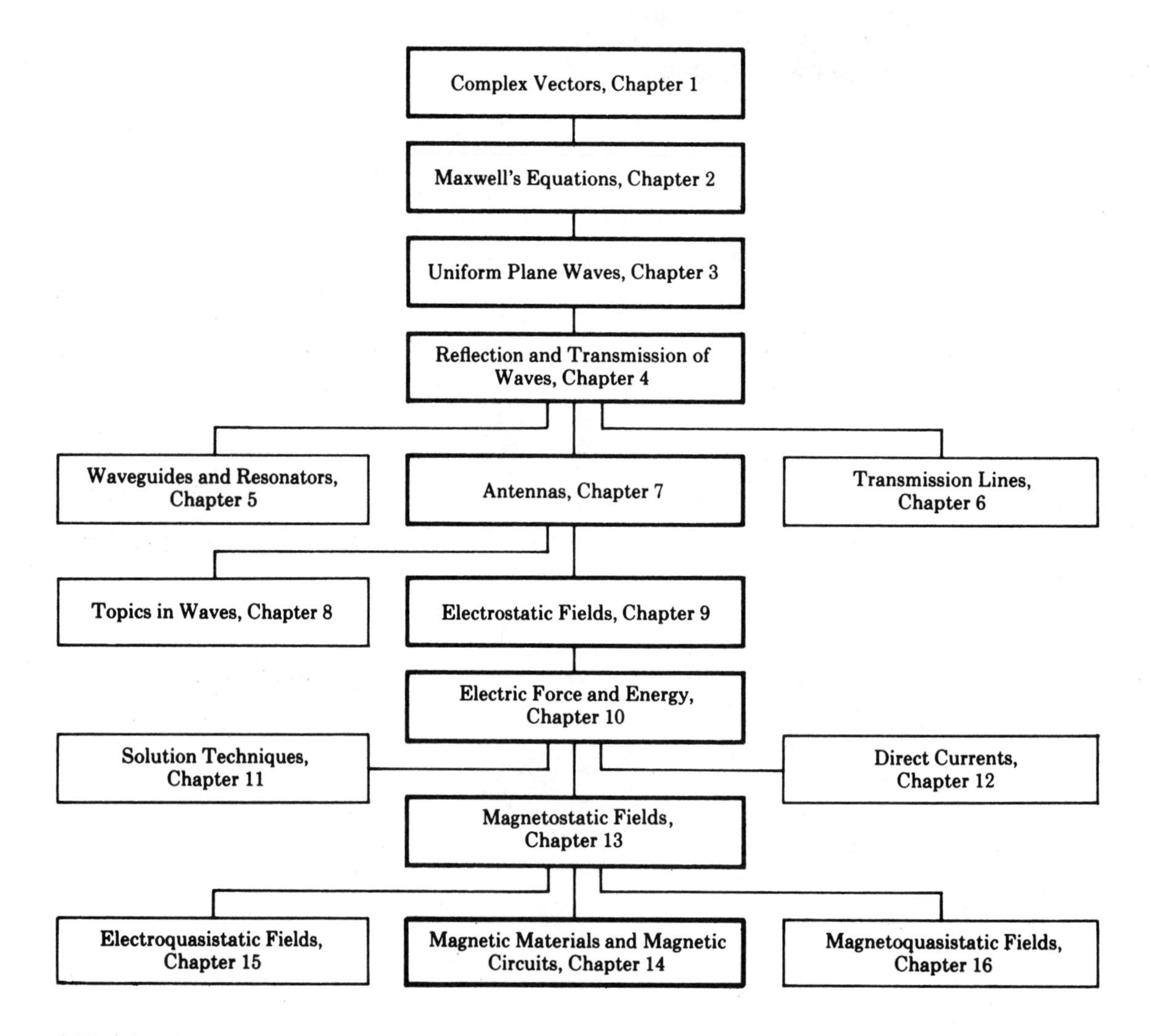

* The nine chapters in the central column may be covered in a one-semester course.

CONTENTS

FOREWORD TO THE THIRD EDITION

In the third edition, we continue to incorporate more examples of the application of electromagnetic fields and waves. A few introductory discussions of the safety of human exposure to electromagnetic radiation have been added to address the conerns over this important topic. The progress of some applications described in this book, such as in the areas of magnetic disk memory and optical fibers, has been updated.

This edition contains several new features. A summary has been added at the end of each chapter to review for readers what they have learned in that chapter. Comments have been added to all examples to explain their purposes. Available computer codes are recommended as study aids to learn the material covered in each chapter whenever such codes are available. These codes are called CAEME software. It was developed with support from the National Science Foundation and IEEE. Because these codes are continuously being revised, recommendations for their use are given in the Instructor's Manual.

The authors are indebted to their colleagues and reviewers for suggesting ways to improve this book: David Jackson, Richard Liu, Stuart Long, Jeffery Williams, and Donald Wilton of the University of Houston; and Hsiu Han, Kit Lai, Lifang Wang, Murat Veysoglu, and Eric Yang of Massachusetts Institute of Technology. The reviewers of the third edition are:

I. M. Besieris
Virginia Polytechnic Institute and State University

Lynn Carpenter
Pennsylvania State University

Hollace L. Cox, Jr.
University of Louisville

S. T. Hsieh
Tulane University

David R. Jackson
University of Houston

R. Janaswamy
U.S. Naval Postgraduate School

Hung-Mou Lee
U.S. Naval Postgraduate School

Stuart A. Long
University of Houston

T. C. Poon
Virginia Polytechnic Institute and State University
Keith A. Ross
South Dakota School of Mines and Technology
A. Safaai-Jazi
Virginia Polytechnic Institute and State University
Surendra Singh
The University of Tulsa
Glen S. Smith
Georgia Institute of Technology
Robert D. Strattan
The University of Tulsa

FOREWORD TO THE SECOND EDITION

Applied electromagnetism is an exciting subject to study and teach. New applications are continuously being generated. There seem to be endless possibilities for newly invented electromagnetic devices to meet the challenge of modern technology. This book has been revised to incorporate more applied examples that emphasize the usefulness and importance of the subject of electromagnetism which, we believe, is the cornerstone of the electrical engineering discipline.

In this new edition, a section covering electromagnetic radiation power and radiation pressure has been added. As exploration and exploitation of outer space have intensified, an understanding of this subject becomes quite desirable. Other new application examples in this edition include the tails of comets, very-long-baseline arrays, triboelectricity, laser printing, and thermomagnetic copying. Many new problems have been added and answers to the odd-numbered problems are provided in Appendix E. In addition, many minor revisions have been made and some sections have been completely rewritten.

The authors are indebted to their colleagues, students, and reviewers for pointing out errors and ambiguities in the original edition and for suggesting ways to improve the clarity of presentation of certain topics. In this regard, we would like to acknowledge the contributions of David Jackson, Stuart Long, Soon Poh, Ann Tulintseff, Donald Wilton, Eric Yang, and the following second edition reviewers: Donald Herrick, Environmental Research Institute of Michigan; Paul McIsaac, Cornell University; Paul Steffes, Georgia Institute of Technology; and A. K. Chan, Texas A & M University.

PREFACE TO THE FIRST EDITION

This book is intended as an introduction to applied electromagnetism for undergraduate students. It contains enough material for a two-semester or two-quarter course. For a one-semester course, we recommend covering the nine chapters outlined in the flowchart on page iv.

Applied Electromagnetism presents the theory of electromagnetism along with many examples of applications. Another salient feature of this book is that dynamic electromagnetic fields are presented before the discussion of static fields. Advanced mathematics is avoided so that the only required mathematical background is knowledge of calculus and some elementary awareness of the vector del operators.

Electromagnetic theory taught at the undergraduate level has usually been presented in the following sequence: (a) introduction to vector analysis to prepare for the use of the del operators in Maxwell's equations; (b) treatment of static fields, energy, forces, and, in some schools, quasi-static fields as well; and (c) study of electromagnetic waves by using the complete set of Maxwell's equations and discussion of transmission lines, waveguides, antennas, and, in physics-oriented courses, the dynamics of particles in electromagnetic fields.

We feel that this sequence of presentation has become undesirable. The historical treatment, starting from electrostatics and moving to electromagnetic waves, was largely due to the development and application of electromagnetic machinery. The importance of power and of machinery notwithstanding, a new era, centered on applications of electromagnetic waves to lasers, communications, optics, and remote sensing, has now begun. We think that a presentation starting with waves and specializing to statics and quasi-statics is both more logical and simpler. It is more logical because, when a subject reaches maturity, a deductive approach is always superior to an inductive approach. It is simpler because the Lorentz force law is not needed in treating waves. Also, with this presentation the complication of cylindrical and spherical coordinates can be deferred to a later stage when the students have become familiar with the concepts.

Possibly because of the influence of most textbooks, electromagnetic theory has been taught from a cultural point of view rather than from a point of view that connects it to possible applications to other scientific and engineering disciplines. Electromagnetic theory is undoubtedly a fundamental subject in electrical engineering and is certainly extremely elegant in structure and uniquely rigorous in formulation. Some students are attracted by the beauty of the theory and may decide to devote their professional lives to this subject. However, many other students, whose aptitudes and tastes do not incline them to mathematical abstraction, are discouraged, and they shy away from this theory because it is hard for them to see its applications to their fields of interest. We feel that the cultural value of electromagnetic theory is enhanced when students have learned the wide spectrum of applications of electromagnetism to modern science and engineering. As we have mentioned, the conventional sequence of presentation is unnecessary. Studying statics is not a necessary step toward understanding electromagnetic waves. All students can be highly motivated if they find that a subject is relevant to their daily experience and their chosen fields. With this fact in mind, we present electromagnetism in a text that stresses applications and at the same time preserves the unique and coherent features of Maxwell's theory.

Chapter 1 reviews the necessary mathematical background. Chapter 2 is the foundation of the book. Chapters 3 and 4 treat fundamental aspects of electromagnetic waves. A one-semester course can omit Chapters 5, 6, and 8. For instance, to understand 5 and 6, one must have the background presented in Chapter 4. But Chapter 7 also follows from Chapter 4, and the omission of 5 and 6 does not affect the continuity. Chapter 9 begins our treatment of statics. Again, a one-semester course may omit the side blocks in the flow chart—namely, Chapters 11, 12, 15, and 16.

Acknowledgments

This book is based on the lecture notes used by the authors at the University of Houston. We are indebted to our students and colleagues who have made valuable suggestions and comments leading to many important changes. In this regard, we are particularly grateful to Professors Stuart A. Long and Louis D. Smullin. We also wish to acknowledge the valuable comments of the manuscript reviewers: Andrew Blanchard, Texas A & M University; Kenneth A. Connor, Rensselaer Polytechnic University; Henry Domingos, Clarkson College of Technology; Charles E. Smith, University of Mississippi; Glenn S. Smith, Georgia Institute of Technology; Sedki Riad, Virginia Polytechnic Institute and State University; Leung Tsang, Texas A & M University; and Piergiorgio Uslenghi, University of Illinois at Chicago Circle. Proofreading of the final manuscript was done independently by Frank S. C. Huang, Sompongse Toomsawasdi, Soon Y. Poh, Tarek M. Habashy, Jay K. Lee, and S. L. Chuang. The original manuscript and several revised versions were typed by Cindy Kopf, Micki Maes, and Diana Phillips. Finally and most importantly, we wish to acknowledge that this book could have never been written without the understanding and encouragement of our wives, Wei Liu Shen and Wen-yuan Yu Kong.

LCS and JAK

NOTATIONS, SYMBOLS, AND UNITS

Real Scalars: Real scalars are indicated by italic type or Greek letters—for example, a or ρ.

Complex Scalars: Complex scalars are indicated by roman type or bold Greek letters—for example, r or $\boldsymbol{\rho}_v$.

Real Vectors: Real vectors are indicated by boldface italic type—for example, $\boldsymbol{B}$.

Unit Vectors: Unit vectors—that is, vectors with magnitude equal to unity—are indicated by the symbol ˆ above a quantity and boldface italic type or Greek letters—for example, $\hat{\boldsymbol{x}}$.

Complex Vectors: Complex vectors are indicated by boldface type—for example, **J**.

Time-Harmonic Scalars and Vectors: Time-harmonic quantities are always represented by cosine functions. For example,

$$V(t) = a \cos(\omega t + \phi)$$

The phasor notation of $V(t)$ given above is

$$V = a \exp(j\phi) \exp(j\omega t).$$

The factor $\exp(j\omega t)$ is usually omitted in writing. Note that most electrical engineering books use $\exp(j\omega t)$.

This book uses SI Units (the International System of Units). Appendix A lists frequently used symbols and their units.

1 COMPLEX VECTORS

Physical quantities are usually described mathematically by real variables of space and time and frequently by vector quantities. The reader may remember that vector quantities were introduced in physics to quantify force, velocity, and acceleration. For instance, the velocity of wind can be characterized by a vector, with the vector's direction describing the wind direction and its length describing the magnitude of the wind speed. The large class of physical quantities that vary periodically with time are called **time-harmonic**. For instance, household electricity varies at 60 Hz (cycles per second), and the output of a helium-neon gas laser is 474×10^{12} Hz. In mathematical manipulations, the time-harmonic real quantities are conveniently represented by complex variables.

In this chapter we shall explain in detail how to represent a time-harmonic real physical quantity by a complex one, especially when the physical quantity is a vector. For review purposes, we include complex algebra, time-harmonic scalars, and real vectors in the first three sections. In sections 1.4 and 1.5 we introduce the complex-vector representation for time-harmonic vectors and discuss the time-average values of their products.

1.1 COMPLEX ALGEBRA

A complex number c is represented by $\mathrm{c} = a + jb$, where a is the **real** part and b is the **imaginary** part of c, both are real numbers, j is the imaginary number defined by $j^2 = -1$. On a complex plane, c is represented by the arrow OA as shown in Figure 1.1a. Sometimes it is convenient to write c in **phasor form**:

$$\mathrm{c} = a + jb = |\mathrm{c}|e^{j\phi} = |\mathrm{c}|\cos\phi + j|\mathrm{c}|\sin\phi \tag{1.1}$$

where

$$|\mathrm{c}| = \sqrt{a^2 + b^2} \tag{1.2}$$

is called the **magnitude** or the **absolute value*** of c and where

$$\phi = \tan^{-1}\left(\frac{b}{a}\right) \tag{1.3}$$

in radians is called the **phase angle** or simply the **phase**** of c.

---

* Some mathematicians like to call it **modulus** of c.

** Also called **argument** of c.

Equality

Two complex numbers are equal if and only if their real parts and imaginary parts are both equal. For example, if c $= a + jb$, h $= f + jg$ and c = h, then $a = f$ and $b = g$.

Addition and Subtraction

Let c $= a + jb$ and h $= f + jg$; then

$$\text{c} + \text{h} = (a + f) + j(b + g)$$

and

$$\text{c} - \text{h} = (a - f) + j(b - g)$$

Addition and subtraction of two complex numbers are represented graphically by the diagrams in Figures 1.1b and 1.1c.

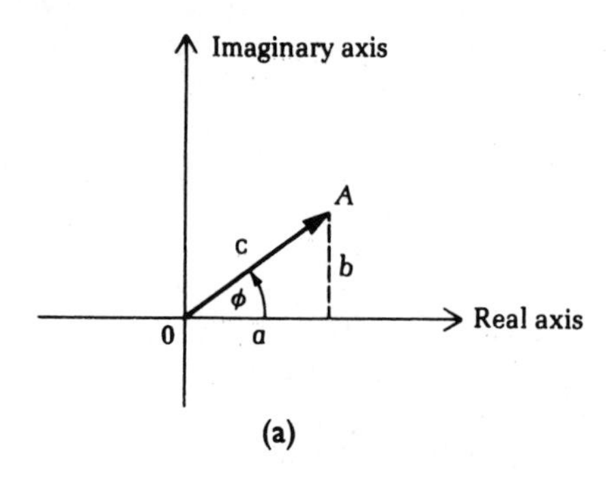

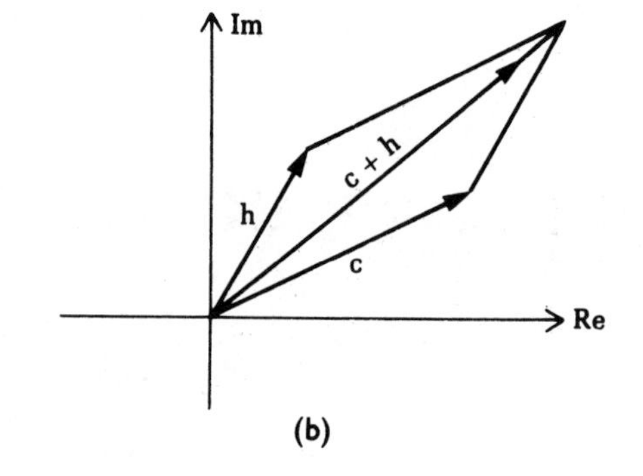

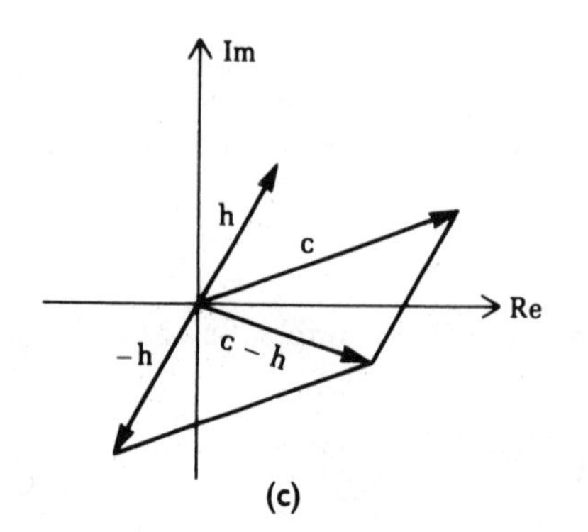

Figure 1.1 Graphic representation of complex numbers.

Multiplication and Division

$$\mathrm{ch} = (a + jb)(f + jg) = (af - bg) + j(bf + ag)$$

$$\frac{\mathrm{c}}{\mathrm{h}} = \frac{a + jb}{f + jg} = \frac{(a + jb)(f - jg)}{(f + jg)(f - jg)} = \frac{(af + bg)}{(f^2 + g^2)} + j\frac{(bf - ag)}{(f^2 + g^2)}$$

Note that complex numbers can be multiplied and divided readily if the complex numbers are expressed in phasor form. Let

$$\mathrm{c} = |\mathrm{c}|e^{j\phi_1} \quad \text{and} \quad \mathrm{h} = |\mathrm{h}|e^{j\phi_2}$$

then

$$\mathrm{ch} = |\mathrm{c}||\mathrm{h}|e^{j(\phi_1 + \phi_2)}$$

$$\frac{\mathrm{c}}{\mathrm{h}} = \frac{|\mathrm{c}|}{|\mathrm{h}|}e^{j(\phi_1 - \phi_2)}$$

Example 1.1 *This example shows numerical results of adding, subtracting, multiplying, and dividing between two complex numbers.*

Let $\mathrm{x} = 1 + j2$ and $\mathrm{y} = -2 - j$; then $\mathrm{x} + \mathrm{y} = -1 + j$, and $\mathrm{x} - \mathrm{y} = 3 + j3$.

Also,

$$\mathrm{xy} = (1 + j2)(-2 - j) = (-2 + 2) + j(-4 - 1) = -j5$$

and

$$\frac{\mathrm{x}}{\mathrm{y}} = \frac{1 + j2}{-2 - j} = \frac{(1 + j2)(-2 + j)}{(-2 - j)(-2 + j)} = \frac{-4 - j3}{5} = -0.8 - j0.6$$

The multiplication and division can be carried out in different ways:

$$\mathrm{x} = 2.236\, e^{j1.107}$$

$$\mathrm{y} = 2.236\, e^{-j2.678}$$

Thus,

$$\mathrm{xy} = 5\, e^{-j1.571} = -j5$$

and

$$\frac{\mathrm{x}}{\mathrm{y}} = e^{j3.785} = -e^{j0.643} = -0.8 - j0.6$$

Complex Conjugate

Let $c = a + jb$; then the **complex conjugate** of c is denoted by c^*, and $c^* = a - jb$. It can be shown that cc^* is always real and positive unless $c = 0$, in which case cc^* is equal to zero. See Problem 1.4.

Square Root of a Complex Number

Let $z = 1 + j2$. In phasor form

$$z = 2.236\, e^{j1.107}$$

The square root of z is readily obtained:

$$\sqrt{z} = 1.495\, e^{j0.554}$$

Note that the complex number $1 + j2$ can also be expressed as

$$1 + j2 = 2.236\, e^{j(1.107 + 2n\pi)}$$

where n is an arbitrary integer. If we let $n = 0$, we obtain the above result. If $n = 1$, we obtain

$$\sqrt{z} = 1.495\, e^{j(0.554 + \pi)}$$

The equation above is the second solution. However, this case has two and only two solutions.

In fact, if $z^m = c$, where m is an integer, there are exactly m distinct solutions for z.

1.2 COMPLEX REPRESENTATION OF TIME-HARMONIC SCALARS

Consider a time-harmonic real physical quantity $V(t)$ that varies sinusoidally with time.

$$V(t) = V_0 \cos(\omega t + \phi) \tag{1.4}$$

where V_0 is called the **amplitude**, ω the **angular frequency**, and ϕ the **phase** of $V(t)$. Note that $\omega = 2\pi f$, where f is called the **frequency** of $V(t)$. Figure 1.2 shows $V(t)$ as a function of time t.

We can also write $V(t)$ as follows:

$$\boxed{V(t) = \mathrm{Re}\{\mathrm{V}\, e^{j\omega t}\}} \quad \textbf{(conversion between real time and its phasor notation)} \tag{1.5}$$

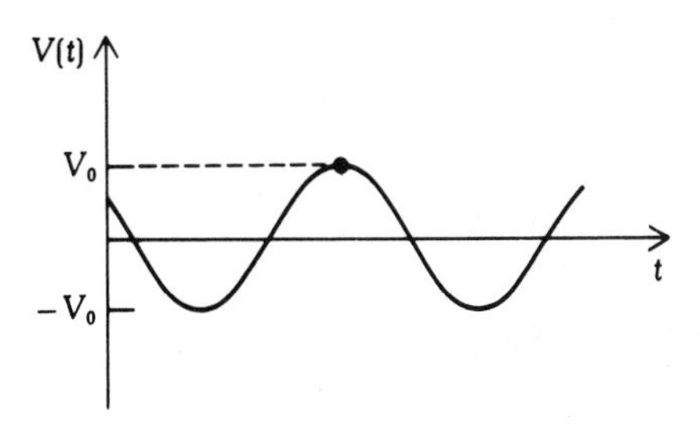

Figure 1.2 A time-harmonic function $V(t)$.

where Re{ } means taking the real part of the quantity in the braces { } and $\mathrm{V} = V_0 e^{j\phi}$. For simplicity, the symbol $\mathrm{Re}\{(\ )e^{j\omega t}\}$ can be omitted in writing. Thus

$$V(t) \leftrightarrow \text{(is equivalent to) } \mathrm{V}$$

The complex quantity V is called a **phasor** because it is a complex number presenting a time-harmonic physical quantity. Notice that $V(t)$ is a function of time t, whereas V is no longer a function of t. Whereas $V(t)$ is a real quantity, V is now complex.

Let

$$U(t) = U_0 \cos(\omega t + \phi)$$

Using the above rule of equivalence, we find

$$U(t) \leftrightarrow \mathrm{U} = U_0 e^{j\phi}$$

It can easily be proved that

$$V(t) + U(t) \leftrightarrow \mathrm{V} + \mathrm{U}$$

It can also be shown that

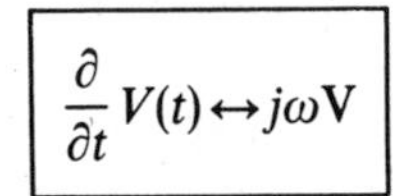

$$\frac{\partial}{\partial t} V(t) \leftrightarrow j\omega \mathrm{V}$$

(conversion of time derivative to phasor notation)

because

$$\frac{\partial V(t)}{dt} = -\omega V_0 \sin(\omega t + \phi) = \mathrm{Re}\{j\omega V_0 e^{j\phi} e^{j\omega t}\}$$

Therefore, $j\omega$ can replace the time derivative $\partial/\partial t$ in the complex representation of time-harmonic quantities.

The rule of equivalence for time-harmonic quantities should not be carried too far. For example,

$$V(t)U(t) \not\leftrightarrow \text{(is not equivalent to) } \mathrm{VU}$$

The rule of equivalence applies only to addition, subtraction, and time derivatives of time-harmonic quantities having the same frequency.

Example 1.2 *This example shows how to represent a real time-harmonic function by a complex number (phasor).*

Let $V(t) = 100 \cos(120\pi t + 60°)$ volts; then the phasor notation of $V(t)$ is

$$\mathrm{V} = 100\, e^{j(\pi/3)} = 50 + j86.6 \text{ volts}$$

Example 1.3 *This example shows how to convert a phasor notation to the corresponding time-harmonic function.*

The current along a conductor is a function of the distance z and is given by

$$\mathrm{I(z)} = j21.2 \cos(kz) \text{ amperes}$$

The frequency is 100 MHz. What is the current as a function of z and t?

Solution

$$I(z, t) = \mathrm{Re}\{\mathrm{I(z)}\, e^{j\omega t}\} = \mathrm{Re}\{j21.2 \cos(kz)\, e^{j\omega t}\}$$
$$= -21.2 \cos(kz) \sin(\omega t)$$

where $\omega = 2\pi \times 10^8$.

Example 1.4 *This example shows that the sum of two time-harmonic functions of the same frequency may be represented by a single phasor.*

What is the phasor notation of $V(t) = \cos[120\pi t + (\pi/3)] + \sin(120\pi t)$?

Solution We note that the phasor notation of $\cos[120\pi t + (\pi/3)]$ is $\mathrm{V}_1 = e^{j\pi/3}$. Similarly, the phasor notation of $\sin(120\pi t) = \cos[120\pi t - (\pi/2)]$ is $\mathrm{V}_2 = e^{-j\pi/2}$. Thus, the phasor notation of $V(t)$ is

$$\mathrm{V}_1 + \mathrm{V}_2 = e^{j\pi/3} + e^{-j\pi/2} = 0.5 - j0.134$$

Example 1.5 *This example reminds you that the phasor notation cannot be applied to functions with different frequencies.*

What is the phasor notation of $V(t) = \cos(120\pi t) + \cos(240\pi t)$?

Solution $V(t)$ has no phasor notation because it represents a sum of two signals of different frequencies.

Example 1.6 *This example reminds you that the phasor notation cannot be applied to the product of two time-harmonic functions, even if they are of the same frequency.*

Does the signal $V(t) = \cos(\omega t + \pi/3)\sin(\omega t)$ have a phasor notation?

Solution The phasor notation of $\cos(\omega t + \pi/3)$ is $e^{j\pi/3}$, and the phasor notation of $\sin(\omega t)$ is $e^{-j\pi/2}$ or $-j$. However, the product of $\cos(\omega t + \pi/3)\sin(\omega t)$ cannot be represented by the product of the two phasors. We note that

$$V(t) = \cos\left(\omega t + \frac{\pi}{3}\right)\sin(\omega t) = \frac{1}{2}\left[\sin\left(2\omega t + \frac{\pi}{3}\right) - \sin\left(\frac{\pi}{3}\right)\right]$$

This signal represents a sum of two signals, one with a frequency of 2ω and the other with a frequency of zero.

1.3 REAL VECTORS

Some physical quantities must be characterized by both a magnitude and a direction. Examples are the force acting on a particle, the velocity of a rocket, and an electric or magnetic field. These quantities are **vectors**.

A three-dimensional coordinate system has three basic vectors. In a rectangular coordinate system, they are $\hat{x}$, $\hat{y}$, and $\hat{z}$. These vectors are called **unit vectors**. Unit vectors have magnitudes equal to unity. They are dimensionless. Their only function is to indicate direction. The directions of $\hat{x}$, $\hat{y}$, and $\hat{z}$ are defined as

$\hat{x}$ = a unit vector in the direction of increasing x

$\hat{y}$ = a unit vector in the direction of increasing y

$\hat{z}$ = a unit vector in the direction of increasing z

These unit vectors are illustrated in Figure 1.3a.

To represent a vector $\boldsymbol{V}$ in the rectangular coordinate system, we draw an arrow in the direction of the vector, as shown in Figure 1.3b. The length OA is equal to the magnitude of the vector.

Note that we use $\boldsymbol{V}$ to symbolize a vector and V to symbolize a complex number. The vector $\boldsymbol{V}$ is written as

$$\boldsymbol{V} = V_x\hat{x} + V_y\hat{y} + V_z\hat{z} \tag{1.6}$$

where V_x, V_y, and V_z are the projections of the arrow OA on the x, y, and z axes, respectively. V_x, V_y, and V_z can be positive, negative, or zero, depending on whether the projection of the arrow is pointing in the positive direction of the axis, is pointing in the negative direction of the axis, or is simply zero.

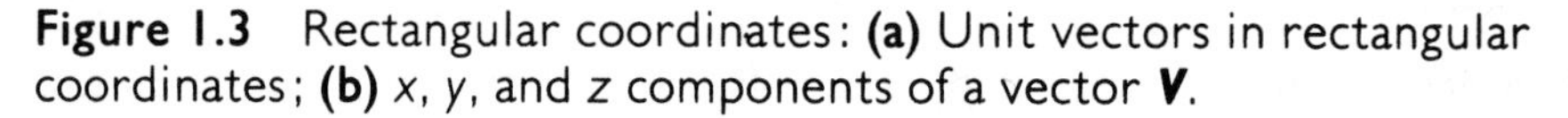
Figure 1.3 Rectangular coordinates: **(a)** Unit vectors in rectangular coordinates; **(b)** x, y, and z components of a vector $\boldsymbol{V}$.

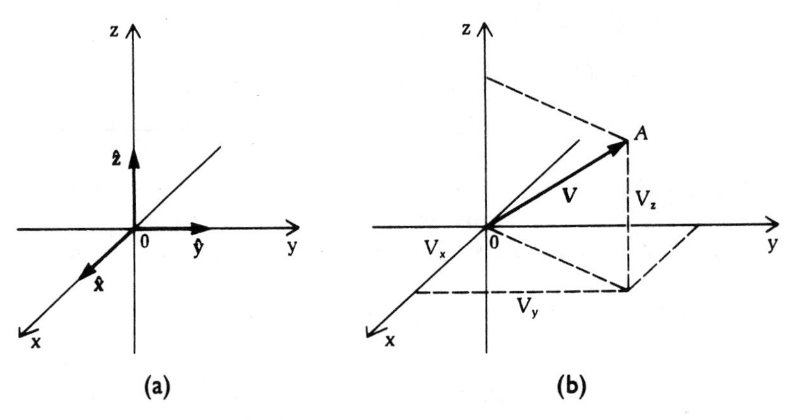

The magnitude of V is denoted by $|V|$, and

$$|V| = \sqrt{V_x^2 + V_y^2 + V_z^2} \tag{1.7}$$

If all three components of a vector are equal to zero, we call the vector **zero vector,** $\boldsymbol{O}$, or simply 0.

Equality

Two vectors are equal if and only if their x, y, and z components are respectively equal. For example, if $V = V_x\hat{x} + V_y\hat{y} + V_z\hat{z}$, $U = U_x\hat{x} + U_y\hat{y} + U_z\hat{z}$, and $V = U$, then $V_x = U_x$, $V_y = U_y$, and $V_z = U_z$.

Addition and Subtraction

Let

$$V = V_x\hat{x} + V_y\hat{y} + V_z\hat{z} \quad \text{and} \quad U = U_x\hat{x} + U_y\hat{y} + U_z\hat{z}$$

Then

$$V + U = (V_x + U_x)\hat{x} + (V_y + U_y)\hat{y} + (V_z + U_z)\hat{z} \tag{1.8}$$

$$V - U = (V_x - U_x)\hat{x} + (V_y - U_y)\hat{y} + (V_z - U_z)\hat{z} \tag{1.9}$$

Addition and subtraction of two vectors can be represented graphically by the diagrams shown in Figures 1.4a and 1.4b. Note the similarity between Figures 1.1 and 1.4. Also note that Figure 1.1 shows complex planes, whereas Figure 1.4 shows three-dimensional real spaces.

Figure 1.4 (a) Addition of two vectors: $\boldsymbol{V} + \boldsymbol{U}$; (b) Subtraction of two vectors: $\boldsymbol{V} - \boldsymbol{U}$.

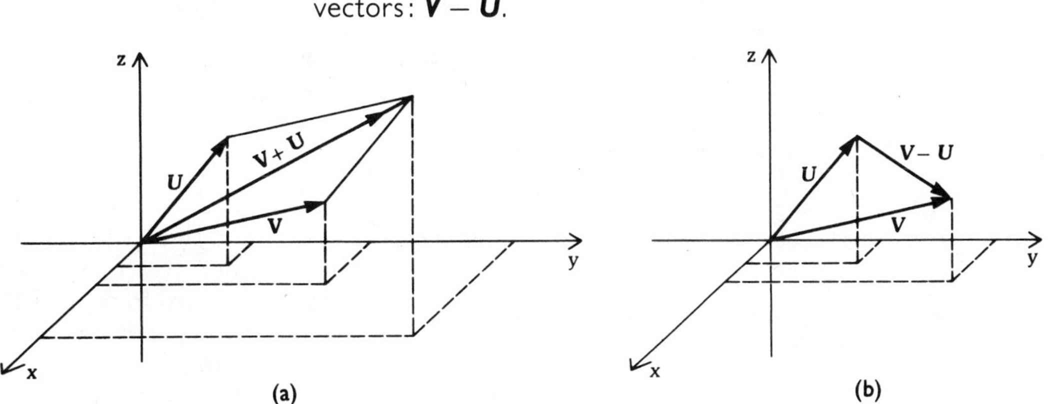

Example 1.7

This example shows how to write physical quantities (velocities of an airplane and the wind) in terms of vectors. The relative velocity between two moving objects can be obtained in terms of vector subtraction.

An airplane travels in a northwesterly direction at 375 km/h relative to the ground. There is a westerly wind of 150 km/h relative to the ground. How fast and in what direction would the plane travel with no wind?

Solution Let $\hat{x}$ be east and $\hat{y}$ be north; then $\boldsymbol{V} = -\hat{x}(375/\sqrt{2}) + \hat{y}(375/\sqrt{2})$, which is the airplane velocity relative to the ground. $\boldsymbol{W} = \hat{x}150$ is the wind velocity.

Let $\boldsymbol{U}$ be the velocity of the airplane relative to the wind; then $\boldsymbol{V} = \boldsymbol{U} + \boldsymbol{W}$. Therefore,

$$\boldsymbol{U} = \boldsymbol{V} - \boldsymbol{W}$$

which is also the airplane velocity if there is no wind.

Multiplication—Cross and Dot Products

The dot product (or scalar product) of two vectors $\boldsymbol{V}$ and $\boldsymbol{U}$ is defined as

$$\boxed{\boldsymbol{V} \cdot \boldsymbol{U} = V_x U_x + V_y U_y + V_z U_z} \quad \textbf{(dot product)} \qquad \textbf{(1.10a)}$$

Note that the result is a scalar.

An equivalent definition of the dot product is

$$\boxed{V \cdot U = |V||U|\cos\theta} \quad \textbf{(equivalent definition of dot product)} \tag{1.10b}$$

where θ is defined in Figure 1.5.

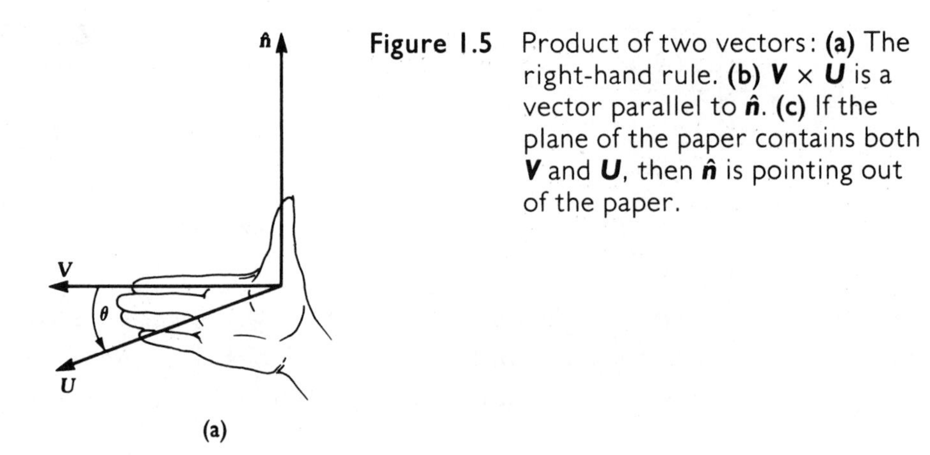

Figure 1.5 Product of two vectors: **(a)** The right-hand rule. **(b)** $\boldsymbol{V} \times \boldsymbol{U}$ is a vector parallel to $\hat{\boldsymbol{n}}$. **(c)** If the plane of the paper contains both $\boldsymbol{V}$ and $\boldsymbol{U}$, then $\hat{\boldsymbol{n}}$ is pointing out of the paper.

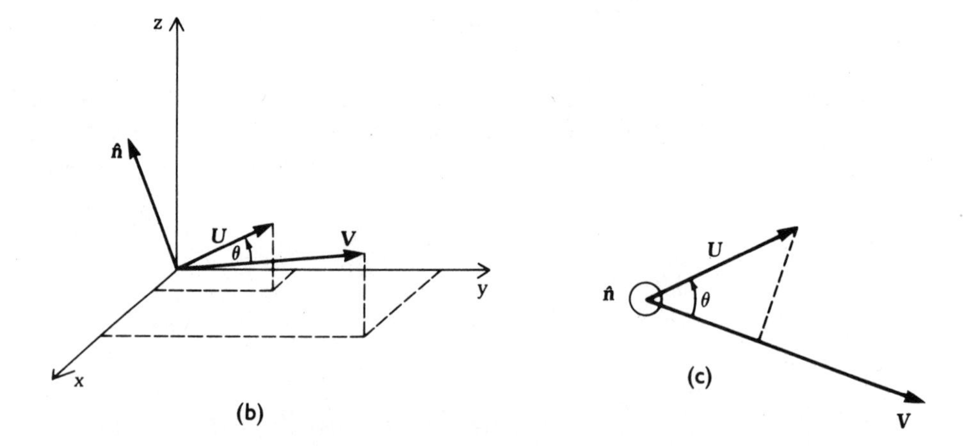

From Figure 1.5(c) and Equation (1.10b), we see that the scalar product of V and U is equal to the magnitude of V times the projection of U on V; or equivalently, the magnitude of U times the projection of V on U.

The dot product is commutative and distributive. That is,

$$U \cdot V = V \cdot U \quad \textbf{(commutative)} \tag{1.11a}$$

$$A \cdot (B + C) = A \cdot B + A \cdot C \quad \textbf{(distributive)} \tag{1.11b}$$

We can prove (1.11a) by using either (1.10a) or (1.10b). The distributive law may be proven by writing $\boldsymbol{A}$, $\boldsymbol{B}$, and $\boldsymbol{C}$ in terms of their x, y, and z components and by showing that the left-hand side of (1.11b) is equal to the right-hand side (see Problem 1.18).

The cross product between two vectors $\boldsymbol{V}$ and $\boldsymbol{U}$ is sometimes referred to as the vector product of $\boldsymbol{V}$ and $\boldsymbol{U}$ because the result is a vector:

$$\boldsymbol{V} \times \boldsymbol{U} = (V_y U_z - V_z U_y)\hat{\boldsymbol{x}} + (V_z U_x - V_x U_z)\hat{\boldsymbol{y}} + (V_x U_y - V_y U_x)\hat{\boldsymbol{z}} \tag{1.12}$$

To remember the above formula more easily, we can put it in a determinant form as follows:

$$\boldsymbol{V} \times \boldsymbol{U} = \begin{vmatrix} \hat{\boldsymbol{x}} & \hat{\boldsymbol{y}} & \hat{\boldsymbol{z}} \\ V_x & V_y & V_z \\ U_x & U_y & U_z \end{vmatrix} = \hat{\boldsymbol{x}} \begin{vmatrix} V_y & V_z \\ U_y & U_z \end{vmatrix} - \hat{\boldsymbol{y}} \begin{vmatrix} V_x & V_z \\ U_x & U_z \end{vmatrix} + \hat{\boldsymbol{z}} \begin{vmatrix} V_x & V_y \\ U_x & U_y \end{vmatrix}$$

(cross product) (1.13)

Identical results can be obtained after expanding the two-by-two determinants.

Note that an equivalent definition of the cross product is

$$\boldsymbol{V} \times \boldsymbol{U} = \hat{\boldsymbol{n}}|\boldsymbol{V}||\boldsymbol{U}|\sin\theta$$ **(equivalent definition of cross product) (1.14)**

where $\hat{\boldsymbol{n}}$ is a vector normal to the plane that contains both $\boldsymbol{V}$ and $\boldsymbol{U}$. The direction of $\hat{\boldsymbol{n}}$ is also determined by the right-hand rule: Hold your right hand as shown in Figure 1.5. Let your fingers point in the direction from $\boldsymbol{V}$ to $\boldsymbol{U}$. Your thumb is then pointing in the $\hat{\boldsymbol{n}}$ direction.

Consider the situation of a force $\boldsymbol{F}$ newton applied at point B of a rod pivoted at point A, as shown in Figure 1.6. In physics, we learn that this force produces a torque $\boldsymbol{T}$. The magnitude of the torque is $F\ell \sin\theta$, where ℓ is the length of the rod between A and B. The direction of the torque is such that it tends to rotate the rod in the counterclockwise direction. The above description can be conveniently summed up by the following cross product:

$$\boldsymbol{T} = \ell\hat{\boldsymbol{r}} \times \boldsymbol{F}$$

Note that a rotation in the counterclockwise direction is represented by a vector pointing out of the paper. Using the right-hand rule, the fingers follow the rotation while the thumb points in the direction of the vector that represents the rotation.

Figure 1.6

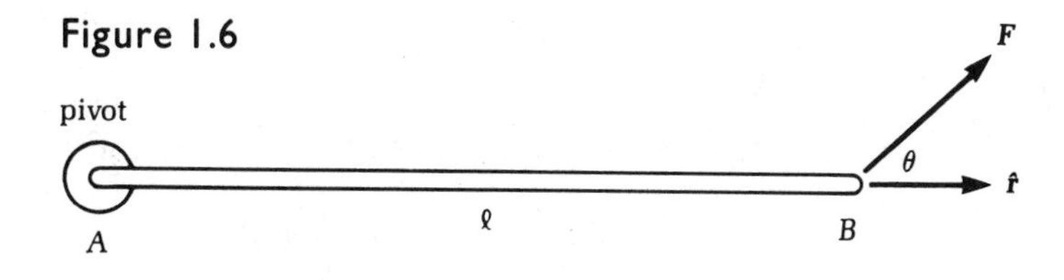

The cross product is *not* commutative because

$$\boldsymbol{V} \times \boldsymbol{U} = -\boldsymbol{U} \times \boldsymbol{V}$$

The above product equation is derived directly from the definition of the vector product, that is, (1.12) or (1.14).

The cross product is distributive; that is,

$$\boldsymbol{A} \times (\boldsymbol{B} + \boldsymbol{C}) = \boldsymbol{A} \times \boldsymbol{B} + \boldsymbol{A} \times \boldsymbol{C}$$

However, the cross product is *not* associative. That is,

$$\boldsymbol{A} \times (\boldsymbol{B} \times \boldsymbol{C}) \neq (\boldsymbol{A} \times \boldsymbol{B}) \times \boldsymbol{C}$$

In fact,

$$\boldsymbol{A} \times (\boldsymbol{B} \times \boldsymbol{C}) = \boldsymbol{B}(\boldsymbol{A} \cdot \boldsymbol{C}) - \boldsymbol{C}(\boldsymbol{A} \cdot \boldsymbol{B}) \tag{1.15}$$

The above equation may be derived by expanding $\boldsymbol{A}$, $\boldsymbol{B}$, and $\boldsymbol{C}$ vectors in x, y, and z components (see Problem 1.21).

We are defining these operations between vectors because they can neatly express some physical laws. Note that we have not defined division of two vectors or square root of a vector.

Example 1.8 *This example shows the derivation of the law of cosines for a triangle.*

Let A, B, and C be the three sides of a triangle, and let the angle between A and B be θ (see Figure 1.7). Find C^2 in terms of A, B, and θ.

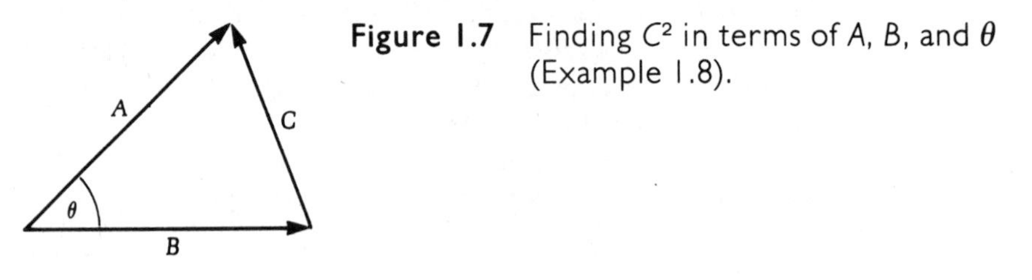

Figure 1.7 Finding C^2 in terms of A, B, and θ (Example 1.8).

Solution Defining three vectors $\boldsymbol{A}$, $\boldsymbol{B}$, and $\boldsymbol{C}$, which coincide with the three sides of the triangle as shown in the figure, we see that $\boldsymbol{C} = \boldsymbol{A} - \boldsymbol{B}$. Thus,

$$C^2 = (\boldsymbol{A} - \boldsymbol{B}) \cdot (\boldsymbol{A} - \boldsymbol{B}) = A^2 + B^2 - 2\boldsymbol{A} \cdot \boldsymbol{B} = A^2 + B^2 - 2AB \cos \theta$$

Example 1.9 *This example consists of numerical examples of vector algebra.*

Let $\boldsymbol{V} = \hat{x} + 2\hat{y} + 3\hat{z}$ and $\boldsymbol{U} = -2\hat{x} + 3\hat{y} - \hat{z}$; then

$$V + U = -\hat{x} + 5\hat{y} + 2\hat{z}$$

$$V - U = 3\hat{x} - \hat{y} + 4\hat{z}$$

$$V \cdot U = -2 + 6 - 3 = 1$$

and

$$V \times U = \begin{vmatrix} \hat{x} & \hat{y} & \hat{z} \\ 1 & 2 & 3 \\ -2 & 3 & -1 \end{vmatrix} = \hat{x}\begin{vmatrix} 2 & 3 \\ 3 & -1 \end{vmatrix} - \hat{y}\begin{vmatrix} 1 & 3 \\ -2 & -1 \end{vmatrix} + \hat{z}\begin{vmatrix} 1 & 2 \\ -2 & 3 \end{vmatrix}$$

$$= -11\hat{x} - 5\hat{y} + 7\hat{z}$$

Example 1.10 *This example presents a good way to find the angle between two vectors.*

Using the values of V and U given in the previous example, find the angle θ between them.

Solution

$$V \cdot U = |V||U| \cos \theta$$

$$|V| = (1 + 4 + 9)^{1/2} = 3.742$$

$$|U| = (4 + 9 + 1)^{1/2} = 3.742$$

Therefore,

$$V \cdot U = 14 \cos \theta$$

On the other hand, the previous example shows that

$$V \cdot U = 1$$

Thus,

$$\cos \theta = \tfrac{1}{14}$$

or

$$\theta = \cos^{-1}(\tfrac{1}{14}) = 1.499 \text{ rad} = 85.9^\circ$$

Example 1.11 *This example shows that, using the scalar product, we can check whether two vectors are perpendicular to each other in space and, using the cross product, we can check whether they are parallel.*

Let $V = \hat{x} + b\hat{y} + c\hat{z}$ and $U = -\hat{x} + 3\hat{y} - 8\hat{z}$. What are the values of b and c that will make (a) $V \parallel U$ and (b) $V \perp U$?

Solution (a) $V \parallel U$ implies $V \times U = 0$.

$$V \times U = \hat{x}(-8b - 3c) - \hat{y}(-8 + c) + \hat{z}(3 + b) = 0$$

A zero vector means that each of its components is zero:

$$-8b - 3c = 0$$
$$-8 + c = 0$$
$$3 + b = 0$$

The solution $b = -3$ and $c = 8$ satisfies all three equations.
(b) $V \perp U$ implies $V \cdot U = 0$.

$$V \cdot U = -1 + 3b - 8c = 0$$

Thus, the vector $V = \hat{x} + b\hat{y} + [(3b - 1)/8]\hat{z}$ is perpendicular to U for any value of b.

1.4 COMPLEX VECTORS

We have discussed the method of phasor representation of time-harmonic scalars. This method can readily be applied to time-harmonic vectors. A vector's three components, each of which is a scalar, can be individually transformed into phasor notation as shown below.

$$\begin{aligned} V(t) &= \hat{x}V_x \cos(\omega t + \phi_x) + \hat{y}V_y \cos(\omega t + \phi_y) + \hat{z}V_z \cos(\omega t + \phi_z) \\ &= \mathrm{Re}\{[\hat{x}V_x \exp(j\phi_x) + \hat{y}V_y \exp(j\phi_y) + \hat{z}V_z \exp(j\phi_z)] \exp(j\omega t)\} \end{aligned} \quad \textbf{(1.16)}$$

The equivalent phasor notation for $V(t)$ is the above expression with the expressions Re and $e^{j\omega t}$ omitted.

$$V(t) \leftrightarrow \mathbf{V} = \hat{x}V_x \exp(j\phi_x) + \hat{y}V_y \exp(j\phi_y) + \hat{z}V_z \exp(j\phi_z) \quad \textbf{(1.17)}$$

The complex vector $\mathbf{V}$ is not an explicit function of time. Remember that $\mathbf{V}$ is implicitly a vector function of time and that the conversion rule is as follows:

$$\boxed{V(t) = \mathrm{Re}\{\mathbf{V}\, e^{j\omega t}\}} \quad \textbf{(conversion between real vector and complex vector)} \quad \textbf{(1.18)}$$

Example 1.12 *This example provides numerical examples of complex vector algebra.*

Let $\mathbf{C} = j\hat{x} + (1 + j)\hat{y} + (3 - j4)\hat{z}$ and $\mathbf{D} = (2 - j3)\hat{x} + j2\hat{y} - 2\hat{z}$; then

$$\mathbf{C} + \mathbf{D} = (2 - j2)\hat{x} + (1 + j3)\hat{y} + (1 - j4)\hat{z}$$

$$\mathbf{C} - \mathbf{D} = (-2 + j4)\hat{x} + (1 - j)\hat{y} + (5 - j4)\hat{z}$$

$$\mathbf{C} \cdot \mathbf{D} = (3 + j2) + (-2 + j2) + (-6 + j8) = -5 + j12$$

and

$$\mathbf{C} \times \mathbf{D} = (-10 - j8)\hat{x} + (-6 - j15)\hat{y} + (-7 + j)\hat{z}$$

Example 1.13 *This example shows numerical examples of vector operation involving the conjugate of a complex vector.*

Using the values of $\mathbf{C}$ and $\mathbf{D}$ in the previous example, find $\frac{1}{2}\operatorname{Re}\{\mathbf{C} \times \mathbf{D}^*\}$ and $\mathbf{D} \cdot \mathbf{D}^*$.

Solution

$$\mathbf{C} \times \mathbf{D}^* = \begin{bmatrix} \hat{x} & \hat{y} & \hat{z} \\ j & 1 + j & 3 - j4 \\ 2 + j3 & -j2 & -2 \end{bmatrix} = \hat{x}(6 + j4) + \hat{y}(18 + j3) + \hat{z}(3 - j5)$$

$$\tfrac{1}{2}\operatorname{Re}\{\mathbf{C} \times \mathbf{D}^*\} = 3\hat{x} + 9\hat{y} + 1.5\hat{z}$$

$$\mathbf{D} \cdot \mathbf{D}^* = 13 + 4 + 4 = 21$$

Example 1.14 *This example shows the conversion from complex to real-time expression.*

Find $\boldsymbol{C}(t)$ and $\boldsymbol{D}(t)$ in terms of ω, using the values given in Example 1.12.

Solution

$$\begin{aligned} \boldsymbol{C}(t) &= \operatorname{Re}\{\mathbf{C}\, e^{j\omega t}\} \\ &= -\sin(\omega t)\hat{x} + [\cos(\omega t) - \sin(\omega t)]\hat{y} + [3\cos(\omega t) + 4\sin(\omega t)]\hat{z} \end{aligned}$$

An alternative solution is the following:

$$\begin{aligned} \boldsymbol{C}(t) &= \operatorname{Re}\{[e^{j\pi/2}\hat{x} + \sqrt{2}\, e^{j\pi/4}\hat{y} + 5\, e^{-j0.93}\hat{z}]\, e^{j\omega t}\} \\ &= \hat{x}\cos(\omega t + \tfrac{1}{2}\pi) + \hat{y}\sqrt{2}\cos(\omega t + \tfrac{1}{4}\pi) + \hat{z}5\cos(\omega t - 0.93) \end{aligned}$$

Similarly,

$$\begin{aligned} \boldsymbol{D}(t) &= \hat{x}[2\cos(\omega t) + 3\sin(\omega t)] + \hat{y}[-2\sin(\omega t)] + \hat{z}[-2\cos(\omega t)] \\ &= \hat{x}3.61\cos(\omega t - 0.98) + \hat{y}2\cos(\omega t + \tfrac{1}{2}\pi) + \hat{z}2\cos(\omega t + \pi) \end{aligned}$$

Example 1.15 *This example illustrates the fact that a complex vector representing a time-harmonic real vector usually does not have a constant direction in real-time domain. The direction of the vector varies with time periodically.*

Let $\mathbf{A} = \hat{x} + j\hat{y}$, and sketch the motion of the tip of the vector $A(t)$ as a function of time.

Solution $A(t) = \hat{x}\cos(\omega t) - \hat{y}\sin(\omega t)$. When $\omega t = 0$, the vector is pointing in the $\hat{x}$ direction, and, when $\omega t = \pi/2$, the vector is pointing in the $-\hat{y}$ direction, and so on. In fact, the vector rotates clockwise as seen from the positive z axis. The tip traces a circle of radius equal to unity with angular frequency ω, as shown in Figure 1.8.

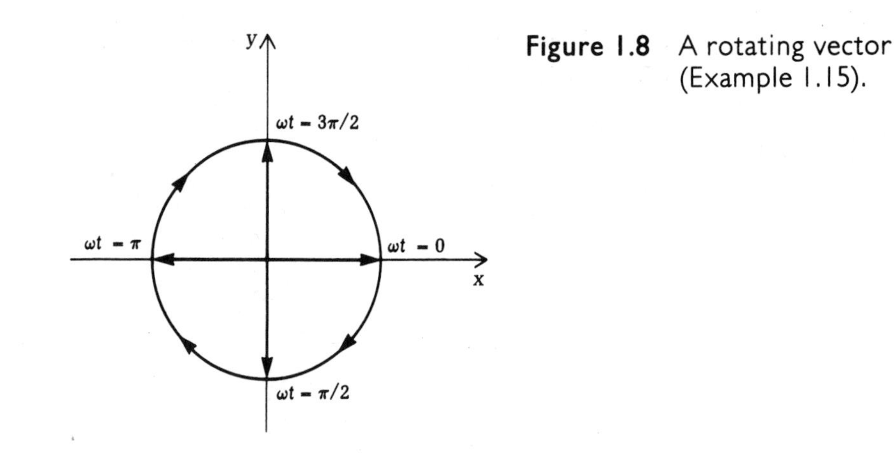

Figure 1.8 A rotating vector (Example 1.15).

Example 1.16 *In Example 1.11, we checked whether two real vectors are perpendicular or parallel by finding the dot and the cross products. Can we do the same for complex vectors? This example gives you the answer to this question.*

Let $\mathbf{A} = \hat{x} + j\hat{y}$ and $\mathbf{B} = j(\hat{x} + j\hat{y})$. Show that $\mathbf{A} \times \mathbf{B} = 0$ but that $A(t) \times B(t)$ is not zero.

Solution The first part, that $\mathbf{A} \times \mathbf{B} = 0$, can be shown by carrying out the cross product directly. Then, since $A(t) = \hat{x}\cos(\omega t) - \hat{y}\sin(\omega t)$ and $B(t) = -\hat{x}\sin(\omega t) - \hat{y}\cos(\omega t)$,

$$A(t) \times B(t) = \hat{z}[-\cos^2(\omega t) - \sin^2(\omega t)] = -\hat{z} \neq 0$$

Thus, *when the cross product of two complex vectors is zero, their time-domain counterparts are not necessarily parallel at all times.*

1.5 TIME AVERAGES

The time-average value of a time-harmonic quantity is always zero. As seen from equations (1.4) or (1.16), the time-average values for $V(t)$ and $\boldsymbol{V}(t)$ are defined as $\langle V(t)\rangle$ and $\langle \boldsymbol{V}(t)\rangle$, respectively, where

$$\langle V(t)\rangle \equiv \frac{1}{T}\int_0^T dt\, V_0 \cos(\omega t + \phi) = 0$$

$$\langle \boldsymbol{V}(t)\rangle \equiv \frac{1}{T}\int_0^T dt[\hat{\boldsymbol{x}} V_x \cos(\omega t + \phi_x) + \hat{\boldsymbol{y}} V_y \cos(\omega t + \phi_y) + \hat{\boldsymbol{z}} V_z \cos(\omega t + \phi_z)]$$
$$= 0$$

where $T = 1/f = 2\pi/\omega$ represents a period in time. However, the time-average value for the product of two time-harmonic quantities is not always zero. For instance,

$$\langle V^2(t)\rangle = \frac{1}{T}\int_0^T dt\, V_0^2 \cos^2(\omega t + \phi)$$
$$= \frac{1}{T}\int_0^T dt\, V_0^2[\tfrac{1}{2} + \tfrac{1}{2}\cos(2\omega t + 2\phi)]$$
$$= \frac{V_0^2}{2}$$

which is not zero.

In many instances we are interested in time-average quantities like power delivered to a motor or power lost to heat. We all know that customers pay for the time-average power they draw from the power company. It is therefore important to know how to compute the time-average values of time-harmonic physical quantities.

Consider complex vectors $\mathbf{A} = \boldsymbol{A}_R + j\mathbf{A}_I$ and $\mathbf{B} = \boldsymbol{B}_R + j\boldsymbol{B}_I$, where $\boldsymbol{A}_R$ and $\boldsymbol{B}_R$ denote the real parts and $\boldsymbol{A}_I$ and $\boldsymbol{B}_I$ their imaginary parts. Their time-domain counterparts can be written as

$$\boldsymbol{A}(t) = \text{Re}\{\mathbf{A}\, e^{j\omega t}\} = \boldsymbol{A}_R \cos\omega t - \boldsymbol{A}_I \sin\omega t$$

$$\boldsymbol{B}(t) = \text{Re}\{\mathbf{B}\, e^{j\omega t}\} = \boldsymbol{B}_R \cos\omega t - \boldsymbol{B}_I \sin\omega t$$

The cross product of $\boldsymbol{A}(t)$ and $\boldsymbol{B}(t)$ then becomes

$$\boldsymbol{A}(t) \times \boldsymbol{B}(t) = \boldsymbol{A}_R \times \boldsymbol{B}_R \cos^2\omega t + \boldsymbol{A}_I \times \boldsymbol{B}_I \sin^2\omega t$$
$$- \tfrac{1}{2}(\boldsymbol{A}_R \times \boldsymbol{B}_I + \boldsymbol{A}_I \times \boldsymbol{B}_R)\sin 2\omega t$$

The time-average value of $\boldsymbol{A}(t) \times \boldsymbol{B}(t)$ is then

$$\langle \boldsymbol{A}(t) \times \boldsymbol{B}(t)\rangle = \tfrac{1}{2}(\boldsymbol{A}_R \times \boldsymbol{B}_R + \boldsymbol{A}_I \times \boldsymbol{B}_I)$$

If we take the cross product of **A** and **B***, we then get

$$\mathbf{A} \times \mathbf{B}^* = \boldsymbol{A}_R \times \boldsymbol{B}_R + \boldsymbol{A}_I \times \boldsymbol{B}_I + j(\boldsymbol{A}_I \times \boldsymbol{B}_R - \boldsymbol{A}_R \times \boldsymbol{B}_I)$$

We thus see that

$$\boxed{\langle \boldsymbol{A}(t) \times \boldsymbol{B}(t) \rangle = \tfrac{1}{2}\,\mathrm{Re}\,\{\mathbf{A} \times \mathbf{B}^*\}} \quad \textbf{(time-average rule)} \qquad \textbf{(1.19)}$$

The time-average value for the cross product of $\boldsymbol{A}(t)$ and $\boldsymbol{B}(t)$ is equal to half the real part of the cross product of **A** and the complex conjugate of **B**, **B***. This rule will be very important in later chapters when we encounter the concept of Poynting's power-density vector.

The dot product of $\boldsymbol{A}(t)$ and $\boldsymbol{B}(t)$ is

$$\begin{aligned}\boldsymbol{A}(t) \cdot \boldsymbol{B}(t) = {} & \boldsymbol{A}_R \cdot \boldsymbol{B}_R \cos^2 \omega t + \boldsymbol{A}_I \cdot \boldsymbol{B}_I \sin^2 \omega t \\ & - (\boldsymbol{A}_R \cdot \boldsymbol{B}_I + \boldsymbol{A}_I \cdot \boldsymbol{B}_R) \sin \omega t \cos \omega t\end{aligned}$$

The time-average value of $\boldsymbol{A}(t) \cdot \boldsymbol{B}(t)$ is then

$$\langle \boldsymbol{A}(t) \cdot \boldsymbol{B}(t) \rangle = \tfrac{1}{2}(\boldsymbol{A}_R \cdot \boldsymbol{B}_R + \boldsymbol{A}_I \cdot \boldsymbol{B}_I)$$

If we take the dot product of **A** and **B***, we get

$$\mathbf{A} \cdot \mathbf{B}^* = \boldsymbol{A}_R \cdot \boldsymbol{B}_R + \boldsymbol{A}_I \cdot \boldsymbol{B}_I + j(\boldsymbol{A}_I \cdot \boldsymbol{B}_R - \boldsymbol{A}_R \cdot \boldsymbol{B}_I)$$

Therefore, we can write

$$\langle \boldsymbol{A}(t) \cdot \boldsymbol{B}(t) \rangle = \tfrac{1}{2}\,\mathrm{Re}\,(\mathbf{A} \cdot \mathbf{B}^*) \qquad \textbf{(1.20)}$$

Example 1.17 *This example shows shortcuts to computing the time averages of the cross product and the scalar product of two time-harmonic vectors. We can obtain the answer from their phasor notations without having to carry out any integration.*

Find $\langle \boldsymbol{C}(t) \times \boldsymbol{D}(t) \rangle$ and $\langle \boldsymbol{C}(t) \cdot \boldsymbol{D}(t) \rangle$, where **C** and **D** are given in Example 1.12.

Solution The normal way to compute the time averages is to carry out the cross and dot products first, and then integrate over the period. The shortcut is to use Equations (1.19) and (1.20). The phasor expressions of $\boldsymbol{C}(t)$ and $\boldsymbol{D}(t)$ are given in Example 1.12. Thus,

$$\begin{aligned}\langle \boldsymbol{C}(t) \times \boldsymbol{D}(t) \rangle &= \tfrac{1}{2}\,\mathrm{Re}\,(\mathbf{C} \times \mathbf{D}^*) \\ &= 3\hat{x} + 9\hat{y} + 1.5\hat{z} \\ \langle \boldsymbol{C}(t) \cdot \boldsymbol{D}(t) \rangle &= \tfrac{1}{2}\,\mathrm{Re}\,(\mathbf{C} \cdot \mathbf{D}^*) \\ &= -3.5\end{aligned}$$

SUMMARY

You should have learned the following:

Complex Numbers

- phasor notation
- addition, subtraction, multiplication, and division
- square root

Time-Harmonic Scalar and Vector Functions

- how to represent them with complex numbers and vectors
- that complex expressions cannot be used if the function contains more than one frequency component or if it is the product of two time-harmonic functions of the same frequency
- how to convert complex numbers and vectors back to real-time expressions

Vectors

- addition and subtraction
- dot and cross products
- how to find the angle between two real vectors

Time Averages

- a quick way to compute time averages of dot and cross products of two time-harmonic vectors

---

Problems

1.1 Let $a = 8 + j2$ and $b = -3 + j$. Calculate (a) $a + b$, (b) $a - b$, (c) ab, and (d) a/b. Give the answer in real and imaginary parts.

1.2 Repeat (c) and (d) in Problem 1.1 with the answer given in phasor form.

1.3 Find the real part, the imaginary part, and the magnitude of $e^{j\omega t}$, where ω and t are real numbers.

1.4 Let c be a complex number. Show that the following statements are always true.
(a) $(c + c^*)$ is real.
(b) $(c - c^*)$ is imaginary.
(c) c/c^* has a magnitude equal to 1.
(d) cc^* is always real and positive unless $c = 0$, in which case $cc^* = 0$.

1.5 Consider the equation $z^2 = 1 + j$. Find two values of z that satisfy this equation.

1.6 Let a be a real number, and let $|a| \ll 1$. Show that the square root of $(1 + ja)$ is approximately equal to $\pm(1 + ja/2)$.

1.7 Let a be a positive real number, and let $a \gg 1$. Show that the square root of $(1 + ja)$ is approximately equal to $\pm(1 + j)(a/2)^{1/2}$.

1.8 Obtain the phasor notation of the following time-harmonic functions (if possible):

(a) $V(t) = 6\cos(\omega t + \pi/4)$
(b) $I(t) = -8\sin(\omega t)$
(c) $A(t) = 3\sin(\omega t) - 2\cos(\omega t)$
(d) $C(t) = 6\cos(120\pi t) - \pi/2)$
(e) $D(t) = 1 - \cos(\omega t)$
(f) $U(t) = \sin(\omega t + \pi/3)\sin(\omega t + \pi/6)$

1.9 Obtain $C(t)$ in terms of ω from the following phasors: (a) $\mathrm{C} = 1 + j$, (b) $\mathrm{C} = 4\exp(j0.8)$, and (c) $\mathrm{C} = 3\exp(j\pi/2) + 4\exp(j0.8)$.

1.10 Show that, if $\mathrm{V} = r + jx$ and $\mathrm{U} = g + jy$, then $V(t)U(t) \neq \mathrm{Re}\{\mathrm{VU}\, e^{j\omega t}\}$. Find the correct expression for $V(t)U(t)$ in terms of r, x, g, y, and ωt.

1.11 Let $\boldsymbol{A} = -8\hat{x} + 9\hat{y} - \hat{z}$ and $\boldsymbol{B} = 2\hat{x} - 4\hat{y} + 3\hat{z}$. Find (a) $\boldsymbol{A} + \boldsymbol{B}$, (b) $\boldsymbol{A} - \boldsymbol{B}$, (c) $\boldsymbol{A} \cdot \boldsymbol{B}$, and (d) $\boldsymbol{A} \times \boldsymbol{B}$.

1.12 Find the angle between the $\boldsymbol{A}$ and $\boldsymbol{B}$ given in Problem 1.11.

1.13 Show that $\boldsymbol{V}$ is always perpendicular to $\boldsymbol{V} \times \boldsymbol{U}$. Hint: Use the expressions of $\boldsymbol{V} \times \boldsymbol{U}$ in Equation (1.13) and find the dot product of $\boldsymbol{V}$ and $\boldsymbol{V} \times \boldsymbol{U}$.

1.14 In Example 1.7, find $|\boldsymbol{U}|$, the speed of the airplane relative to the wind.

1.15 Find the vector $\boldsymbol{C}$ that is parallel to $\boldsymbol{A} = 5\hat{x} - 8\hat{y} + 2\hat{z}$ and has a magnitude equal to 1.

1.16 Find a unit vector $\hat{\boldsymbol{n}}$ that points in the same direction as an arrow drawn from point A to point B where the rectangular coordinates of A and B are $(1, 0, 2)$ and $(-1, 3, -2)$, respectively.

1.17 Show that the definition of the dot product $\boldsymbol{V} \cdot \boldsymbol{U}$ given by Equation (1.10a) is equivalent to that given by (1.10b). To simplify the algebra, you may choose the coordinates so that the x axis is along $\boldsymbol{V}$ and the z axis is perpendicular to both $\boldsymbol{V}$ and $\boldsymbol{U}$. In other words, let $\boldsymbol{V} = a\hat{x}$ and $\boldsymbol{U} = b\hat{x} + c\hat{y}$.

1.18 Prove Equation (1.11b) using the approach suggested in the text.

1.19 Show that the definition of the cross product $\boldsymbol{V} \times \boldsymbol{U}$ given by Equation (1.12) is equivalent to that given by Equation (1.14). To simplify the algebra, do what is suggested in Problem 1.17.

1.20 Figure P1.20 shows that a vector $\boldsymbol{V}$ is along the x axis and a vector $\boldsymbol{U}$ is on the x-y plane forming a 135° angle with the x axis. The magnitudes of these vectors are v and u, respectively. Use Equation (1.14) to express $\boldsymbol{V} \times \boldsymbol{U}$ in terms v and u. Now, we can also say that

Figure P1.20

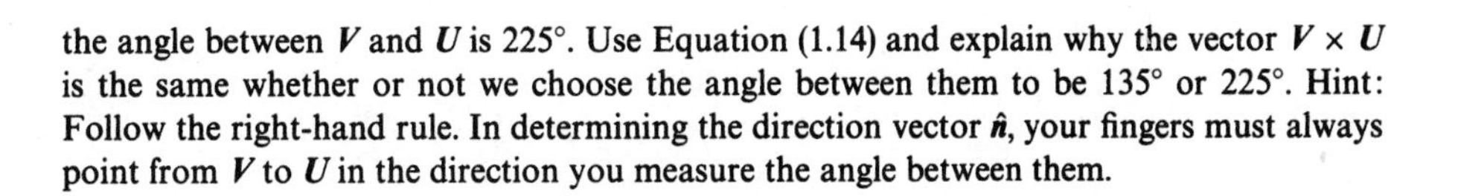
the angle between $\boldsymbol{V}$ and $\boldsymbol{U}$ is 225°. Use Equation (1.14) and explain why the vector $\boldsymbol{V} \times \boldsymbol{U}$ is the same whether or not we choose the angle between them to be 135° or 225°. Hint: Follow the right-hand rule. In determining the direction vector $\hat{\boldsymbol{n}}$, your fingers must always point from $\boldsymbol{V}$ to $\boldsymbol{U}$ in the direction you measure the angle between them.

1.21 Prove Equation (1.15) using the approach suggested in the text.

1.22 Find the phasor notations of the following time-harmonic vectors:

(a) $\boldsymbol{V}(t) = 3\cos(\omega t)\hat{x} + 4\sin(\omega t)\hat{y} + \hat{z}\cos(\omega t + \pi/2)$
(b) $\boldsymbol{E}(t) = [3\cos(\omega t) + 4\sin(\omega t)]\hat{x} + 8[\cos(\omega t) - \sin(\omega t)]\hat{z}$
(c) $\boldsymbol{H}(t) = 0.5\cos(kz - \omega t)\hat{x}$

1.23 Find the phasor notation of the following vector:

$$\boldsymbol{C}(t) = (\partial/\partial t)\boldsymbol{E}(t)$$

where $\boldsymbol{E}$ is given in Problem 1.22(b).

1.24 From the following complex vectors, find $\boldsymbol{C}(t)$ in terms of ωt: (a) $\mathbf{C} = \hat{x} - j\hat{y}$, (b) $\mathbf{C} = j(\hat{x} - j\hat{y})$, and (c) $\mathbf{C} = \exp(-jkz)\hat{x} + j\exp(jkz)\hat{y}$.

1.25 Let $\mathbf{A} = \hat{x} + j\hat{y} + (1 + j2)\hat{z}$, and let $\mathbf{B} = -\hat{x} - (1 - j2)\hat{y} + j\hat{z}$. Find (a) $\mathbf{A} + \mathbf{B}$, (b) $\mathbf{A} - \mathbf{B}$, (c) $\mathbf{A} \cdot \mathbf{B}$, and (d) $\mathbf{A} \times \mathbf{B}$.

1.26 Find $\mathbf{A} \cdot \mathbf{A}^*$ and $\text{Re}\{\mathbf{A} \times \mathbf{B}^*\}$ for the values of $\mathbf{A}$ and $\mathbf{B}$ given in Problem 1.25.

1.27 Consider the two vectors $\mathbf{A} = \hat{x} + j\hat{y}$ and $\mathbf{B} = \hat{x} + j\hat{y}$. (They are actually the same vector.) Are $\boldsymbol{A}(t)$ and $\boldsymbol{B}(t)$ parallel to each other or perpendicular to each other?

1.28 Calculate $\mathbf{A} \cdot \mathbf{B}$, given $\mathbf{A} = \hat{x} + j2\hat{y}$ and $\mathbf{B} = 2\hat{x} + j\hat{y}$. Are $\boldsymbol{A}(t)$ and $\boldsymbol{B}(t)$ perpendicular at all times?

1.29 Trace the tip of the vector $\mathcal{A}(t)$, where (a) $\mathbf{A} = \hat{x} - j\hat{y}$ and (b) $\mathbf{A} = 4\hat{x} + j3\hat{y}$.

1.30 Write a computer program to trace the tip of a two-dimensional vector $\mathcal{A}(t)$ where $\mathbf{A} = (a + jb)\hat{x} + (c + jd)\hat{y}$. The program should take a, b, c, and d as input parameters and then trace the tip of $\mathcal{A}(t)$ as a function of time on the screen or on a plotter. Solving this problem will give you practice with conversion of a complex vector to a real-time vector. You will see that the real vector represents a time-varying vector with the tip tracing a straight line, a circle, or an ellipse.

2 MAXWELL'S EQUATIONS

In the study of electromagnetics, we are concerned with four vector quantities called **electromagnetic fields**:

$\boldsymbol{E}$ = electric field strength	(volts per meter)
$\boldsymbol{D}$ = electric flux density	(coulombs per square meter)
$\boldsymbol{H}$ = magnetic field strength	(amperes per meter)
$\boldsymbol{B}$ = magnetic flux density	(webers per square meter) or (teslas)

These are vectors, and they are functions of space and time. These electromagnetic fields are always present. Some of them come from natural sources such as the sun, lightning, and stars. Others, such as radio waves and laser lights, are generated by artificial means. Electromagnetic fields have a wide range of applications: from communication to medical treatment, from environmental sensing to energy generation. In this book we shall study a number of such applications of electromagnetic fields as well as the fundamental theory.

2.1 MAXWELL'S EQUATIONS

The fundamental theory of electromagnetic fields is based on Maxwell's equations. These equations govern the electromagnetic fields $\boldsymbol{E}$, $\boldsymbol{D}$, $\boldsymbol{H}$, and $\boldsymbol{B}$:

$$\nabla \times \boldsymbol{E} = -\frac{\partial \boldsymbol{B}}{\partial t} \qquad \textbf{(Faraday's law of induction)} \tag{2.1}$$

$$\nabla \times \boldsymbol{H} = \boldsymbol{J} + \frac{\partial \boldsymbol{D}}{\partial t} \qquad \textbf{(Ampère's law)} \tag{2.2}$$

$$\nabla \cdot \boldsymbol{B} = 0 \qquad \textbf{(magnetic Gauss' law)} \tag{2.3}$$

$$\nabla \cdot \boldsymbol{D} = \rho_v \qquad \textbf{(electric Gauss' law)} \tag{2.4}$$

where

$\boldsymbol{J}$ = electric current density	(amperes per square meter)
ρ_v = electric charge density	(coulombs per square meter)

are the sources generating the electromagnetic fields. Equations (2.1)–(2.4) express the physical laws governing the $\boldsymbol{E}$, $\boldsymbol{D}$, $\boldsymbol{H}$, and $\boldsymbol{B}$ fields and the sources $\boldsymbol{J}$ and ρ_v at every point in space at all times. So far there has been no experimental evidence of an electromagnetic field that does not satisfy all four of Maxwell's equations.

The reader is expected to accept with an open mind the physical laws postulated in Maxwell's equations. These four equations, together with the Lorentz force law discussed in Section 2.4, are the only postulates that will be needed to explain the electromagnetic phenomena and to discuss the applications given in this book. The reader may be surprised to find out how much information can be obtained from these five fundamental equations.

Maxwell's first equation is Faraday's law of induction, discovered experimentally by Michael Faraday (1791–1867). Maxwell's second equation refers to Ampère's law, although the original law of André Marie Ampère (1775–1836) did not contain the **displacement current** ($\partial \boldsymbol{D}/\partial t$) term. James Clerk Maxwell (1831–1879) first proposed the addition of this term to the **conduction** or **convection current** $\boldsymbol{J}$ in Ampère's law. The introduction of this extra term was very significant because it made it possible to predict the existence of electromagnetic waves. Their existence was later demonstrated by Heinrich Rudolf Hertz (1857–1894).* Equations (2.3) and (2.4) are known respectively as magnetic Gauss' law and electric Gauss' law.

The ∇ Operator

The symbol ∇ appears in all of Maxwell's equations. This symbol represents a vector partial-differentiation operator. The operation $\nabla \times \boldsymbol{A}$ is called the **curl of** $\boldsymbol{A}$, and the operation $\nabla \cdot \boldsymbol{A}$ is called the **divergence of** $\boldsymbol{A}$. In rectangular coordinates, suppose that

$$\boldsymbol{A} = \hat{x}A_x(x, y, z) + \hat{y}A_y(x, y, z) + \hat{z}A_z(x, y, z)$$

then

$$\nabla \times \boldsymbol{A} = \hat{x}\left(\frac{\partial A_z}{\partial y} - \frac{\partial A_y}{\partial z}\right) + \hat{y}\left(\frac{\partial A_x}{\partial z} - \frac{\partial A_z}{\partial x}\right) + \hat{z}\left(\frac{\partial A_y}{\partial x} - \frac{\partial A_x}{\partial y}\right) \quad \textbf{(curl of } A\textbf{)} \qquad (2.5)$$

$$\nabla \cdot \boldsymbol{A} = \frac{\partial A_x}{\partial x} + \frac{\partial A_y}{\partial y} + \frac{\partial A_z}{\partial z} \quad \textbf{(divergence of } A\textbf{)} \qquad (2.6)$$

* An interesting history of the theoretical and experimental development of electromagnetic science appears in a series of articles by C. Susskind, "The Early History of Electronics," *IEEE Spectrum*, August 1968, December 1968, April 1969, August 1969, April 1970, and September 1970.

Note that $\nabla \cdot \boldsymbol{A}$ is a scalar and $\nabla \times \boldsymbol{A}$ is a vector. In the rectangular coordinates, we can treat the operator ∇ as a vector:

$$\boxed{\nabla = \hat{x}\frac{\partial}{\partial x} + \hat{y}\frac{\partial}{\partial y} + \hat{z}\frac{\partial}{\partial z}} \qquad \textbf{(}\boldsymbol{\nabla}\textbf{ operator)} \tag{2.7}$$

where $\hat{x}$, $\hat{y}$, and $\hat{z}$ denote unit vectors along the x, y, and z axes. Thus, $\nabla \cdot \boldsymbol{A}$ is the scalar product of the vectors ∇ and $\boldsymbol{A}$, and the cross product of ∇ and $\boldsymbol{A}$ is $\nabla \times \boldsymbol{A}$. However, we must note that the vector operator ∇ is somewhat different from the ordinary vectors in that the former must operate on some functions. For ordinary vectors $\boldsymbol{A}$ and $\boldsymbol{B}$,

$$\boldsymbol{A} \times \boldsymbol{B} = -\boldsymbol{B} \times \boldsymbol{A} \qquad \text{and} \qquad \boldsymbol{A} \cdot \boldsymbol{B} = \boldsymbol{B} \cdot \boldsymbol{A}$$

But for ∇ operators,

$$\nabla \times \boldsymbol{A} \neq -\boldsymbol{A} \times \nabla$$

$$\nabla \cdot \boldsymbol{A} \neq \boldsymbol{A} \cdot \nabla$$

The right-hand sides of the above equations are not defined.

It is useful to introduce yet another operation involving the ∇ operator. Let $\phi(x, y, z)$ be a scalar function of the coordinates; then the following operation is called the **gradient of ϕ**:

$$\boxed{\nabla\phi = \hat{x}\frac{\partial\phi}{\partial x} + \hat{y}\frac{\partial\phi}{\partial y} + \hat{z}\frac{\partial\phi}{\partial z}} \qquad \textbf{(gradient of } \boldsymbol{\phi}\textbf{)} \tag{2.8}$$

At this point, we must caution students that treating the ∇ operator as a vector in the definition of curl, divergence, and gradient is only permissible in rectangular coordinates. In other coordinate systems, such as spherical or cylindrical coordinates, the explicit form of curl, divergence, and gradient will be more complicated, and the ∇ operator will no longer be able to be regarded as an ordinary vector operator. Cylindrical and spherical coordinates will be introduced in Chapters 5 and 7, respectively.

We list some of the useful vector identities that can be proved using Equations (2.5)–(2.8). In the following equations, $\boldsymbol{a}$ and $\boldsymbol{b}$ are arbitrary (real or complex) vector functions that have continuous partial derivatives.

$$\nabla \times (\nabla \times \boldsymbol{a}) \equiv \nabla(\nabla \cdot \boldsymbol{a}) - \nabla^2\boldsymbol{a} \tag{2.9}$$

where

$$\nabla^2 = \frac{\partial^2}{\partial x^2} + \frac{\partial^2}{\partial y^2} + \frac{\partial^2}{\partial z^2} \tag{2.10}$$

which is called the **Laplacian operator**. It is also true that

$$\nabla \cdot (\nabla \times \boldsymbol{a}) \equiv 0 \tag{2.11a}$$

$$\nabla \times \nabla \phi \equiv 0 \tag{2.11b}$$

and

$$\nabla \cdot (\boldsymbol{a} \times \boldsymbol{b}) \equiv \boldsymbol{b} \cdot (\nabla \times \boldsymbol{a}) - \boldsymbol{a} \cdot (\nabla \times \boldsymbol{b}) \tag{2.12}$$

Both sides of (2.12) are scalars.

Interpretation of $\nabla \times A$

Let us consider the $\hat{x}$ component of $\nabla \times A$ around a small rectangle on the y-z plane, as shown in Figure 2.1a. From (2.5) we obtain

$$(\nabla \times A) \cdot \hat{x} = \frac{\partial A_z}{\partial y} - \frac{\partial A_y}{\partial z}$$

By the definition of partial differentiation, we have

$$\frac{\partial A_z}{\partial y} = \frac{(A_z \text{ on } b) - (A_z \text{ on } d)}{\Delta y}$$

$$\frac{\partial A_y}{\partial z} = \frac{(A_y \text{ on } c) - (A_y \text{ on } a)}{\Delta z}$$

Thus,

$$(\nabla \times A) \cdot \hat{x} = \frac{(A_y \text{ on } a)\Delta y + (A_z \text{ on } b)\Delta z + (-A_y \text{ on } c)\Delta y + (-A_2 \text{ on } d)\Delta z}{\Delta y \, \Delta z}$$

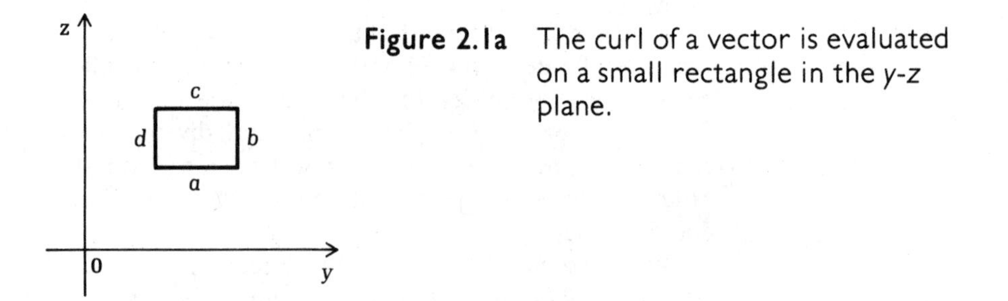

Figure 2.1a The curl of a vector is evaluated on a small rectangle in the y-z plane.

We see that, to obtain the curl of A in the $\hat{x}$ direction, we go along the field of A around a small rectangle of the y-z plane in the same direction as our right-hand fingers with thumb pointing in the $\hat{x}$ direction, as shown in Figure 2.1b. Notice that the small rectangle must be a differential area of infinitesimal size, such that $\Delta y \to 0$ and $\Delta z \to 0$ assuming that the field components A_y and A_z are constant on the boundary of the rectangle.

From the preceding discussion, we see that, in the limit that the area of the

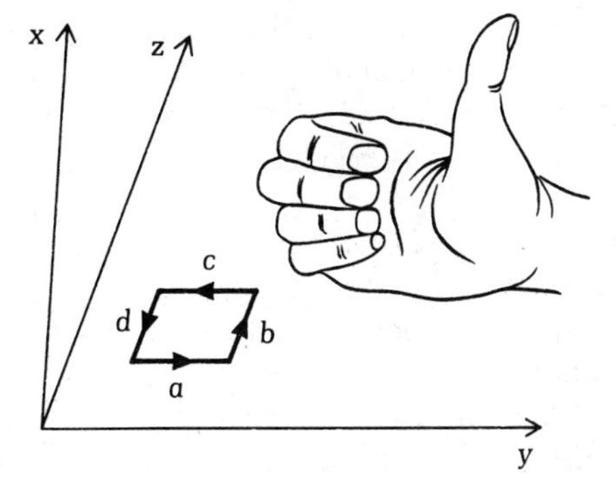

Figure 2.1b

loop becomes infinitesimally small,

$$(\nabla \times \boldsymbol{A}) \cdot \hat{x} = \frac{\int_{abcd} \boldsymbol{A} \cdot d\ell}{\Delta s} \tag{2.13}$$

That is, the x component of $\nabla \times \boldsymbol{A}$ is equal to the integration of $\boldsymbol{A} \cdot d\boldsymbol{l}$ along a closed loop a-b-c-d on the y-z plane, divided by Δs, the area of that loop. The direction of integration is determined by the right-hand rule, as shown in Figure 2.1b.

Faraday Induction

Equation (2.13) is a vector identity and is valid for any vector. Let us substitute $\boldsymbol{A}$ by the electric field $\boldsymbol{E}$; then we obtain

$$\int_{abcd} \boldsymbol{E} \cdot d\ell = [(\nabla \times \boldsymbol{E}) \cdot \hat{x}]\,\Delta s$$

The left-hand side of the above equation is the integration of the electric field along a closed loop, and the result is in terms of volts. This is the so-called **electromotive force**, or **emf**. The right-hand side, according to (2.1), is equal to

$$-\frac{\partial}{\partial t}[\boldsymbol{B} \cdot \hat{x}]\,\Delta s$$

which is equal to the rate of decrease of the magnetic flux going through the a-b-c-d loop. Putting them together, we obtain

$$\int_{abcd} \boldsymbol{E} \cdot d\ell = -\frac{\partial}{\partial t}[\boldsymbol{B} \cdot \hat{x}]\,\Delta s \tag{2.14}$$

Equation (2.14) states that the emf induced in a closed loop *a-b-c-d* is equal to the rate of decrease of the total magnetic flux going through the loop. The phenomenon that a changing magnetic flux in a circuit induces voltage in that circuit is called **Faraday induction.** Faraday's law and its applications will be studied in detail in Chapter 16.

We can place the loop *a-b-c-d* on the *x-y* plane or on the *z-x* plane and obtain similar Faraday's induction laws for each case. Therefore, (2.1) is really a succinct statement of three Faraday's induction laws, one for each orientation of the loop in the *x*, *y*, and *z* direction, respectively.

In summary, $\nabla \times \boldsymbol{E}$ has three components. The x component is proportional to the integration of the vector $\boldsymbol{E}$ along a small loop oriented in the x direction; Equation (2.13) shows this relationship. The y component of $\nabla \times \boldsymbol{E}$ is proportional to the integration of the vector $\boldsymbol{E}$ along a small loop oriented in the y direction, and so on for the z component.

Ampère's Law

The second Maxwell's equation involves $\nabla \times \boldsymbol{H}$. If we substitute $\boldsymbol{A}$ in (2.13) by $\boldsymbol{H}$, we obtain

$$\int_{abcd} \boldsymbol{H} \cdot d\ell = [(\nabla \times \boldsymbol{H}) \cdot \hat{x}]\,\Delta s \tag{2.15}$$

For the steady-state case for which the time derivative is zero, (2.2) becomes

$$\nabla \times \boldsymbol{H} = \boldsymbol{J}$$

Substituting the above equation in (2.15) yields

$$\int_{abcd} \boldsymbol{H} \cdot d\ell = [\boldsymbol{J} \cdot \hat{x}]\,\Delta s = \Delta I \tag{2.16}$$

Equation (2.16) states that integration of the magnetic field strength $\boldsymbol{H}$ along a small loop is equal to the total current flowing through that loop. In Chapter 13 we will prove that the above statement is true even for a loop that is not so small. This equation is called **Ampère's law.**

The original Ampère's law was for the steady-state case, as the $\partial \boldsymbol{D}/\partial t$ term in (2.2) did not appear in the equation. It was Maxwell who first proposed the addition of that term. When it is added, (2.16) becomes

$$\int_{abcd} \boldsymbol{H} \cdot d\ell = \left[\boldsymbol{J} \cdot \hat{x} + \frac{\partial}{\partial t} \boldsymbol{D} \cdot \hat{x}\right] \Delta s = \Delta I + \Delta I_{\mathrm{d}} \tag{2.17}$$

where the ΔI_{d} term is called the displacement current flowing through the loop *a-b-c-d*. In the circuit theory, we learned that the current flowing through a capacitor is given by $C\,dV/dt$, where C is the capacitance. This capacitor current is actually the displacement current given in (2.17).

Interpretation of $\nabla \cdot \boldsymbol{A}$

Consider a vector function $\boldsymbol{A}$ that has an $\hat{x}$ component only. From (2.6) we obtain

$$\nabla \cdot \boldsymbol{A} = \frac{\partial A_x}{\partial x}$$

Referring to Figure 2.2 and to the definition of a partial differentiation, we have

$$\frac{\partial A_x}{\partial x} = \frac{(A_x \text{ at right}) - (A_x \text{ at left})}{\Delta x}$$

$$\nabla \cdot \boldsymbol{A} = \frac{(A_x \text{ at right})\,\Delta y\,\Delta z - (A_x \text{ at left})\,\Delta y\,\Delta z}{\text{volume}}$$

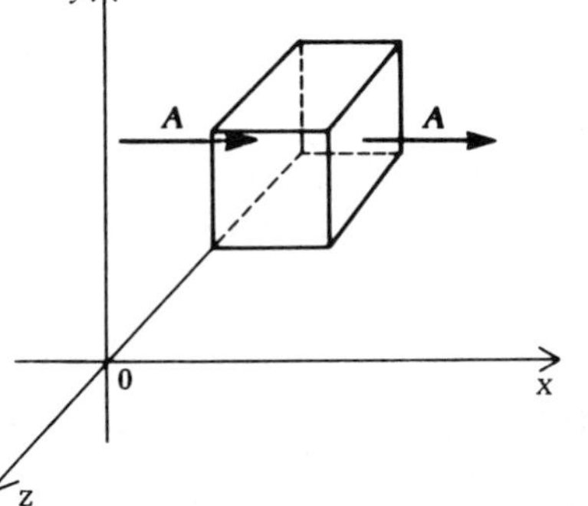

Figure 2.2 Definition of divergence.

The first term in the numerator of the above equation is the flow of $\boldsymbol{A}$ out of a small rectangular block through surface area $\Delta y\,\Delta z$ on the right, and the second term is the flow of $\boldsymbol{A}$ into the block through surface area $\Delta y\,\Delta z$ on the left. Similar results will be obtained for a vector of $\boldsymbol{A}$ that has all three components. Thus, the divergence of a vector $\boldsymbol{A}$ is equal to the net flow of $\boldsymbol{A}$ out of the small block divided by the block's volume. Notice that the block must be a differential volume of infinitesimal size, such that $\Delta x \to 0$, $\Delta y \to 0$, and $\Delta z \to 0$ assuming that $\boldsymbol{A}$ on the differential surfaces is constant.

From the preceding discussion, we see that, in the limit that the rectangular block shown in Figure 2.2 becomes infinitesimally small, the divergence of $\boldsymbol{A}$ may be expressed as

$$\nabla \cdot \boldsymbol{A} = \frac{\int_{\text{block}} \boldsymbol{A} \cdot d\boldsymbol{s}}{\Delta v} \tag{2.18}$$

That is, $\nabla \cdot \boldsymbol{A}$ is equal to the integration of $\boldsymbol{A} \cdot d\boldsymbol{s}$ over the surface of the block, divided by Δv, the volume of that block. Note that the $d\boldsymbol{s}$ vector always points away from the volume. Consequently, $\boldsymbol{A} \cdot d\boldsymbol{s}$ represents the "outflow" of the vector $\boldsymbol{A}$. We can say that $\nabla \cdot \boldsymbol{A}$ represents the net outflow of the vector $\boldsymbol{A}$ per unit volume.

Gauss' Law

Equation (2.3) is the **magnetic Gauss' law**. It states that the total outflow of the magnetic field $\boldsymbol{B}$ from any infinitesimal volume is always equal to zero. This means that the magnetic field can neither originate from any point in space nor terminate at any point. A few magnetic field lines inside and outside a permanent magnet are sketched in Figure 2.3 to illustrate this point. We can see that if we follow a magnetic field line, it will be a continuous loop with no beginning or end.

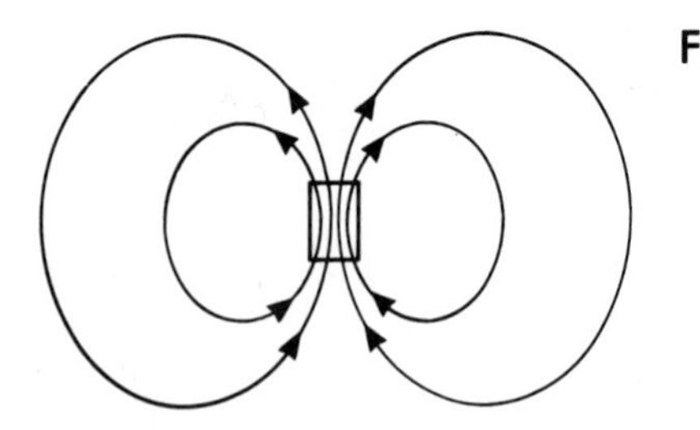

Figure 2.3 Magnetic field lines due to a permanent magnet: $\nabla \cdot \mathbf{B} = 0$ implies that $\mathbf{B}$ lines are continuous with no beginning or end.

Equation (2.18) is a vector identity and is valid for any vector. Substituting $\boldsymbol{D}$ for $\boldsymbol{A}$ in that equation yields

$$\int_{\text{block}} \boldsymbol{D} \cdot d\boldsymbol{s} = [\nabla \cdot \boldsymbol{D}]\, \Delta v = \rho_v\, \Delta v \tag{2.19}$$

where the last equality comes from (2.4).

Equation (2.4), or equivalently (2.19), is the **electric Gauss' law**. It states that the total outflow of the electric field $\boldsymbol{D}$ from any infinitesimal volume is equal to the total electric charge in that volume. Unlike the magnetic field, the electric field can originate from a point in space or terminate at a point. Since the divergence of $\boldsymbol{D}$ represents the outflow of $\boldsymbol{D}$, the $\boldsymbol{D}$ field lines originate from the point where a positive charge is located and terminate at the point where a negative charge is located. Figure 2.4 illustrates some electric field lines near a positive charge and two negative charges.

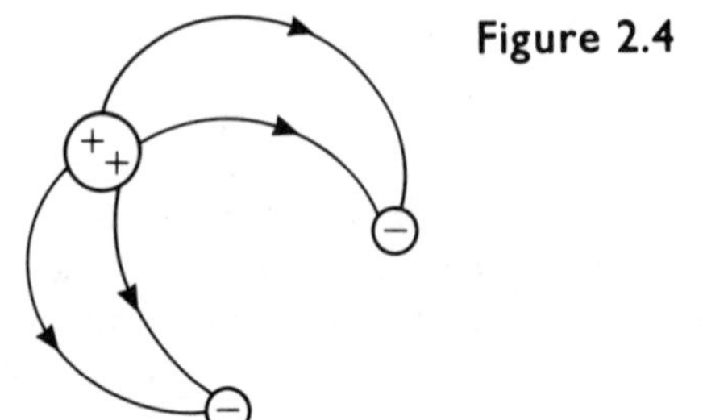

Figure 2.4 Electric field lines due to some electric charges: $\mathbf{E}$ lines originate from the positive charge and end at the negative charges.

Conservation Law

Considering the vector identity (2.11), we dot-multiply both sides of (2.2) by ∇, interchange the time-space derivatives, and obtain

$$\nabla \cdot \boldsymbol{J} + \frac{\partial}{\partial t} \nabla \cdot \boldsymbol{D} = 0$$

Using (2.4), we obtain the continuity law for current and charge densities.

$$\boxed{\nabla \cdot \boldsymbol{J} + \frac{\partial \rho_v}{\partial t} = 0} \quad \text{(law of conservation of electric charge)} \tag{2.20}$$

We recognize that (2.20) is a direct consequence of Maxwell's equations. Considering that all Maxwell's equations for the field vectors, for the current density $\boldsymbol{J}$, and for the charge density, ρ_v are valid at all times and at all places, what is the physical meaning of (2.20)?

If we move the second term to the right-hand side of (2.20), we have

$$\nabla \cdot \boldsymbol{J} = -\frac{\partial \rho_v}{\partial t}$$

The left-hand side of the equation is the "flow of electric current out of a differential volume," and the right-hand side is the "rate of decrease of electric charge in the volume." The equation simply says that the rate of transfer of electric charge out of any differential volume is equal to the rate of decrease of total electric charge in that volume. This statement implies that electric charge is conserved—it can neither be created nor be destroyed. This law is also known as the **law of conservation of electric charge**.

By a similar process, we may dot-multiply (2.1) by ∇ and find

$$\frac{\partial}{\partial t} \nabla \cdot \boldsymbol{B} = 0$$

which states that $\nabla \cdot \boldsymbol{B}$ is a constant and is independent of time. The magnetic Gauss' law (2.3) tells us that this constant is zero.

Static Fields

The electromagnetic fields are generally functions of space and time. In the special case in which they are time-invariant, by setting the time derivative equal to zero, we obtain from (2.1) and (2.4)

$$\nabla \times \boldsymbol{E} = 0 \tag{2.21a}$$

$$\nabla \cdot \boldsymbol{D} = \rho_v \tag{2.21b}$$

Equations (2.21a) and (2.21b) govern the behavior of electrostatic fields $\boldsymbol{E}$ and $\boldsymbol{D}$. The equations governing the behavior of the magnetostatic fields $\boldsymbol{H}$ and $\boldsymbol{B}$ are derived from (2.2) and (2.3) by setting the time derivative equal to zero:

$$\nabla \times \boldsymbol{H} = \boldsymbol{J} \tag{2.22a}$$

$$\nabla \cdot \boldsymbol{B} = 0 \tag{2.22b}$$

Notice that, although time-varying electric and magnetic fields are interrelated because they must satisfy the coupled partial differential equations (2.1)–(2.4), the electric fields $\boldsymbol{E}$ and $\boldsymbol{D}$ and the magnetic fields $\boldsymbol{H}$ and $\boldsymbol{B}$ are decoupled when we set the time derivatives equal to zero. The electrostatic fields satisfy equations (2.21), and the magnetostatic fields satisfy equations (2.22). The electrostatic fields are generated by charge distributions ρ_v, and the magnetostatic fields are generated by current densities $\boldsymbol{J}$. The interdependence between $\boldsymbol{J}$ and ρ_v also breaks down in the static case because the law of conservation of charge (2.20) simply requires that $\nabla \cdot \boldsymbol{J} = 0$. For this reason, it is clear that the existence of the electrostatic fields is independent of the existence of the magnetostatic fields.

Dynamic Electromagnetic Fields

The independence of electric and magnetic fields does not hold for time-varying fields. Consider a region free of currents and charges. In Figure 2.5 we assume that $\boldsymbol{J}$ and ρ_v are localized in a limited region V_0 and that the space outside V_0, is source-free. In the source-free region, Maxwell's equations are

$$\nabla \times \boldsymbol{E} = -\frac{\partial \boldsymbol{B}}{\partial t} \tag{2.23a}$$

$$\nabla \times \boldsymbol{H} = \frac{\partial \boldsymbol{D}}{\partial t} \tag{2.23b}$$

$$\nabla \cdot \boldsymbol{B} = 0 \tag{2.23c}$$

$$\nabla \cdot \boldsymbol{D} = 0 \tag{2.23d}$$

We see that in the source-free region V a time-varying magnetic field is always accompanied by an electric field and that a time-varying $\boldsymbol{D}$ field is always accompanied by a magnetic field $\boldsymbol{H}$. For example, a radio antenna generates radio waves that consist of both electric and magnetic fields.

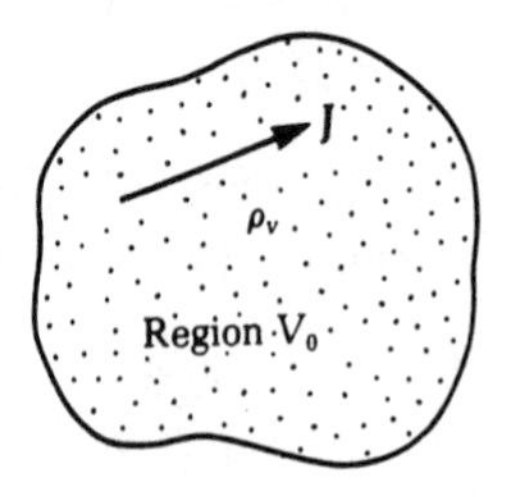

Figure 2.5 Region V_0 contains sources ρ_v and J, whereas region V is source-free.

Example 2.1 *This example shows how to carry out ∇ operations on some specific functions.*

Let $\boldsymbol{A} = \hat{y} \sin x$; $\phi = \sin(x)\sin(2y)$; find $\nabla \times \boldsymbol{A}$, $\nabla \cdot \boldsymbol{A}$, and $\nabla\phi$.

Solution

$$\nabla \times \boldsymbol{A} = \hat{z} \cos x$$

$$\nabla \cdot \boldsymbol{A} = 0$$

$$\nabla\phi = \hat{x} \cos(x)\sin(2y) + \hat{y}2 \sin(x)\cos(2y)$$

Example 2.2 *Remember that **E** and **B** fields are, in general, related. When one is given, the other can usually be found.*

If $\boldsymbol{E} = \hat{x} \cos(x)$, is $\boldsymbol{B}$ varying with time?

Solution From (2.1)

$$-\frac{\partial \boldsymbol{B}}{\partial t} = \nabla \times \boldsymbol{E} = 0$$

Therefore, $\boldsymbol{B}$ is not varying with time.

Example 2.3 *Remember that Maxwell's equations have to be satisfied everywhere at all times.*

Can the following $\boldsymbol{B}$ field exist in the region defined by $0 \le x \le 1$ and by $0 \le y \le 1$? $\boldsymbol{B} = \hat{x} \sin(x) + \hat{y} \sin(y)$.

Solution Because $\nabla \cdot \boldsymbol{B} = \cos(x) + \cos(y)$, which is not identically zero in the region, it violates Maxwell's equation (2.3) and hence cannot actually exist.

Example 2.4 *It is said that the electrostatic fields are generated by electric charges. It is possible, however, that **D** field is not zero in a region but $\rho_v = 0$. There is no contradiction in these statements because the field may be produced by charges on the boundary surface of the region. The **volume** charge density ρ_v does not include the **surface** charge density.*

In a region where $\boldsymbol{D} = \cos(x)\sinh(y)\hat{x} + \sin(x)\cosh(y)\hat{y}$ what is the charge density ρ_v?

Solution According to (2.4),

$$\rho_v = \nabla \cdot \boldsymbol{D} = -\sin(x)\sinh(y) + \sin(x)\sinh(y) = 0$$

2.2 CONSTITUTIVE RELATIONS

In solving the electromagnetic-field problems, we assume that the sources $\boldsymbol{J}$ and ρ_v are given and that they satisfy the continuity equation (2.20):

$$\nabla \cdot \boldsymbol{J} + \frac{\partial \rho_v}{\partial t} = 0$$

Before we begin solving Maxwell's equations with the given source distributions, we must first ask the following two questions:

(a) How many scalar equations are there in Maxwell's equations?

There are three scalar equations for each curl equation and one for each divergence equation. However, the divergence equations are derivable from the continuity equation and the curl equations. Taking the divergence of (2.2) and using the vector identity $\nabla \cdot (\nabla \times \boldsymbol{H}) = 0$, we find

$$\frac{\partial}{\partial t}(\nabla \cdot \boldsymbol{D}) + \nabla \cdot \boldsymbol{J} = 0$$

Considering (2.20), we see that

$$\frac{\partial}{\partial t}(\nabla \cdot \boldsymbol{D} - \rho_v) = 0$$

Thus, (2.4) is not independent of (2.2). Similarly, after interchanging time and space derivatives, taking the divergence of (2.1) yields

$$\frac{\partial}{\partial t} \nabla \cdot \boldsymbol{B} = 0$$

Thus, the divergence of $\boldsymbol{B}$ is a constant and is independent of time. The constant is always zero. Equation (2.3) is then a consequence of (2.1). Thus, we have a total of six independent equations.

(b) How many independent scalar-field variables are there in Maxwell's equations?

There are 12, one for each component of $\boldsymbol{E}$, $\boldsymbol{H}$, $\boldsymbol{D}$, and $\boldsymbol{B}$. Therefore, we see that the set of Maxwell's equations is not sufficient to solve for the 12 unknowns. We need six more scalar equations, called the constitutive relations.

Physically, the constitutive relations provide information about the environment in which electromagnetic fields occur—for example, free space, water, or plasma media. Mathematically, we may characterize as follows a simple medium with a **permittivity** ϵ and a **permeability** μ:

$$\boxed{D = \epsilon E} \quad \textbf{(definition of permittivity)} \tag{2.24}$$

$$\boxed{B = \mu H} \quad \textbf{(definition of permeability)} \tag{2.25}$$

Equations (2.24) and (2.25) are the constitutive relations for the simple medium. For free space, $\mu = \mu_0 = 4\pi \times 10^{-7}$ H/m, and $\epsilon = \epsilon_0 = 8.85 \times 10^{-12}$ F/m. In general, ϵ and μ can be functions of many parameters. When ϵ or μ is a function of the frequency, we call the medium dispersive.

Water is an example of a dispersive medium. The H_2O molecule tends to line up with the $\boldsymbol{E}$ field. Quantitatively, the ability of the water molecules to line up with the $\boldsymbol{E}$ field is described by a dimensionless electric **susceptibility** χ_e. The $\boldsymbol{D}$ field is then composed of a free-space part $\epsilon_0 \boldsymbol{E}$ and a material part characterized by a dipole moment density $\boldsymbol{P}$, which is related to χ_e and $\boldsymbol{E}$ by

$$\boldsymbol{P} = \chi_e \epsilon_0 \boldsymbol{E} \tag{2.26a}$$

The dipole moment density P induced by the $\boldsymbol{E}$ field may be thought of as composed of many individual dipoles:*

$$P = Nqd \tag{2.26b}$$

where N is the number of dipoles per unit volume and d is the distance between the $-q$ and the $+q$ charges. The direction of $\boldsymbol{P}$ points from the negative charge to the positive charge. The $\boldsymbol{D}$ field is the direct sum of $\epsilon_0 \boldsymbol{E}$, and $\boldsymbol{P}$:

$$\boldsymbol{D} = \epsilon_0 \boldsymbol{E} + \boldsymbol{P} = \epsilon_0(1 + \chi_e)\boldsymbol{E} \tag{2.26c}$$

and thus

$$\epsilon = (1 + \chi_e)\epsilon_0 \tag{2.26d}$$

At low frequencies, the H_2O molecules can align with the slowly varying field relatively easily, and χ_e is large. At high frequencies, their response to the rapidly varying field becomes more difficult, and χ_e is small. At very low frequencies, liquid water has $\epsilon \sim 80\epsilon_0$, whereas at optical frequencies $\epsilon \sim 2\epsilon_0$. As we shall discuss in Section 2.4, another example of a dispersive medium is an electron plasma.

---

* A dipole is a pair of positive and negative charges of equal magnitude separated by a small distance d. Section 10.1 discusses dipoles further.

2.3 MAXWELL'S EQUATIONS FOR TIME-HARMONIC FIELDS

The electromagnetic-field vectors and the electric-current and electric-charge densities discussed in the previous section are real functions of space and time. In this section, we shall be concerned with time-harmonic fields at a **single frequency**. We assume that all field vectors and all currents and charge densities vary cosinusoidally with time at a single angular frequency ω. Then, we can write, for example, the $\hat{x}$ component of the real $\boldsymbol{E}$ vector E_x in the following form:

$$E_x(x, y, z, t) = E_1(x, y, z)\cos(\omega t + \phi_1) = \text{Re}\{E_1(x, y, z)\, e^{j\phi_1}\, e^{j\omega t}\} \qquad \textbf{(2.27a)}$$

Now let

$$\mathrm{E}_x(x, y, z) = E_1(x, y, z)\, e^{j\phi_1} \qquad \textbf{(2.27b)}$$

The complex function $\mathrm{E}_x(x, y, z)$, together with the frequency ω, contains all necessary information about the original function $E_x(x, y, z, t)$. As discussed in detail in Chapter 1, we call $\mathrm{E}_x(x, y, z)$ the complex representation of $E_x(x, y, z, t)$.

The same reasoning applies to the $\hat{y}$ and $\hat{z}$ components of the $\mathbf{E}$ vector. In summary, we can write

$$\boldsymbol{E}(x, y, z, t) \leftrightarrow \mathbf{E}(x, y, z)$$

The $\mathbf{E}$ vector is now a complex vector, each component of which possesses both real and imaginary parts. Similar expressions apply to all other field vectors. As discussed in Chapter 1, the time derivatives can be represented by $j\omega$. Thus,

$$\frac{\partial}{\partial t}\boldsymbol{B}(x, y, z, t) \leftrightarrow j\omega\mathbf{B}(x, y, z)$$

and similarly for $\partial \boldsymbol{D}/\partial t$. With these complex notations for the time-harmonic quantities, Maxwell's equations become

$$\nabla \times \mathbf{E} = -j\omega\mathbf{B} \qquad \textbf{(2.28a)}$$

$$\nabla \times \mathbf{H} = j\omega\mathbf{D} + \mathbf{J} \qquad \textbf{(2.28b)}$$

$$\nabla \cdot \mathbf{B} = 0 \qquad \textbf{(2.28c)}$$

$$\nabla \cdot \mathbf{D} = \rho_v \qquad \textbf{(2.28d)}$$

(Maxwell's equations for time-harmonic fields)

Notice that all field vectors are now complex quantities independent of time. To obtain the real space-time expression for a quantity given in the complex notation, we merely have to multiply the complex quantity by $e^{j\omega t}$ and then take the real part. In the time-harmonic case, the equation of conservation of charge (2.20) becomes $\nabla \cdot \mathbf{J} + j\omega\rho_v = 0$. It can be put in another form:

$$\nabla \cdot \mathbf{J} = -j\omega\rho_v \tag{2.29}$$

The physical quantity that the left-hand side of the equation represents is the total time-harmonic current flowing out of a unit volume, and the term ρ_v at the right-hand side represents the amount of time-harmonic charge in that volume.

2.4 LORENTZ FORCE LAW

The Lorentz force law* specifies the force acting on a particle with charge q when it is moving with velocity $\boldsymbol{v}$ in an electromagnetic field characterized by $\boldsymbol{E}$ and $\boldsymbol{B}$:

$$\boxed{\boldsymbol{F} = q(\boldsymbol{E} + \boldsymbol{v} \times \boldsymbol{B})} \quad \textbf{(Lorentz force law)} \tag{2.30}$$

The unit for the force $\boldsymbol{F}$ is newton, and the unit for the charge q is coulomb. The Lorentz force law is the link of electromagnetic fields to mechanics because Newton's laws of mechanics also govern the force. Because charge particles such as electrons and positively charged nuclei are building blocks of material media, we can also regard the Lorentz force law as describing the action of the field on media. The action of media on the field is described by Maxwell's equations and the constitutive relations because the former govern the behavior of the electromagnetic fields $\boldsymbol{E}$, $\boldsymbol{B}$, $\boldsymbol{D}$, and $\boldsymbol{H}$, and the latter prescribe the media by relating $\boldsymbol{D}$ to $\boldsymbol{E}$ and $\boldsymbol{B}$ to $\boldsymbol{H}$.

In the first eight chapters of this book, we discuss electromagnetic-wave phenomena that do not require the use of the Lorentz force law. However, in the following example, we shall make use of this law to find the constitutive relations of a plasma medium.

Plasma

Plasma is an ionized gas consisting of positively charged ions and negatively charged electrons that are both free to move. It is quite different from un-ionized gas, in which charged particles are bound together and in which only a limited displacement between the charged particles can occur under the influence of the external $\boldsymbol{E}$ field. In plasma, external excitations free electrons from the positive nuclei of gas molecules. Plasma exists naturally in what we call the **ionosphere**. The ionosphere is a region 80 to 120 km above the surface of the earth. It is where ultraviolet light from the sun ionizes air molecules.

---

* In honor of Hendrik Antoon Lorentz (1853–1928) of The Netherlands.

When an electric field is applied to plasma, the charged particles move according to Newton's law. Because the mass of the electron is much smaller than that of the ion, we can ignore motions of ions. For a typical electron in a time-harmonic **E** field,

$$\mathbf{F} = -e\mathbf{E}$$

where $e = 1.6 \times 10^{-19}$ C. Let $\mathbf{x}$ be the displacement of the electron away from the positive ion. Newton's second law states that

$$\mathbf{F} = m\frac{d^2\mathbf{x}}{dt^2} = -m\omega^2\mathbf{x}$$

where $m = 9.1 \times 10^{-31}$ kg is the mass of the electron and ω is the angular frequency of the time-harmonic **E** field. According to (2.26b), the dipole moment density **P** is as follows:

$$\mathbf{P} = -Ne\mathbf{x}$$

where N is the number of electrons in the plasma per unit volume. Eliminating $\mathbf{x}$ from the above equations, we obtain

$$\mathbf{P} = \frac{-Ne^2}{m\omega^2}\mathbf{E}$$

Because **D** is composed of the free-space part $\epsilon_0\mathbf{E}$ and the material part **P**, we find

$$\mathbf{D} = \epsilon_0\mathbf{E} + \mathbf{P} = \epsilon_0\mathbf{E} - \frac{Ne^2}{m\omega^2}\mathbf{E} = \epsilon_0\left\{1 - \frac{\omega_p^2}{\omega^2}\right\}\mathbf{E} \tag{2.31a}$$

where

$$\omega_p = \sqrt{\frac{Ne^2}{m\epsilon_0}} \tag{2.31b}$$

is called the **plasma frequency**. Thus, the plasma can be modeled as a dielectric material with permittivity given by

$$\epsilon = \epsilon_0\left\{1 - \frac{\omega_p^2}{\omega^2}\right\} \tag{2.32}$$

Electron density in the ionosphere varies with height and with the time of day. The typical daytime ionosphere contains 10^{12} electrons/m^3, corresponding to $\omega_p = 5.64 \times 10^7$ rad/s or $f_p = 9$ MHz. Thus for electromagnetic fields of frequencies much higher than 9 MHz, the ionosphere is not much different from free space. However, for those of lower frequencies, the ionosphere is a medium with negative permittivity. We shall study the consequences of this negative permittivity in Chapter 4.

2.5 POYNTING'S THEOREM

In Section 2.4, we learned the relation between electromagnetic fields and physical force. Through the Lorentz force law, we can determine how much force a charged particle will experience when it enters a region where electromagnetic fields exist. In this section, we shall establish a relation between electromagnetic fields and another familiar quantity: energy or power.

The equation that will lead to a power equation may be derived from (2.1) and (2.2) by dot-multiplying (2.2) with $\boldsymbol{E}$ and then subtracting it from (2.1) after the latter is dot-multiplied with $\boldsymbol{H}$. Using the vector identity $\boldsymbol{H} \cdot \nabla \times \boldsymbol{E} - \boldsymbol{E} \cdot \nabla \times \boldsymbol{H} = \nabla \cdot (\boldsymbol{E} \times \boldsymbol{H})$, we find

$$\nabla \cdot (\boldsymbol{E} \times \boldsymbol{H}) = -\boldsymbol{H} \cdot \frac{\partial \boldsymbol{B}}{\partial t} - \boldsymbol{E} \cdot \frac{\partial \boldsymbol{D}}{\partial t} - \boldsymbol{J} \cdot \boldsymbol{E}$$

Using the constitutive relations $\boldsymbol{B} = \mu \boldsymbol{H}$ and $\boldsymbol{D} = \epsilon \boldsymbol{E}$ and noticing that $\boldsymbol{H} \cdot (\partial \boldsymbol{H}/\partial t) = (\partial/\partial t)[(1/2)\boldsymbol{H} \cdot \boldsymbol{H}]$, we can write

$$\nabla \cdot (\boldsymbol{E} \times \boldsymbol{H}) + \frac{\partial}{\partial t}\left(\frac{1}{2}\mu \boldsymbol{H} \cdot \boldsymbol{H}\right) + \frac{\partial}{\partial t}\left(\frac{1}{2}\epsilon \boldsymbol{E} \cdot \boldsymbol{E}\right) + \boldsymbol{J} \cdot \boldsymbol{E} = 0 \qquad \textbf{(2.33)}$$

Note that each term in the above equation has the unit watts/m^3. The term $\boldsymbol{J} \cdot \boldsymbol{E}$ is easily recognized as the ohmic loss because $\boldsymbol{J}$ is the current density in A/m^2 and $\boldsymbol{E}$ is the electric field in V/m. $\boldsymbol{J} \cdot \boldsymbol{E}$ is the power lost per unit volume by the electric field to heat.

The term $\frac{1}{2}\epsilon \boldsymbol{E} \cdot \boldsymbol{E}$ is identified as the stored electric energy per unit volume:

$$\boxed{U_E = \tfrac{1}{2}\epsilon \boldsymbol{E} \cdot \boldsymbol{E}} \quad \textbf{(stored electric energy)} \qquad \textbf{(2.34)}$$

The term $\frac{1}{2}\mu \boldsymbol{H} \cdot \boldsymbol{H}$ is identified as the stored magnetic energy per unit volume:

$$\boxed{U_H = \tfrac{1}{2}\mu \boldsymbol{H} \cdot \boldsymbol{H}} \quad \textbf{(stored magnetic energy)} \qquad \textbf{(2.35)}$$

Let us define the **Poynting vector*** $\boldsymbol{S}$ where

$$\boxed{\boldsymbol{S} = \boldsymbol{E} \times \boldsymbol{H}} \quad \textbf{(Poynting vector)} \qquad \textbf{(2.36)}$$

The unit of $\boldsymbol{S}$ is watts/m^2, which is power density. The first term in (2.33) is $\nabla \cdot \boldsymbol{S}$. Remembering that the divergence of a vector represents outflow of the vector across a small volume, we identify $\nabla \cdot \boldsymbol{S}$ as the outflow of electromagnetic power.

---

* In honor of the British physicist John Henry Poynting (1852–1914).

With the preceding terms identified, (2.33) can be put in words:

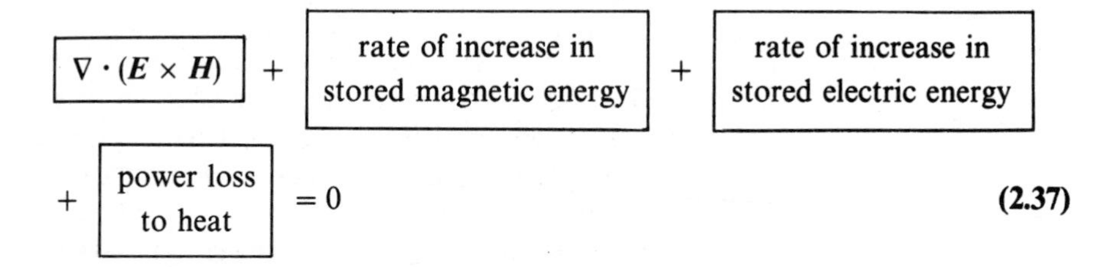

(2.37)

Equation (2.37) is a statement of conservation of power. It states that the sum of electromagnetic power flowing out of a volume, the rate of increase of stored magnetic energy and electric energy in that volume, and the power lost to ohmic heat must equal to zero.

We have learned two things from the preceding derivation. First, the Poynting vector $\boldsymbol{S} = \boldsymbol{E} \times \boldsymbol{H}$ represents flow of electromagnetic power per unit area. Second, Maxwell's equations are consistent with the law of conservation of energy.

For time-harmonic electromagnetic fields, the **time-average Poynting's vector** $\langle \boldsymbol{S} \rangle$ is defined as the average of the time-domain Poynting vector $\boldsymbol{S}(x, y, z, t)$ over a period $T = 2\pi/\omega$.

$$\langle \boldsymbol{S} \rangle = \frac{1}{2\pi} \int_0^{2\pi} d(\omega t) \boldsymbol{E}(x, y, z, t) \times \boldsymbol{H}(x, y, z, t)$$

From Section 1.5, we conclude that

$$\boxed{\langle \boldsymbol{S} \rangle = \tfrac{1}{2} \operatorname{Re}\{\mathbf{E} \times \mathbf{H}^*\}} \quad \textbf{(time-average Poynting vector)} \qquad \textbf{(2.38)}$$

where $\mathbf{E}$ and $\mathbf{H}$ are the complex counterparts of the $\boldsymbol{E}$ and $\boldsymbol{H}$ fields.

This formula is useful for computing time-average electromagnetic power flow.

Example 2.5 *This example shows that, if the electric field is known, then the corresponding magnetic field is determined by Maxwell's equations.*

If $\mathbf{E} = \hat{x} E_0 e^{-jkz}$ in a region, find the corresponding $\mathbf{B}$. (k is a constant.)

Solution According to (2.28a),

$$\mathbf{B} = \frac{1}{-j\omega} \nabla \times \mathbf{E} = \hat{y} E_0 \frac{k}{\omega} e^{-jkz}$$

Note that this $\mathbf{B}$ field satisfies (2.28c).

Example 2.6 *This example shows how to obtain explicit expressions of the time-domain electromagnetic fields, the Poynting vector, the time-average Poynting vector, the stored electric and magnetic energy, and the time average of the stored electric and magnetic energy.*

With **E** given in the previous example, find $\boldsymbol{E}(t)$, $\boldsymbol{H}(t)$, $\boldsymbol{S}(t)$, and $\langle \boldsymbol{S} \rangle$ assuming that both μ and ϵ are real constants. Also find the electric energy U_E, the magnetic energy U_H, and their time-average values.

Solution

$$\boldsymbol{E}(t) = \text{Re}\{\hat{x}E_0 e^{-jkz} e^{j\omega t}\} = \hat{x}E_0 \cos(\omega t - kz)$$

$$\boldsymbol{H}(t) = \text{Re}\left\{\hat{y}\frac{k}{\omega\mu} E_0 e^{-jkz} e^{j\omega t}\right\} = \hat{y}\frac{k}{\omega\mu} E_0 \cos(\omega t - kz)$$

$$\boldsymbol{S}(t) = \boldsymbol{E} \times \boldsymbol{H} = \hat{z}\frac{k}{\omega\mu} E_0^2 \cos^2(\omega t - kz)$$

$$\langle \boldsymbol{S} \rangle = \frac{1}{2}\text{Re}\{\mathbf{E} \times \mathbf{H}^*\} = \hat{z}\frac{k}{2\omega\mu} E_0^2$$

$$U_E = \frac{\epsilon E_0^2}{2} \cos^2(\omega t - kz)$$

$$U_H = \frac{k^2 E_0^2}{2\omega^2\mu} \cos^2(\omega t - kz)$$

$$\langle U_E \rangle = \frac{\epsilon}{4} E_0^2$$

$$\langle U_H \rangle = \frac{k^2}{4\omega^2\mu} E_0^2$$

SUMMARY

1. The electric fields ($\boldsymbol{E}$ and $\boldsymbol{D}$) and the magnetic fields ($\boldsymbol{B}$ and $\boldsymbol{H}$) satisfy a set of equations called Maxwell's equations.
2. Maxwell's equations are a collection of experimentally obtained laws that have been tested thoroughly and found to be true everywhere at all times.
3. Maxwell's equations involve the partial differential operators curl ($\nabla \times$) and divergence ($\nabla \cdot$). Interpretation of these operators were discussed.
4. The electric fields are related by the constitutive relation $\boldsymbol{D} = \epsilon \boldsymbol{E}$, where ϵ is called the permittivity of the medium.
5. The magnetic fields are related by the constitutive relation $\boldsymbol{B} = \mu \boldsymbol{H}$, where μ is called the permeability of the medium.
6. The Lorentz force law gives the force acting on a charged particle in electromagnetic fields.

7. The flow of electromagnetic power is determined by the cross product of $\boldsymbol{E}$ and $\boldsymbol{H}$, or the Poynting vector.
8. The stored electric energy, the stored magnetic energy, and the ohmic loss were given.

Problems

2.1 Let $A = 5\hat{x} + 6yz\hat{y} + x^3\hat{z}$; find $\nabla \times A$ and $\nabla \cdot A$.

2.2 Let $\phi = xyz$; find $\nabla\phi$ and $\nabla \cdot \nabla\phi$.

2.3 Let $\boldsymbol{a} = a_1\hat{x} + a_2\hat{y} + a_3\hat{z}$ and $\boldsymbol{b} = b_1\hat{x} + b_2\hat{y} + b_3\hat{z}$. Show that Equations (2.9), (2.11a), and (2.12) are true.

2.4 Show that $\nabla \times (\boldsymbol{a} + \boldsymbol{b}) = \nabla \times \boldsymbol{a} + \nabla \times \boldsymbol{b}$ and $\nabla \cdot (\boldsymbol{a} + \boldsymbol{b}) = \nabla \cdot \boldsymbol{a} + \nabla \cdot \boldsymbol{b}$.

2.5 Show that $\nabla(\Phi_1\Phi_2) = \Phi_1\nabla\Phi_2 + \Phi_2\nabla\Phi_1$ and that $\nabla \cdot (\Phi A) = A \cdot \nabla\Phi + \Phi\nabla \cdot A$.

2.6 Show that $\nabla \times (\Phi A) = \nabla\Phi \times A + \Phi\nabla \times A$.

2.7
(a) It is known that the vector $\boldsymbol{a}$ is equal to zero at one point. Does that imply that $\nabla \times \boldsymbol{a} = 0$ at that point? Give a counterexample if your answer is no.
(b) Does $\boldsymbol{E} = 0$ on a line always imply $\nabla \times \boldsymbol{E} = 0$ on that line? Give a counterexample if the answer is no.
(c) It is found that the $\boldsymbol{E}$ field is zero on a surface. Does it follow that $\partial\boldsymbol{B}/\partial t = 0$ on that surface?

2.8 If $\nabla \cdot A = \nabla \cdot B$, is $A = B$? If yes, explain briefly. If no, give a counterexample.

2.9 If $\nabla \times A = \nabla \times B$, is $A = B$? If yes, explain briefly. If no, give a counterexample.

2.10 The vector A on the front surface of a small cube is equal to $8\hat{x}$, as shown in Figure P2.10.

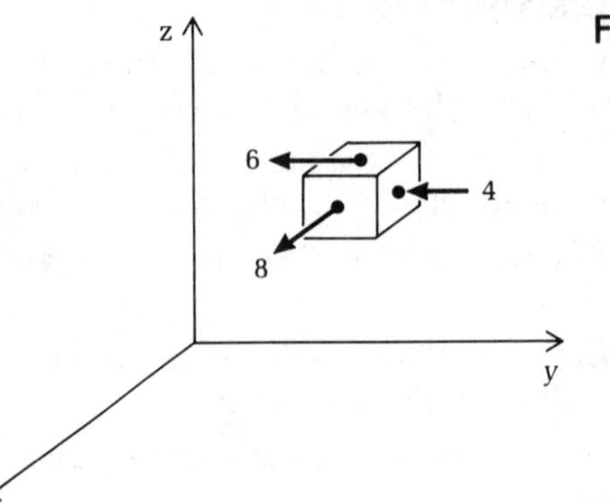

Figure P2.10

Its direction and magnitude on two other surfaces are also indicated. The vector $\boldsymbol{A}$ is equal to zero on the remaining three surfaces of the cube. Assume that the volume of the cube is unity; find $\nabla \cdot \boldsymbol{A}$.

2.11 The direction and magnitude of the vector $\boldsymbol{A}$ on the four sides of a small square are indicated in Figure P2.11. Assume that the area of the square is unity; find the z component of $\nabla \times \boldsymbol{A}$.

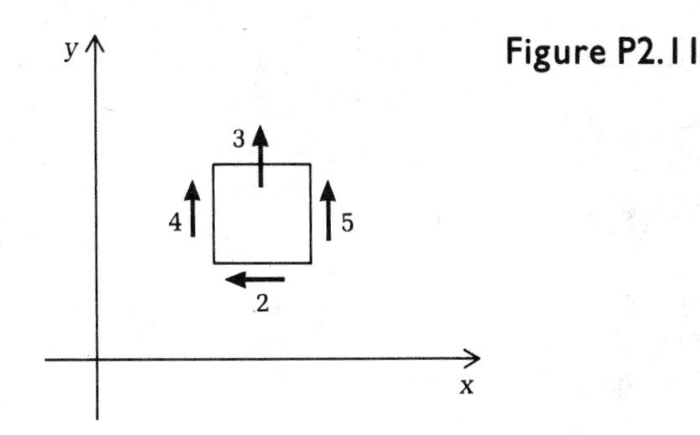

Figure P2.11

2.12 In a source-free region, $\boldsymbol{H} = z\hat{y} + y\hat{z}$. Does $\boldsymbol{D}$ vary with time?

2.13 What is the charge density in a region where $\boldsymbol{D} = 2x\hat{x}$?

2.14 Find the magnetic field $\boldsymbol{B}(y, t)$ associated with the electric field $\boldsymbol{E}(y, t)$ given as follows:

$$\boldsymbol{E}(y, t) = \hat{x}0.3 \cos(\omega t + ky)$$

where ω and k are constants.

2.15 Express k in terms of the magnetic permeability and dielectric permittivity of the medium when the electromagnetic fields are given in Problem 2.14 in a source-free region.

2.16 Let $\boldsymbol{E}_1$, $\boldsymbol{B}_1$, $\boldsymbol{H}_1$, and $\boldsymbol{D}_1$ satisfy equations (2.1)–(2.4) with given $\boldsymbol{J}_1$ and ρ_{v1}. Let also $\boldsymbol{E}_2$, $\boldsymbol{B}_2$, $\boldsymbol{H}_2$, and $\boldsymbol{D}_2$ satisfy equations (2.1)–(2.4) with given $\boldsymbol{J}_2$ and ρ_{v2}. What are the electromagnetic fields due to a current $\boldsymbol{J}_t$ and charge ρ_{vt} where $\boldsymbol{J}_t = \boldsymbol{J}_1 + \boldsymbol{J}_2$ and $\rho_{vt} = \rho_{v1} + \rho_{v2}$? You must show that your proposed solution satisfies Maxwell's equations. What is the appropriate name for the theorem you have just proved?

2.17 Show that equations (2.28c) and (2.28d) can be derived from equations (2.28a), (2.28b), and the conservation equation (2.29).

2.18 To represent time-harmonic fields, most physics books use the factor $e^{-i\omega t}$ instead of $e^{j\omega t}$, which most electrical engineering books use. For a time-harmonic real function $A(x, y, z, t) = a(x, y, z) \cos(\omega t + \phi)$, find the phasor notation that corresponds to the physicists' convention. What is the corresponding conversion rule by which phasors can be transformed back to the real-time expression?

2.19 Refer to Problem 2.18 about the notation $e^{-i\omega t}$ adopted in most physics books. Write the time-harmonic Maxwell's equations using that notation.

2.20 What is the range of effective permittivity of the ionosphere at AM broadcasting frequencies? Use the following data: $N = 10^{12}\ \mathrm{m}^{-3}$ and $f = 500$ kHz to 1 MHz.

2.21 Show that the dimension of each term of equation (2.33) is watts per cubic meter.

2.22 Indicate in watts, meters, and joules the dimensions of the following quantities: (a) $\boldsymbol{E} \cdot \boldsymbol{D}$, (b) $\boldsymbol{H} \cdot \boldsymbol{B}$, and (c) $\boldsymbol{S}$.

2.23 Let $\mathbf{E} = (\hat{x} + j\hat{y})e^{-jz}$ and $\mathbf{H} = (\hat{y} - j\hat{x})e^{i-jz}$. Find $\boldsymbol{S}$ in terms of z and ωt and find $\langle \boldsymbol{S} \rangle$.

2.24 Show that $\boldsymbol{S} \neq \mathrm{Re}\{\mathbf{E} \times \mathbf{H}\, e^{j\omega t}\}$.

2.25 Show that $\boldsymbol{S} \neq \mathrm{Re}\{\mathbf{E}\, e^{j\omega t} \times \mathbf{H}\, e^{j\omega t}\}$.

2.26 Compare the energy stored in a cubic region one meter on a side which has a uniform $\boldsymbol{E}$ field of 10^4 V/m to the energy stored in a similar region with a uniform $\boldsymbol{B}$ field of 10^4 G. (One G $= 10^{-4}\ \mathrm{Wb/m^2}$). The medium is air.

2.27 Repeat Problem 2.26 for the case where the medium is water instead of air. Use $\epsilon = 80\epsilon_0$ and $\mu = \mu_0$ for water.

3 UNIFORM PLANE WAVES

The known spectrum of electromagnetic waves covers a wide range of frequencies. Radio waves, television signals, radar beams, visible light, X rays, and gamma rays are examples of electromagnetic waves. In free space, these waves all propagate with the same velocity, 3×10^8 m/s. They obey the governing laws described by Maxwell's equations. Figure 3.1 shows the spectrum of the electromagnetic wave.

For convenience, the spectrum is divided into regions. These divisions are not intended to be precise, and each region is also subdivided. For instance, in the visible range, the spectrum is further divided into red, orange, yellow, green, blue, purple, and violet, in order of increasing frequency.

The previous chapter introduced Maxwell's equations and constitutive equations. In this chapter, we shall use these equations to study the behavior of electromagnetic plane waves. The plane wave is an idealization of most electromagnetic waves, and it characterizes almost all the important properties of the waves generated from most electromagnetic sources. In addition to studying plane waves in free space, we shall also investigate their behavior in dissipative media.

3.1 ELECTROMAGNETIC SOURCES

Electromagnetic sources are everywhere. The sun, the stars, lightning, and tornados are natural sources of electromagnetic waves. In fact, every terrestrial object reflects part of the sun's electromagnetic radiation, absorbs part of it, and then continuously reradiates some of it. There are also artificial electromagnetic sources such as nuclear explosions, electronic circuits with vacuum tubes or transistors, microwave diodes, lasers, masers, magnetrons in microwave ovens, radio antennas, and so on. In the following paragraphs we discuss some natural and some man-made electromagnetic sources. Chapter 7 will investigate in detail the radiation characteristics of various antennas.

Radio Waves from the Sun

Table 1 lists the power spectrum of the solar radiation the earth receives in the 50 MHz–50 GHz range.* The data gives the average quantity received from the "quiet sun." During a solar "flare," the power density at some frequencies may be 1200 times higher!

---

* G. Abetti, *Solar Research*, New York: Macmillan, 1963, p. 131.

Figure 3.1 Electromagnetic spectrum.

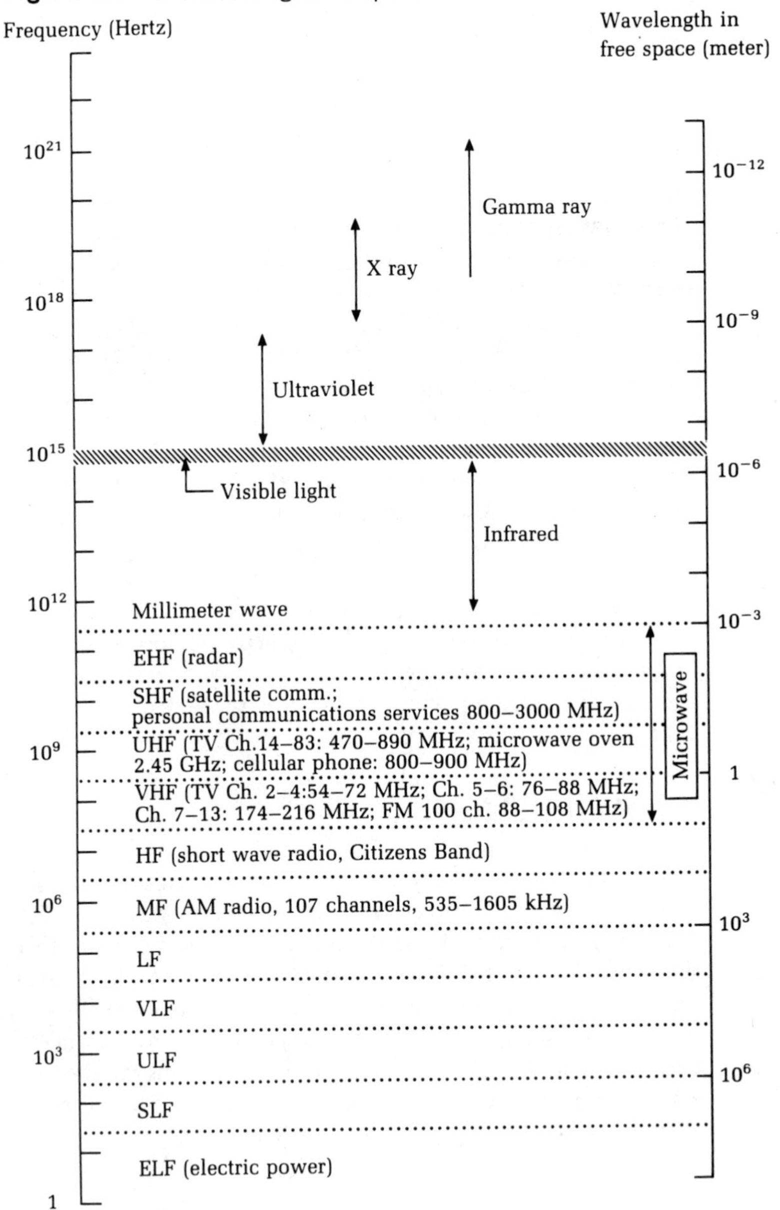

Table 1 Solar Radiation Measured on Earth.

Frequency	Power density
	(10^{-22} W/m²-Hz)
50 MHz	0.6
1 GHz	35.0
50 GHz	3400.0

Solar Energy

The total power density of the solar radiation measured on the earth—that is, the power density of the entire frequency spectrum, is approximately 1.4 kW/m^2. The radius of the earth is 6367 km. Thus,

$$P = \pi r^2 \times 1.4 \times 10^3 = 1.78 \times 10^{11} \text{ MW}$$

The total solar energy incident on the earth is

$$\text{Energy} = 1.78 \times 10^{17} \times 24 \times 60 \times 60 = 1.54 \times 10^{22} \text{ joules per day}$$

In comparison, the world total primary energy output (oil, gas, coal, hydroelectric power, and nuclear power) in 1990 was 1.0×10^{18} joules per day.

Extraterrestrial Sources

Life on the earth depends on energy from the sun received through radiation. The sun radiates a wide spectrum of electromagnetic energy from gamma and X rays through ultraviolet and visible light to infrared and radio waves. The observable stars are sources of electromagnetic waves in the visible (light) region of the spectrum. Many other stars cannot be observed by conventional telescopes. **Radio stars** are heavenly bodies that emit primarily radio waves. In 1967, scientists discovered regular radio pulses coming from deep in the universe. Speculation was rife that some civilization outside our solar system might be trying to send radio signals to communicate with earthlings. However, these radio pulses were found to originate from spinning radio stars that are now called **pulsars** (pulsating stars). Radio waves from pulsars were once called LGM signals, that is, signals from the Little Green Man. **Quasars**, an acronym for "quasi-stellar radio sources," were discovered in 1963. They are radio objects with tremendous radiation power.

Remote Sensing

There are two methods for remote sensing of the earth with microwaves; active and passive.

Active microwave remote sensing uses a transmitter and a receiver. This pair is installed in aircraft or spacecraft. The microwave transmitting antenna emits microwaves and directs them toward a target on the ground, and the receiver acquires the signal reflected from the ground target. Because different surfaces have different reflection characteristics, the nature of the target—for example, rock, soil, calm ocean, or turbulent ocean—can be determined by processing the recorded data.

The passive remote sensing devices take advantage of the fact that different surfaces emit different spectra of electromagnetic waves, including microwaves. A receiver is installed on the sensing aircraft or spacecraft. Focused on the earth's surface, the antenna can record the intensity and the frequency of the target's microwave emissions. To gather information on the state of an unknown and remote surface, the collected data are processed and then sometimes compared with the data collected from surfaces of known properties.

Radio Interferences from Thunderstorms

We shall explain why the noise in a car radio increases appreciably in or near a thunderstorm.

For a receiver that is located approximately 2 km from a typical lightning stroke and that has a bandwidth equal to 1 kHz, the field intensity produced by the lightning is approximately 0.4 V/m at 100 kHz, and this field intensity decreases linearly at a rate of 20 dB per decade of increase in frequency.* Thus, for a car radio tuned to an AM station of 1 MHz, the noise caused by lightning has an intensity of about 40 mV/m, and this intensity is strong enough to interfere with the signal. (The Federal Communications Commission of the United States requires a minimum of 25 mV/m field intensity for AM stations covering the commercial area of a city.)

3.2 UNIFORM PLANE WAVES IN FREE SPACE

Suppose that electromagnetic fields are generated in free space by sources $\boldsymbol{J}$ and ρ_v in a localized region; then, for electromagnetic fields outside this region, we have $\boldsymbol{J} = 0$, $\rho_v = 0$, and

$$\nabla \times \mathbf{E} = -j\omega\mu_0 \mathbf{H} \tag{3.1}$$

$$\nabla \times \mathbf{H} = j\omega\epsilon_0 \mathbf{E} \tag{3.2}$$

$$\nabla \cdot \mathbf{H} = 0 \tag{3.3}$$

$$\nabla \cdot \mathbf{E} = 0 \tag{3.4}$$

where we use the free-space constitutive relations $\mathbf{D} = \epsilon_0 \mathbf{E}$ and $\mathbf{B} = \mu_0 \mathbf{H}$. We now proceed to derive wave equations for the **E** or **H** field and to examine some simple solutions.

---

* A. Kimpara, "Electromagnetic energy radiated from lightning," in S. C. Coroniti, *Problems of Atmospheric and Space Electricity*, Amsterdam: Elsevier Publication Co., 1965, pp. 352–365. A. D. Watt and E. L. Maxwell, "Characteristics of atmospheric noise from 1 to 100 KC." *Proc. IRE*, 45 (1957): 787.

Taking the curl of (3.1) and substituting (3.2), we obtain

$$\nabla \times (\nabla \times \mathbf{E}) = \omega^2 \mu_0 \epsilon_0 \mathbf{E}$$

We use the vector identity (2.9), that is, $\nabla \times (\nabla \times \mathbf{E}) = \nabla(\nabla \cdot \mathbf{E}) - \nabla^2 \mathbf{E}$, and (3.4) to obtain a **wave equation** for **E**.

$$\nabla^2 \mathbf{E} + \omega^2 \mu_0 \epsilon_0 \mathbf{E} = 0 \tag{3.5a}$$

We can derive a similar equation for **H**.

The wave equation is a vector second-order differential equation. We consider the simple solution where the **E** field is parallel to the x axis and is a function of z coordinate only. In that case, the wave equation (3.5a) becomes

$$\frac{\partial^2 \mathrm{E}_x}{\partial z^2} + \omega^2 \mu_0 \epsilon_0 \mathrm{E}_x = 0 \tag{3.5b}$$

A solution to the above differential equation is

$$\boxed{\mathbf{E} = \hat{x} E_0 e^{-jkz}} \quad \textbf{(electric field of a uniform plane wave)} \tag{3.6}$$

Substituting (3.6) in the wave equation (3.5) yields

$$(-k^2 + \omega^2 \mu_0 \epsilon_0) E_0 = 0$$

Since E_0 is the amplitude of **E**, it is not zero and we have

$$\boxed{k^2 = \omega^2 \mu_0 \epsilon_0} \quad \textbf{(}\boldsymbol{k}\textbf{ is the wavenumber)} \tag{3.7}$$

which is called the **dispersion relation**. Equation (3.6) with k given by (3.7) constitutes a solution to the wave equation (3.5).

What does the electric field look like in space and time? Let us transform the solution in (3.6) to real space and time by the rule

$$\boldsymbol{E}(z, t) = \mathrm{Re}\,\{\mathbf{E}\, e^{j\omega t}\} = \hat{x} E_0 \cos(\omega t - kz) \tag{3.8}$$

There are two ways to examine the solution in (3.8). First we observe at one particular point in space, for example, at $z = 0$. In Figure 3.2 we plot the x component of $\boldsymbol{E}$ at $z = 0$:

$$E_x = E_0 \cos \omega t$$

We find that E_x is periodic in time, with period T determined by $\omega T = 2\pi$. The frequency $f = 1/T = \omega/2\pi$. We call $\omega = 2\pi f$ the angular frequency of the wave.

Next we imagine that we are taking a series of pictures of the solution. In other words, we try to see the spatial variation of $\boldsymbol{E}$ at successive times, as shown in Figure 3.3. We observe that at $z_0 = \lambda$ such that $k\lambda = 2\pi$ the function repeats

Figure 3.2 The x component of the electric field at $z = 0$ as a function of time [Equation (3.8)].

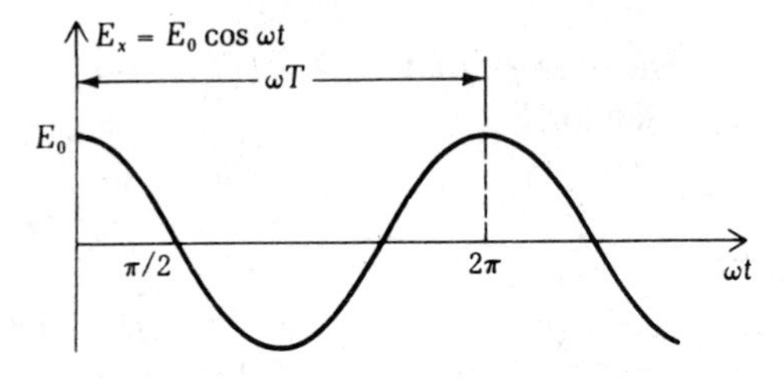

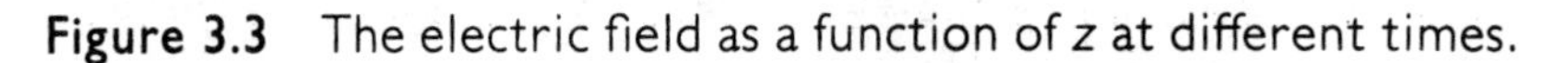
Figure 3.3 The electric field as a function of z at different times.

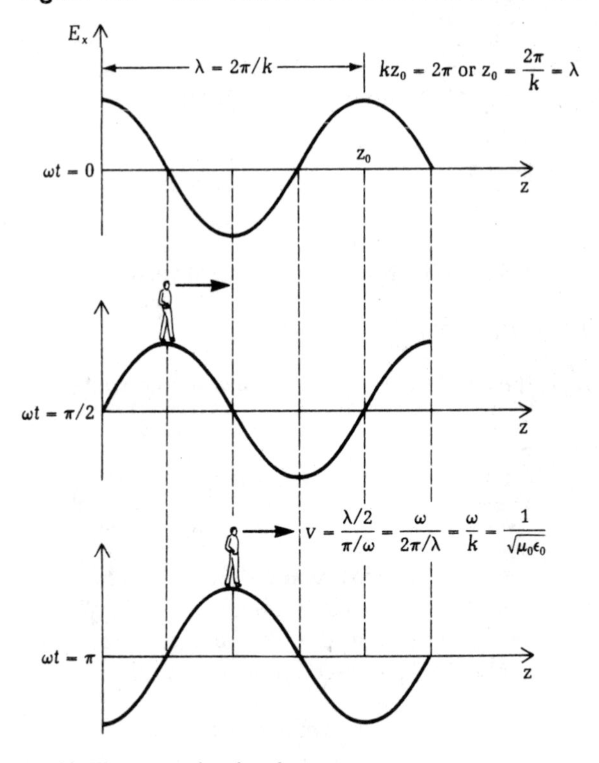

itself. The quantity λ, where

$$\boxed{\lambda = \frac{2\pi}{k}} \quad \textbf{(wavelength)} \tag{3.9}$$

is called the **wavelength**. The number of wavelengths contained in a spatial distance of 2π is given by

$$k = \frac{2\pi}{\lambda}$$

Thus, k is called the **wavenumber**.

The sketch shows how the solution, which can now be recognized as a sinusoidal wave, progresses with time. Imagine that we ride along with the wave. We ask, "at what velocity shall we move in order to keep up with the wave?" Mathematically, answering this question requires that $\cos(\omega t - kz)$ be a constant or that the phase of the wave be a constant. That is,

$$\omega t - kz = \text{a constant}$$

So the velocity of propagation is given by

$$\frac{dz}{dt} = v = \frac{\omega}{k}$$

From (3.7), we obtain

$$\boxed{v = \frac{\omega}{k} = \frac{1}{\sqrt{\mu_0 \epsilon_0}}} \quad \textbf{(velocity)} \qquad \textbf{(3.10)}$$

which is the velocity of light in free space.

The magnetic field **H** of the wave can be determined from either (3.1) or (3.2). We find

$$\boxed{\mathbf{H} = \hat{y}\sqrt{\frac{\epsilon_0}{\mu_0}}\, E_0 e^{-jkz}} \quad \textbf{(H field of a uniform plane wave)} \qquad \textbf{(3.11)}$$

or in the time domain

$$\boldsymbol{H} = \hat{y}\sqrt{\frac{\epsilon_0}{\mu_0}}\, E_0 \cos(\omega t - kz)$$

The space-time variation of $\boldsymbol{H}$ is similar to that of $\boldsymbol{E}$ as shown in Figures 3.2 and 3.3. Their amplitudes are related by the factor $(\mu_0/\epsilon_0)^{1/2}$, which is known as the characteristic impedance, or **intrinsic impedance**, of the free space (η).

$$\boxed{\eta = \sqrt{\frac{\mu_0}{\epsilon_0}}} \quad \textbf{(intrinsic impedance)} \qquad \textbf{(3.12)}$$

Thus, the wave has the electric field $\boldsymbol{E}$ in the $\hat{x}$ direction and the magnetic field $\boldsymbol{H}$ in the $\hat{y}$ direction, and it propagates in the $\hat{z}$ direction.

The Poynting vector is

$$\boldsymbol{S} = \hat{z}\,\frac{E_0^2}{\eta}\cos^2(\omega t - kz)$$

Notice that in all the above discussions, the wave solution in (3.6) or equivalently in (3.8) is independent of x and y coordinates. In other words for an observer anywhere in the x-y plane with the same value for z, the phenomena are the same. The constant-phase front is defined by setting the phase in (3.6) equal to a constant. We have

$$kz = C$$

where C is a constant defining a plane perpendicular to the z axis at $z = C/k$. We call waves whose phase fronts are planes plane waves. Thus, the wave solution in (3.6) or equivalently in (3.8) represents a **plane wave**. The amplitudes of $\boldsymbol{E}$ and $\boldsymbol{H}$ on a given constant-phase plane also assume uniform values. A plane wave with uniform amplitudes over its constant-phase planes is called a **uniform plane wave**.

The term *plane wave* is an idealized simplification. The output of a laser beam, for instance, has a beam width of several hundred thousand wavelengths. For practical purposes, it can be approximated by a uniform plane wave. The radiated field from a remote antenna is another example of a uniform plane wave. Strictly speaking, most waves from antennas are spherical rather than plane waves. But if the observation point is limited to a small region far from the source, we can always approximate these waves by uniform plane waves. Thus, in most cases, radio or television antennas can be considered to be intercepting uniform plane waves.

Example 3.1 *This example shows that the speed of an electromagnetic wave in free space is equal to the speed of light. It should be, because light is an electromagnetic wave.*

Calculate the velocity of a uniform plane wave in free space.

Solution The velocity of the uniform plane electromagnetic wave in free space is given by (3.10). The numerical values of the permittivity ϵ_0 and the permeability μ_0 of free space are

$$\epsilon_0 \approx \frac{1}{36\pi} \times 10^{-9} \text{ F/m}$$

$$\mu_0 = 4\pi \times 10^{-7} \text{ H/m}$$

Substituting these values in (3.10) yields $v = 3 \times 10^8$ m/s. The intrinsic impedance that relates $\boldsymbol{E}$ and $\boldsymbol{H}$ is given by $\eta = (\mu_0/\epsilon_0)^{1/2} \approx 120\pi \approx 377\,\Omega$.

Note that the velocity is independent of the frequency of the wave. A 10 GHz electromagnetic wave sent from an earth station to the moon will travel at the same speed as a laser light traveling in space. For historical reasons, the constant 3×10^8 m/s is called the speed of light in vacuum.*

Example 3.2 *Remember that the orientation of the **E** field of a uniform plane electromagnetic wave is perpendicular to the **H** field of that wave and that both are perpendicular to the direction from which the wave propagates.*

A uniform electromagnetic wave is traveling at an angle θ with respect to the z axis, as shown in Figure 3.4. The $\boldsymbol{E}$ field is pointing out of the paper (in the $\hat{y}$ direction). What is the direction of the $\boldsymbol{H}$ field?

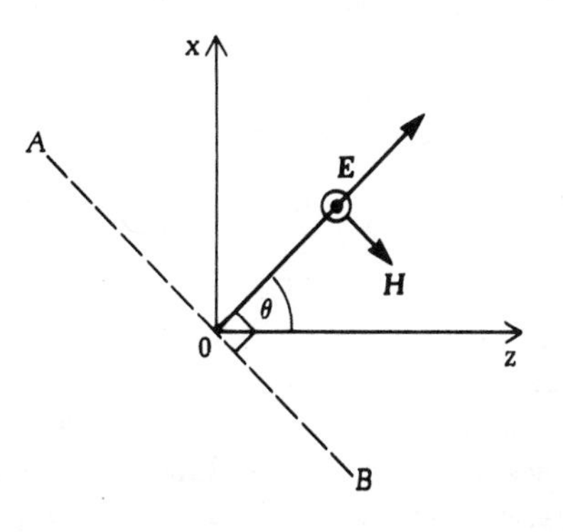

Figure 3.4 A plane wave traveling in a direction at an angle to the z axis (Example 3.2).

Solution Because $\boldsymbol{E}$, $\boldsymbol{H}$, and the direction of wave propagation are perpendicular to one another, in this case the $\boldsymbol{H}$ field is parallel to the line AOB shown in Figure 3.4. Furthermore, $\langle \boldsymbol{S} \rangle$ must be in the direction of wave propagation. This requires that the $\boldsymbol{H}$ field points in the OB direction. Mathematically speaking, $\boldsymbol{H}$ is the direction of the unit vector $\hat{\boldsymbol{u}}$ where $\hat{\boldsymbol{u}} = \hat{z} \sin \theta - \hat{x} \cos \theta$.

Human Exposure to RF Electromagnetic Radiation

There are two kinds of radiation. The first kind are the alpha and the beta particles given out by radioactive materials; electromagnetic waves or photons are the second kind. In this book we are concerned with only electromagnetic waves.

People have been subject to electromagnetic (EM) radiation since the beginning of humankind. Once EM radiation came only from natural sources such as the sun and thunderstorms. Today we are subject to additional EM radiation from artificial sources. At the low end of the frequency spectrum (60 or 50 Hz) are the EM fields generated by electric power lines and small and large appliances. At the high end is nuclear radiation consisting of gamma rays and X rays. In

* A more accurate value of the speed of light in vacuum is 2.998×10^8 m/s, which implies that the value of ϵ_0 given above is only approximate. However, the value of μ_0 is exact.

between are the so-called radio-frequency (RF) EM waves that carry everything from AM and FM radio and television broadcasts, ham and citizen band radios, cordless and cellular phones, and personal communication devices. The term *RF* is layperson's language used to describe the frequency range between a few kilohertz to several hundred gigahertz. Therefore, radars for air traffic controls or for automobile speed checks, microwave ovens, computers, and other electronic products are also radiating or leaking RF EM waves, although they are not associated with radios.

Very high-energy electromagnetic waves, such as gamma rays or X rays, are called ionizing radiation because they ionize molecules in their paths. Uncontrolled exposure to large amounts of these waves is known to cause sickness and even death in humans. However, natural radiation from the sun is the source of life on earth, and harnessed artificial radiation has been used medically to save lives.

The biological effects of non-ionizing RF electromagnetic waves are not well understood at this time, despite numerous studies on the subject. There is no proof that exposure to low-frequency EM fields from power lines will cause sickness in humans. However, some studies have found a weak statistical correlation between occurrence of leukemia* and the length of exposure time to electric power lines. As for human exposure to RF electromagnetic fields, the IEEE has proposed safety limits based on more than eight years of studies by more than 100 engineers, biophysicists, and cancer researchers. These limits are shown in Figure 3.5.** At frequencies higher than 100 MHz, the human exposure limits are in terms of power density, or magnitude of the time-average Poynting vector. At lower frequencies, the limits are expressed separately in terms of the ***E*** field and ***H*** field, because each field has different biological effects.

Although the IEEE standard shown in Figure 3.5 is the accepted exposure limit at the present time, some experts recommend practicing "prudent avoidance." That is, avoid exposure to electromagnetic radiation if it can be accomplished with small investment of money or effort.

Example 3.3 *If we know the strength of the* ***E*** *field, we can compute the power density associated with an electromagnetic wave. We can then refer to the IEEE guideline to find safety limits of human exposure to radio-frequency electromagnetic radiation.*

Broadcasting stations typically maintain approximately 25 mV/m field strength in a city. Compute the time-average power density and see whether it satisfies the safety limit set by the IEEE.

* P. Thomas, "Power Struggle," *Harvard Health Letter*, Vol. 18, no. 9 (1993): p 1.

** M. Fischetti, "The cellular phone scare," *IEEE Spectrum*, June, 1993, p. 43. The reader should refer to the latest guidelines, because the data in Figure 3.5 may be modified when new studies progress. Contact IEEE Service Center for further information.

Figure 3.5 IEEE safety limits for human exposure to RF electromagnetic fields.

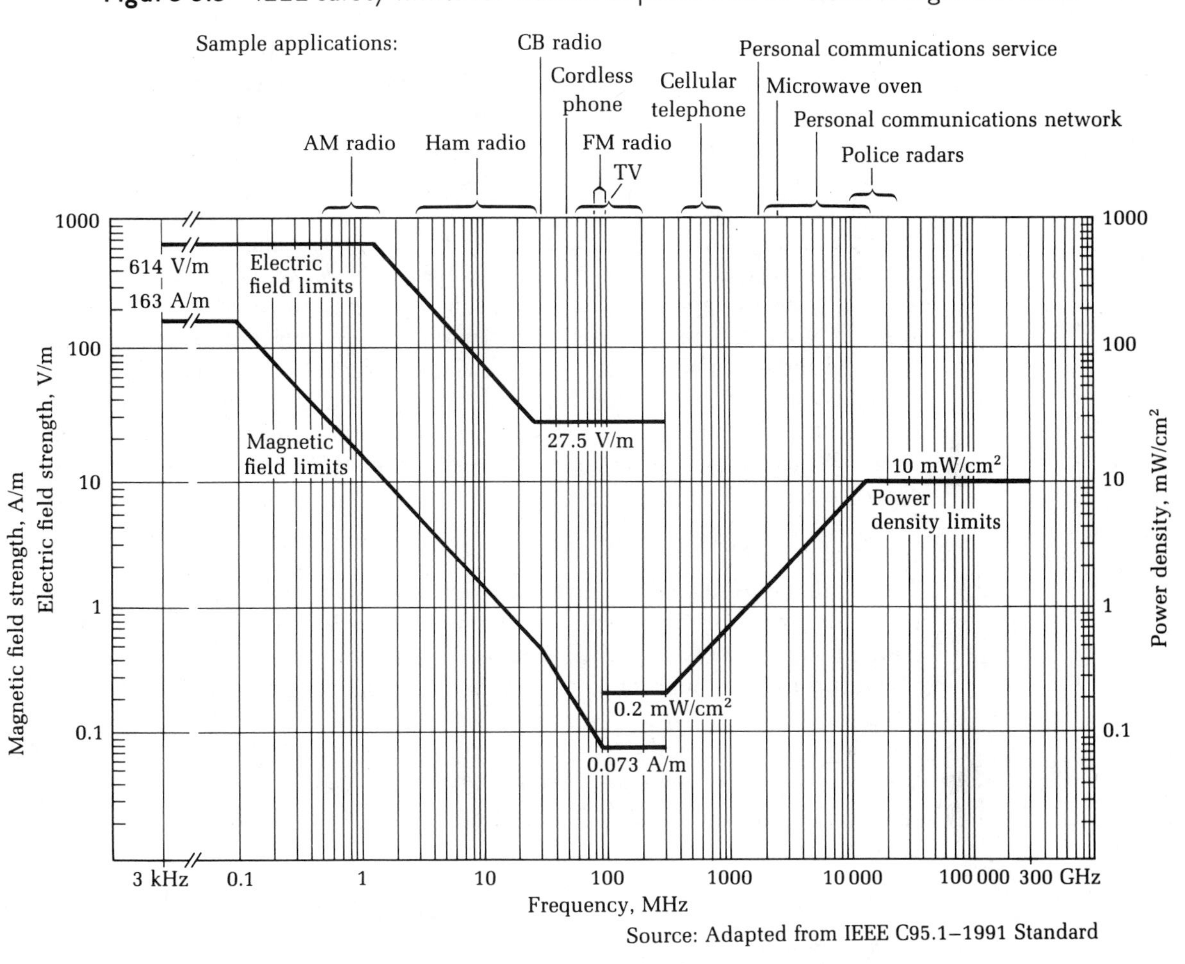

Source: Adapted from IEEE C95.1–1991 Standard

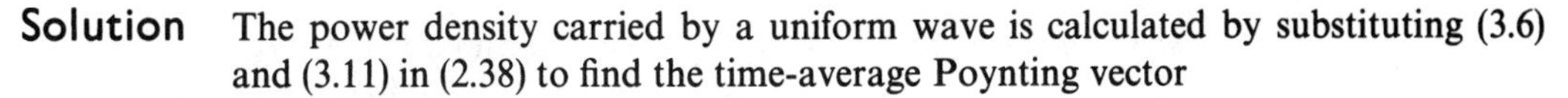

Solution The power density carried by a uniform wave is calculated by substituting (3.6) and (3.11) in (2.38) to find the time-average Poynting vector

$$\langle \boldsymbol{S} \rangle = \frac{1}{2}\mathrm{Re}\{\mathbf{E} \times \mathbf{H}^*\} = \hat{z}\,\frac{1}{2}\sqrt{\frac{\epsilon_0}{\mu_0}}\,|E_0|^2$$

Substituting $E_0 = 0.025$ and the free space ϵ_0 and μ_0 into the above equation yields

$$\langle S_z \rangle = \frac{(0.025)^2}{2 \times 120\pi}\ \mathrm{W/m^2}$$
$$= 0.83\ \mu\mathrm{W/m^2}$$
$$= 0.83 \times 10^{-10}\ \mathrm{W/cm^2}$$

This power density is well below the limits set by the IEEE in the entire RF range.

3.3 POLARIZATION

At a fixed point in space, the $\boldsymbol{E}$ vector of a time-harmonic electromagnetic wave varies sinusoidally with time. The **polarization** of the wave is described by the **locus of the tip of the $\boldsymbol{E}$ vector as time progresses.** If the locus is a straight line the wave is said to be **linearly polarized.** It is **circularly polarized** if the locus is a circle and **elliptically polarized** if the locus is an ellipse. An electromagnetic wave, for example, sunlight or lamplight, may also be randomly polarized. In such cases, the wave is **unpolarized.** An unpolarized wave can be regarded as a wave containing many linearly polarized waves with their polarization randomly oriented in space. A wave can also be **partially polarized**, such as skylight or light reflected from the surface of an object— that is, glare. A partially polarized wave can be thought of as a mixture of polarized waves and unpolarized waves.

The uniform plane wave discussed in the previous section is linearly polarized. We can show this by taking the real part of (3.6) multiplied by $e^{j\omega t}$ to find $\boldsymbol{E}(z, t)$:

$$\boldsymbol{E}(z, t) = E_0 \cos(\omega t - kz)\hat{x}$$

Tracing the tip of the vector $\boldsymbol{E}$ at any point z will show that the tip always stays on the x axis with maximum displacement E_0. We thus conclude that the uniform plane wave is linearly polarized.

Next consider a uniform plane wave with the following electric-field vector:

$$\mathbf{E} = \hat{x} a\, e^{-j(kz - \phi_a)} + \hat{y} b\, e^{-j(kz - \phi_b)} \tag{3.13}$$

We can easily show that substituting (3.13) into the wave equation (3.5) yields the same dispersion relation (3.7). The real time-space $\boldsymbol{E}$ vector in (3.13) has x and y components:

$$E_x = a \cos(\omega t - kz + \phi_a) \tag{3.14a}$$

$$E_y = b \cos(\omega t - kz + \phi_b) \tag{3.14b}$$

where a and b are real constants. To determine the locus of the tip of $\boldsymbol{E}$ in the x-y plane as a function of time at any z, we can eliminate the variable $(\omega t - kz)$ from (3.14a) and (3.14b) to obtain an equation for E_x and E_y. Consider the following cases.

Linear Polarization

$$\phi = \phi_b - \phi_a = 0 \quad \text{or} \quad \pi$$

When this relation holds between the phases E_x and E_y, we obtain

$$E_y = \pm\left(\frac{b}{a}\right)E_x \tag{3.15}$$

This result is a straight line with slope $\pm b/a$. The plus sign applies to the case $\phi = 0$, and the minus sign to $\phi = \pi$. The former case is shown in Figure 3.6a.

Figure 3.6 Polarization of an electric field: **(a)** Linear polarization. **(b)** Circular polarization (left-hand circular polarization if wave propagates in $\hat{\mathbf{z}}$ direction). **(c)** Circular polarization (right-hand if wave propagates in $\hat{\mathbf{z}}$ direction). **(d)** Elliptical polarization.

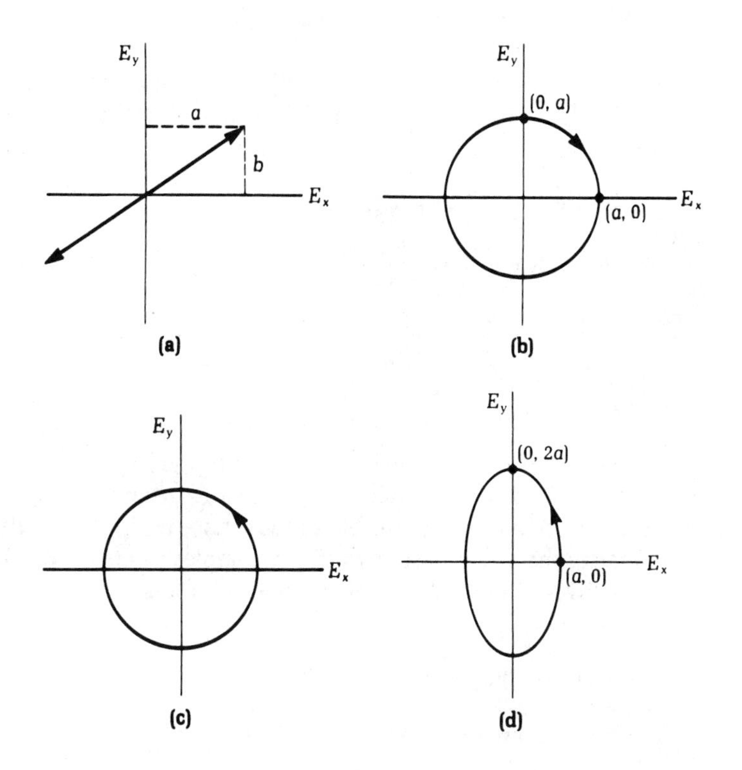

Circular Polarization

$$\phi = \phi_b - \phi_a = \pm \frac{\pi}{2} \qquad \text{and} \qquad A = \frac{b}{a} = 1$$

Consider the case $\phi = \pi/2$ and $A = 1$. Equations (3.14) become

$$E_x = a \cos(\omega t - kz + \phi_a) \tag{3.16a}$$

$$E_y = -a \sin(\omega t - kz + \phi_a) \tag{3.16b}$$

Elimination of t yields

$$E_x^2 + E_y^2 = a^2 \tag{3.17}$$

As shown in Figure 3.6b, this result is a circle in the E_x-E_y plane, and the circle's radius is equal to a. Note that the tip of $\boldsymbol{E}$ moves clockwise along the circle as time progresses. If we use left-hand fingers to follow the tip's motion, the thumb will point in the direction of wave propagation. We call this wave left-hand circularly polarized. Figure 3.6c shows that the wave will be right-hand circularly polarized when $\phi = -\pi/2$ and $A = 1$.

Elliptical Polarization

The wave represented by (3.13) is elliptically polarized if it is neither linearly nor circularly polarized. Consider the case $\phi = -\pi/2$ and $A = b/a = 2$. Equations (3.14) become

$$E_x = a\cos(\omega t - kz + \phi_a) \tag{3.18a}$$

$$E_y = 2a\sin(\omega t - kz + \phi_a) \tag{3.18b}$$

Eliminating t yields

$$\left(\frac{E_x}{a}\right)^2 + \left(\frac{E_y}{2a}\right)^2 = 1 \tag{3.19}$$

As shown in Figure 3.6d, this result is an ellipse.

For other ϕ and A values, the wave is generally elliptically polarized. To predict the shape and sense of rotation for the tip of the electric field, we first summarize the above cases by using a complex plane on which each point can be marked by a magnitude A and a phase ϕ (Figure 3.7). Let the complex electric field be

$$\mathbf{E} = (\hat{x}\mathrm{E}_x + \hat{y}\mathrm{E}_y)e^{-jkz} \tag{3.20}$$

Then A and ϕ are determined by the following equation:

$$\frac{\mathrm{E}_y}{\mathrm{E}_x} = A\,e^{j\phi} \tag{3.21}$$

The linearly polarized waves correspond to the ratio $\mathrm{E}_y/\mathrm{E}_x$ being located on the real axis, where $\phi = 0$ or π. For the circularly polarized waves, this ratio is located at $A = 1$ and $\phi = \pm\pi/2$. Furthermore, if the ratio is located on the upper half of the plane, the corresponding wave is left-hand elliptically polarized. Similarly, if the ratio (3.21) is located on the lower half of the plane, the corresponding wave is right-hand elliptically polarized. For instance, if $A = 1$ and $\phi = \pi/4$, the wave is left-hand elliptically polarized with the ellipse slanted toward the upper right corner, as shown in Figure 3.7. As A approaches infinity, the wave becomes linearly polarized in the y direction, since the complex $\mathbf{E}$ field has a vanishing x component.

Figure 3.7 Polarization chart: polarization is determined by the complex ratio of E_y/E_x. The wave propagates in the $\hat{z}$ direction.

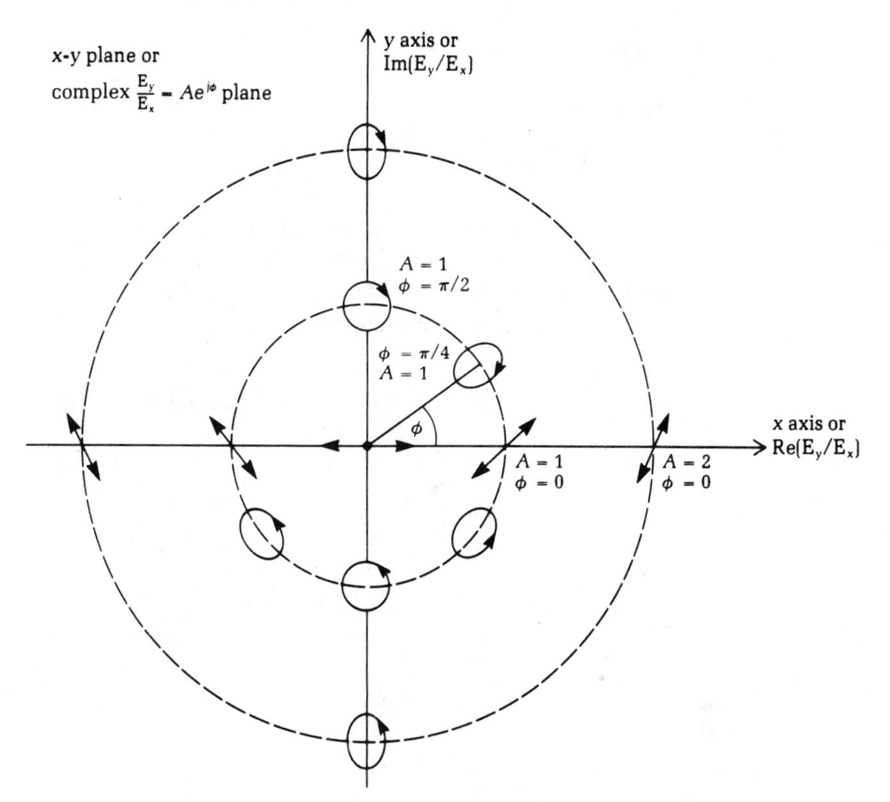

Example 3.4 ***The polarization of an electromagnetic wave is determined by the locus of the tip of the E field vector. In this example, we trace the tip of the H field vector. For a uniform plane wave, the polarization type (linear, circular, or elliptical) of the E field and that of the H field are the same, but they are pointing in perpendicular directions.***

Find the $\boldsymbol{H}$ field associated with the $\boldsymbol{E}$ field in (3.13). When $a = b$ and $\phi_b - \phi_a = \pi/2$, the tip of the $\boldsymbol{E}$ field given by (3.13) traces a circle as shown in Figure 3.6b. Find the locus of the tip of the corresponding $\boldsymbol{H}$ field.

Solution Maxwell's equation (3.1) shows that the $\boldsymbol{H}$ field corresponding to the $\boldsymbol{E}$ field given by (3.13) is

$$\boldsymbol{H} = -\hat{x}\,\frac{b}{\eta}\,e^{j\phi_b}\,e^{-jkz} + \hat{y}\,\frac{a}{\eta}\,e^{j\phi_a}\,e^{-jkz}$$

Using the above result, we can express the time-domain $\boldsymbol{H}$ field as follows:

$$H_x(z, t) = \frac{a}{\eta} \sin(\omega t - kz + \phi_a)$$

$$H_y(z, t) = \frac{a}{\eta} \cos(\omega t - kz + \phi_a)$$

Thus,

$$H_x^2 + H_y^2 = \left(\frac{a}{\eta}\right)^2$$

The locus of the tip of the $\boldsymbol{H}$ field is a circle. Comparing with (3.16), we see that at $kz = \phi_a$ and $\omega t = 0$, $\boldsymbol{E}$ is in the $\hat{x}$ direction and $\boldsymbol{H}$ is in the $\hat{y}$ direction. As time progresses, both vectors rotate in the same direction and remain perpendicular at all times. This situation is shown in Figure 3.8.

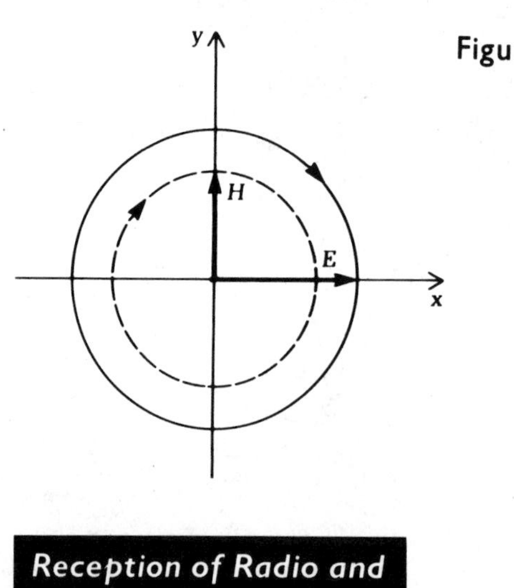

Figure 3.8 **E** and **H** fields of a circularly polarized wave.

Reception of Radio and Television Signals

As Chapter 7 will discuss, the electromagnetic waves radiated by AM stations always have the $\boldsymbol{E}$ field perpendicular to the ground and parallel to the antenna towers. Listeners' antennas are oriented parallel to the $\boldsymbol{E}$ field for maximum reception, as we can see from many vertically oriented wire antennas. For television broadcasting, however, the $\boldsymbol{E}$ field is horizontal. Thus, the television receiving antennas are oriented so that their wires are parallel to the ground and perpendicular to the direction from which the signal comes. Therefore, rooftop television antennas should be oriented this way to receive the maximum signal. Most of the FM stations in the United States broadcast with circular polarization. To receive the FM broadcast, the receiving wire-antenna may be oriented in any direction as long as the antenna is on the plane perpendicular to the direction from which the signal is coming. This is because the antenna will always intercept the electric field that is rotating around a circular path in that plane.

Eye Detection of Polarized Light

Sun navigation was first observed in 1911.* It was found that some species of ants, horseshoe crabs, honeybees, and so on, are sensitive to polarized light. As long as there is a small patch of blue sky, the animals can navigate as well as ever. The sky polarization depends on the angle between the sun's rays and a particular point in the sky and on an observer's line of sight to the same point. We see from Figure 3.9 that, because of scattering from air molecules in the sky, light is completely unpolarized when an observer looks directly toward the sun and becomes more linearly polarized when the observer looks at other parts of the sky farther away from the direct path of the sun's rays.

Figure 3.9 Polarization of skylight.

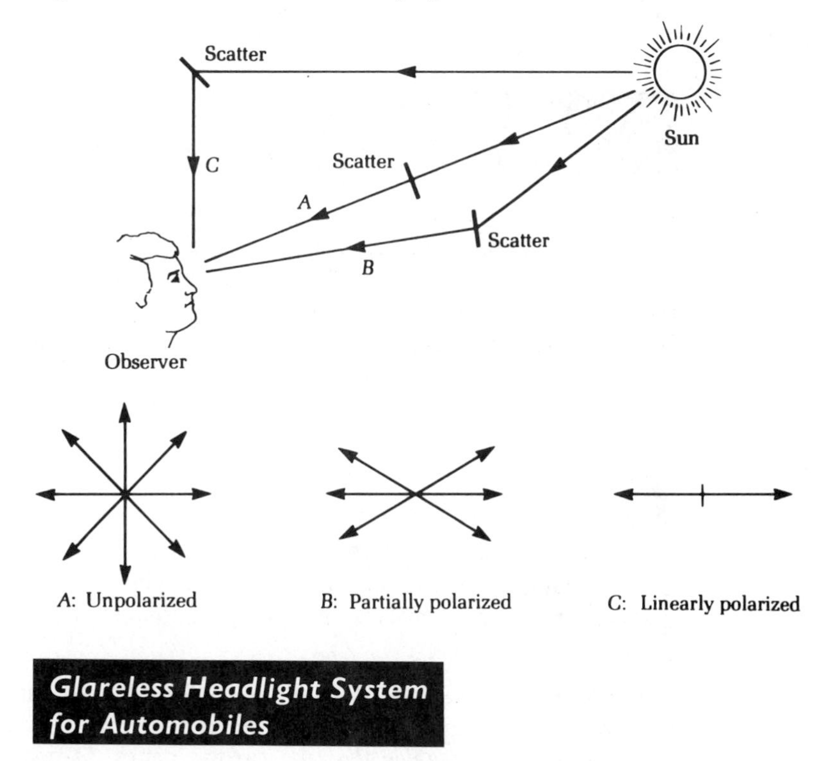

Glareless Headlight System for Automobiles

Future automobile-industry safety standards may require that every automobile be equipped with a glareless headlight system. The glareless headlight system has headlights polarized with their axes at 45° to the horizontal plane and a windshield made of glass with the same direction of polarization. If every car had such

* T. H. Waterman, "Polarized light and animal navigation," *Scientific American*, July 1955, p. 88.

a system, the headlight glare from oncoming traffic would be drastically reduced. At the same time, the driver's visibility would not be affected because only light of the same polarization passes through the polarized glass, which absorbs or reflects light of other polarizations.

Stereoscopic Pictures with Polarized Light

A three-dimensional impression of an object is conveyed to our brain because each of our eyes has a slightly different view of an object. This fact is used to make stereoscopic pictures. Using two polarizations in making a picture can provide the two different views of our two eyes. When we look at the specially photographed picture through a pair of glasses with different polarizations, a three-dimensional effect is made to appear.

Communication System Using Orthogonal Polarizations

Because of the great demand for telecommunication, the Federal Communications Commission allocates specific frequency bands to users in the United States. To increase their capacity within the fixed frequency band allocated to them, domestic satellite communication systems will use two orthogonal polarizations. These orthogonal polarizations could double the capacity of a system that ordinarily uses only single polarization.

3.4 PLANE WAVES IN DISSIPATIVE MEDIA

In the discussion of constitutive relations, we have omitted one important class of media—namely, conductors. A conductor is characterized by a conductivity σ and is governed by Ohm's law. For isotropic conductors, Ohm's law states that

$$\mathbf{J}_c = \sigma \mathbf{E} \tag{3.22}$$

where $\mathbf{J}_c$ denotes the conduction current. The unit of conductivity σ is siemens per meter, commonly known as mhos per meter. From Ampère's law, $\nabla \times \mathbf{H} = j\omega\mathbf{D} + \mathbf{J}$, we see that the current density $\mathbf{J}$ can in fact embody two kinds of currents—namely, the source current $\mathbf{J}_0$ and the conduction current $\mathbf{J}_c$, $\mathbf{J} = \mathbf{J}_c + \mathbf{J}_0$. The source current $\mathbf{J}_0$ can be either a conduction or a convection current. Remember that for an isotropic dielectric $\mathbf{D} = \epsilon\mathbf{E}$. It is instructive to see that, in a conducting medium, Ampère's law become

$$\nabla \times \mathbf{H} = j\omega\left(\epsilon - j\frac{\sigma}{\omega}\right)\mathbf{E} + \mathbf{J}_0$$

Thus, if we define

$$\boxed{\boldsymbol{\epsilon} = \epsilon - j\frac{\sigma}{\omega}} \quad \textbf{(complex permittivity)} \tag{3.23}$$

the conductivity becomes an imaginary part of the **complex permittivity $\boldsymbol{\epsilon}$.**

Maxwell's equations in a conducting medium devoid of any source can be written as

$$\nabla \times \mathbf{E} = -j\omega\mu\mathbf{H} \tag{3.24a}$$

$$\nabla \times \mathbf{H} = j\omega\boldsymbol{\epsilon}\mathbf{E} \tag{3.24b}$$

$$\nabla \cdot \mathbf{H} = 0 \tag{3.24c}$$

$$\nabla \cdot \mathbf{E} = 0 \tag{3.24d}$$

As we have seen, for a conducting medium $\boldsymbol{\epsilon}$ is complex, as shown in (3.23). As in the case of free space, we can derive a wave equation for **E** by taking the curl of (3.24a) and using (3.24d). We find

$$(\nabla^2 + \omega^2\mu\boldsymbol{\epsilon})\mathbf{E} = 0$$

The set of plane-wave fields is still a solution of Maxwell's equations. Namely,

$$\mathbf{E} = \hat{x}E_0 e^{-j\mathrm{k}z} \tag{3.25a}$$

$$\mathbf{H} = \hat{y}\left(\frac{E_0}{\eta}\right)e^{-j\mathrm{k}z} \tag{3.25b}$$

(electromagnetic fields of a uniform plane wave in a dissipative medium)

where

$$\boxed{\mathrm{k}^2 = \omega^2\mu\boldsymbol{\epsilon}} \quad \textbf{(k is complex wavenumber)} \tag{3.26a}$$

is the dispersion relation derived from the wave equation and

$$\boxed{\eta = \sqrt{\frac{\mu}{\boldsymbol{\epsilon}}}} \quad \textbf{(complex intrinsic impedance)} \tag{3.26b}$$

is the intrinsic impedance of the isotropic media. Note that k and η are now complex numbers.

Because k and η are complex, we can separate k into real and imaginary parts and express η in phasor notation:

$$\mathrm{k} = k_R - jk_I \quad \textbf{(complex wavenumber and} \tag{3.27a}$$

$$\eta = |\eta| e^{j\phi} \quad \textbf{intrinsic impedance)} \tag{3.27b}$$

where k_R is the real part of k and k_I is the negative of the imaginary part of k. Both k_R and k_I are assumed to be real and positive. We shall see that the minus sign in front of k_I will give the correct physical result for positive k_I values. Substitution of (3.27) in (3.25) yields

$$\mathbf{E} = \hat{x} E_0 e^{-k_I z} e^{-jk_R z} = \hat{x} \mathrm{E}_x \tag{3.28a}$$

$$\mathbf{H} = \hat{y} \left[\frac{E_0}{|\eta|} \right] e^{-k_I z} e^{-jk_R z} e^{-j\phi} \tag{3.28b}$$

The instantaneous value is $E_x(z, t) = \mathrm{Re}(\mathrm{E}_x e^{j\omega t})$, and from (3.28) we find

$$E_x = E_0 e^{-k_I z} \cos(\omega t - k_R z) \tag{3.29}$$

The above equation represents a wave traveling in the $\hat{z}$ direction with a velocity equal to v, where

$$v = \frac{\omega}{k_R}$$

As the wave travels, the amplitude is attenuated exponentially at the rate k_I nepers per meter. Notice that if we use a positive sign in front of k_I in (3.27), the wave inside the conductor will grow instead of decay, a fact contrary to physics.

We now define a **penetration depth** d_p such that, when $k_I z = k_I d_p = 1$, the amplitude of the electric field shown in (3.29) will decay to $1/e$ of its value at $z = 0$. Thus, the penetration depth is as follows:

$$d_p = \frac{1}{k_I} \quad \textbf{(penetration depth)}$$

For conducting media, we have

$$\mathrm{k} = \omega \sqrt{\mu\epsilon} \left[1 - j \frac{\sigma}{\omega\epsilon} \right]^{1/2} = k_R - jk_I \tag{3.30}$$

where $\sigma/\omega\epsilon$ is called the **loss tangent** of the conducting media.

Waves in Non-Dissipative Media

In materials where the effective conductivity is zero, then the wave number k and the intrinsic impedance η are real numbers. The attenuation constant k_I is zero so that the **E** and **H** fields of the uniform plane wave do not decay in a non-dissipative medium. The EM wave in such a medium behaves very much like the wave in free space. The speed of the wave is given by $1/\sqrt{\mu\epsilon}$.

Waves in Slightly Conducting Media

A slightly conducting medium is one for which $\sigma/\omega\epsilon \ll 1$. The value of k in (3.30) can be approximated by

$$\mathrm{k} = \omega\sqrt{\mu\epsilon\left(1 - j\frac{\sigma}{\omega\epsilon}\right)} \approx \omega\sqrt{\mu\epsilon}\left(1 - j\frac{\sigma}{2\omega\epsilon}\right)$$

(See Problem 1.6.) We find

$$k_R = \omega\sqrt{\mu\epsilon} \tag{3.31a}$$

$$k_I = \frac{\sigma}{2}\sqrt{\frac{\mu}{\epsilon}} \tag{3.31b}$$

Thus, the wave propagates in the $\hat{z}$ direction with k_R as the propagating constant and decays exponentially in the same direction at the rate of k_I nepers per meter. After entering a slightly conducting medium and traveling a penetration depth

$$d_p = \frac{2}{\sigma}\sqrt{\frac{\epsilon}{\mu}} \tag{3.31c}$$

the amplitude of the wave will decay to $1/e$ times its original value.

Penetration Depth of Ice

Ice is a slightly conducting medium with conductivity $\sigma \approx 10^{-6}$ mho/m and $\epsilon \approx 3.2\epsilon_0$. The loss tangent $= 10^{-6}/(2\pi f \times 3.2 \times 8.85 \times 10^{-12}) = 5.6 \times 10^3/f$, which is a small number for frequency f in megahertz range or above. The penetration depth is found to be $d_p = 9.5$ km. These facts suggest that the megahertz frequency range is useful for probing ice thickness, and that range has indeed been useful in some glacier studies. During the Apollo era, the lunar subsurface was also probed with the megahertz frequencies because the lunar surface possesses very low conductivities at these frequencies. It is perhaps surprising to find that d_p is apparently independent of frequency, as given in (3.31c). As a matter of fact, when frequency reaches gigahertz range, scattering effects from air bubbles in ice become so important that the simple conductivity description is no longer suitable for ice.

Waves in Highly Conducting Media

A highly conducting medium is also called a good conductor, for which $\sigma/\omega\epsilon \gg 1$. In this case, the wavenumber k can be approximated by

$$k \simeq \omega\sqrt{\mu\epsilon}\left(-j\frac{\sigma}{\omega\epsilon}\right)^{1/2} = \sqrt{\omega\mu\left(\frac{\sigma}{2}\right)}(1-j) \tag{3.32a}$$

(See Problem 1.7.) Therefore, the penetration depth becomes

$$d_p = \sqrt{\frac{2}{\omega\mu\sigma}} \equiv \delta \tag{3.32b}$$

The symbol δ signifies that d_p is so small it is better called the **skin depth** $\boldsymbol{\delta}$. The terms **skin depth** and **penetration depth** are used interchangeably for highly conducting media, where most of the conduction current concentrates on the surface and very little flows inside the conductor. This phenomenon is called the **skin effect**.

Perfect Conductors

A perfect conductor is an idealized material in which no electric field is allowed to exist. Examining Ohm's law

$$\mathbf{J} = \sigma\mathbf{E}$$

we see that, if **E** is zero, then $\sigma \to \infty$ in order to have finite current. It follows from (3.32b) that the skin depth for a perfect conductor is zero.

For practical purposes, ordinary metals such as copper, aluminum, gold, silver, etc., can be regarded as perfect conductors in solving electromagnetic wave problems. At room temperature, the conductivity for copper is $\sigma = 5.7 \times 10^7$ mho/m (or S/m). At very low temperatures (several Kelvins), some metals exhibit a property called **superconductivity**. The conductivity of a superconducting lead at 4.2 K has been found to be greater than 2.7×10^{20} mho/m for direct currents.*

In 1986 physicists K. A. Müller and J. G. Bednorz realized that ceramics composed of lanthanum, barium, copper, and oxygen became superconductors when cooled to a temperature near 40° K. Shortly thereafter, a research group led by C. W. Chu at the University of Houston and his former student, M. K. Wu, at the University of Alabama discovered that a chemical compound lost all resistivity at 92° K. The material is $Y_1Ba_2Cu_3O_7$, now known as 1-2-3 (so named because of the ratio of yttrium, barium, and copper atoms in the compound). Since then, other superconducting oxides have been discovered. These new

* T. Van Duzer and C. W. Turner, "Superconductivity, new roles for an old discovery," *IEEE Spectrum*, December 1972, p. 53. D. J. Quinn III and W. B. Ittner III, "Resistance in superconductor," *J. Appl. Phys.*, 33 (1962): 748.

materials are called **high-temperature superconductors** because the superconductivity occurs at relatively high Tc, the critical temperature at which superconductivity occurs. The metallic superconductors have to be cooled by liquid helium to a few degrees above absolute zero in order to reach the superconducting state. Newly developed high-temperature superconductors can be immersed in relatively inexpensive liquid nitrogen, which maintains at 77° K. High-temperature superconducting materials have now been developed with the critical temperature equal to 125° K. The hope is that some day material will be found to be superconducting at room temperature.*

Example 3.5 *This example shows that a material may be a good conductor at low frequencies but becomes a not-so-good conductor at higher frequencies. The ω factor in the denominator of (3.23) plays an important role in determining whether a material is a good or a poor conductor.*

Seawater can be characterized by $\sigma = 4$ mho/m, $\epsilon = 81\epsilon_0$, and $\mu = \mu_0$. Find the complex ϵ for seawater at 60 Hz, 1 MHz, and 100 MHz.

Solution

$$\epsilon = 81\epsilon_0 - j\frac{\sigma}{\omega}$$

$$= 81\epsilon_0\left(1 - j\frac{\sigma}{81\omega\epsilon_0}\right)$$

$$= \begin{cases} 81\epsilon_0(1 - j1.48 \times 10^7) & \text{at 60 Hz} \\ 81\epsilon_0(1 - j8.9 \times 10^2) & \text{at 1 MHz} \\ 81\epsilon_0(1 - j8.9) & \text{at 100 MHz} \end{cases}$$

The ratio of the imaginary part of ϵ to its real part represents the relative amplitude of the conduction current to the displacement current in the medium. For frequencies below 100 MHz, this ratio for seawater is large, and hence seawater is regarded as a good conductor up to 100 MHz.

Example 3.6 *This example shows how to obtain the attenuation rate of an electromagnetic wave in a good conductor. There are two ways to compute the skin depth. The first is to use a computer and the second is to use the approximate formula (3.32).*

---

* R. J. Cava, "Superconductors beyond 1-2-3," *Scientific American*, August 1990, p. 42; and M. Cyrot and D. Pavuna, *Introduction to Superconductivity and High-Tc Materials*, World Scientific Publishing Co.: Singapore, 1992.

A 100 Hz electromagnetic wave is propagating vertically down into seawater, which is characterized by $\mu = \mu_0$, $\epsilon = 81\epsilon_0$, and $\sigma = 4$ mho/m. The electric-field intensity measured just beneath the surface of the seawater is 1 V/m. What is the intensity of E at a depth of 100 m?

Solution We first compute the complex permittivity ϵ using (3.23).

$$\epsilon = 81 \times 8.8542 \times 10^{-12} - j\frac{4}{2\pi \times 100}$$

$$= 0.71719020 \times 10^{-9} - j0.63661977 \times 10^{-2}$$

Then the complex wave number k is computed from (3.26a):

$$\mathrm{k} = 2\pi \times 100(4\pi \times 10^{-7} \times \epsilon)^{1/2}$$

$$= 0.039738355 - j0.039738351$$

The EM wave attenuates like $E_0 \exp(-0.039738351z)$. Thus, at a depth of 100 m, the intensity of E is approximately

$$|E| = (1 \text{ V/m})\exp(-3.9738) = 0.00188 \text{ V/m}$$

We can obtain an approximate solution as follows. Since the imaginary part of ϵ is much greater than the real part (in absolute value), seawater at 100 Hz is a good conductor. We use (3.32a) and obtain

$$\mathrm{k} = (1 - j)(\pi \times 100 \times 4 \times \pi \times 10^{-7} \times 4)^{1/2}$$

$$= 0.039738353 - j0.039738353$$

The difference between the approximate k and exact k is in the last digit. Since the value of ϵ_0 is given only for five significant digits, the difference between the exact and the approximate values of k is insignificant for this case.

Example 3.7 *This example shows how to compute the time-average Poynting vector of an EM wave in a dissipative medium.*

In the previous example, what is the power density just beneath the surface of the seawater?

Solution Using (3.28) yields

$$\langle \boldsymbol{S} \rangle = \frac{1}{2}\mathrm{Re}\left\{\frac{E_0^2}{|\boldsymbol{\eta}|}(e^{-2k_I z}e^{j\phi})\right\}\hat{z} = \frac{E_0^2}{2|\boldsymbol{\eta}|}e^{-2k_I z}\cos(\phi)\hat{z}$$

Now

$$\boldsymbol{\epsilon} = \epsilon - j\frac{\sigma}{\omega}$$

$$= 81 \times 8.8542 \times 10^{-12} - j\frac{4}{2\pi \times 100}$$

$$\approx -j6.37 \times 10^{-3}$$

$$\boldsymbol{\eta} = \sqrt{\frac{\mu}{\boldsymbol{\epsilon}}}$$

$$= \sqrt{4\pi \times \frac{10^{-7}}{-j6.37 \times 10^{-3}}}$$

$$= 1.4 \times 10^{-2}(\sqrt{j})$$

Thus,

$$|\boldsymbol{\eta}| = 1.4 \times 10^{-2}$$

$$\phi = 45°$$

At $z = 0$, $|\mathbf{E}| = 1$ V/m, so that

$$\langle \boldsymbol{S} \rangle = \hat{z}25.3 \text{ W/m}^2$$

Radio Communication Between Submarines

The main difficulty in radio communication between submarines is the high attenuation of electromagnetic waves propagating in the ocean. The relative permittivity of seawater is approximately 79, and its average conductivity is approximately 4 mho/m. From (3.30), the attenuation constant k_I is computed as a function of the frequency, and the result is shown in Figure 3.10. The attenuation is expressed in decibels per meter. The attenuation dB is defined as

$$\text{dB} = 20\log_{10}(\text{ratio of amplitudes})$$

Here the ratio is the amplitude of the **E** field at z and that at $z + 1$ meters. According to (3.28a),

$$\text{dB} = 20\log_{10}[\exp(k_I)]$$

$$= 8.686k_I$$

Figure 3.10 shows that the attenuation increases rapidly as the frequency is varied from low frequencies to higher frequencies. For example, after traveling 100 m, an EM wave in seawater is attenuated approximately 11 dB at 10 Hz, 35 dB at 100 Hz, and 109 dB at 1 kHz. Although the loss of signal strength can be reduced by using low frequencies, the rate of message transmission is also reduced. At an operating frequency of 50 Hz, for example, transmitting a single word may take several hours.

Figure 3.10 Attenuation of electromagnetic waves in seawater: the abscissa is logarithm, to the base in 10, of the frequency in hertz.

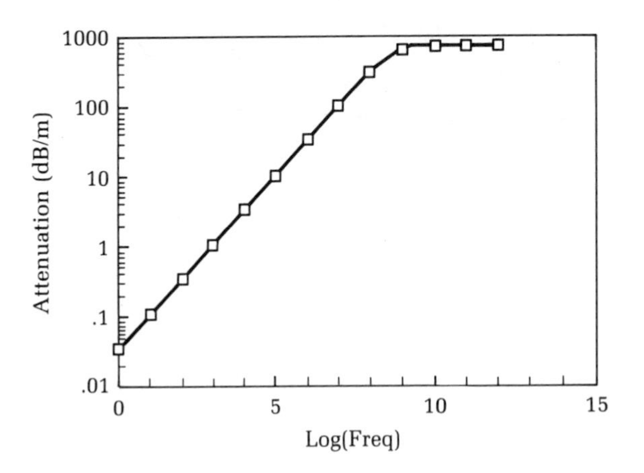

Example 3.8 *Absorbing EM waves by dissipative media can be used to our advantage. One such application is shielding an area against radio interference, as demonstrated in this example.*

To shield a room from radio interference, the room must be enclosed in a layer of copper five skin-depths thick. If the frequency to be shielded against is 10 kHz to 100 MHz, what should be the thickness of the copper (in millimeters)?

Solution For copper, $\mu = \mu_0$, $\epsilon = \epsilon_0$, and $\sigma = 5.8 \times 10^7$ mho/m. At $f = 10^4$ Hz, we have

$$\frac{\sigma}{\omega\epsilon} = \frac{5.8 \times 10^7 \times 36\pi \times 10^9}{2\pi \times 10^4} \gg 1$$

Thus, copper is a good conductor, and (3.32b) can be used to calculate its skin depth:

$$d_p = \frac{1}{\sqrt{\pi \times 10^4 \times 4\pi \times 10^{-7} \times 5.8 \times 10^7}} = 0.66 \text{ mm}$$

Thus, the copper layer should be 3.3 mm thick.

In fact, copper is a good conductor even at the highest microwave frequency (10^{12} Hz). From (3.32b), we can see that d_p is inversely proportional to the square root of frequency. The skin depth of copper is smaller than 0.66 mm for frequencies higher than 10 kHz, or the 3.3 mm-thick copper layer is more than five skin depths for frequencies higher than 10 kHz. Therefore, the shielding designed for 10 kHz will be adequate to block most of the electromagnetic waves with frequencies higher than 10 kHz.

Permittivities of Various Materials

The permittivities of a wide variety of materials at different frequencies are listed in the literature.* We know that the dielectric constant of almost any material depends on frequency. Take water, for example: its dielectric constant is approximately $80\epsilon_0$ at frequencies below VHF but is reduced to $1.75\epsilon_0$ at optical frequencies. The loss of many materials is caused not only by conductivity but also by the "friction" among polarized molecules.

Let us consider distilled water, which contains very few free electrons. The conductivity and the permittivity of distilled water at 25° C are

$$\sigma = 1.7 \times 10^{-4} \text{ mho/m}$$

$$\epsilon = 78\epsilon_0$$

Using these values in (3.23) with $f = 3$ GHz gives

$$\epsilon = 78\epsilon_0(1 - j1.3 \times 10^{-5}) \text{ (conduction loss only)}$$

But in reality the complex permittivity of distilled water at 3 GHz is

$$\epsilon = 76.7\epsilon_0(1 - j0.157) \text{ (measured)}$$

Adding salt to distilled water greatly increases its conductivity. For example, with 0.1 molal of salt added to water, the conductivity and the permittivity become:

$$\sigma = 1.04 \text{ mho/m}$$

$$\epsilon = 78.2\epsilon_0$$

If we substitute these values into (3.23), with $f = 3$ GHz, we obtain

$$\epsilon = 78.2\epsilon_0(1 - j0.08) \text{ (conduction loss only)}$$

In reality,

$$\epsilon = 75.5\epsilon_0(1 - j0.24) \text{ (measured)}$$

Thus, the electromagnetic loss in saline water at 3 GHz is due both to the conduction loss and to the dielectric loss.

To include all loss factors in the complex permittivity, we write

$$\epsilon = \epsilon' - j\epsilon''$$

where ϵ'' accounts for all losses including that due to conductivity σ/ω. The **loss tangent** $\tan\delta$ is defined as

$$\boxed{\tan\delta = \frac{\epsilon''}{\epsilon'}} \quad \textbf{(loss tangent)} \qquad \textbf{(3.33)}$$

* A. R. Von Hippel, *Dielectric Materials and Applications*, Cambridge: MIT Press, 1954.

The real part of the permittivity ϵ' and the loss tangent of some material are shown as follows:

Freshly fallen snow:	$\epsilon' = 1.20\epsilon_0$,	$\tan\delta = 3 \times 10^{-4}$,	$f = 3$ GHz
Bottom round steak:	$\epsilon' = 40\epsilon_0$,	$\tan\delta = 0.3$,	$f = 3$ GHz
Polystyrene foam:	$\epsilon' = 1.03\epsilon_0$,	$\tan\delta = 0.3 \times 10^{-4}$	$f = 3$ GHz

Microwave Ovens

In cooking, traditional ovens use either electric or gas heat to raise the temperature of the food. They rely on heat conduction to transfer the heat from the outside surface into the food. This process of heat transfer is slow, and the distribution of heating energy between the outside and the inside of the food is often uneven.

A microwave oven converts 60 Hz electric power into microwave power, which is then irradiated into the food. Most meat and vegetables are lossy media at microwave frequencies. While the microwave power penetrates the medium, part of the electromagnetic power is lost to the medium as heat, which instantly raises the temperature of the medium. Note that the loss tangent for steak is quite high at microwave frequencies. This high loss-tangent is what enables a microwave oven (operating frequency 2.45 GHz) to cook a steak. Also note that the dielectric constant of polystyrene foam is very close to that of free space. Microwave experiments often use this material because it is "microwave transparent." This microwave transparency also explains why microwaves in a microwave oven can enter and heat a hamburger contained in a polystyrene-foam box without burning the box.

To estimate the depth of the microwave penetration, we use the permittivity for bottom round steak $\epsilon = 40(1 - j0.3)\epsilon_0$ to calculate the complex wave number k. We find that at 3 GHz

$$k = 402 - j59$$

The penetration depth d_p is equal to $1/k_I = 1.70$ cm. The microwave power heating the inside of a steak 0.85 cm from the surface is roughly 37% of that heating the surface. More important, the power reaches the inside as soon as the oven is turned on. Because of its fast heating capability, the microwave oven is also widely used in industry for cooking and for thawing frozen food.

3.5 RADIATION POWER AND RADIATION PRESSURE

A uniform plane wave carries electromagnetic power. The power density is obtained from the Poynting vector. For a uniform wave in free space with the following electromagnetic fields

$$\mathbf{E} = \hat{x}E_0 e^{-jkz} \tag{3.34a}$$

$$\mathbf{H} = \hat{y}(E_0/\eta) e^{-jkz} \tag{3.34b}$$

the time-averaging Poynting vector $\langle S \rangle$ is, according to (2.38),

$$\langle S \rangle = \hat{z}(1/2)(E_0^2/\eta) \tag{3.35}$$

Equation (3.35) gives the time-average power density, in watts per square meter, carried by a uniform plane wave in free space with the electromagnetic fields given by (3.34). The direction of the Poynting vector is in the direction of wave propagation. This electromagnetic power is sometimes referred to as the **radiation power** of the uniform plane wave. Note that because free space is not a dissipative medium, the power does not attenuate as the wave propagates.

Considering now a uniform plane wave in a dissipative medium. The electromagnetic fields are expressed as

$$\mathbf{E} = \hat{x}E_0 e^{-jkz} \tag{3.36a}$$

$$\mathbf{H} = \hat{y}(E_0/\eta) e^{-jkz} \tag{3.36b}$$

The above expressions are similar to those given in (3.34), except that k and η are complex numbers in (3.36). The time-average Poynting vector is

$$\langle S \rangle = \hat{z}(1/2)(E_0^2/|\eta|) \exp(-2k_I z) \cos(\phi) \tag{3.37}$$

where k_I and ϕ are defined in (3.27). Note that for a uniform plane wave propagating in a dissipative medium, the radiation power is attenuated exponentially. The attenuation is due to the fact that the medium absorbs part of the electromagnetic power.

When a uniform plane wave encounters a flat object, part of the wave is reflected and part is transmitted. The reflection and transmission of uniform plane waves will be discussed in Chapter 4. In this section, however, it will be shown that when a uniform plane wave impinges on a flat surface, it also exerts a force on the surface. This force is referred to as the **radiation pressure**.

Consider a uniform plane wave in free space with electromagnetic fields given in (3.34). Assume that the wave encounters an object. For simplicity, the surface of the object is assumed to be flat and perpendicular to the z axis, as shown in Figure 3.11. It is further assumed that the electromagnetic properties of the object are only slightly different from those of the free space so that reflection of the wave may be neglected. In other words, the object is assumed to be a perfect absorber of the electromagnetic wave.

On the surface of the flat object, the electric field is given by (3.34a). For a charge q in the atomic structure of the object, it will experience an electric force F_e, where, according to (2.30),

$$F_e = \hat{x}qE_0 \cos(\omega t - kz) \tag{3.38}$$

Note that the time-domain expression is used. This force causes the charge to move. But this charge is also subject to atomic forces which tend to restrict its

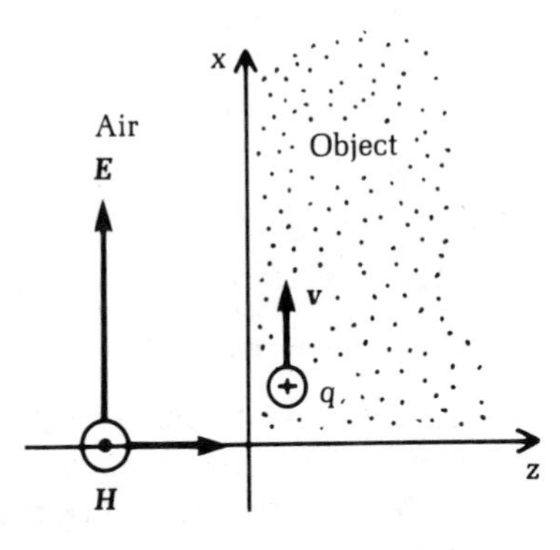

Figure 3.11 An electromagnetic wave impinges from air to a flat object. It exerts a pressure on the surface of the object.

motion. In the steady state, the charge moves at a velocity which is proportional to the electric force F_e. That is,

$$v = \hat{x} dqE_0 \cos(\omega t - kz) \tag{3.39}$$

The constant d is a damping factor. Note that in the above equation, the *velocity* instead of the *acceleration* of the charge q is proportional to the force. This is because the charge in question is in the atomic structure of the object and is bound by atomic forces. Actually, Equation (3.39) is similar to Ohm's law, which states that the current in a medium is linearly proportional to the electric field. That is,

$$J = \sigma E$$

It is well known that Ohm's law is valid in most common materials.

The magnetic force acting on the moving charge is then

$$\begin{aligned} F_m &= qv \times B \\ &= \hat{z}(\mu dq^2 E_0^2/\eta)\cos^2(\omega t - kz) \end{aligned} \tag{3.40}$$

Note that while F_e and v change signs periodically, F_m always remains positive and directed in $\hat{z}$.

Let U_q be the work done on the charge by the electric field. Then

$$\begin{aligned} dU_q/dt &= F_e \cdot v \\ &= dq^2 E_0^2 \cos^2(\omega t - kz) \end{aligned} \tag{3.41}$$

The magnetic field does not do work on q because the magnetic force is always perpendicular to v.

Comparing (3.40) and (3.41) leads to the following equation:

$$dU_q/dt = \hat{z} \cdot F_m \eta/\mu \tag{3.42}$$

The above equation is for a charge on the surface of an object.

Multiplying the above equation by an area A, the left-hand side becomes

$$dU_q A/dt = S_z A \tag{3.43a}$$

where S_z is the z component of the Poynting vector, because the work done to the charges must come from the electromagnetic wave. The right-hand side of (3.42), after multiplying by the area A, becomes

$$\hat{z} \cdot \boldsymbol{F}_{\mathrm{m}} A\eta/\mu = PA\eta/\mu = PA/(\mu\epsilon)^{1/2} \tag{3.43b}$$

where P is the pressure on those charges. Equating (3.43a) and (3.43b) yields

$$\begin{aligned} P &= (\mu\epsilon)^{1/2} S_z \\ &= \epsilon E_0^2 \cos^2(\omega t - kz) \end{aligned} \tag{3.44}$$

The quantity P is referred to as the **radiation pressure**. Notice that the radiation pressure is always positive and directed toward the object illuminated by the EM wave. The time-average radiation pressure is given as follows:

$$\langle \boldsymbol{P} \rangle = \hat{z}(1/2)\epsilon E_0^2 = (1/c)\langle \boldsymbol{S} \rangle \qquad \text{N/m}^2 \tag{3.45}$$

Equations (3.44) and (3.45) are valid only for the case of a uniform plane wave in free space impinging on a flat surface of a perfect absorber. When the object is a perfect reflector, the radiation pressure is twice the value given in (3.44). For arbitrarily shaped objects, the radiation pressure is expressed in terms of Maxwell stress tensor.* Interested readers should refer to the reference cited for further details.

Example 3.9 *The radiation pressure of an electromagnetic wave is usually too weak for humans to feel. This example gives the order of magnitude of the radiation pressure due to solar radiation.*

What is the total force due to solar radiation pressure acting on a person sunbathing on a beach? Compare the strength of this force to the gravitational force.

Solution We mentioned in Section 3.1 that the solar radiation power measured on the earth is approximately 1.4 kW/m^2. Substituting this number in (3.45) yields

$$\begin{aligned} \langle P \rangle &= 1.4 \times 10^3/3 \times 10^8 \\ &= 4.7 \times 10^{-6} \text{ N/m}^2 \end{aligned}$$

Assume that if a person's cross-sectional area is approximately 0.6 m^2, then the total time-average solar radiation pressure force on this person is

$$F = 2.8 \times 10^{-6} \text{ newton}$$

The gravitation force on a person weighing 70 kg on earth is 686 newtons. Therefore that person cannot feel the force due to solar radiation.

---

* J. D. Jackson, *Classical Electrodynamics*, 2nd edition, New York: John Wiley & Sons, 1975, Section 6.8.

Example 3.10 *Although the solar radiation pressure cannot be felt by humans on earth, it can be utilized in space, as is shown in this example.*

Consider a place in the solar system where the gravitational force on an object is mainly due to the sun. A space "sailboat" is to be designed such that the solar radiation pressure on its sail could counter the sun's gravitational force. Assume that the total weight of the "boat" is 1000 kg. What is the surface area of the sail?

Solution The gravitational force between two masses of m_1 and m_2 separated by a distance r is

$$F_g = 6.67 \times 10^{-11}(m_1 m_2/r^2) \tag{3.46}$$

Assuming that the sun radiates an equal amount of electromagnetic power in all directions, the solar radiation power density at a distance r from the sun can be expressed as follows

$$\langle S \rangle = 1.4 \times 10^3 (r_e^2/r^2),$$

where r_e is the distance between earth and sun. If A is the surface area of the sail, then the total radiation force on the sail is

$$F_r = 1.4 \times 10^3 (A/c)(r_e^2/r^2) \tag{3.47}$$

Equating (3.46) and (3.47) and using the following constants:

$$m_1 = \text{mass of the "sail boat"} = 10^3 \text{ kg}$$

$$m_2 = \text{mass of the sun} = 1.99 \times 10^{30} \text{ kg}$$

$$c = \text{velocity of electromagnetic waves in free space}$$

$$= 3 \times 10^8 \text{ m/s}$$

$$r_e = \text{mean distance between earth and sun} = 1.5 \times 10^{11} \text{ m}$$

the surface area A is

$$A = 1.26 \times 10^6 \text{ m}^2$$

The National Aeronautics and Space Administration has plans to build a similar "space sailboat." The sail is to be made of thin polymer film coated with aluminum only a few atoms thick. The above calculation assumes that the sail is a perfect absorber. For aluminum-coated film it is more accurate to assume that the film totally reflects the electromagnetic wave. In this case, the radiation pressure is twice as great and the surface area needed is only half as large.

The Tails of Comets

Comets are generally nebulous celestial bodies with small masses, and most of them travel along elliptical paths in the solar system. More than 1600 comets have been recorded so far, but the actual number of comets may be as many as 100 billion. A comet is usually a conglomeration with an icy nucleus made of water, methane, ammonia, and dust particles, held together by gravitational attraction. The nucleus is approximately 1 to 100 km in diameter.

When a comet approaches the sun, ices in its nucleus begin to vaporize and the nucleus is surrounded by escaping gases that form a sphere of the order 10^4 to 10^5 km in diameter. This sphere, which scatters the sunlight and is quite visible, is called the head, or the coma, of the comet. When the comet is close to the sun, gases and dust particles in the coma ultimately form the comet's tail, which in some cases extends 100 million kilometers (Figure 3.12). It is because of this long shining tail that the spectacular space visitor is given the name "comet," which in Greek means "hairy star."

Figure 3.12 Halley's comet (photographed in 1986 near Houston, Texas, by Mark V. Rowe).

Comet sightings have been recorded in the history of humankind for more than two thousand years. An ancient Chinese drawing made between 246–177 B.C. shows 27 different kinds of comets (Figure 3.13). Modern observation of the comet has confirmed that comet tails vary greatly in appearance from comet to comet, and the tail of each individual comet may change from time to time.

What is the mechanism that causes the formation of a comet's tail? It has been noted that all comet tails point away from the sun regardless of whether the comet is approaching or moving away from the sun. This observation provides strong evidence that there are forces due to the sun that are greater than the gravitational force between the comet and the sun. Otherwise, comet tails would point toward the sun. The current explanation of comet tails is as follows. There are neutral fine dust particles and ionized gas molecules in a comet's coma. The **radiation pressure** of the sunlight pushes these neutral fine dust particles away

Figure 3.13 Various comet shapes drawn on silk found in China. These figures were painted between 246 to 177 B.C. Below the figures are Chinese names for these comets.

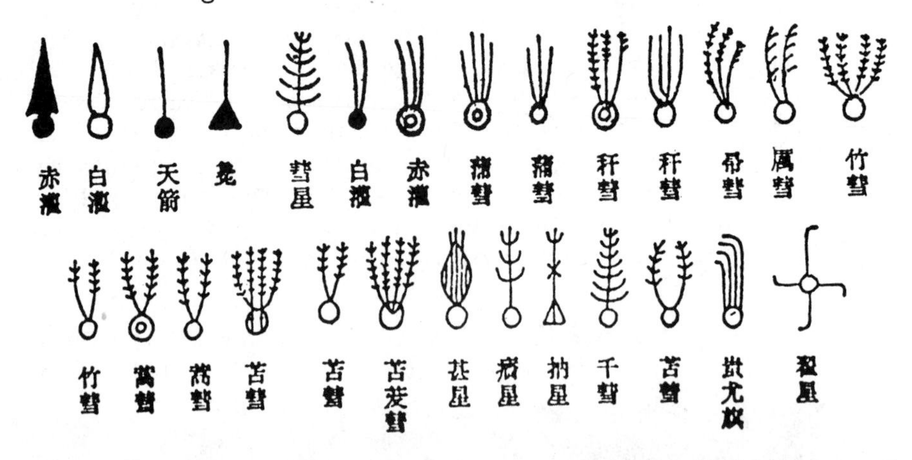

from the sun, and the latter form a tail called a Type II tail. Furthermore, the ionized gases form a neutral plasma consisting of ions and electrons. The plasma forms a Type I tail because of the **solar wind**. Solar wind is the flow of protons and electrons issuing from the sun and moving outward at a speed up to 1000 km/s. These fast-moving particles react with the comet's plasma and form the plasma tail.*

SUMMARY

1. Electromagnetic waves come from the sun and stars, thunderstorms and tornados, power lines and microwave ovens, and television stations and cellular phones. The frequency ranges from zero to 10^{19} Hz. Visible light is an EM wave.
2. From Maxwell's equations in free space, we can derive a simple solution where the **E** field is given by (3.6) and the **H** field by (3.11). This set of solutions is interpreted physically as a wave. The wavelength λ and the wave number k are given in terms of the frequency of the wave, the dielectric permittivity ϵ, and the magnetic permeability μ of the free space.
3. The factor e^{-jkz} represents a wave propagating in the positive z direction, and e^{+jkz} represents a wave propagating in the negative z direction.
4. The uniform plane wave in free space has the following properties:
   (a) Its **E** field and **H** field are perpendicular to each other, and both are perpendicular to the direction of wave propagation.

---

* L. F. Biermann and R. Lust, "The tails of comets," *Scientific American*, Vol. 199, October 1958, pp. 44–50.

E. N. Parker, "The solar wind," *Scientific American*, Vol. 210, no. 4, April 1964, pp. 66–76.

(b) Its **E** field and **H** field are proportional to each other by a constant called the intrinsic impedance η. They are in phase.

(c) The speed of a uniform plane wave is determined by $1/\sqrt{\mu_0 \epsilon_0}$, which is equal to the speed of the light in vacuum.

(d) The **E** and **H** fields do not decay as the wave propagates.

5. The uniform plane wave carries an EM power. The density of the power is determined by the Poynting vector.

6. IEEE has issued a guideline giving the safety limit of human exposure to EM radiation in the frequency range 3 kHz to 300 GHz.

7. The polarization of an EM wave is determined by the locus of the tip of the $\boldsymbol{E}$ field vector as a function of time. If the locus traces a straight line, a circle, or an ellipse, then the polarization is linear, circular, or elliptical, respectively. Otherwise, it is randomly or partially polarized.

8. The mathematical expression of a uniform plane wave in a dissipative medium is identical to that in the free space. The difference is that the wave number k and the intrinsic impedance η are complex numbers in a dissipative medium.

9. In a non-dissipative medium, the wave number k and the intrinsic impedance η are real. A uniform EM wave in a non-dissipative medium has the following properties:

   (a) Its **E** field and **H** field are perpendicular to each other, and both are perpendicular to the direction of wave propagation.

   (b) Its **E** field and **H** field are proportional to each other by a constant called the intrinsic impedance η. They are in phase.

   (c) The speed of a uniform plane wave is determined by $1/\sqrt{\mu\epsilon}$.

   (d) The **E** and **H** fields do not decay as the wave propagates.

10. The uniform plane wave in a dissipative medium has the following properties:

   (a) Its **E** field and **H** field are perpendicular to each other, and both are perpendicular to the direction of wave propagation.

   (b) Its **E** field and **H** field are proportional to each other by a constant called the intrinsic impedance η. Since η is generally complex, usually **E** and **H** are not in phase.

   (c) The phase velocity of an EM wave is determined by ω/k_R.

   (d) Both the **E** and the **H** fields decay like $e^{-k_I z}$. The skin depth or the penetration depth is defined as $1/k_I$. Thus the fields decay by a factor e^{-1} or 8.686 dB for every skin depth the wave travels.

11. When an EM wave impinges on an absorbing material, it exerts a radiation pressure on the material. The pressure is given by (3.44) or (3.45).

Problems

3.1 Estimate the power density of electromagnetic radiation from the sun received on earth in the same frequency band as that of the VHF television channel 2 (54–60 MHz).

3.2 The closest star to us is α Centauri, which is approximately 4.33 light-years distant from the sun. The farthest planet is Pluto, which is 6×10^9 km from the sun. Is α Centauri or Pluto closer to the sun? A light-year is a unit of length that is the distance a light wave covers in one year.

3.3 An electromagnetic pulse is sent from an earth station to the moon, and the reflected pulse is received 2.56 s later. How far is the moon from the earth? (An electromagnetic pulse consists of a wide spectrum of electromagnetic waves at different frequencies.)

3.4 Find the SI units of the following quantities associated with a uniform electromagnetic wave: (a) ω, (b) k, (c) f, (d) T, and (e) λ.

3.5 A helium-neon laser emits light at a wavelength 6.328×10^{-7} m in air. Calculate its frequency, period, and wave number.

3.6 Figure P3.6 shows a dipole antenna. It is very effective in receiving television signals when its length is approximately equal to one-half the signal wavelength. What are approximate antenna lengths for receiving signals for the following: (a) Channel 2 ($f = 57$ MHz) and (b) Channel 13 ($f = 213$ MHz)?

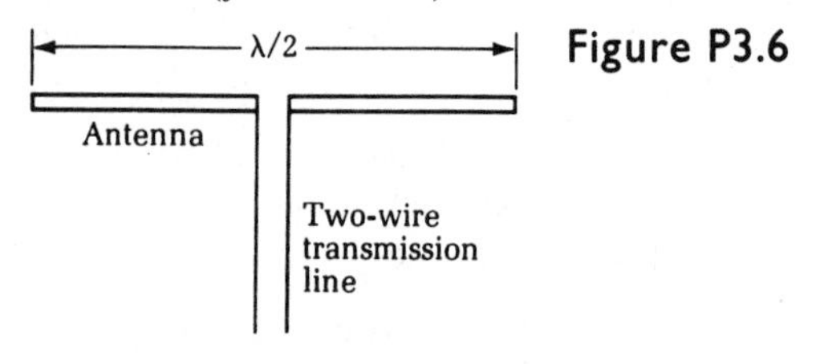

Figure P3.6

3.7 The following set of electromagnetic fields satisfies the time-harmonic Maxwell's equations in free space:

$$\mathbf{E} = E_0 e^{+jkz} \hat{x}$$

and

$$\mathbf{H} = H_0 e^{+jkz} \hat{y}$$

Express H_0 and k in terms of E_0 and ϵ_0 and μ_0.

3.8 Do the fields in the previous problem represent a uniform plane wave? In what direction does the wave travel? Find its velocity and determine the time-average Poynting vector $\langle S \rangle$.

3.9 Study the following **E** field in a source-free region:

$$\mathbf{E} = \hat{x} E_0 e^{-jkx}$$

Does it satisfy Maxwell's equations? If so, find the k and the **H** field. If not, explain why not.

3.10 Let $\mathbf{H} = 5e^{+j8y}\hat{x}$ and the medium be free space. Use (3.2) to find the corresponding $\mathbf{E}$ field. Express your answer in terms of ω, ϵ, etc. What is the direction of propagation of the wave and what is the frequency?

3.11 Find the polarization (linear, circular, or elliptical and left-hand or right-hand) of the following fields:

(a) $\mathbf{E} = (j\hat{x} + \hat{y})e^{-jkz}$
(b) $\mathbf{E} = [(1 + j)\hat{y} + (1 - j)\hat{z}]e^{-jkx}$
(c) $\mathbf{E} = [(2 + j)\hat{x} + (3 - j)\hat{z}]e^{-jky}$
(d) $\mathbf{E} = (j\hat{x} + j2\hat{y})e^{+jkz}$

3.12 Show that, if $a = b$ and $\phi_a - \phi_b = \pi/4$, the wave is elliptically polarized. [Refer to Equation (3.13).] Do not try to obtain an analytical expression for the locus. Just obtain a pair of parametric equations similar to (3.14), calculate E_x and E_y at ten points ($\omega t = 0, 10°, \ldots, 90°$), and sketch the locus.

3.13 Show that an elliptically polarized wave can be decomposed into two circularly polarized waves, one left-handed and the other right-handed. Hint: Let

$$\mathbf{E} = (\mathrm{a}\hat{x} + \mathrm{b}\hat{y})e^{-jkz}$$

where a and b are, in general, complex numbers. Then, let

$$\mathbf{E} = (\mathrm{a}'\hat{x} + j\mathrm{a}'\hat{y})e^{-jkz} + (\mathrm{b}'\hat{x} - j\mathrm{b}'\hat{y})e^{-jkz}$$

and solve for a′ and b′ in terms of a and b.

3.14 Show that a linearly polarized wave can be decomposed into two circularly polarized waves.

3.15 A dipole antenna is in the x-y plane and makes a 45° angle to the x axis. A receiver attached to the antenna is calibrated to read directly the component of the $\boldsymbol{E}$ field that is parallel to the dipole. What are the readings when the fields are those given in (a)–(d) of Problem 3.11?

3.16 An electromagnetic wave in a vacuum has frequency f_0, wavelength λ_0, wavenumber k_0, and velocity v_0. When it enters a dielectric medium characterized by μ_0 and $\epsilon = 4\epsilon_0$, what are the f, λ, k, and v of the wave in this medium?

3.17 A uniform plane wave propagates in a non-dissipative medium in the positive z direction. The frequency of the wave is 20 MHz. A probe located at $z = 0$ measures the phase of the wave to be 98°. An identical probe located at $z = 2$ m measures the phase to be $-15°$. What is the relative dielectric constant of the medium?

3.18 Aluminum has $\epsilon = \epsilon_0$, $\mu = \mu_0$, and $\sigma = 3.54 \times 10^7$ mho/m. If an antenna for UHF reception is made of wood coated with a layer of aluminum, and if its thickness ought to be five times greater than the skin depth of the aluminum at that frequency, determine the thickness of the aluminum layer. Is ordinary aluminum foil thick enough for that purpose? Use $f = 1$ GHz. Ordinary aluminum foil is approximately 1/1000 in. thick.

3.19 Calculate the attenuation rate and skin depth of earth for a uniform plane wave of 10 MHz. Take the following data for the earth: $\mu = \mu_0$, $\epsilon = 4\epsilon_0$, and $\sigma = 10^{-4}$ mho/m.

3.20 Find the power density in earth where the field intensity is 1 V/m. Use the data in Problem 3.19.

3.21 Assume that a medium has $\epsilon = 10\epsilon_0$, $\sigma = 0.1$ mho/m, and $\mu = \mu_0$. A uniform electromagnetic wave of 100 MHz propagates in this medium in the $\hat{x}$ direction. Find the ratio of the electric field at $x = 0$ to that at $x = 0.1$ m. Also find the phase difference of the electric field between these points.

3.22 Suppose that an airplane uses a radar to measure its altitude. Let the frequency of the radar be 3 GHz. Suppose further that the ground is covered with a meter of hard-packed snow.

(a) What is the difference between the apparent altitude measured by the radar and the true altitude?

(b) How much attenuation in dB does the radar signal suffer because of the snow? Consider only the attenuation of the wave in the snow, and neglect the effect of snow on the reflection at air-snow and at snow-ground interfaces. Refer to Figure P3.22. Use $\epsilon = 1.5\epsilon_0$ and $\tan\delta = 9 \times 10^{-4}$ for hard-packed snow at 3 GHz.

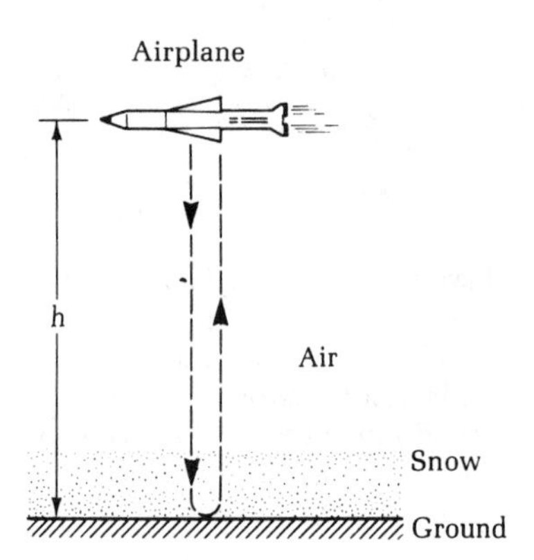

Figure P3.22

3.23 The following data are given for a uniform plane wave in a dissipative medium:

(i) amplitude of E_x at $z = 0$ is 1 V/m,
(ii) phase of E_x at $z = 0$ is zero,
(iii) $k = 0.5 - j0.5$ (1/m),
(iv) direction of propagation is in $\hat{z}$,
(v) intrinsic impedance of the medium is $1 + j$ ohms

(a) Find the phasor expression for E_x as a function of z.
(b) Find the phasor expression for **H** as a function of z.
(c) Sketch the time-domain E_x at $z = 0$ and at $z = 2$ m, as functions of ωt.
(d) Sketch the time-domain $\boldsymbol{H}$ fields at $z = 0$ and $z = 2$ m as functions of ωt.

3.24 Assume that the field in air is given by Problem 3.11(a); what is the radiation pressure on an object when the wave hits it? Assume that the object is a perfect absorber. You first need to find the corresponding **H** field in air.

3.25 Consider that a small space vehicle with 100 kg of mass is located in outer space where the gravitational field is negligible and the fuel has been exhausted. A searchlight of 1 kW is turned on, with hopes that the vehicle will gain some speed. How much speed will it finally gain if the searchlight can last 48 hours? Hint: The light wave carries radiation pressure, and there is a reaction force on the source of the light.

3.26 An ice particle of radius a is r distance away from the sun. The gravitational force acting on the particle is given by (3.46). The ice particle's mass can be obtained from its volume and its density, which is assumed to be 1 gram/cm^3. The ice particle is also subject to radiation pressure, which is given in (3.47). The force acting on the ice particle due to radiation pressure is approximately equal to the cross-sectional area of the particle times the radiation pressure. Show that, when the particle's radius is less than a critical value, the radiation force will be greater than the gravitational force, and this critical radius is independent of r, the distance from the sun. As a result, all particles with radii smaller than this critical radius tend to be blown out of the solar system. Find the value of this critical radius.

4 REFLECTION AND TRANSMISSION OF WAVES

We have stated that electromagnetic fields in a medium must satisfy Maxwell's equations and the medium's constitutive relations. We have also shown that uniform plane waves can exist in an unbounded medium and that the velocity, wave number, and amplitude of the wave depend on the characteristics of the medium.

We shall now consider the reflection and transmission of electromagnetic waves at a boundary surface separating two different media. This situation is shown in Figure 4.1. The electromagnetic fields in both medium 1 and medium 2 satisfy Maxwell's equations and their respective constitutive relations. At the boundary surface, the derivatives do not exist, and the differential equations break down. However, if we write in the forms of differences, Maxwell's equations are applicable at the boundary surfaces as well. We shall make use of these forms to derive the boundary conditions and thus to supplement the differential descriptions whenever a boundary surface is encountered. Applying the boundary conditions to a plane interface, we shall then treat the problem of reflection and transmission of a plane wave from a boundary separating two different media, and we shall study the applications of the results.

4.1 BOUNDARY CONDITIONS

Consider a ribbonlike area across the boundary (Figure 4.1). The longer side has length ℓ and the shorter side length w. We choose our coordinates such that x is tangent to the boundary, y is parallel to the normal vector $\hat{\boldsymbol{n}}$, and z is pointing out of the paper.

Take z components of Ampère's law (2.28b), and we have

$$\frac{\partial \mathrm{H}_y}{\partial x} - \frac{\partial \mathrm{H}_x}{\partial y} = \mathrm{J}_z + j\omega \mathrm{D}_z$$

Figure 4.1 A region containing two media with different characteristics. The unit vector normal to the boundary pointing from medium 2 to medium 1 is denoted $\hat{\boldsymbol{n}}$. If the boundary is curved, $\hat{\boldsymbol{n}}$ is a function of position.

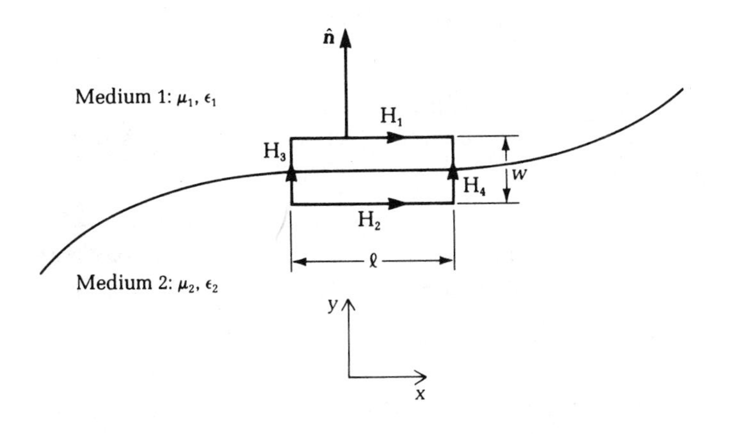

Or, approximately,

$$\frac{\mathbf{H}_4 - \mathbf{H}_3}{\ell} - \frac{\mathbf{H}_1 - \mathbf{H}_2}{w} = \mathbf{J}_z + j\omega \mathbf{D}_z$$

or

$$\mathbf{H}_2 - \mathbf{H}_1 = \mathbf{J}_z w + \left(j\omega \mathbf{D}_z + \frac{\mathbf{H}_3 - \mathbf{H}_4}{\ell} \right) w \tag{4.1}$$

We now let the ribbon area shrink to a point in a manner such that w goes to zero before ℓ does.

We also recognize the fact that in producing electromagnetic waves, with, for instance, an antenna made of perfect conductors, the current flows on the conductor surface because the conductor's skin depth is zero. We call such a current a **surface current**; it flows along the surface with an infinitesimal depth w and a volume current density $\mathbf{J}_v$ such that $\mathbf{J}_v w$ is a constant. Thus, we formally define a surface-current density to be

$$\mathbf{J}_s = \lim_{w \to 0} \mathbf{J}_v w \tag{4.2}$$

Note that, although current densities may be infinite at the boundary, the field vectors are finite everywhere.

Taking the limit as $w \to 0$, we obtain from (4.1)

$$(\mathbf{H}_2 - \mathbf{H}_1) = \mathbf{J}_s$$

or, in general,

$$\boxed{\hat{\boldsymbol{n}} \times (\mathbf{H}_1 - \mathbf{H}_2) = \mathbf{J}_s} \quad \textbf{(boundary condition for H field)} \tag{4.3}$$

Remember that $\mathbf{H}_2$ is the magnetic field on the side of medium 2 and $\mathbf{H}_1$ is the magnetic field on the side of medium 1.

Applying the above argument to Faraday's law (2.28a), we obtain

$$\boxed{\hat{n} \times (\mathbf{E}_1 - \mathbf{E}_2) = 0} \quad \textbf{(boundary condition for E field)} \tag{4.4}$$

because there is no source term.

The boundary conditions derived above can be summarized in the following statement:

> The tangential electric field **E** is continuous across the boundary surface. The discontinuity in the tangential magnetic field **H** is equal to the surface current $\mathbf{J}_s$.

Notice that, for ordinary materials with finite conductivity, the skin depth δ is not equal to zero and that, from the definition of $\mathbf{J}_s$ in (4.2), we know that the surface current $\mathbf{J}_s$ is zero in materials with finite conductivity, even though the volume current $\mathbf{J}_v$ is not zero. $\mathbf{J}_s$ may only exist on the surface of perfect conductors. We can then deduce the following two statements:

1. For two media having finite conductivities, both tangential electric field **E** and magnetic field **H** are continuous across the boundary.
2. On the surface of a perfect conductor, the tangential field **E** is zero, and the surface current $\mathbf{J}_s = \hat{n} \times \mathbf{H}$, where $\hat{n}$ is the unit vector normal to the conductor's surface.

The last statement is deduced from the fact that a perfect conductor allows no electromagnetic field to exist inside ($\mathbf{E}_2 = 0$, $\mathbf{H}_2 = 0$).

Let us now consider the "pill box" regions shown in Figure 4.2 and apply

Figure 4.2 Boundary condition of the vector **D**.

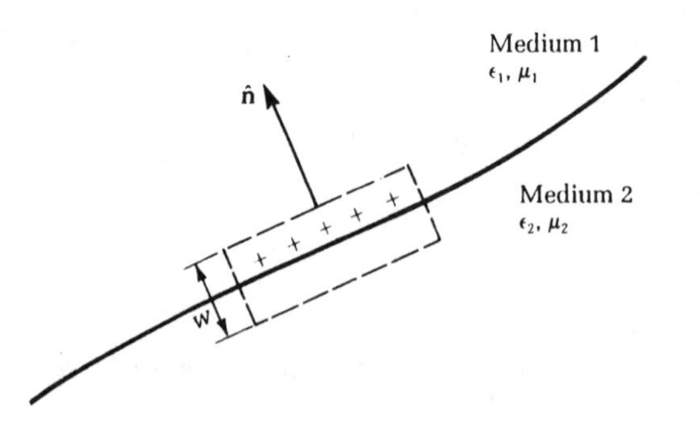

(2.28d) to this volume. According to the physical meaning of the divergence operator discussed in Section 2.1, (2.28d) states that the total D flux out of a volume is equal to the total charge in the volume. In the present case,

$$\text{flux of } D \text{ out of the box} = (D_{1n} - D_{2n}) \cdot \text{area}$$

$$\text{total charge in the box} = \rho_s \cdot \text{area}$$

where ρ_s is the surface charge density (coulombs per square meter) on the boundary. Notice that the flux of D out of the box from the side walls of the pill box is proportional to the thickness of the box w, and it has been neglected. Therefore, (2.28d) yields the following boundary condition:

$$\boxed{D_{1n} - D_{2n} = \rho_s} \quad \textbf{(boundary condition for D field)} \qquad \textbf{(4.5)}$$

Similarly, from (2.28c), we obtain

$$\boxed{B_{1n} - B_{2n} = 0} \quad \textbf{(boundary condition for B field)} \qquad \textbf{(4.6)}$$

The above boundary conditions can be summarized in the following statement:

> The normal component **B** is continuous across the boundary surface. The discontinuity in the normal component of **D** is equal to the surface charge density ρ_s.

Example 4.1

*This example shows that, from the boundary conditions, we can find the surface current on a perfect conductor from the **H** field, and vice versa.*

A region contains a perfectly conducting half-space and air, as shown in Figure 4.3. We know that the surface current $\mathbf{J}_s$ on the surface of the perfect conductor is $\mathbf{J}_s = \hat{x}2$ amperes per meter. What is the tangential **H** field in air just above the perfect conductor?

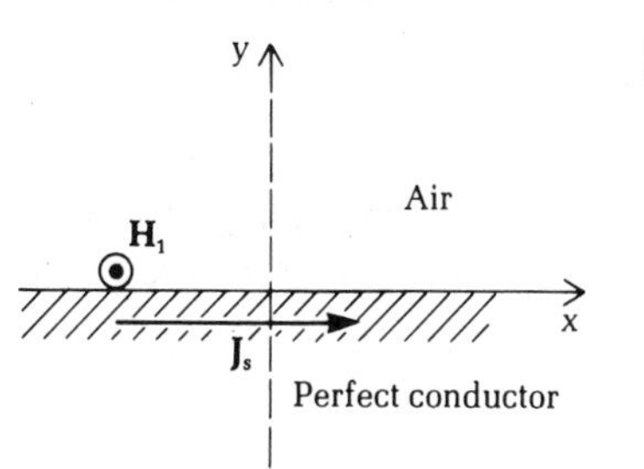

Figure 4.3 **H** in air and $\mathbf{J}_s$ on the surface of the conductor are related by the boundary condition (Example 4.1).

Solution We solve this problem by using (4.3), which gives

$$\hat{n} \times (\mathbf{H}_1 - \mathbf{H}_2) = \hat{x}2$$

Now $\hat{n}$ is equal to $\hat{y}$ and $\mathbf{H}_2$ is identically zero because medium 2 is a perfect conductor. Consequently,

$$\hat{y} \times \mathbf{H}_1 = \hat{x}2$$

Thus,

$$\mathbf{H}_1 \text{ tangential} = \hat{z}2 \text{ amperes per meter at } y = 0^+$$

Example 4.2 *From the boundary conditions, we can find the surface charge on a perfect conductor from the **D** field, and vice versa.*

The electric field on the surface of a perfect conductor is equal to 2 V/m. The conductor is immersed in water, as shown in Figure 4.4. What is the surface-charge density on the conductor?

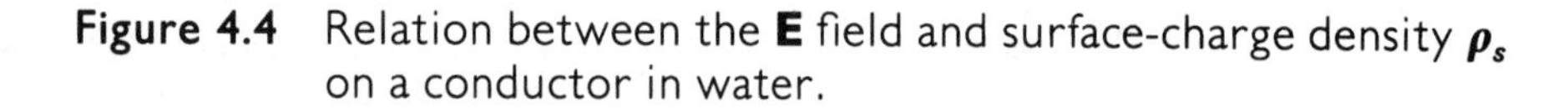

Figure 4.4 Relation between the **E** field and surface-charge density ρ_s on a conductor in water.

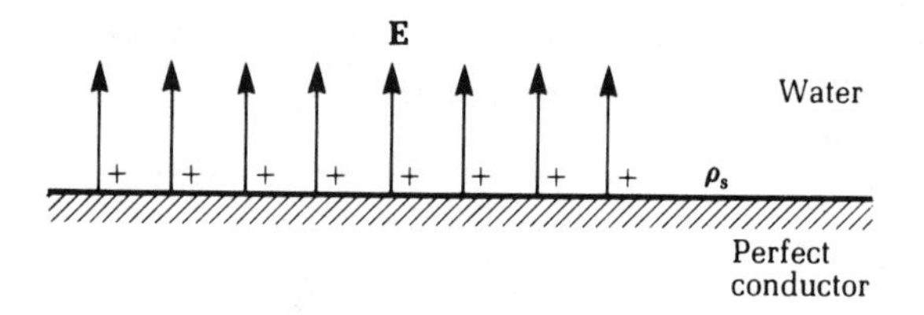

Solution By definition, the field in a perfect conductor is equal to zero. Thus, $D_{2n} = 0$ in (4.5) and we obtain

$$\rho_s = D_{1n} = \epsilon_1 E_{1n} = 80 \times \frac{1}{36\pi} \times 10^{-9} \times 2 = 1.41 \times 10^{-9} \text{ C/m}^2$$

4.2 REFLECTION AND TRANSMISSION AT A DIELECTRIC INTERFACE

First consider a uniform plane wave propagating in a direction other than the $\hat{z}$ direction. Let the wave be linearly polarized, with the electric field vector in the $\hat{y}$ direction. We write

$$\mathbf{E} = \hat{y}E_0 e^{-jk_x x - jk_z z} \tag{4.7}$$

This wave is propagating in the positive $\hat{x}$ direction and positive $\hat{z}$ direction as shown in Figure 4.5. A wave vector $\boldsymbol{k}$ is defined by

$$\boldsymbol{k} = \hat{x}k_x + \hat{z}k_z \quad \textbf{(definition of wave vector)} \tag{4.8}$$

Figure 4.5 A uniform plane wave is propagating in the **k** direction.

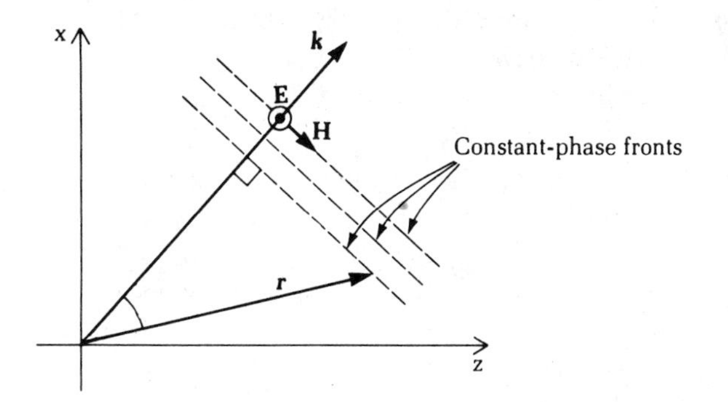

To specify the observation points, we define a position vector

$$\boldsymbol{r} = \hat{x}x + \hat{z}z \tag{4.9}$$

We can see from (4.7) that

$$\mathbf{E} = \hat{y}E_0 e^{-j\boldsymbol{k}\cdot\boldsymbol{r}} \tag{4.10}$$

The phase of the wave will be a constant when

$$\boldsymbol{k} \cdot \boldsymbol{r} = \text{constant}$$

This defines a constant-phase front, which is a plane perpendicular to $\boldsymbol{k}$. Note that for all observation points $\boldsymbol{r}$ on this phase front, $\boldsymbol{k} \cdot \boldsymbol{r}$ is a constant. The wave is called a plane wave because all its constant-phase fronts are planes. The plane wave is also called a uniform plane wave because the electric field assumes uniform amplitudes along the constant-phase fronts. The magnetic field **H** associated with the electric field is calculated from Faraday's law

$$\mathbf{H} = -\frac{1}{j\omega\mu}\nabla \times \mathbf{E} = (-\hat{x}k_z + \hat{z}k_x)\frac{E_0}{\omega\mu}e^{-jk_x x - jk_z z}$$

We see that **H** is perpendicular to both **E** and $\boldsymbol{k}$ as $\mathbf{H} \cdot \mathbf{E} = 0$ and $\boldsymbol{k} \cdot \mathbf{H} = 0$.

Furthermore,

$$\mathbf{E} \times \mathbf{H}^* = (\hat{x}k_x + \hat{z}k_z)\frac{|E_0|^2}{\omega\mu}$$

which is in the direction of $\boldsymbol{k}$. Thus, as indicated in Figure 4.5, the magnetic-field vector $\mathbf{H}$ is such that the Poynting vector is pointing in the wave vector $\boldsymbol{k}$ direction.

We now let this uniform plane wave impinge upon a plane dielectric interface, as shown in Figure 4.6. When the incident wave impinges on the boundary at an oblique angle, the normal of the boundary and the incident ray form a plane called the **plane of incidence**. The $\mathbf{E}$ field of the incident wave may be polarized perpendicular or parallel to the plane of incidence.

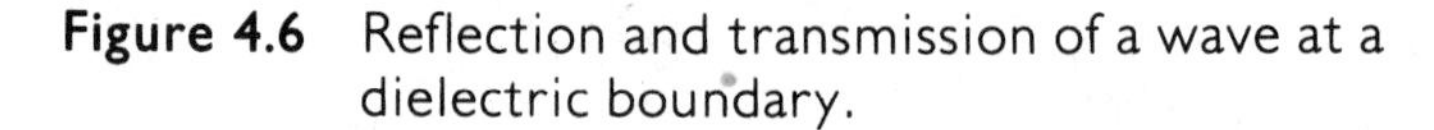

Figure 4.6 Reflection and transmission of a wave at a dielectric boundary.

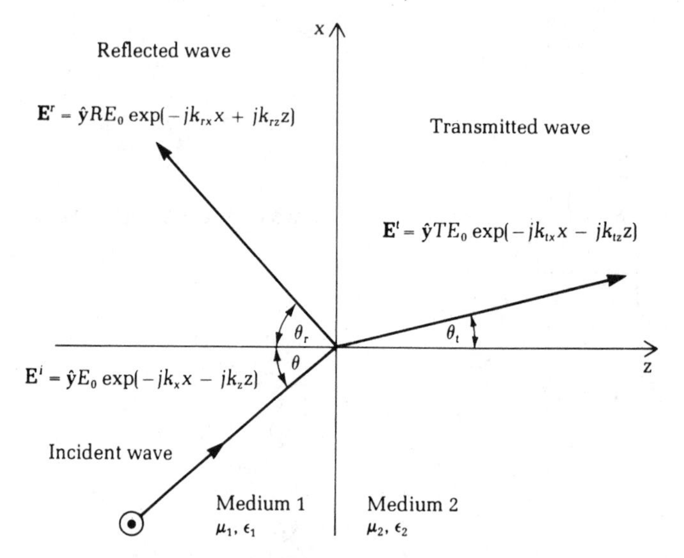

We now consider a **perpendicularly** polarized incident wave. The mathematical expressions of the $\mathbf{E}$ and $\mathbf{H}$ fields associated with the incident wave are given below.

$$\mathbf{E}^i = \hat{y}E_0 e^{-jk_x x - jk_z z} \tag{4.11}$$

$$\mathbf{H}^i = (-\hat{x}k_2 + \hat{z}k_x)\frac{E_0}{\omega\mu_1} e^{-jk_x x - jk_z z} \tag{4.12}$$

The reflected wave is given by

$$\mathbf{E}^r = \hat{y}R_I E_0 e^{-jk_{rx} x + jk_{rz} z} \tag{4.13}$$

$$\mathbf{H}^r = (+\hat{x}k_{rz} + \hat{z}k_{rx})\frac{R_I E_0}{\omega\mu_1} e^{-jk_{rx} x + jk_{rz} z} \tag{4.14}$$

where R_I is the reflection coefficient for the perpendicularly polarized wave. The wave vector for the reflected wave is $\boldsymbol{k}_r = \hat{x}k_{rx} - \hat{z}k_{rz}$, with a negative sign for the $\hat{z}$ component to signify the fact that the reflected wave propagates in the positive $\hat{x}$ direction and the *negative* $\hat{z}$ *direction.* For the transmitted wave, we have

$$\mathbf{E}^t = \hat{y}T_I E_0 e^{-jk_{tx}x - jk_{tz}z} \tag{4.15}$$

$$\mathbf{H}^t = (-\hat{x}k_{tz} + \hat{z}k_{tx})\frac{T_I E_0}{\omega\mu_2} e^{-jk_{tx}x - jk_{tz}z} \tag{4.16}$$

where T_I is the transmission coefficient. Assume that neither of the two media under consideration is a perfect conductor, so that the surface current $\boldsymbol{J}_s = 0$. Then, the boundary conditions (4.4) and (4.3) require that both the tangential electric-field and magnetic-field components be continuous at $z = 0$. We thus have

$$e^{-jk_x x} + R_I e^{-jk_{rx}x} = T_I e^{-jk_{tx}x} \tag{4.17}$$

$$-\frac{k_z}{\omega\mu_1} e^{-jk_x x} + \frac{k_{rz}}{\omega\mu_1} R_I e^{-jk_{rx}x} = -\frac{k_{tz}}{\omega\mu_2} T_I e^{-jk_{tx}x} \tag{4.18}$$

It is important to note that the incident-wave field as well as the reflected-wave field are everywhere in medium 1 and that the above equations must be satisfied for *all values of* x. The consequence is that

$$\boxed{k_x = k_{rx} = k_{tx}} \qquad \textbf{(phase matching condition)} \tag{4.19}$$

According to (4.19), the tangential components of the three wave vectors $\boldsymbol{k}$, $\boldsymbol{k}_r$, and $\boldsymbol{k}_t$ are equal. This condition is known as the **phase matching condition.**

We can obtain the magnitudes of the three wave vectors by substituting the solutions for $\mathbf{E}^i$, $\mathbf{E}^r$, and $\mathbf{E}^t$ into the wave equation

$$(\nabla^2 + \omega^2\mu_1\epsilon_1)\begin{Bmatrix}\mathbf{E}^i \\ \mathbf{E}^r\end{Bmatrix} = 0$$

and

$$(\nabla^2 + \omega^2\mu_2\epsilon_2)\mathbf{E}^t = 0$$

We find

$$k_x^2 + k_z^2 = \omega^2\mu_1\epsilon_1 = k_1^2 \tag{4.20a}$$

$$k_{rx}^2 + k_{rz}^2 = \omega^2\mu_1\epsilon_1 = k_1^2 \tag{4.20b}$$

and

$$k_{tx}^2 + k_{tz}^2 = \omega^2\mu_2\epsilon_2 = k_2^2 \tag{4.21}$$

From the phase matching condition $k_x = k_{rx} = k_{tx}$ and from (4.20a) and (4.20b), we find that $k_z = k_{rz}$. In Figure 4.6 we show that the z components of $\boldsymbol{k}_r$ and $\boldsymbol{k}_i$ have the same magnitude, although they point in opposite directions.

Using the phase-matching condition (4.19) and the fact that $k_{rz} = k_z$ in (4.17) and (4.18), we find that the following two equations:

$$1 + R_I = T_I$$

$$1 - R_I = \frac{\mu_1 k_{tz}}{\mu_2 k_z} T_I$$

which gives rise to the solution for R_I and T_I with

$$R_I = \frac{\mu_2 k_z - \mu_1 k_{tz}}{\mu_2 k_z + \mu_1 k_{tz}} \quad \textbf{(reflection coefficient for perpendicularly polarized wave)} \tag{4.22}$$

$$T_I = \frac{2\mu_2 k_z}{\mu_2 k_z + \mu_1 k_{tz}} \quad \textbf{(transmission coefficient for perpendicularly polarized wave)} \tag{4.23}$$

Equations (4.22) and (4.23) are known respectively as the Fresnel reflection and transmission coefficients for the perpendicular polarization.

Referring to the angle of incidence θ, the angle of reflection θ_r, and the angle of transmission θ_t, we see that

$$k_x = k_1 \sin\theta$$

$$k_{rx} = k_1 \sin\theta_r$$

$$k_{tx} = k_2 \sin\theta_t$$

Substituting these relations into the phase-matching condition (4.19), we find

$$k_1 \sin\theta_r = k_1 \sin\theta = k_2 \sin\theta_t$$

The first equal sign states that $\theta_r = \theta$—that is, the angle of reflection is equal to the angle of incidence. To relate the second equality to a familiar law in optics, we define the refractive indices

$$n_1 = c\sqrt{\mu_1 \epsilon_1} = \frac{c}{\omega} k_1 \tag{4.24a}$$

$$n_2 = c\sqrt{\mu_2 \epsilon_2} = \frac{c}{\omega} k_2 \tag{4.24b}$$

where $c = 1/\sqrt{\mu_0 \epsilon_0} = 3 \times 10^8$ m/s (the speed of light in a vacuum). Thus, the phase-matching condition of $k_{tx} = k_x$ gives rise to

$$n_1 \sin\theta = n_2 \sin\theta_t \quad \textbf{(Snell's law)} \tag{4.25}$$

which is the well-known **Snell's law**, or law of refraction.

The phase-matching conditions in (4.19) can be represented graphically. In Figure 4.7a we illustrate the case for which $n_1 < n_2$. The semicircle representing the magnitude of k_1 in medium 1 is smaller than that representing the magnitude of k_2 in medium 2. An incident wave with $\boldsymbol{k}_i$ gives rise to $\boldsymbol{k}_r$ and $\boldsymbol{k}_t$, whose x components are equal to that of $\boldsymbol{k}_i$. In Figure 4.7b we show the case where $n_1 > n_2$. For the incident angles greater than θ_c, k_x is larger than the magnitude of k_2. In that case,

$$k_{tz}^2 = k_2^2 - k_x^2 < 0$$

or

$$k_{tz} = \pm j\alpha$$

with $\alpha = \sqrt{k_x^2 - k_2^2}$ being a positive real number. We choose $k_{tz} = -j\alpha$ so that the transmitted electric field

$$\mathbf{E}^t = \hat{y} T E_0 e^{-\alpha z} e^{-jk_x x} \tag{4.26}$$

is decaying exponentially in the $+\hat{z}$ direction.

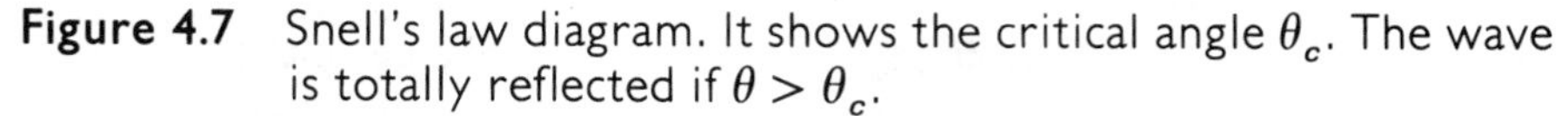

Figure 4.7 Snell's law diagram. It shows the critical angle θ_c. The wave is totally reflected if $\theta > \theta_c$.

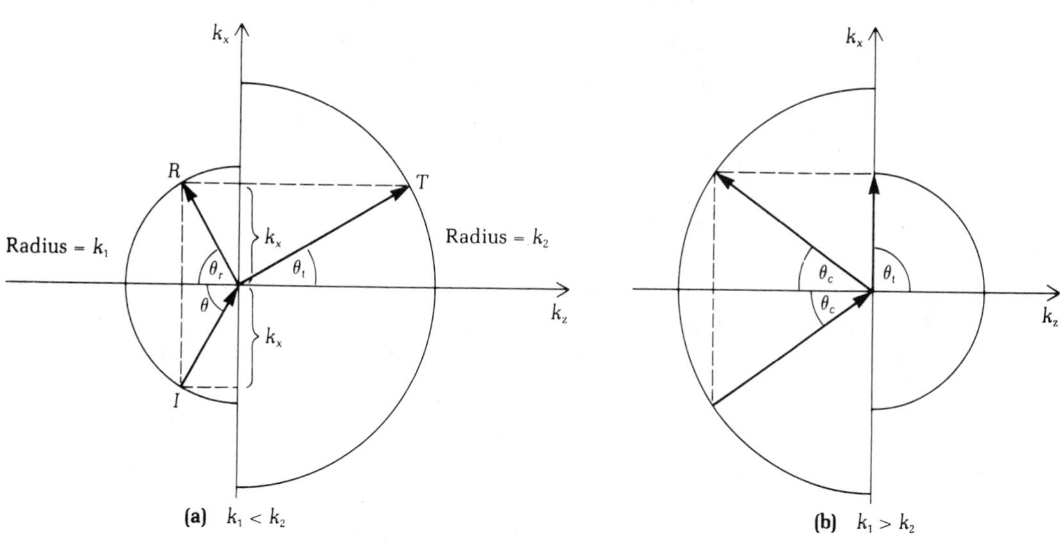

We multiply (4.26) by $e^{j\omega t}$ and take its real part to obtain

$$\boldsymbol{E}^t(\boldsymbol{r}, t) = \hat{y}\,|T|\,E_0 e^{-\alpha z} \cos(\omega t - k_x x + \theta_0)$$

where θ_0 is the phase angle of T. The wave thus propagates in the $\hat{x}$ direction, the constant-phase fronts of which are planes. On these constant-phase fronts, however, the magnitude of $|\boldsymbol{E}^t|$ decays exponentially as $e^{-\alpha z}$. We call this wave a nonuniform plane wave because, even though the phase front is a plane, the wave amplitude on the plane phase front is not uniform. Because the magnitude decays exponentially away from the interface and because the wave is propagating along the interface, the wave is also called a **surface wave.**

Surface waves occur when the incident angle is larger than the **critical angle** θ_c—that is, surface waves are generated when

$$k_x > k_2 \qquad \text{or} \qquad \theta > \theta_c$$

The critical angle θ_c is given as follows:

$$\boxed{\theta_c = \sin^{-1}\frac{k_2}{k_1}} \qquad \textbf{(critical angle)} \tag{4.27}$$

It is important to remember that the critical angle occurs only when the wave is incident from a medium with k_1 larger than k_2 of the other medium. We will see later (Example 4.4) that the wave is totally reflected when the incident angle θ is greater than the critical angle θ_c.

Example 4.3

We want to emphasize that light can be totally reflected without using a mirror. Total reflection occurs when the light is incident at an angle greater than the critical angle from a denser medium (higher dielectric constant) to a less dense medium. The magnitude of the reflection coefficient given by (4.22) is equal to unity when that happens. The formal proof is given in Example 4.4.

For an isotropic light source located under water, only light rays within a cone of θ_c can be refracted to the air, as shown in Figure 4.8. The permittivity of the

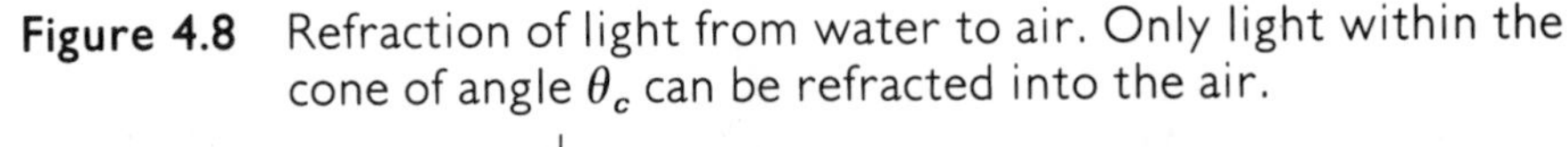

Figure 4.8 Refraction of light from water to air. Only light within the cone of angle θ_c can be refracted into the air.

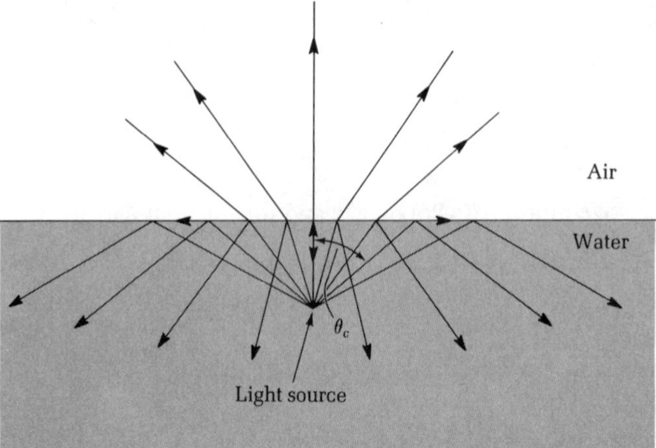

water at optical frequencies is $1.77\epsilon_0$. Calculate the value of the critical angle.

Solution Using (4.27), we obtain

$$\theta_c = \sin^{-1}\left(\frac{1}{\sqrt{1.77}}\right) = 49^\circ$$

Parallel Polarization

An incident wave of arbitrary polarization can always be decomposed into two waves having perpendicular and parallel polarizations. The electric field of the perpendicularly polarized wave is perpendicular to the plane of incidence, and the parallel-polarized wave's electric-field vector is parallel to that plane. We now consider a parallel-polarized incident wave. Because the plane of incidence now contains both the incident **k** vector and the electric-field vector **E** and because the magnetic-field vector is perpendicular to both **k** and **E**, we conclude that the **H** vector for parallel polarization must be perpendicular to the plane of incidence.

This situation is shown in Figure 4.9. The mathematical expressions of the incident, reflected, and transmitted waves can be written as

$$\mathbf{H}^i = \hat{y} H_0 e^{-jk_x x - jk_z z} \tag{4.28}$$

$$\mathbf{E}^i = (\hat{x} k_z - \hat{z} k_x) \frac{H_0}{\omega \epsilon_1} e^{-jk_x x - jk_z z} \tag{4.29}$$

$$\mathbf{H}^r = \hat{y} R_{II} H_0 e^{-jk_{rx} x + jk_{rz} z} \tag{4.30}$$

$$\mathbf{E}^r = (-\hat{x} k_{rz} - \hat{z} k_{rx}) \frac{R_{II} H_0}{\omega \epsilon_1} e^{-jk_{rx} x + jk_{rz} z} \tag{4.31}$$

$$\mathbf{H}^t = \hat{y} T_{II} H_0 e^{-jk_{tx} x - jk_{tz} z} \tag{4.32}$$

$$\mathbf{E}^t = (\hat{x} k_{tz} - \hat{z} k_{tx}) \frac{T_{II} H_0}{\omega \epsilon_2} e^{-jk_{tx} x - jk_{tz} z} \tag{4.33}$$

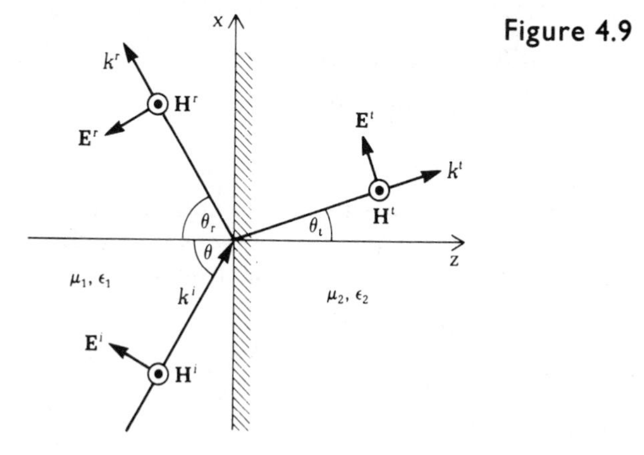

Figure 4.9 Reflection and transmission of a parallel-polarized plane wave.

where R_{II} and T_{II} are, respectively, the reflection and transmission coefficients for the magnetic-field vector for the parallelly polarized waves.

Note that the mathematical expressions for parallel-polarized waves are identical to those for the perpendicularly polarized waves if **H** is replaced by **E**, **E** by $-$**H**, μ by ϵ, and ϵ by μ. This observation enables us to write the reflection and transmission coefficients for this case directly from (4.22) and (4.23).

$$R_{II} = \frac{\epsilon_2 k_z - \epsilon_1 k_{tz}}{\epsilon_2 k_z + \epsilon_1 k_{tz}} \quad \text{(reflection coefficient for parallel polarization)} \tag{4.34}$$

$$T_{II} = \frac{2\epsilon_2 k_z}{\epsilon_2 k_z + \epsilon_1 k_{tz}} \quad \text{(transmission coefficient for parallel polarization)} \tag{4.35}$$

These are known as the Fresnel reflection and transmission coefficients for parallel polarization. These equations can be derived by applying the boundary conditions. The $\boldsymbol{k}$ vectors are related in the same way as they are for perpendicular polarization.

Brewster Angle

Consider that the two media shown in Figure 4.9 are lossless and have equal permeability—that is, $\mu_1 = \mu_2$. It follows that for parallel polarization there is always an angle θ_b such that, if $\theta = \theta_b$, the wave is totally transmitted and the reflection coefficient R_{II} is zero. For perpendicular polarization, however, there is no angle at which the incident wave will be totally transmitted. These statements can be verified by examining (4.22) and (4.34).

Let $\mu_1 = \mu_2$; then from (4.22) we find that, if $R_I = 0$, we must have $k_z = k_{tz}$. In terms of incident angle θ and transmitted angle θ_t,

$$\omega\sqrt{\mu_1\epsilon_1} \cos\theta = \omega\sqrt{\mu_1\epsilon_2} \cos\theta_t$$

By phase-matching conditions, we also have

$$\omega\sqrt{\mu_1\epsilon_1} \sin\theta = \omega\sqrt{\mu_1\epsilon_2} \sin\theta_t$$

To satisfy the above two equations, we must have $\theta_t = \theta$ and $\epsilon_2 = \epsilon_1$. In other words, the perpendicularly polarized wave is totally transmitted only if the two media are actually the same medium. But from (4.34) we seen that $R_{II} = 0$ gives

$$\omega\sqrt{\mu_1\epsilon_2} \cos\theta_b = \omega\sqrt{\mu_1\epsilon_1} \cos\theta_t$$

and the phase-matching condition gives

$$\omega\sqrt{\mu_1\epsilon_1} \sin\theta_b = \omega\sqrt{\mu_1\epsilon_2} \sin\theta_t$$

Solving the above two equations, we find $\theta_t + \theta_b = \pi/2$ and

$$\boxed{\theta_b = \tan^{-1}\sqrt{\frac{\epsilon_2}{\epsilon_1}}} \qquad \textbf{(Brewster angle)} \qquad (4.36)$$

where the incident angle θ_b is called the **Brewster angle.**

If a wave is arbitrarily polarized and is incident on the boundary of the two dielectric media at the Brewster angle, the reflected wave will contain only the perpendicular polarization because the parallel-polarized component of the wave is totally transmitted. For this reason, the Brewster angle is also called the **polarization angle.**

Reflection of Unpolarized Light from Dielectrics

An unpolarized light beam in air strikes at the surface of a piece of dielectric material. Suppose that the surface of the dielectric is horizontal and that the dielectric constant is $2.25\epsilon_0$ (glass). We are interested in knowing the composition of the reflected ray.

The incident ray may be divided into two equal components, one horizontally polarized (perpendicularly polarized) and the other polarized in the plane of incidence (parallel-polarized). R_I denotes the reflection coefficient of the former and R_{II} that of the latter. The values of $|R_I|^2$ and $|R_{II}|^2$ are calculated from (4.22) and (4.34) and plotted as functions of the incident angle θ, as shown in Figure 4.10 (p. 98). We can see that in general $|R_I|^2$ is greater than $|R_{II}|^2$. As a result, the reflected wave has a greater amount of light polarized in the horizontal direction than polarized in the other direction.

Brewster Window

A gas laser is often composed of a tube containing gas fitted with a Brewster-angle window and external mirrors, as shown in Figure 4.11. Two mirrors reflect the light back and forth in such a manner that a standing wave is sustained in the "cavity." The open cavity is said to be at "resonance." Part of the light is transmitted through the mirrors and becomes the output of the laser.

The light travels only along the z axis except in the Brewster windows. The beam may be decomposed into two components with two different polarizations. One component of the wave is polarized in the $\hat{x}$ direction, the other in the $\hat{y}$ direction. The $\hat{x}$-polarized wave is parallel-polarized with respect to the dielectric slab called the "window." The normal of the window makes an angle equal to the Brewster angle, so that the $\hat{x}$-polarized wave is totally transmitted through the

Figure 4.10 Reflected power as a function of incident angle. The material is glass with $\epsilon = 2.25\epsilon_0$. The Brewster angle is 56°.

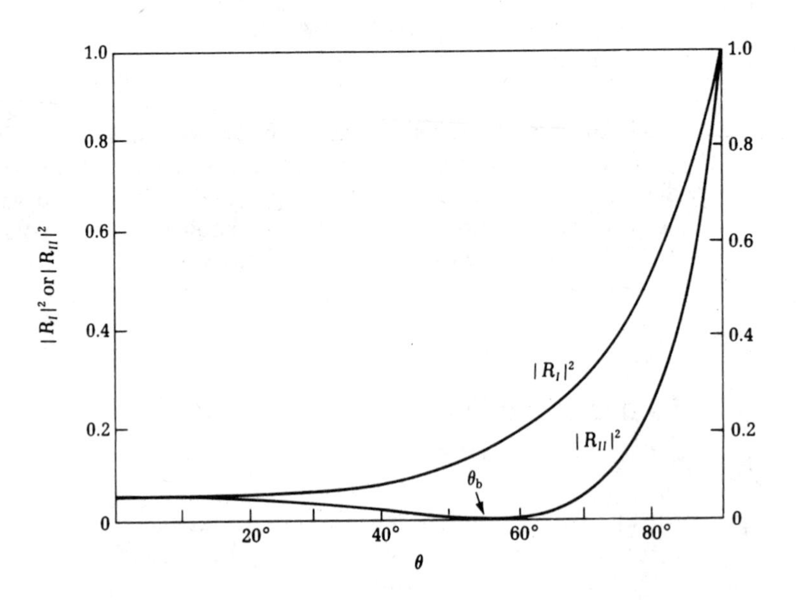

Figure 4.11 A gas laser with Brewster windows. The windows are transparent to the $\hat{\mathbf{x}}$-polarized wave but reflect the $\hat{\mathbf{y}}$-polarized wave. Thus, only the $\hat{\mathbf{x}}$-polarized wave is resonant in the cavity, and the output of the laser is polarized linearly in the $\hat{\mathbf{x}}$ direction.

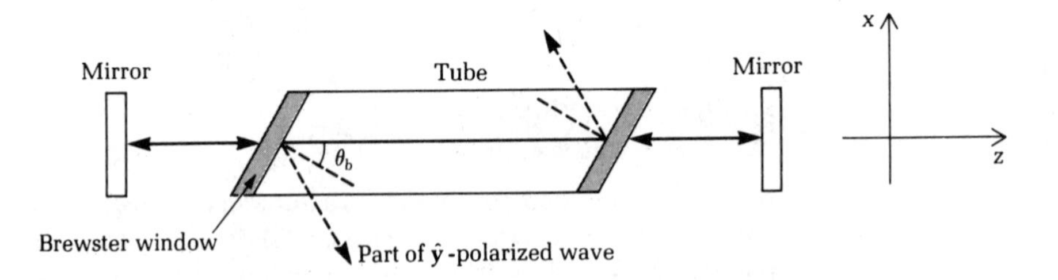

window. We say that the Brewster window is transparent to the $\hat{x}$-polarized wave. On the other hand, the $\hat{y}$-polarized wave is perpendicularly polarized with respect to the window. Thus, part of its energy is reflected and part is transmitted. Because of the reflection loss at the Brewster windows, the $\hat{y}$-polarized wave will not be resonant in the cavity. As a result, the output of the laser is linearly polarized in the $\hat{x}$ direction.

Power Conservation

The time-average Poynting power density for the perpendicular polarization can be calculated from (4.11)–(4.16). Notice that, by phase-matching condition (4.19), $k_{tx} = k_{rx} = k_x = k_1 \sin\theta$ and $k_{rz} = k_z = k_1 \cos\theta$ are all real numbers but $k_{tz} = (k_2^2 - k_x^2)^{1/2}$ can become imaginary when $k_2 < k_1 \sin\theta$ at incidence angles larger than the critical angle. We thus obtain, at $z = 0$,

$$\langle S^i \rangle = \frac{1}{2}\text{Re}\{\mathbf{E}^i \times (\mathbf{H}^i)^*\} = \frac{|E_0|^2}{2\omega\mu_1}\{\hat{x}k_x + \hat{z}k_z\} \tag{4.37a}$$

$$\langle S^r \rangle = \frac{1}{2}\text{Re}\{\mathbf{E}^r \times (\mathbf{H}^r)^*\} = \frac{|E_0|^2}{2\omega\mu_1}\{\hat{x}k_x - \hat{z}k_z\}|\text{R}_I|^2 \tag{4.37b}$$

$$\langle S^t \rangle = \frac{1}{2}\text{Re}\{\mathbf{E}^t \times (\mathbf{H}^t)^*\} = \frac{|E_0|^2}{2\omega\mu_2}\text{Re}\{\hat{x}k_x + \hat{z}k_{tz}^*\}|\text{T}_I|^2 \tag{4.37c}$$

Similarly, for the parallel polarization, we calculate from (4.28)–(4.33),

$$\langle S^i \rangle = \frac{|H_0|^2}{2\omega\epsilon_1}\{\hat{x}k_x + \hat{z}k_z\} \tag{4.38a}$$

$$\langle S^r \rangle = \frac{|H_0|^2}{2\omega\epsilon_1}\{\hat{x}k_x - \hat{z}k_z\}|\text{R}_{II}|^2 \tag{4.38b}$$

$$\langle S^t \rangle = \frac{|H_0|^2}{2\omega}\text{Re}\left\{(\hat{x}k_x + \hat{z}k_{tz})\frac{1}{\epsilon_2}\right\}|\text{T}_{II}|^2 \tag{4.38c}$$

To observe power conservation, we take a sample volume across the boundary surface, as shown in Figure 4.12. Clearly, the $\hat{x}$ components of the Poynting

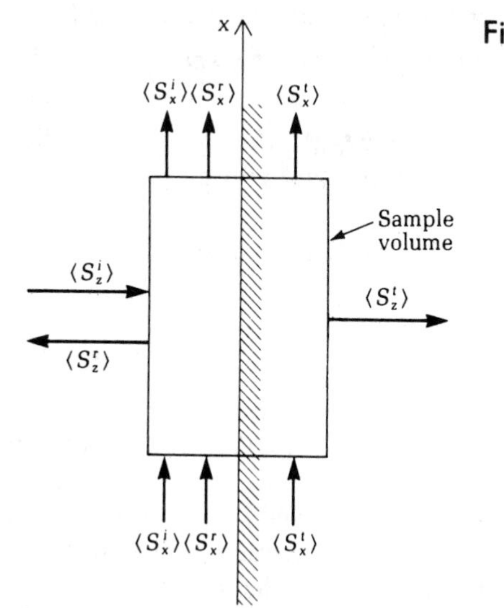

Figure 4.12 Conservation of power on the boundary.

vectors are continuous in and out of the sample volume. For the $\hat{z}$ components, we denote P_r the percentage of power reflected from the boundary and P_t the percentage of power transmitted to the second medium. Then from (4.37a) and (4.37b), or from (4.38a) and (4.38b),

$$P_r = \frac{\langle S_z^r \rangle}{\langle S_z^i \rangle} = |\mathrm{R}|^2 \tag{4.39}$$

where R is either the reflection coefficient R_I or R_{II}. The transmitted power P_{tI} for perpendicular polarization can be obtained using (4.37a) and (4.37c):

$$P_{tI} = \frac{\langle S_z^t \rangle}{\langle S_z^i \rangle} = |T_I|^2 \operatorname{Re}\left\{\frac{k_{tz}^* \mu_1}{k_z \mu_2}\right\} \tag{4.40}$$

Similarly, the transmitted power for parallel polarization is obtained from (4.38a) and (4.38c):

$$P_{tII} = \frac{\langle S_z^t \rangle}{\langle S_z^i \rangle} = |\mathrm{T}_{II}|^2 \operatorname{Re}\left\{\frac{k_{tz}^* \epsilon_1}{k_z \epsilon_2}\right\} \tag{4.41}$$

When the expressions for T_I and T_{II} given in (4.23) and (4.35) are substituted in (4.40) and (4.41), it can be proved that

$$P_r + P_t = 1 \tag{4.42}$$

The above equation is valid for both perpendicular and parallel polarizations. It is consistent with the law of conservation of energy.

Example 4.4 *This example illustrates that, when the incident angle is greater than the critical angle, the magnitude of the reflection coefficient given by (4.22) or (4.34) is equal to unity. That is, total reflection occurs for both polarizations.*

Show that, at incident angles $\theta > \theta_c$, the reflection coefficients for both perpendicular and parallel polarizations have unit magnitude.

Solution For $\theta > \theta_c$, $k_{tz} = -j\alpha_t$, the reflection coefficients become

$$\mathrm{R} = \frac{1 + ju}{1 - ju} = e^{j2\tan^{-1}u}$$

where $u = \mu_1\alpha_t/\mu_2 k_z$ for perpendicular polarization and $u = \epsilon_1\alpha_t/\epsilon_2 k_z$ for parallel polarization. The reflection coefficient is seen to have unit magnitude and a phase shift of $2\tan^{-1} u$, which is known as the Goos-Hanschen shift. The Goos-Hanschen shift can in fact be observed with a beam of light incident at $\theta > \theta_c$. The reflected beam is seen to be laterally shifted instead of reflected from the same position as the incident beam at the interface.

Reflection from the Ionosphere

Before the era of satellite technology, the ionosphere was principally used as a reflecting surface that enabled long-distance communication with radio waves. The ionosphere can be characterized as a plasma medium. Its permittivity was derived in Chapter 2, which gives

$$\epsilon = \epsilon_0\left(1 - \frac{\omega_p^2}{\omega^2}\right)$$

where ω_p is the plasma frequency. For the earth's ionosphere, we may take approximately $\omega_p = 2\pi \times 9 \times 10^6$ Hz.

With $\omega = \omega_p/\sqrt{5}$, consider a wave incident upon the ionosphere at 45°, as shown in Figure 4.13. The permittivity is then $\epsilon = -4\epsilon_0$. We express all wave numbers in terms of the free-space wave number k_0.

$$k_1 = k_0$$

$$k_2 = \omega\sqrt{\mu_0(-4\epsilon_0)} = -j2k_0$$

$$k_x = k_0 \sin 45° = 0.707k_0$$

$$k_z = k_0 \cos 45° = 0.707k_0$$

$$k_{tz} = \sqrt{-4k_0^2 - 0.5k_0^2} = -j2.12k_0$$

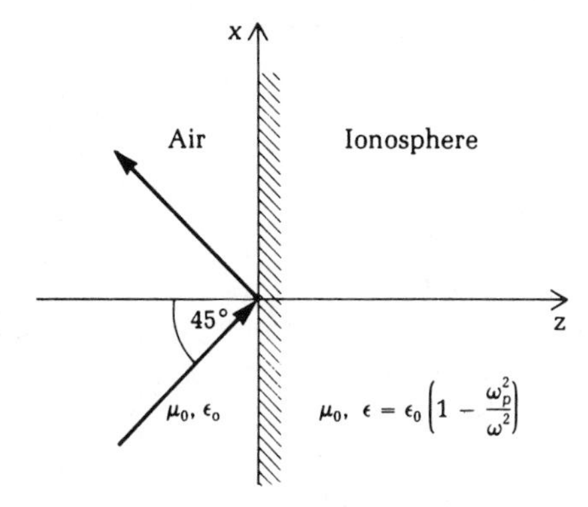

Figure 4.13 Reflection of an electromagnetic wave from the ionosphere.

Note that minus signs in the expressions for k_2 and k_{tz} make the wave attenuate in the $+\hat{z}$ direction. Using (4.22) and (4.34), we find

$$R_I = \frac{0.707 + j2.12}{0.707 - j2.12} = 1.0\, e^{j143°}$$

$$R_{II} = \frac{-4 \times 0.707 + j2.12}{-4 \times 0.707 - j2.12} = 1.0\, e^{j286°}$$

Note that the reflection coefficients have amplitudes equal to unity regardless of their polarization. However, the phase shift of the reflected wave will depend on the type of polarization.

Reflection of Electromagnetic Waves from Skin

In Chapter 3, we learned the safety limit of human exposure to EM radiation. The power density limit shown in Figure 3.5 is the incident EM power. How much of that power enters the human body depends on a number of factors, such as the incident angle, the frequency, and the polarization of the wave. Here we shall give an approximate estimate of the percentage of the reflected and the transmitted power when an EM wave impinges on a human body.

Table 4.1* lists the dielectric constant and conductivity of biological materials. Consider that a 915 MHz EM wave impinges on skin tissue at a 30° angle. Let us assume that polarization is parallel to the plane of incidence. To find the reflection and transmission coefficients, we first note from Table 4.1 that, for skin at 915 MHz, the relative dielectric constant is 51 and the conductivity is 1.6 mho/m. We then use (3.7) to find k_1.

Table 4.1 Electrical Properties of Biological Materials.

	Muscle, skin		Fat, bone	
Frequency (MHz)	ϵ relative	σ mho/m	ϵ relative	σ mho/m
100	71.7	0.889	7.45	0.0191–0.0759
915	51	1.60	5.6	0.0556–0.147
3000	46	2.26	5.5	0.110–0.234

$$k_1 = 2\pi \times 915 \times 10^6 \sqrt{\mu_0 \epsilon_0} = 19.18$$

Since

$$\epsilon_2 = \epsilon_0\left(51 - j\frac{1.6}{\omega\epsilon_0}\right) = (51 - j31.43)\epsilon_0$$

* Quoted from C. C. Johnson and A. W. Guy, "Nonionizing electromagnetic wave effects in biological materials and systems," *Proceedings of IEEE*, Vol. 60, no. 6 (1972): 692.

we have, from (3.26a),

$$k_2 = \omega\sqrt{\mu_0 \epsilon_2} = 142.8 - j40.47$$

Now,

$$k_x = k_1 \sin(30^\circ) = 0.5k_1$$

$$k_z = k_1 \cos(30^\circ) = 0.866k_1$$

$$k_{tz} = \sqrt{k_2^2 - k_x^2} = 142.5 - j40.56$$

Substituting the above numbers in (4.34) yields

$$R_{II} = 0.750\ e^{-j4.7^\circ}$$

Therefore, according to (4.39), the percentage of power reflected is 56.3%, and 43.7% of power is transmitted to the skin.

4.3 STANDING WAVES

When an electromagnetic wave impinges on a boundary, some of it is reflected and some transmitted, as shown in Figure 4.6. The reflected wave interacts with the incident wave and creates an interference pattern, or a **standing-wave** pattern. The sum of the incident and reflected waves results in a standing wave. Note that in the second medium, medium 2 shown in Figure 4.6, there is only one wave—the transmitted wave. Therefore, no standing wave exists in medium 2.

To see what a standing wave is, let us first consider a simple case of reflection of an EM wave by a perfectly conducting plane. Ordinary metals like copper and aluminum have such high conductivities that they can be considered as perfect conductors in many practical situations. But theoretically, a perfect conductor is a medium with infinite conductivity.

Because the complex permittivity of a medium is defined as

$$\epsilon_2 = \epsilon_2 - j\frac{\sigma_2}{\omega}$$

[see (3.23)], the perfect conductor can also be regarded as a medium with infinite permittivity. Substituting $\epsilon_2 \to \infty$ and $k_{tz} \sim \omega\sqrt{\mu\epsilon_2} \to \infty$ in the formulas for the reflection coefficients obtained in the preceding section, we obtain

$$R_I = -1 \qquad \textbf{(4.43a)}$$

$$R_{II} = 1 \qquad \textbf{(4.43b)}$$

In this section, we shall study special cases involving waves reflected from perfectly conducting planes.

Normal Incidence of a Plane Wave on a Perfect Conductor

Consider a perfectly conducting half-space, as shown in Figure 4.14. A uniform plane electromagnetic wave of the following form impinges normally on the boundary:

$$\mathbf{E}^i = \hat{x}E_0 e^{-jkz} \tag{4.44a}$$

$$\mathbf{H}^i = \hat{y}\left(\frac{E_0}{\eta_0}\right)e^{-jkz} \tag{4.44b}$$

Figure 4.14 Reflection of a wave from a perfect conductor (normal incidence): **(a)** Reflection of a normally incident wave from a perfect conductor. **(b)** Standing-wave pattern of the E field. **(c)** Standing-wave pattern of the H field.

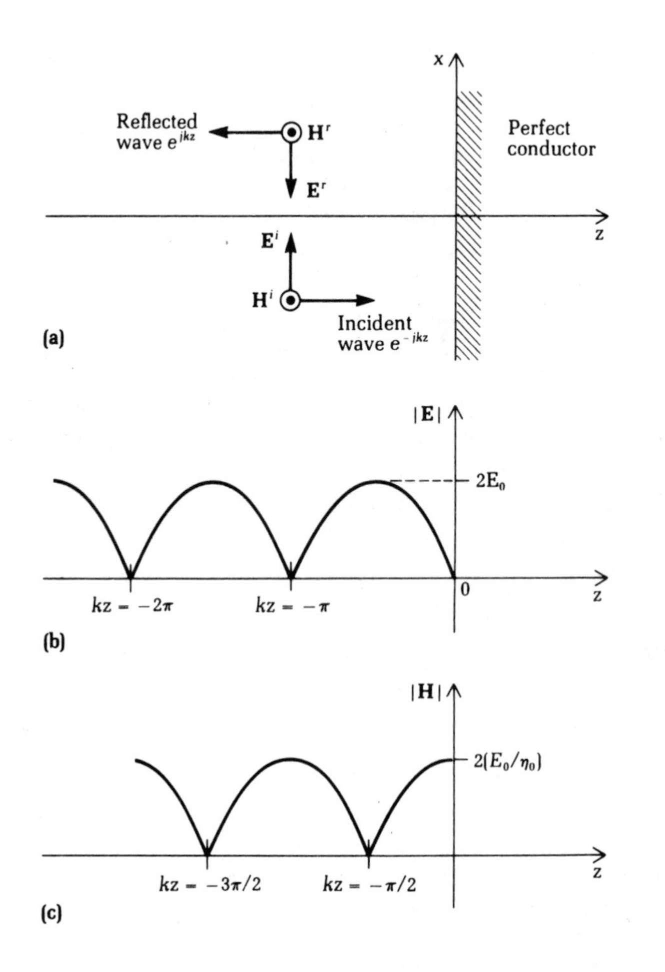

The present case is the special case of a parallel-polarized wave incident at $\theta = 0$, as shown in Figure 4.9.

According to (4.43b), the reflected wave is given by:

$$\mathbf{E}^r = -\hat{x}E_0\, e^{jkz} \tag{4.45a}$$

$$\mathbf{H}^r = \hat{y}\left(\frac{E_0}{\eta_0}\right) e^{jkz} \tag{4.45b}$$

The total electromagnetic field in medium 1 is the sum of the incident and the reflected waves:

$$\mathbf{E} = \hat{x}E_0(e^{-jkz} - e^{jkz}) = -\hat{x}2jE_0 \sin kz \tag{4.46a}$$

$$\mathbf{H} = \hat{y}\frac{E_0}{\eta_0}(e^{-jkz} + e^{jkz}) = \hat{y}\left(\frac{2E_0}{\eta_0}\right)\cos kz \tag{4.46b}$$

The boundary condition (4.3) states that at $z = 0$,

$$\mathbf{J}_s = -(\hat{z}) \times \mathbf{H} = \hat{x}\frac{2E_0}{\eta_0} \tag{4.47}$$

This is the surface current induced on the surface of the perfect conductor because of the incident wave.

The instantaneous values for $\boldsymbol{E}$ and $\boldsymbol{H}$ are obtained from the rule $\boldsymbol{E}(z, t) = \text{Re}\{\mathbf{E}\, e^{j\omega t}\}$ and from the corresponding rule for $\boldsymbol{H}(z, t)$. We find

$$\boldsymbol{E} = \hat{x}2E_0 \sin kz \sin \omega t \tag{4.48}$$

$$\boldsymbol{H} = \hat{y}2\left(\frac{E_0}{\eta_0}\right)\cos kz \cos \omega t \tag{4.49}$$

The field patterns for $|\mathbf{E}|$ and $|\mathbf{H}|$ are shown in Figures 4.14b and 4.14c. These patterns are called **standing-wave patterns** because the waveform does not shift in space as time progresses. For example, as time progresses, a meter located at $kz = -\pi/2$ will detect an $\mathbf{E}$ field with amplitude equal to $2E_0$, as seen from Figure 4.14b or calculated from Equation (4.48). For a detector located at $kz = -\pi/4$, the amplitude reads $1.414E_0$. The detector located at $kz = -\pi$ will measure zero $\mathbf{E}$ field. The phase of the $\mathbf{E}$ field, according to (4.46a), is 90° for $-\pi < kz < 0$, $-90°$ for $-2\pi < kz < -\pi$, 90° for $-3\pi < kz < -2\pi$, and so on. In other words, the phase is either 90° or $-90°$. This situation is very different from the propagating wave in air, where the phase of the field progresses continuously with distance and the amplitude is a constant. Since the wave with its electromagnetic fields expressed in (4.46) does not seem to go anywhere, it is called a **standing wave**.

From (4.46) we see that the $\mathbf{E}$ and $\mathbf{H}$ fields are out of phase by 90°, and their standing-wave patterns are displaced by $\lambda/4$ from each other. That is, as seen from Figures 4.14b and 4.14c, the $\mathbf{E}$ field is maximum where the $\mathbf{H}$ field is minimum, and vice versa. Note that two adjacent nulls of either the $\mathbf{E}$ or $\mathbf{H}$ field are separated by $kd = \pi$, or $d = \lambda/2$.

In summary, a standing wave is the result of the interference of two waves that are of the same frequency and are propagating in opposite directions. A standing wave is formed in front of a metal plate when a uniform plane impinges normally on the plate.

Fabry-Perot Resonators

We see from Figure 4.14 that, if we place conducting plates at locations where the electric field is zero and $z = -n\pi/k$, the fields will not be disturbed. The solution (4.46) will still be valid. In a parallel plate region, the standing electromagnetic wave as given by (4.46) constitutes a simple model for the Fabry-Perot resonator (Figure 4.15), which is used for laser cavities at optical frequencies and in electron-beam devices, such as microwave tubes at millimeter and submillimeter wavelengths. The resonant frequency is determined by the plate separation d, where $d = n\pi/k$ and n is an integer, which gives

$$f = \frac{\omega}{2\pi} = \frac{ck}{2\pi} = \frac{nc}{2d} \tag{4.50}$$

We have assumed that the medium inside the cavity is free space, and c is the speed of light.

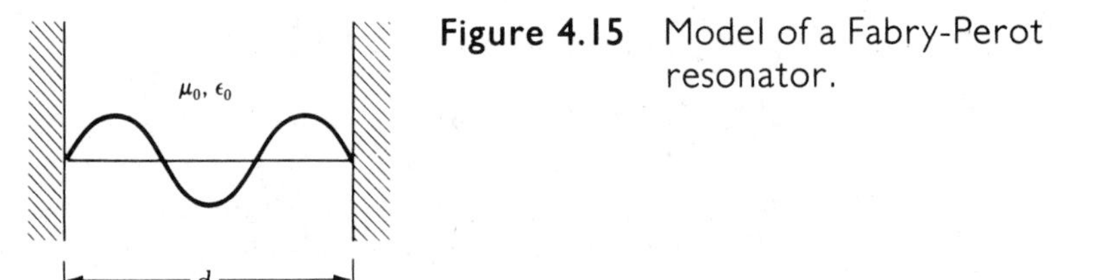

Figure 4.15 Model of a Fabry-Perot resonator.

Oblique Incidence of a Plane Wave on a Perfect Conductor

Consider a perfectly conducting plane, as shown in Figure 4.16. A perpendicularly polarized incident wave impinges at the boundary surface. This situation is similar to the one shown in Figure 4.6 except for the fact that in the present case the second medium is a perfect conductor.

The incident fields are expressed as follows:

$$\mathbf{E}^i = \hat{y}E_0\, e^{-jk_x x - jk_z z} \tag{4.51a}$$

According to (4.43a), the reflection coefficient is equal to -1. Consequently the reflected fields are given as follows:

$$\mathbf{E}^r = -\hat{y}E_0\, e^{-jk_x x + jk_z z} \tag{4.51b}$$

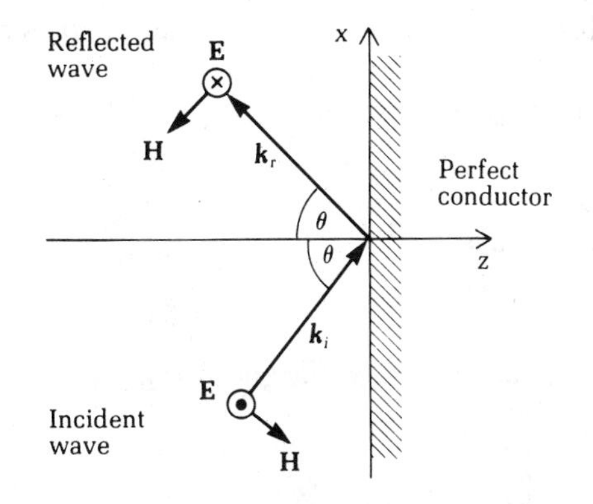

Figure 4.16 Reflection of a wave from a perfect conductor (oblique incidence).

This result is obtained directly from (4.13). The total field in medium 1 is then the sum of the incident and the reflected waves:

$$\mathbf{E} = \hat{y}(-2jE_0)\sin k_z z e^{-jk_x x} \tag{4.51c}$$

Notice that this wave is a standing wave in the $\hat{z}$ direction, but that it propagates in the $\hat{x}$ direction as manifested by the factor $e^{-jk_x x}$. Furthermore, the electric field vanishes not only at $z = 0$ but also at $z = -n\pi/k_z$, as shown in Figure 4.17.

Figure 4.17 Standing-wave patterns for the total field arising from reflection of a wave from a perfect conductor with oblique incidence and perpendicular polarization.

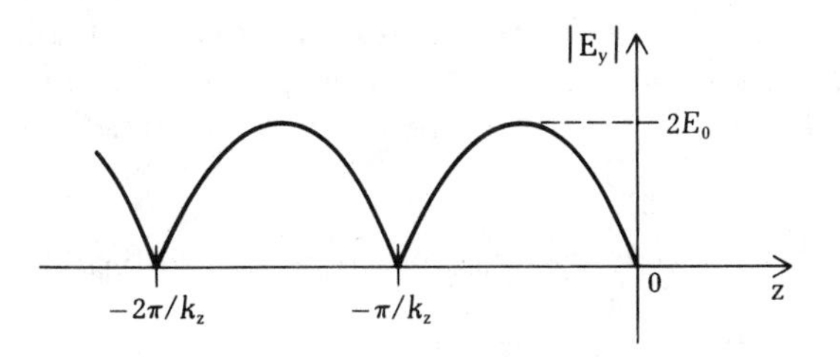

Since $k_z = k_1 \cos\theta$, two adjacent nulls of either the **E** or **H** field are separated by $k_z d = \pi$, or $d = \lambda/(2\cos\theta)$.

Standing Wave in Front of a Dielectric Medium

Consider that a uniform plane EM wave impinges on a plane dielectric interface. Assume that medium 1 is air and medium 2 is soil characterized by $\epsilon = 10\epsilon_0$, $\sigma = 0.01$ mho/m, and $\mu = \mu_0$. The frequency is 50 MHz. The polarization is per-

perpendicular, like the case shown in Figure 4.6. Let the incident angle θ be equal to zero.

The total **E** field in medium 1 is then

$$\mathbf{E}_1 = \hat{y}(E_0 e^{-jk_{1z}} + R_I e^{+jk_{1z}}) \tag{4.52a}$$

and the **E** field in medium 2 is

$$\mathbf{E}_2 = \hat{y}\mathrm{T}_I \mathrm{E}_0 e^{-jk_{2z}} \tag{4.52b}$$

The reflection and the transmission coefficients are given by (4.22) and (4.23).

$$k_z = k_1 \cos\theta = k_1 = 1.048$$

$$\mathrm{k}_{tz} = \sqrt{k_2^2 - k_x^2} = k_2 = 3.365 - j0.587$$

$$\mathrm{R}_I = 0.537e^{j173.4^\circ}$$

$$\mathrm{T}_I = 0.471e^{j7.6^\circ}$$

Substituting the R_I value in (4.52a) and then taking the absolute value, we obtain

$$|\mathrm{E}_{1y}| = E_0 |1 + 0.537e^{+j(2k_{1z}+173.4^\circ)}|$$

Similarly, substituting the T_I and k_2 values in (4.52b) and taking the absolute value, we have

$$|\mathrm{E}_{2y}| = 0.471E_0 e^{-0.587z}$$

The **E** field strength is plotted in Figure 4.18. We can see that there is a standing wave in medium 1 due to the interference between the incident and the reflected waves. Because the reflection is not total, the minimum field strength is not zero, as it is in the case shown in Figure 4.14. There is no standing wave in medium 2. The **E** field decays exponentially because medium 2 is dissipative.

Figure 4.18 Electric field strength as a function of distance.

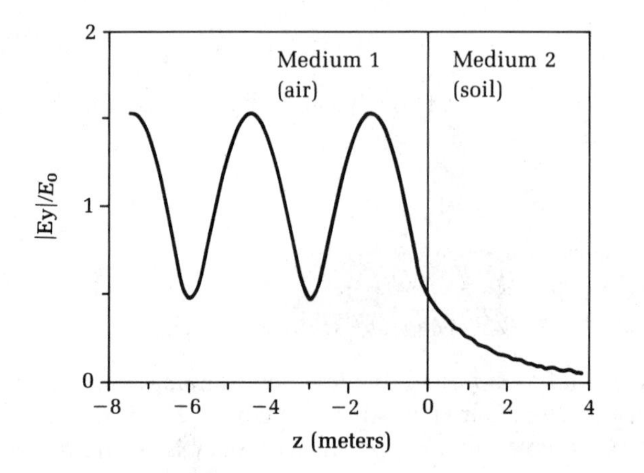

The maximum **E** field strength in medium 1 occurs at z_{max}, where

$$2k_1 z_{max} + 173.4° = -2n\pi$$

where n is an integer. Therefore, the first maximum corresponds to $n = 0$ and is located at

$$z_{max} = -(173.4\pi/180)/(2k_1) = -1.44 \text{ m}$$

and the maximum field strength is $1.537E_0$.

The minimum **E** field strength in medium 1 occurs at z_{min} where

$$2k_1 z_{min} + 173.4° = -(2n + 1)\pi$$

where n is an integer. Therefore the first minimum corresponds to $n = 0$ and is located at

$$z_{min} = -(\pi + 173.4\pi/180)/(2k_1) = -2.94 \text{ m}$$

and the minimum field strength is $0.463E_0$.

Microwave Reflectors

The reflection of an electromagnetic beam from a finite conducting plate is shown in Figure 4.19. We assume that the cross section of the beam is large compared with its wavelength. Thus, near the center of the beam the wave can be regarded as a uniform plane wave. Near point I, only the incident wave is present. Near point R, only the reflected wave is present. In the triangle region, ABC, both incident and reflected waves are present, and an interference pattern is generated.

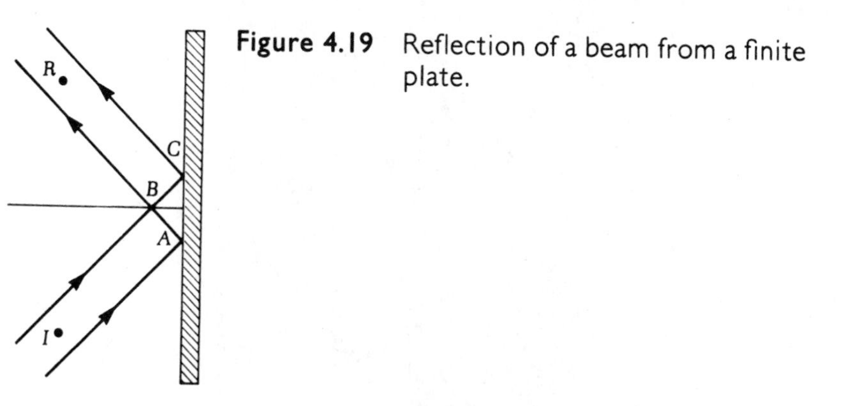

Figure 4.19 Reflection of a beam from a finite plate.

The situation is quite complicated near the edges of the beams where the expression of the uniform plane wave is not applicable. Moreover, the conducting plate must be large compared with both the wavelength and the cross section of the beam. Otherwise, the effect of the edges of the plate and even of its back surface must be taken into account.

Reflecting plates are used in microwave relay stations (see Figure 4.20) or at

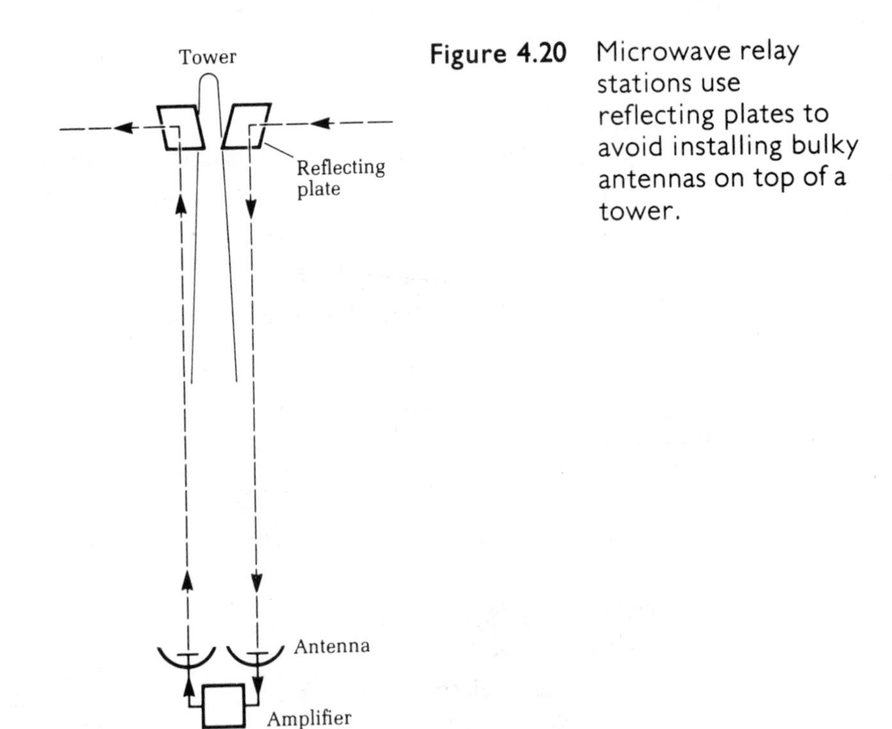

Figure 4.20 Microwave relay stations use reflecting plates to avoid installing bulky antennas on top of a tower.

Figure 4.21 Air traffic control tower at the Houston Intercontinental Airport. A reflector on the tower directs the wave to and from the dish antenna on the ground.

airport control towers (see Figure 4.21). With this arrangement, the antennas need not be installed at the top of a tower.

Reflection by Grating of Parallel Wires

The surface current $\mathbf{J}_s$ on the surface of a conducting plate induced by an electromagnetic field may be found by applying the boundary condition (4.3). For the case of oblique incidence with perpendicular polarization, as shown in Figure 4.16, we have

$$\mathbf{J}_s = -\hat{z} \times \mathbf{H}|_{z=0} \tag{4.53}$$

From (4.51c) and Maxwell's equations, one can derive the corresponding $\mathbf{H}$ field:

$$\mathbf{H}|_{z=0} = -\hat{x}\frac{2E_0}{\eta}\cos\theta e^{-jkx\sin\theta} \tag{4.54}$$

Thus,

$$\mathbf{J}_s = \hat{y}\left(\frac{2E_0}{\eta}\right)\cos\theta e^{-jkx\sin\theta} \tag{4.55}$$

Note that the current flows in the $\hat{y}$ direction and that no current flows in the $\hat{x}$ direction. In fact, if the conducting plate is replaced by a grating of parallel conducting wires arranged in the $\hat{y}$ direction, these wires also serve as a reflector that is as effective as a solid conducting plate. Experiments have found that grates are effective when the spacing of the wire in the grate is much smaller than the wavelength of the wave. Grates are used instead of conducting plates to reflect linearly polarized electromagnetic waves because they reduce weight, save

Figure 4.22 To receive linearly polarized electromagnetic waves, wire gratings may replace metal plates for reflector antennas.

material, and decrease resistance to wind. Based on these considerations, some reflectors use wires to replace metal dishes for transmitting and receiving electromagnetic waves. An example of such a structure is shown in Figure 4.22.

SUMMARY

1. Boundary conditions: the tangential **E** and **H** fields are continuous, and the normal **D** and **B** are continuous across a boundary of two media if neither one is a perfect conductor.
2. If one of the media is a perfect conductor (e.g., medium 2), then $\mathbf{E}_1$ and $\mathbf{D}_1$ are perpendicular to the boundary; and $\mathbf{D}_1$ is proportional to the surface charge on the conductor. Also, $\mathbf{H}_1$ and $\mathbf{B}_1$ are parallel to the boundary, and $\mathbf{H}_1$ is proportional to the surface current on the conductor.
3. Reflection and transmission coefficients of a uniform plane wave impinging on a planar boundary can be solved by using the boundary conditions. One set of coefficients is for parallel polarization, the other set for perpendicular polarization.
4. The reflected angle is always equal to the incident angle. The transmitted angle and the incident angle are related by Snell's law. These results are independent of polarization.
5. If medium 2 is dissipative, then let $k_x = k_1 \sin\theta$, $k_z = k_1 \cos\theta$, $k_{tx} = k_x$, and $k_{tz} = \sqrt{k_2^2 - k_x^2}$.
6. Total reflection occurs when
   (a) both media are non-dissipative
   (b) $k_1 > k_2$
   (c) incident angle is greater than the critical angle
   It is independent of polarization.
7. Total transmission occurs when
   (a) both media are non-dissipative
   (b) incident angle is equal to the Brewster angle
   (c) the polarization is parallel (if $\mu_1 = \mu_2$)
8. Standing waves are the result of interference between the incident and the reflected waves.
9. The maxima and minima of the standing-wave pattern are determined by the reflection coefficient. Their positions are determined by the phase of the reflection coefficient. Adjacent maximum and minimum are separated by $\lambda/(2\cos\theta)$, where λ is the wavelength and θ the incident angle.

Problems

4.1 The **E** field measured in air just above a glass plate is equal to 2 V/m in magnitude and is direct at 45° away from the boundary, as shown in Figure P4.1. The magnitude of the **E**

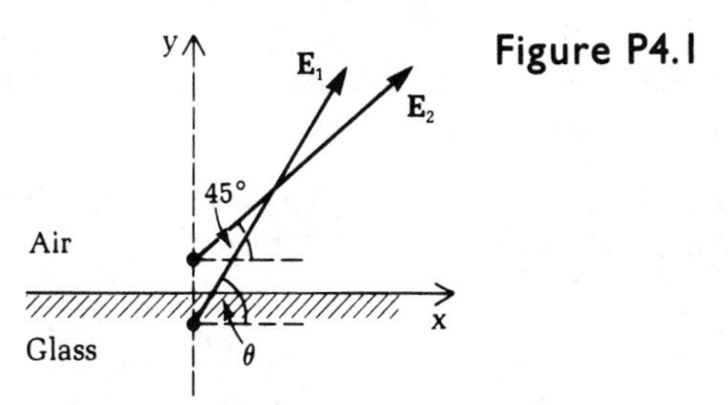

Figure P4.1

field measured just below the boundary is equal to 3 V/m. Find the angle θ for the **E** field in the glass just below the boundary.

4.2 The **H** field in air just above a perfect conductor is given by

$$\mathbf{H}_1 = 3\hat{x} + 4\hat{z} \text{ amperes per meter}$$

as shown in Figure P4.2. Find the surface current $\mathbf{J}_s$ on the surface of the perfect conductor. The conductor occupies the space $y < 0$.

Figure P4.2

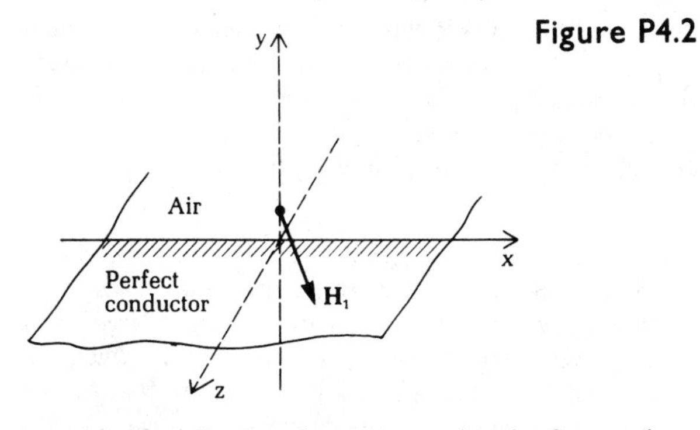

4.3 Match the following descriptions with the figures shown in Figure P4.3. Fields are near the interface but on opposite sides of the boundary.

Figure P4.3

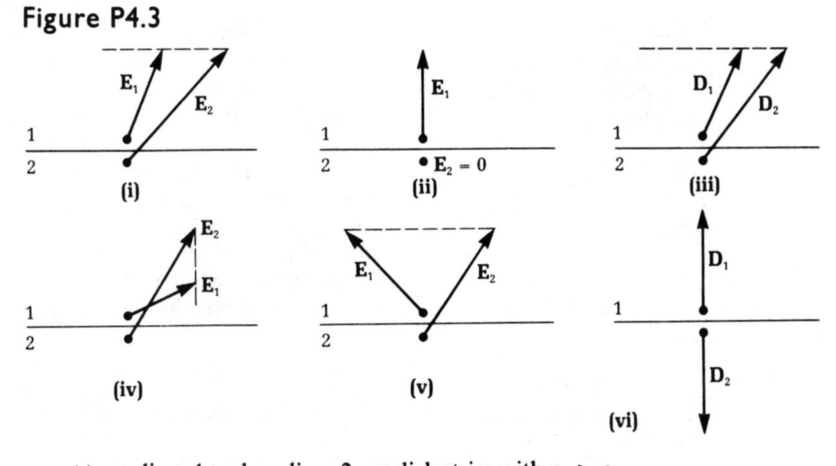

(a) medium 1 and medium 2 are dielectrics with $\epsilon_1 > \epsilon_2$
(b) medium 1 and medium 2 are dielectrics with $\epsilon_1 < \epsilon_2$
(c) impossible

(d) impossible
(e) there is a positive surface charge on the boundary between two dielectrics
(f) medium 2 is a perfect conductor

4.4 In the three-media configuration shown in Figure P4.4, the wave numbers are k_1, k_2, and k_3. Find the transmission angle in medium 3 in terms of θ_1 and the wave numbers. Assume all k's are real.

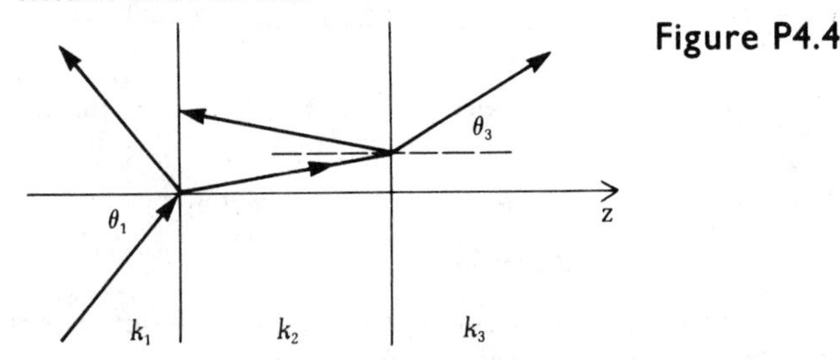

Figure P4.4

4.5 A small bug is found in the "Jurassic Park" ruin embedded at the middle of a rectangular block of amber ($\epsilon_r = 1.55$). The top surface of the amber block is a square of 5 cm by 5 cm and the height is 2.5 cm. Is it possible to cover a portion of the top surface of the block so that the bug will not be seen at any viewing angle from the top? If so, find the shape and the minimum area of the cover. Hint: Consider conditions for total reflection, and neglect multiple internal reflections.

4.6 A parallel-polarized wave is incident from medium 1 on the plane boundary between medium 1 and medium 2. Both media are dielectrics with $\mu_1 = \mu_2 = \mu_0$ and real permittivities ϵ_1 and ϵ_2. We know that, when the incident angle is larger than the critical angle θ_c, no time-average power is transferred to medium 2. Also, when the incident angle is equal to the Brewster angle θ_b, the reflected power is zero. Now imagine a situation in which the Brewster angle is greater than the critical angle. A wave incident at the Brewster angle will not be reflected, because the incident angle is equal to θ_b, nor will it be transmitted, because the incident angle is greater than θ_c. Is this situation possible? Why?

4.7 Solid-state lasers (ruby or glass) are often fabricated of rods with the ends bevelled at the Brewster angle. Let $\epsilon = 2\epsilon_0$ for the rod. Sketch the proper arrangement of the external mirrors and their angles. Indicate the bevelled angle of the glass rod. What is the polarization of the output of the laser beam? (See Figure P4.7.)

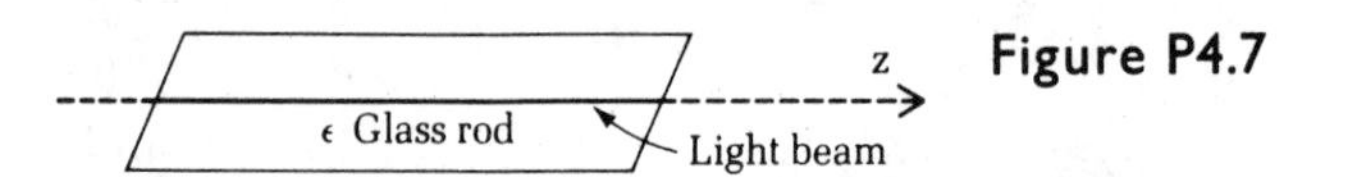

Figure P4.7

4.8 A perpendicularly polarized electromagnetic wave impinges from medium 1 (characterized by $\mu_1 = \mu_0$ and $\epsilon_1 = 4\epsilon_0$) to medium 2 (characterized by $\mu_2 = \mu_0$ and $\epsilon_2 = \epsilon_0$). This situation is shown in Figure P4.8.

(a) What is the critical angle?
(b) Let the incident angle be 60°; find k_x and k_z in terms of $k_0 = \omega\sqrt{\mu_0 \epsilon_0}$.
(c) Find k_{tz} in terms of k_0.
(d) In the second medium, find the distance z_0 at which the field strength is $1/e$ of that at $z = 0^+$.
(e) Find $|\mathrm{R}_I|$ and the phase shift arg (R_I).

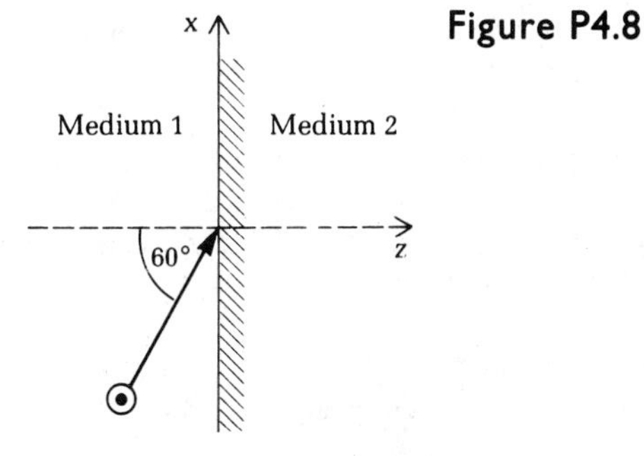

Figure P4.8

4.9 A uniform plane wave in air impinges normally on a dielectric wall. The magnitude of the total **E** field measured in front of the wall is shown in Figure P4.9.

(a) What is the permittivity of the dielectric wall? Assume $\mu_2 = \mu_0$.
(b) What is the frequency of the wave?

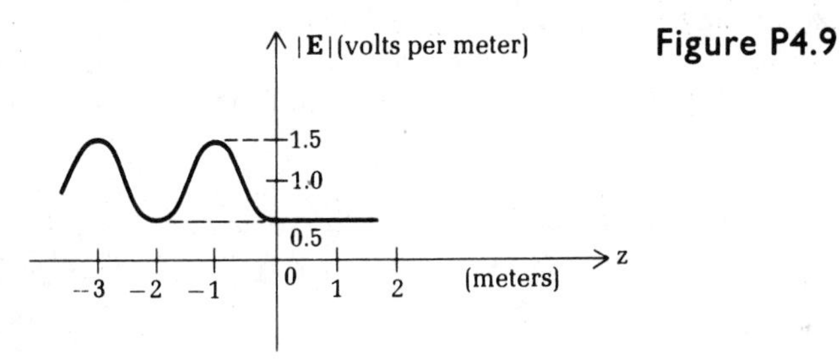

Figure P4.9

4.10 A uniform plane wave in air impinges on a lossless dielectric material at a 45° angle, as shown in Figure P4.10. The transmitted wave propagates in a 30° direction with respect to the normal. The frequency is 300 MHz.

(a) Find ϵ_2 in terms of ϵ_0.
(b) Find the reflection coefficient R_{II}.
(c) Obtain the mathematical expressions for the incident **E** field, the reflected **E** field, and the transmitted **E** field.
(d) In both media, sketch the standing-wave pattern of $|E_{x\,\text{total}}|$ as a function of z.

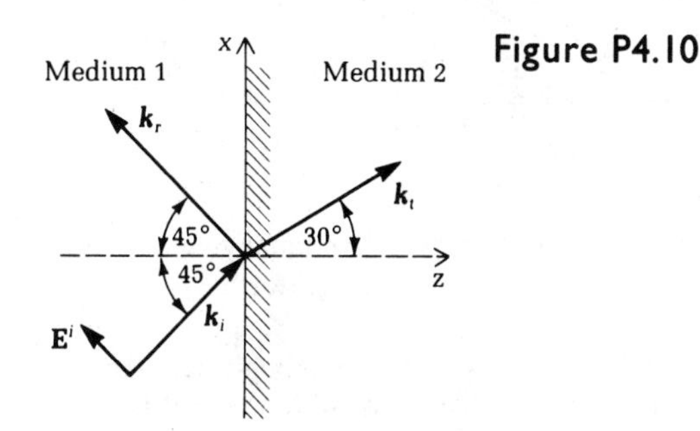

Figure P4.10

4.11 For two media with $\mu_1 \neq \mu_2$ and $\epsilon_1 \neq \epsilon_2$, find the Brewster angle for both the perpendicular polarization and the parallel polarization.

4.12 If a wire antenna is attached parallel to the metallic surface of a vehicle and is insulated from the surface by a thin layer of dielectric material with a thickness approximately equal to 1 mm, would it receive an AM signal ($f = 1$ MHz)? Hint: Wire antenna interacts only with the E field in the direction of the wire.

4.13 It is known that the transmitting antenna of an FM station is located in the direction perpendicular to a metallic plate, as shown in Figure 4.14. The frequency of the signal is 94 MHz.

(a) Where should a receiving antenna be placed to receive maximum signal? The antenna is a dipole that interacts with the E field.
(b) If the amplitude of the incident E field is 1 V/m, what is the amplitude of the E field at this optimum position?

4.14 It is found that by placing a conducting plate 0.8 m behind a dipole antenna, the received signal coming from the normal direction is twice as strong as the incident field. What is the frequency of the signal? What would the strength of the total E field be if the frequency of the wave is changed to 98 MHz while the antenna is still placed 0.8 m from the plate?

4.15 What would the E field be if the receiving system in Problem 4.13 were to detect a wave coming in at an angle 10° off the normal? Assume that all other parameters remain the same.

4.16 Consider a 90° "corner reflector" shown in Figure P4.16. It is made of two conducting plates placed perpendicularly to each other. A uniform plane wave with $\mathbf{E} = \hat{z}E_0 \exp(jkx \cos\theta + jky \sin\theta)$ impinges on the structure at an angle θ. Show that the total electric field is $\mathbf{E} = -\hat{z}4E_0 \sin(kx \cos\theta)\sin(ky \sin\theta)$. Hint: The field is the sum of four waves with four $\boldsymbol{k}$-vectors shown in Figure P4.16.

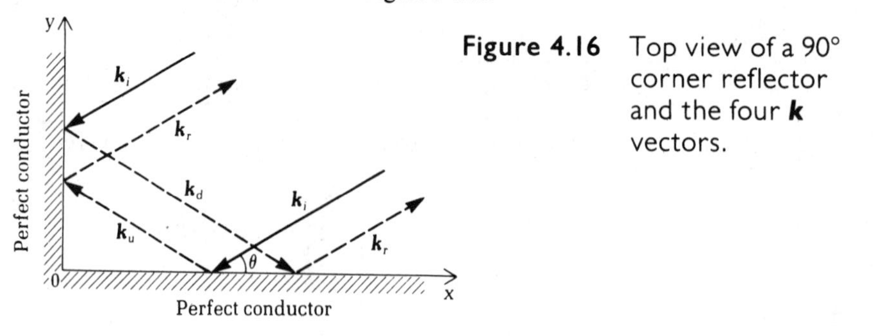

Figure 4.16 Top view of a 90° corner reflector and the four **k** vectors.

4.17 For a parallel-polarized uniform plane wave impinging on a perfect conductor at an angle θ, find the electric and magnetic fields **E** and **H** for the incident and for the reflected waves.

4.18 Match the following descriptions to the standing-wave patterns shown in Figure P4.18. The incident wave in medium 1 has an amplitude equal to 1 V/m. Note: There are three

Figure P4.18

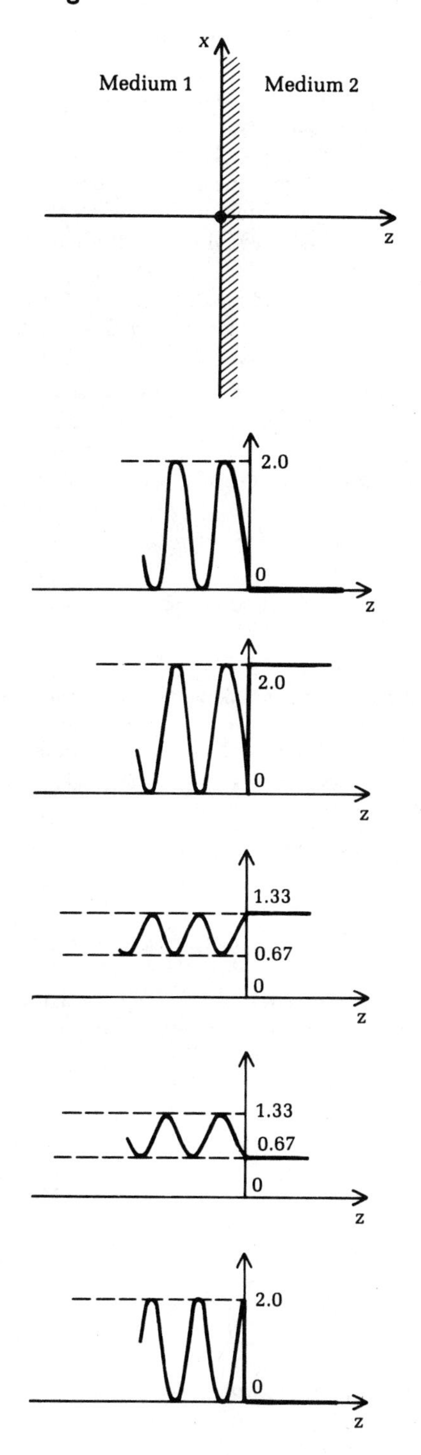
x
Medium 1
Medium 2
z
2.0
0
z
2.0
0
z
1.33
0.67
0
z
1.33
0.67
0
z
2.0
0
z

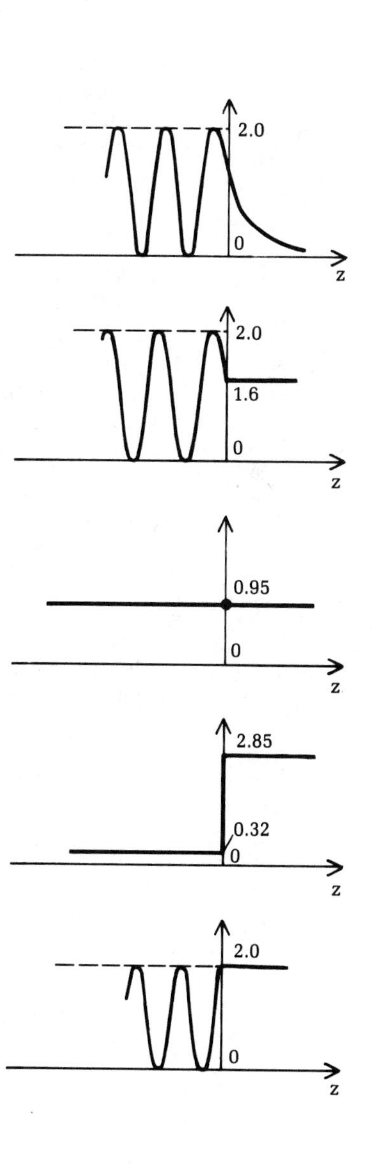
2.0
0
z
2.0
1.6
0
z
0.95
0
z
2.85
0.32
0
z
2.0
0
z

patterns that do not fit any of the following descriptions. Cross out these patterns.

(i) Plot of $|E_{y\ \text{total}}|$, with medium 1 being air, medium 2 having $\epsilon_2 = 4\epsilon_0$ and $\mu_2 = \mu_0$. Normal incidence.
(ii) Plot of $|E_{y\ \text{total}}|$, with medium 1 being characterized by $\epsilon_1 = 4\epsilon_0$ and $\mu_1 = \mu_0$, and medium 2 being air. Normal incidence.
(iii) Plot of $|E_{y\ \text{total}}|$, with medium 1 being characterized by $\epsilon_1 = 4\epsilon_0$ and $\mu_1 = \mu_0$, and medium 2 being air. The incidence angle is greater than the critical angle.
(iv) Plot of $|E_{x\ \text{total}}|$, incidence angle is equal to the Brewster angle.
(v) Plot of $|E_{z\ \text{total}}|$, incidence angle is equal to the Brewster angle. ϵ_1 is greater than ϵ_2.
(vi) Plot of $|E_{y\ \text{total}}|$. Medium 1 is air and medium 2 is a perfect conductor.
(vii) Plot of $|H_{y\ \text{total}}|(\eta_1)$. Medium 1 is air and medium 2 is a perfect conductor.

4.19 Use the formula given in Problem 4.16 for the total electric field. Find the optimum position of a dipole antenna placed in front of the 90° corner reflector. The θ angle of the incident wave is 30°. The frequency is 100 MHz. Express the position in x-y coordinates in meters. What is the "gain" of this receiving antenna? Gain is defined as follows:

$$\text{Gain} = 20 \log_{10}\left|\frac{E_a}{E_0}\right| \text{(dB)}$$

where E_a is the E field at the antenna position and E_0 is the field strength of the incident wave.

4.20 Consider the reflection from the ionosphere shown in Figure 4.13. Find the total **E** fields in both media, assuming perpendicular polarization with the incident **E** field strength equal to E_0. Sketch $|E_y|$ as a function of z. What are the minimum and the maximum values of $|E_y|$ in medium 1? Where is the first minimum located?

4.21 Review the "reflection from skin" topic present in Section 4.3. Find the percentage of the reflected power for the case of perpendicular polarization.

4.22 Review the "standing wave in front of a dielectric medium" topic in Section 4.4. Repeat the computation for 20 MHz. Find the reflection and the transmission coefficients, $|\mathbf{E}_y|$ in both media, and plot the latter as a function of z. Locate the first z_{max} and z_{min}. Find the maximum and the minimum values of $|\mathbf{E}_y|$ in medium 1.

5 WAVEGUIDES AND RESONATORS

In Chapter 3 we learned that electromagnetic waves can propagate in an unbounded medium. The path of the wave is straight, and its intensity is uniform on the transverse plane. We shall now consider the interesting problem of transmitting electromagnetic power from one point to another in such a way that the wave intensity is limited to a finite cross section (that is, in such a way that the intensity is not uniform on the transverse plane) and in such a way that the wave can be guided along a curved path. The structure that transmits electromagnetic waves in this manner is called a **waveguide** or a **transmission line**.

The wires that we see hanging over huge towers in the countryside are transmission lines guiding low-frequency (60 Hz) electromagnetic power from generating plants to distribution stations. Buried underground or underwater are coaxial lines that transmit hundreds of thousands of telephone conversations across land and sea by electromagnetic waves of high frequency (up to several gigahertz). Recent developments in fiberglass have aided the creation of communication systems at optical frequencies. These systems provide small physical volume, large channel capacity, and broad bandwidth. Fiber optics is a rapidly developing field with many applications in data transmission, data processing, photoelectric devices, photocopying, and biomedical instrumentation like fiberscopes.

5.1 PARALLEL-PLATE WAVEGUIDES

Figure 5.1 shows a parallel-plate waveguide. We assume that both plates are perfect conductors and that the plate dimension w in the $\hat{y}$ direction is very large so no field vectors have any y-dependence—that is, $\partial/\partial y = 0$. This latter assumption is physically justified when the fields are confined between the plates and when consequently there are negligible fringing fields outside the edges at $y = 0$ and $y = w$.

Using the fact that $\partial/\partial y = 0$ in Maxwell's equations, we expand them in rectangular coordinates. From (2.28a) we have

$$-\frac{\partial}{\partial z}\mathrm{E}_y = -j\omega\mu\mathrm{H}_x \tag{5.1a}$$

Figure 5.1 A parallel-plate waveguide.

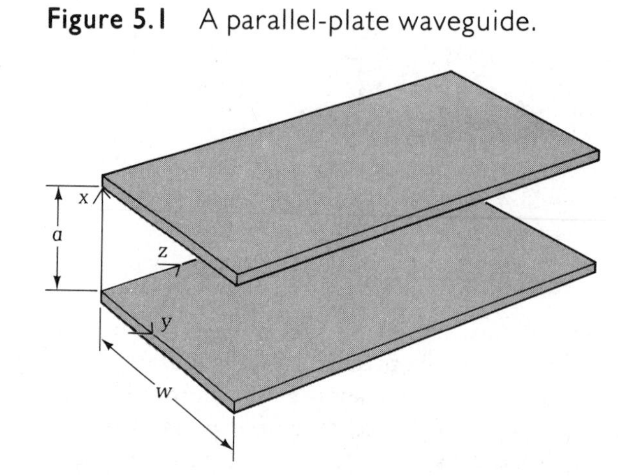

$$\frac{\partial}{\partial z}E_x - \frac{\partial}{\partial x}E_z = -j\omega\mu H_y \tag{5.1b}$$

$$\frac{\partial}{\partial x}E_y = -j\omega\mu H_z \tag{5.1c}$$

and from $\nabla \times \mathbf{H} = j\omega\epsilon\mathbf{E}$, (2.28b)

$$-\frac{\partial}{\partial z}H_y = j\omega\epsilon E_x \tag{5.2a}$$

$$\frac{\partial}{\partial z}H_x - \frac{\partial}{\partial x}H_z = j\omega\epsilon E_y \tag{5.2b}$$

$$\frac{\partial}{\partial x}H_y = j\omega\epsilon E_z \tag{5.2c}$$

Substituting (5.1a) and (5.1c) in (5.2b), we obtain the wave equation for E_y

$$\left(\frac{\partial^2}{\partial x^2} + \frac{\partial^2}{\partial z^2} + \omega^2\mu\epsilon\right)E_y = 0 \tag{5.3}$$

Substituting (5.2a) and (5.2c) in (5.1b) gives us a similar equation for H_y.

$$\left(\frac{\partial^2}{\partial x^2} + \frac{\partial^2}{\partial z^2} + \omega^2\mu\epsilon\right)H_y = 0 \tag{5.4}$$

It is very important to note that (5.3), (5.1a), and (5.1c) consist of the field components E_y, H_x, and H_z while (5.4), (5.2a), and (5.2c) consist of the field components H_y, E_x, and E_z. The former set characterizes the TE (transverse electric) waves, and the latter set characterizes the TM (transverse magnetic) waves.

TE Waves in Parallel-Plate Waveguides

The electric field E_y for TE waves in parallel-plate waveguides is determined from (5.3), (5.1a), and (5.1c), subject to the boundary condition of $E_y = 0$ at $x = 0$ and $x = a$. The solution is expected to represent waves propagating in the $\hat{z}$ direction. Assuming that the wave is propagating in the positive $\hat{z}$ direction, we write

$$E_y = E_0 \sin k_x x e^{-jk_z z} \tag{5.5}$$

Substituting in (5.3) yields the dispersion relation

$$k_x^2 + k_z^2 = \omega^2 \mu \epsilon = k^2 \tag{5.6}$$

Clearly, (5.5) satisfies the boundary condition of $E_y = 0$ at $x = 0$. In order to satisfy the boundary of $E_y = 0$ at $x = a$, we require

$$k_x a = m\pi \tag{5.7}$$

where m is any integer except zero. When $m = 0$, we see that $k_x = 0$ and consequently from (5.5) that $E_y = 0$. Equation (5.7) is known as the guidance condition, and it is a direct consequence of the boundary condition.

The solution (5.5) with k_x given in (5.7) and k_z given in (5.6) can be expressed in another form by decomposing the sine into exponents. We write

$$E_y = \frac{jE_0}{2}[\exp(-jk_x x - jk_z z) - \exp(jk_x x - jk_z z)] \tag{5.8}$$

The first term represents a plane wave propagating in the positive $\hat{x}$ and positive $\hat{z}$ directions. The second term represents a plane wave in the negative $\hat{x}$ and positive $\hat{z}$ directions (Figure 5.2). The sign for E_y is different for the two plane-wave components, and this difference indicates that the reflection coefficient $R = -1$. We see from the guidance condition that $2k_x a = 2m\pi$, meaning that for a round trip the phase of the wave in the $\hat{x}$ direction $(2k_x a)$ adds up to $2m\pi$, representing constructive interference. In terms of the directions of the guided-wave components, the guidance condition is satisfied only for those plane waves shown in Figure 5.3.

The propagation constant in the $\hat{z}$ direction is k_z, which is obtained from (5.6) and (5.7).

$$k_z = \left[\omega^2 \mu \epsilon - \left(\frac{m\pi}{a}\right)^2\right]^{1/2} = \omega\sqrt{\mu\epsilon}\left[1 - \left(\frac{m\lambda}{2a}\right)^2\right]^{1/2} \tag{5.9}$$

Figure 5.2 A TE wave propagating in a parallel-plate waveguide.

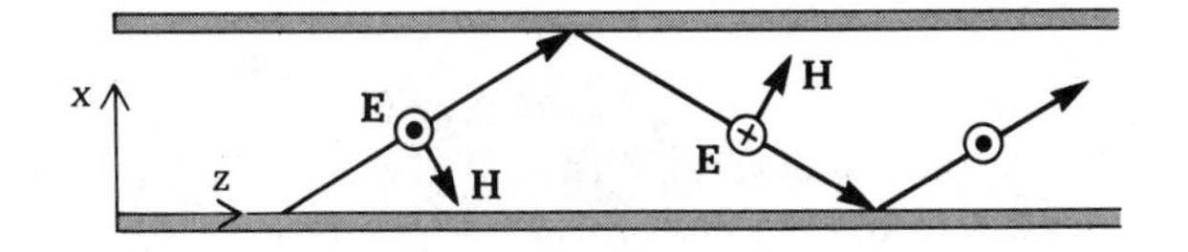

Figure 5.3 The bouncing wave shown in Figure 5.2 can propagate along a parallel-plate waveguide only if k_x is equal to $m\pi/a$, $m = 1, 2, 3, \ldots$.

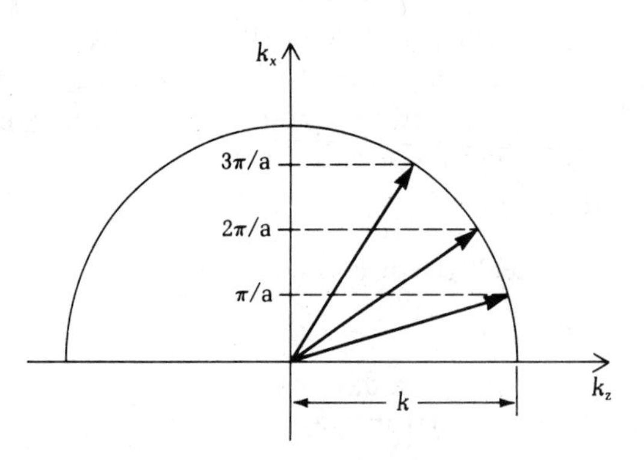

where we used the fact that $k^2 = \omega^2\mu\epsilon = (2\pi/\lambda)^2$ with λ representing the wavelength. The guided wave propagates with the phase velocity

$$v_p = \frac{\omega}{k_z} = \frac{1}{\sqrt{\mu\epsilon}}\left[1 - \left(\frac{m\lambda}{2a}\right)^2\right]^{-1/2} \tag{5.10}$$

We observed from (5.9) that k_z will be imaginary when

$$\omega\sqrt{\mu\epsilon} < \frac{m\pi}{a}$$

or when

$$\lambda > \frac{2a}{m}$$

If k_z becomes imaginary, the wave will attenuate exponentially in the $\hat{z}$ direction, and the velocity for the guided wave will become imaginary. The frequency at which this happens is called the **cutoff frequency**. It is given by

$$\boxed{f_c = \frac{\omega_c}{2\pi} = \frac{m}{2a\sqrt{\mu\epsilon}}} \qquad \textbf{(cutoff frequency of TE}_m\textbf{ mode)}$$

and the corresponding wavelength

$$\boxed{\lambda_c = \frac{2a}{m}} \qquad \textbf{(cutoff wavelength of TE}_m\textbf{ mode)} \tag{5.11}$$

is called the **cutoff wavelength**. We call the guided-wave field with the above cutoff frequencies and wavelengths the **TE$_m$ modes**.

TM Waves in Parallel-Plate Waveguides

The magnetic field H_y for TM waves in parallel-plate waveguides is a solution of (5.4), which gives

$$H_y = (A \sin k_x x + H_0 \cos k_x x)e^{-jk_z z} \tag{5.12a}$$

From (5.2a) and (5.2c), we find that for the electric field

$$E_x = \frac{k_z}{\omega\epsilon}(A \sin k_x x + H_0 \cos k_x x)e^{-jk_z z} \tag{5.12b}$$

$$E_z = \frac{k_x}{j\omega\epsilon}(A \cos k_x x - H_0 \sin k_x x)e^{-jk_z z} \tag{5.12c}$$

To satisfy the boundary conditions of $E_z = 0$ at $x = 0$, we see that $A = 0$. The boundary condition of $E_z = 0$ at $x = a$ then gives the guidance condition

$$k_x a = m\pi$$

which is identical to the boundary condition (5.7) for TE waves. Figure 5.4 shows a TM wave propagating in the parallel-plate waveguide. This picture is similar to Figure 5.2.

Figure 5.4 A TM wave propagating in a parallel-plate waveguide.

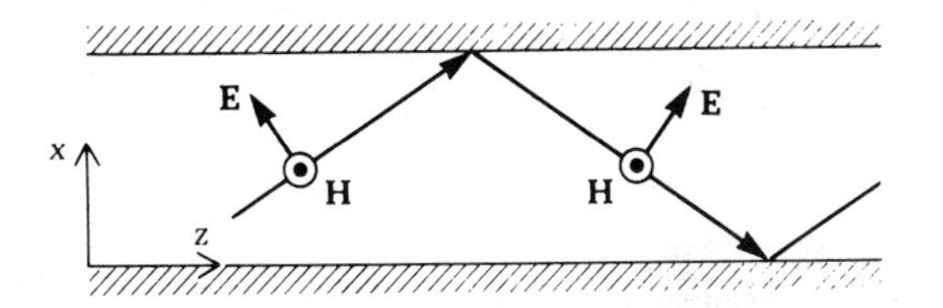

There is, however, a significant difference between the TE and the TM waves inside a parallel-plate waveguide. For the TM wave, we can now have $m = 0$, in which case, the guidance condition gives

$$k_x = 0$$

the dispersion relation yields

$$k_z = k = \omega\sqrt{\mu\epsilon}$$

and the field solutions become

$$H_y = H_0 e^{-jkz} \tag{5.13a}$$

$$E_x = \sqrt{\frac{\mu}{\epsilon}} H_0 e^{-jkz} = E_0 e^{-jkz} \tag{5.13b}$$

(electromagnetic fields of TM_0 or TEM mode)

This mode is the TM_0 mode, which is better known as the TEM mode, whose cutoff frequency is zero.

We can see that the TEM mode inside the parallel-plate waveguide is a "chopped" plane wave propagating in the $\hat{z}$ direction (Figure 5.5). The boundary conditions for the magnetic fields give us the surface currents.

$$\mathbf{J}_s = \hat{n} \times \mathbf{H} = \pm \hat{x} \times \hat{y} H_0 e^{-jkz} = \pm \hat{z} \frac{E_0}{\eta} e^{-jkz} \tag{5.13c}$$

where + and − signs denote the plates at $x = 0$ and a. The plates on which surface charges are generated interrupt the electric-field vectors. The surface charges are found to be

$$\rho_s = \pm \hat{x} \cdot \epsilon \mathbf{E} = \pm \epsilon E_0 e^{-jkz} \tag{5.13d}$$

on the plates at $x = 0$ and a with the plus and minus signs, respectively. Because the electric-field vector is perpendicular to the plate surfaces, the boundary condition of zero tangential electric field on the plate surfaces is satisfied.

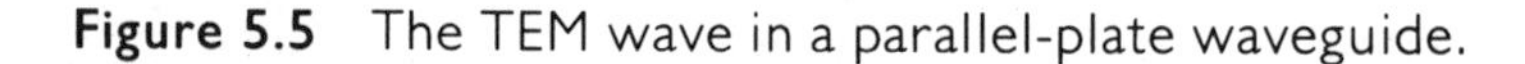

Figure 5.5 The TEM wave in a parallel-plate waveguide.

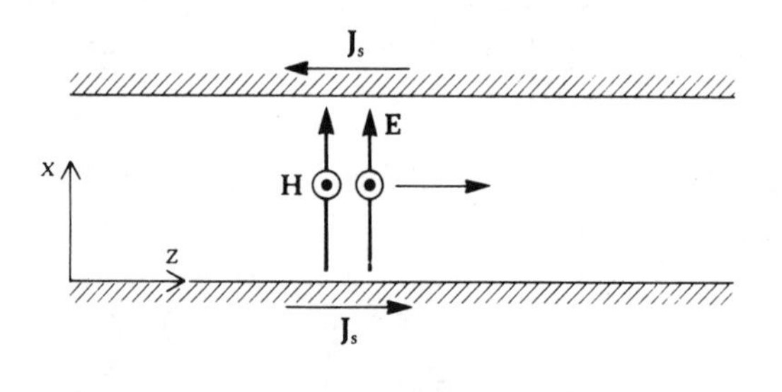

Microstrips or Striplines

Figure 5.6 shows a metallic strip supported by a layer of dielectric substrate over a conducting ground plate. This strip is called the stripline or microstrip. The electromagnetic wave that can propagate in this line is approximately given by (5.13). Strictly speaking, however, the analysis given in this section is only valid when the conducting plates are infinitely wide in the $\hat{y}$ direction. In practice, the upper plate's width is always finite. As long as the width of the strip is much

Figure 5.6 Stripline or microstrip: a variation of the parallel-plate waveguide.

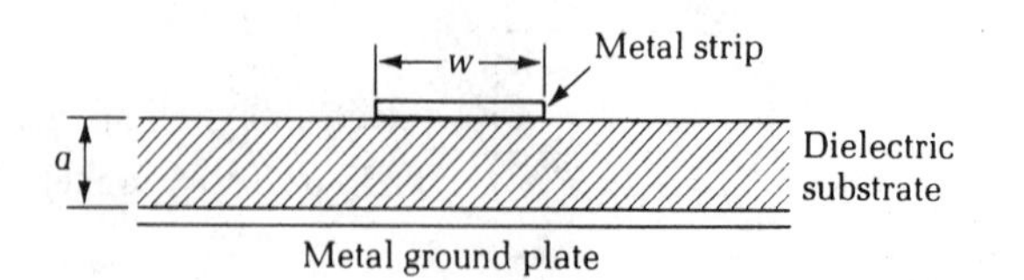

greater than the spacing between the upper and the lower conductors, the dominant electromagnetic fields in the stripline are approximately given by (5.13). Striplines or microstrips are widely used in microwave integrated circuits (MIC's) because they can easily be fabricated on dielectric substrate by printed-circuit techniques.

Ionosphere-Earth Waveguide

The ionosphere is a region composed of several layers of plasma, with the lowest layer approximately 80 km above the surface of the earth. At low frequencies, say 1 MHz, both the earth and the ionosphere are good conductors. Thus, the ionosphere and the earth form a "natural" parallel-plate waveguide for low-frequency electromagnetic waves. This fact explains why low-frequency signals may be guided to follow the curvature of the earth and to propagate at great distances. At higher frequencies, the dielectric constant of the ionosphere approaches that of free space, and hence high-frequency signals propagate through the ionosphere without being reflected back to the earth. It follows that transmission of a high-frequency signal, such as a television signal, relies on line-of-sight direct transmission.

Parallel Plates Made of Imperfect Conductors

When parallel-plate conductors are made of metals like copper or aluminum, rigorously speaking, the uniform plane wave given by (5.13) cannot be generated. Nevertheless, it is a good approximation to assume that the wave is a slightly modified uniform plane wave. The perturbation method that leads to an estimate of the effect of the finite conductivity on the strength of the **E** field is outlined below.

If the wave given by (5.13a) and (5.13b) is to exist, the surface current on the plates should be given by (5.13c), that is,

$$\mathbf{J}_s = \hat{n} \times \mathbf{H} = \mp \hat{z} H_0 e^{-jkz} \tag{5.13e}$$

on the top and bottom plates. Consequently, a tangential **E** field in the $\hat{z}$ direction must exist to support the current. In good conductors, the current must be distributed in depth rather than concentrated on the surface. The distribution of the current for $x \leq 0$ is given by

$$\mathbf{J} = J_0 e^{+jk_2 x} e^{-jkz} \hat{z} = \hat{z} \sigma E_2 e^{-jkz} e^{jk_2 x} \tag{5.14}$$

where E_2 is the magnitude of the electric field inside the conductor and $k_2 = k_{2R} - jk_{2I}$ is the complex wave number of the conductor. A good approximation would be to assume the $\mathbf{J}_s$ given by (5.13e) and $\mathbf{J}$ given by (5.14) are related by the following equation:

$$\mathbf{J}_s = \int_{-\infty}^{0} \mathbf{J}\, dx = \hat{z}\frac{\sigma E_2}{j\mathrm{k}_2} e^{-jkz}$$

Considering (5.13c) and noting the $H_0 = E_0/\eta$, we obtain

$$|E_2/E_0| = \frac{|\mathrm{k}_2|}{\sigma\eta} \approx \frac{\sqrt{\omega\mu\sigma}}{\sigma\eta} = \sqrt{\frac{\omega\epsilon}{\sigma}} \tag{5.15}$$

where we assume $\sigma/\omega\epsilon \gg 1$. It follows that E_2 is only a fraction of E_0. Then, the boundary condition requires that the tangential E field be continuous; thus, the E field between the plates must be modified so that its tangential components on the surface of the plates are equal to E_2, rather than equal to zero. The actual E field configuration in a parallel-plate waveguide of finite conductivity is shown in Figure 5.7a.

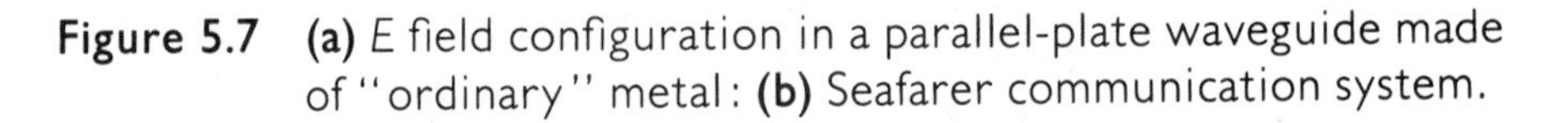

Figure 5.7 **(a)** E field configuration in a parallel-plate waveguide made of "ordinary" metal: **(b)** Seafarer communication system.

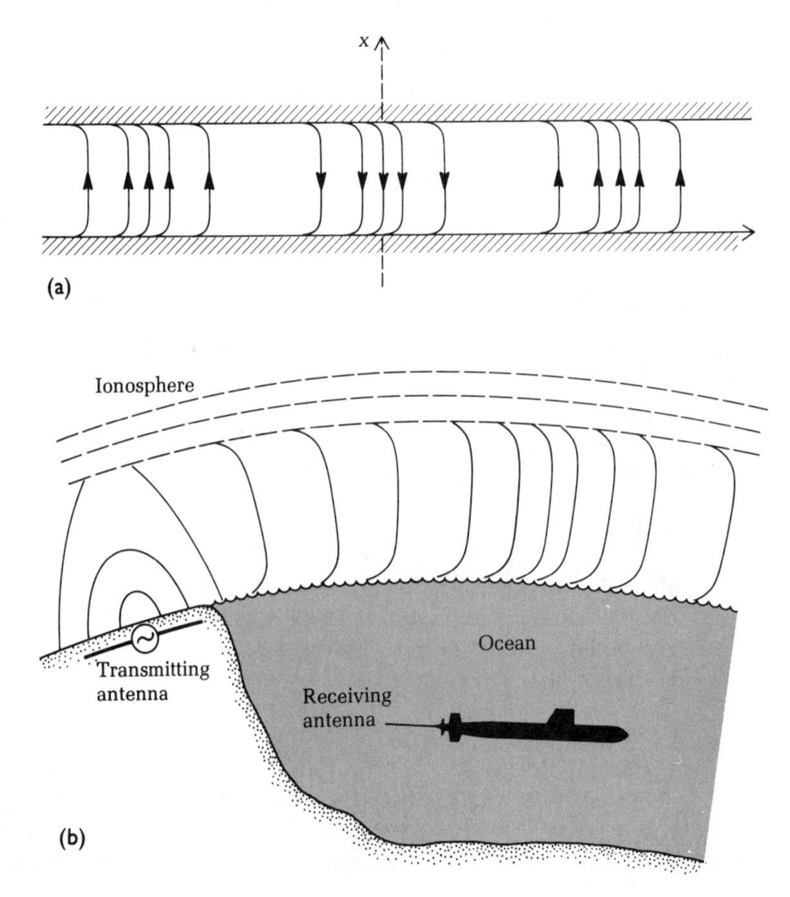

Project Seafarer

Project Seafarer* is a proposed global communication network for transmitting command and control messages to the United States strategic forces, including Polaris and Trident submarines that may be more than 200 m under the surface of the ocean. To increase the ability of electromagnetic waves to penetrate the conducting seawater, and extremely low frequency (ELF) must be used. The frequency band employed is between 45 Hz and 75 Hz. Because of the choice of frequency, the physical dimension of the transmitting antenna will be enormous. The Seafarer project proposed by the Navy calls for embedding cables 2–3 m underground over a 10,360 km^2 area (approximately 4,000 mi^2) in the central United States.

As sketched in Figure 5.7b, the horizontal antennas generate a vertically polarized E field, and the electromagnetic wave propagates in the ionosphere-earth waveguide.** Because of the finite conductivity of the seawater (4 mho/m), the E field is slightly tilted in the horizontal direction.

Using (5.15) and taking $f = 45$ Hz, we find that the ratio of the horizontal component of the E field on the surface of the ocean to the vertical component of the E field in the air is approximately 2.5×10^{-5}. Notice that the horizontally polarized electromagnetic wave that "leaks" into the ocean must attenuate about 0.23 dB/m before it finally reaches a submarine. Hence, an estimated power of more than 14 MW is needed to feed the transmitting antenna!

The location of the transmitting antenna will be determined primarily by the conductivity of the earth in the selected area. High conductivity of the earth near the antenna would effectively reduce both the size of the antenna and its efficiency. Possible sites for the antenna are northern Wisconsin, northern Michigan, and central Texas, where the conductivity is comparatively low, about 10^{-4} mho/m as compared with 10^{-3} mho/m for "ordinary" dry earth.

Example 5.1 *This example illustrates the total time-average power being transmitted along a parallel-plate waveguide in TEM mode. It is expressed in terms of the E field strength, the dimension of the waveguide, and the characteristics of the medium between the plates.*

Calculate the electromagnetic power transmitted by the TEM mode along a parallel-plate waveguide. The fields are given in (5.13). The waveguide is shown in Figure 5.1, and the medium between the plates is air.

---

* The project was originally called Project Sanguine.

** A more accurate model for the ELF propagation associated with the Seafarer communication system says that the earth-ionosphere forms a concentric spherical shell rather than a parallel-plate waveguide. See J. R. Wait, "Project Sanguine," *Science* 178 (October 1972): 272.

Solution We first calculate the time-average Poynting vector $\langle S \rangle$:

$$\langle S \rangle = \frac{1}{2}\text{Re}\,[\mathbf{E} \times \mathbf{H}^*] = \frac{E_0^2}{2\eta}\hat{z}$$

Note that k and η are real because the medium is air.

The total power P is obtained by integrating $\langle S \rangle$ over the cross section of the waveguide:

$$P = \int \langle S \rangle \cdot \hat{z}\, dA = \frac{E_0^2 wa}{2\eta} \text{ (watts)}$$

Example 5.2 *The maximum power that can be propagated in a waveguide is determined by the maximum E field that the medium can endure without suffering a breakdown.*

What is the maximum power that can be propagated in TEM mode in an air-filled parallel-plate waveguide of cross section $w = 5$ cm and $a = 1$ cm without causing breakdown? The breakdown E in air is approximately 2×10^6 V/m. Use a safety factor of 10 so that the maximum E in the waveguide is less than 1/10 of the breakdown E. Assume that the plates are perfectly conducting so the ohmic loss on the plates may be neglected.

Solution The maximum E field in the parallel-plate waveguide is E_0, according to (5.13). Thus, we let

$$E_0 = 0.1 \times 2 \times 10^6 = 2 \times 10^5$$

The maximum power P is obtained by use of the result calculated in the previous example:

$$P = \frac{(2 \times 10^5)^2 \times 10^{-2} \times 10^{-2} \times 5}{2 \times 120\pi} = 26.5 \text{ kW}$$

5.2 RECTANGULAR WAVEGUIDES AND RESONATORS

Consider the TE waves in the parallel-plate waveguides given by (5.5) and shown in Figure 5.2. The electric-field vector is in the $\hat{y}$ direction and has no components in the $\hat{x}$ and $\hat{z}$ directions. We can put conducting plates at $y = 0$ and $y = b$ to form a rectangular waveguide as shown in Figure 5.8. Surface charges will be induced on the newly placed plates. From (5.5) these surface charges can

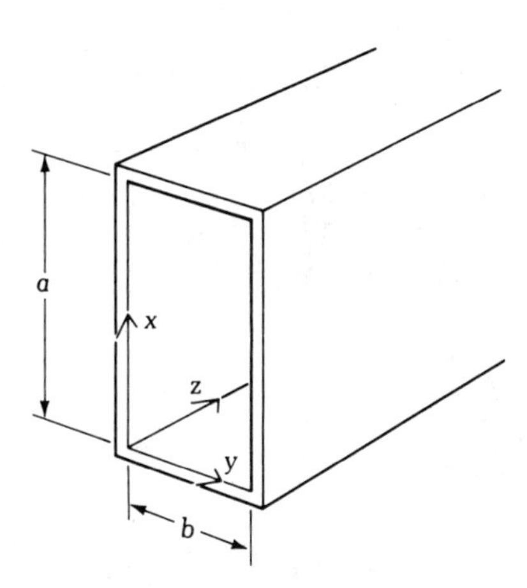

Figure 5.8 A rectangular waveguide.

be determined to be

$$\rho_s = \pm\hat{y}\cdot\epsilon\mathbf{E} = \pm\epsilon \mathrm{E}_y = \pm\epsilon \mathrm{E}_0 \sin k_x x e^{-jk_z z}$$

The electromagnetic fields inside the rectangular waveguide are the same as those in the parallel-plate waveguides.

$$\mathbf{E} = \hat{y}E_0 \sin k_x x e^{-jk_z z} \tag{5.16a}$$

$$\mathbf{H} = \frac{1}{-j\omega\mu}\nabla\times\mathbf{E} = E_0\left(\hat{x}\frac{-k_z}{\omega\mu}\sin k_x x + \hat{z}\frac{jk_x}{\omega\mu}\cos k_x x\right)e^{-jk_z z} \tag{5.16b}$$

where

$$k_x = \frac{m\pi}{a}$$

The TE_m modes in parallel-plate waveguides are known as TE_{m0} modes inside rectangular waveguides, where the second index 0 signifies that there is no variation in the $\hat{y}$ direction. If we relabeled the x axis y and the y axis x, then the TE_{m0} modes would also be relabeled TE_{0m} modes. There are other permissible solutions inside the rectangular waveguides. However, the most useful mode of propagation is the TE_{10} mode, which is called the fundamental mode of the rectangular waveguide.

We now consider the TM waves in the parallel-plate waveguides given by (5.12) and shown in Figure 5.4. The expressions for the fields are repeated as follows:

$$\mathrm{H}_y = H_0 \cos\frac{m\pi x}{a}e^{-jk_z z}$$

$$\mathrm{E}_x = \frac{k_z}{\omega\epsilon}H_0 \cos\frac{m\pi x}{a}e^{-jk_z z} \tag{5.17}$$

$$\mathrm{E}_z = \frac{jk_x}{\omega\epsilon}H_0 \sin\frac{m\pi x}{a}e^{-jk_z z}$$

If we now place perfectly conducting plates at $y = 0$ and $y = b$, we find that the tangential electric-field components E_x and E_z are not zero. The above solutions must be altered in order to satisfy the additional boundary conditions at $y = 0$ and $y = b$. There must now be variations in the $\hat{y}$ direction.

The complete solution for the TM waves in a rectangular waveguide is rather involved, and it takes the following forms:

$$E_z = E_0 \sin\frac{m\pi x}{a}\sin\frac{n\pi y}{b}e^{-jk_z z}$$

$$E_x = \frac{-jk_z k_x}{\omega^2\mu\epsilon - k_z^2}E_0 \cos\frac{m\pi x}{a}\sin\frac{n\pi y}{b}e^{-jk_z z}$$

$$E_y = \frac{-jk_z k_y}{\omega^2\mu\epsilon - k_z^2}E_0 \sin\frac{m\pi x}{a}\cos\frac{n\pi y}{b}e^{-jk_z z} \tag{5.18}$$

$$H_x = \frac{j\omega\epsilon k_y}{\omega^2\mu\epsilon - k_z^2}E_0 \sin\frac{m\pi x}{a}\cos\frac{n\pi y}{b}e^{-jk_z z}$$

$$H_y = \frac{-j\omega\epsilon k_x}{\omega^2\mu\epsilon - k_z^2}E_0 \cos\frac{m\pi x}{a}\sin\frac{n\pi y}{b}e^{-jk_z z}$$

We can see that all the boundary conditions are satisfied because at $x = 0, a$, we have $E_z = E_y = 0$ and at $y = 0, b$, we have $E_z = E_x = 0$.

The dispersion relations for the TM_{mn} modes are obtained by substituting the above solutions into the wave equation. We find

$$k_y = \frac{n\pi}{b}$$

$$k_z^2 = \omega^2\mu\epsilon - \left(\frac{m\pi}{a}\right)^2 - \left(\frac{n\pi}{b}\right)^2 \tag{5.19}$$

A **guide wavelength** λ_g is defined by letting $k_z = 2\pi/\lambda_g$, which gives

$$\lambda_g = \frac{2\pi}{\sqrt{\omega^2\mu\epsilon - \left(\frac{m\pi}{a}\right)^2 - \left(\frac{n\pi}{b}\right)^2}} \tag{5.20}$$

This wavelength will be used in Chapter 6 when we model a rectangular waveguide carrying the TE_{10} mode with a transmission line. Cutoff occurs as $k_z = 0$, and it yields

$$f_c = \frac{1}{2\sqrt{\mu\epsilon}}\sqrt{\left(\frac{m}{a}\right)^2 + \left(\frac{n}{b}\right)^2} \quad \text{(cutoff frequency of } TE_{mn} \text{ and } TM_{mn} \text{ modes)} \tag{5.21}$$

Notice that for TM_{mn} modes neither m nor n can be zero or else the field solutions would be zero.

To be complete, we also write down the exact solutions for the TE_{mn} modes.

$$H_z = H_0 \cos\frac{m\pi x}{a}\cos\frac{n\pi y}{b} e^{-jk_z z}$$

$$H_x = \frac{jk_z k_x}{\omega^2\mu\epsilon - k_z^2} H_0 \sin\frac{m\pi x}{a}\cos\frac{n\pi y}{b} e^{-jk_z z}$$

$$H_y = \frac{jk_z k_y}{\omega^2\mu\epsilon - k_z^2} H_0 \cos\frac{m\pi x}{a}\sin\frac{n\pi y}{b} e^{-jk_z z} \tag{5.22}$$

$$E_x = \frac{j\omega\mu k_y}{\omega^2\mu\epsilon - k_z^2} H_0 \cos\frac{m\pi x}{a}\sin\frac{n\pi y}{b} e^{-jk_z z}$$

$$E_y = \frac{-j\omega\mu k_x}{\omega^2\mu\epsilon - k_z^2} H_0 \sin\frac{m\pi x}{a}\cos\frac{n\pi y}{b} e^{-jk_z z}$$

The dispersion relations for the TE_{mn} modes are identical to those for the TM_{mn} modes. As $n = 0$, we find the expressions for the TE_{m0} modes identical to those given by (5.16). Notice that for TE_{mn} modes, either m or n may be zero but they cannot both be equal to zero.

X-Band Waveguide

A standard X-band waveguide has the following dimensions: $a = 2.29$ cm, and $b = 1.02$ cm. The medium inside the waveguide is air.

The cutoff frequency of the TE_{mn} and TM_{mn} modes can be calculated from (5.21). By using these data, the following table of cutoff frequencies for the first few modes can be obtained.

n \ m	0	1	2	3
0		TE_{10} 6.55 GHz	TE_{20} 13.10 GHz	TE_{30} 19.65 GHz
1	TE_{01} 14.71 GHz	TE_{11}, TM_{11} 16.10 GHz	TE_{21}, TM_{21} 19.69 GHz	TE_{31}, TM_{31} 24.54 GHz
2	TE_{02} 29.41 GHz	TE_{12}, TM_{12} 30.13 GHz	TE_{22}, TM_{22} 32.20 GHz	TE_{32}, TM_{32} 35.37 GHz

We see that the TE_{10} mode can propagate when the operating frequency is higher than 6.55 GHz. Above 13.1 GHz, the TE_{20} mode can also propagate. If the operating frequency is higher than 14.71 GHz, three modes can exist simultaneously in the waveguide: TE_{10}, TE_{20}, and TE_{01}. No TM modes can exist unless the frequency is higher than 16.1 GHz. To ensure one-mode operation, this waveguide should be used only in the frequency range 6.55 to 13.1 GHz. Note also that TE_{00} mode corresponds to the trivial solution that $\mathbf{E} = \mathbf{H} = 0$, and consequently it is of no practical use.

The TE_{10} Mode

The discussion given above shows that in order to ensure the one-mode operation, a rectangular waveguide should be operated in the TE_{10} mode. The electromagnetic fields of the TE_{10} mode can be obtained by substituting $m = 1$ and $n = 0$ in (5.22). It is seen that only E_y, H_z, and H_x are present in the TE_{10} mode. If the maximum electric field of the TE_{10} mode is expressed as E_0, then the electromagnetic fields of the TE_{10} mode can be expressed in terms of E_0 as follows

$$\begin{aligned} E_y &= E_0 \sin\frac{\pi x}{a} e^{-jk_z z} \\ H_x &= -\frac{k_z}{\omega\mu} E_0 \sin\frac{\pi x}{a} e^{-jk_z z} \\ H_z &= j\frac{\pi/a}{\omega\mu} E_0 \cos\frac{\pi x}{a} e^{-jk_z z} \end{aligned} \tag{5.23}$$

where

$$k_z = \sqrt{\omega^2\mu\epsilon - \left(\frac{\pi}{a}\right)^2}$$

Attenuation of the TE_{10} Mode in Rectangular Waveguides

Table 1 lists the loss characteristics of some commercial waveguides to show the order of magnitude of the ohmic loss caused by imperfect conducting walls for TE_{10}-mode operation.

Table 1 Loss Characteristics of Rectangular Waveguides with TE_{10} Mode.

Cross Section (cm)	Material	Frequency	Attenuation (dB/100 m)
2.850 × 1.262	brass	10 GHz	14
0.711 × 0.355	silver	40 GHz	49

Example 5.3 *This example shows the total time-average power being transmitted along a rectangular waveguide in TE_{10} mode. It is expressed in terms of the E field strength, the dimension of the waveguide, and the characteristics of the medium inside.*

Calculate the time-average power transmitted in TE_{10} mode along the rectangular waveguide shown in Figure 5.8.

Solution The time-averaging Poynting vector $\langle S \rangle$ is

$$\langle S \rangle = \frac{1}{2}\text{Re}\{\mathbf{E} \times \mathbf{H}^*\}$$

$$= \frac{1}{2}\text{Re}\left\{E_0 \sin\frac{\pi x}{a}\hat{y}\left[\frac{-E_0 k_z}{\omega\mu}\sin\frac{\pi x}{a}\hat{x} - \frac{jE_0 \pi}{\omega a\mu}\cos\frac{\pi x}{a}\hat{z}\right]\right\} \tag{5.24a}$$

$$= \frac{1}{2}\text{Re}\left\{\hat{z}\frac{E_0^2 k_z}{\omega\mu}\sin^2\frac{\pi x}{a} - \hat{x}\frac{jE_0^2 \pi}{\omega a\mu}\sin\frac{\pi x}{a}\cos\frac{\pi x}{a}\right\}$$

$$= \frac{E_0^2 k_z}{2\omega\mu}\sin^2\frac{\pi x}{a}\hat{z}$$

Integration over the cross section of the waveguide yields the total transmitted power P:

$$P = \int_0^a dx \int_0^b dy \frac{E_0^2 k_z}{2\omega\mu}\sin^2\frac{\pi x}{a} = \frac{E_0^2 abk_z}{4\omega\mu} \tag{5.24b}$$

Example 5.4 *The TE_{10} mode of a rectangular waveguide has the lowest cutoff frequency. This example shows that the next higher cutoff frequency is either the TE_{01} or the TE_{20} mode, depending on whether b is greater than a/2. It also shows that if b = a/2, the time-average power being transmitted in the waveguide is maximized without sacrificing the operating frequency bandwidth.*

For the waveguide shown in Figure 5.8, with what value of b is the operating frequency band of the waveguide in the TE_{10} mode maximum?

Solution Equation (5.21) gives the cutoff frequency of the TE_{10} mode. The next mode is either the TE_{01} or the TE_{20} mode, depending on whether b is greater than $a/2$.

To be more specific, the cutoff frequencies of the TE_{10}, TE_{01}, and TE_{20} modes are:

$$f_c(TE_{10}) = \frac{1}{2\sqrt{\mu\epsilon}}\frac{1}{a}$$

$$f_c(TE_{01}) = \frac{1}{2\sqrt{\mu\epsilon}}\frac{1}{b}$$

$$f_c(TE_{20}) = \frac{1}{2\sqrt{\mu\epsilon}}\frac{2}{a}$$

If $a/2 < b < a$, then

$$f_c(TE_{10}) < f_c(TE_{01}) < f_c(TE_{20})$$

and the operating frequency band should be limited to

$$\frac{1}{2a\sqrt{\mu\epsilon}} < f < \frac{1}{2b\sqrt{\mu\epsilon}} < \frac{1}{a\sqrt{\mu\epsilon}} \tag{5.25}$$

On the other hand, if $b < a/2$, then

$$f_c(TE_{10}) < f_c(TE_{20}) < f_c(TE_{01})$$

and the operating frequency band should be limited to

$$\frac{1}{2a\sqrt{\mu\epsilon}} < f < \frac{1}{a\sqrt{\mu\epsilon}} < \frac{1}{2b\sqrt{\mu\epsilon}} \tag{5.26}$$

Comparing (5.25) and (5.26) reveals that maximum operating frequency is obtained with b less than $a/2$. The frequency band is given by (5.26).

Equation (5.24) shows that the time-average power transmitted by the waveguide is directly proportional to b. Thus, b is usually set equal to $a/2$ so that power is maximized without sacrifice of the operating frequency bandwidth.

Example 5.5 *This example shows a practical design problem concerning a rectangular waveguide.*

Design a rectangular waveguide to transmit 10 GHz electromagnetic waves, and satisfy the following specifications: (a) the 10 GHz frequency is at the middle of the frequency band of the waveguide, and (b) $b = a/2$.

Solution The operating frequency band of a rectangular waveguide with $b = a/2$ is given in (5.26):

$$\frac{3 \times 10^8}{2a} < f < \frac{3 \times 10^8}{a}$$

Let the middle frequency be 10 GHz, and we have

$$10 \times 10^9 = \frac{1}{2}\left(\frac{3 \times 10^8}{2a} + \frac{3 \times 10^8}{a}\right)$$

Therefore,

$$a = 2.25 \text{ cm} \qquad b = 1.125 \text{ cm}$$

An air-filled rectangular waveguide of 2.25×1.125 cm will satisfy the design specifications.

Example 5.6 *This example gives a numerical value of the maximum power permitted in a rectangular waveguide. Note that it is concerned only with the possibility of the breakdown of air. The ohmic loss, which causes heating problems, is not considered.*

What is the maximum time-average power the waveguide discussed in the previous example can transmit without causing breakdown in air? The breakdown E of air is 2×10^6 V/m. Use a safety factor of 10 so that the maximum E field anywhere in the waveguide is less than 2×10^5 V/m. The frequency is 10 GHz. Neglect ohmic loss.

Solution The maximum E field exists at the central line ($x = a/2$) of the waveguide, and its maximum amplitude is E_0. Setting E_0 equal to 2×10^5 and using (5.24), we find

$$P_{\max} = 5.0 \text{ kW}$$

Note that Z_{TE} is calculated to be 505.8 Ω.

Rectangular Cavity Resonators

Consider the parallel-plate waveguide in Figure 5.1. Placing conducting plates at $y = 0$ and $y = b$, we form a rectangular waveguide as shown in Figure 5.8. The electric field of the TE_{m0} mode in the rectangular waveguide is given by (5.16a):

$$\mathbf{E} = \hat{y} E_0 \sin k_x x \, e^{-jk_z z} \tag{5.27}$$

with

$$k_x = \frac{m\pi}{a}$$

Closing the rectangular waveguide by placing conducting plates at $z = 0$ and $z = d$, we have a rectangular cavity resonator as shown in Figure 5.9. Inside the cavity, we not only have waves propagating in the positive $\hat{z}$ direction, as indicated by (5.27), but we must also have waves propagating in the negative $\hat{z}$ direc-

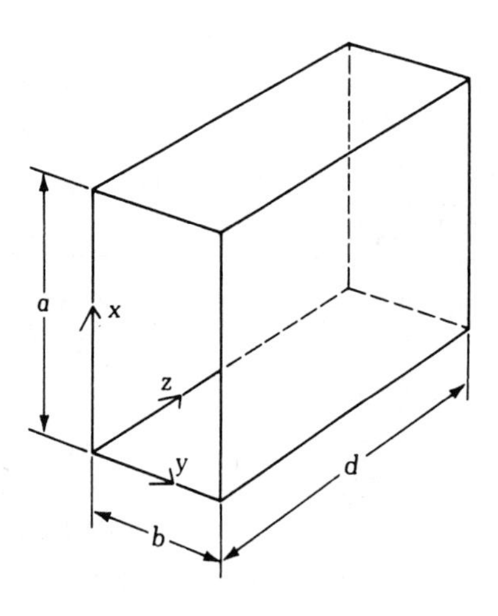

Figure 5.9 A rectangular cavity resonator.

tion as a result of reflections at the newly placed conducting plates, as indicated in Figure 5.10. We write

$$\mathbf{E} = \hat{y}E_0 \sin k_x x(e^{-jk_z z} + Ae^{jk_z z}) \tag{5.28}$$

Figure 5.10 Fields in a rectangular cavity resonator: **(a)** The **E** and **H** fields. **(b)** The forward- and backward-bouncing waves in the cavity.

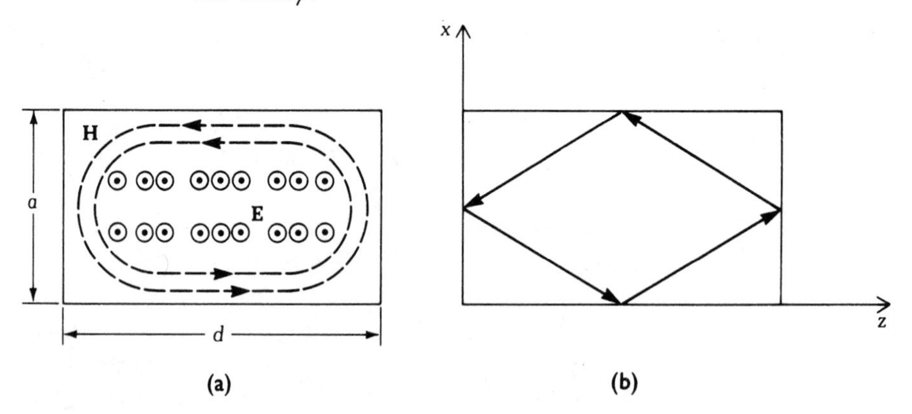

where the constant A is determined by the boundary conditions at $z = 0$ and $z = d$. The boundary conditions require that E_y be zero. Letting $E_y = 0$ at $z = 0$, we find $A = -1$. Letting $E_y = 0$ at $z = d$, we find

$$k_z = \frac{p\pi}{d}$$

where p is an integer. From the dispersion relation we see that

$$k = \sqrt{k_x^2 + k_z^2} = \left[\left(\frac{m\pi}{a}\right)^2 + \left(\frac{p\pi}{d}\right)^2\right]^{1/2}$$

Thus, the angular frequency $\omega = k(\mu\epsilon)^{-1/2}$ is fixed for a given pair of m and p. We call

$$f_{mp} = \frac{1}{2\sqrt{\mu\epsilon}}\left[\left(\frac{m}{a}\right)^2 + \left(\frac{p}{d}\right)^2\right]^{1/2} \tag{5.29}$$

the resonant frequency of the TE_{m0p} mode. The field expressions for the TE_{m0p} mode take the form

$$\mathbf{E} = \hat{y}E_{mp}\sin\frac{m\pi x}{a}\sin\frac{p\pi z}{d} \tag{5.30a}$$

$$\begin{aligned}\mathbf{H} &= \frac{1}{-j\omega\mu}\nabla\times\mathbf{E}\\ &= \hat{x}E_{mp}\frac{p\pi}{j\omega\mu d}\sin\frac{m\pi x}{a}\cos\frac{p\pi z}{d} - \hat{z}E_{mp}\frac{m\pi}{j\omega\mu a}\cos\frac{m\pi x}{a}\sin\frac{p\pi z}{d}\end{aligned} \tag{5.30b}$$

Clearly, the boundary conditions for tangential electric fields at the conducting surfaces are satisfied.

The electric field for the TE_{101} mode may be written as follows:

$$\text{E}_x = \text{E}_z = 0$$

$$\text{E}_y = E_0\sin\frac{\pi x}{a}\sin\frac{\pi z}{d} = \frac{-E_0}{4}(e^{-j(\pi/a)x} - e^{j(\pi/a)x})(e^{-j(\pi/d)z} - e^{j(\pi/d)z})$$

The bouncing-wave picture for the resonator mode is illustrated in Figure 5.10.

As a numerical example, let $a = d = 1.4142$ cm. We find the resonant frequency

$$f_{101} = 15 \text{ GHz}$$

for the TE_{101} mode.

Frequency Meter

A microwave cavity may be used as a frequency meter for measuring the unknown frequency of a signal. Figure 5.11 shows the arrangement of this experi-

Figure 5.11 **(a)** When the cavity is not resonant with the signal, it has little effect on wave propagation. **(b)** When the cavity is at resonance, it absorbs power from the waveguide, so the detector reading dips.

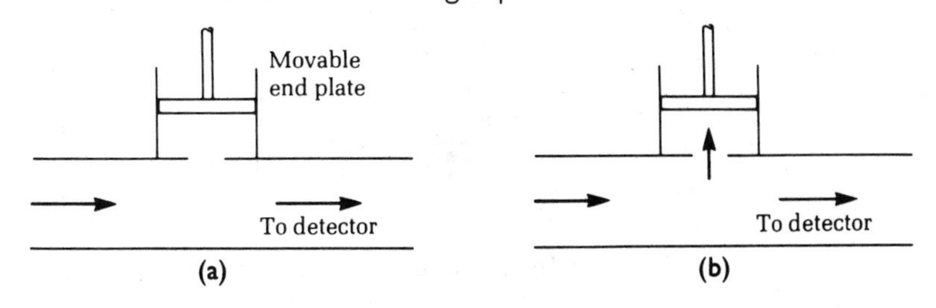

ment. The cavity is coupled to a waveguide through a small opening. The end plate of the cavity is moved by a micrometer. When the cavity is not resonant with the signal frequency, it has very little effect on the propagation of the wave. When the end plate is so adjusted that the resonant frequency of the cavity is equal to the frequency of the signal, the cavity absorbs some of the power to maintain its resonance. Consequently, the detector reading registers a dip. The position of the end plate is read on the micrometer. It is then compared with a calibrated table for the corresponding frequency.

5.3 DIELECTRIC SLAB WAVEGUIDES

Total reflection of an interface from an optically dense medium to one that is less dense provides another ideal way to guide waves. Dielectric waveguides in integrated optical circuitry and optical fibers of circular cross sections have found important applications at optical frequencies. In this section we shall not analyze guidance by cylindrical optical fibers because of the relative complexity of the mathematics required. Instead, we shall concentrate on the guidance of an electromagnetic wave by a dielectric slab. The qualitative conclusions obtained from this example are applicable to propagation of electromagnetic waves along cylindrical structures.

Consider the structure shown in Figure 5.12. It has dielectric constant ϵ_1, whereas the dielectric constant of the medium surrounding the slab is ϵ_2. The permeabilities are equal, $\mu_1 = \mu_0$. We are looking for wave solutions with the **E** field polarized along the $\hat{y}$ direction.

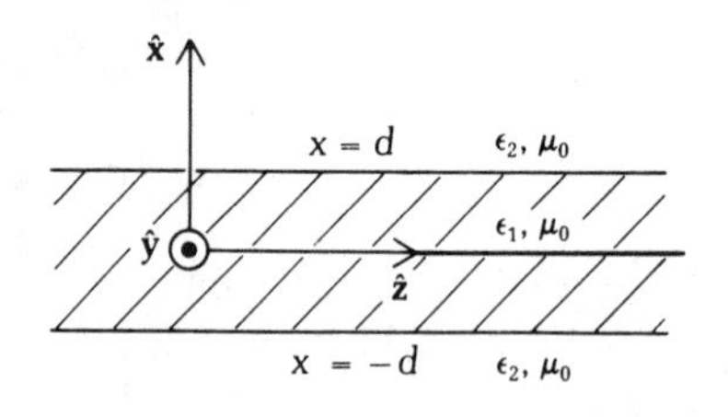

Figure 5.12 Wave propagating in a dielectric slab.

By analogy with parallel-plate metallic waveguides, we expect guidance to occur by reflection of a TE wave

$$\exp(jk_{1x}x - jk_z z)$$

off the bottom interface. This reflection produces a reflected wave

$$\exp(-jk_{1x}x - jk_z z)$$

of equal magnitude, if indeed the condition for total internal reflection is satisfied.

A standing wave has to result in a waveguide with the pattern

$$\cos k_{1x} x e^{-jk_z z}$$

or

$$\sin k_{1x} x e^{-jk_z z}$$

The former can be called a symmetric solution, the latter an antisymmetric solution.

Let us therefore look for a self-consistent solution that matches all the boundary conditions. We start with an assumed symmetric $\mathbf{E}$ field inside the dielectric slab as follows:

$$\mathbf{E} = \hat{y} E_1 \cos k_{1x} x e^{-jk_z z} \qquad \text{for } |x| < d \tag{5.31}$$

The magnetic field accompanying $\mathbf{E}$ of (5.31) is obtained from Faraday's law

$$\nabla \times \mathbf{E} = -j\omega\mu_0 \mathbf{H}$$

Substituting (5.31) into the above equation yields

$$\left.\begin{aligned} H_x &= -\frac{k_z}{\omega\mu_0} E_1 \cos k_{1x} x e^{-jk_z z} \qquad &(5.32a) \\ H_z &= -\frac{jk_{1x}}{\omega\mu_0} E_1 \sin k_{1x} x e^{-jk_z z} \qquad &(5.32b) \end{aligned}\right\} \text{for } |x| \le d$$

Outside the dielectric slab, the $\mathbf{E}$ field must decay away from the slab if indeed the wave is guided by the structure

$$\mathbf{E} = \hat{y} E_0 e^{\pm \alpha_x x} e^{-jk_z z} \qquad \text{for } |x| > d \tag{5.33}$$

The + and − signs correspond to the lower and the upper region, respectively. The constant E_0 must be the same on both sides of the slab because we are looking for an $\mathbf{E}$ field that is a symmetric function of x.

The $\mathbf{H}$ field is again obtained from Faraday's law:

$$\left.\begin{aligned} H_x &= -\frac{k_z}{\omega\mu_0} E_0 e^{\mp \alpha_x x} e^{-jk_z z} \qquad &(5.34a) \\ H_z &= \mp \frac{j\alpha_x}{\omega\mu_0} E_0 e^{\mp \alpha_x x} e^{-jk_z z} \qquad &(5.34b) \end{aligned}\right\} \text{for } |x| \ge d$$

Matching of the tangential $\mathbf{E}$ and $\mathbf{H}$ fields at $x = d$ gives

$$E_1 \cos k_{1x} d = E_0 e^{-\alpha_x d} \tag{5.35a}$$

and

$$-\frac{jk_{1x}}{\omega\mu_0} E_1 \sin k_{1x} d = -\frac{j\alpha_x}{\omega\mu_0} E_0 e^{-\alpha_x d} \tag{5.35b}$$

The matching of boundary conditions at $x = -d$ reproduces the above equations because we were consistent in setting up a symmetric E field in the region $|x| > d$.

Combining (5.35a) and (5.35b) gives the guidance relation

$$k_{1x}d \tan k_{1x}d = \alpha_x d \tag{5.36}$$

This is the equation which selects the propagation constant k_z of the mode. Indeed, both k_{1x} and α_x may be related to k_z by the dispersion relations in the two media.

$$k_{1x}^2 + k_z^2 = \omega^2 \mu_0 \epsilon_1 \tag{5.37}$$

$$-\alpha_x^2 + k_z^2 = \omega^2 \mu_0 \epsilon_2 \tag{5.38}$$

We may eliminate k_z from (5.37) and (5.38) by subtracting the two equations. The resulting expression gives α_x in terms of k_{1x}.

$$k_{1x}^2 + \alpha_x^2 = \omega^2 \mu_0(\epsilon_1 - \epsilon_2) \tag{5.39}$$

Eliminating α_x from (5.36) and (5.39), we find

$$k_{1x}^2 \sec^2 k_{1x}d = \omega^2 \mu_0(\epsilon_1 - \epsilon_2) \tag{5.40}$$

We see from (5.40) that, as $\omega \to \infty$, $k_{1x}d$ approaches $(2n-1)\pi/2$, where n is an integer characterizing the TE_n mode. When $k_{1x}d$ approaches this constant and $\omega\sqrt{\mu\epsilon_1}$ goes to infinity, we find from (5.37) that

$$k_z \approx \omega\sqrt{\mu_0 \epsilon_1}$$

The propagation constant is equal to that of the dielectric medium, and the wave is concentrated inside the slab.

One may similarly prove that k_z approaches $\omega\sqrt{\mu_0\epsilon_2}$ when ω either approaches zero for the dominant mode or approaches the cutoff frequency for any higher order mode. Consider the dominant mode as an example. From (5.40) we see that $k_{1x}d$ approaches zero as $\omega \to 0$. Therefore, $\sec k_{1x}d \approx 1$ and (5.40) gives

$$k_{1x} \approx \omega\sqrt{\mu_0(\epsilon_1 - \epsilon_2)}$$

Introducing this expression into (5.37), we find

$$k_z \approx \omega\sqrt{\mu_0 \epsilon_1}$$

Thus, the propagation constant is equal to that of the surrounding medium, and a substantial part of the wave's energy is guided along the outside of the slab. This propagation is the characteristic of a guided wave approaching cutoff inside a dielectric medium. Cutoff occurs when α_x becomes imaginary and the wave is no longer evanescent outside the slab. We set $\alpha_x = 0$ in (5.38) and conclude that $k_z = \omega(\mu_0\epsilon_2)^{1/2}$ for a guided wave approaching cutoff.

The cutoff frequencies of the TE_n modes can be determined from (5.39)–(5.40) by setting $\alpha_x = 0$. We find

$$\sec^2 k_{1x}d = 1$$

which yields

$$k_{1x}d = n\pi$$

Substituting into (5.40), we find the angular cutoff frequency

$$\omega_c = \frac{n\pi}{d\sqrt{\mu_0(\epsilon_1 - \epsilon_2)}}$$

For the fundamental TE_0 mode, $\omega_c = 0$. Letting $d = 3 \times 10^{-6}$ m and $\epsilon_1 = 1.47\epsilon_0$ and $\epsilon_2 = 1.46\epsilon_0$, we find the cutoff frequency of the next higher mode TE_1 to be

$$f_c = \frac{3 \times 10^8}{2 \times 3 \times 10^{-6} \times 10^{-1}} = 5 \times 10^{14} \text{ Hz}$$

which is in the visible range. Thus, the guided wave inside the dielectric slab can have single-mode propagation with the TE_0 mode up to 500 THz.

As shown in Figure 5.13, the guidance of the wave inside the slab can be interpreted as bouncing waves with an incident angle larger than the critical

Figure 5.13 A dielectric waveguide.

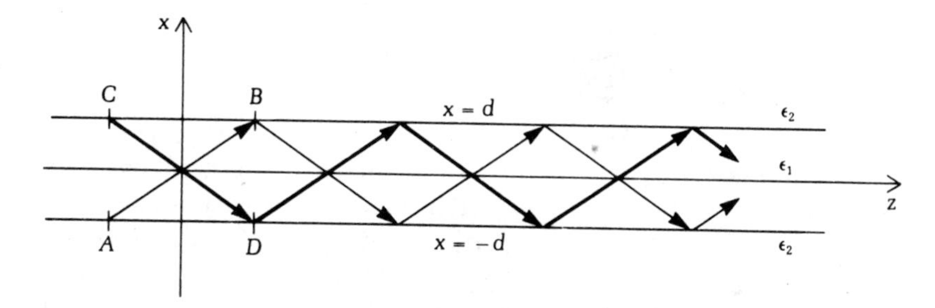

angle of reflection at the boundary surface. We notice that at total reflection the Fresnel reflection coefficient takes the following form:

$$\mathbf{R}_I = \frac{1 + j\dfrac{\alpha_x}{k_{1x}}}{1 - j\dfrac{\alpha_x}{k_{1x}}} = e^{j\psi}$$

where

$$\psi = 2\tan^{-1}\frac{\alpha_x}{k_{1x}} \tag{5.41}$$

is the phase shift. We now try to cast the guidance condition (5.36) in a form containing ψ. We find

$$n\pi + k_{1x}d = \frac{\psi}{2} \tag{5.42}$$

Taking the tangent of the above equation reduces to (5.36). Equation (5.42) can be written as

$$\psi - 2k_{1x}d = 2n\pi \tag{5.43}$$

The interpretation is as follows. Refer to Figure 5.13 in which two bouncing waves AB and CD are shown. The phase functions of these waves are:

$$\text{phase of } AB \text{ wave} = -k_{1x}x - k_z z$$

$$\text{phase of } CD \text{ wave} = k_{1x}x - k_z z$$

On the upper boundary where $x = d$, the phase of the AB wave is $-k_{1x}d - k_z z$ and that of the CD wave is $k_{1x}d - k_z z$. After reflection of the AB wave a phase shift ψ is added. If the phase of this reflected wave of AB is equal to or differs by $2n\pi$ from that of the unreflected wave of CD, a constructive interference is obtained. Mathematically speaking, constructive interference occurs if the following condition holds:

$$-k_{1x}d - k_z z + \psi = k_{1x}d - k_z z + 2n\pi$$

The above equation reduces to (5.43). Therefore the guidance relation may be obtained by satisfying the condition for constructive interference.

Optical Fibers

Optical fibers are dielectric waveguides used to transmit electromagnetic waves with frequencies close to or in the visible light spectrum. Thin fibers made of glass were first proposed in 1966 for long-distance communications. At that time, the attenuation rate of the light waves in the fiber was approximately 1000 dB/km. In less than twenty years, the rate was reduced to 0.2 dB/km when the fiber material was purified. It is now possible to transmit 1 Gbit/s with repeaters every 100 km. The commonly used wavelengths are 1.3 and 1.55 μm because of their low losses in the fiber material.

Because the wavelength of the light is short, there may exist many modes that can propagate in the dielectric waveguide when its thickness or diameter is many wavelengths long. Those optical waveguides are called multimode fibers. If one makes the size of the fiber smaller and fabricates it in such a way that its dielectric constant is varied, like the profile shown in Figure 5.14, it is possible to obtain a fiber that allows only one propagating mode. These new waveguides are called the single-mode or mono-mode optical fibers. Multimode waveguide is not preferred because there always exists the possibility of power in the primary mode being converted to other modes and thus unable to be recovered at the receiver end.

The diameter of a typical single-mode fiber is 7 μm, with 42 μm cladding surrounding it. The main material is glass, or SiO_2. Variation of the dielectric constant is achieved by dopping the glass with germanium, phosphorous, and boron oxides.*

* For details of single-mode or multimode optical fibers, see S. Ungar, *Fibre Optics*, John Wiley and Sons, West Sussex: England, 1990; or E. G. Neumann, *Single-Mode Fibers*, Springer-Verlag, Berlin: Germany, 1988.

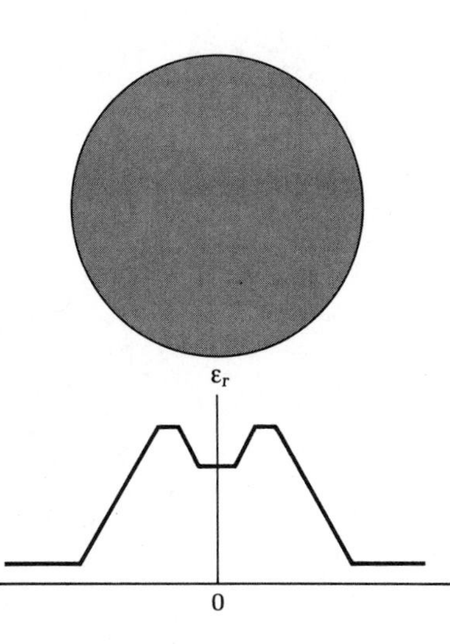

Figure 5.14 Variation of the dielectric constant in the radial direction in a single-mode optical fiber.

5.4 COAXIAL LINES

Coaxial lines are probably the most commonly used transmission lines. The essential parts of a coaxial line are an inner conductor of radius a and an outer conductor of inner radius b, insulated by a dielectric layer of permittivity ϵ, as shown in Figure 5.15. To discuss the solutions inside coaxial lines, we need to employ the cylindrical coordinate system.

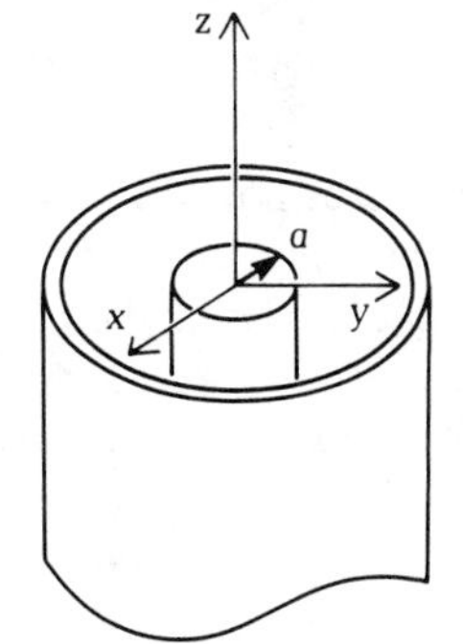

Figure 5.15 A coaxial line.

Cylindrical Coordinates

In cylindrical coordinates, a point P in space is represented by three coordinates ρ, ϕ, and z, as shown in Figure 5.16, where

ρ = its distance from the z axis, or length OA
ϕ = the angle between OA and the $\hat{x}$ axis
z = its distance from the x-y plane

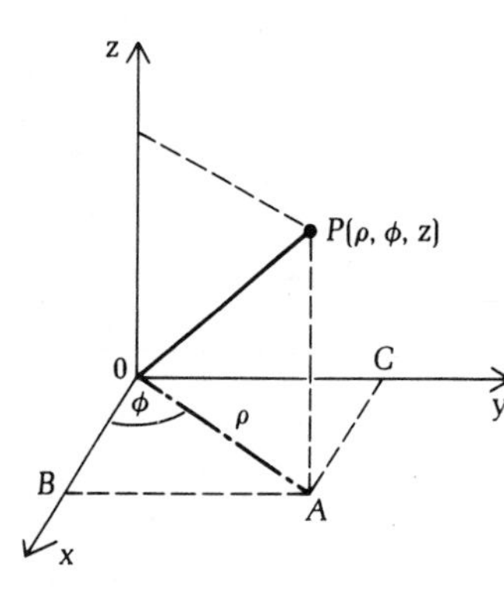

Figure 5.16 The cylindrical coordinate system.

In cylindrical coordinates, vector directions can also be expressed in terms of three unit vectors $\hat{\rho}$, $\hat{\phi}$, and $\hat{z}$, where

$\hat{\rho}$ = the unit vector in the direction of increasing ρ
$\hat{\phi}$ = the unit vector in the direction of increasing ϕ
$\hat{z}$ = the unit vector in the direction of increasing z

It is very important to bear in mind that $\hat{\rho}$ and ϕ depend on the location of the vector. For instance, at point B, $\hat{\rho}$ points in the $\hat{x}$ direction, but at C it points in the $\hat{y}$ direction.

The ∇ Operator in Cylindrical Coordinates

The ∇ operator in cylindrical coordinates does not have the simple appearance it has in rectangular coordinates. The reason for this difference is that the unit vectors $\hat{\rho}$ and $\hat{\phi}$ do not point in fixed directions and the coordinates ρ, ϕ, and z do not even have the same units; ρ and z are in meters, but ϕ is in radians. The divergence and the curl of a vector $\boldsymbol{A}$ take the following forms:

$$\nabla \cdot \boldsymbol{A} \equiv \frac{1}{\rho}\frac{\partial}{\partial \rho}(\rho A_\rho) + \frac{1}{\rho}\frac{\partial}{\partial \phi}A_\phi + \frac{\partial}{\partial z}A_z \tag{5.44}$$

$$\begin{aligned}\nabla \times \boldsymbol{A} &\equiv \hat{\rho}\left(\frac{1}{\rho}\frac{\partial}{\partial \phi}A_z - \frac{\partial}{\partial \rho}A_\phi\right) + \hat{\phi}\left(\frac{\partial}{\partial z}A_\rho - \frac{\partial}{\partial z}A_z\right) \\ &\quad + \hat{z}\frac{1}{\rho}\left[\frac{\partial}{\partial \rho}(\rho A_\phi) - \frac{\partial}{\partial \phi}A_\rho\right] \\ &= \frac{1}{\rho}\begin{vmatrix} \hat{\rho} & \rho\hat{\phi} & \hat{z} \\ \dfrac{\partial}{\partial \rho} & \dfrac{\partial}{\partial \phi} & \dfrac{\partial}{\partial z} \\ A_\rho & \rho A_\phi & A_z \end{vmatrix}\end{aligned} \tag{5.45}$$

The gradient and the Laplacian of a scalar ψ take the following forms

$$\nabla\psi = \left(\hat{\rho}\frac{\partial}{\partial\rho} + \hat{\phi}\frac{1}{\rho}\frac{\partial}{\partial\phi} + \hat{z}\frac{\partial}{\partial z}\right)\psi \tag{5.46}$$

$$\nabla^2\psi = \frac{1}{\rho}\frac{\partial}{\partial\rho}\left(\rho\frac{\partial}{\partial\rho}\psi\right) + \frac{1}{\rho^2}\frac{\partial^2}{\partial\phi^2}\psi + \frac{\partial^2}{\partial z^2}\psi \tag{5.47}$$

The TEM Mode in a Coaxial Line

To search for the solutions for the fundamental mode in the coaxial line, we consider the case when the inner radius a is very close to the outer radius b. We can imagine that the coaxial line shown in Figure 5.15 is cut along the $\hat{x}$ axis and unfolded into a parallel strip (Figure 5.17). From our experience with parallel-

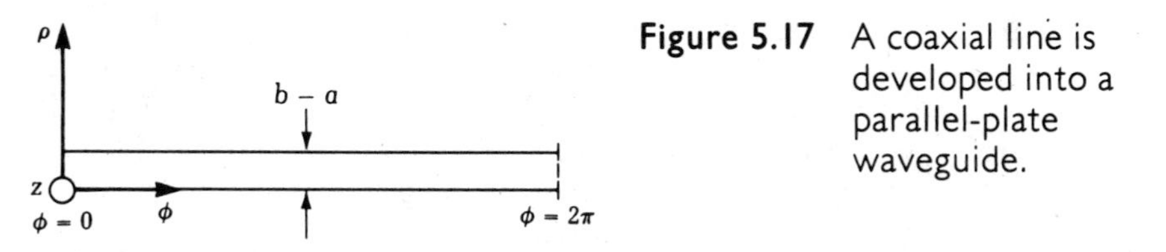

Figure 5.17 A coaxial line is developed into a parallel-plate waveguide.

plate waveguides, we know that the fundamental mode has **E** in the $\hat{\rho}$ direction and **H** in the $\hat{\phi}$ direction and that both are independent of ϕ. The solution for E_ρ and H_ϕ for waves propagating in the $\hat{z}$ direction takes the following form:

$$\mathbf{E} = \hat{\rho}\frac{V_0}{\rho}e^{-jkz} \qquad \textbf{(TEM mode in a coaxial line)} \tag{5.48a}$$

$$\mathbf{H} = \hat{\phi}\frac{V_0}{\eta\rho}e^{-jkz} \tag{5.48b}$$

where the propagation constant k and the proportionality constant η are given by

$$k = \omega\sqrt{\mu\epsilon} \tag{5.49a}$$

$$\eta = \sqrt{\frac{\mu}{\epsilon}} \qquad \textbf{(propagation constant and proportionality constant of the TEM mode in a coaxial line)} \tag{5.49b}$$

Direct substitution shows that the TEM mode given by (5.48) satisfies Maxwell's equations and the boundary conditions. Examining boundary conditions shows that the surface current on the inner conductor is

$$\mathbf{J}_s = \frac{V_0}{\eta a}e^{-jkz}\hat{z} \tag{5.50a}$$

The total current on the surface of the inner conductor is

$$\mathbf{I}_s = 2\pi a \mathbf{J}_s = \frac{2\pi V_0}{\eta} e^{-jkz}\hat{z} \tag{5.50b}$$

The total current on the inner surface of the outer conductor is equal to that on the inner conductor with the opposite sign. Because the electric and the magnetic fields are both transverse to the direction of wave propagation, the particular set of fields given by (5.48) is called the transverse electromagnetic (TEM) mode of the coaxial line.

Attenuation of the TEM Wave in Coaxial Lines

In the analysis given in this section, the conductors are assumed to be perfect conductors. In practice, the coaxial lines are made of aluminum or brass. These metals have high, but finite, conductivity. The E field of the TEM mode in the coaxial line has a slight longitudinal component. Some of the power leaks into the conducting surfaces and is dissipated as heat. Thus, the signal in the line is attenuated as it propagates.

Table 2 lists the loss characteristic of a typical commercial coaxial line. The inner conductor of the line is made of bare copper of 0.284 cm ID and 0.411 cm OD; the outer conductor is made of aluminum of 1.143 cm ID and 1.270 cm OD; and between them in polyethylene foam with $\epsilon = 1.5\epsilon_0$.

Miniature coaxial lines made of superconductors may greatly reduce the attenuation. For example, a superconducting coaxial line of 1.04 mm OD and 0.45 mm ID with polyethylene insulation has only 0.1 dB/100 m attenuation at 1 GHz.*

Table 2 Attenuation Characteristic of a Coaxial Line.

Frequency	10 MHz	100 MHz	1 GHz	10 GHz
Attenuation (dB/100 m)	0.72	2.62	9.8	49

Measurement of Wavelength

Consider a coaxial line that extends from $z = 0$ to $z = -\infty$. At the $z = 0$ end, the line is short-circuited by a conducting annular ring, as shown in Figure 5.18. The total electromagnetic fields in the coaxial line consist of the TEM mode in the $\hat{z}$

* N. S. Nahman, "Miniature superconducting coaxial transmission lines," *Proc. IEEE* 61 (1973): 76.

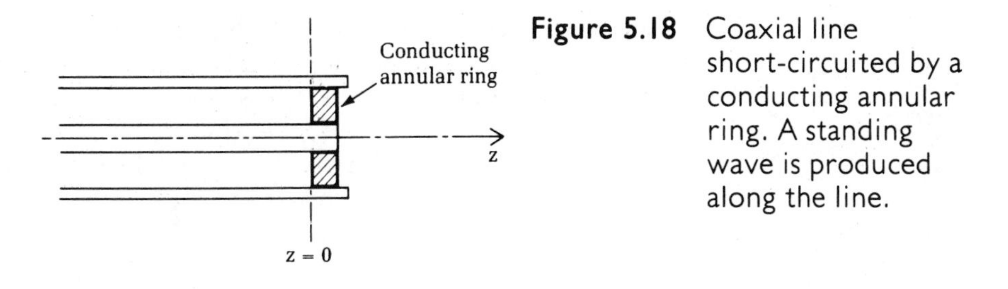

Figure 5.18 Coaxial line short-circuited by a conducting annular ring. A standing wave is produced along the line.

direction (incident wave) and the TEM mode in the $-\hat{z}$ direction (reflected wave). The reflected fields are given by

$$\mathbf{E}^r = \hat{\rho}\frac{V_1}{\rho}e^{jkz} \tag{5.51a}$$

$$\mathbf{H}^r = \hat{\phi}\frac{-V_1}{\eta\rho}e^{jkz} \tag{5.51b}$$

where k and η are given by (5.49). The total fields may then be expressed as

$$\mathbf{E} = \hat{\rho}\left(\frac{V_0}{\rho}e^{-jkz} + \frac{V_1}{\rho}e^{jkz}\right) \tag{5.52a}$$

$$\mathbf{H} = \hat{\phi}\left(\frac{V_0}{\rho\eta}e^{-jkz} - \frac{V_1}{\rho\eta}e^{jkz}\right) \tag{5.52b}$$

where the constants V_0 and V_1 are related by the boundary conditions. The boundary conditions require, among other things, that the tangential **E** field be zero on the surface of the annular ring. For this reason, we know that $V_1 = -V_0$, so that the **E** and **H** fields take the following form:

$$\mathbf{E} = \hat{\rho}\frac{-j2V_0}{\rho}\sin(kz) \tag{5.53a}$$

$$\mathbf{H} = \hat{\phi}\frac{2V_0}{\eta\rho}\cos(kz) \tag{5.53b}$$

Thus, if a probe is moving at constant ρ along the z axis, it will record a standing-wave pattern similar to that resulting from a uniform plane wave normally incident on a perfectly conducting plate. The wavelength of the electromagnetic wave can then be measured by measuring the distance between the minima.

Slotted Coaxial Lines

Many microwave experiments use slotted coaxial lines to measure the standing-wave pattern on the line. To record the pattern, a slot must be cut on the outer conductor so that a probe can be inserted into the space between the conductors and can be moved along the axis of the line. Notice that the longitudinal slot on the outer conductor will interfere very little with the fields because the current on the inner surface of the outer conductors flows only in the $\hat{z}$ direction. A slot cut in the $\hat{\phi}$ direction, for example, would cause considerable leakage of electromagnetic power because it would prevent the flow of current along $\hat{z}$, a flow that is essential to maintain the TEM mode in the line.

Example 5.7 *This example shows how to carry out the divergence and curl operations in cylindrical coordinates.*

Find $\nabla \cdot \mathbf{A}$ and $\nabla \times \mathbf{A}$ when $\mathbf{A} = \hat{\rho}(1/\rho)e^{-jkz}$ in cylindrical coordinates. Assume $\rho \neq 0$.

Solution Using (5.44), we obtain

$$\nabla \cdot \mathbf{A} = \frac{1}{\rho}\frac{\partial}{\partial \rho}\left(\rho \frac{1}{\rho} e^{-jkz}\right) = 0$$

Using (5.45), we obtain

$$\nabla \times \mathbf{A} = \hat{\phi}\frac{\partial A_\rho}{\partial z} = \hat{\phi}\frac{-jk}{\rho}e^{-jkz}$$

Example 5.8 *In this example, the time-average Poynting vector in a coaxial line and the total time-average power is found in terms of the line parameters.*

Calculate the total time-average electromagnetic power transmitted along a coaxial line when the fields are given by (5.48).

Solution The time-average Poynting vector $\langle S \rangle$ is given by

$$\langle S \rangle = \frac{1}{2}\text{Re}[\mathbf{E} \times \mathbf{H}^*] = \frac{1}{2}\text{Re}\left\{\frac{V_0}{\rho}e^{-jkz}\hat{\rho} \times \hat{\phi}\frac{V_0}{\eta\rho}e^{jkz}\right\} = \hat{z}\frac{V_0^2}{2\eta\rho^2}$$

Therefore, $P = (\pi V_0^2/\eta)\ln(b/a)$.

SUMMARY

1. Electromagnetic waves can propagate between two parallel conducting plates. These waves can be categorized into TEM, TE_m, and TM_m modes.
2. The TE_m or the TM_m mode can propagate in a parallel-plate waveguide only if the operating frequency is higher than the cutoff frequency of that mode.

The cutoff frequency of the TEM mode is zero, and consequently EM waves of all frequencies can propagate in TEM mode in the parallel-plate waveguide.

3. Electromagnetic waves can propagate in rectangular conducting pipes. These waves can be categorized into TE_{mn} and TM_{mn} modes.
4. The TE_{mn} or the TM_{mn} mode can propagate in a rectangular waveguide only if the operating frequency is higher than the cutoff frequency of that mode. The TEM mode cannot propagate in any rectangular waveguide.
5. The TE_{10} mode has the lowest cutoff frequency. The next higher cutoff frequency is that of the TE_{20} mode if $b < a/2$; or that of the TE_{01} mode if $a/2 < b$. Note that $a \times b$ denotes the cross-sectional dimensions of the rectangular waveguide and the wider side is denoted "a."
6. Generally speaking, electromagnetic fields do not exist inside an enclosed metal cavity. However, at certain frequencies known as the resonant frequencies, EM fields can exist in a metal cavity.
7. Electromagnetic waves can be guided in a dielectric slab. Tiny glass fibers can be used to guide EM waves at optical frequencies.
8. The TEM mode in a coaxial line has the E field in the radial direction and the H field in the azimuthal direction. The cutoff frequency of the TEM mode is zero.

Problems

5.1 Show that the complete fields of the TE wave in a parallel-plate waveguide are given as follows:

$$E_y = E_0 \sin k_x x e^{-jk_z z}$$

$$H_x = \frac{-k_z E_0}{\omega\mu} \sin k_x x e^{-jk_z z}$$

$$H_z = \frac{jk_x E_0}{\omega\mu} \cos k_x x e^{-jk_z z}$$

$$E_x = 0; \quad E_z = 0; \quad H_y = 0$$

5.2 Show that the complete fields of the TM wave in a parallel-plate waveguide are given as follows:

$$E_x = H_0 \frac{k_z}{\omega\epsilon} \cos k_x x e^{-jk_z z}$$

$$E_z = H_0 \frac{jk_x}{\omega\epsilon} \sin k_x x e^{-jk_z z}$$

$$H_y = H_0 \cos k_x x e^{-jk_z z}$$

$$H_x = H_z = E_y = 0$$

5.3 What is the lowest frequency of an electromagnetic wave that can be propagated in the TE mode in the earth-ionosphere waveguide? Model the latter as two perfectly conducting parallel plates separated by 80 km.

5.4 Find the surface-charge density ρ_s on the upper and the lower plates of a parallel-plate waveguide for (a) the TE_m mode, and (b) the TM_m mode.

5.5 Find the mathematical expressions for a TEM wave in a parallel-plate waveguide that propagates in the $-\hat{z}$ direction (see Figure 5.1). Sketch the parallel-plate waveguide, and indicate the directions of **E**, **H**, and $\mathbf{J}_s$.

5.6 A microstrip line has the dimensions $a = 0.15$ cm and $w = 0.71$ cm, and the permittivity of the substrate is $\epsilon = 2.5\epsilon_0$, $\mu = \mu_0$, $\sigma = 0$. Estimate the time-average power that is transmitted by the line when $|\mathbf{E}| = 10^4$ V/m.

5.7 The breakdown voltage of the dielectric substrate used in the stripline described in Problem 5.6 is 2×10^7 V/m. Use a safety factor of 10 so that $|\mathbf{E}|$ is less than 2×10^6 V/m everywhere in the line. Find the maximum time-average power that the stripline can transmit. Neglect the ohmic loss.

5.8 With the fields in a rectangular waveguide, find the surface current $\mathbf{J}_s$ on the top ($y = b$) of the waveguide. If we want to cut a slot along z, where should the slot be cut in order to minimize the disturbance it will cause? Assume that only the TE_{10} mode exists in the waveguide.

5.9 Show that, if the wavelength of an electromagnetic wave in an unbounded medium characterized by μ and ϵ is greater than $2a$, then this wave cannot propagate in the rectangular waveguide (shown in Figure 5.8) with the dielectric inside the waveguide also characterized by μ and ϵ.

5.10 An AM radio in an automobile cannot receive any signal when the car is inside a tunnel. Why? Let us assume that the tunnel is the Lincoln Tunnel, which was built in 1939 under the Hudson River in New York. Figure P5.10 shows a cross section of the Lincoln Tunnel.*

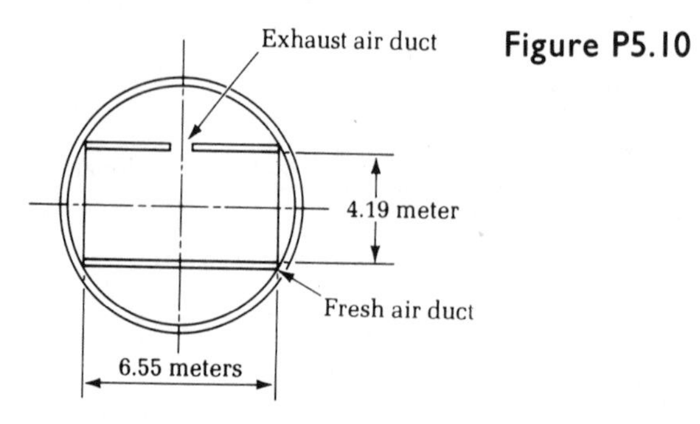

Figure P5.10

* G. E. Sandstrom, *Tunnels*, Holt, Rinehart & Winston: New York, 1963, p. 242.

5.11 Find the frequency ranges for TE_{10} mode operation for those rectangular waveguides listed in Table 1.

5.12 Design an air-filled rectangular waveguide to be used for transmission of electromagnetic power at 2.45 GHz. This frequency should be at the middle of the operating frequency band. The design should also allow maximum power transfer without sacrificing the operating bandwidth. Find the maximum power the waveguide can transmit. Use a safety factor of 10. Neglect ohmic loss. The breakdown E in air is assumed to be 2×10^6 V/m.

5.13 Repeat Problem 5.12, but assume that a dielectric material is used to fill the waveguide. The material is characterized by $\epsilon = 2.50\epsilon_0$, $\mu = \mu_0$, and $\sigma = 0$. The breakdown E field in the dielectric is 10^7 V/m.

5.14 Consider the size of a rectangular waveguide to explain why it is not used to transmit electromagnetic waves in the VHF range. (Take $f = 100$ MHz.)

5.15 The lowest three cutoff frequencies of a rectangular waveguide are: TE_{10} (7 GHz), TE_{20} (14 GHz), and TE_{01} (16 GHz). If the waveguide is connected to a 15 GHz source of electromagnetic wave, then which of the following statements is true?

(a) There are two possible propagating modes: TE_{10} and TE_{20}.
(b) There is only one propagating mode: TE_{01}.
(c) There is only one propagating mode: TE_{10}.
(d) There is only one propagating mode: TE_{20}.

5.16 An air-filled rectangular waveguide is designed to transmit signals in the TE_{10} mode in the frequency band 5 GHz to 10 GHz. The waveguide is then filled with a material with relative dielectric constant equal to 4. In which of the following frequency bands can this waveguide be used to transmit a signal?

(a) 20 GHz to 40 GHz
(b) 10 GHz to 20 GHz
(c) 1.25 GHz to 2.5 GHz
(d) 2.5 GHz to 5 GHz

5.17 The electromagnetic fields associated with the TE_{10} mode propagating in the $\hat{z}$ direction are given by (5.23). Find the electromagnetic fields associated with the TE_{10} mode propagating in the $-\hat{z}$ direction, with maximum electric field equal to E_1.

5.18 Consider a rectangular waveguide shown in Figure P5.18. For the region $z < 0$, the medium is air and for $z > 0$ the medium is characterized by ϵ_2 and μ_2. A TE_{10} mode with

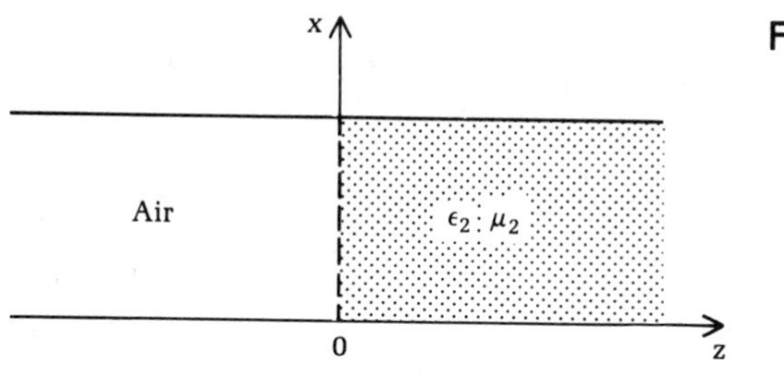

Figure P5.18

maximum E field equal to E_0 impinges on the boundary from the left. The result is that some power is reflected and some is transmitted. Assume that the reflected wave is also TE_{10}, with maximum E field equal to E_1, and the transmitted wave is TE_{10} mode with maximum E field equal to E_2. Find the ratio E_1/E_0 in terms of a, ω, ϵ_0, μ_0, ϵ_2, and μ_2.

5.19 The corner reflector studied in Problem 4.16 requires the solution

$$\mathbf{E} = -\hat{z}4E_0 \sin(kx \cos\theta)\sin(ky \sin\theta)$$

Show that although the coordinates are different this solution is in fact the resonator mode that we studied in Section 5.2. Placing conducting plates at $x = a$ and $y = b$ to form a cavity resonator as shown in Problem 4.16, what are the restrictions on the incident angle θ?

5.20
(a) Find the real-time expression of the fields of the TE_{101} mode in the rectangular cavity shown in Figure 5.9.
(b) Find the total stored electric energy in the cavity as a function of time. Find the corresponding total stored magnetic energy.
(c) Show that energy is stored alternatingly in electric and in magnetic fields, that the maximum stored electric energy is equal to the maximum stored magnetic energy, and that the total stored electromagnetic energy in the cavity is a constant independent of time. Note that these properties are similar to those of the low-frequency LC resonant circuits.

5.21 Find the lowest resonant frequency of the TE_{101} mode in an air-filled rectangular cavity measuring $2 \times 3 \times 5$ cm^3. Note that there are three different choices for designating the z axis and that these result in three different TE_{101} modes.

5.22 Electromagnetic waves in air with wavelengths ranging from 1 to 10 mm are called millimeter waves. Millimeter waves may be guided by dielectric slabs. Consider a dielectric slab with $\epsilon_1 = 10\epsilon_0$ and $\epsilon_2 = \epsilon_0$, as shown in Figure 5.12. What should its thickness be in order that only the TE_0 mode may be excited for frequencies up to 300 GHz?

5.23 Use direct substitution into Maxwell's equations to show that the fields given by (5.48) are solutions of Maxwell's equations in cylindrical coordinates.

5.24 Use the formulas of divergence and curl in cylindrical coordinates to prove that $\nabla \cdot \nabla \times A = 0$ for any vector A.

5.25 Find the rectangular coordinates of a point P where the cylindrical coordinates are $\rho = 1$, $\phi = 30°$, and $z = 2$.

5.26 Find the cylindrical coordinates of a point Q where the rectangular coordinates are x, y, and z.

5.27 Show that the differential volume in the cylindrical coordinates is $\rho\, d\rho\, d\phi\, dz$.

5.28 To convert a vector expressed in cylindrical components into the same in rectangular components, or vice versa, it is convenient to prepare a table for dot products between unit vectors in these coordinate systems. For example, $\hat{x} \cdot \hat{\rho} = \cos\phi$, as shown in the following table. Complete the table.

Dot Products Between Cylindrical and Rectangular Unit Vectors

	$\hat{\rho}$	$\hat{\phi}$	$\hat{z}$
$\hat{x}$	$\cos\phi$		
$\hat{y}$			
$\hat{z}$	0		

5.29 Use the above table to find the rectangular components of the following vector located at $\rho = 2$, $\phi = 30°$, and $z = -3$:

$$A = 8\hat{\rho} + 4\hat{\phi} - 3\hat{z}$$

5.30 What is the maximum time-average power a coaxial line can transmit without causing breakdown? Assume that the coaxial line is air-filled and that the breakdown E of the air is 2×10^6 V/m. Use a safety factor of 10 so that the maximum E field anywhere in the line does not exceed 2×10^5 V/m. The dimension of the line is $2a = 0.411$ cm and $2b = 1.143$ cm. Neglect ohmic loss.

5.31 Consider the coaxial line shown in Figure P5.31. Half of the line ($z < 0$) is filled with air, and half of it ($z > 0$) is filled with a material characterized by ϵ_1 and μ_1. The electromagnetic wave incident from the left has the following fields:

$$\mathbf{E}^i = \hat{\rho}\frac{V_0}{\rho}e^{-jk_0 z}$$

$$\mathbf{H}^i = \hat{\phi}\frac{V_0}{\eta_0 \rho}e^{-jk_0 z}$$

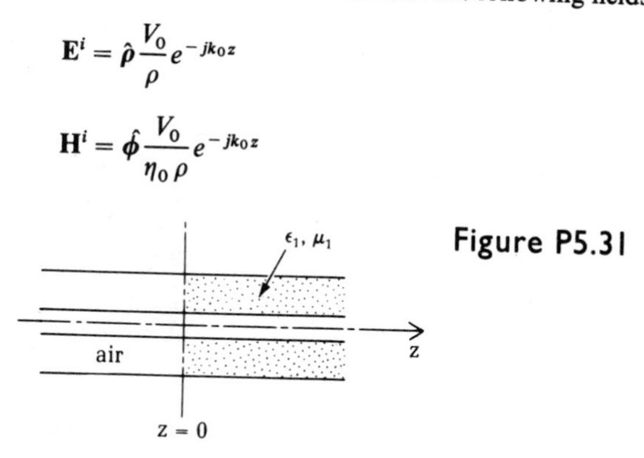

Figure P5.31

The field of the reflected wave may be expressed as follows:

$$\mathbf{E}^r = \hat{\rho}\frac{V'_0}{\rho}e^{jk_0 z}$$

$$\mathbf{H}^r = \hat{\phi}\frac{-V'_0}{\eta_0 \rho}e^{jk_0 z}$$

(a) Write down the fields of the transmitted wave in $z > 0$. What wave number k should be used?

(b) Find V_0' and the amplitude of the transmitted fields in terms of V_0, η_1, and η_0 by matching the boundary conditions at $z = 0$. Compare your result with the reflection and transmission coefficients obtained in Chapter 4 for waves reflected from dielectric boundaries.

5.32 Fill in the blanks for three waveguides discussed in this chapter.

Waveguide	Parallel Plate	Coaxial Line	Rectangular Waveguide
Dimension	width $= w$, height $= a$	Radii: a and b	cross sec: $a \times b$
Medium	air	air	air
Basic mode	TEM	TEM	TE_{10}
Cutoff frequency			
E field lines in x-y plane			
H field lines in x-y plane			
Is velocity of wave independent of frequency? Yes or No			
If breakdown occurs, where would it start?			

6 TRANSMISSION LINES

Most readers have probably been exposed to circuit theory, the study of voltages and currents. Circuit elements are defined by different relationships between V (voltage) and I (current). For example, a resistor has the relationship $V = RI$; an inductor has the relationship $V = L\,dI/dt$ or $V = j\omega LI$; and a capacitor has the relationship $I = C\,dV/dt$ or $I = j\omega CV$. The use of the voltage-current-impedance concept is convenient and descriptive in many engineering applications. In this chapter we shall use circuit parameters to describe the analysis and the application of transmission lines. Transmission lines are modeled after waveguiding structures, like parallel-plate waveguides, coaxial lines, and rectangular waveguides. The field descriptions for these guiding structures are transformed into voltage concepts with impedance and admittance properly defined. The reader may find it interesting that these circuit elements bear little physical resemblance to their low-frequency counterparts governed by Kirchhoff's voltage and current laws. We shall introduce Smith charts to study the properties of transmission lines, and we shall also discuss propagation of pulses on the line and digital-to-analog conversions.

6.1 CONCEPT OF VOLTAGE, CURRENT, AND IMPEDANCE

Chapter 5 analyzes several types of waveguides and gives the electromagnetic fields associated with the propagating waves. From the viewpoint of electromagnetic theory, the solution is complete because we can readily obtain other physical quantities such as power, stored energy, and so on once the information about the fields is available.

Although the description of power transmission in waveguides is satisfactory and complete using the concept of electromagnetic fields, it is sometimes convenient to describe power transmission with parameters of low-frequency networks, like voltages, currents, and impedances. For this purpose, we now introduce rules that enable us to transform the field quantities into network parameters.

Transformation Rules for Transmission Lines and Waveguides

(a) $$\boxed{\text{Voltage } V(z) = \alpha_1 \int_{C_t} \mathbf{E} \cdot d\ell}$$ **(definition of voltage in a transmission line)**

where α_1 is a proportional constant, and C_t is an integration path transverse to z.

(b) $$\boxed{\text{Current } I(z) = \alpha_2 \oint_{C_0} \mathbf{H} \cdot d\ell}$$ **(definition of current in a transmission line)**

where α_2 is a proportional constant, and C_0 is a closed contour of integration.

The power relationship must hold

(c) $$\boxed{\frac{1}{2}\text{Re}\,[V(z)I^*(z)] = \int_A dA\,\hat{z} \cdot \frac{1}{2}\text{Re}\,[\mathbf{E} \times \mathbf{H}^*]}$$ **(power equation)**

where A is the cross-sectional area of the line or waveguide.

The circuit parameters V and I are defined in the above equations. We see that the voltage is proportional to the line integral of the **E** field and the current is proportional to the closed line integral (or contour integral) of the **H** field. Because the product of V and I represents power and the integral of the Poynting vector also represents the same power, definitions for V and I must be consistent with the Poynting theorem. Therefore, the above relationship (c) must be imposed.

Choice of the integration path C_t or C_0 depends on the particular transmission line or waveguide for which the circuit parameters are to be defined. In this section we shall define the circuit parameters for the three most common waveguiding structures discussed in Chapter 5, namely, the parallel-plate waveguide, the coaxial line, and the rectangular waveguide.

Circuit Parameters of Parallel-Plate Waveguides

In a parallel-plate waveguide or stripline of width w and separation a, the electromagnetic fields are approximately as follows:

$$\mathbf{E} = \hat{x} E_0 e^{-jkz} \quad \textbf{(6.1a)}$$

$$\mathbf{H} = \hat{y} \frac{E_0}{\eta} e^{-jkz} \quad \textbf{(6.1b)}$$

We can define the voltage and the current of this waveguide with the transformation rule—that is,

$$V(z) = \alpha_1 \int_0^a \mathbf{E} \cdot \hat{x}\, dx = \alpha_1 E_0\, a e^{-jkz} \tag{6.2a}$$

$$I(z) = \alpha_2 \oint_{C_0} \mathbf{H} \cdot ds = \alpha_2 \frac{E_0}{\eta} w e^{-jkz} \tag{6.2b}$$

where C_0 is a contour around the lower plate, as shown in Figure 6.1a. Notice that in Equation (6.2a) we define $V(z)$ by using the plate at $x = a$ as reference potential—namely, $V(z) = V(z, x = 0) - V(z, x = a)$. The total time-average power transmitted is given by

$$P_t = \int_A \frac{1}{2} \mathrm{Re}\,[\mathbf{E} \times \mathbf{H}^*] \cdot \hat{z}\, dA = \frac{E_0^2\, wa}{2\eta}$$

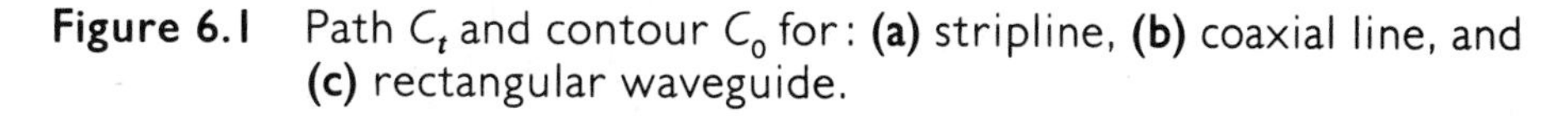
Figure 6.1 Path C_t and contour C_0 for: **(a)** stripline, **(b)** coaxial line, and **(c)** rectangular waveguide.

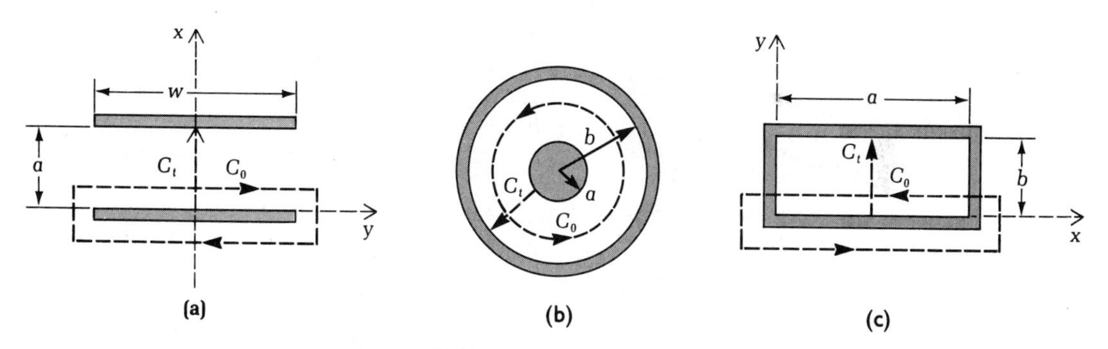

According to transformation rule (c), this result should be equated to $(1/2)\,\mathrm{Re}\,[V(z)I^*(z)]$. Thus, we find that $\alpha_1\alpha_2 = 1$. By convention, we choose $\alpha_1 = 1, \alpha_2 = 1$ so that

$$V(z) = E_0\, a e^{-jkz} \tag{6.3a}$$

$$I(z) = \frac{E_0\, w}{\eta} e^{-jkz} \tag{6.3b}$$

The **characteristic impedance** of a transmission line or of a waveguide is defined as the ratio of voltage and current on a line with no reflected waves. Thus, for the parallel-plate waveguide, the characteristic impedance is defined as

$$Z_0 = \frac{V(z)}{I(z)} = \eta \frac{a}{w} \text{ ohms} \tag{6.4}$$

Circuit Parameters of Coaxial Lines

The TEM mode inside a coaxial line has the following fields:

$$\mathrm{E}_\rho = \frac{V_0}{\rho} e^{-jkz} \tag{6.5a}$$

$$\mathrm{H}_\phi = \frac{V_0}{\eta\rho} e^{-jkz} \tag{6.5b}$$

Using the transformation rule (refer to Figure 6.1b), we define

$$\mathrm{V}(z) = \alpha_1 \int_a^b d\rho \mathrm{E}_\rho = \alpha_1 V_0 \ln\frac{b}{a} e^{-jkz} \tag{6.6a}$$

$$\mathrm{I}(z) = \alpha_2 \oint_{C_0} \rho \, d\phi \mathrm{H}_\phi = \alpha_2 \frac{2\pi V_0}{\eta} e^{-jkz} \tag{6.6b}$$

The total time-average power transmitted along the line is given by

$$P_t = 2\pi \int_a^b \rho \, d\rho \frac{1}{2} \frac{|V_0|^2}{\eta\rho^2} = \frac{\pi |V_0|^2}{\eta} \ln\frac{b}{a}$$

To satisfy transformation rule (c), it can be shown that $\alpha_1\alpha_2 = 1$. By convention, we choose $\alpha_1 = 1$ and $\alpha_2 = 1$.

The characteristic impedance of the coaxial line is then

$$Z_0 = \frac{\eta}{2\pi} \ln\frac{b}{a} \tag{6.7}$$

Commercial coaxial lines are available with Z_0 specified (for example, a 50 Ω or a 75 Ω line).

Circuit Parameters of Rectangular Waveguides

For the TE_{10} mode in a rectangular waveguide, the electromagnetic fields are

$$\mathrm{E}_y = E_0 \sin\frac{\pi x}{a} e^{-jk_z z} \tag{6.8a}$$

$$\mathrm{H}_x = -\frac{k_z}{\omega\mu} E_0 \sin\frac{\pi x}{a} e^{-jk_z z} \tag{6.8b}$$

$$\mathrm{H}_z = j\frac{\pi/a}{\omega\mu} E_0 \cos\frac{\pi x}{a} e^{-jk_z z} \tag{6.8c}$$

$$k_z = \sqrt{\omega^2\mu\epsilon - \left(\frac{\pi}{a}\right)^2} \tag{6.8d}$$

By using transformation rule (a), the voltage may be defined as

$$V(z) = \alpha_1 \int_0^b dy E_y |_{x=a/2} = \alpha_1 E_0 b e^{-jk_z z} \tag{6.9a}$$

with the current defined as

$$I(z) = \alpha_2 \oint_{C_0} dx \mathbf{H} \cdot \hat{x} = \frac{2\alpha_2 E_0 a k_z}{\pi \omega \mu} e^{-jk_z z} \tag{6.9b}$$

where C_0 is shown in Figure 6.1c. The factors α_1 and α_2 are chosen in such a way that the power relationship will hold. The time-average power transmitted along the waveguide may be obtained by integrating Poynting's vector over the cross-sectional area of the waveguide. The result is

$$P_t = \frac{1}{2} \frac{k_z}{\omega \mu} |E_0|^2 \int_0^b dy \int_0^a dx \sin^2 \frac{\pi x}{a} = \frac{ab}{4} \frac{k_z}{\omega \mu} |E_0|^2$$

On the other hand,

$$\frac{1}{2} \mathrm{Re}[V(z) I^*(z)] = \alpha_1 \alpha_2 \frac{abk_z}{\pi \omega \mu} |E_0|^2$$

Thus, according to transformation rule (c), we have $\alpha_1 \alpha_2 = \pi/4$. The characteristic impedance of the waveguide is

$$Z_0 = \frac{\alpha_1 b \pi \omega \mu}{2 \alpha_2 a k_z} \tag{6.10}$$

Solving for α_1 and α_2 yields

$$\alpha_1 = \left(\frac{a k_z Z_0}{2 b \omega \mu} \right)^{1/2}$$

$$\alpha_2 = \left(\frac{\pi^2 b \omega \mu}{8 a k_z Z_0} \right)^{1/2}$$

To summarize, for rectangular waveguides we have

$$V(z) = E_0 \left(\frac{abk_z Z_0}{2 \omega \mu} \right)^{1/2} e^{-jk_z z} \tag{6.11a}$$

$$I(z) = E_0 \left(\frac{abk_z}{2 \omega \mu Z_0} \right)^{1/2} e^{-jk_z z} \tag{6.11b}$$

Traditionally, the characteristic impedance for the TE_{10} mode in the rectangular waveguide is set equal to unity, that is, $Z_0 = 1$.

Example 6.1 *This example demonstrates how to design a stripline to meet the given specifications.*

Design a stripline with characteristic impedance equal to 200 Ω, and maximum time-average power equal to 100 kW. The medium between the plates is characterized by $\mu = \mu_0$, $\epsilon = 2.5\epsilon_0$, $\sigma = 0$. The breakdown E of the dielectric is 2×10^7 V/m. Use a safety factor of 10 so that the maximum E in the line does not exceed 2×10^6 V/m. Neglect ohmic loss.

Solution

$$\eta = \sqrt{\frac{\mu}{\epsilon}} = 238\ \Omega$$

$$Z_0 = 200 = \frac{238 \cdot a}{w}$$

$$P_t = 10^5 = \frac{E_0^2 aw}{2 \times 238}$$

$$E_0 = 2 \times 10^6$$

Solving the above equations yields $a = 3.16$ mm, $w = 3.76$ mm. The design guarantees that when 100 kW power is applied to the line, the dielectric material inside the line will not break down. Note that the ohmic loss due to the plates is not considered. Therefore, the maximum power of 100 kW should not be applied to the line continuously. It may be applied on and off with a small duty cycle so that the ohmic loss in the line does not significantly heat up the plates.

Example 6.2 *This example shows again how to define the voltage and the current in a transmission line in terms of* **E** *and* **H** *fields.*

In a stripline, a wave with the following fields propagates in the negative $\hat{z}$ direction:

$$\mathbf{E} = \hat{x}E' e^{jkz} \tag{6.12a}$$

$$\mathbf{H} = -\hat{y}\frac{E'}{\eta}e^{jkz} \tag{6.12b}$$

Find the corresponding V(z) and I(z).

Solution Because V(z) is proportional to the **E** field and (6.12a) and (6.1a) are similar, we have

$$\text{V}(z) = E'ae^{jkz} \tag{6.13a}$$

The current is proportional to the **H** field, and (6.12b) and (6.1b) differ only by a sign. Thus,

$$I(z) = -\frac{E'w}{\eta} e^{jkz} \quad \textbf{(6.13b)}$$

Example 6.3 *This example reiterates that characteristic impedance of a transmission line is a function of the material in the line and the dimension of the line.*

The inner conductor of a coaxial line is made of copper of 0.284 cm ID and 0.411 cm OD; the outer conductor is made of aluminum of 1.143 cm ID and 1.270 cm OD; and between the inner and outer conductors is polyethylene with $\epsilon = 1.5\epsilon_0$, $\mu = \mu_0$, and $\sigma = 0$. What is the characteristic impedance of this coaxial line?

Solution

$$\eta = \sqrt{\frac{\mu}{\epsilon}} = \sqrt{\frac{\mu_0}{1.5\epsilon_0}} = 308\ \Omega$$

Using (6.7), we have

$$Z_0 = \frac{308}{2\pi} \ln \frac{1.143}{0.411} = 50.1\ \Omega$$

Example 6.4 *This example illustrates how to convert the voltage reading in a transmission line to* ***E*** *field strength in the line.*

The maximum voltage in the coaxial line described in Example 6.3 is 100 V. What is the maximum $|\mathbf{E}|$ in that line?

Solution

$$V_0 \ln\left(\frac{b}{a}\right) = 100$$

$$E_{max} = \frac{V_0}{a} = \frac{100}{a \ln(b/a)} = 4.76 \times 10^4\ \text{V/m}$$

6.2 TRANSMISSION-LINE EQUATIONS

Maxwell's curl equations can be cast in the standard form of transmission-line equations in terms of V and I. For TEM waves in a parallel-plate waveguide of width w and separation a (Figure 6.1a). the two curl equations become

$$\frac{d}{dz}\mathrm{E}_x a = -j\omega\mu\frac{a}{w}\mathrm{H}_y w$$

$$\frac{d}{dz}\mathrm{H}_y w = -j\omega\epsilon\frac{w}{a}\mathrm{E}_x a$$

Using (6.3), we obtain

$$\frac{d\mathrm{V}}{dz} = -j\omega L\mathrm{I} \tag{6.14a}$$

$$\frac{d\mathrm{I}}{dz} = -j\omega C\mathrm{V} \tag{6.14b}$$

where we define

$$L = \frac{\mu a}{w} \equiv \text{inductance per unit length, henrys per meter}$$

$$C = \frac{\epsilon w}{a} \equiv \text{capacitance per unit length, farads per meter}$$

Later chapters will discuss the inductance and the capacitance for different geometries under the static and quasi-static approximations. The two equations in (6.14a) and (6.14b) are the **transmission-line equations.**

Eliminating I from (6.14a) and (6.14b), we obtain a wave equation for the voltage V

$$\frac{d^2\mathrm{V}}{dz^2} = -\omega^2 LC\mathrm{V} \tag{6.15}$$

V has two solutions, one with the dependence $\exp(-j\omega\sqrt{LC}z)$, the other with the dependence $\exp(+j\omega\sqrt{LC}z)$. Each of the two solutions has its own integration constant as multiplier. Introducing two constants V_+ and V_- so that

$$\mathrm{V} = V_+ \mathrm{e}^{-jkz} - V_- e^{+jkz} \tag{6.16}$$

where

$$k = \omega\sqrt{LC}$$

we have written the general solution for the voltage V(z). The amplitude V_+ will always be associated with the exponential e^{-jkz}, which represents a wave traveling in the positive $\hat{z}$ direction, and V_- will always be associated with e^{+jkz}, which represents a wave traveling in the negative $\hat{z}$ direction.

The solution (6.16) is associated with a definite solution for the current. Using (6.14a) we find

$$\mathrm{I} = Y_0(V_+ e^{-jkz} - V_- e^{+jkz}) \tag{6.17}$$

where $Y_0 = \sqrt{C/L} = \sqrt{\epsilon/\mu}\, w/a$ is the characteristic admittance of the transmission line and

$$Z_0 = \frac{1}{Y_0} = \sqrt{\frac{L}{C}} \tag{6.18}$$

is its characteristic impedance.

TEM Waves in a Coaxial Transmission Line

As with the propagation of TEM waves along a parallel-plate transmission line, we can develop the transmission-line equations for TEM waves in a coaxial line and show that the fields can be analogously represented by current and voltage equations. From (6.6), because $\alpha_1 = \alpha_2 = 1$, we find

$$\mathrm{E}_\rho = \frac{\mathrm{V}(z)}{\ln(b/a)}\frac{1}{\rho}$$

$$\mathrm{H}_\phi = \frac{\mathrm{I}(z)}{2\pi}\frac{1}{\rho}$$

Using cylindrical coordinates, the two Maxwell's curl equations become

$$\frac{1}{\ln(b/a)}\frac{d\mathrm{V}}{dz} = -j\omega\mu\frac{\mathrm{I}}{2\pi}$$

$$\frac{1}{2\pi}\frac{d\mathrm{I}}{dz} = -j\frac{\omega\epsilon}{\ln(b/a)}\mathrm{V}$$

These two equations may be cast in the form of (6.14) by defining

$$L \equiv \frac{\mu\ln(b/a)}{2\pi} \text{ henrys per meter} \tag{6.19}$$

$$C \equiv \frac{2\pi\epsilon}{\ln(b/a)} \text{ farads per meter} \tag{6.20}$$

These equations are the inductance and capacitance per unit length, respectively. They will be derived again in Chapters 10 and 13 in relation to electrostatics and magnetostatics.

For the transmission-line equations, we find the same solutions as (6.16) and (6.17), with the characteristic impedance

$$Z_0 = \frac{1}{Y_0} = \sqrt{\frac{L}{C}} = \sqrt{\frac{\mu}{\epsilon}\frac{\ln(b/a)}{2\pi}}$$

TE_{10} Mode in Rectangular Waveguides

For the TE waves in rectangular waveguides, Maxwell's curl equations yield

$$\frac{\partial}{\partial z}E_y = j\omega\mu H_x$$

$$\frac{\partial}{\partial x}E_y = -j\omega\mu H_z$$

$$\frac{\partial}{\partial z}H_x - \frac{\partial}{\partial x}H_z = j\omega\epsilon E_y$$

which after eliminating H_z, can be written as

$$\frac{d}{dz}E_y = j\omega\mu H_x \tag{6.21a}$$

$$\frac{d}{dz}H_x = j\frac{k_z^2}{\omega\mu}E_y \tag{6.21b}$$

Considering (6.9), we can write the field components E_y and H_x for the TE_{10} mode in terms of V(z) and I(z) as follows:

$$E_y = E_0 \sin\frac{\pi x}{a}e^{-jk_z z} = \frac{1}{\alpha_1 b}V(z)\sin\frac{\pi x}{a}$$

$$H_x = -\frac{k_z}{\omega\mu}E_0 \sin\frac{\pi x}{a}e^{-jk_z z} = -\frac{\pi}{2\alpha_2 a}I(z)\sin\frac{\pi x}{a}$$

Substituting into (6.21) and using (6.10), we find the transmission-line equations (6.14) with

$$L = \mu\frac{\alpha_1 \pi b}{\alpha_2 2a} = k_z\frac{Z_0}{\omega}$$

$$C = \frac{k_z^2}{\omega^2\mu}\frac{\alpha_2 2a}{\alpha_1 \pi b} = \frac{k_z}{\omega Z_0}$$

We choose the characteristic impedance to be $Z_0 = 1$. The dispersion relation is

$$\omega^2 LC = k_z^2 = \omega^2\mu\epsilon - \left(\frac{\pi}{a}\right)^2$$

The propagation constant is

$$k_z = \sqrt{\omega^2\mu\epsilon - \left(\frac{\pi}{a}\right)^2}$$

The corresponding guide wavelength λ_g is

$$\lambda_g = \frac{2\pi}{\sqrt{\omega^2\mu\epsilon - (\pi/a)^2}}$$

In subsequent developments in the transmission-line theory, the propagation constant is written as k and the wavelength as λ in conformity with the TEM modes in two-conductor transmission lines. *When applied to the rectangular waveguides carrying the TE_{10} mode, k stands for k_z and λ for λ_g.*

6.3 IMPEDANCES

Figure 6.2 depicts a section of a transmission line of characteristic impedance Z_0 terminated with a load impedance Z_L. We wish to find the relation between a generated wave propagating to the right and any reflected wave propagating to the left that results from the termination Z_L.

Figure 6.2 A transmission line with characteristic impedance Z_0 is terminated with a load impedance Z_L.

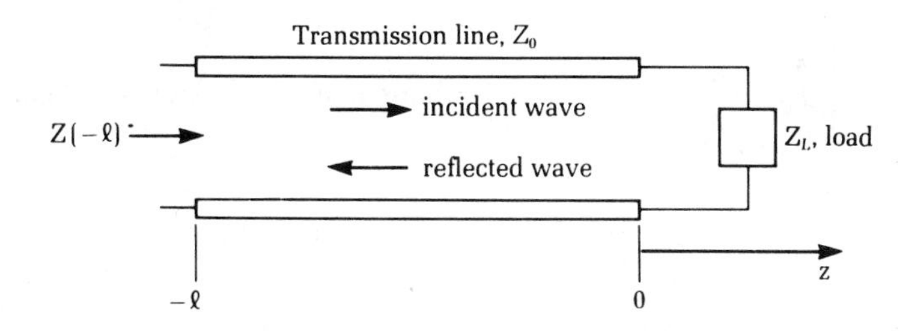

We conventionally place our origin, $z = 0$, at the load. The voltage on the line may be written as

$$V(z) = V_+ e^{-jkz} + V_- e^{jkz}$$

$$\boxed{V(z) = V_+(e^{-jkz} + \Gamma_L e^{jkz})} \quad \textbf{(voltage on a transmission line)} \qquad \textbf{(6.22)}$$

where Γ_L is defined as the **reflection coefficient**:

$$\boxed{\Gamma_L = \frac{V_-}{V_+}} \quad \textbf{(definition of reflection coefficient)}$$

The current I is found from (6.17)

$$\boxed{I(z) = \frac{1}{Z_0} V_+ (e^{-jkz} - \Gamma_L e^{jkz})} \quad \text{(current in a transmission line)} \tag{6.23}$$

The impedance Z(z) at any point z on the line is defined as

$$\boxed{Z(z) \equiv \frac{V(z)}{I(z)} = Z_0 \frac{e^{-jkz} + \Gamma_L e^{jkz}}{e^{-jkz} - \Gamma_L e^{jkz}}} \quad \text{(impedance function in a transmission line)} \tag{6.24}$$

At $z = 0$, the impedance Z(0) is equal to the load impedance Z_L. We obtain the relation

$$\boxed{Z_L = Z_0 \frac{1 + \Gamma_L}{1 - \Gamma_L}} \quad \text{(load impedance in terms of reflection coefficient)} \tag{6.25}$$

which specifies the reflection coefficient

$$\boxed{\Gamma_L = \frac{Z_L - Z_0}{Z_L + Z_0}} \quad \text{(reflection coefficient in terms of load impedance)} \tag{6.26}$$

We now determine the reflection coefficient for various load impedances.

Case 1
Short circuit, $Z_L = 0$. We find from (6.26)

$$\Gamma_L = -1$$

Thus $V_- = -V_+$. The voltage changes sign upon reflection and propagates back along the line.

Case 2
Open circuit, $Z_L = \infty$. We find from (6.26)

$$\Gamma_L = 1$$

In this case the voltage is not inverted because $V_- = V_+$.

Case 3
Matched line $Z_L = Z_0$. We find from (6.26)

$$\Gamma_L = 0 \quad \text{or} \quad V_- = 0$$

There is no reflected wave because $V_- = 0$.

Let us now ask what impedance we find at some point to the left of the load—for example, $z = -\ell$. We find

$$
\begin{aligned}
Z(z = -\ell) &= \frac{V(-\ell)}{I(-\ell)} = Z_0 \frac{e^{jk\ell} + \Gamma_L e^{-jk\ell}}{e^{jk\ell} - \Gamma_L e^{-jk\ell}} \\
&= Z_0 \frac{(Z_L + Z_0)e^{jk\ell} + (Z_L - Z_0)e^{-jk\ell}}{(Z_L + Z_0)e^{jk\ell} - (Z_L - Z_0)e^{-jk\ell}} \\
&= Z_0 \frac{Z_L 2 \cos k\ell + Z_0 2j \sin k\ell}{Z_L 2j \sin k\ell + Z_0 2 \cos k\ell} \\
&= Z_0 \frac{Z_L + jZ_0 \tan k\ell}{Z_0 + jZ_L \tan k\ell}
\end{aligned} \tag{6.27}
$$

Thus, in general, as we move away from the load, the impedance will vary periodically. We shall now examine specific cases.

Case I

Open-circuit load ($Z_L \to +\infty$)

$$Z(-\ell) = \frac{Z_0}{j \tan k\ell}$$

Notice that in the limit of $k\ell \ll 1$, which implies low frequency or a line that is small compared to the wavelength,

$$Z(-\ell) \approx \frac{Z_0}{jk\ell} = \frac{\sqrt{L/C}}{j\omega\sqrt{LC}\ell} = \frac{1}{j\omega C\ell}$$

and our system looks like a capacitor!

In general, we can write $Z = R + jX$. For the open-circuit load, $R = 0$ and $X = -Z_0 \cot k\ell$. Thus, we need only plot X as a function of $k\ell$, as shown in Figure 6.3. It is essential to note that the line impedance repeats itself for every

Figure 6.3 Reactance of an open-circuited transmission line as a function of its length.

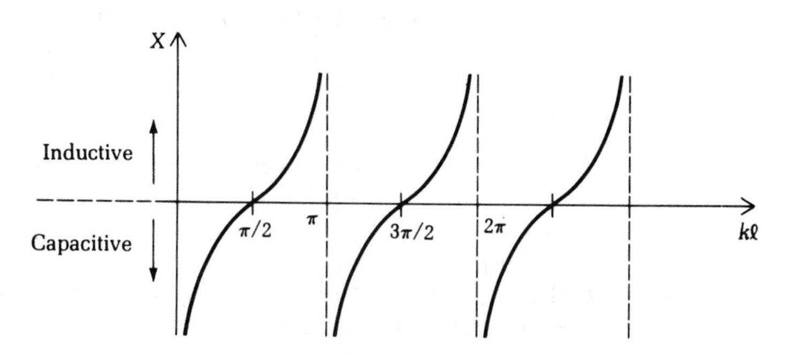

$k\ell = \pi$—namely, $\ell = \lambda/2$. Note also that as $k\ell = \pi/2$—namely, $\ell = \lambda/4$—the open-circuit impedance $Z \to \infty$ will become a short-circuit impedance with $Z = 0$.

Case 2
Short circuit ($Z_L = 0$). We obtain

$$Z(-\ell) = jZ_0 \tan k\ell$$

Plotting X versus $k\ell$, we obtain a curve identical to the one above shifted by $\pi/2$. Again, we see that the line of impedance repeats itself every $\ell = \lambda/2$. Figure 6.4 shows this repetition.

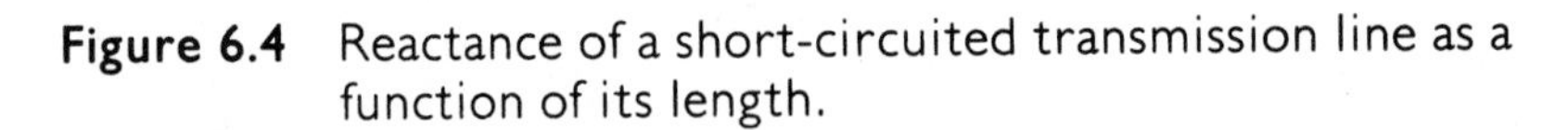

Figure 6.4 Reactance of a short-circuited transmission line as a function of its length.

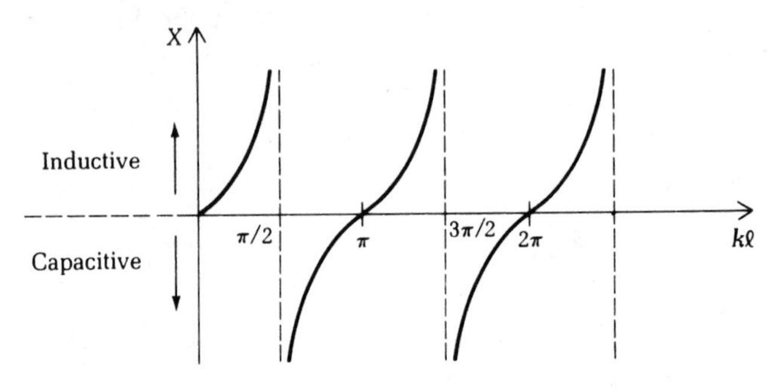

Case 3
Matched line ($Z_L = Z_0$). We see that $Z(-\ell) = Z_0$—that is, the line is matched everywhere.

6.4 GENERALIZED REFLECTION COEFFICIENT AND SMITH CHART

We have found that the impedance depends on the position z on the transmission line. We now define a **generalized reflection coefficient** that is also a function of z. Let $\Gamma(z)$ be the ratio of the reflected wave to the incident wave at z. From (6.22), we have

$$\boxed{\Gamma(z) = \frac{V_- e^{+jkz}}{V_+ e^{-jkz}} = \Gamma_L e^{+2jkz}} \quad \text{(definition of generalized reflection coefficient)} \qquad (6.28)$$

If we plot $\Gamma(z)$ on the complex plane, then as we move from $z = 0$ toward the generator at the left, as shown in Figure 6.5a, the reflection coefficient $\Gamma(z)$ traces

Figure 6.5 Definition of the generalized reflection coefficient. **(a)** $\mathbf{\Gamma}(z)$ on the transmission line. **(b)** The phasor diagram of $\mathbf{\Gamma}(z)$.

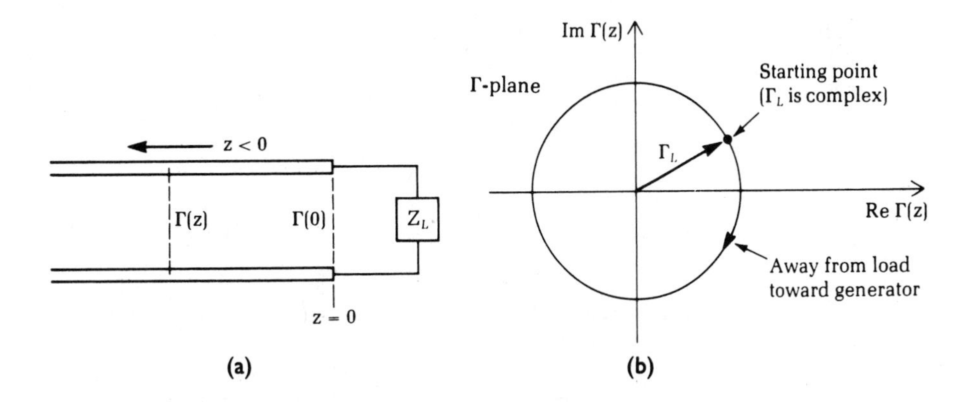

out a circle, as shown in Figure 6.5b. Notice that $\mathbf{\Gamma}$ returns to its starting point on the complex plane when $2kz = 2\pi$, or $z = \lambda/2$.

V(z) can be expressed in terms of $\mathbf{\Gamma}(z)$. Let the incident voltage V_+ be a real number; then

$$V(z) = V_+ e^{-jkz}[1 + \mathbf{\Gamma}(z)]$$

Or,

$$\boxed{|V(z)| = V_+ |1 + \mathbf{\Gamma}(z)|} \quad \textbf{(amplitude of voltage as function of } z\textbf{)}$$

The relation between $|V(z)|$ and $\mathbf{\Gamma}(z)$ can be demonstrated graphically.

Figure 6.6 shows a phasor diagram relating $\mathbf{\Gamma}(z)$ and $|V(z)|$. From Figure 6.6, we can plot $|V|$ versus z, as shown in Figure 6.7. Notice that

$$\boxed{V_{\text{max}} = (1 + |\mathbf{\Gamma}_L|)V_+} \quad \textbf{(maximum voltage)}$$

and

$$\boxed{V_{\text{min}} = (1 - |\mathbf{\Gamma}_L|)V_+} \quad \textbf{(minimum voltage)}$$

We define the ratio

$$\boxed{\frac{V_{\text{max}}}{V_{\text{min}}} = \frac{1 + |\mathbf{\Gamma}_L|}{1 - |\mathbf{\Gamma}_L|} = \text{VSWR}} \quad \textbf{(definition of VSWR)} \qquad \textbf{(6.29)}$$

Figure 6.6 Phasor diagram of $\mathbf{\Gamma}$ and $|V(z)|$, which is proportional to $|1 + \mathbf{\Gamma}|$.

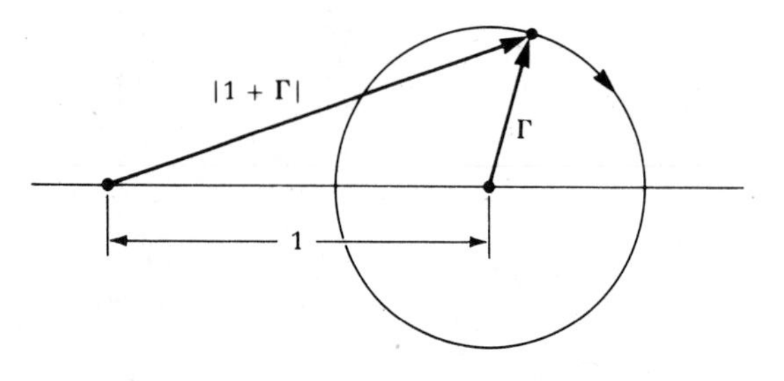

Figure 6.7 Standing-wave pattern of $|V(z)|$ on a transmission line.

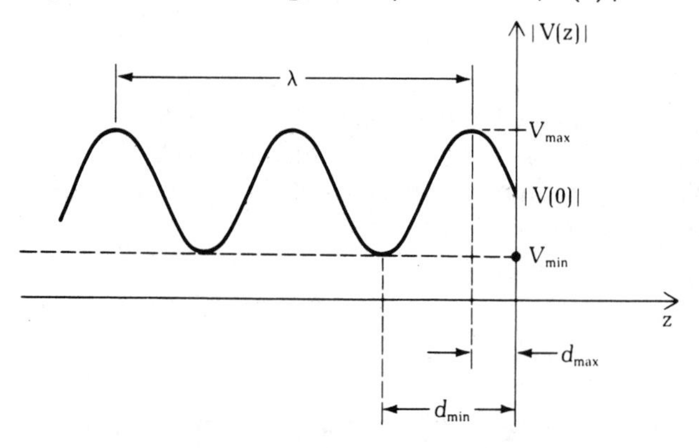

as the voltage standing-wave ratio (VSWR). From (6.29), we obtain

$$\boxed{|\mathbf{\Gamma}_L| = \frac{\text{VSWR} - 1}{\text{VSWR} + 1}} \qquad \textbf{(amplitude of } \mathbf{\Gamma}_L \textbf{ is related to VSWR)} \qquad \textbf{(6.30)}$$

Note that the positions of minimum and maximum voltages are related to the phase of $\mathbf{\Gamma}_L$. Let ϕ denote the phase of $\mathbf{\Gamma}_L$; then we have

$$\boxed{\begin{aligned} 2kd_{\text{max}} &= \phi \\ 2kd_{\text{min}} &= \phi + \pi \end{aligned}} \qquad \textbf{(phase of } \mathbf{\Gamma}_L \textbf{ is related to } \boldsymbol{d_{\text{min}}} \textbf{ and } \boldsymbol{d_{\text{max}}}\textbf{)} \qquad \begin{aligned} &\textbf{(6.31a)} \\ &\textbf{(6.31b)} \end{aligned}$$

Let us again consider some special cases.

Case I
Open circuit ($\mathbf{\Gamma}_L = 1\,e^{j0}$). The corresponding $\mathbf{\Gamma}(z)$ diagram is shown in Figure

6.8a, and the standing-wave pattern of the voltage is shown in Figure 6.8b. Also,

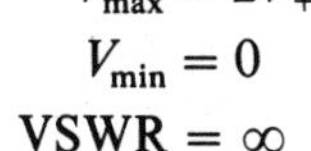

$$V_{max} = 2V_+$$
$$V_{min} = 0$$
$$\text{VSWR} = \infty$$

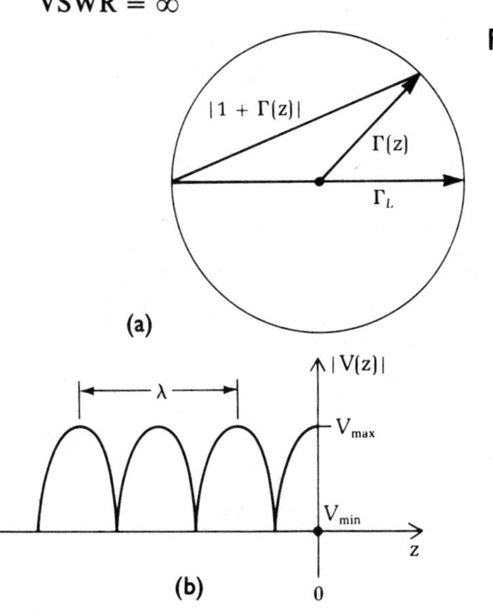

Figure 6.8 Open circuit: **(a)** The phasor diagram for $\mathbf{\Gamma}(z)$. **(b)** The standing-wave pattern for $|V(z)|$.

Case 2
Short circuit ($\mathbf{\Gamma}_L = -1$). Figure 6.9a and Figure 6.9b, respectively, show the $\mathbf{\Gamma}(z)$ diagram and the voltage standing-wave pattern. In this case,

$$V_{max} = 2V_+$$
$$V_{min} = 0$$
$$\text{VSWR} = \infty$$

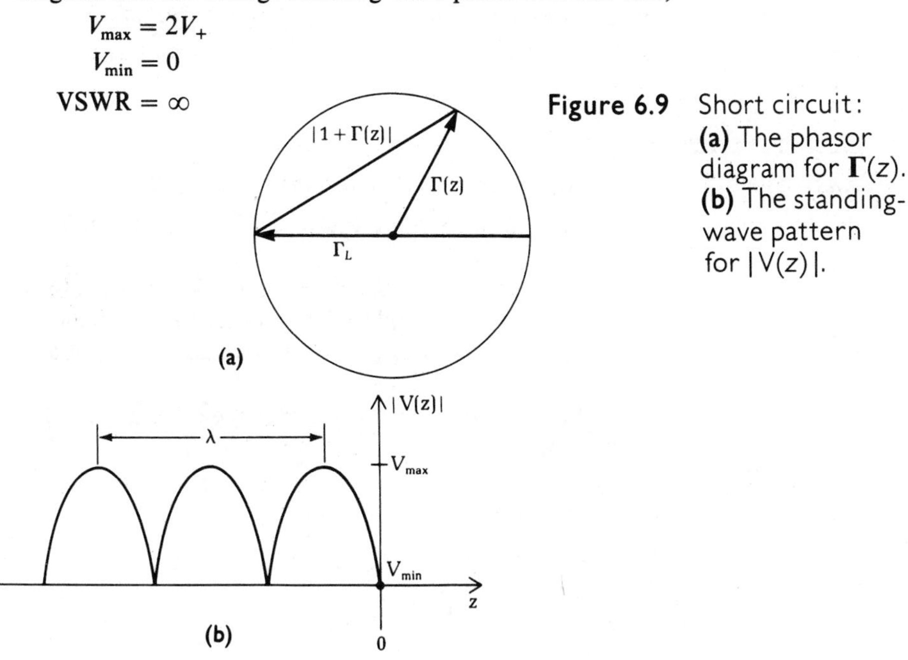

Figure 6.9 Short circuit: **(a)** The phasor diagram for $\mathbf{\Gamma}(z)$. **(b)** The standing-wave pattern for $|V(z)|$.

Note that the difference between the present and the preceding case is that V_{min} and V_{max} occur at different places, as seen from Figures 6.8 and 6.9.

Case 3

Matched load ($\mathbf{\Gamma}_L = 0$). Figure 6.10a and Figure 6.10b, respectively, show the $\mathbf{\Gamma}(z)$ diagram and the voltage standing-wave pattern. In this case,

$$V_{max} = V_+$$
$$V_{min} = V_+$$
$$\text{VSWR} = 1$$

Figure 6.10 Matched load: **(a)** The phasor diagram for $\mathbf{\Gamma}(z)$. **(b)** The standing-wave pattern for $|V(z)|$. This curve is flat because there is no reflected wave.

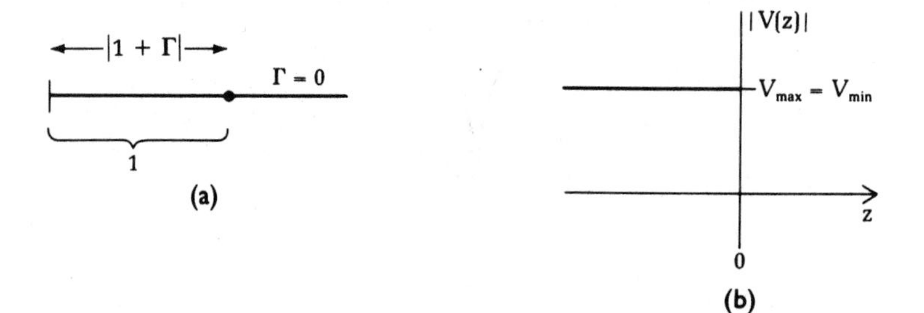

From the three cases discussed above, we can see that, as the load is varied, the VSWR varies between 1 and infinity. When there is a perfect match, VSWR = 1, and all the power is delivered to the load without any reflection.

Example 6.5

VSWR is a measure of reflection. Smaller VSWR corresponds to lower reflection and better matching of the load to the transmission line.

A two-wire line terminates with a television set *A*. The VSWR measured on the line is 5.8. When *A* is replaced by a television set *B*, the VSWR is 1.5. We want to calculate what percentage of power available to sets *A* and *B* is reflected.

Solution From Equation (6.30), we know that the magnitude of the reflection coefficient $\mathbf{\Gamma}$ for set *A* is

$$|\mathbf{\Gamma}| = \frac{5.8 - 1}{5.8 + 1} = 0.706$$

The incident power P_+ (power available to the television set) is given by

$$P_+ = \frac{|V_+|^2}{2Z_0}$$

The reflected power is

$$P_- = \frac{|V_-|^2}{2Z_0}$$

Therefore,

$$\frac{P_-}{P_+} = \left|\frac{V_-}{V_+}\right|^2 = |\Gamma|^2 = 0.5$$

A similar procedure will yield the following result for set B:

$$\frac{P_-}{P_+} = 0.04$$

Thus, television set A reflects 50% of the power that is available to it, whereas set B reflects only 4%. Set B is said to *be better matched to the transmission line* than set A.

Example 6.6 *This example shows that the load impedance can be calculated from the standing-wave pattern. It illustrates the basic technique of high-frequency impedance measurement.*

Figure 6.11 shows the measured $|V(z)|$ along a transmission line. Find the impedance of the load. The characteristic impedance of the line is 50 Ω.

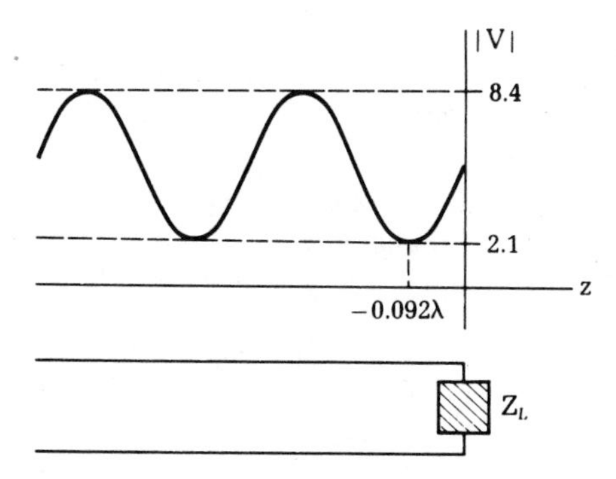

Figure 6.11 Load impedance is determined from the standing-wave pattern (Example 6.6).

Solution From Figure 6.11, we find

$$\text{VSWR} = \frac{8.4}{2.1} = 4.0$$

From (6.30), we obtain

$$|\Gamma_L| = 0.60$$

The voltage minimum is located at $z = -0.092\lambda$; thus, using (6.31b).

$$\phi = 2(0.092 \times 2\pi) - \pi$$

We now know the reflection coefficient Γ_L:

$$\Gamma_L = 0.6e^{j\phi} = 0.6e^{-j1.99} = -0.24 - j0.55$$

Using (6.25), we can calculate Z_L:

$$Z_L = Z_0\left(\frac{1+\Gamma_L}{1-\Gamma_L}\right) = \left(\frac{0.76 - j0.55}{1.24 + j0.55}\right) \times 50 = (0.348 - j0.597) \times 50$$

$$= 17.4 - j30\,\Omega$$

Smith Chart

Note that, wherever $\Gamma(z)$ is known, the corresponding impedance $Z(z)$ can be immediately calculated.

$$Z(z) = \frac{V(z)}{I(z)} = Z_0\frac{1+\Gamma(z)}{1-\Gamma(z)}$$

The normalized impedance is defined as

$$Z_n(z) = \frac{Z(z)}{Z_0} = \frac{1+\Gamma(z)}{1-\Gamma(z)} \qquad \textbf{(definition of normalized impedance)} \qquad \textbf{(6.32)}$$

When every point on the Γ plane is associated with a pair of complex numbers signifying the corresponding impedance, the result is a **Smith chart**. It is a graphical representation of (6.32).

To see how a Smith chart is constructed, we begin by finding the loci of constant Z_n on the complex Γ plane. Let

$$Z_n = r_n + jx_n$$

$$\Gamma = \Gamma_r + j\Gamma_i$$

We have

$$r_n + jx_n = \frac{1 + \Gamma_r + j\Gamma_i}{1 - \Gamma_r - j\Gamma_i} = \frac{1 - \Gamma_r^2 - \Gamma_i^2 + j2\Gamma_i}{(1-\Gamma_r)^2 + \Gamma_i^2} \qquad \textbf{(6.33)}$$

We find the constant-r_n locus by equating the real parts of the above equation:

$$\left(\Gamma_r - 1 + \frac{1}{1+r_n}\right)^2 + \Gamma_i^2 = \left(\frac{1}{1+r_n}\right)^2 \qquad \textbf{(6.34)}$$

On the complex Γ plane, this equation represents circles centered at $\Gamma_r = r_n/(1 + r_n)$ and $\Gamma_i = 0$ with radius $1/(1 + r_n)$ (see Figure 6.12).

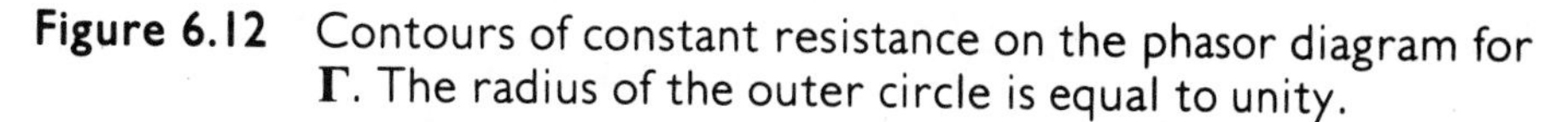

Figure 6.12 Contours of constant resistance on the phasor diagram for Γ. The radius of the outer circle is equal to unity.

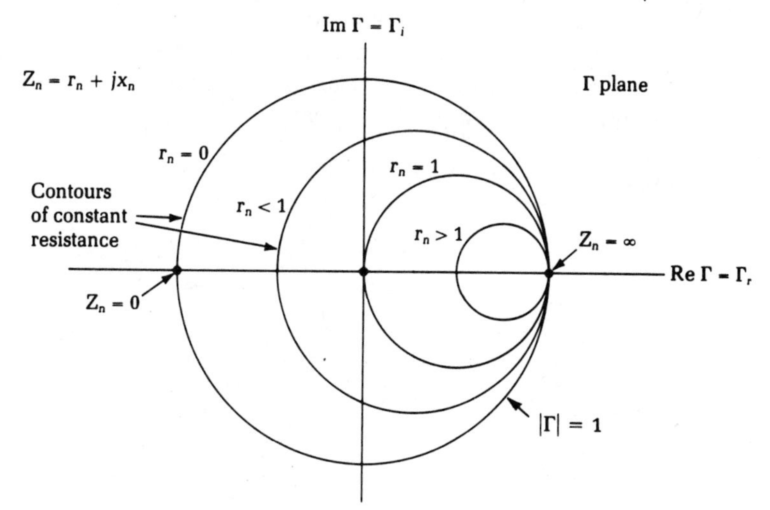

Equating the imaginary parts, we obtain the locus for constant x_n.

$$(\Gamma_r - 1)^2 + \left(\Gamma_i - \frac{1}{x_n}\right)^2 = \left(\frac{1}{x_n}\right)^2 \tag{6.35}$$

On the complex Γ plane, the above equation represents circles centered at $\Gamma_r = 1$ and $\Gamma_i = 1/x_n$ with radius $1/|x_n|$ (see Figure 6.13).

Figure 6.13 Contours of constant reactance on the phasor diagram for Γ.

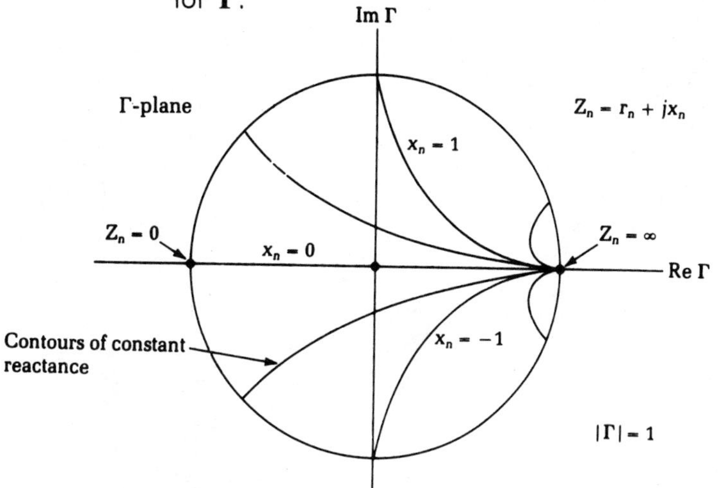

For positive values of r_n, we can easily show that the magnitude of $\mathbf{\Gamma}$ is always smaller than unity.

$$|\mathbf{\Gamma}| = \left|\frac{\mathbf{Z}_n - 1}{\mathbf{Z}_n + 1}\right| = \left[\frac{(r_n - 1)^2 + x_n^2}{(r_n + 1)^2 + x_n^2}\right]^{1/2} < 1$$

We can construct a chart of constant r_n and x_n on the $\mathbf{\Gamma}$ plane within the circle of $|\mathbf{\Gamma}| \leq 1$. This chart is known as the **Smith chart**. A typical Smith chart is shown in Figure 6.14. Note that normalized impedances $\mathbf{Z}_n$ and admittances $\mathbf{Y}_n$ must always be used with the Smith chart.

Figure 6.14 Smith chart.

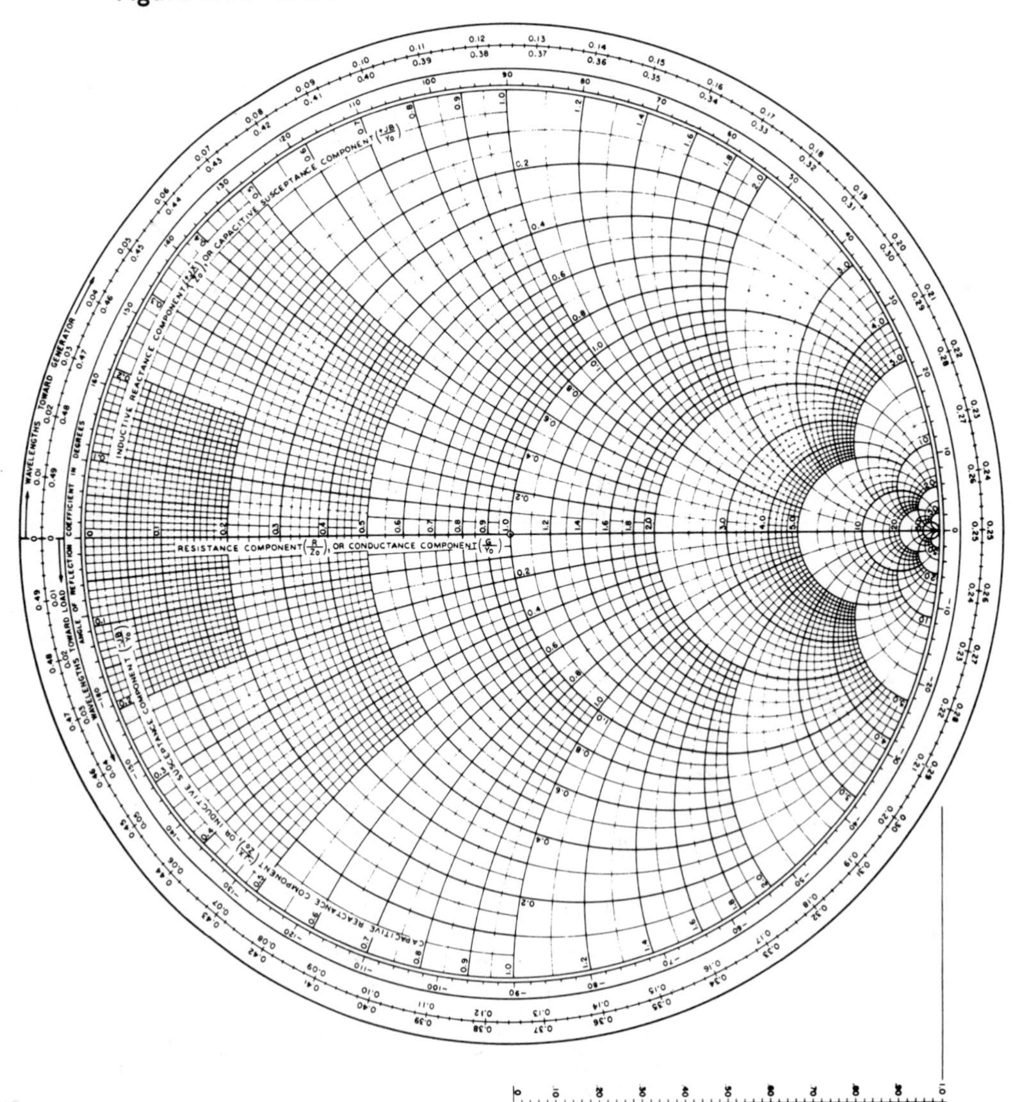

Basic Smith Chart Applications

Application 1: Calculation of $\mathbf{\Gamma}_L$ from Z_{Ln} (Normalized Load Impedance)
For example, a normalized load impedance of $Z_{Ln} = 1 + j2$ corresponds to $\mathbf{\Gamma}_L = 0.707\angle 45°$. Figure 6.15 shows this result.

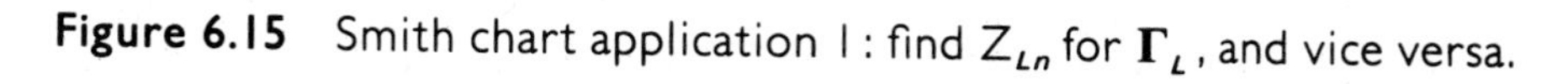
Figure 6.15 Smith chart application 1: find Z_{Ln} for $\mathbf{\Gamma}_L$, and vice versa.

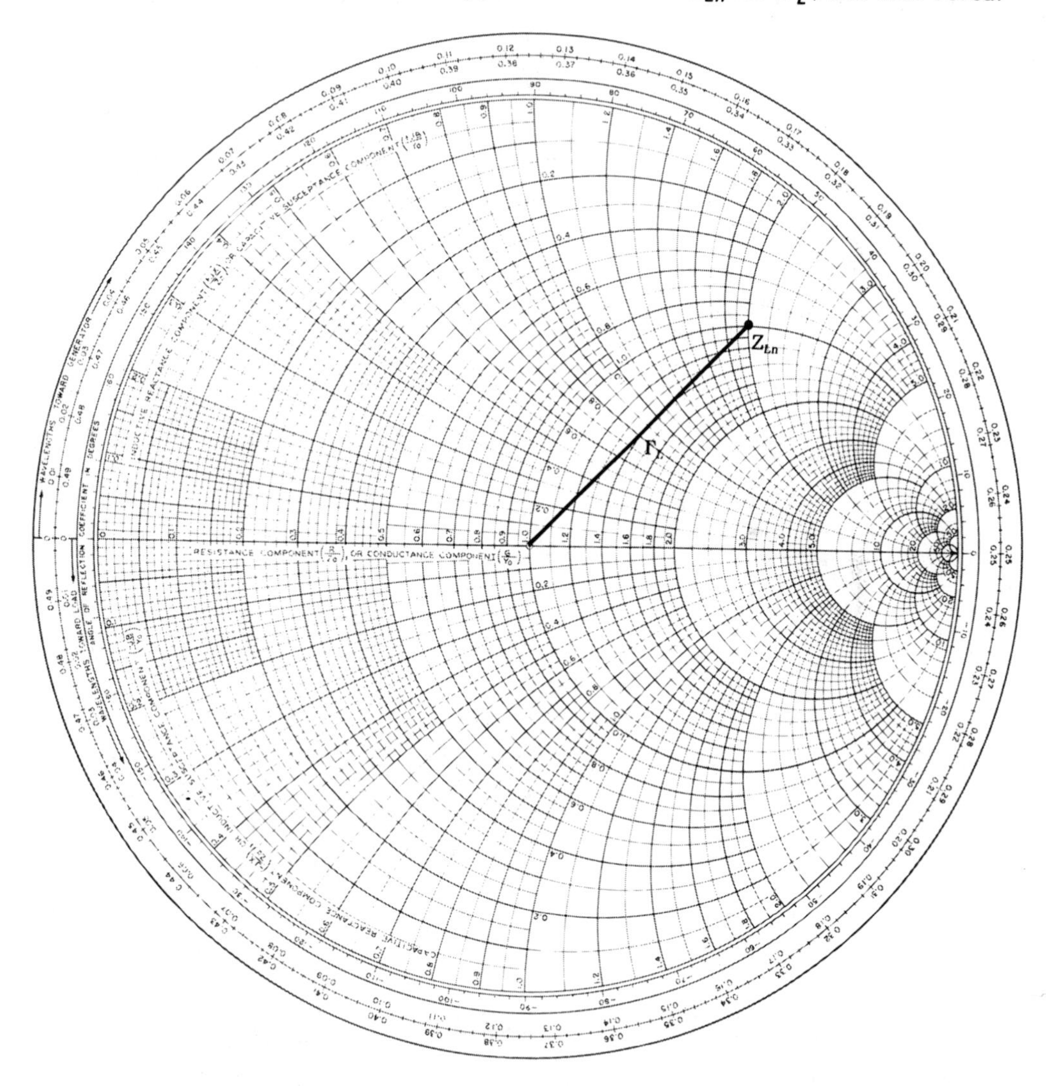

Application 2: Calculation of $Z_n(z)$ from Z_{Ln}
Given the load impedance Z_{Ln}, we can read $\mathbf{\Gamma}_L$ from the chart as described above. The quantity $\mathbf{\Gamma}_L e^{j2kz}$ is simply a point on the circle that is centered at the origin of the polar (phasor) coordinates and that has radius $|\mathbf{\Gamma}_L|$. For example,

the load impedance of $1 + j2$ is transformed into an impedance of $1 - j2$ at $2kz = -\pi/2$. In Figure 6.16 the circle with radius OB is the circle $\mathbf{\Gamma}_L e^{j2kz}$. The point $2kz = -\pi/2$ corresponds to the point $\lambda/8$ away from the load toward the generator. The position of the load is indicated by OC, or 0.188λ. Adding 0.125λ to this value yields 0.313λ or line OD. Line OD intersects the circle at E. The impedance at E reads $1 - j2$. Note that all impedances are normalized.

Figure 6.16 Smith chart applications 2–5.

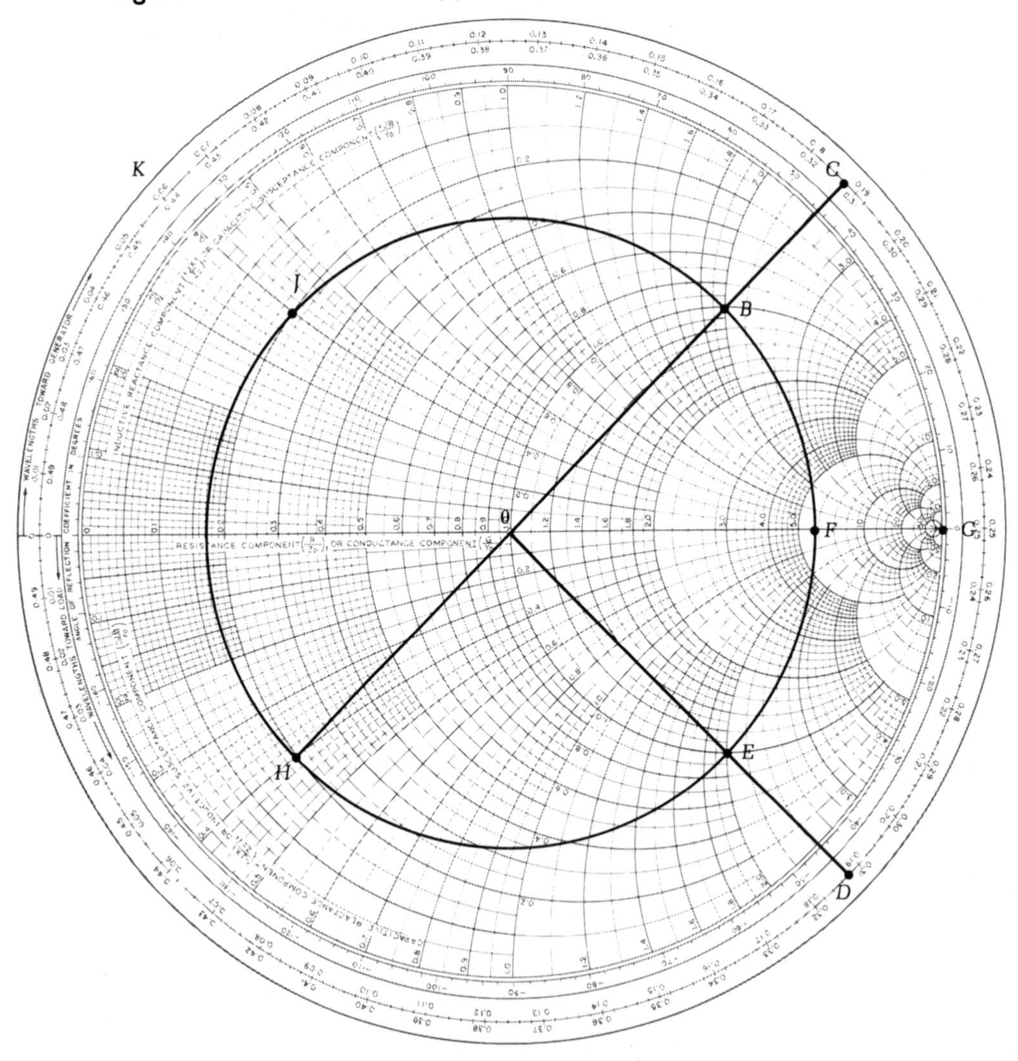

Application 3: Calculation of the Position of V_{max} and V_{min}

The circle of $\mathbf{\Gamma}_L e^{j2kz}$ mentioned above intersects the $\phi = 0$ line of the Smith chart. At the point of intersection, the combined phase angle of $\mathbf{\Gamma}_L e^{j2kz}$, which is equal to $\phi_0 + 2kz$, is zero, which is the condition for maximum voltage. Thus, the angle between the original $\mathbf{\Gamma}_L$ and the line $\phi = 0$ represents the position of the voltage maximum relative to the load position measured in radians. Similarly, the angle between the line $\phi = \pi$ and the original $\mathbf{\Gamma}_L$ represents the position of the voltage minimum relative to the load position.

For example, for a load impedance of $1 + j2$, the load is located at B in the Smith chart shown in Figure 6.16. The position of the load is marked by C or 0.188λ. The maximum voltage is at F with the position indicated by G or 0.25λ. Thus, the maximum voltage is located at $(0.25 - 0.188)\lambda$ or 0.062λ away from the load toward the generator side.

Application 4: Calculation of the VSWR

When read in the scale of the circular-arc coordinates, the point of intersection of the positive real Γ axis and the Γ circle mentioned above marks the value of the VSWR, according to (6.29) and (6.32). For example, the VSWR on a transmission line with load impedance $1 + j2$ is equal to 5.8, which is the impedance read at point F in Figure 6.16.

Application 5: Calculation of Admittance from Impedance

The admittance $Y_n(z)$ is defined as the inverse of $Z_n(z)$. Note that in (6.24) the denominator and the numerator differ only by a sign. Thus, for every increase of $2kz = \pi$, these two interchange values, and thus the value of $Z_n(z)$ is the inverse of the original value. We may conclude that $Z_n(z)$ and $Y_n(z)$ stay on the same circle of radius $|\mathbf{\Gamma}|$ but are opposite to each other. For example, a load impedance of $1 + j2$ is equivalent to an admittance of $0.2 - j0.4$. In other words, the value of the impedance read at H in Figure 6.16 is equal to the value of admittance at the load. The load impedance is read at B, and the load admittance is read at H.

When we rotate on the $\mathbf{\Gamma}$ plane from B to E, which corresponds to $\lambda/8$ from the load, the normalized impedance reads $1 - j2$ at point E. The corresponding admittance $0.2 + j0.4$ is read at point J, which is also $\lambda/8$ from the point H registering the load admittance. It thus follows that the Smith chart is equally useful as an admittance chart.

All of these five applications are reversible. For example, Application 1 can be used to find Z_{Ln} from $\mathbf{\Gamma}_L$. In measuring an unknown load impedance on a transmission line, the VSWR and the position of $V_{\min}$ are usually recorded. The load impedance can be obtained by reversing these applications. For example, for an unknown load that produces a VSWR of 4.0 on the transmission line with $V_{\min}$ at 0.2λ from the load, we can find the quantity $|\mathbf{\Gamma}|$ by first using Application 4 in the reverse order, with the result that $|\mathbf{\Gamma}| = 0.6$. In Figure 6.17 the circle that passes through the point A (where $Z_n = 4.0 + j0$) is the circle with $|\mathbf{\Gamma}| = 0.6$. Point B in Figure 6.17 represents the minimum voltage point. Point C is reached tracing back on the circle of $|\mathbf{\Gamma}| = 0.6$ counterclockwise by 0.2λ. The load impedance is read at C, yielding $Z_{Ln} = 1.65 - j1.80$.

The last example shows the essence of the impedance concept in microwave circuits. The "load" mentioned in the example can be either a horn antenna attached at the end of a rectangular waveguide or simply a rectangular waveguide left open-ended at $z = 0$. Note that the open-ended waveguide does not

Figure 6.17 Finding Z_{Ln} from VSWR and position of voltage minimum.

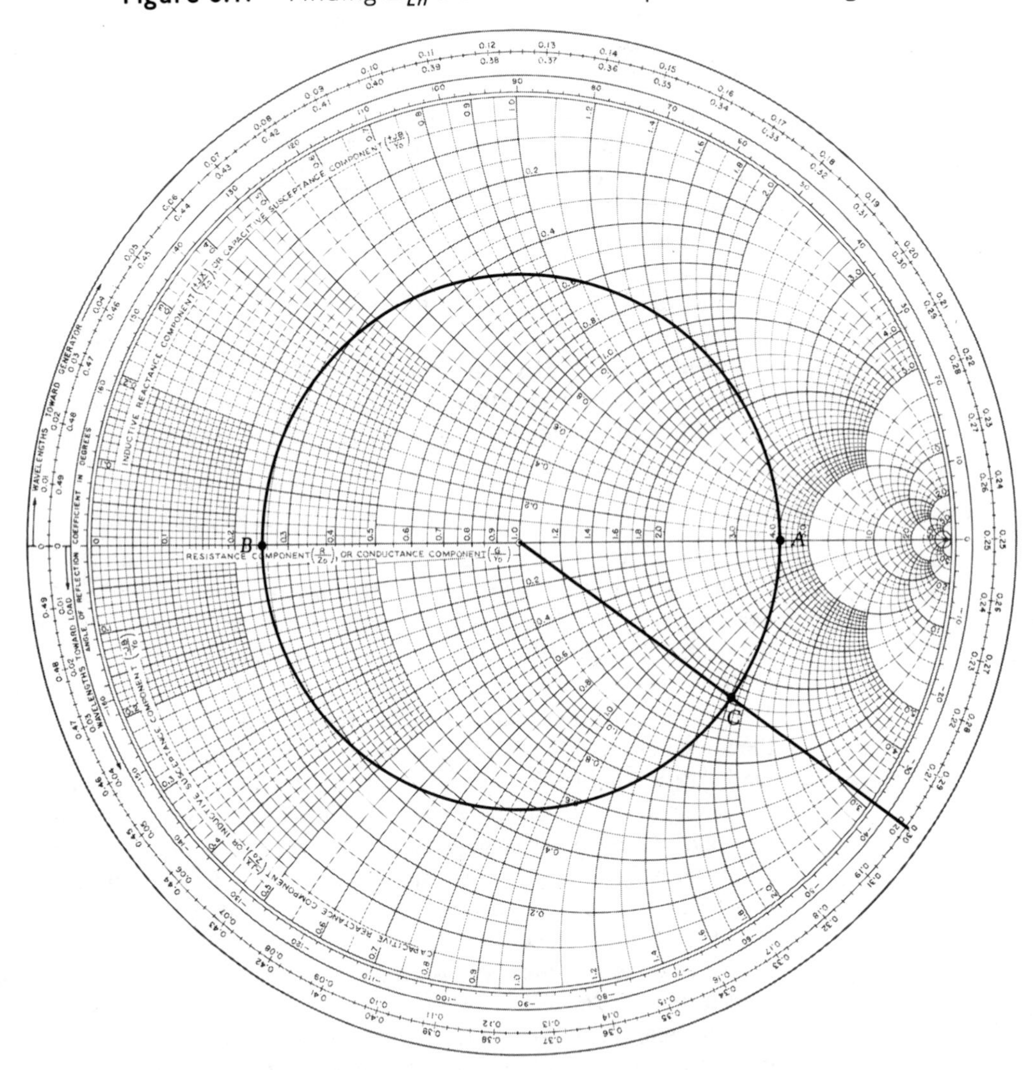

necessarily have an impedance of infinity. How do we describe the effects of the horn?

In the last example, the open-ended waveguide produces a standing-wave ratio (of the E field) of 4.0 with the minimum E field at 0.2λ from the horn. Or, using circuit terminology, we can say that the effect of the horn is to terminate the transmission line with the load of $1.65 - j1.80$. Because the value of the impedance has a real part, the load absorbs power from the line. In reality, of course, the horn absorbs power from the line on one end and radiates the power on the other end. The fact that load impedance is different from unity means that the horn and the line (the waveguide) are mismatched. The following paragraph discusses a technique of impedance matching that uses the Smith chart.

Impedance Matching by Shunt Admittance

The physical meaning of the reflection coefficient $\mathbf{\Gamma}$ is that the quantity $\mathbf{\Gamma\Gamma}^* = |\mathbf{\Gamma}|^2$ is the power reflected from the load back to the generator relative to the incident power. Normally, it is desirable to connect a load to the transmission line with least reflection. $\mathbf{\Gamma}$ is zero if the load impedance is equal to the characteristic impedance of the line. Under this condition, the transmission line is said to have a matched load. A matched load absorbs all power that is incident on it.

Consider a line with a normalized load impedance of $2 + j2$. The reflection coefficient $\mathbf{\Gamma}_L$ equals $0.62 \angle 30°$. Thus 38% or $(0.62)^2$ of the incident power is reflected back toward the generator. To tune the load for a perfect match, one might consider adding a series reactance somewhere along the line so that the combined impedance would be unity. Such a scheme is not practical because it is difficult to obtain a series impedance on a microwave transmission line. Remember that the transmission line under consideration can be a coaxial line or a rectangular waveguide, for either of which series impedance cannot be produced easily. However, as we shall show, it is relatively easy to construct a shunt susceptance for a transmission line.

The admittance of the load in this example equals $0.25 - j0.25$. A shunt admittance of $j0.25$ could be added parallel to the load to make the total admittance equal 0.25. This admittance corresponds to a $\mathbf{\Gamma}_L$ of $0.6 \angle 0°$, which is only slightly smaller than the original $|\mathbf{\Gamma}_L|$ without the shunt admittance. A better match may be obtained if a shunt admittance is added at a point where the admittance equals $1 + jb$; then, addition of a shunt admittance of $-jb$ will make the resulting admittance equal unity. To determine the point of insertion, recall that the admittance trajectory of the line as a function of z is a circle with radius $|\mathbf{\Gamma}_L|$ on the Γ plane. This circle on the Smith chart intersects the circular arc representing $Y_n = 1$, and, therefore, at the point of intersection the admittance is $1 + jb$. In the present case, we have two intersection points: $1 + j1.57$ and $1 - j1.57$. The relative position of these two points are 0.219λ and 0.363λ away from the load, respectively. Thus, for a transmission line with a load impedance of $2 + j2$, a perfect match of the load can be achieved if a shunt admittance of $-j1.57$ is added to the transmission line at 0.219λ from the load or if a shunt admittance of $j1.57$ is added to 0.363λ from the load. Figure 6.18 illustrates this procedure.

Figure 6.19a shows the voltage and the current standing-wave patterns on a transmission line with a load of $2 + j2$ impedance. The incident voltage V_+ is one volt. Figure 6.19b shows the pattern on the line after the shunt admittance $-j1.57$ has been added to the line at $z = -0.219\lambda$. Note that to the left of the shunt there is no standing wave because the line is perfectly matched. A standing wave still exists between the load and the shunt. Magnitudes of V and I between the load and the shunt are greater than those of the untuned line by a factor equal to 1/0.78 for the same amount of incident power. As a result, the load absorbs a greater amount of power.

Figure 6.18 Impedance matching using shunt admittance.

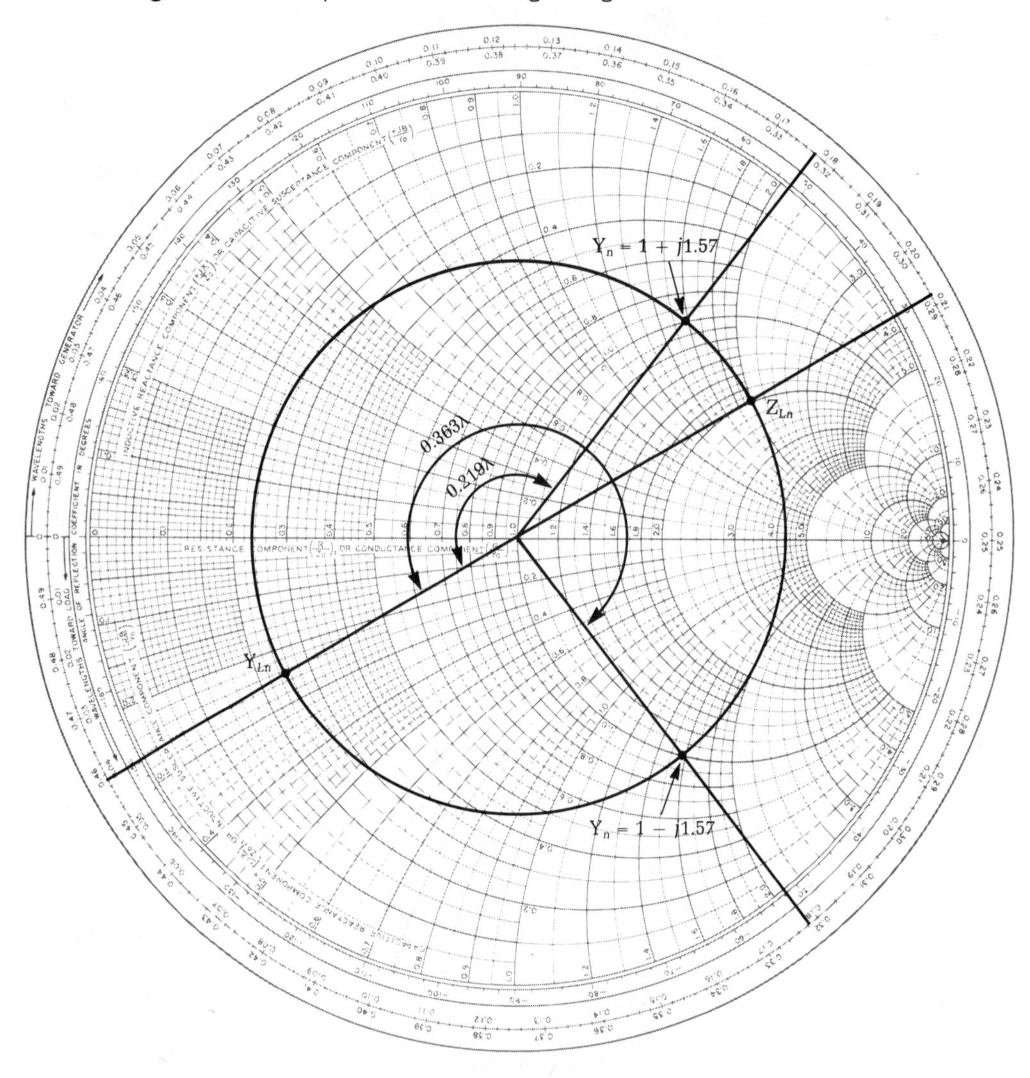

Microwave Inductors and Capacitors

In the preceding discussion we have seen that a mismatched line may be matched by adding a reactive element parallel to the load at proper positions on the transmission line. What do these microwave inductors or capacitors look like? How do we make an inductor with, say, a $-j1.57$ admittance? The questions do not have just one answer.

To produce on a transmission line an effect equivalent to a shunt susceptance, we must consider the actual transmission line in use. For a coaxial or microstrip line, a short-circuited section of a line similar to the main line can be attached

Figure 6.19 **(a)** Voltage and current standing-wave patterns on a transmission line with a load impedance $2 + j2$.
(b) Standing-wave pattern after $-j1.57$ shunt admittance has been added at $z = -0.219\lambda$. I is the current on the main line.

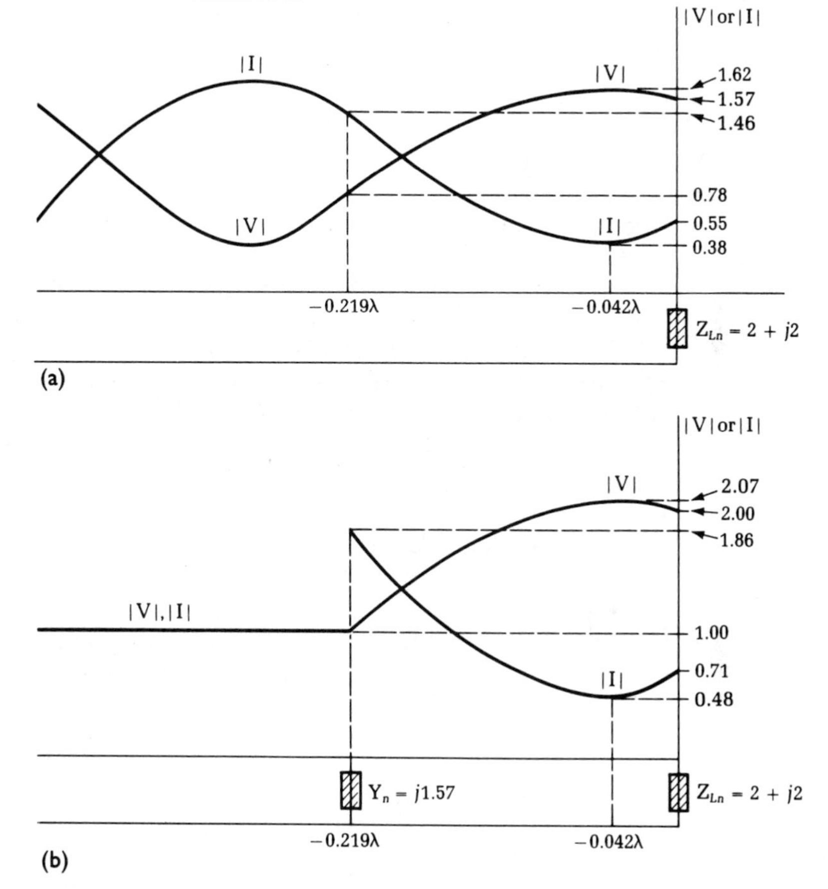

perpendicular to the main line, as shown in Figure 6.20. The reason that the shunted line should be kept perpendicular to the main line is that in such an arrangement the current on the shunt section will have a minimum effect on the field associated with the main line and vice versa, so that the shunt section and the main line can be considered as separate parts.

The susceptance of a short-circuited line of length ℓ is, according to (6.32),

$$Y_n(-\ell) = \frac{1 - \Gamma_L e^{-j2k\ell}}{1 + \Gamma_L e^{-j2k\ell}} = \frac{1}{Z_n(-\ell)}$$

The reflection coefficient Γ_L equals -1, from (6.26). The expression can then be simplified as

$$Y_n(-\ell) = -j \cot(k\ell) \tag{6.36}$$

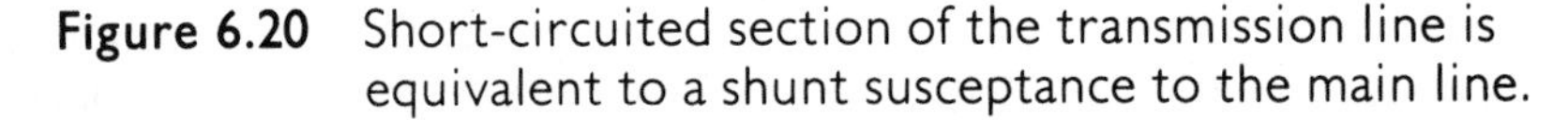

Figure 6.20 Short-circuited section of the transmission line is equivalent to a shunt susceptance to the main line.

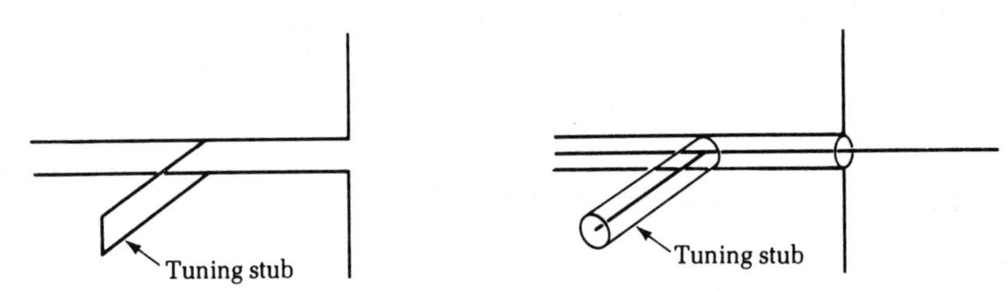

The end plate or wire can be moved back and forth on the shunt section to get a wide range of susceptance values. For the example of this section, a susceptance of $-j1.57$ can be obtained by a short-circuited line 0.09λ long, which is readily determined from the Smith chart.

For rectangular waveguides, a shunt susceptance may be obtained by inserting a metal iris, as shown in Figure 6.21. In Figure 6.21a the iris is inductive, and the susceptance is given by the formula*

$$b_0 = \frac{-\lambda_g}{a} \cot^2 \left(\frac{\pi d}{2a} \right) \tag{6.37}$$

where λ_g is the guide wavelength.

Figure 6.21 **(a)** Inductive and **(b)** capacitive irises for rectangular waveguides. Shaded areas represent the metal surface of the irises.

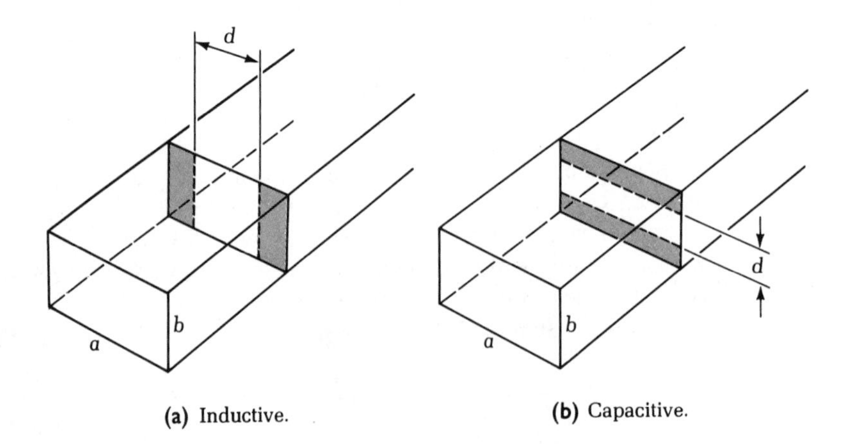

(a) Inductive. **(b)** Capacitive.

* G. G. Montgomery, R. H. Dicke, and E. Purcell, *Principles of Microwave Circuits*, Radiation Laboratory Series No. 8, New York: McGraw-Hill, 1948, pp. 165–166.

In Figure 6.21b the iris is capacitive, and the susceptance is given by

$$b_0 = \frac{4b}{\lambda_g} \ln\left[\csc\left(\frac{\pi d}{2b}\right)\right] \tag{6.38}$$

Such devices cannot be moved freely. An alternative scheme is to use metal screws of small diameter introduced from the broad side of the waveguide. The net effect is a shunt capacitive susceptance. Several screws can be placed on the waveguide to vary the position of the shunt susceptance for impedance matching.

Example 6.7 *This example shows that a segment of a short-circuited line can be used as a reactance or a susceptance.*

For a short-circuited 50 Ω transmission line of length ℓ, the input admittance at the other end is $j0.058$ siemens. Find the length ℓ in λ.

Solution The load admittance Y_{Ln} of a short-circuited line is infinity. Its position on the Smith chart is shown in Figure 6.22. The reflection coefficient is -1 and the $|\Gamma|$ circle is the unit circle that forms the boundary of the Smith chart.

The input admittance $Y = j0.058$ siemens has a normalized value of $j0.058/0.02 = j2.9$. The position is shown in Figure 6.22. The distance ℓ is read from the distance scale, with the following result:

$$\ell = 0.250\lambda + 0.197\lambda = 0.447\lambda$$

Example 6.8 *This example illustrates the load-matching technique. A mismatched load (an array of 10 antennas) becomes matched after a stub tuner of the right length is placed at the right place.*

A phased-array radar has 10 antenna elements that are matched to their feed lines, as sketched in Figure 6.23 (p. 187). All transmission lines are 50-Ω coaxial lines filled with a dielectric of $\epsilon = 2.3\epsilon_0$ and $\mu = \mu_0$. For the stub tuner on the main line, find the position ℓ_1 and length ℓ_2 that will maximize the power to the antenna array. The operating frequency is 1 GHz.

Solution As seen from the main feed line, the 10 antennas represent a load impedance of 5 Ω or a normalized load admittance of 10. The corresponding position of Y_{Ln} is shown in the Smith chart of Figure 6.24 (p.188).

From the Smith chart, the reflection coefficient is found to be equal -0.82. Thus, if there is no stub tuner, 67% of the power will be reflected back to the generator.

Figure 6.22 Use of Smith chart to find reactance of a short-circuited line.

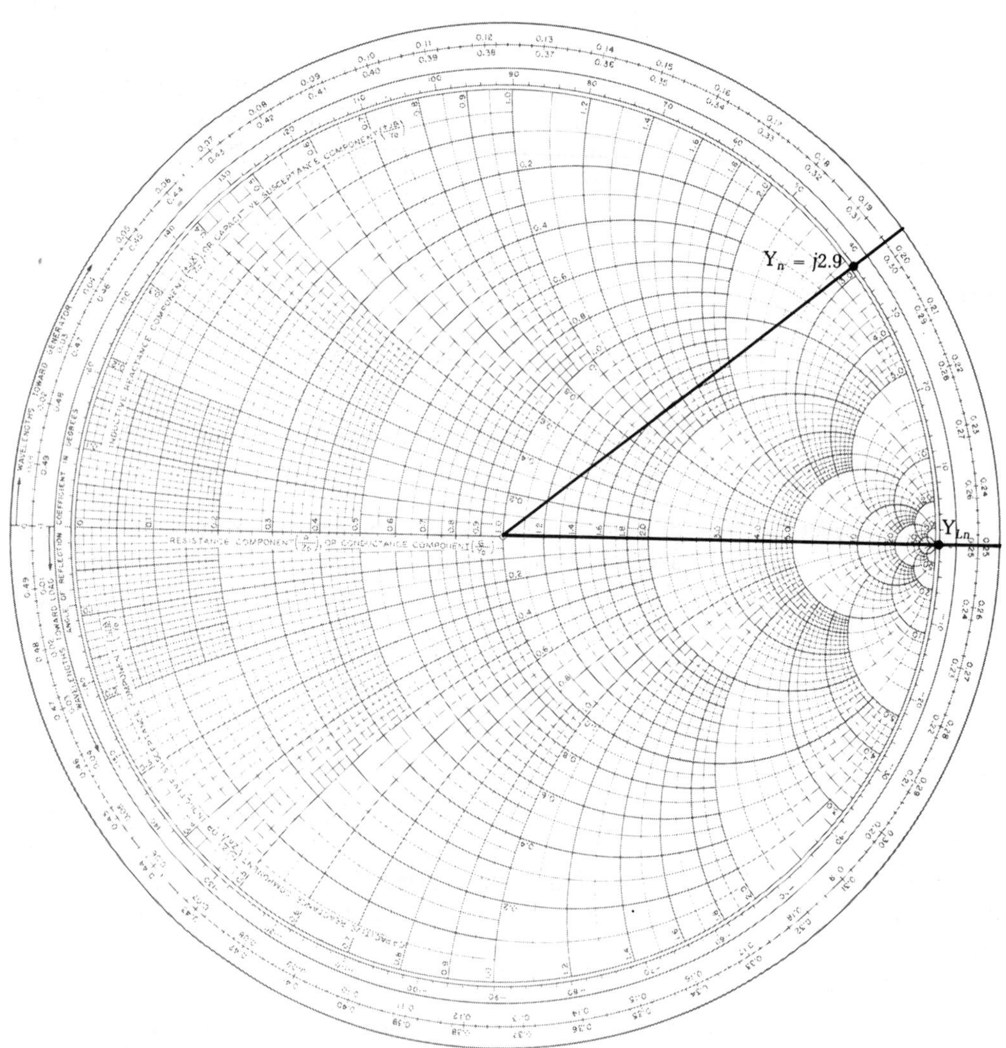

The $|\Gamma| = 0.82$ circle intersects the $Y_n = 1 + jb$ circle at points A and B. At point A, the admittance is $1 - j2.9$. Thus, if a stub tuner is put at this position with an input admittance equal to $j2.9$, the net admittance will be unity, and the main line will be perfectly matched. From Figure 6.24, we find that A is $(0.298 - 0.25)\lambda$ or 0.048λ away from the load. The frequency is 1 GHz, so that the guided wavelength in the coaxial line is given by

$$\lambda = \frac{3 \times 10^8}{10^9\sqrt{2.3}} = 19.8 \times 10^{-2} \text{ m} \tag{6.39}$$

Figure 6.23 Impedance matching for a ten-element antenna array (Example 6.8).

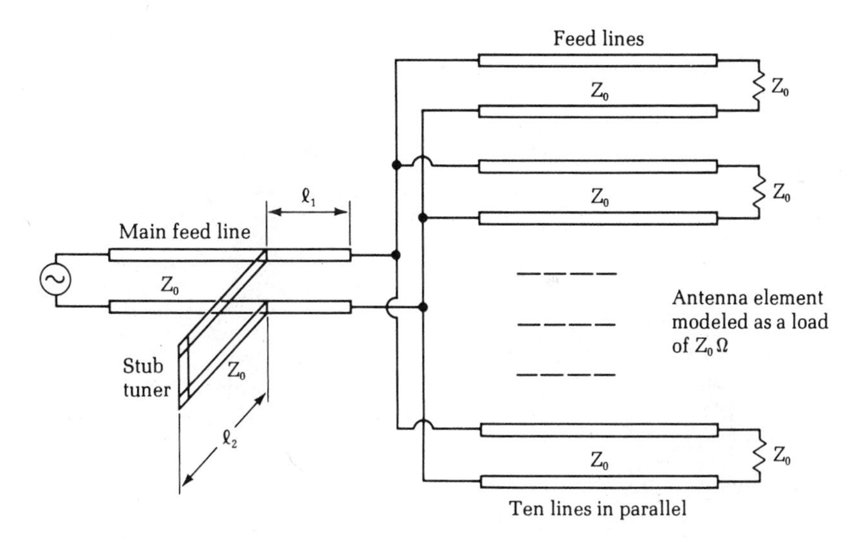

Thus, the stub tuner should be placed 0.95 cm from the load.

From the result obtained in Example 6.7, to obtain an input admittance of $j2.9$, a short-circuited line of 0.447λ must be used. In the present case, the length of the stub tuner should be 8.85 cm. In conclusion, maximum power to these antennas may be obtained if $\ell_1 = 0.95$ cm and $\ell_2 = 8.85$ cm.

6.5 TRANSIENTS ON TRANSMISSION LINES

Throughout our discussion, we have assumed single operating frequency or monochromatic wave. In some applications, however, pulsed signals rather than single-frequency signals will be encountered. An example is furnished by the problem of sending a 0.1 ns (10^{-10} s) pulse through a microstripline in a computer circuit. Another example is the effect of lightning on computers and power systems. Lightning generates intense electromagnetic pulses radiating in all directions. It can cause temporary short circuits in power transmission lines.

A pulse can be regarded as a result of superposition of waves of many frequencies. A pulse-generating electromagnetic source is mathematically equivalent to a set of sources generating a spectrum of electromagnetic waves. When a pulse is coupled to a transmission line, the propagation characteristics of the different frequency components are likely to be different. Each frequency component must be treated individually, and the results must be summed up at the end to find the overall effect of the transmission line on the pulse.

Figure 6.24 Matching impedance of phase array to main line (Example 6.8).

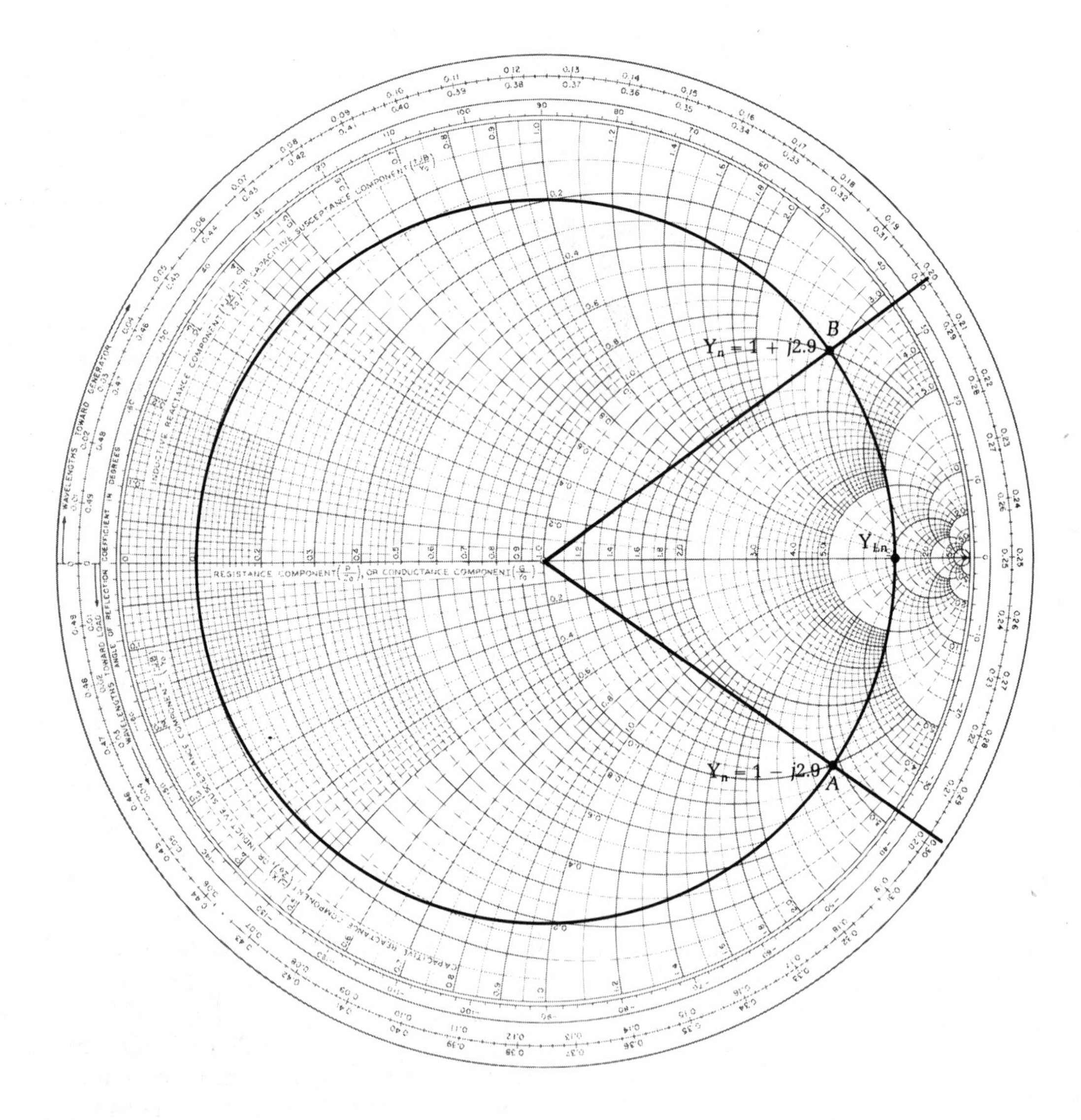

An electromagnetic pulse passing through a transmission line or through a waveguide will be distorted in shape when any of the following conditions exist: (a) the velocity of propagation of the wave on the line is not a constant but is a function of frequency, (b) the attenuation constant is a function of frequency, or (c) for some frequency components in the spectrum, more than one propagation mode exists in the transmission line.

Transmission of a Pulse on an Ideal Transmission Line

On an ideal transmission line, we assume that the velocity of the wave is a constant equal to $(\mu\epsilon)^{-1/2}$ and independent of frequency. The wave suffers no attenuation, so that the characteristic impedance Z_0 of the line is a real number.* These assumptions are approximately valid for TEM waves on two-wire lines, microstrips, and coaxial lines.

Consider a line that has length ℓ and characteristic impedance Z_0, that is terminated with a resistive load of R_L, and that is driven by a pulse generator at $z = 0$. Figure 6.25a shows this situation. The internal impedance of the generator equals R_g. Suppose that at $t = 0$ the pulse generator sends out a single pulse voltage V_0 and width Δ, as shown in Figure 6.25b. We assume that Δ is much less than ℓ/v, which is the time it takes for the pulse to travel the entire length of the transmission line. Here, v is the velocity of the wave.

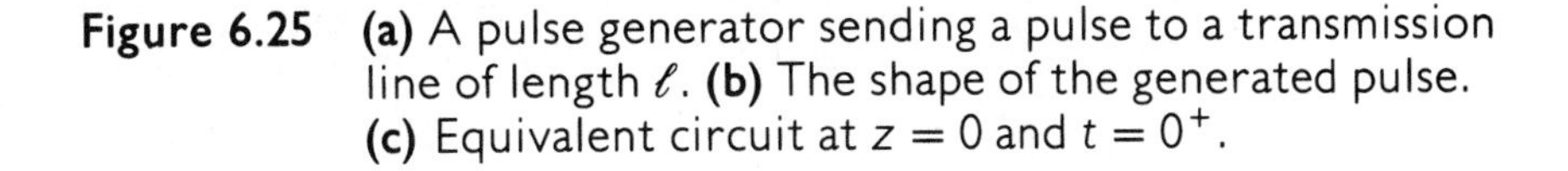

Figure 6.25 **(a)** A pulse generator sending a pulse to a transmission line of length ℓ. **(b)** The shape of the generated pulse. **(c)** Equivalent circuit at $z = 0$ and $t = 0^+$.

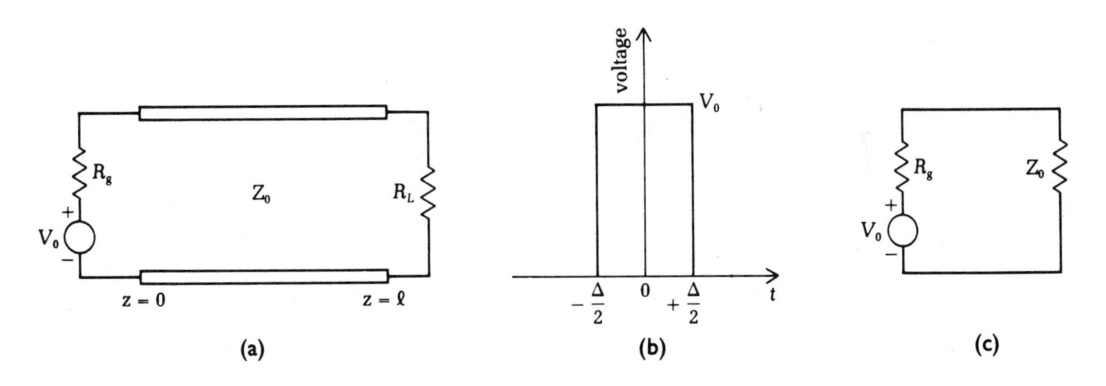

The pulse is reflected at the load at $t = \ell/v$. Reflection occurs because the load is not perfectly matched to the line. The reflected wave arrives at $z = 0$ at $t = 2\ell/v$. This pulse is reflected back toward the load. Reflection continues until the energy of the pulse is entirely absorbed by the two resistors.

At the beginning, the wave cannot tell whether or not the line is matched at the other end. It follows that the impedance looking toward the load at $z = 0$ is simply Z_0, the characteristic impedance of the line. The equivalent circuit at

---

* Because we are considering time-domain quantities, all variables and constants are real numbers. However, we shall follow the convention of using the term *characteristic impedance* and the symbol Z_0 despite the fact that the term impedance usually implies time-harmonic variations.

$z = 0$ and $t = 0^+$ is as shown in Figure 6.25c. Clearly, the voltage of the wave that appears in the transmission line is given by

$$V_+ = V_0 \frac{Z_0}{Z_0 + R_g} \tag{6.40a}$$

The corresponding current is given by

$$I_+ = V_+/Z_0 \tag{6.40b}$$

The pulse with voltage V_+ and current I_+ travels along the line with velocity v, and at $t = \ell/v$ it arrives at the load. A reflection takes place, and the reflection coefficient is given by (6.26):

$$\Gamma_1 = \frac{R_L - Z_0}{R_L + Z_0} \tag{6.41}$$

The reflected pulse has a voltage $\Gamma_1 V_+$ and a current $-\Gamma_1 I_+$ [see (6.17)] and travels with a velocity v toward the generator. At $t = 2\ell/v$, the pulse arrives at the generator, where the second reflection takes place. The pulse is reflected to travel back to the load again with a voltage $V_+ \Gamma_1 \Gamma_2$ and a current $I_+ \Gamma_1 \Gamma_2$, where

$$\Gamma_2 = \frac{R_g - Z_0}{R_g + Z_0} \tag{6.42}$$

This process continues indefinitely.

Digital-to-Analog Converter

In a pulse code modulation (PCM) system, an analog signal is periodically sampled, and each sample is converted to binary form. The binary representation is then transmitted as a sequence of pulses, with each pulse having the value of either one or zero. Normally, the least significant digit is transmitted first. For example, in a three-digit code, the analog sample is digitized into one of eight levels. Figure 6.26 shows three pulse trains that correspond to the binary repre-

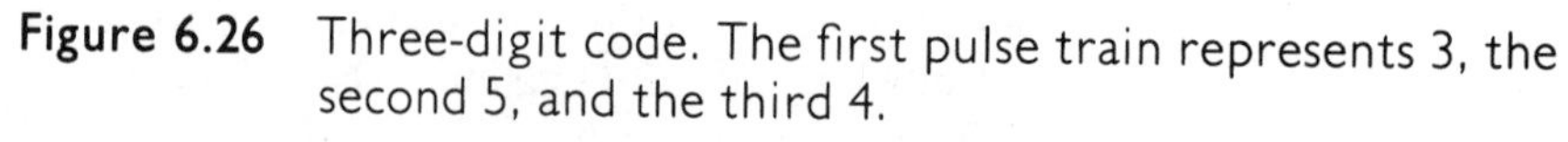

Figure 6.26 Three-digit code. The first pulse train represents 3, the second 5, and the third 4.

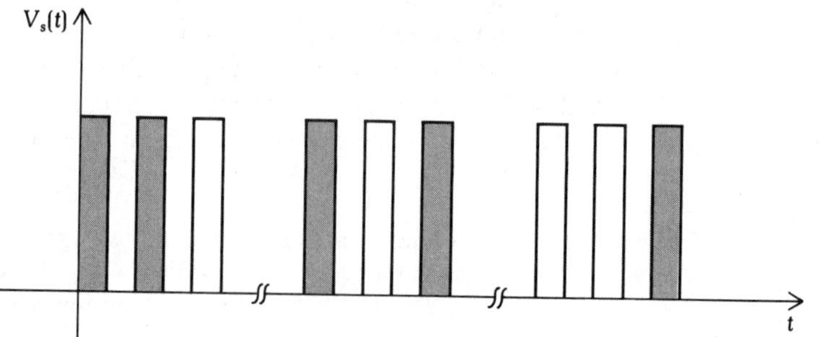

sentation of a sequence of sampled values, three, five, and four. The time between trains is often much greater than the duration of each pulse train, so that several different signals can be transmitted in the same channel.

At the receiver, each pulse train must be converted back into analog form. Thus, if a three-digit code is used and the binary signal of a train of three pulses of values a_1, a_2, and a_3 (where a_1, a_2, and a_3 are either 1 or 0) is received, a scheme must be devised to reconstruct the decimal equivalent $a_3(2^2) + a_2(2^1) + a_1(2^0)$ (to within a constant multiplier). We shall discuss a possible digital-to-analog converter (D-A converter) that uses the reflection from a mismatched transmission line for digital-to-analog conversion.

Part A

For the network shown in Figure 6.27a (p. 192), let us determine $h(t)$, the voltage at the *midpoint* of the transmission line, when the input voltage is a single unit pulse, as shown in Figure 6.27b. Assume $\Delta \ll \ell/v$, where v is the velocity of the wave on the line.

At $t = 0$, the voltage appearing at the transmission line at $z = 0$ is a pulse with amplitude f_+, where

$$f_+ = \frac{Z_0}{R + Z_0}$$

This pulse travels along the line and is reflected by the load at $t = T$, where $T = \ell/v$. The reflected wave is a pulse with amplitude f_-, where $f_- = f_+ \Gamma$ and $\Gamma = (R - Z_0)/(R + Z_0)$. At $t = 2T$, the reflected pulse arrives at $z = 0$, and a secondary reflected wave is produced, which travels in the positive z direction with amplitude $f_+ \Gamma^2$. This process continues indefinitely. The voltage at the midpoint is sketched in Figure 6.27c.

Part B

Consider the input voltage shown in Figure 6.28, where the pulse heights are a_1, a_2, and a_3, which can be either 1 or 0. Assume $T = \ell/v$. Let us find $V_0(t)$, the voltage at the midpoint of the transmission line.

By superposition, because the input is

$$V_s(t) = a_1 f(t) + a_2 f(t - T) + a_3 f(t - 2T)$$

where $f(t)$ is a single pulse, as shown in Figure 6.27b, the response at the midpoint should be

$$V_0(t) = a_1 h(t) + a_2 h(t - T) + a_3 h(t - 2T)$$

Part C

For the excitation given in Part B (Figure 6.28), we shall study the value of R that should be used and the time at which V_0 should be sampled in order that the sampled output be proportional to the decimal equivalent of the binary signal represented by the pulse shown in Figure 6.28.

Figure 6.27 **(a)** Transmission line with load resistance R. The internal impedance of the source is also R. **(b)** Input signal (a unit pulse). **(c)** Voltage at the midpoint.

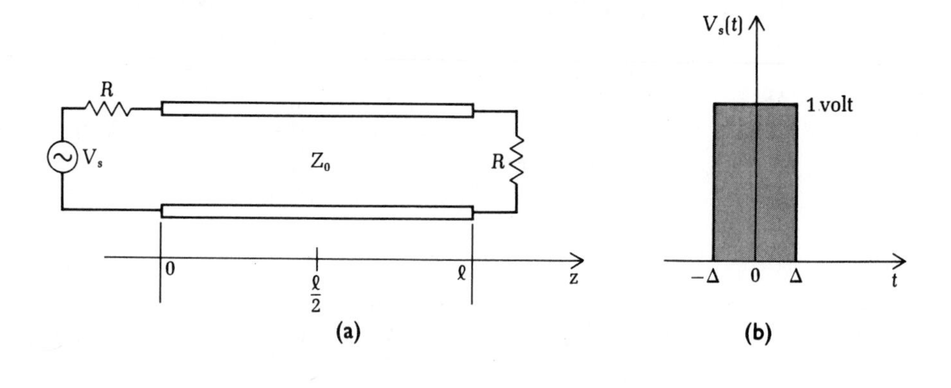

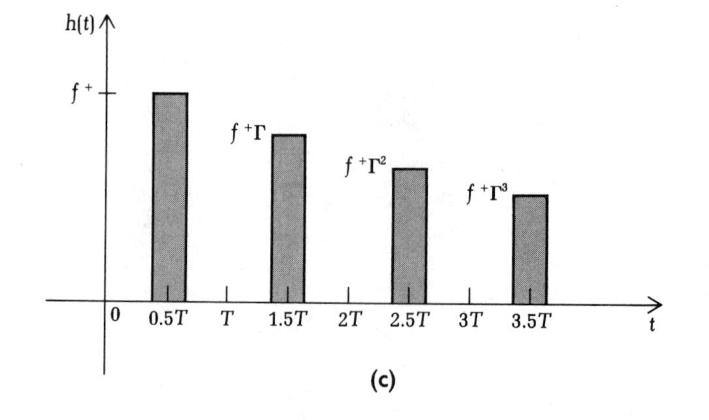

Figure 6.28 The input signal is a train of pulses of amplitudes a_1, a_2, and a_3, which may be either 1 or 0.

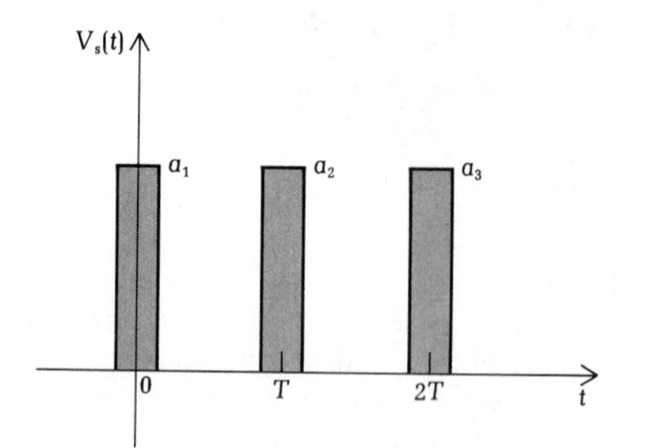

For the excitation in Part B, V_0 must be sampled after $t = 2.5T$ because it takes at least $2.5T$ for the sampler, which is located at $z = (1/2)\ell$, to detect the arrival of the third pulse. At $t = [n + (1/2)]T$, with $n \geq 2$, the voltage at $z = (1/2)\ell$ is

$$V_0[(n + 1/2)T] = f_+ (a_1\Gamma^n + a_2\Gamma^{n-1} + a_3\Gamma^{n-2})$$

or

$$V_0[(n + 1/2T] = f_+ \Gamma^{n-2}(a_3 + \Gamma a_2 + \Gamma^2 a_1)$$

For V_0 to be proportional to the decimal equivalence of $a_3(2^2) + a_2(2^1) + a_1(2^0)$, we must sample it at $t = (n + 1/2)T$, with $n \geq 2$, and

$$1:\Gamma:\Gamma^2 = 2^2:2^1:2^0$$

In other words, the reflection coefficient Γ must be equal to 1/2. From (6.25), we know that $R = 3Z_0$.

In summary, if $R = 3Z_0$, a sampler located at the midpoint of the transmission line should sample the voltage at $t = (n + 1/2)T$, where n is any integer greater than one. The analog output of the sampled voltage is directly proportional to the digital value of the signal train.

Example 6.9 *When a pulse propagates in a line that is not matched at both ends, it will be bounced back and forth from these ends. The amplitude of the pulse will decrease each time it is reflected from an end. This example illustrates this phenomenon.*

For the circuit shown in Figure 6.25a, $Z_0 = 50\ \Omega$, $R_L = 20\ \Omega$, $R_g = 30\ \Omega$, $\ell = 2$ m, $v = 1.0 \times 10^8$ m/s, $\Delta = 10^{-9}$ s, $V_0 = 1$ V. Plot the voltage and current at $z = \ell/2$ as functions of time.

Solution Using (6.40)–(6.42), we find

$$V_+ = \frac{50}{50 + 30} \times 1 = 0.625 \text{ V}$$

$$I_+ = \frac{0.625}{50} = 12.5 \text{ mA}$$

$$\Gamma_1 = \frac{20 - 50}{20 + 50} = -0.429$$

$$\Gamma_2 = \frac{30 - 50}{30 + 50} = -0.250$$

$$\ell/v = 2 \times 10^{-8} \text{ s} = 20 \text{ ns}$$

The incident pulse arrives at the midpoint at $t = 10$ ns with voltage 0.625 V and current 12.5 mA. The reflected pulse arrives 20 ns later with voltage -0.268 V and current 5.36 mA. The pulse reflected by R_g arrives at $t = 50$ ns with voltage 0.067 V and current 1.34 mA, and so on. These pulses are shown in Figure 6.29. Note that according to (6.23) the voltage and the current have the same sign when the wave propagates in the positive $\hat{z}$ direction and opposite signs when the wave propagates in the negative $\hat{z}$ direction.

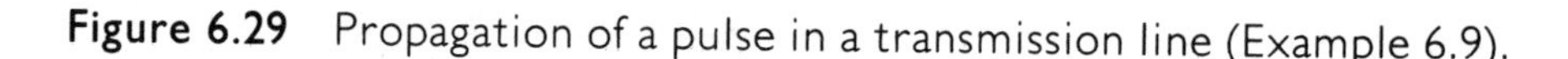

Figure 6.29 Propagation of a pulse in a transmission line (Example 6.9).

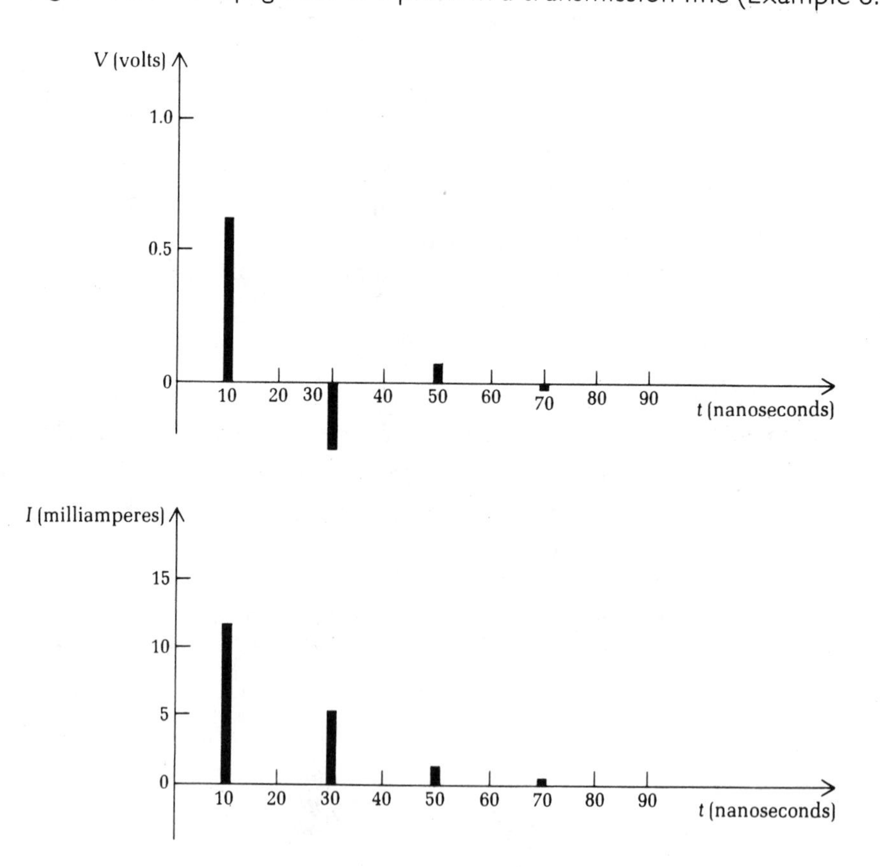

Example 6.10 ***This example shows how to compute how much energy carried by a pulse is absorbed in the load.***

For the circuit shown in Figure 6.25a and with the circuit parameters given in Example 6.9, calculate the percentage of energy absorbed by the load based on the energy delivered by the generator.

Solution The energy delivered by the generator to the system shown in Figure 6.25a is given by

$$E_g = VI\Delta = V_0\left(\frac{V_0}{Z_0 + R_g}\right)\Delta$$

The amount of energy that appears originally on the transmission line is E_+, which is given by

$$E_+ = E_g\frac{Z_0}{Z_0 + R_g}$$

Part of this energy is reflected by the load, and the reflected energy is $E_+(\Gamma_1)^2$. Thus, during the first reflection, the load absorbs $E_+(1 - \Gamma_1^2)$.

The energy carried by the second incident wave to the load is given by

$$E_+(\Gamma_1\Gamma_2)^2$$

Again the load absorbs $(1 - \Gamma_1^2)$ and reflects Γ_1^2 of the incident energy back to the generator side. This process continues indefinitely. The total energy absorbed by the load is expressed by an infinite summation:

$$\begin{aligned} E_L &= (1 - \Gamma_1^2)E_+(1 + \Gamma_1^2\Gamma_2^2 + \Gamma_1^4\Gamma_2^4 + \cdots) \\ &= (1 - \Gamma_1^2)E_g\frac{Z_0}{Z_0 + R_g}\frac{1}{1 - (\Gamma_1\Gamma_2)^2} \end{aligned}$$

The last term in the above equation is obtained by summing the infinite series $(1 + x + x^2 + x^3 + \cdots) = 1/(1 - x)$ for $|x| < 1$.

The ratio of E_L/E_g is given by

$$\frac{E_L}{E_g} = \frac{Z_0}{Z_0 + R_g}\frac{1 - \Gamma_1^2}{1 - (\Gamma_1\Gamma_2)^2}$$

Using the data obtained in the previous example, we obtain:

$$\frac{E_L}{E_g} = 52\%$$

Example 6.11 *This example gives the specification of the pulse that might someday be used in the A-D converter circuit discussed in this section.*

To make the scheme practical, what should be the width of each of the pulses in the A-D converter circuit discussed in this section? Assume that the practical length of the transmission line is 10 cm and that $v = 10^8$ m/s.

Solution The pulses must be separated by a time interval T where

$$T = \frac{\ell}{v} = \frac{10^{-1}}{10^8} = 10^{-9}\ \text{s} = 1\ \text{ns}$$

The width 2Δ of each pulse must be much less than T. Taking "much less" to mean "ten times less," we have

$$2\Delta = 0.1 \text{ ns} = 100 \text{ ps (picoseconds)}$$

Thus, the width of the pulse must be less than 100 ps. At present, the shortest switching time has been achieved using the Josephson junction:* 34 ps.

Distortion of a Pulse by Dispersion

A transmission line is frequency-dispersive to a particular mode if the velocity of that propagating mode is a function of the frequency—that is, if $v = \omega/k$ is not a constant. To demonstrate that a dispersive line will change the shape of a multi-frequency signal as it propagates along the line, let us consider as an example a wave of only two frequency components. Suppose that the propagating mode is A and that the propagation constant β versus the frequency is given by a function shown in Figure 6.30a. Let one of the components of the E field of mode A be given at the starting point $z = 0$ as follows:

$$E(0, t) = \text{Re}\,[e^{j\omega_1 t} + e^{j2\omega_1 t}] \tag{6.43}$$

According to Figure 6.30a the propagation constant associated with frequency ω_1 is equal to 1 and that associated with frequency $2\omega_1$ is equal to 1.5. The field component at z is then given by

$$E(z, t) = \text{Re}\,[e^{-jz}e^{j\omega_1 t} + e^{-j1.5z}e^{j2\omega_1 t}] \tag{6.44}$$

It is obvious that at an arbitrary point z the two frequency components are out of phase. For comparison, the original wave shape at $z = 0$ is shown in Figure 6.30b, and the wave shape at $z = \pi$ is shown in Figure 6.30c. Evidently there has been considerable distortion in the waveform. When a wave is in the form of a pulse that has a continuous spectrum, the foregoing procedure can still be used to find the waveform at any point along the line, except that the summation process is replaced by integration.

For a TE_{10} mode in a rectangular waveguide, the velocity is indeed a function of frequency. We therefore conclude that rectangular waveguides are not wide-band devices. Nevertheless, they serve a rather useful purpose of transmitting narrow-band signals arising from high operating frequencies. At 10 GHz operating frequency, a narrow frequency band of 1% still yields a useful bandwidth of 100 MHz.

One might be misled by the fact that the parallel-plate waveguide and the coaxial line are wide-band transmission lines because for TEM mode operation the velocity is a constant independent of the operating frequency. May we then conclude that these lines are ideal for transmitting, say, nanosecond pulses in

* See "Josephson switch: Faster than transistor," *IEEE Spectrum*, April 1973, p. 87.

Figure 6.30 **(a)** ω-β diagram of a transmission line. **(b)** Original waveform. **(c)** Waveform at $z = \pi$.

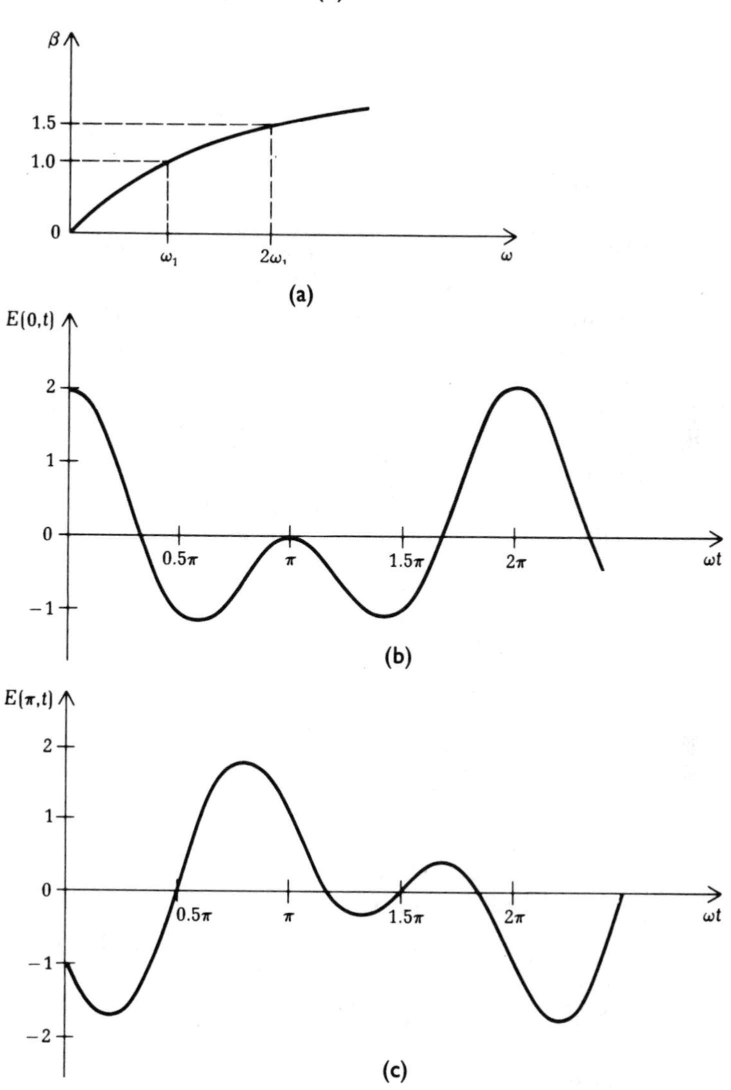

computer circuits? To answer this question, we need to point out quickly that, although the velocity is a constant for the TEM mode in these lines, other propagation modes also exist at high frequencies. High frequencies are always associated with a narrow pulse, and the velocities of those higher modes are not independent of frequency. For a narrow pulse propagating in a coaxial line, for example, the pulse may initially be propagated in the TEM mode without dispersion. If, however, the surfaces of the line have imperfections, or if the line bends, or if anything makes the line deviate from the ideal model, the higher frequency components will readily convert into higher-order modes, for which the transmission line is then dispersive. As for the microstriplines, it is only approximately true that the velocity of the TEM mode is independent of frequency. A more accurate analysis has shown that although β is close to k, it is not strictly a linear function of the frequency.*

Multiple-Mode Problem

In the discussion of rectangular waveguides, we mentioned that, if the operating frequency is too high, many higher-order modes can exist in addition to the fundamental TE_{10} mode. Although we did not mention the existence of higher-order modes in our discussion of coaxial lines, they do exist when the operating frequency exceeds a certain limit.

In coaxial lines, the first higher-order mode appears when $\lambda = 2b$. Higher operating frequency will yield more higher-order modes. The rule is that, to avoid multiple-mode propagation, the frequency spectrum of the signal must not contain frequencies higher than the cutoff frequency of the lowest higher-order mode.

Radiation Problem

Now let us consider a signal propagating in a transmission line, such as the microstripline that is not completely shielded by conducting walls. If the spectrum contains frequency components for which the cross-sectional area of the line is comparable with the wavelength, not only will the multiple-mode problem occur but the line will also become a radiator—that is, the transmission line will radiate some of the energy of the high-frequency components. Chapter 7 discusses this radiation phenomenon.

Transients on an Ideal Transmission Line

The preceding topic was concerned with propagation of "narrow" pulses on a transmission line. By "narrow" we mean that the width of the pulse Δ is much less than ℓ/v, where ℓ is the length of the line and v the propagation velocity of the pulse. Digital signals transmitted in cables or computer circuits generally satisfy this condition.

We shall now consider a different case, one in which the width of the pulse is very wide. The limiting case is a step-function voltage applied to a transmission

* R. Mittra and T. Itoh, "A new technique for the analysis of the dispersion characteristics of microstrip lines," *IEEE Trans.* MTT-19, no. 1 (January 1971): 47–56.

line of finite length. The short-circuiting of a power line due to lightning is one of many practical examples of such a case.

Consider a lossless transmission line with characteristic impedance Z_0, connected with a load resistance R_L and a generator resistance R_g with a constant voltage source V_0 (Figure 6.31). The boundary conditions at $z = 0$ and $z = \ell$ are, according to Kirchhoff's voltage law,

$$V_0 = V + R_g I \quad \text{at} \quad z = 0 \tag{6.45a}$$

$$V = R_L I \quad \text{at} \quad z = \ell \tag{6.45b}$$

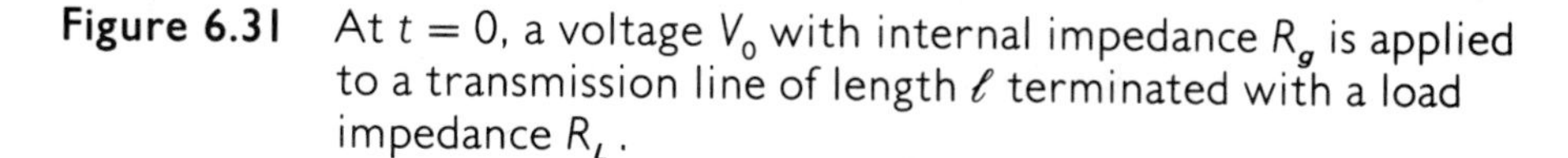
Figure 6.31 At $t = 0$, a voltage V_0 with internal impedance R_g is applied to a transmission line of length ℓ terminated with a load impedance R_L.

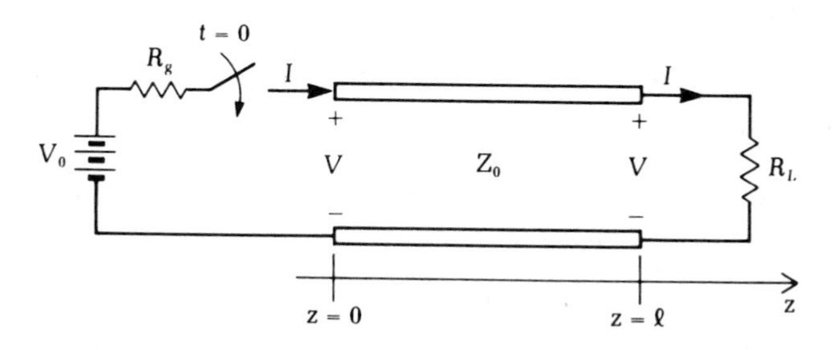

Let the switch be closed at $t = 0$ and a positive traveling wave initiated at $z = 0$. At $t = \ell/v$, where v is the velocity of the wave on the line, the positive traveling wave will be reflected by the load resistance, R_L, and at $t = 2\ell/v$ it will be reflected again by the generator resistance R_g, and so on. As $t \to \infty$, steady state will be reached, and the voltage on the line will simply become $R_L V_0/(R_L + R_g)$. The buildup of the voltage on the line from time zero to infinity is the subject of our studies.

First, consider the time interval $0 \le t \le \ell/v$. At $t = 0$, the boundary condition at $z = 0$ is given by (6.45a). A positive traveling voltage V_+ and the accompanying current $I_+ = V_+/Z_0$ is generated, and we have

$$V_0 = V_+ + R_g I_+$$

which gives

$$V_+ = \frac{Z_0}{R_g + Z_0} V_0 \tag{6.46}$$

Figure 6.32a illustrates this result.

Second, consider the time interval $\ell/v \le t < 2\ell/v$. At $t = \ell/v$, the boundary condition at $z = \ell$ is given by (6.45b). A negative traveling voltage V_- and the accompanying current $I_- = -V_-/Z_0$ is generated, and we have

$$(V_+ + V_-) = R_L(I_+ + I_-) = \frac{R_L}{Z_0}(V_+ - V_-)$$

which gives

$$V_- = \frac{R_L/Z_0 - 1}{R_L/Z_0 + 1} V_+ = \Gamma_L V_+ \tag{6.47}$$

where Γ_L is the reflection coefficient at the load. Figure 6.32b illustrates this result.

Third, consider the time interval $2\ell/v \leq t < 3\ell/v$. At $t = 2\ell/v$, the boundary condition at $z = 0$ is again given by (6.45a). A new positive traveling wave V'_+ and its accompanying current $I'_+ = V'_+/Z_0$ is generated, and we find

$$V_0 = (V_+ + V_- + V'_+) + \frac{R_g}{Z_0}(V_+ - V_- + V'_+)$$

which gives, considering (6.46) and (6.47),

$$V'_+ = \frac{R_g/Z_0 - 1}{R_g/Z_0 + 1} V_- = \Gamma_g V_- \tag{6.48}$$

where Γ_g is the reflection coefficient at the generator end. Figure 6.32c illustrates this result.

Figure 6.32 Transient voltage waves on the transmission line shown in Figure 6.31.

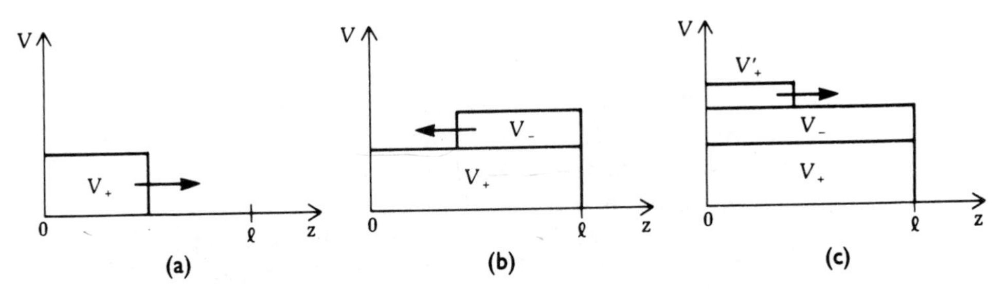

Summing up the voltages on the line for $t \to \infty$, we obtain

$$\begin{aligned} V &= V_+ + V_- + V'_+ + V'_- + V''_+ + V''_- + \cdots \\ &= V_+(1 + \Gamma_L + \Gamma_g\Gamma_L + \Gamma_g\Gamma_L^2 + \Gamma_g^2\Gamma_L^2 + \Gamma_g^2\Gamma_L^3 + \cdots) \\ &= V_+[(1 + \Gamma_L)(1 + \Gamma_g\Gamma_L + \Gamma_g^2\Gamma_L^2 + \cdots)] \\ &= V_+(1 + \Gamma_L)\frac{1}{1 - \Gamma_g\Gamma_L} = V_0\frac{R_L}{R_L + R_g} \end{aligned}$$

This result is the steady-state voltage on the line and is what we set out to prove.

Reflection Diagrams

The preceding analysis of transients on a transmission line can be illustrated graphically, as shown in Figure 6.33. The abscissa in Figure 6.33 denotes the transmission line of length ℓ. The ordinate represents time. The incident voltage

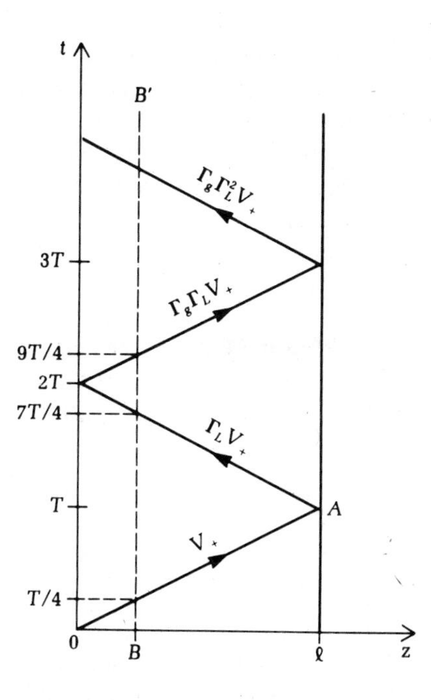

Figure 6.33 The voltage reflection diagram of the transmission line shown in Figure 6.31.

wave V_+, the amplitude of which is given by (6.46), starts at $z = 0$ at $t = 0$, propagates down the line , and arrives at $z = \ell$ at $t = \ell/v$. This wave is represented by the line OA. The reflected wave, with amplitude given by $V_+ \Gamma_L$, is represented by the line starting at A and terminating at $t = 2T$, $z = 0$. Subsequent reflected waves are represented by zigzag lines.

To find the voltage as a function of time at any point $z = z_0$ on the transmission line, one can draw a vertical line that passes the point $z = z_0$. The voltage at $z = z_0$ can be found by simply adding all reflected waves that cross this vertical line up to that particular time. For example, Figure 6.34 shows the voltage at $z = \ell/4$. Between $0 < t < T/4$, none of those zigzag lines crosses the vertical line BB'. Therefore, the voltage is zero during that time period. The line BB' intersects V_+ at $t = T/4$, $\Gamma_L V_+$ at $t = 7T/4$, $\Gamma_g \Gamma_L V_+$ at $9T/4$, etc. Consequently, the voltage has discontinuities at these times, as shown in Figure 6.34.

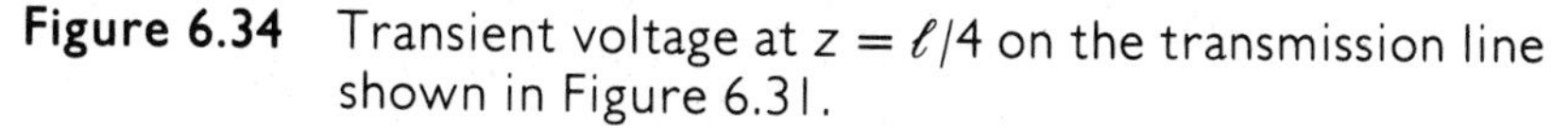
Figure 6.34 Transient voltage at $z = \ell/4$ on the transmission line shown in Figure 6.31.

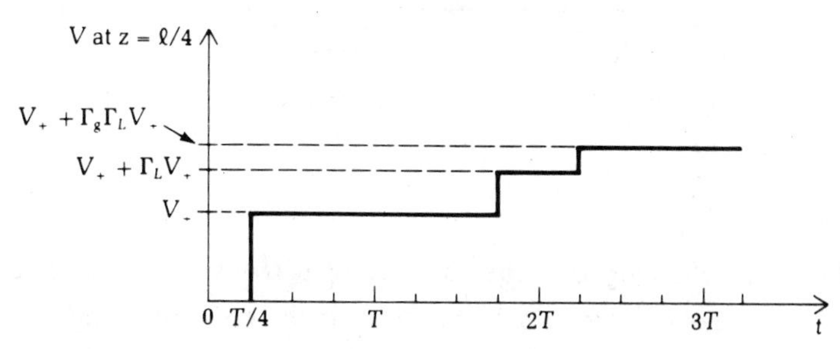

Figure 6.33 is called the **voltage reflection diagram** of the transmission line. To study the transient **current** on the transmission line, one can similarly construct a corresponding **current reflection diagram**. Figure 6.35 shows the latter for the transmission line shown in Figure 6.31. Note that the current wave traveling in the negative $\hat{z}$ direction is opposite in sign from the accompanying voltage wave. The voltage waves and the current waves are of the same sign when they are traveling in the positive $\hat{z}$ direction. Figure 6.36 shows the current at $z = \ell/4$ as a function of time.

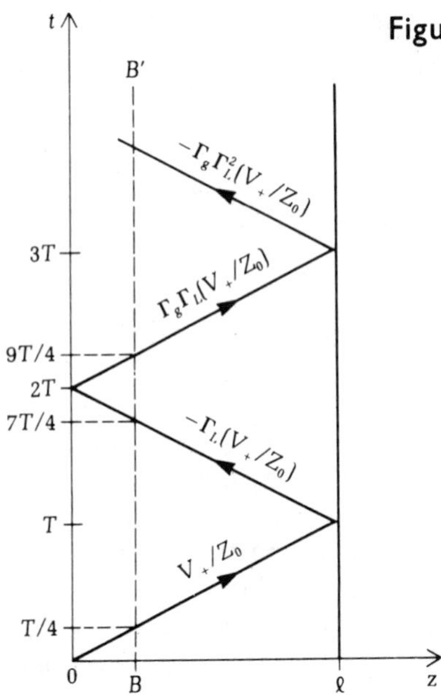

Figure 6.35 The current reflection diagram of the transmission line shown in Figure 6.31.

Figure 6.36 Transient current at $z = \ell/4$ on the transmission line shown in Figure 6.31.

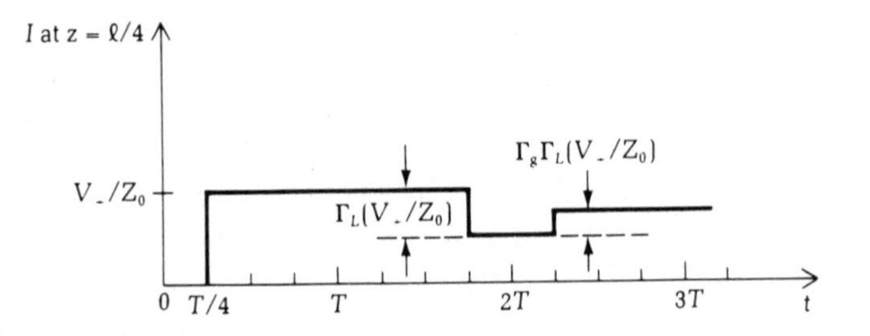

Example 6.12 *The following is a special case of a transmission line terminated at one end by an open circuit and matched at the generator end.*

Consider the transmission line shown in Figure 6.37a. Notice that the line is open-circuited at the load. Draw the voltage reflection diagram. What is the voltage V(z) for $t > 2\ell/v$?

Solution From (6.46) we find

$$V_+ = \tfrac{1}{2}V_0$$

The reflection coefficients Γ_L and Γ_g are

$$\Gamma_L = 1$$

$$\Gamma_g = 0$$

Figure 6.37b shows the voltage reflection diagram. It is clear from the diagram that the line reaches the steady state at $t = 2\ell/v$. Therefore, V(z) = V_0 for $t > 2\ell/v$.

Figure 6.37 **(a)** A transmission line with an open circuit. **(b)** The voltage reflection diagram (Example 6.12).

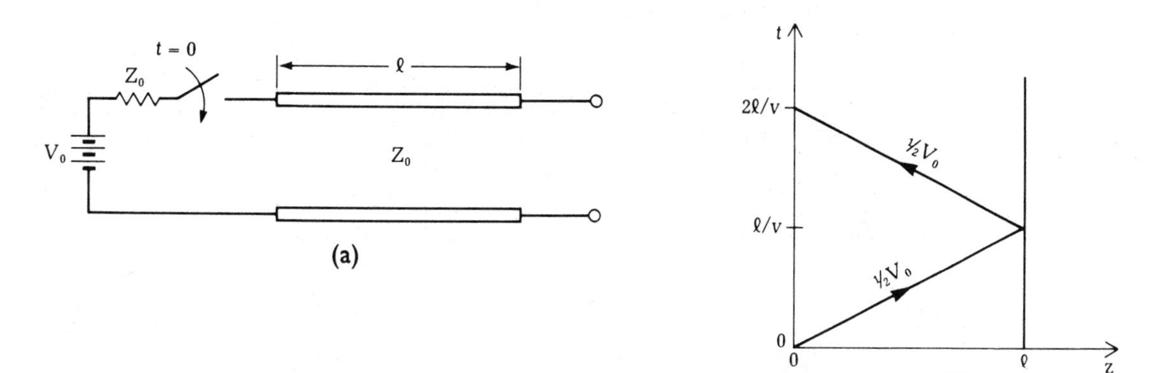

Transients on a Line with Reactive Loading

We shall now study the transient response of a transmission line that is terminated not by a resistor but by either a capacitor or an inductor. Let us first consider a transmission line with a capacitive load, as shown in Figure 6.38a (p. 205).

The boundary condition at the load is as follows:

$$I_L = C\frac{dV_L}{dt} \tag{6.49}$$

At $t = T = \ell/v$, the V_+ wave arrives at the load, and at the same time the V_- wave is generated. The voltage across the capacitor is the sum of V_+ and V_-:

$$V_L = V_+ + V_- \tag{6.50}$$

The load current I_L is the sum of I_+ and I_-:

$$I_L = I_+ + I_-$$

Using the relationships $I_+ = V_+/Z_0$ and $I_- = -V_-/Z_0$, we obtain

$$I_L = (V_+ - V_-)/Z_0 \tag{6.51}$$

Combining (6.50) and (6.51) to eliminate V_- yields

$$2V_+ = V_L + Z_0 I_L \tag{6.52}$$

Substituting (6.52) into (6.49) to eliminate I_L, we obtain a differential equation for V_L:

$$C\frac{dV_L}{dt} + \frac{1}{Z_0}V_L = \frac{2V_+}{Z_0} \tag{6.53}$$

The solution of (6.53) is

$$V_L = 2V_+ + Ae^{-t/Z_0C}$$

where A is an arbitrary constant. Assuming that the capacitor is not charged until the voltage wave V_+ arrives, we have

$$V_L(t = T) = 0 = 2V_+ + Ae^{-T/Z_0C} \tag{6.54}$$

The above equation determines the constant A. Substituting the value A found from (6.54), we find the voltage across the capacitor for $t > T$:

$$V_L = 2V_+[1 - e^{-(t-T)/Z_0C}] \tag{6.55}$$

The reflected wave V_- can then be found from (6.55) and (6.50):

$$V_- = V_L - V_+ = \frac{V_0}{2}[1 - 2e^{-(t-T)/Z_0C}] \tag{6.56}$$

$V_-(t)$ is shown in Figure 6.38b. Note that $V_+ = V_0/2$.

This reflected wave travels toward the source with velocity v. Because the source is matched, no reflection is originated at this end of the line.

The voltage $V(z)$ on the line in the time period $T < t < 2T$ is shown in Figure 6.38c. Note that $z_0 = \ell - (t - T)v$.

When the transmission line is terminated with an inductor, as shown in Figure 6.39a, the boundary condition at the load is then given by

$$V_L = L\frac{dI_L}{dt} \tag{6.57}$$

We can find the reflected wave by carrying out an analysis similar to the one for a transmission line with a capacitor. The result is as follows:

$$V_- = V_L - V_+ = V_0(e^{-Z_0(t-T)/L} - 1/2) \tag{6.58}$$

Figure 6.39b shows the voltage $V(z)$ on the line during the time period $T < t < 2T$.

Figure 6.38 **(a)** A transmission line with capacitive load. **(b)** The reflected wave originated at the capacitor. **(c)** The voltage on the line for $T < t < 2T$.

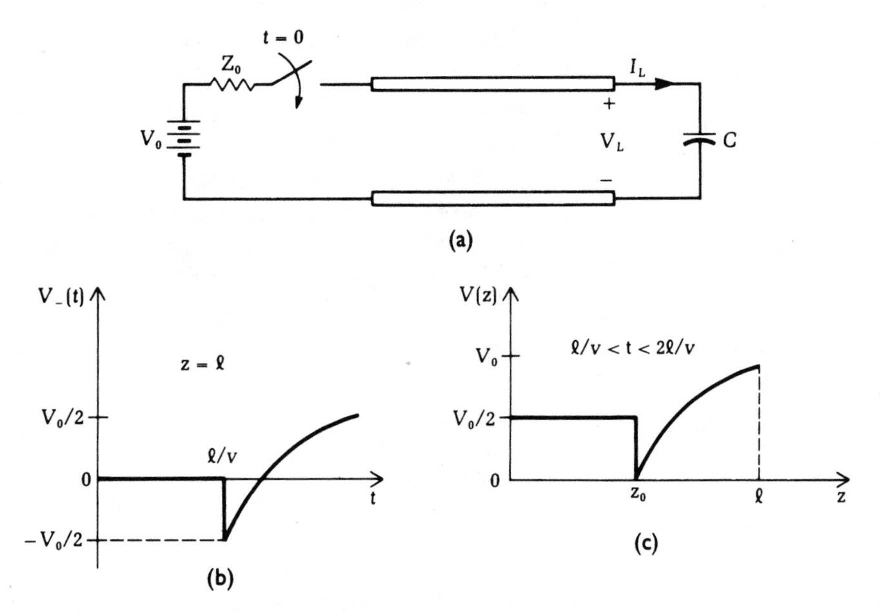

Figure 6.39 **(a)** A transmission line with inductive loading. **(b)** Voltage $V(z)$ on the line for $T < t < 2T$.

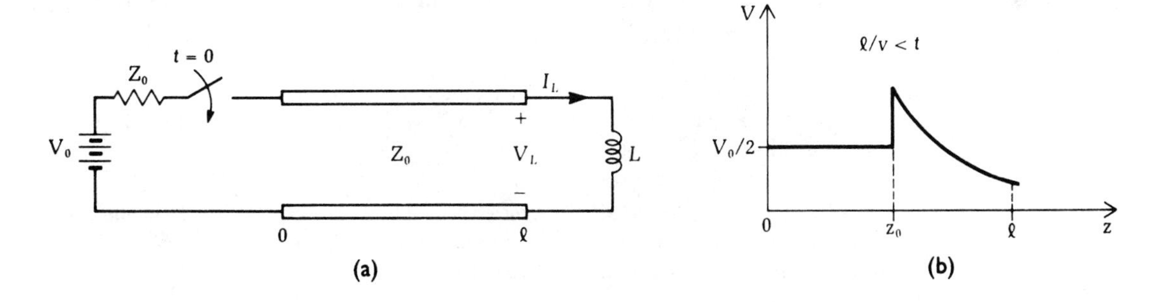

SUMMARY

1. Propagation of electromagnetic waves in waveguides and coaxial lines can be represented by propagation of equivalent voltages and currents in a transmission line.
2. The equivalent voltage is proportional to the integration of the E field along a path on the transverse plane of the waveguide or the coaxial line.
3. The equivalent current is proportional to the contour integration of the H field along a closed path on the transverse plane of the waveguide or the coaxial line.

4. The equivalent voltage and current must satisfy the power relationship.
5. In a transmission line, the reflection coefficient can be expressed in terms of the load impedance, and vice versa.
6. A general reflection coefficient is defined. It is a function of z. Likewise, a general impedance function of z is defined.
7. The Smith chart gives a graphical, one-to-one relationship between the reflection coefficient and the impedance function. That is, if one is given, the other can be read from the Smith chart.
8. A standing wave exists in a transmission line terminated by an unmatched load. The VSWR in the line and the position of maximum and minimum voltages can be found graphically by using the Smith chart.
9. An unmatched transmission line can be matched by adding a shunt susceptance at certain place in the line. The position and the value of the susceptance can be found by using the Smith chart.
10. Propagation of pulses in a transmission line can be analyzed using a similar principle, provided that the velocity of the EM wave and the characteristic impedance of the line do not vary significantly in the frequency spectrum of the pulse.

Problems

6.1 What is the voltage in the stripline discussed in Example 6.1 when the time-average power being transmitted is 10 kW?

6.2 Consider the coaxial line discussed in Example 6.3. Calculate the maximum time-average power that may be transmitted in the line. Use the breakdown $E = 2 \times 10^7$ V/m and a safety factor of 10.

6.3 Two coaxial lines have equal characteristic impedances: 50 Ω. Both are air-filled. The first line has a power capacity of 1 MW, and the second line's capacity is 1 kW. Find the ratios a_1/a_2 and b_1/b_2. Consider only the breakdown E field.

6.4 Use (6.1b) and the boundary condition (4.3) to obtain the surface-current density $\mathbf{J}_s$ on the lower plate of the parallel-plate waveguide. Then calculate the total current on the lower plate. Compare the current with the definition of $I(z)$ given by (6.3b).

6.5 Find the surface-current density $\mathbf{J}_s$ on the inner conductor of a coaxial line. Then calculate the total current on it. Compare the total current with $I(z)$ defined for the coaxial line.

6.6 A transmission line is short-circuited ($Z_L = 0$).
(a) Find the expressions for $|V(z)|$ and $|I(z)|$ as a function of kz, Z_0, and V_+.
(b) Sketch $|V(z)|$ and $|I(z)|$.
(c) Find VSWR on the line.

6.7 Repeat Problem 6.6 for a transmission line with an open circuit at the load ($Z_L = \infty$).

6.8 Repeat Problem 6.6 for a transmission line with a matched load ($Z_L = Z_0$).

6.9 A transmission line is terminated with a normalized load of $0.8 + j1.0$. Calculate (a) the VSWR, (b) the position of the voltage minimum, and (c) the percentage of the incident power that is reflected by the load. Sketch $|V(z)|$ as a function of z/λ.

6.10 Solve the problem discussed in Example 6.6 by using the Smith chart. Find the position of a shunt susceptance that can tune the line to have a perfect match. Determine the value (in mhos) of the shunt susceptance.

6.11 For an open-circuited 50-Ω transmission line of length ℓ, the input impedance at the other end is $j33\,\Omega$. Find the length ℓ (in λ).

6.12 Repeat Problem 6.11 when the line is short-circuited at one end.

6.13 For the first waveguide in Table 1 of Section 5.2, design an iris that will give a $j1.57$ admittance at $f = 8$ GHz.

6.14 From the Smith chart, find Γ_L for the following Z_{Ln}: (a) $1 + j2$, (b) ∞, (c) 0 and (d) $0.55 - j0.38$.

6.15 Use the Smith chart to find Z_{Ln} from the following Γ_L: (a) $0.6e^{j45°}$, (b) -0.3, and (c) 0.

6.16 For a load impedance of $0.4 - j0.5$, find the location of the first voltage minimum and the first voltage maximum at the load end.

6.17 From the Smith chart, find the admittances for the following impedances: (a) $Z_n = 0.3 - j0.6$ and (b) $Z_n = 5 + j3$.

6.18 To measure the impedance of an antenna, we measure the standing wave on the transmission line to which the antenna is connected. The standing-wave pattern is shown in Figure P6.18. Find the impedance of the antenna using the Smith chart. The characteristic impedance of the transmission line is 50 ohms.

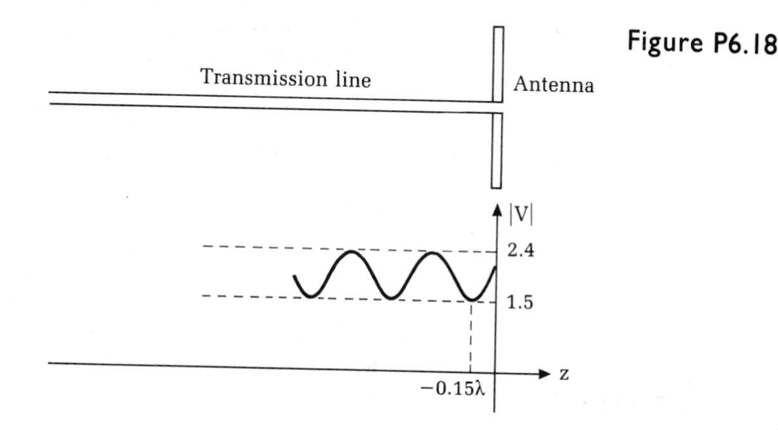

Figure P6.18

6.19 A transmission line is matched by a shunt susceptance of $Y_{ns} = j0.9$ placed 0.15λ from the load toward the generator. Find the normalized load impedance. Hint: Use the Smith chart.

6.20 A transmission line is terminated with a normalized impedance $Z_{Ln} = 2 + j2$, as shown in Figure 6.19a. The incident $V_+ = 1.0$, and the characteristic impedance of the line is 1.0. Show that $V_{max} = I_{max} = 1.62$, $|V(0)| = 1.57$, $|V(-0.219\lambda)| = 0.78$, $V_{min} = I_{min} = 0.38$, $|I(0)| = 0.55$, and $|I(-0.219\lambda)| = 1.46$.

6.21 A shunt admittance of $Y_{Ln} = -j1.57$ is added to the transmission line that is terminated by a load $Z_{Ln} = 2 + j2$, as shown in Figure 6.19b. The position of the shunt is $0.219\lambda_g$ from the load, so that the line is perfectly matched. Let $V_+ = 1.0$ and $Z_0 = 1.0$ and show that $V_{max} = 2.07$, $|V(0)| = 2.00$, $I_{max} = 1.86$, $|I(0)| = 0.71$, and $I_{min} = 0.48$.

6.22 In Example 6.8, find another set of solutions of ℓ_1 and ℓ_2 (in centimeters).

6.23 If the load impedance of a transmission line is purely reactive, that is, $Z_{Ln} = ja$, where a is a real number, can it be matched by adding a shunt susceptance somewhere in the line? Why?

6.24 For the circuit shown in Figure 6.25a, let $Z_0 = 50\ \Omega$, $R_L = 70\ \Omega$, $R_g = 50\ \Omega$, $\ell = 2$ m, $v = 10^8$ m/s, $\Delta = 10^{-9}$ s, and $V_0 = 1$. Plot the voltage and current at $z = \ell/2$ as a function of time.

6.25 Calculate the percentage of energy generated by the pulse generator that is absorbed by the load in the circuit of Problem 6.24.

6.26 For a four-digit code system, design a D-A converter similar to that discussed in Section 6.5 using the transmission line shown in Figure 6.27a. Specify the value of R, the location of the sampler, and the time that a sample should be taken.

6.27 In plotting Figure 6.32, it is implicitly assumed that $R_L > Z_0$ and that $R_g > Z_0$, so that both Γ_L and Γ_g are positive numbers. Sketch a similar diagram for the case in which $R_L = 0.5Z_0$ and $R_g = 0.5Z_0$.

6.28 Draw the voltage and the current reflection diagrams for the transmission line that is short-circuited, as shown in Figure P6.28. Plot V and I as functions of time at $z = \ell/2$.

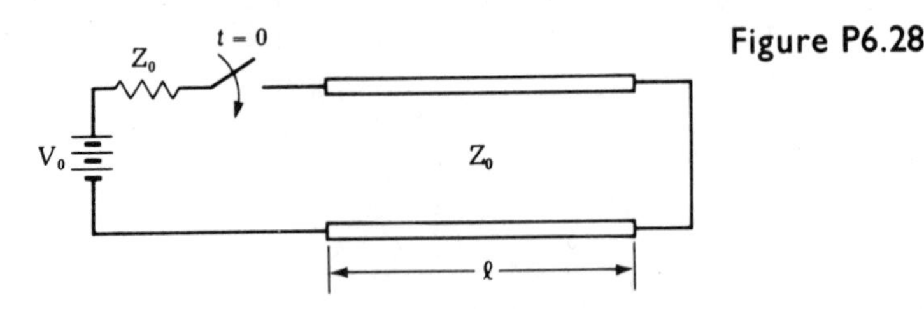

Figure P6.28

6.29 Draw the voltage and the current reflection diagrams for the transmission line that is perfectly matched, as shown in Figure P6.29. Plot V and I as functions of time at $z = \ell/2$.

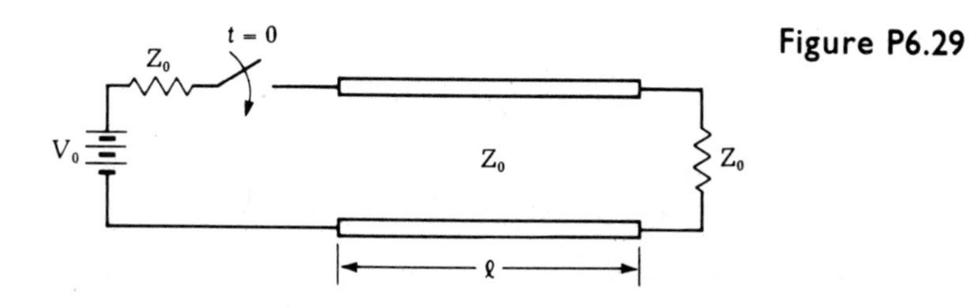

Figure P6.29

6.30 Refer to Figure 6.31, and let $R_g = 2Z_0$ and $R_L = 0.5Z_0$. Draw the voltage reflection diagram for $0 < t < 6T$, and plot V at $z = 3\ell/4$ for $0 < t < 6T$.

6.31 Refer to Figure 6.31, and let $R_g = 2Z_0$ and $R_L = 0.5Z_0$. Draw the current reflection diagram for $0 < t < 6T$, and plot I at $z = 3\ell/4$ for $0 < t < 6T$.

6.32 Refer to Figure 6.38a, and obtain an expression for I_-. Sketch $I_-(t)$ and $I(z)$ versus z for the time period $T < t < 2T$. The sketch should be similar to Figure 6.38b and Figure 6.38c.

6.33 Derive Equation (6.58).

7 ANTENNAS

Humans have invented devices that produce electromagnetic energy for communication, sensing and detection, and other needs. Whatever the purposes, a system must transmit the electromagnetic energy in a desirable manner. Transmission lines and waveguides are examples of such systems. Another way to transmit electromagnetic energy is to send it through the air, as in radio or television broadcasting, or by point-to-point microwave links. For broadcasting systems, the electromagnetic signal must cover a wide area; but, for point-to-point communication, the signal is narrowly limited. The device for radiating and receiving radio waves is an antenna. In this chapter we shall study a few simple antenna systems. The purpose is to obtain some ideas about designing or using antennas so that they will radiate or receive electromagnetic waves in the manner appropriate for certain special applications.

7.1 VECTOR AND SCALAR POTENTIAL FUNCTIONS

To send electromagnetic waves into a medium, one must maintain oscillating currents somewhere in the medium. This fact is evident from Maxwell's equations (2.28a) and (2.28b), which are repeated below:

$$\nabla \times \mathbf{E} = -j\omega\mu\mathbf{H}$$

$$\nabla \times \mathbf{H} = \mathbf{J} + j\omega\epsilon\mathbf{E}$$

When the current $\mathbf{J}$ is known, the above equations can be solved for $\mathbf{E}$ and $\mathbf{H}$. These Maxwell's equations can also be solved through two auxiliary functions called the vector and the scalar potentials $\mathbf{A}$ and Φ. We will first express the electromagnetic fields by these potential functions, then obtain partial differential equations for $\mathbf{A}$ and Φ, and finally find solutions of $\mathbf{A}$ and Φ, from which we can calculate $\mathbf{E}$ and $\mathbf{H}$. The introduction of the potential functions will also facilitate the solution of the Maxwell's equations, as we shall demonstrate in this section.

The **vector potential A** is defined as

$$\boxed{\mathbf{B} = \nabla \times \mathbf{A}} \quad \textbf{(definition of } \bar{\mathbf{A}}\textbf{)} \tag{7.1}$$

Note that Maxwell's equation $\nabla \cdot \mathbf{B} = 0$ is now automatically satisfied because the divergence of a curl of a vector is identically equal to zero [see (2.11a)].

Substituting (7.1) in Faraday's law $\nabla \times \mathbf{E} = -j\omega\mathbf{B}$ yields

$$\nabla \times (\mathbf{E} + j\omega\mathbf{A}) = 0 \tag{7.2}$$

Because the curl of the gradient of a scalar is identically equal to zero, we choose the following equation to define the **scalar potential** Φ:

$$\boxed{\mathbf{E} = -j\omega\mathbf{A} - \nabla\Phi} \quad \textbf{(definition of } \Phi\textbf{)} \tag{7.3}$$

The choice of the sign for Φ follows the convention used in static-electric potential.

Thus far, two of the four Maxwell's equations are satisfied. Equation (7.1) does not uniquely define $\mathbf{A}$. Consider $\mathbf{A}' = \mathbf{A} + \nabla\psi$, where ψ is any scalar function; then, $\mathbf{B} = \nabla \times \mathbf{A} = \nabla \times \mathbf{A}'$ [see (2.11b)]. Thus, both $\mathbf{A}$ and $\mathbf{A}'$ give the same $\mathbf{B}$. To define $\mathbf{A}$ uniquely, we also have to specify $\nabla \cdot \mathbf{A}$, which is done with the so-called **Lorentz condition**:

$$\boxed{\nabla \cdot \mathbf{A} + j\omega\epsilon\Phi = 0} \quad \textbf{(Lorentz condition)} \tag{7.4}$$

The Lorentz condition is chosen in accordance with the principle of relativity.*

Substituting (7.1) and (7.3) in (2.28b) yields

$$\nabla \times (\nabla \times \mathbf{A}) = j\omega\mu\epsilon(-j\omega\mathbf{A} - \nabla\Phi) + \mu\mathbf{J}$$

Noting the vector identity $\nabla \times (\nabla \times \mathbf{A}) = \nabla(\nabla \cdot \mathbf{A}) - \nabla^2\mathbf{A}$ and applying the Lorentz condition (7.4), we obtain

$$\boxed{\nabla^2\mathbf{A} + \omega^2\mu\epsilon\mathbf{A} = -\mu\mathbf{J}} \quad \textbf{(differential equation for } \bar{\mathbf{A}}\textbf{)} \tag{7.5}$$

Substituting (7.3) in $\nabla \cdot \mathbf{E} = \rho_v/\epsilon$ and applying the Lorentz condition, we find

$$\boxed{\nabla^2\Phi + \omega^2\mu\epsilon\Phi = \frac{-\rho_v}{\epsilon}} \quad \textbf{(differential equation for } \Phi\textbf{)} \tag{7.6}$$

Equations (7.5) and (7.6) contain four scalar equations, which are known as the inhomogeneous Helmholtz wave equations. The mathematical procedure leading to the solution of the Helmholtz equation is complicated and lies beyond the scope of this book.** However, the results are relatively simple and have wide

* J. A. Kong, *Theory of Electromagnetic Waves*, New York: Wiley, 1975, chaps. 2 and 7.

** R. F. Harrington, *Time Harmonic Electromagnetic Fields*, New York: McGraw-Hill, 1962, p. 80.

applications. Thus, we shall state the results in the following two equations and fully explore their applications in this and later chapters.

$$\mathbf{A}(\boldsymbol{r}) = \frac{\mu}{4\pi} \iiint_v dV' \frac{\mathbf{J}(\boldsymbol{r}')e^{-jk|\boldsymbol{r}-\boldsymbol{r}'|}}{|\boldsymbol{r}-\boldsymbol{r}'|} \tag{7.7}$$

(solutions for $\bar{\mathbf{A}}$ and Φ)

$$\Phi(\boldsymbol{r}) = \frac{1}{4\pi\epsilon} \iiint_v dV' \frac{\rho_v(\boldsymbol{r}')e^{-jk|\boldsymbol{r}-\boldsymbol{r}'|}}{|\boldsymbol{r}-\boldsymbol{r}'|} \tag{7.8}$$

where $k = \omega\sqrt{\mu\epsilon}$, $\boldsymbol{r}$ is the vector indicating the position of the potentials, $\boldsymbol{r}'$ is the position vector of the sources, and $|\boldsymbol{r} - \boldsymbol{r}'|$ is the distance between the observation point $\boldsymbol{r}$ and the source point $\boldsymbol{r}'$. The integration extends over all points $\boldsymbol{r}'$ where the source is not zero (Figure 7.1).

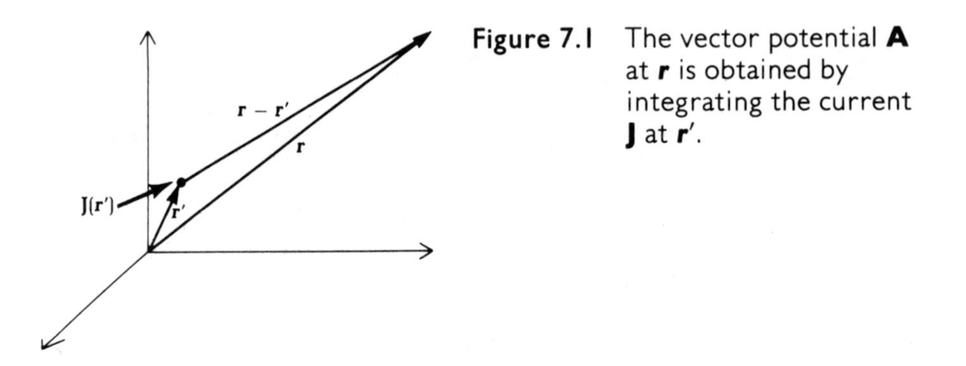

Figure 7.1 The vector potential **A** at $\boldsymbol{r}$ is obtained by integrating the current **J** at $\boldsymbol{r}'$.

We have now shown that if the sources are specified, instead of solving the original Maxwell's equations, we can calculate the vector potential **A** and the scalar potential Φ by using (7.7) and (7.8). We can then obtain the electric and magnetic fields from (7.3) and (7.1), respectively.

7.2 FIELDS OF AN INFINITESIMAL ANTENNA

An infinitesimal antenna is an extremely short and thin wire driven by a current source, usually located at its center. Assume the antenna has an infinitesimal length Δz, a cross-sectional area ΔA, and a constant current I. Its current density $\mathbf{J}(\boldsymbol{r}')$ (amperes per square meter) multiplied by ΔA is equal to I. Because the antenna source occupies an infinitesimal volume, the differential volume $\Delta V' = \Delta A\, \Delta z$ multiplied by $\mathbf{J}(\boldsymbol{r}')$ gives

$$\Delta V' \mathbf{J}(\boldsymbol{r}') = \mathrm{I}\, \Delta z \hat{z} \tag{7.9}$$

Placing the infinitesimal antenna at the origin, we may set $\boldsymbol{r}' = 0$ and obtain from (7.7)

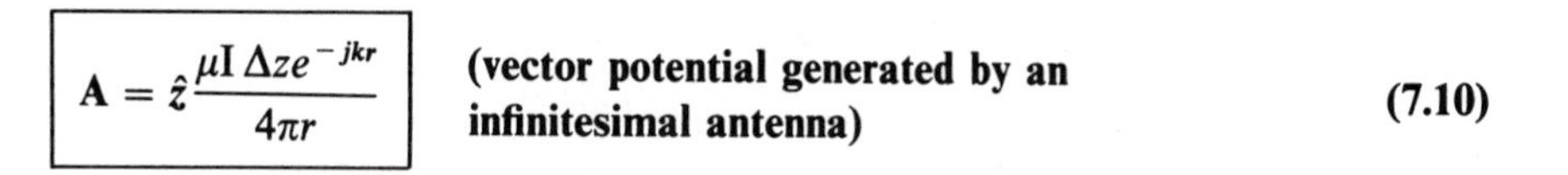

$$\mathbf{A} = \hat{z}\frac{\mu I \,\Delta z e^{-jkr}}{4\pi r} \qquad \textbf{(vector potential generated by an infinitesimal antenna)} \qquad (7.10)$$

To calculate the magnetic field from (7.1), we first resort to a discussion of **spherical coordinate systems.**

In spherical coordinates, a point P is space is represented by three coordinates, r, θ, and ϕ, as shown in Figure 7.2, where

$r =$ the length OP $\theta =$ the angle between the z axis and OP $\phi =$ the angle between the x axis and OA, which is the projection of OP on the x-y plane	**(spherical coordinates)**

Figure 7.2 Spherical coordinates.

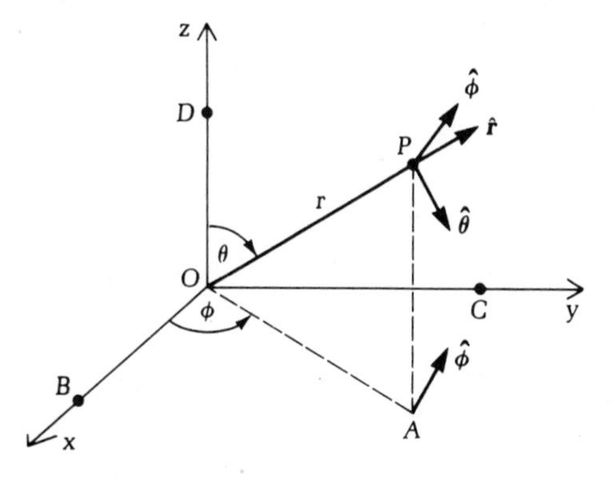

Directions of a vector can also be expressed in spherical coordinates by three unit vectors $\hat{\boldsymbol{r}}$, $\hat{\boldsymbol{\theta}}$, and $\hat{\boldsymbol{\phi}}$, where

$\hat{\boldsymbol{r}} =$ the unit vector in the direction of increasing r $\hat{\boldsymbol{\theta}} =$ the unit vector in the direction of increasing θ $\hat{\boldsymbol{\phi}} =$ the unit vector in the direction of increasing ϕ	**(unit vectors in spherical coordinates)**

Bear in mind that all of the three unit vectors in spherical coordinates depend on the location of the vector. For instance, at point B, $\hat{\boldsymbol{r}} = \hat{\boldsymbol{x}}$, $\hat{\boldsymbol{\theta}} = -\hat{\boldsymbol{z}}$, and $\hat{\boldsymbol{\phi}} = \hat{\boldsymbol{y}}$; at point C, $\hat{\boldsymbol{r}} = \hat{\boldsymbol{y}}$, $\hat{\boldsymbol{\theta}} = -\hat{\boldsymbol{z}}$, and $\hat{\boldsymbol{\phi}} = -\hat{\boldsymbol{x}}$, and at point D, $\hat{\boldsymbol{r}} = \hat{\boldsymbol{z}}$, and $\hat{\boldsymbol{\theta}}$ and $\hat{\boldsymbol{\phi}}$ are not defined. The divergence and the curl operator on a vector A in the spherical coordinate system are given as follows:

$$\nabla \cdot A \equiv \frac{1}{r^2}\frac{\partial}{\partial r}(r^2 A_r) + \frac{1}{r \sin\theta}\frac{\partial}{\partial\theta}(\sin\theta A_\theta) + \frac{1}{r\sin\theta}\frac{\partial A_\phi}{\partial\phi} \tag{7.11}$$

$$\begin{aligned}\nabla \times A &\equiv \frac{1}{r\sin\theta}\left[\frac{\partial}{\partial\theta}(A_\phi \sin\theta) - \frac{\partial A_\theta}{\partial\phi}\right]\hat{r} \\ &\quad + \frac{1}{r}\left[\frac{1}{\sin\theta}\frac{\partial A_r}{\partial\phi} - \frac{\partial}{\partial r}(rA_\phi)\right]\hat{\theta} + \frac{1}{r}\left[\frac{\partial}{\partial r}(rA_\theta) - \frac{\partial A_r}{\partial\theta}\right]\hat{\phi} \\ &= \frac{1}{r^2\sin\theta}\begin{vmatrix} \hat{r} & r\hat{\theta} & r\sin\theta\hat{\phi} \\ \dfrac{\partial}{\partial r} & \dfrac{\partial}{\partial\theta} & \dfrac{\partial}{\partial\phi} \\ A_r & rA_\theta & r\sin\theta A_\phi \end{vmatrix}\end{aligned} \tag{7.12}$$

The gradient and Laplacian of scalar function ψ takes the following forms:

$$\nabla\psi = \hat{r}\frac{\partial}{\partial r}\psi + \hat{\theta}\frac{1}{r}\frac{\partial}{\partial\theta}\psi + \hat{\phi}\frac{1}{r\sin\theta}\frac{\partial}{\partial\phi}\psi$$

$$\nabla^2\psi = \frac{1}{r^2}\frac{\partial}{\partial r}\left(r^2\frac{\partial\psi}{\partial r}\right) + \frac{1}{r^2\sin\theta}\frac{\partial}{\partial\theta}\left(\sin\theta\frac{\partial\psi}{\partial\theta}\right) + \frac{1}{r^2\sin^2\theta}\frac{\partial^2\psi}{\partial\phi^2}$$

We now apply (7.12) to evaluate $\mathbf{B}$ from (7.1). First, we note that $\hat{z} = \hat{r}\cos\theta - \hat{\theta}\sin\theta$. We obtain

$$\mathbf{H} = \frac{1}{\mu}\nabla \times \mathbf{A} = \hat{\phi}\frac{jk\mathrm{I}\,\Delta z e^{-jkr}}{4\pi r}\left(1 + \frac{1}{jkr}\right)\sin\theta \tag{7.13}$$

The corresponding electric field $\mathbf{E}$ is calculated from Ampère's law. Outside the dipole source,

$$\begin{aligned}\mathbf{E} &= \frac{1}{j\omega\epsilon}\nabla \times \mathbf{H} = \sqrt{\frac{\mu}{\epsilon}}\frac{jk\mathrm{I}\,\Delta z e^{-jkr}}{4\pi r} \\ &\quad \times \left\{\hat{r}\left[\frac{1}{jkr} + \frac{1}{(jkr)^2}\right]2\cos\theta + \hat{\theta}\left[1 + \frac{1}{jkr} + \frac{1}{(jkr)^2}\right]\sin\theta\right\}\end{aligned} \tag{7.14}$$

Thus, the magnetic field has only $\hat{\phi}$ components circulating around the dipole, whereas the electric field lies in planes containing the dipole axis and possesses no $\hat{\phi}$ component.

Hertzian Dipoles

The small antenna that we just described has a constant-current distribution over the dipole length Δz. To see how this distribution is possible, consider two charge reservoirs separated by Δz. At one instant of time, say $t = 0$, one of the two reservoirs is full of positive charge $+q$, and the other has an equal amount of negative charge $-q$ (Figure 7.3). At $t = T/4$, the charges from the two reservoirs neutralize each other as a result of the constant current flow I. The current flow exchanges the polarities of the two charge distributions after a half cycle, and the process continues at the angular frequency $\omega = 2\pi/T$. This model constitutes an oscillating dipole with a dipole moment defined by

$$p = q\,\Delta z \tag{7.15}$$

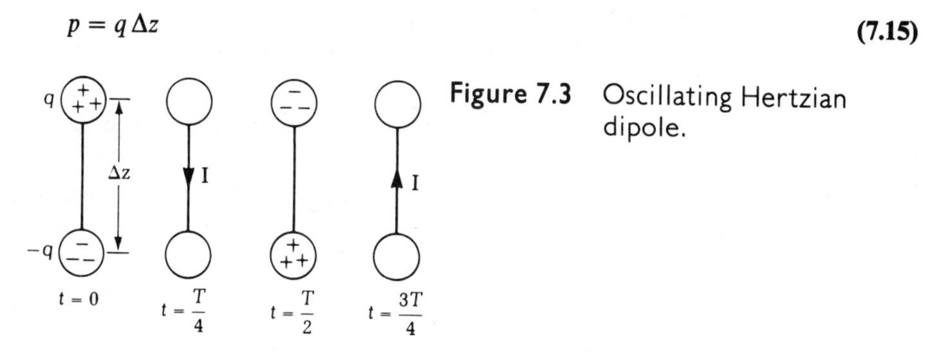

Figure 7.3 Oscillating Hertzian dipole.

Time derivative of p constitutes the current $\mathrm{I}\,\Delta z$:

$$\mathrm{I}\,\Delta z = j\omega p \tag{7.16}$$

This oscillating dipole is known as the Hertzian dipole. It is the simplest radiator, and longer or more complex antennas can be thought of as the superposition of many Hertzian dipoles.

Radiation Fields of a Hertzian Dipole Antenna

Radiation fields occur at great distances from the Hertzian dipole such that $kr = 2\pi r/\lambda \gg 1$, with only terms involving $1/kr$ dominant. The radiation fields of the Hertzian dipole antenna are

$$\mathbf{H} = \hat{\boldsymbol{\phi}}\frac{jk\mathrm{I}\,\Delta z e^{-jkr}}{4\pi r}\sin\theta \tag{7.17a}$$

$$\mathbf{E} = \hat{\boldsymbol{\theta}}\sqrt{\frac{\mu}{\epsilon}}\frac{jk\mathrm{I}\,\Delta z\, e^{-jkr}}{4\pi r}\sin\theta \tag{7.17b}$$

(radiation fields generated by a Hertzian dipole antenna)

At great distances from the antenna, the electromagnetic fields resemble those of a uniform plane wave. Comparing (7.17) with the uniform plane wave discussed in Chapter 3, we notice the following similarities and differences.

Similarities:

1. The **E** field, the **H** field, and the propagation path are perpendicular.

2. **E** and **H** are in phase and their magnitudes are related by the intrinsic impedance of the medium, $\eta = \sqrt{\mu/\epsilon}$.

Differences:

1. **E** and **H** decay as $1/r$ for a dipole field but remain constant for a uniform plane wave.
2. Field intensity is a function of θ for a dipole but is a constant for a uniform plane wave.
3. The constant phase surface is spherical for a dipole field but is planar for a uniform plane wave.
4. The wave radiated by the dipole travels in the *radial* direction with a velocity given by ω/k, whereas a uniform plane wave travels with the same velocity in a *fixed* direction.

Radiation Patterns

In the radiation field, the magnitude of $|\mathbf{E}|$ can be plotted at a constant radius r.

$$|E_\theta| = \sqrt{\frac{\mu}{\epsilon}}\frac{k|I|\Delta z}{4\pi r}|\sin\theta|$$

Thus, $|E_\theta|$ is a maximum at $\theta = \pi/2$ and zero at $\theta = 0$.

Figure 7.4a shows a 3-d plot of $|E_\theta|$, normalized so that the maximum value (on the $\theta = 90°$ plane) is equal to 1 V/m. Figure 7.4b shows the cross-sectional view of the field pattern at the x-z plane. The plots are called **radiation field patterns**. The radiation field pattern of a Hertzian dipole can be visualized as a doughnut without a hole.

The time-average power density radiated by the infinitesimal antenna in the radiation zone is readily calculated from

$$\langle S\rangle = \frac{1}{2}\text{Re}[\mathbf{E}\times\mathbf{H}^*] = \hat{r}\frac{1}{2}\sqrt{\frac{\mu}{\epsilon}}|H_\phi|^2 = \hat{r}\frac{\eta}{2}\left(\frac{k|I|\Delta z}{4\pi r}\right)^2\sin^2\theta \qquad (7.18)$$

(time-average Poynting generated by the dipole antenna)

Figure 7.5 plots the magnitude of $\langle S_r\rangle$ on the x-z and x-y planes. These patterns are known as the radiation power patterns.

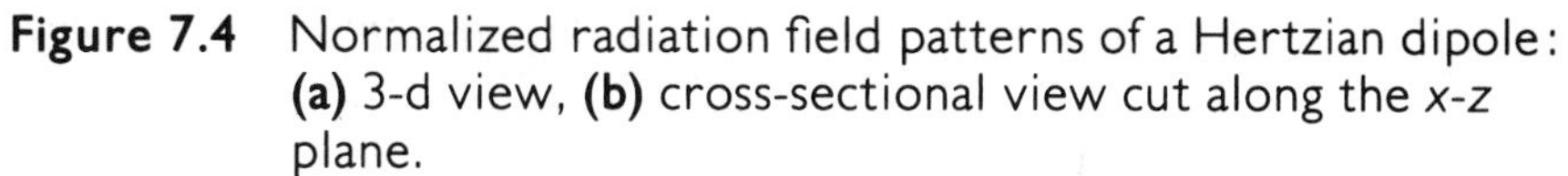

Figure 7.4 Normalized radiation field patterns of a Hertzian dipole: **(a)** 3-d view, **(b)** cross-sectional view cut along the x-z plane.

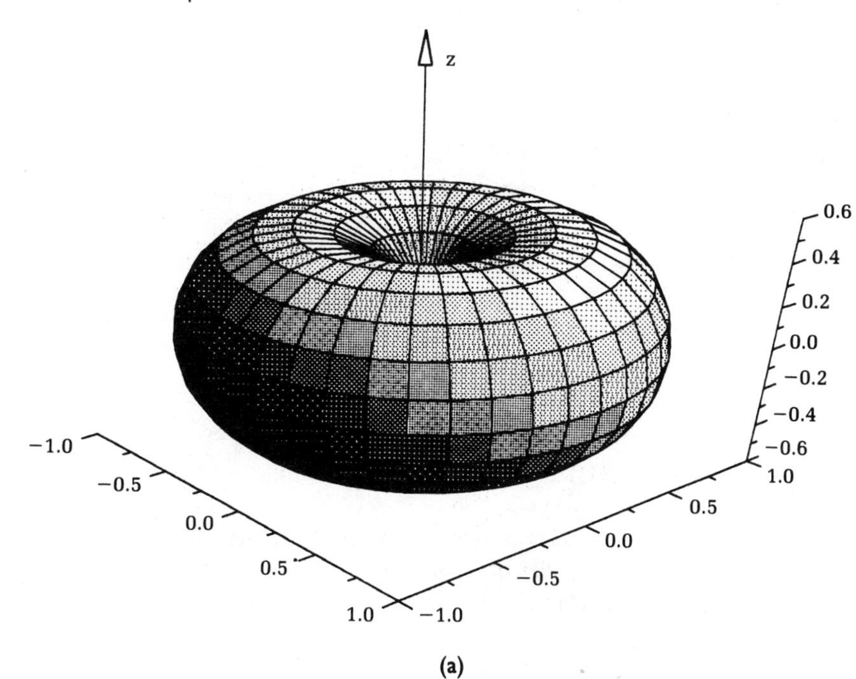

(a)

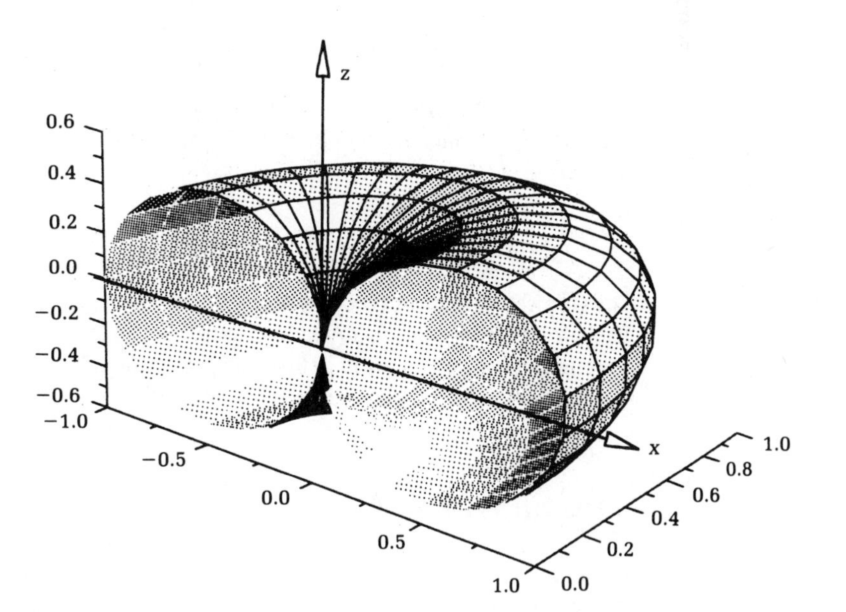

(b)

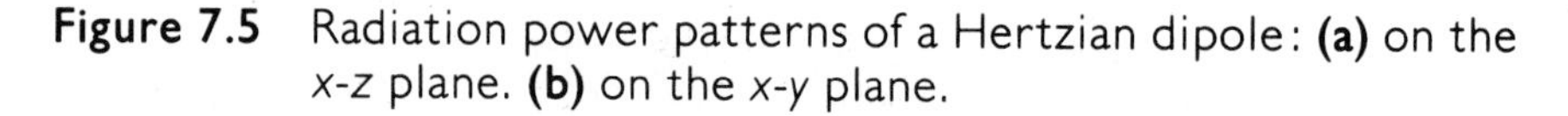

Figure 7.5 Radiation power patterns of a Hertzian dipole: **(a)** on the *x-z* plane. **(b)** on the *x-y* plane.

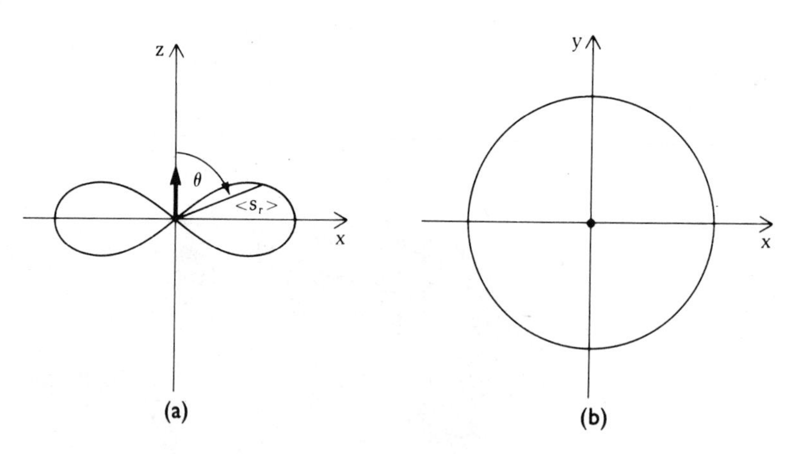

Antenna Gain

The total power radiated by the antenna is calculated by integrating the Poynting power density over the surface of the sphere at r. The integration surface element is $r^2 \sin\theta \, d\theta \, d\phi$. We find from (7.18) the total radiated power

$$\begin{aligned} P &= \int_0^{2\pi} d\phi \int_0^{\pi} d\theta r^2 \sin\theta \langle S_r \rangle \\ &= \frac{\eta}{2} \left| \frac{kI\,\Delta z}{4\pi} \right|^2 2\pi \int_0^{\pi} d\theta \sin^3\theta = \frac{4\pi}{3} \eta \left| \frac{kI\,\Delta z}{4\pi} \right|^2 \end{aligned} \tag{7.19}$$

The antenna **directive gain** $D(\theta, \phi)$ is defined as the ratio of the Poynting power density $\langle S_r \rangle$, which is a function of angle, over P divided by the area of the sphere:

$$D(\theta, \phi) = \frac{\langle S_r \rangle}{P/4\pi r^2} = \frac{3}{2} \sin^2\theta$$

Thus, in the direction $\theta = \pi/2$, the gain is 1.5, whereas at $\theta = 0$ the gain is zero. The plot of $D(\theta, \phi)$, as shown in Figure 7.6, is known as the gain pattern. The **directivity** of an antenna is defined as the value of the gain in the direction of its maximum value. Thus, the directivity of a Hertzian dipole is 1.5.

Capacitor-Plate Antennas

Rigorously, a short current-carrying wire cannot be regarded as a small dipole with constant current. The current on a short wire must fall to zero at both ends of the wire. One structure that carries almost uniform current is the capacitor-

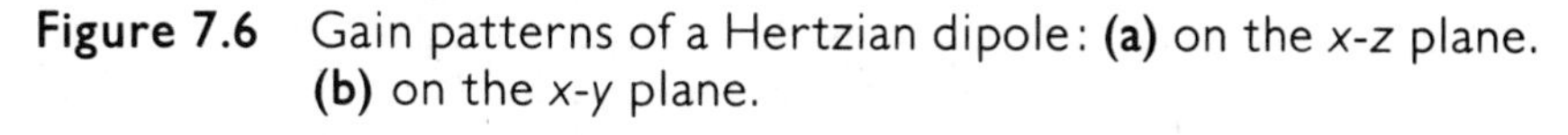
Figure 7.6 Gain patterns of a Hertzian dipole: **(a)** on the x-z plane. **(b)** on the x-y plane.

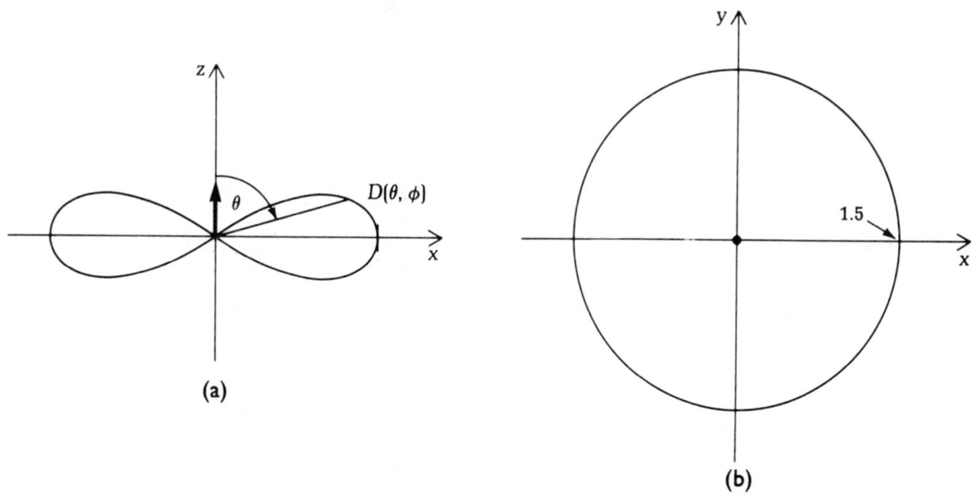

plate antenna, a short wire with two plates attached to its ends, as shown in Figure 7.7. The current on the wire is almost uniform, and the currents on the plates are radial and flow in opposite directions, so that they do not contribute significantly to the radiation fields. Therefore, for all practical purposes the capacitor-plate antenna can be regarded as a Hertzian dipole.

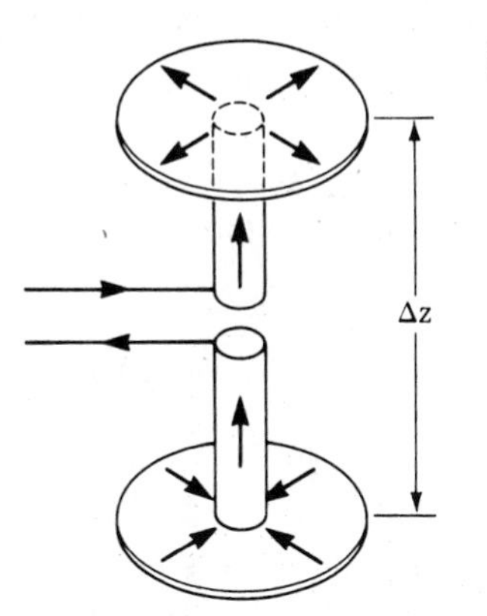

Figure 7.7 Capacitor-plate antenna. The plates serve as charge reservoirs so that a constant current can be maintained on the vertical section of the antenna.

Short Dipole Antennas

As mentioned earlier, the current on a short wire must fall to zero at both ends of the wire. Thus, a good approximation of the current on a short dipole antenna is a triangular distribution—that is, the current decreases linearly from the center toward both ends, as shown in Figure 7.8a. Equations (7.13) and (7.14) give an

Figure 7.8 These three antennas produce approximately the same electromagnetic fields: **(a)** short dipole, **(b)** Hertzian dipole, and **(c)** capacitor-plate antenna.

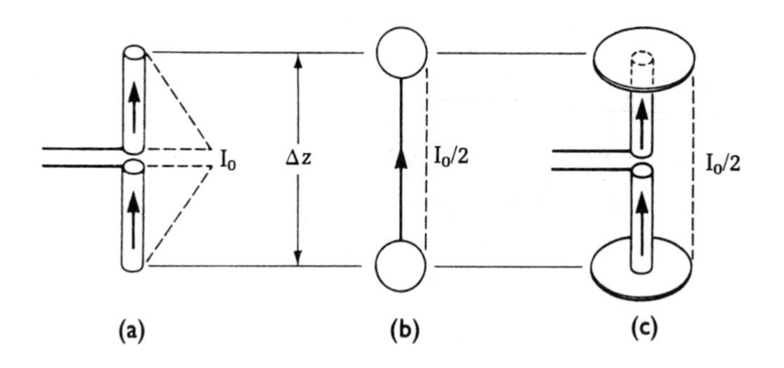

approximation of the electromagnetic fields produced by this antenna, provided that the I in these equations is replaced by 0.5I. In other words, the short dipole antenna shown in Figure 7.8a is equivalent to the short Hertzian dipole or the short capacitor-plate antenna shown in Figures 7.8b and 7.8c, respectively. The latter two carry half of the current at the center than does the short dipole. Analytical proof of this statement is given in the next section.

Example 7.1 *If the length and the radius of a wire antenna are much smaller than the wavelength, then (7.19) can be used to compute its total radiated power.*

Calculate the total power radiated by a small capacitor-plate antenna. Assume that the antenna transmits AM radio frequency of 1 MHz, $\Delta z = 1$ m, the radius of the wire $= 0.3$ cm, and $I = 1$ A.

Solution We first calculate the wavelength of the signal to see whether (7.19) is applicable.

$$\lambda = \frac{3 \times 10^8}{10^6} = 300 \text{ m}$$

Thus, the antenna is $\lambda/300$ long, and the radius of the brass wire is $1 \times 10^{-5}\lambda$. Both are much shorter than a wavelength, and consequently (7.19) is applicable.

$$P_t = \frac{120\pi \times (2\pi/300)^2}{12\pi} = 4.39 \text{ mW}$$

Example 7.2 *This example shows the difference and similarity between a short dipole and a capacitor-plate antenna. They have similar radiation patterns, but the* *radiated powers are different because the current on the short dipole tapers off toward the ends.*

The capacitor-plate antenna in the previous example is to be replaced by a short dipole. The frequency is still 1 MHz; the current at the center of the short dipole is 1 A; the radius of the wire is 0.3 cm; and the radiated power needs to be the same as that of the capacitor-plate antenna—that is, 4.39 mW. How long should the short dipole antenna be?

Solution The radiated power of a short dipole antenna is given by (7.19), with $I = 1/2$ A. Thus, to have the same power, Δz must be doubled—that is, the short dipole should be 2 m long, which is $\lambda/150$ long and still much shorter than λ.

Example 7.3

This example illustrates that the electric field of a dipole antenna varies with the distance r as well as with the angle θ.

Assume that the transmitting antenna of a station is a small vertical dipole on the surface of the ground. A person with a receiver is 5 km from the transmitter. What are the minimum distances (horizontal and vertical) that this person must move to detect a 3-dB change in the signal intensity?

Solution If the person has moved on the ground and directly away from the antenna, then using (7.17b) with $\theta = 90°$, we have

$$\left|\frac{E(r')}{E(5000)}\right| = \frac{5000}{r'}$$

To convert into dB, we take the logarithm of the ratio to base 10 and multiply the result by 20:

$$20 \log_{10}\left(\frac{5000}{r'}\right) = -3 \text{ dB}$$

Solve for r', and the result is $r' = 7070$ m.

If the person has moved vertically up h meters to detect a -3-dB signal difference, we have

$$20 \log_{10}\left|\frac{5000 \sin \theta'}{r' \sin 90°}\right| = -3$$

where $r' = (5000^2 + h^2)^{1/2}$ and $\sin \theta' = 5000/r'$.

Solving for h we obtain $h = 3211$ m. Note that the direction of the E field at a height h is not parallel to the E field at the ground. The direction $\hat{\boldsymbol{\theta}}$ at the ground is different from that at height h.

Electromagnetic Radiation from Cellular Telephones

Cellular telephones transmit and receive signals to and from substations that are located throughout a city. The substation that receives the clearest signal from the cellular phone puts the message into the local or long-distance telephone network. The personal communication services (PCS) network is similar to the cellular phone system. It provides both voice and data communications and is designed for worldwide coverage. The 800 to 3000 MHz frequency band has been assigned for these communication devices.*

Because cellular phones or PCS units must communicate with substations that are typically located several kilometers away, the radiation from these devices must be powerful enough to ensure signal clarity. A typical device radiates about 0.1 to 1.0 W. This level of power and the proximity of the antenna to the head has brought up safety concerns.**

Here we estimate the radiation power density from the antenna on a cellular phone, as shown in Figure 7.9(a). Figure 7.9(b) shows the model of our computation. The antenna is modeled as a short dipole, and the Poynting vector in the y direction is computed on the x-y plane y meters away from the antenna. The Poynting vector is computed instead of the E or H field because the safety guidelines shown in Figure 3.5 are expressed in terms of power at high frequencies.

The H and E fields of an infinitesimal antenna are given by Equations (7.13) and (7.14). The corresponding radiation fields are given by (7.17). To compute the Poynting vector, we can use either (7.13)–(7.14) or (7.17) and obtain the same results (see Problem 7.8). The Poynting vector is given by (7.18) and the total radiated power is given by (7.19). Since the total radiated power usually is provided by the phone manufacturer, we express the Poynting vector in terms of the total radiated power P:

$$\langle S \rangle = \frac{3P}{8\pi r^2} (\sin\theta)^2\, \hat{r}$$

The y component of the Poynting vector is

$$\langle S_y \rangle = \frac{3P}{8\pi r^2} (\sin\theta)^3 \sin\phi$$

To facilitate the plotting of $\langle S_y \rangle$ on the x-z plane at a fixed y location, the above expression is transformed into rectangular coordinates:

$$\langle S_y \rangle = \frac{3Py(x^2 + y^2)}{8\pi r^5}$$

where $r = (x^2 + y^2 + z^2)^{1/2}$. Figure 7.9(c) shows the plot of $\langle S_y \rangle$ on the x-z plane

---

* B. Z. Kobb, "Personal wireless," *IEEE Spectrum*, June 1993, p. 20.

** M. Fischetti, "The cellular phone scare," *IEEE Spectrum*, June 1993, p. 43.

Figure 7.9 Electromagnetic radiation from a cellular telephone. **(a)** A cellular phone. **(b)** The antenna on the cellular phone is modeled as a short dipole and $\langle S_y \rangle$, the y component of the time-average Poynting vector, is evaluated on a y = constant plane. **(c)** Plot of $\langle S_y \rangle$ on $y = 5$ cm plane.

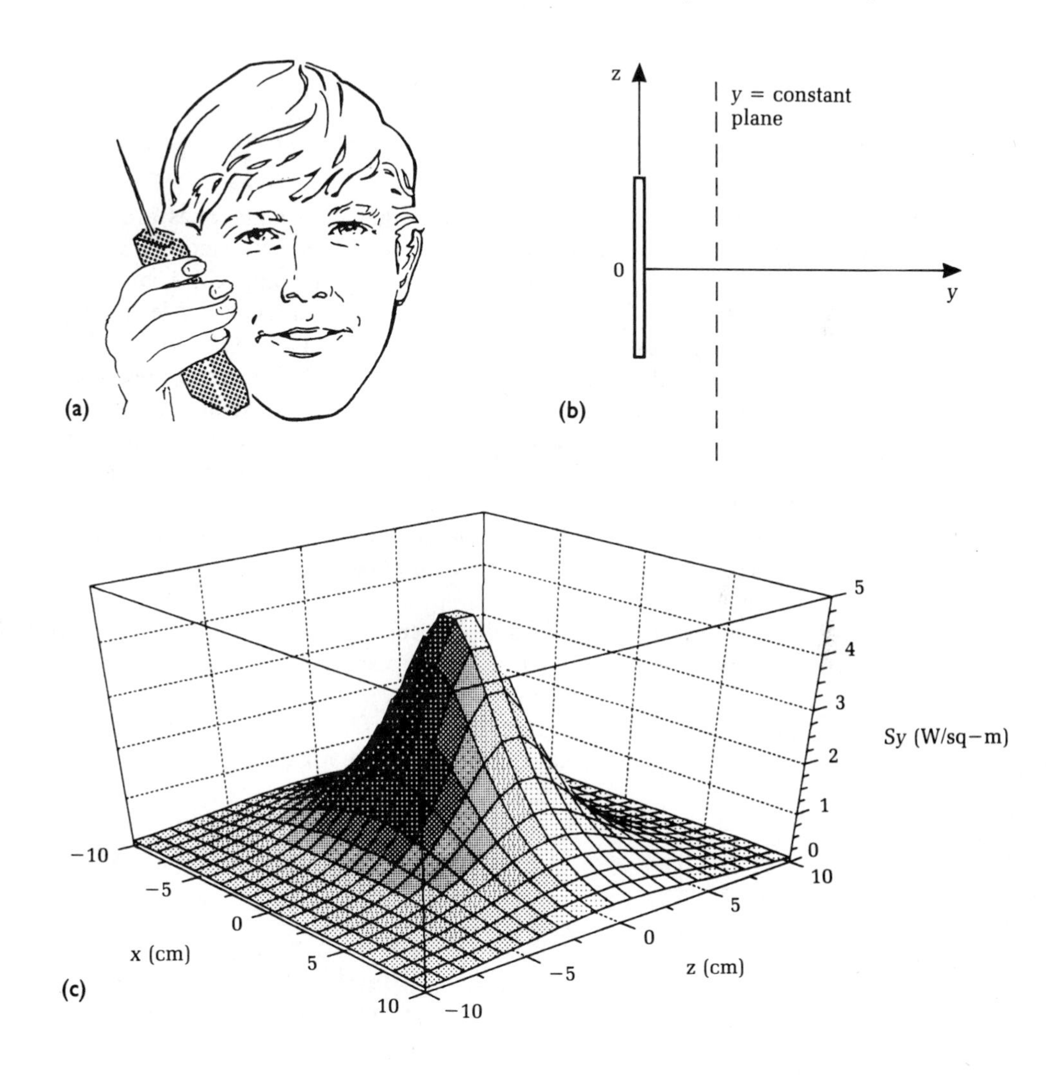

at $y = 5$ cm. It is assumed that the total radiated power of the cellular phone is 0.1 W.

From Figure 7.9(c), we see that the peak power density is approximately 4.8 W/m^2, or 0.48 mW/cm^2. It is just at the level recommended by the IEEE guidelines shown in Figure 3.5. It should be noted that several approximations and assumptions are made here. First, the antenna is assumed to be a short dipole. Second, it is assumed that the radiation power pattern is the same as that in free space. In reality, of course, a user's head interacts with the antenna and the pattern is different from the free-space pattern assumed here. Third, the radiated power is assumed to be 0.1 W and the spacing between the antenna and the head is 5 cm. All these factors affect the computed result. In practice, measurement of the radiation power should be carried out to ensure that the actual exposure intensity satisfies the IEEE guidelines.

7.3 LINEAR ANTENNAS

In the previous section we found that longer antennas are required if radiated power is to be increased. However, if the antenna length is greater than 0.1λ, the formula (7.19) for the radiated power is not applicable, and neither is the field equation (7.17). In practice, linear dipole antennas of length equal to 0.5λ are commonly used at VHF or higher frequencies. They are usually referred to as half-wave dipoles. The impedance of a half-wave dipole matches well with common transmission lines. In this section, we shall discuss the radiation fields produced by a dipole longer than 0.1λ, so that the result will be applicable to the half-wave dipole. Note that, although the radius of the antenna a does not appear in any of the field equations obtained in the previous section, it is implicitly assumed that the radius should be much smaller than a wavelength. For practical purposes, the restriction is usually taken to mean $a < 0.01\lambda$, which this section also assumes.

Figure 7.10a shows a dipole antenna that has a generator attached at $z = 0$, its lower end at $z = -h_1$, and its upper end at $z = h_2$. The radius of the dipole is

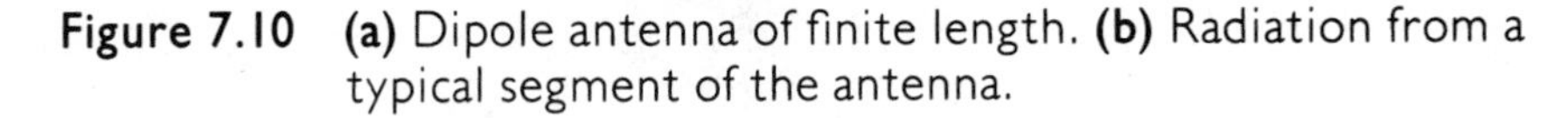
Figure 7.10 **(a)** Dipole antenna of finite length. **(b)** Radiation from a typical segment of the antenna.

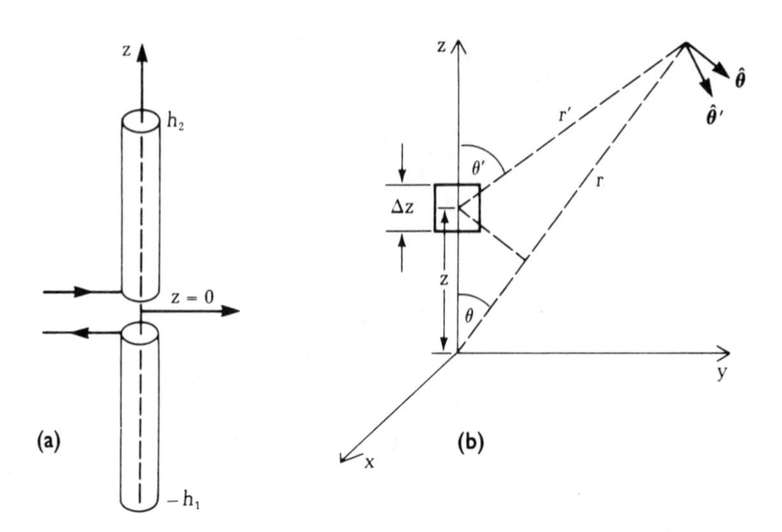

a. The total length $h_1 + h_2$ is comparable to a wavelength. As a consequence, the current on the antenna will be a function of z, rather than a constant. The current distribution is noted by $I(z)$.

To calculate the radiation fields, the antenna is divided into many infinitesimal elements. These elements are smaller than 0.1λ, so that radiation fields attributable to each element are obtained by using (7.17). For a typical segment Δz, as shown in Figure 7.10b, the electric field arising from this section is given by

$$\Delta\bar{\mathbf{E}} = \hat{\boldsymbol{\theta}}' \frac{j\mathrm{I}k\,\Delta z\eta e^{-jkr'} \sin\theta'}{4\pi r'} \tag{7.20}$$

At distances far away from the antenna, where r is much greater than the length of the antenna, the following approximations hold:

$$\theta \approx \theta' \tag{7.21a}$$

$$r' \approx r - z\cos\theta \tag{7.21b}$$

Using these approximations, we may write the electric field caused by typical segment of the antenna as

$$\Delta\mathbf{E} = \hat{\boldsymbol{\theta}} \frac{j\mathrm{I}k\,\Delta z\eta e^{-jkr} e^{jkz\cos\theta}}{4\pi r} \sin\theta \tag{7.22}$$

The total field is obtained by integrating over the source distribution which yields

$$\mathbf{E} = \hat{\boldsymbol{\theta}} \frac{jk\eta e^{-jkr}}{4\pi r} \sin\theta\, \mathrm{U}(\theta) \tag{7.23a}$$

(radiation fields of a linear antenna)

$$\mathbf{H} = \hat{\boldsymbol{\phi}} \frac{jke^{-jkr}}{4\pi r} \sin\theta\, \mathrm{U}(\theta) \tag{7.23b}$$

where

$$\mathrm{U}(\theta) = \int_{-h_1}^{h_2} \mathrm{I}(z) \exp(jkz\cos\theta)\, dz \tag{7.24}$$

(definition of the $\mathrm{U}(\theta)$ function)

Note that the second term in (7.21b) is neglected in the denominator of (7.22) but is kept in the exponential because $\exp(jkz\cos\theta)$ changes with z. In fact, if $z\cos\theta$ changes from 0 to $\lambda/2$, $\exp(jkz\cos\theta)$ changes from 1 to -1, which makes the integrand change sign accordingly and significantly affects the value of the integral.

Short Dipole Antennas

For a short dipole, it is a good approximation to assume that

$$I(z) = \begin{cases} I_0\left(1 - \dfrac{z}{h_2}\right) & \text{for } z \geq 0 \\ I_0\left(1 + \dfrac{z}{h_1}\right) & \text{for } z < 0 \end{cases}$$

Using (7.24) and $k(h_1 + h_2) \ll 1$, we get

$$U(\theta) \approx \int_{-h_1}^{h_2} I(z)\, dz = \frac{1}{2} I_0(h_1 + h_2)$$

Thus, the radiation fields of a short dipole antenna are the same as those of a short capacitor-plate antenna or a Hertzian dipole when in (7.17)

$$\Delta z = (h_1 + h_2) \quad \text{and} \quad I = \frac{1}{2} I_0$$

If both antennas are equally long and have equal current at driving points, the capacitor-plate antenna radiates four times as much power as does the dipole.

Center-Driven Half-Wave Dipoles

A center-driven half-wave dipole is a dipole antenna with $h_1 = h_2 = h = \lambda/4$. It may be the most commonly used radiating element. The current on this antenna is assumed to be cosinusoidal—that is,

$$I(z) = I_0 \cos(kz) \tag{7.25}$$

The half-wave dipole can be thought of as made from a two-wire transmission line. The transmission line is open-circuited at one end. The current on the transmission line is cosinusoidal, as shown in Figure 7.11a. Then, the transmission line is folded outward at $1/4\lambda$ from the end to form a half-wave dipole, as shown in Figure 7.11b. We assume that the current on the $1/4\lambda$ section is not affected by the folding of the transmission line. Note that the current on the transmission line must be modified because the load is now not infinity. It has a resistive part, which accounts for the radiation. Nevertheless, (7.25) is a good approximation for calculating radiation power. By using this current, $U(\theta)$ may be calculated from (7.24):

$$U(\theta) = \frac{I_0}{k} \int_{-\pi/2}^{\pi/2} d(kz) \cos(kz) \exp(jkz \cos\theta) = \frac{2I_0 \cos(\frac{1}{2}\pi \cos\theta)}{k \sin^2\theta} \tag{7.26}$$

The time-average Poynting vector is given by

$$\langle \boldsymbol{S} \rangle = \frac{I_0^2 \eta \cos^2(\frac{1}{2}\pi \cos\theta)}{8\pi^2 r^2 \sin^2\theta} \hat{\boldsymbol{r}} \tag{7.27a}$$

Figure 7.11 (a) Current distribution on an open-circuited, two-wire line. (b) Approximate current distribution on a half-wave dipole antenna.

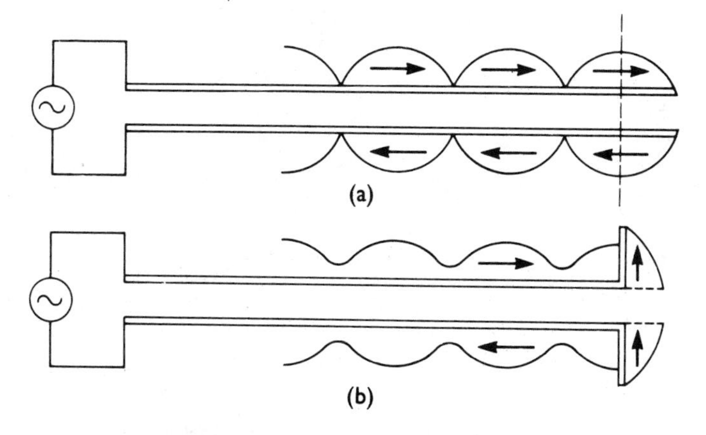

and the total radiated power is given by

$$P_t = \frac{I_0^2 \eta \, \text{Cin}(2\pi)}{8\pi} \tag{7.27b}$$

where $I_0 = |I|$ and $\text{Cin}(z)$ is the cosine integral for which numerical tables are available.* For $I_0 = 1$ A, the total power radiated by the half-wave dipole is 36.6 W [$\text{Cin}(2\pi) = 2.4376$].

The radiation patterns of an infinitesimal dipole, a capacitor-plate antenna, and a short dipole are almost the same when normalized, as shown in Figures 7.4–7.6. Figures 7.12a and 7.12b (p. 228) show the same pattern with a half-wave dipole. Although the radiation intensity of a half-wave dipole is much higher than that of short antennas with equal driving-point currents, the *normalized* radiation patterns of both antennas are similar.

Note that the radiation patterns of these dipoles are functions of θ. They radiate electromagnetic waves more effectively in some directions than in others. Hence, these antennas are called **directional antennas**. Because they are directional in elevation (θ) but not directional in azimuth, these dipole antennas are also called **omnidirectional antennas**.

Center-Driven Dipole Antennas

For a general linear antenna with $h_1 = h_2 = \ell/2$, the current may be written as

$$I(z) = I_0 \sin\left[k\left(\frac{\ell}{2} - |z|\right)\right] \tag{7.28}$$

---

* R. W. P. King, *Theory of Linear Antennas*, Cambridge: Harvard University Press, 1956, p. 560. $\text{Cin}(z) = \int_0^z [(1 - \cos u)/u] \, du$.

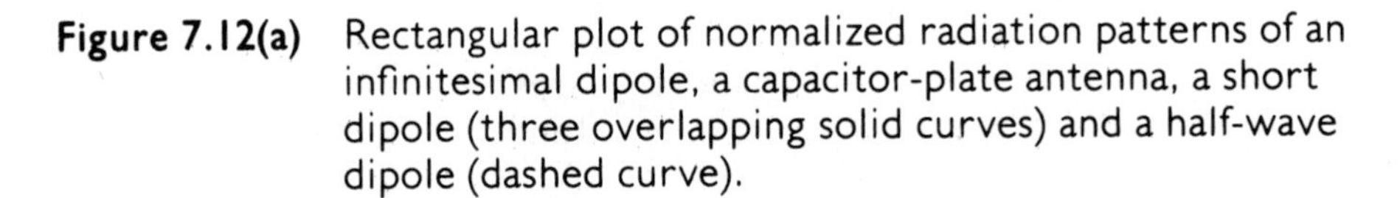
Figure 7.12(a) Rectangular plot of normalized radiation patterns of an infinitesimal dipole, a capacitor-plate antenna, a short dipole (three overlapping solid curves) and a half-wave dipole (dashed curve).

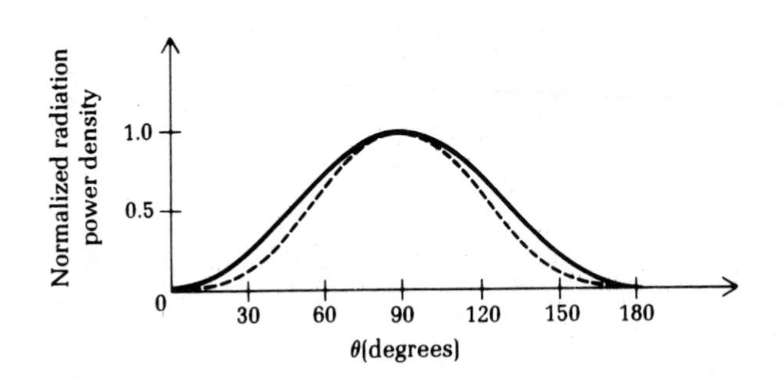

Figure 7.12(b) Polar plot of normalized radiation patterns in **(a)**.

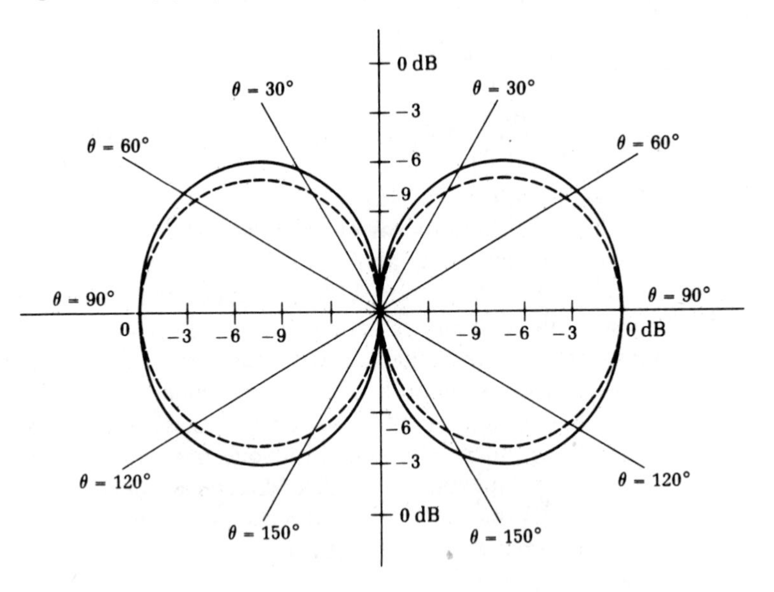

We can see that the current is zero at $z = \pm\ell/2$. Applying (7.24), we calculate

$$U(\theta) = I_0 \int_{-\ell/2}^{\ell/2} dz \sin\left[k\left(\frac{\ell}{2} - |z|\right)\right] \exp(jkz\cos\theta)$$

$$= I_0\left\{\int_{-\ell/2}^{0} dz \sin\left[k\left(\frac{\ell}{2} + z\right)\right] \exp(jkz\cos\theta)\right.$$

$$\left. + \int_{0}^{\ell/2} dz \sin\left[k\left(\frac{\ell}{2} - z\right)\right] \exp(jkz\cos\theta)\right\}$$

We thus find the electric field to be

$$\mathbf{E} = \hat{\theta}\eta \frac{j\mathrm{I}_0 e^{-jkr}}{2\pi r \sin\theta}\left[\cos\left(\frac{k\ell}{2}\cos\theta\right) - \cos\frac{k\ell}{2}\right] \tag{7.29}$$

As $\theta \to 0$, L'Hôpital's rule gives $|\mathrm{E}_\theta| \to 0$. We can plot field patterns for various linear antenna cases.

Case 1
For a half-wavelength dipole, $\ell = \lambda/2$ and $k\ell = \pi$. We have already discussed this case.

Case 2
For a linear antenna one and a half wavelengths long, $\ell = (3/2)\lambda$, we have

$$|\mathrm{E}_\theta| = \frac{\eta \mathrm{I}_0}{2\pi r \sin\theta}\left|\cos\left(\frac{3\pi}{2}\cos\theta\right)\right|$$

Plotting the radiation field pattern, we see that $\mathrm{E}_\theta = 0$ also at $\cos\theta = 1/3$. See Figure 7.13 for the pattern.

Figure 7.13 Radiation pattern of a dipole 1.5λ long.

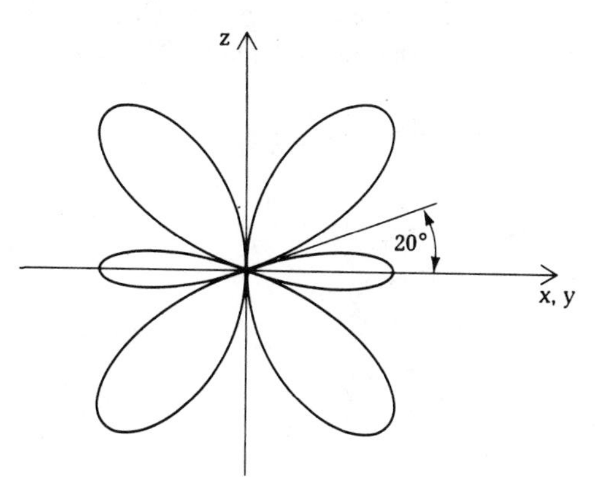

Case 3
For a linear antenna two wavelengths long, $\ell = 2\lambda$, we find

$$|\mathrm{E}_\theta| = \frac{\eta \mathrm{I}_0}{2\pi r \sin\theta}|\cos(2\pi\cos\theta) - 1|$$

The radiation field pattern (Figure 7.14) has nulls existing at $\theta = 0$, $\theta = \pi/2$.

Figure 7.14 Radiation pattern of a two-wavelength long dipole.

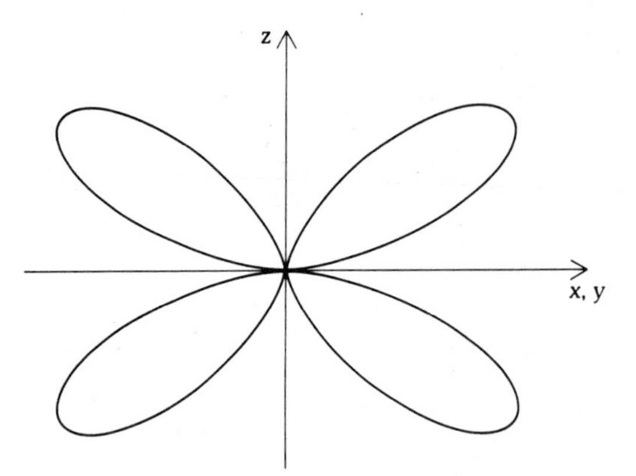

Note that these field patterns are calculated with the assumption that currents on the antennas are given by the simple sinusoidal equation (7.28). More accurate formulas for the currents and for the field patterns are available in the literature.*

Monopole Antennas

In practice, there are several ways of exciting a dipole antenna. In Figure 7.11 the dipole is formed by bending a two-wire transmission line. In Figure 7.15 a coaxial line is terminated at a ground plane with the inner conductor of the coaxial line extended above the ground plane. This construction is a monopole.

Figure 7.15 Monopole antenna fed by a coaxial line.

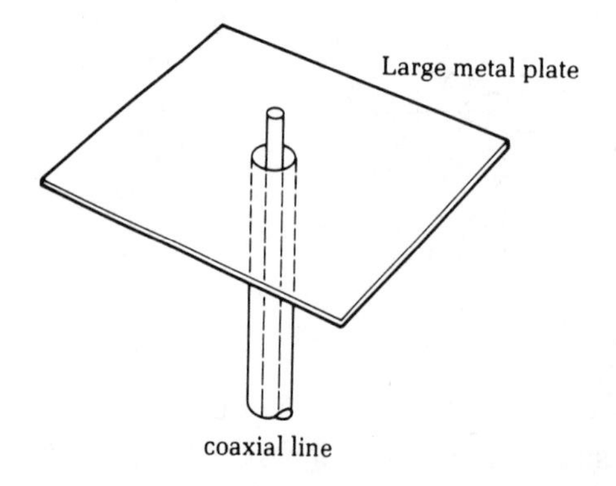

* L. C. Shen, T. T. Wu, and R. W. P. King, "A simple formula of current in dipole antennas," *IEEE Trans.* AP-16, No. 5 (September 1968): 542–547.

L. C. Shen, "The field pattern of a long antenna with multiple excitations," *IEEE Trans.* AP-16, No. 6 (November 1968): 643–646.

The radiation fields of a monopole of height h are approximately equal to those of a center-driven dipole of length $2h$. When the ground plane is infinitely large and infinitely conducting, this relationship is exact, according to image theory, which will be developed in Chapter 11. However, the input impedance of a monopole, defined as the ratio of the voltage and the current at the driving point, is one-half that of a corresponding dipole.

Example 7.4 *This example shows how to relate the E field strength at certain distance from the antenna to the power radiated by it. Also, it reiterates the fact that the monopole antenna radiates half the power as the dipole antenna does when the currents on both antennas are equal.*

According to the United States Federal Communications Commission Radio Broadcast Standards, a minimum signal strength of 25 mV/m must be maintained in the commercial and industrial districts of a city. Suppose that the broadcasting antenna is a quarter-wave monopole over the ground. Calculate the total radiated power that the antenna must produce. Assume that the maximum distance between the antenna and the city limits is 15 km.

Solution From (7.26), we find that

$$U(90°) = \frac{2I_0}{k}$$

Substituting this value into (7.23a) yields

$$|E(90°)| = \frac{2\eta I_0}{(4\pi \times 15{,}000)}$$

Equating, this result to 25 mV yields $I_0 = 6.25$ A.

Then, substituting I_0 in (7.27) times a factor of 0.5 (monopole), we find the answer:

$$P_t = 714 \text{ W}$$

7.4 ARRAY ANTENNAS

An **array antenna** is a system that comprises a number of radiating elements, generally similar, that are arranged and excited to obtain directional patterns. In this section we shall consider a simple case, one in which the radiators are all identical, are all oriented in the $\hat{z}$ direction, are all aligned on the y axis, and are all equally spaced, as shown in Figure 7.16. Furthermore, we assume that the elements are driven by currents with equal magnitudes and with progressive phase shifts—that is, that the current on the first element is I(z), on the second I(z)$e^{j\psi}$, on the third I(z)$e^{i2\psi}$, and so on. An array that has all of these properties is called a **uniform linear array.**

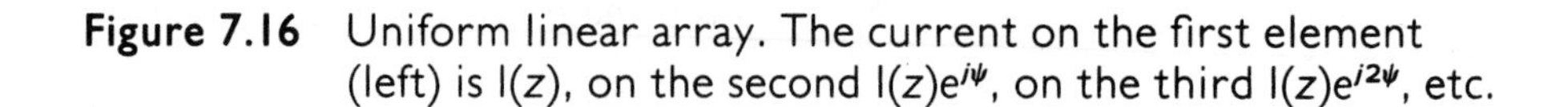

Figure 7.16 Uniform linear array. The current on the first element (left) is I(z), on the second I(z)$e^{j\psi}$, on the third I(z)$e^{j2\psi}$, etc.

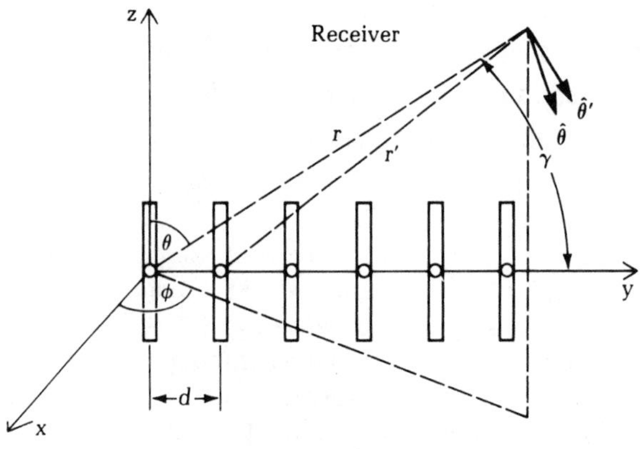

Using the principle of superposition, we can obtain the total E field of the array antenna by adding the field contributed by each individual dipole. The radiation field from element 1, according to (7.23a), is

$$\mathbf{E}_1 = \hat{\boldsymbol{\theta}} \frac{jk\eta e^{-jkr}}{4\pi r} \sin\theta \mathrm{U}(\theta) \tag{7.30}$$

Similarly, the radiation field from the second element is

$$\mathbf{E}_2 = \hat{\boldsymbol{\theta}}' \frac{jk\eta e^{-jkr'}}{4\pi r'} \sin\theta' \, e^{j\psi} \mathrm{U}(\theta') \tag{7.31}$$

Because distances r and r' are approximately equal at distances far from the origin, we use the following approximation:

$$\theta' \approx \theta$$

$$r' \approx r - d(\hat{y} \cdot \hat{r}) = r - d\cos\gamma = r - d\sin\theta\sin\phi \tag{7.32}$$

The angle γ is the angle between the r and the y axis. Substituting (7.32) in the numerator and letting $r' = r$ in the denominator of (7.31), we obtain

$$\mathbf{E}_2 = \hat{\boldsymbol{\theta}} \frac{jk\eta e^{-jkr}}{4\pi r} e^{jkd\sin\theta\sin\phi} \sin\theta e^{j\psi} \mathrm{U}(\theta)$$

$$= \mathbf{E}_1 \exp[j(kd\sin\theta\sin\phi + \psi)] \tag{7.33}$$

Thus, $\mathbf{E}_2$ differs from $\mathbf{E}_1$ only by a phase factor $\exp[j(kd \sin\theta \sin\phi + \psi)]$. The first term in the phase factor is due to the fact that the direct electromagnetic wave from element 2 travels a shorter distance to the receiver than the wave from element 1. The second term is due to the difference in phase between the currents on elements 1 and 2.

Generalizing this procedure, we can calculate the E field of a uniform linear array of N elements as follows:

$$\mathbf{E}_t = \hat{\theta}\frac{jk\eta e^{-jkr}}{4\pi r}\sin\theta \mathrm{U}(\theta)\{1 + \exp[j(\psi + kd\cos\gamma)] + \cdots + \exp[j(N-1)(\psi + kd\cos\gamma)]\} \tag{7.34}$$

where $\cos\gamma = \sin\theta\sin\phi$ and where

$$\mathrm{U}(\theta) = \int_{-h}^{h} \mathrm{I}(z)\exp(jkz\cos\theta)\,dz$$

The terms in the brackets of (7.34) are of the form

$$\sum_{n=0}^{N-1} x^n = \frac{1 - x^N}{1 - x}$$

with the identification $x = \exp[j(\psi + kd\cos\gamma)]$. Thus, the total E field of the array is the product of two terms:

$$\boxed{\mathbf{E}_t = [\mathbf{E}_e(\theta)][\mathrm{F}(\theta, \phi)]} \quad \textbf{(total E field of an array)} \tag{7.35}$$

The first term on the right side of (7.35) is the field of a center-driven dipole, which is given by (7.23a) or (7.30). The second term is called the **array factor,** given by

$$\mathrm{F}(\theta, \phi) = \frac{1 - \exp[jN(\psi + kd\cos\gamma)]}{1 - \exp[j(\psi + kd\cos\gamma)]} \tag{7.36}$$

The array factor is the sum of the phases of the E fields radiated by the elements of the array. It also appears in the expression of the H field of the array. The magnitude of the electric field

$$|\mathrm{E}_\theta| = |\mathrm{E}_e||\mathrm{F}|$$

where

$$|\mathrm{F}| = \left|\frac{1 - \exp[jN(kd\cos\gamma + \psi)]}{1 - \exp[j(kd\cos\gamma + \psi)]}\right|$$

Noting that

$$|1 - e^{jx}| = \left|2j\sin\frac{x}{2}e^{jx/2}\right| = 2\left|\sin\frac{x}{2}\right|$$

we find

$$|\mathrm{F}| = \left| \frac{\sin N\left(\dfrac{kd\cos\gamma + \psi}{2}\right)}{\sin\left(\dfrac{kd\cos\gamma + \psi}{2}\right)} \right| \quad \textbf{(array factor)} \qquad \textbf{(7.37)}$$

where N is the number of elements in the array, d is the distance between two adjacent elements, $\cos\gamma = \sin\theta\sin\phi$, and ψ is the relative phase difference between two adjacent elements (see Figure 7.16).

Note that the maxima of $|\mathrm{F}|$ occur at $kd\cos\gamma + \psi = 0, 2\pi$, etc., and have magnitude of N. Nulls of $|\mathrm{F}|$ occur at

$$N\frac{kd\cos\gamma + \psi}{2} = m\pi$$

where m is an integer not equal to 0, N, $2N$, etc.

Example 7.5 *Various radiation patterns are produced by simply varying the distance between two dipole antennas or the relative phase between the currents on them.*

We now consider the radiation properties in the $\theta = \pi/2$ plane of two dipoles separated by a distance d.

Case I
$d = \lambda/2, \quad kd = \pi, \quad \psi = 0$

Whether the dipole antennas are short dipoles or half-wave dipoles, the radiation pattern on the $\theta = \pi/2$ plane is the same: a constant independent of ϕ. Therefore $\mathbf{E}_e(\theta)$ in (7.35) is a constant independent of ϕ and the variation of $\mathbf{E}_t$ in the x-y plane is determined by $\mathrm{F}(\theta, \phi)$, which is given by (7.37). Now, substituting $N = 2$, $kd = \pi$, $\psi = 0$, and $\cos\gamma = \sin\theta\sin\phi = \sin\phi$ in (7.37) yields

$$|\mathrm{E}_\theta| \sim \left|\cos\left(\frac{\pi}{2}\sin\phi\right)\right|$$

We may easily understand this field pattern (see Figure 7.17). At large distances, where the radiation approximation is valid, the phase fronts look almost like plane waves. We note that in the $\hat{x}$ direction, the two waves will arrive in phase and add to each other. On the other hand, along the $\hat{y}$ direction the two antennas have waves which are 180° out of phase and cancel each other.

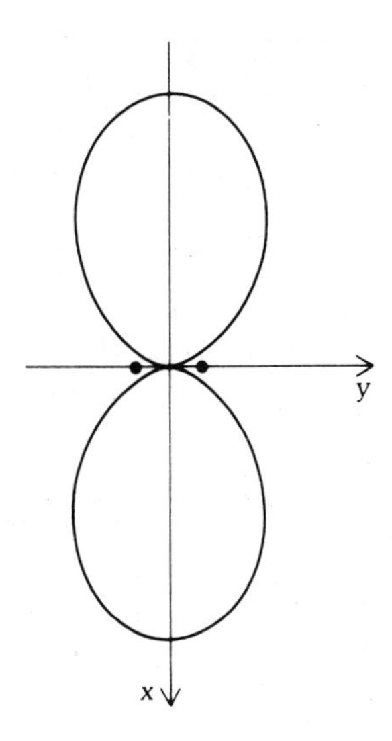

Figure 7.17 Radiation pattern of a two-element array with $d = \lambda/2$, $\psi = 0$.

Case 2
$d = \lambda/2, \psi = \pi$

$$|\mathrm{E}_\theta| \sim \left|\cos\left(\frac{\pi}{2}\sin\phi + \frac{\pi}{2}\right)\right|$$

This pattern may be described using the same physical reasoning as in Case 1. (See Figure 7.18.)

Figure 7.18 Radiation pattern of a two-element array with $d = \lambda/2$, $\psi = \pi$.

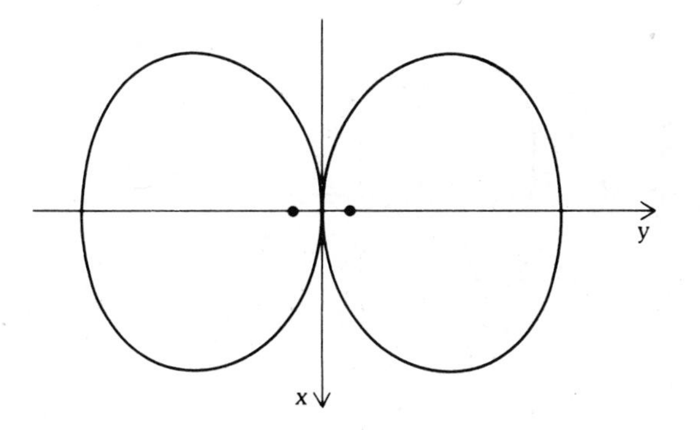

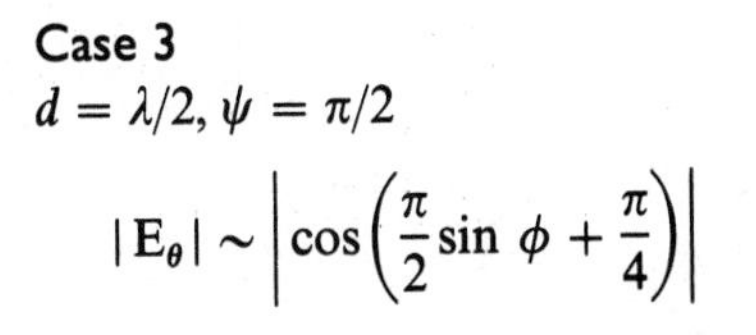

Case 3
$d = \lambda/2, \psi = \pi/2$

$$|\mathrm{E}_\theta| \sim \left|\cos\left(\frac{\pi}{2}\sin\phi + \frac{\pi}{4}\right)\right|$$

It becomes apparent that the direction of the field maximum may be varied merely by changing the antenna phase. (See Figure 7.19.)

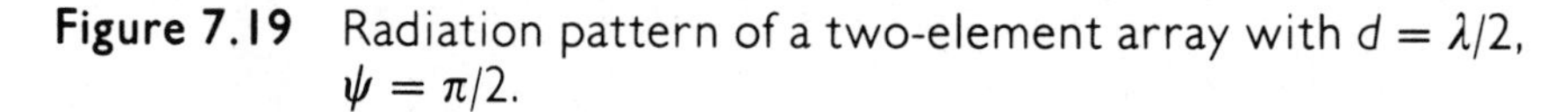

Figure 7.19 Radiation pattern of a two-element array with $d = \lambda/2$, $\psi = \pi/2$.

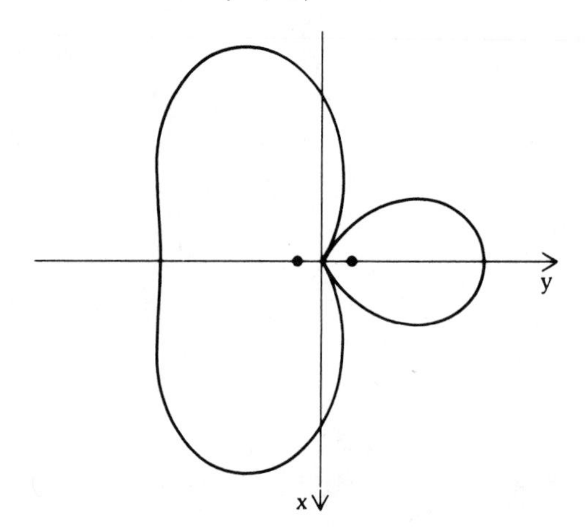

Case 4

$d = \lambda, \psi = 0$

$$|E_\theta| \sim |\cos(\pi \sin \phi)|$$

(See Figure 7.20.)

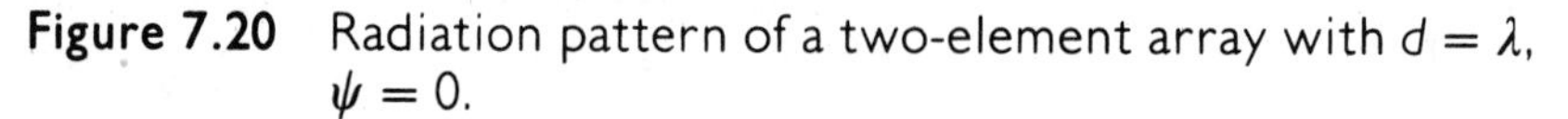

Figure 7.20 Radiation pattern of a two-element array with $d = \lambda$, $\psi = 0$.

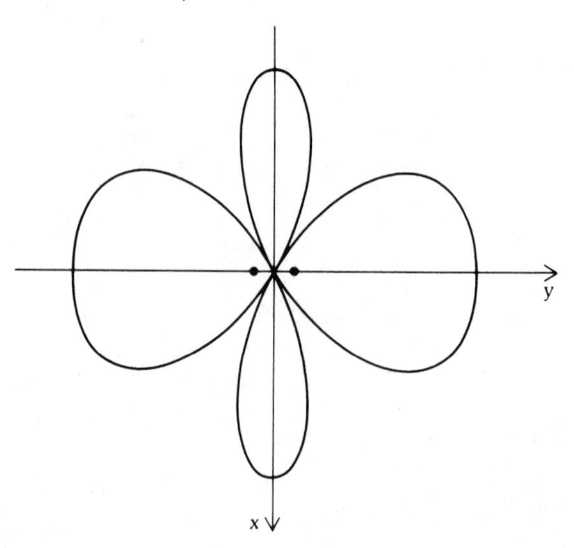

Case 5
$d = \lambda, \psi = \pi$

$$|E_\theta| \sim \left|\cos\left(\pi \sin\phi + \frac{\pi}{2}\right)\right|$$

(See Figure 7.21.)

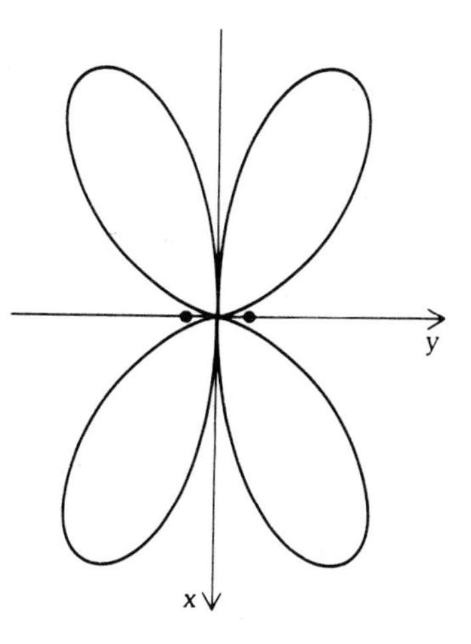

Figure 7.21 Radiation pattern of a two-element array with $d = \lambda$, $\psi = \pi$.

Example 7.6 *As the number of elements in an array is increased, we see from this example that the radiation pattern of the array becomes more complicated.*

Consider a four-element array and let $\psi = 0$. We have on $\theta = 90°$ plane:

$$|E_\theta| \sim \left|\frac{\sin(2kd \sin\phi)}{\sin[(kd/2)\sin\phi]}\right|$$

Solution We may construct a short table of $|E_\theta|$ *vs.* ϕ for $kd = \pi$.

ϕ	$\|E_\theta\|$
0	4
$\pi/6$	0
$\pi/2$	0

The pattern will look like Figure 7.22.
For $kd = 3\pi/2$ (Figure 7.23),

$$|E_\theta| \sim \left|\frac{\sin(3\pi \sin\phi)}{\sin[(3\pi/4)\sin\phi]}\right|$$

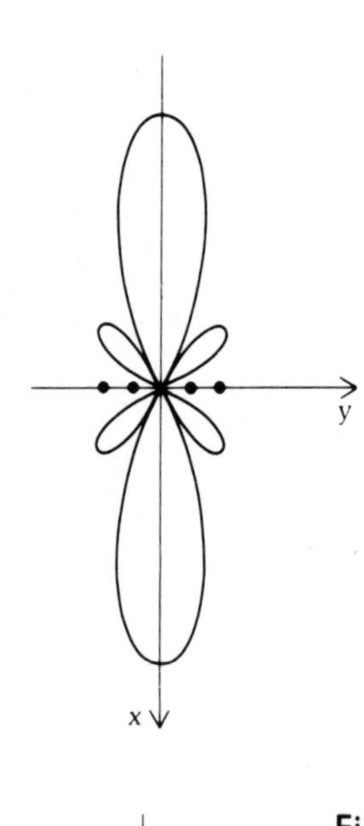

Figure 7.22 Radiation pattern of a four-element array with $d = \lambda/2$, $\psi = 0$.

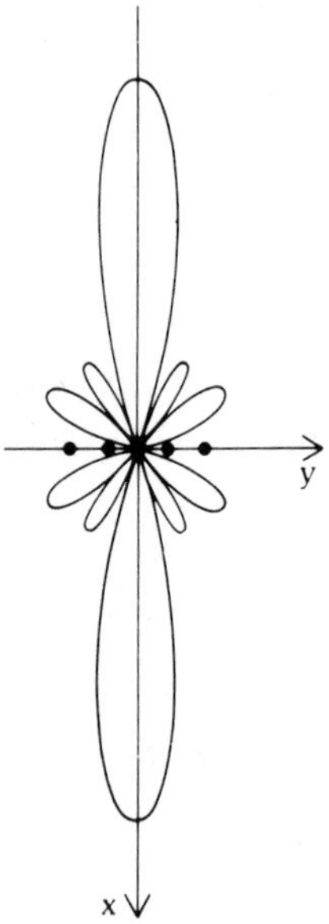

Figure 7.23 Radiation pattern of a four-element array with $d = 0.75\lambda$, $\psi = 0$.

Pattern Multiplication

Consider again the four-element array shown in Figure 7.24a. Suppose we treat elements 1 and 2 as a single unit and likewise dipoles 3 and 4. We denote the unit as $\otimes$, as shown in Figure 7.24b.

The unit has a figure eight radiation pattern similar to that of Figure 7.17. The group of four elements may be replaced by two units separated by λ, as

Figure 7.24 **(a)** A four-element array with $d = \lambda/2$. **(b)** Combine two elements into a unit. **(c)** A four-element array is equivalent to a two-element array consisting of two units like the one shown in (b).

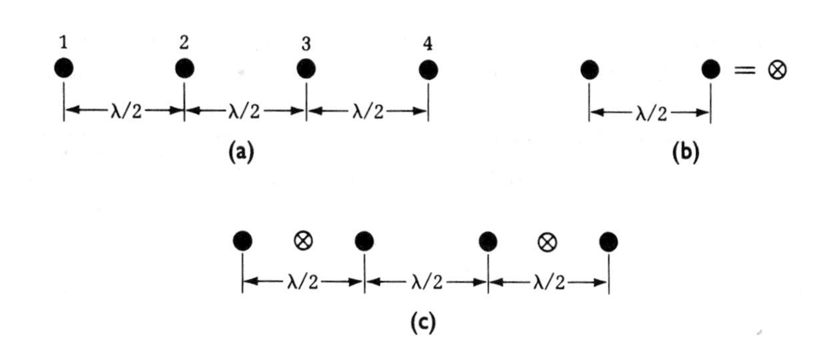

shown in Figure 7.24c. If the unit had an isotropic pattern, the group pattern of two isotropic units separated by λ would be that shown in Figure 7.20, which is reproduced in Figure 7.25. However, each ⊗ element does not have an isotropic pattern. To obtain the true pattern, we must multiply the true radiation pattern that displays the characteristics of each unit by the isotropic group pattern—that is,

$$\text{(group pattern)} \times \text{(unit pattern)} = \text{(resultant pattern)}$$

which is the pattern shown in Figure 7.25. This pattern is identical to that in Figure 7.22. This technique is a quick and useful way of arriving at resultant radiation patterns.

Figure 7.25 Pattern multiplication.

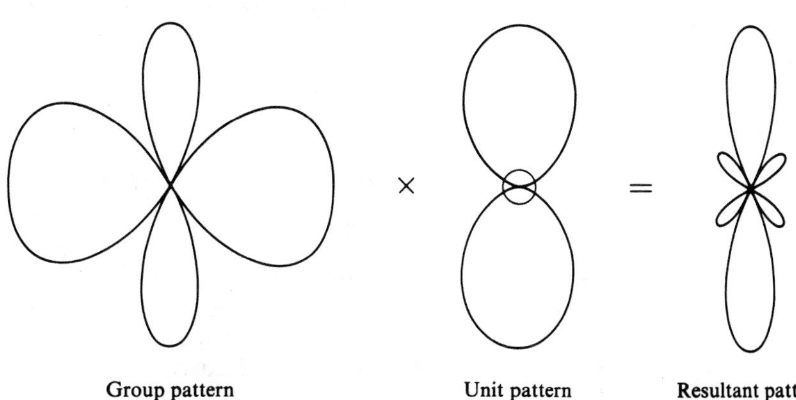

Example 7.7 *This example shows how to apply the pattern multiplication principle to obtain the radiation pattern of a four-element array by multiplying two patterns of two-element arrays.*

Consider a four-element array shown in Figure 7.26a. The currents on the elements are all in phase. Find the array pattern using the multiplication method.

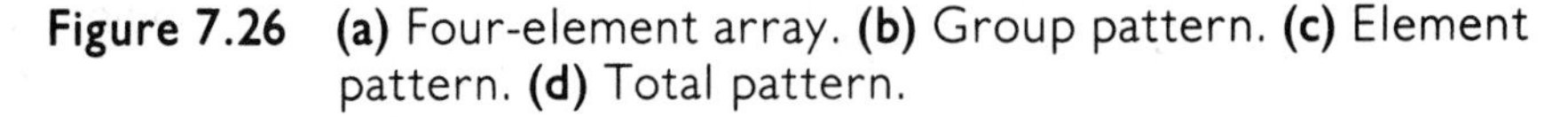

Figure 7.26 **(a)** Four-element array. **(b)** Group pattern. **(c)** Element pattern. **(d)** Total pattern.

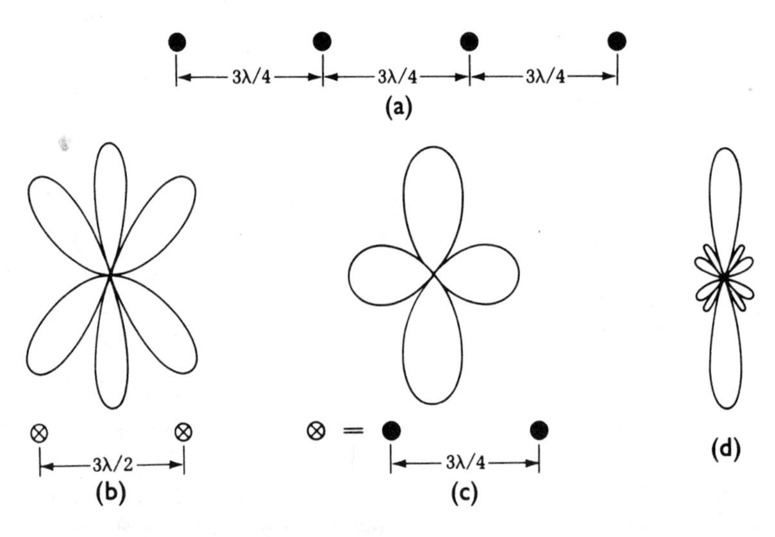

Solution The array shown in Figure 7.26a is equivalent to the array shown in Figure 7.26b with the element of (b) shown in (c). The corresponding patterns of the two-element arrays with $d = 3\lambda/2$ and $3\lambda/4$ are shown in Figure 7.26b and c, respectively. Multiplying these two patterns yields the resultant pattern shown in Figure 7.26d, which is identical to that shown in Figure 7.23.

Binomial Arrays

Many applications require that the antenna field pattern have only major lobes with minimum side lobes. For example, in Figure 7.26, we say that the pattern has two major lobes and eight side lobes. The side lobes in the antenna pattern may confuse radar return and interfere with other communication systems. We shall illustrate a way to obtain a pattern with no side lobes.

Let us begin with a two-element array with the pattern shown in Figure 7.27a.* If we put two such arrays $\lambda/2$ apart, we obtain the resultant field pattern

* An array with this kind of pattern is called a **broadside array**.

Figure 7.27 **(a)** A unit consisting of a two-element array has a "figure 8" pattern. **(b)** Two such units form an array that has a sharper "figure 8" pattern. **(c)** The array in (b) is equivalent to a three-element array with currents equal to 1, 2, 1 units, respectively, on these elements. **(d)** A unit representing the three-element array shown in (c). **(e)** The radiation pattern of the two-unit array. **(f)** The four-element binomial array equivalent to the one shown in (e). **(g)** A five-element binomial array.

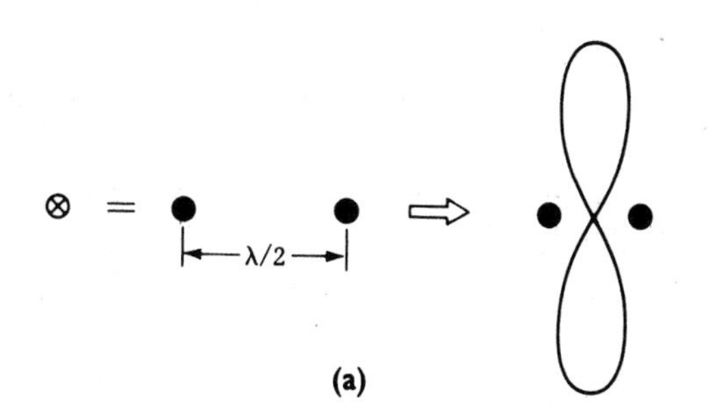

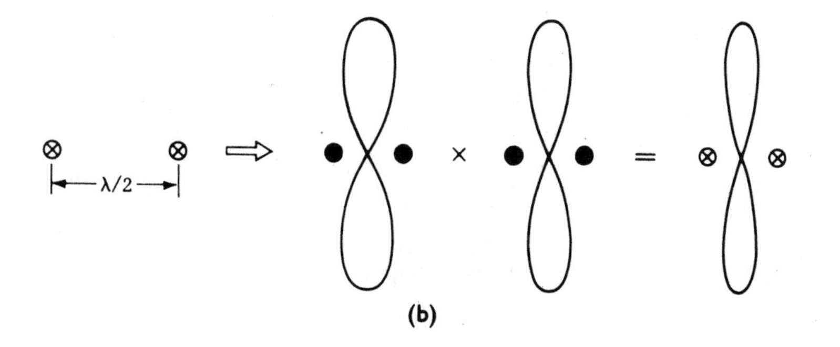

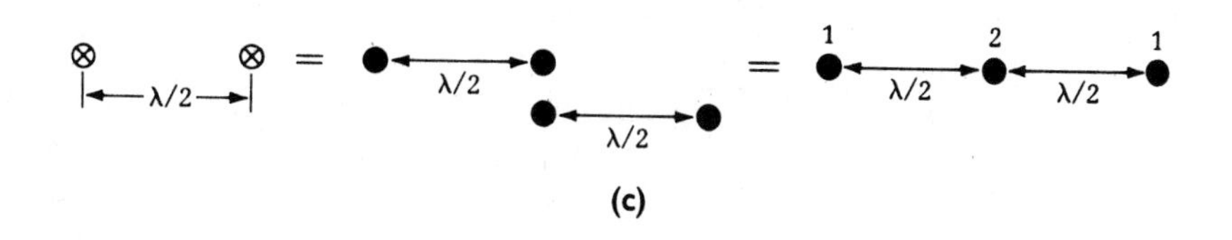

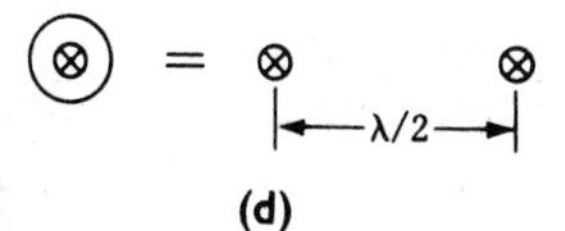

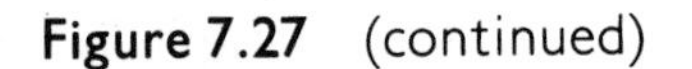
Figure 7.27 (continued)

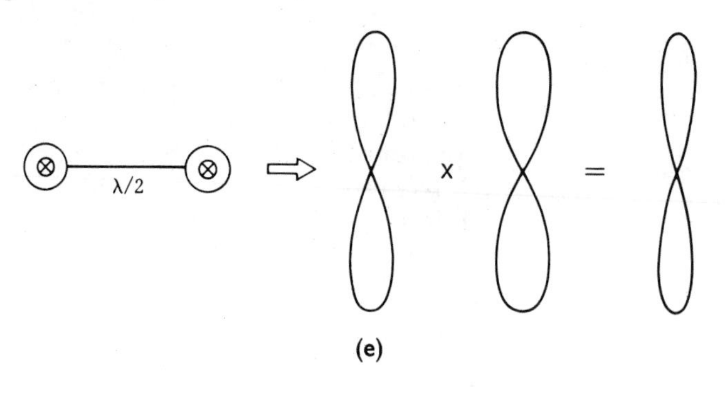

(e)

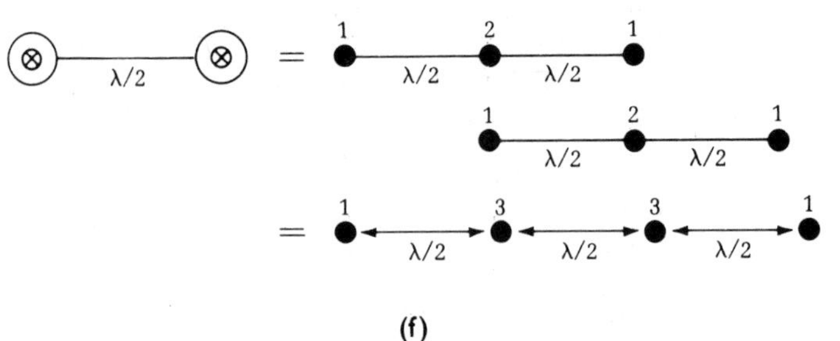

(f)

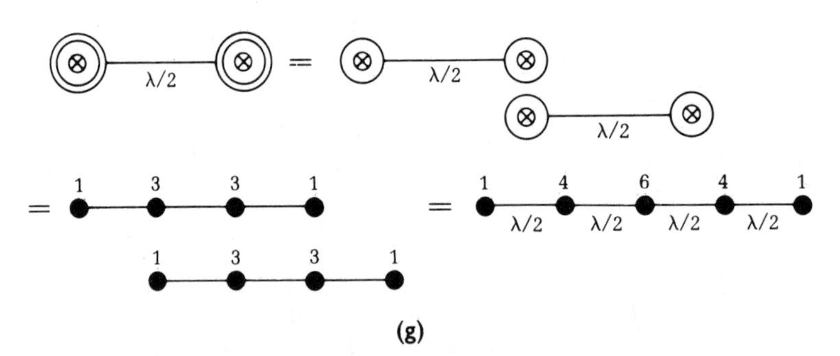

(g)

by the method of pattern multiplication shown in Figure 7.27b. Note that the resulting pattern has no side lobes. The constituents of this array are shown in Figure 7.27c, which has three elements with amplitudes 1, 2, 1, and in-phase currents.

This process can be continued. For example, if we put two such three-element arrays $\lambda/2$ apart, as shown in Figure 7.27d, we obtain a radiation pattern that still has no side lobes, as illustrated in Figure 7.27e. This array consists of four elements with amplitudes 1, 3, 3, 1, respectively, and all currents in phase. Figure 7.27f shows the procedure for finding the current distribution. The next larger array with no side lobes will look like that shown in Figure 7.27g. At this point we can conclude that the array that possesses no side lobes consists of dipoles whose relative amplitudes are given by the binomial coefficients. It is called the **binomial array**.

Radiation from a Transmission Line

A transmission line does not radiate power if it is completely shielded from the outside, as is the coaxial line with both generator and load sealed in a tube. If it is not completely enclosed by conducting walls, a transmission line does radiate power.

For the two-wire line shown in Figure 7.28, with the currents equal in magnitude but opposite in direction, it is reasonable to expect that radiation from both lines cancels out. This cancellation occurs only when the separation between the wires is very small; otherwise, radiation fields arising from the transmission line are appreciable. Here we shall make a quantitative analysis of the radiation of a two-wire transmission line.

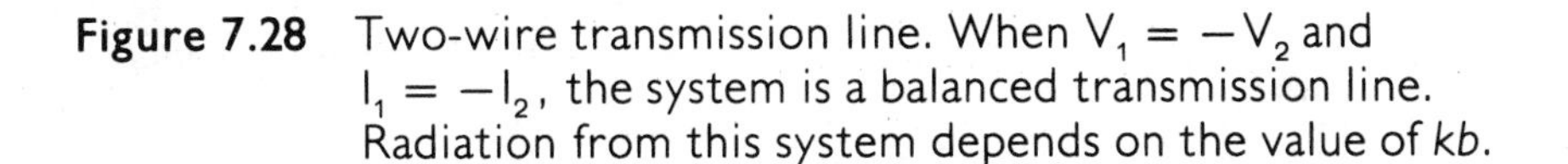

Figure 7.28 Two-wire transmission line. When $V_1 = -V_2$ and $I_1 = -I_2$, the system is a balanced transmission line. Radiation from this system depends on the value of kb.

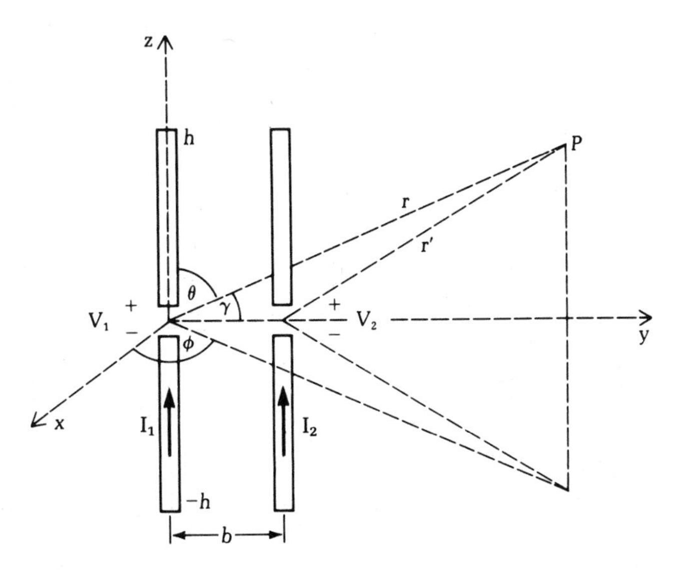

Let us consider the two-wire transmission line shown in Figure 7.28. This system can be considered a two-element array with $I_2 = I_1 e^{j\pi}$. Let us further assume that the transmission line is $1/2\lambda$ long, so that the current on the line is given by

$$I_1(z) = I_0 \cos kz \tag{7.38}$$

Using (7.35), we have

$$\mathbf{E}_t = [\mathbf{E}_e(\theta)][\mathbf{F}(\theta, \phi)] \tag{7.39}$$

where $\mathbf{E}_e(\theta)$ is given by (7.23a) and (7.26) and

$$F(\theta, \phi) = \frac{1 - e^{j2p}}{1 - e^{jp}} = 1 + e^{jp} \tag{7.40}$$

$$p = \pi + kb \sin \theta \sin \phi$$

For a transmission line, the separation b is usually small compared with λ; that is,

$$kb \ll 1$$

Under this condition, the array factor may be approximated:

$$F(\theta, \phi) = 1 - \exp(jkb \sin \theta \sin \phi) \approx -jkb \sin \theta \sin \phi \tag{7.41}$$

The radiation fields are then given by

$$\mathbf{E}_t = \hat{\theta}\frac{jk\eta e^{-jkr}}{4\pi r}(-jkb \sin^2 \theta \sin \phi)U(\theta) \tag{7.42a}$$

$$\mathbf{H}_t = \hat{\phi}\frac{jke^{-jkr}}{4\pi r}(-jkb \sin^2 \theta \sin \phi)U(\theta) \tag{7.42b}$$

$$U(\theta) = \frac{2I_0 \cos(\frac{1}{2}\pi \cos \theta)}{k \sin^2 \theta} \tag{7.43}$$

It can be shown that the total radiated power is given by

$$\boxed{P_t = \frac{\eta}{8\pi}|I_0 kb|^2} \quad \textbf{(radiated power of a 1/4-}\boldsymbol{\lambda}\textbf{ two-wire line)} \tag{7.44}$$

Although this result is for the two-wire line shown in Figure 7.28, with $h = 0.25\lambda$, the qualitative picture is the same for other types of transmission lines.* In general it is true that, to keep the radiated power small, the separation of the lines must be very small compared with the wavelength. As the separation increases, the radiated power increases rapidly.

---

* R. W. P. King and C. W. Harrison, "Transmission-line missile antennas," *IEEE Trans.* AP-8 (January 1960): 88–90.

Josephson Junctions

The device called the Josephson junction* has thus far achieved the shortest switching time: approximately 10 ps (10×10^{-12} s).** A Josephson junction is made of two bulk superconductors separated by a thin film of insulating barrier. A wide-band system is required to cope with a Josephson junction device. In fact, the frequency spectrum of a 10 ps square-wave pulse extends well over 100 GHz. To keep the radiation of a transmission line minimal, the separation between the conductors of the transmission line must be small compared with the wavelength. For the 100 GHz frequency component in the 10 ps pulse, the wavelength is 3 mm. If $kb \ll 1$ means $kb = 0.1$, the separation of the two conductors of the line must be less than 0.05 mm.

Phased Array

A phased antenna array has the phases of its elements controlled electronically so that the radiation pattern can be changed, usually in such a way that the direction of its major radiation scans over a wide angle (Figure 7.29). Unlike conven-

Figure 7.29 Phased array. The phase of the current on each element is controlled electronically to change the radiation pattern rapidly without mechanical movement of the antenna structure.

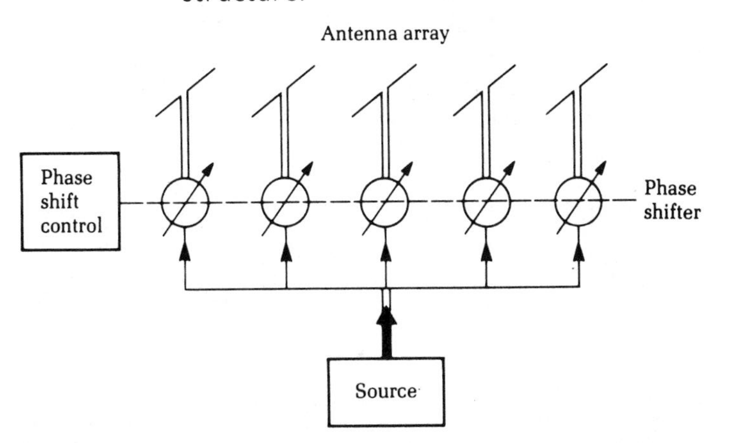

* Named after Brian D. Josephson, co-winner of the 1973 Nobel Prize for Physics for his work in the field of the tunneling effect of superconducting junctions. This work initiated a whole new technology called electronics of superconductors.

** See "Josephson Switch: Faster Than Transistor," in "Focal points—A review of current developments," *IEEE Spectrum*, April 1973, p. 87; H. Hayakawa and S. Kotani, "Josephson Digital Devices," in *Compound and Josephson High-Speed Devices*, ed. T. Misugi and A. Shibatomi, New York: Plenum Press, 1993, p. 255.

tional tracking radars, the mechanical structure of a phased array does not move while it scans.

To underatand the basic operating principle of a phased array, let us study the radiation pattern of a uniform linear array. For simplicity, we consider only the pattern in the horizontal ($\theta = 90°$) plane. Because perpendicular to a dipole axis the radiation pattern is omnidirectional, the array factor determines the pattern on the horizontal plane. According to (7.37), the major radiation always occurs at

$$\psi + kd \sin \phi_{max} = 0, 2\pi, \text{etc.}$$

where ϕ_{max} is the coordinate angle of the major lobe of the pattern. Assume $kd = \pi$, then the following table may be obtained:

ψ	180°	120°	90°	30°	0°
ϕ_{max}	90° −90°	222° −42°	210° −30°	190° −10°	180° 0°

Thus, when the phase shift ψ is controlled and changed electronically, the array antenna may be made to scan rapidly over a wide angle. Figure 7.30 shows the change of field pattern of a six-element array as ψ is varied from 180° to −180°.

By using a two-dimensional array in which elements are distributed over a surface, one may obtain a sharp beam that can perform three-dimensional scanning with high resolution. Phased arrays are used primarily in military radars for detecting ballistic missiles or aircrafts. Figure 7.31 shows a large phased array. It is a solid-state UHF array with two array faces. Each face contains 2677 antenna elements, occupies an area 72.5 ft in diameter, and transmits 585 kW peak power and 145 kW average power. Dipoles are used for array elements. A phased array with several thousand elements usually has a beam as narrow as a few degrees and thus a radar range of several thousand miles.

Figure 7.30 Field patterns of a six-element array with $d = \lambda/2$ and varying phase (below and on the following two pages).

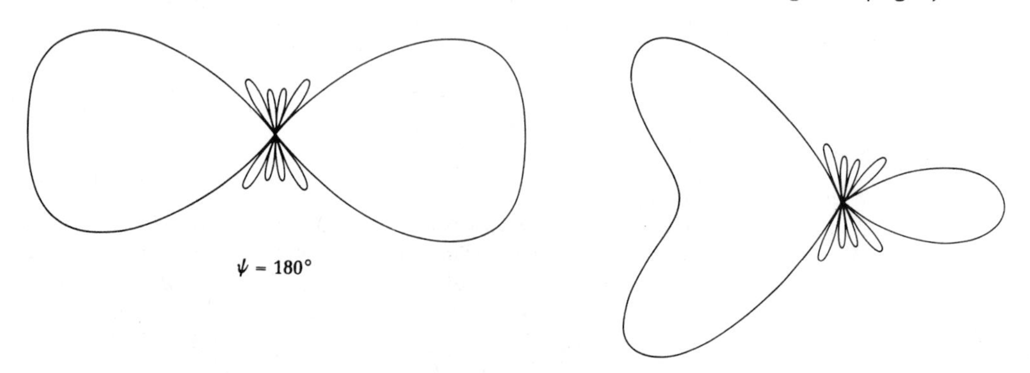

Figure 7.30 (continued)

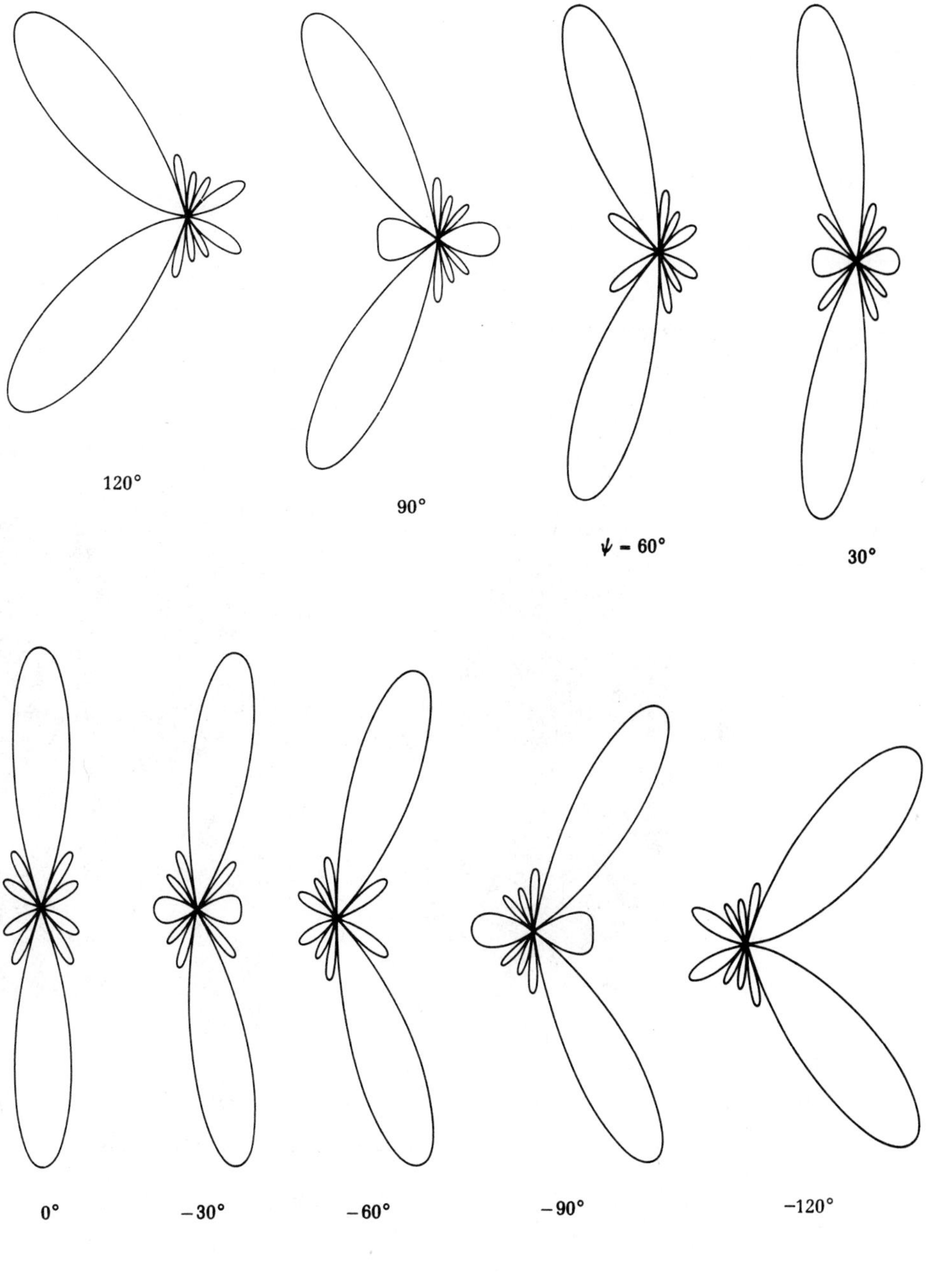

120°   90° $\psi = 60°$ 30°

0°   −30°   −60°   −90°   −120°

Figure 7.30 (continued)

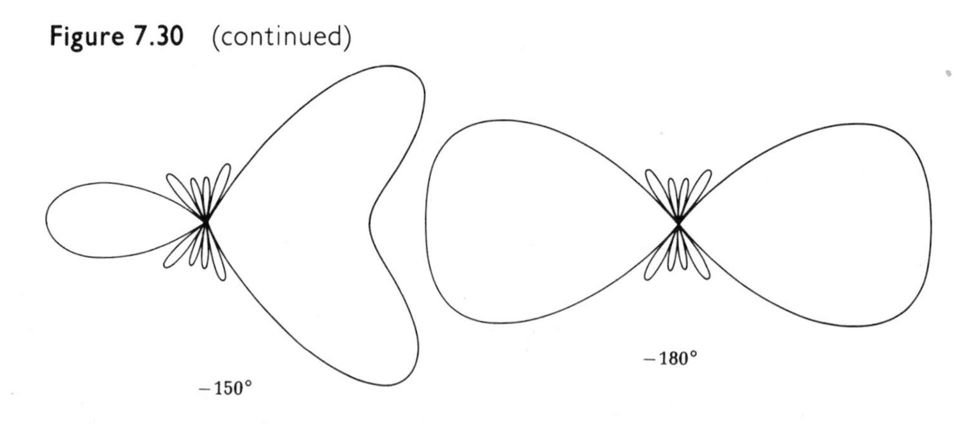

Figure 7.31 A two-face phased array. Each face contains 2677 dipole elements. (Courtesy of U.S. Air Force.)

Example 7.8 *This example shows an application of the array: how to direct the radiation power to cover a specific area and not waste power to other areas.*

A transmitter is located on the coast, 15 km west of a city, as shown in Figure 7.32a. Consider the two cases:

1. A quarter-wavelength monopole antenna is used. To maintain the FCC standard of 25 mV/m field strength in the city, how much radiation power must be provided?

Figure 7.32 **(a)** Design of a two-element array to radiate power in the direction of a city. **(b)** Normalized radiation patterns of a monopole antenna and an array antenna in the horizontal ($\theta = 90°$) plane. The spacing between elements A and B is $1/4\lambda$, and the phase angle of the current on B lags that on A by 90°.

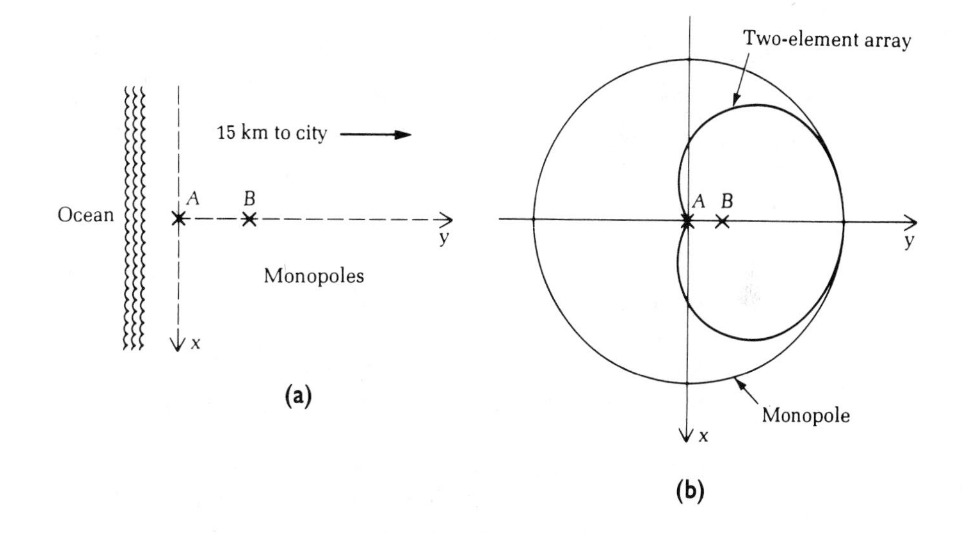

2. An array antenna of two quarter-wave monopoles is used. How much radiation power must be provided to maintain this same field strength?

Solution Example 7.4 answers the first case. The radiated power is 714 W, and the current at the driving point should be 6.25 A. Note that the monopole provides an omnidirectional pattern. In this case, a significant amount of power is wasted in an unpopulated area.

We hope to use the array to direct more power toward the city. This change can be accomplished by spacing the two monopoles $1/4\lambda$ apart and exciting them in such a way that the current on monopole B lags that on monopole A by 90° in phase angle. Note that an electromagnetic wave travels a distance of λ meters in $1/f$ seconds. A wave originating at A and traveling toward B finds itself in phase with the wave originating at B, so that these two waves reinforce each other in the $\hat{y}$ direction. The situation is just the opposite for waves traveling in the direction of the ocean, and as a result they cancel each other.

Mathematically, the total time-average Poynting vector may be obtained by using the expression for the Poynting vector of a dipole given by (7.27a) and the expression for the array factor given by (7.37):

$$\langle \boldsymbol{S} \rangle = \frac{I_0^2 \eta \cos^2(\frac{1}{2}\pi \cos\theta) \cdot \sin^2[\frac{1}{2}\pi(-1 + \sin\theta \sin\phi)]}{8\pi^2 r^2 \sin^2\theta \cdot \sin^2[\frac{1}{4}\pi(-1 + \sin\theta \sin\phi)]} \hat{\boldsymbol{r}}$$

Figure 7.32b shows the normalized antenna pattern in the $\theta = 90^\circ$ plane. In the direction $\theta = \phi = 90^\circ$, the array factor is two. Thus, to maintain the same field strength as that maintained by a single monopole, the current I_0 in the array is equal to half that in a single monopole—that is, for the two-element array, $I_0 =$ 3.13 A.

The total radiated power of the array may be obtained by integration:

$$P_t = r^2 \int_0^{\pi/2} \sin\theta\, d\theta \int_0^{2\pi} d\phi(\langle \boldsymbol{S} \rangle \cdot \hat{r})$$

This integral is analytically integrable because

$$|F(\theta, \phi)|^2 = 4\cos^2\left[\tfrac{1}{4}\pi(1 - \sin\theta \sin\phi)\right]$$
$$= 2[1 + \sin\tfrac{1}{2}(\pi \sin\theta \sin\phi)]$$

so that

$$\int_0^{2\pi} |F(\theta, \phi)|^2\, d\phi = 4\pi$$

Note that for a single monopole $F = 1$ and the integral over ϕ is 2π. The integration over θ is the same for both monopole and array. Bearing in mind that I_0 is half that of the monopole, we conclude that the power radiated by the array is half that radiated by the monopole—that is, $P_t = 357$ W. The array then is more directional than the monopole. In terms of antenna engineering, the **directivity** of the array is 3 dB higher than that of the monopole.

Example 7.9 *The "beam width" of an antenna array is sometimes used to describe the sharpness of the radiation pattern. This example shows how to calculate the "3-dB beam width" of an antenna array.*

A uniform linear array consists of ten half-wave dipoles spaced $1/4\lambda$ apart with all currents in phase. Study the field pattern on the $\theta = 90^\circ$ plane.

(a) Find the directions in which maximum radiation occurs.

(b) Find the 3-dB beam width in the direction of maximum radiation. (The 3-dB beam width is defined as the angle that is between two directions on each side of the maximum radiation direction and in which the radiation power intensity is 3 dB below the maximum intensity.)

Solution Because the pattern on the $\theta = 90^\circ$ plane is what interests us, we only need to examine the array factor given by (7.37). In the present case, $\psi = 0$, $kd = \pi/2$, and $\theta = 90^\circ$. Maximum radiation occurs at $kd \cos\gamma + \psi = 0$ or $\phi = 0$ or π.

Substituting different values of ϕ in (7.37), we obtain the following table:

ϕ	0	0.1	0.15	0.16	0.17	0.18
$\|F\|^2$	100	81.3	62.0	57.9	53.7	49.6

By interpolating, we find that $|F|^2 = 50$ if $\phi = 0.179$ or $10.3°$. Thus, the beam width is $20.5°$.

Very Long Baseline Arrays

As mentioned in Section 3.1, in our universe billions of stars constantly radiate electromagnetic waves in visible as well as in radio frequency range. Optical telescopes were developed to observe light-emitting celestial bodies. Since the 1940s, astronomers have used high-resolution radio antennas, or radio telescopes, to probe the universe. Antenna arrays with individual antennas separated by great distances are used to achieve high resolution.

Consider a two-element array separated by d with zero phase shift. The array factor $|F|$, according to (7.37), is as follows:

$$|F(\theta, \phi)| = 2|\cos[(\tfrac{1}{2})kd \sin(\theta)\sin(\phi)]| \qquad \textbf{(7.45)}$$

The array factor on the $\phi = 90°$ plane with $d = 10\lambda$ is shown in Figure 7.33(a) and the same quantity with $d = 20\lambda$ is shown in Figure 7.33(b). We see that these array factors have many maxima and nulls. They are sometimes called interference, or fringe patterns. When the separation d is increased, more maxima and nulls appear and the angular width of each maximum lobe is decreased.

The above discussion is based on the assumption that the array is a radiating structure. According to the reciprocal theorem, the same field pattern is obtained

Figure 7.33a The array factor of a two-element array with $d = 10\lambda$, $\phi = 90°$.

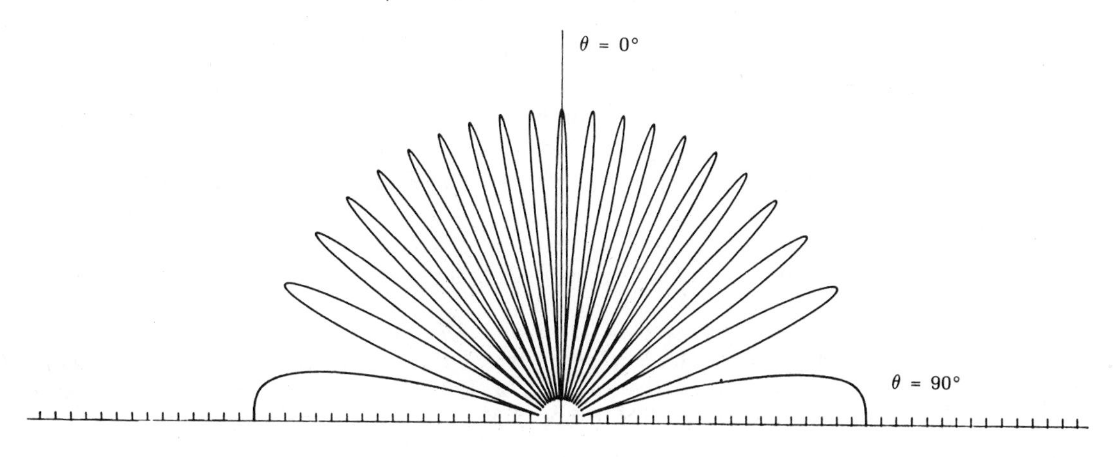

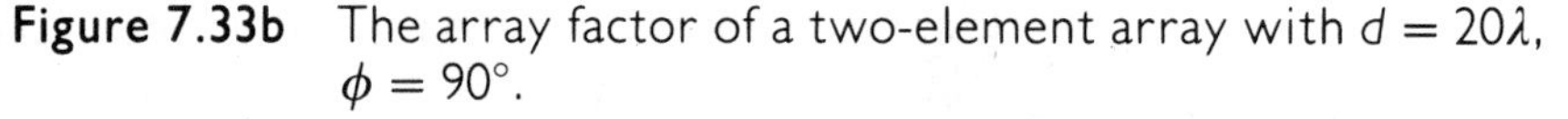
Figure 7.33b The array factor of a two-element array with $d = 20\lambda$, $\phi = 90°$.

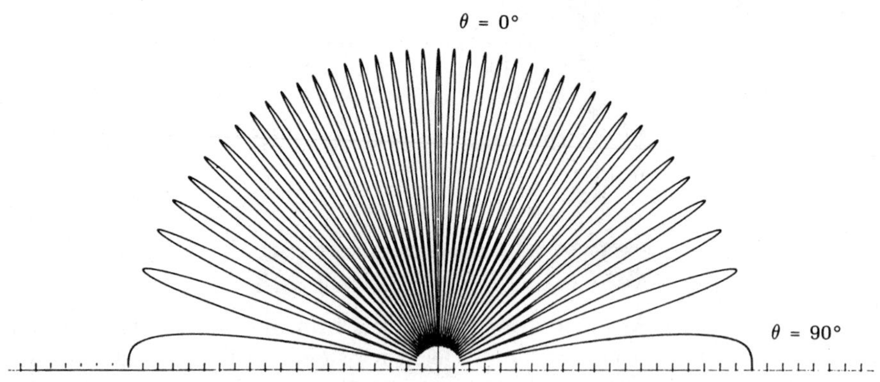

when the array is used as a receiver.* For radio telescopes the separation d is thousands or even millions of wavelengths, resulting in a very narrow fringe pattern.

Let θ_0 be the angle of a null in the $\phi = 90°$ plane. Then,

$$(1/2)kd \sin \theta_0 = (m + 1/2)\pi \tag{7.46}$$

where m is an integer. Let the next null in the increasing θ direction be at $\theta_0 + \Delta$, where Δ is the width of the beam near $\theta = \theta_0$. Then

$$(1/2)kd \sin(\theta_0 + \Delta) = (m + 1 + 1/2)\pi \tag{7.47}$$

Subtracting (7.46) from (7.47) yields

$$(1/2)kd[\sin(\theta_0 + \Delta) - \sin \theta_0] = \pi \tag{7.48}$$

If d is thousands of wavelengths long, then Δ is a small number so that $\sin \Delta \approx \Delta$ and $\cos \Delta \approx 1$. Using these approximate values in (7.48) we find that

$$(1/2)kd[\Delta \cos \theta_0] = \pi$$

Since $k = 2\pi/\lambda$, then

$$\Delta = \lambda/(d \cos \theta_0) \tag{7.49}$$

Antenna arrays have been constructed for astronomical and geophysical applications. They consist of many antennas located throughout the world. These systems are called the very long baseline arrays (VLBA). For example,** an

* D. K. Cheng, *Field and Wave Electromagnetics*, Reading, MA: Addison-Wesley, 1983, Section 11.6.

** C. S. Jacobs and O. J. Sovers, "An assessment of the accuracy of the deep space network extragalactic reference frame," in *Developments in Astrometry and Their Impact on Astrophysics and Geodynamics*, ed. I. I. Mueller and B. Kolaczek, The Netherlands: Kluwer Academic Publishers, p. 173.

antenna located near Goldstone, California, can be electronically linked to another located near Tidbinbilla, Australia, to form a baseline of 10,600 km. The operating frequency is 8.4 GHz. Substituting these values in (7.49) with $\theta_0 = 0$, we have

$$\Delta = 3.4 \times 10^{-9} \text{ radian}$$

This is smaller than the angle spanned by a dime located in New York City when it is viewed from Los Angeles.

In ordinary arrays, individual antenna elements are connected by transmission lines. For the very long baseline array, such physical interconnection of antennas is not practical. Instead, the signal received by each antenna in the array is recorded on magnetic tapes which are later transported to a central facility where the tapes are replayed simultaneously. The key to such processing is a very accurate time standard for all recorded data. At present, time synchronization of the recordings is provided by hydrogen masers, which are accurate within 20 nanoseconds.*

The technique of using VLBA to achieve high resolution is called very long baseline interferometry (VLBI). This technique can also be applied to geophysics. For example, if the array is locked on a distant radio source, then the small change in the spacing between the two widely separated antennas may be observed. The small change in spacing is due to tectonic motion of the earth's crustal plates. An antenna pair located in Westford, Massachusetts, and Wettzell, Germany, was able to detect an increase in the distance between these points at a rate equal to 18 mm per year.**

SUMMARY

1. The electromagnetic fields can be expressed in terms of vector potential **A** and scalar potential Φ. **A** and Φ can be calculated from the current and the charge distributions.
2. The Hertzian dipole is a fundamental source of electromagnetic waves. It is a filament of alternating current of I amperes with length Δz. Its radiation fields are given by (7.17). It is assumed that Δz is much less than the wavelength λ.
3. An antenna usually radiates EM waves with different intensities in different directions. A plot of the radiation field intensity or radiation power intensity as a function of direction is called the radiation pattern of the antenna. The gain of an antenna is a measure of how directional the radiation pattern is.

---

* K. I. Kellermann and A. R. Thompson, "The very long baseline array," *Science*, Vol. 229, No. 4709, July 1985, pp. 123–130.

** W. E. Carter and D. S. Robertson, "Very-long-baseline interferometry applied to geophysics," in *Developments in Astrometry and Their Impact on Astrophysics and Geodynamics*, ed. I. I. Mueller and B. Kolaczek, The Netherlands: Kluwer Academic Publishers, p. 133.

4. The capacitor antenna is a practical realization of the Hertzian dipole. Its radiation fields are the same as those of the Hertzian dipole if both carry the same current and are of the same length.
5. The current on a short dipole is maximum at the center and tapers to zero at both ends. It can be regarded as a Hertzian dipole of the same length carrying $I_0/2$ current, where I_0 is the current at the center of the short dipole.
6. The EM fields due to a general dipole antenna of length $h_1 + h_2$ are similar to those generated by a Hertzian dipole, except that an extra factor $U(\theta)$ appears in their formulas (7.23–7.24).
7. A commonly used antenna is the half-wave dipole for which the total length is $\lambda/2$. Its radiation fields are given by (7.23) with $U(\theta)$ given by (7.26).
8. The electric field of a uniform array antenna is the product of the array factor $F(\theta, \phi)$ and the E field due to a single element of the array.
9. By adjusting the phase shift and the spacing of a uniform array, one may direct the radiation in a desired direction. If the phase shift is carried out electronically, one may direct the radiation in different directions sequentially to scan a portion of the sky rapidly (the phased array).

Problems

7.1 Find the rectangular coordinates of a point P where the spherical coordinates are ($r = 1$, $\theta = 60°$, $\phi = 30°$).

7.2 The rectangular coordinates of a point Q are $(1, 2, -4)$. Find its spherical coordinates.

7.3 Show that $\nabla \cdot \nabla \times \boldsymbol{A} = 0$ in spherical coordinates for any vector $\boldsymbol{A}$.

7.4 Show that the differential spherical surface element is equal to $ds = r^2 \sin\theta \, d\theta \, d\phi$. Hint: Refer to Figure P7.4.

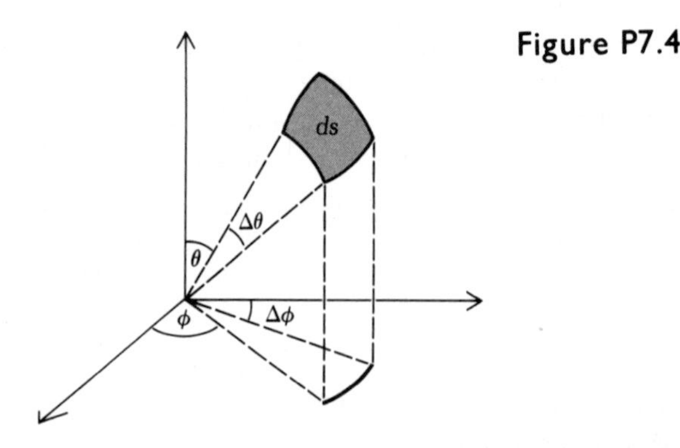

Figure P7.4

7.5 To convert a vector in spherical coordinates to the same rectangular coordinates, it is convenient to prepare a table for dot products between unit vectors in these coordinate systems. For example, $\hat{x} \cdot \hat{r} = \sin\theta \cos\phi$, as indicated in the following table. Complete this table.

·	$\hat{r}$	$\hat{\theta}$	$\hat{\phi}$
$\hat{x}$	$\sin\theta\cos\phi$		
$\hat{y}$			
$\hat{z}$			

7.6 Use the table prepared in the preceding problem to express the following vector A located at ($r = 1$, $\theta = -60°$, $\phi = 45°$) in rectangular coordinates:

$$A = 12\hat{r} + 8\hat{\theta} - 5\hat{\phi}$$

7.7 Show that the distance function $|r - r'|$ that appears in (7.7) and (7.8) can be expressed in spherical coordinates as

$$|r - r'|^2 = r^2 + r'^2 - 2rr' \cos\gamma$$

$$\cos\gamma = \cos\theta\cos\theta' + \sin\theta\sin\theta'\cos(\phi - \phi')$$

where γ is the angle between the vectors r and r' and (r, θ, ϕ) and (r', θ', ϕ') are spherical coordinates of r and r', respectively.
Hint: Express r and r' in x, y, z coordinates using the results obtained in Problem 7.5.

7.8 Calculate the time-average Poynting vector from (7.13) and (7.14). Then calculate it again using (7.17a, b). Show that the results are the same. Assume that the medium is lossless, that is, the wave number k is real.

7.9 The E field at $kr \gg 1$ due to a short dipole oriented in the $\hat{z}$ direction and located at the origin is given as follows:

$$\mathrm{E}_\theta = \frac{10^3 e^{-jkr}}{r} \sin\theta \ (\mathrm{V/m})$$

(a) What is the corresponding H field?
(b) Sketch the radiation field pattern on the x-y plane. Assume $r = 100$ m, and label the scale in units of V/m.
(c) Sketch the radiation field pattern on the x-z plane. Assume $r = 100$ m and label the scale in units of V/m.
(d) Sketch the radiation *power* pattern on the y-z plane. Assume $r = 100$ m, and label the scale in units of W/m^2.

7.10 A vertical receiving dipole antenna at P is 15 km away from a capacitor-plate antenna that is also placed vertically, as shown in Figure P7.10. The receiving antenna measures an E field equal to 10 mV/m. What is the value of E that the same receiving antenna will detect at a height 3 km above P? What must the orientation of the receiving dipole be to obtain a maximum reading? (A maximum reading is obtained if the dipole is parallel to the E field.)

7.11 The power lost on a cylindrical conductor that is Δz long and that carries I amperes of current is given by

$$P_{\mathrm{ohm}} = \tfrac{1}{2} I^2 R_s \Delta z$$

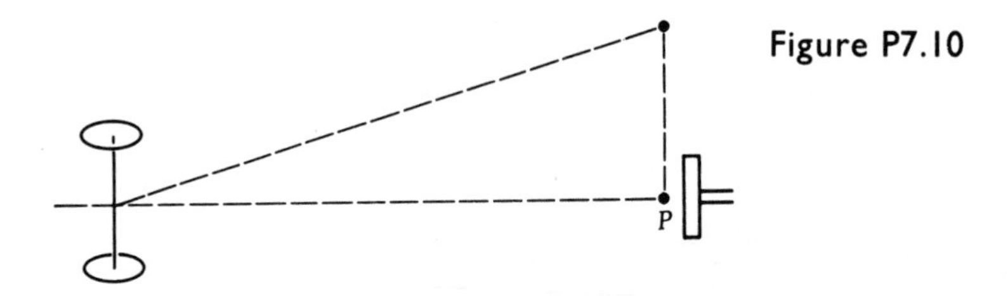

Figure P7.10

where P_{ohm} is the loss due to finite conductivity of the wire, R_s is the surface resistance given by $1/(\sigma d_s 2\pi a)$, and d_s is the skin depth. The efficiency of the antenna is given by

$$\eta_a = \frac{\text{Power radiated}}{\text{Power radiated} + P_{ohm}}$$

Assume that a short antenna of length Δz has an efficiency of 10 percent. Is the efficiency improved by increasing the length to $2\,\Delta z$, while maintaining the same current and, if so, by how much? Assume that the antenna is still a short antenna after its length is increased to $2\,\Delta z$.

7.12 Consider the antenna system consisting of two short dipoles arranged perpendicularly to each other in space, as shown in Figure P7.12. These dipoles are driven by the same amount of power from a common source. However, the current on the $\hat{x}$-oriented dipole has a $-90°$ phase with respect to that on the $\hat{y}$-oriented dipole because of a phase shifter inserted in the transmission line that leads to the former. Find the total radiated electric field on the z axis. Verify that this antenna system radiates a circularly polarized wave in the $\hat{z}$ direction. Is the wave left-hand or right-hand circularly polarized?

Figure P7.12

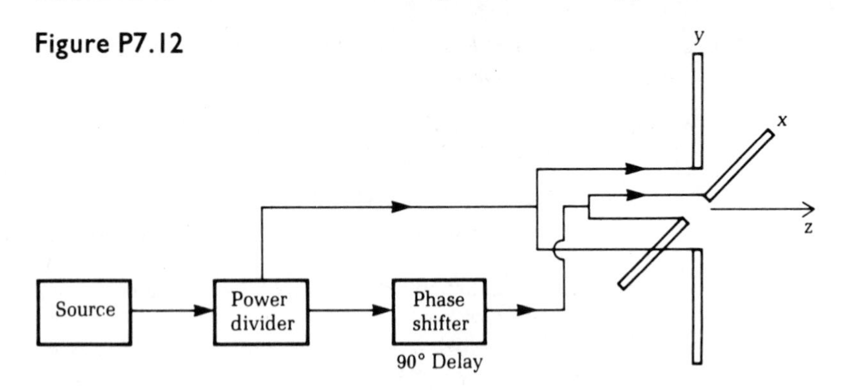

7.13 Find the expression of the total radiated electric field on the x axis that is due to the antenna system discussed in the preceding problem. What is its polarization on the x axis?

7.14 A certain application requires that a field strength of 1 V/m be maintained at a point 1 km from an antenna located in free space. What power must be fed to the antenna if it is (a) an isotropic antenna, (b) a short dipole, and (c) a half-wave dipole? Neglect ohmic loss. An isotropic antenna radiates an equal amount of power in all directions.

7.15 The current at the center of an antenna is 100 A; what is the E field 1 km away from it on the horizontal ($\theta = 90°$) plane at 10 MHz if the antenna is (a) a dipole with $h_1 = h_2 = 0.5$ m, (b) a capacitor-plate antenna with $\Delta z = 1$ m, and (c) a half-wave dipole?

7.16 Find the radiated electric field of a linear antenna that is 3 m long ($\ell = 3$ m) and that operates at 100 MHz in air. Plot its radiation pattern.

7.17 Find the directivity of (a) an isotropic antenna, (b) a capacitor-plate antenna, and (c) a half-wave dipole.

7.18 Consider the two-element array as shown in Figure P7.18 (top view). The antenna elements are identical half-wave dipoles. The current at the center of the first element is 1 A with $\psi = 0$, and that of the second element is 1 A with $\psi = 90°$. If only dipole 1 was present, the field on the x-y plane and 10 km away from the origin would be 1 mV/m. What are the fields at points A, B, C, and D (all are 10 km away from the origin) when both dipole 1 and 2 are present?

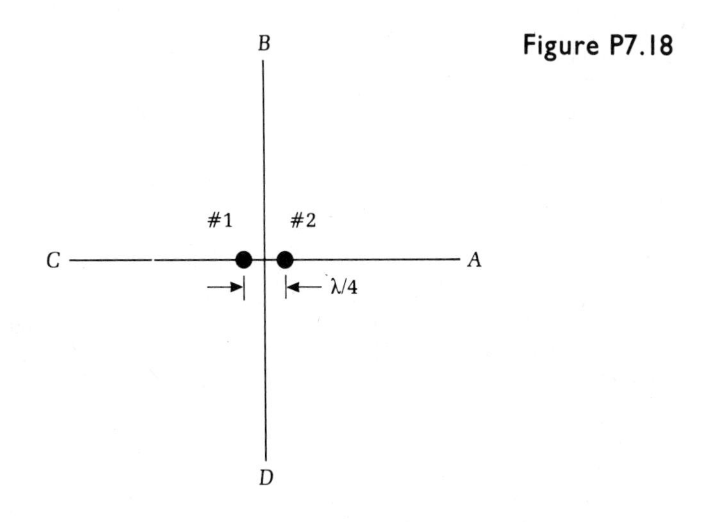

Figure P7.18

7.19 Consider a uniform linear array of two half-wave dipoles that are 1.5 wavelengths apart. The currents on these two dipoles are in phase. Sketch the radiation pattern in the horizontal ($\theta = 90°$) plane. Show clearly the number of lobes in this pattern. Also, estimate the beam width of each of the major lobes. The beam width is the angle between two directions in which the radiation intensity is one-half (-3 dB) the maximum value of the beam.

7.20 Figure 7.33(b) shows the array factor of a two-element array separated by 20λ. Find the beam width (in terms of the angle between two adjacent nulls) of this array factor near $\phi = 90°$ and $\theta = 30°$.

(a) Use the approximate formula given by (7.49).

(b) Find the exact value starting from (7.45).

7.21 Find the directivity of the two-wire transmission line shown in Figure 7.28 with radiation fields given by (7.42).

7.22 Find the field pattern of a two-element array with $d = \lambda/4$ and $\psi = 0$. Sketch the field pattern on the x-y plane.

7.23 Find the field pattern of a four-element array with $d = \lambda/4$ and $\psi = 0$. Sketch the field pattern on the x-y plane. (a) Use (7.37) to obtain the field-pattern formula, and (b) use the result obtained in the preceding problem and in Figure 7.17 and the pattern-multiplication technique.

7.24 Consider a uniform array of four identical short dipoles as shown in Figure P7.24. The field due to the #1 element alone is the same as that given in Problem 7.9. The current on all dipoles are the same except the phases, which are 0° for #1, −90° for #2, −180° for #3, and −270° for #4.

(a) Sketch the radiation field pattern on the x-y plane. Assume $r = 100$ m and label the scale in units of V/m.

(b) Repeat (a) on the y-z plane.

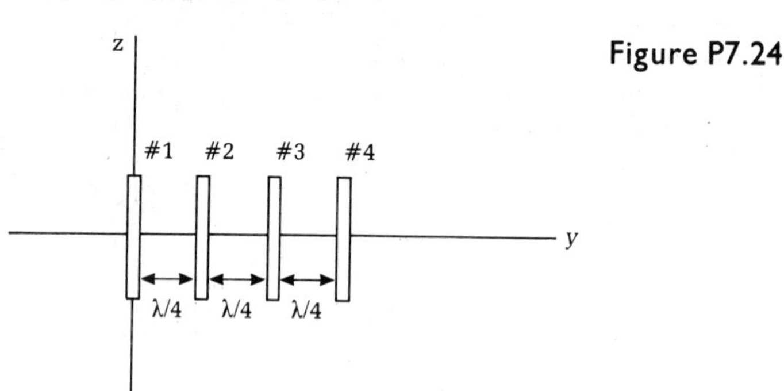

Figure P7.24

7.25 A uniform linear array consists of six short dipoles. The spacing between adjacent elements is $\lambda/4$, as shown in Figure P7.25.

(a) What should the phase shift ψ be, in order to point the maximum radiation in the $\phi = 90°$ (that is, $\hat{y}$) direction?

(b) Suppose that the E field due to the first element (the dipole at far left) is given as follows:

$$E_{\theta e} = \frac{1000}{r} e^{-jkr} \sin\theta$$

Calculate $|E_\theta|$ of the entire array at point A (0, 1000, 0), point B (1000, 0, 0), point C (0, −1000, 0), and point D (−1000, 0, 0), separately. All positions are given in rectangular coordinates in meters. Use the phase shift found in (a).

(c) Sketch the field pattern of the array in the x-y plane.

(d) Sketch the field pattern of the array in the x-z plane.

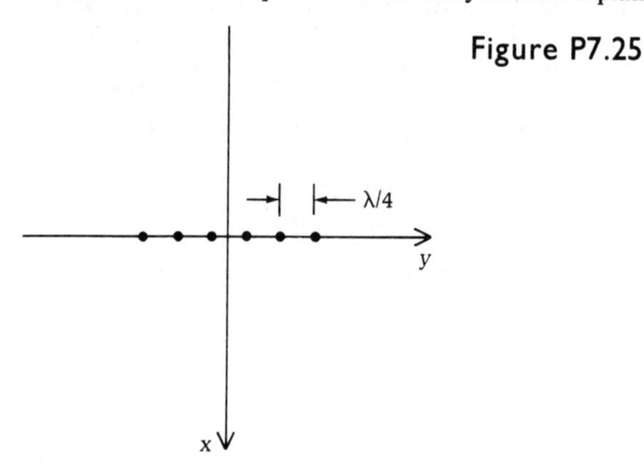

Figure P7.25

7.26 Write a computer program to plot field patterns of a ten-element phased array with $d = \lambda/4$ and varying phases.

8 TOPICS IN WAVES

In this chapter, a few advanced topics in electromagnetic waves are introduced to illustrate more applications. In the preceding chapters, we have seen how an electromagnetic wave is reflected from a planar boundary and how the transmission or refraction occurs. The reader may wonder what happens when an electromagnetic wave encounters a finite object, like an airplane. The answer is that the wave is scattered by the finite object. Solving the problem of the scattering of a wave by an airplane involves advanced mathematics and is beyond the scope of this book. But we can see the basic mechanism by studying the scattering of a wave by a sphere; this is the first topic discussed in this chapter. In fact, the scattering of an electromagnetic wave is the basic reason the sky is blue! A related topic is the case of a plane wave impinging on a plane containing a hole. This is called **diffraction** and is the second topic studied here.

In Chapter 3, we studied the wave propagation in unbounded media. We assumed that the wave was infinite—that is, a plane wave. In practice, however, all waves are finite in extent. Therefore, the third topic covered in this chapter is the **Gaussian beam**. It is an electromagnetic wave for which the field strength is concentrated at the center. A laser or a radar beam is more like a Gaussian beam than like a planar wave.

Until now we have assumed that the frequency of the electromagnetic wave is constant, which implies that the reflection, refraction, and the scattering of waves occur off a stationary reflector or target. What if the reflector moves with respect to the source of the electromagnetic wave? If that happens, the reflected or scattered wave has a different frequency than that of the incident wave. This phenomenon is called the **Doppler effect** and it is discussed in this chapter also.

Finally, you may have realized that the media in which the electromagnetic waves propagate are assumed to be isotropic. That is, their electromagnetic properties are the same no matter in which direction a wave propagates. But some natural and artificial materials are **anisotropic**: they have different properties depending on the orientation of the fields. Propagation of electromagnetic waves in anisotropic media is the final topic covered in this chapter.

8.1 RAYLEIGH SCATTERING

When an electromagnetic wave encounters an object, it is scattered. The scattering process largely depends on the size of the object. In many cases, the scatterer can be modeled as a sphere. The scattering of electromagnetic waves by a dielectric sphere with a radius very much smaller than the wavelength of the wave is called **Rayleigh scattering.**

This section studies the Rayleigh scattering problem. The dielectric constant of the sphere is ϵ and the radius a. The incident electromagnetic wave is a uniform plane wave. The fields of the scattered wave satisfy Maxwell's equations. From (7.13) and (7.14) we know that the following sets of fields form a set of solutions:

$$H_\phi = \frac{A \sin\theta}{\eta r^3}[(jkr) + (jkr)^2]e^{-jkr} \tag{8.1a}$$

$$E_r = \frac{2A \cos\theta}{r^3}(1 + jkr)e^{-jkr} \tag{8.1b}$$

$$E_\theta = \frac{A \sin\theta}{r^3}[1 + (jkr) + (jkr)^2]e^{-jkr} \tag{8.1c}$$

where A is a constant.

Set the origin of the spherical coordinates at the center of the sphere, as shown in Figure 8.1. The incident wave can be expressed as

$$\mathbf{E} = E_o(\hat{\boldsymbol{r}} \cos\theta - \hat{\boldsymbol{\theta}} \sin\theta)e^{-j\mathbf{k}\cdot\boldsymbol{r}} \tag{8.2}$$

Figure 8.1 A dielectric sphere in a uniform plane wave.

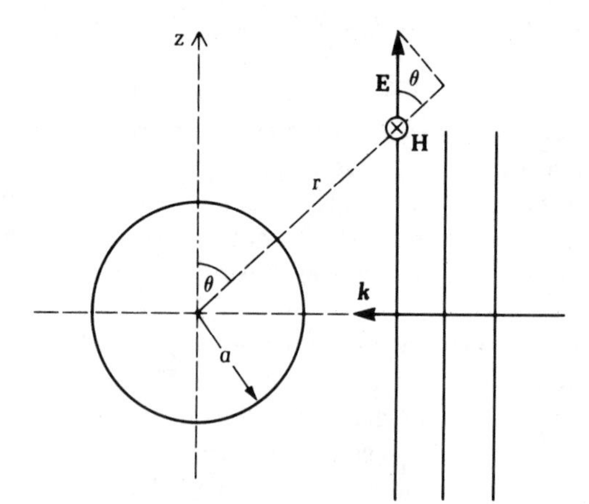

Because the sphere is very small compared with the wavelength, the scattered fields take the following form near the sphere, where $kr \ll 1$:

$$\mathbf{E}_s = \frac{A}{r^3}(\hat{\boldsymbol{r}}\, 2\cos\theta + \hat{\boldsymbol{\theta}} \sin\theta) \tag{8.3}$$

Inside the sphere, an electric field is excited. It has amplitude B and is parallel to the incident field:

$$\mathbf{E}_i = B(\hat{\boldsymbol{r}} \cos\theta - \hat{\boldsymbol{\theta}} \sin\theta) \tag{8.4}$$

Matching the boundary conditions on the **E** and **D** fields, we obtain two equations

$$-E_0 + \frac{A}{a^3} = -B$$

$$E_0 + \frac{2A}{a^3} = \frac{\epsilon}{\epsilon_0} B$$

Note that on the sphere $\boldsymbol{k} \cdot \boldsymbol{r} = ka \ll 1$, so that $e^{-j\boldsymbol{k} \cdot \boldsymbol{r}} \simeq 1$. Solving for A, we have

$$A = \left(\frac{\epsilon - \epsilon_0}{\epsilon + 2\epsilon_0}\right) a^3 E_0 \tag{8.5}$$

Substituting back into (8.1), we obtain the complete solution for the scattered wave. We see that the scattered wave is similar to the field radiated by a small dipole antenna. The scattering mechanism may be regarded as a process in which the incident wave induces in the sphere a dipole moment that oscillates at the same frequency as the incident wave. This oscillating dipole acts in every respect as a dipole antenna that reradiates the incident power in all directions.

Consider the field of the scattered wave at a distance far from the sphere, where $kr \gg 1$. According to (8.1), the scattered fields become

$$\mathrm{H}_\phi = -\frac{E_0(\epsilon - \epsilon_0)k^2 a^3}{\eta(\epsilon + 2\epsilon_0)r} \sin\theta e^{-jkr} = \frac{1}{\eta}\mathrm{E}_\theta \tag{8.6}$$

Figure 8.2 shows a plot of the E_θ given in (8.6). This plot is called the scattering-field pattern of the small dielectric sphere. Note that electromagnetic waves do not scatter uniformly in all directions. In fact, in Rayleigh scattering, the scattered fields are strongest in the forward and in the backward directions.

Figure 8.2 Scattering-field pattern of a small dielectric sphere. The incident wave is polarized in the $\hat{\mathbf{z}}$ direction. The scattered fields in the OA, OB, and OC directions are polarized in the directions of the arrows. The strengths of the scattered fields in these directions are proportional to the lengths OA, OB, and OC, respectively.

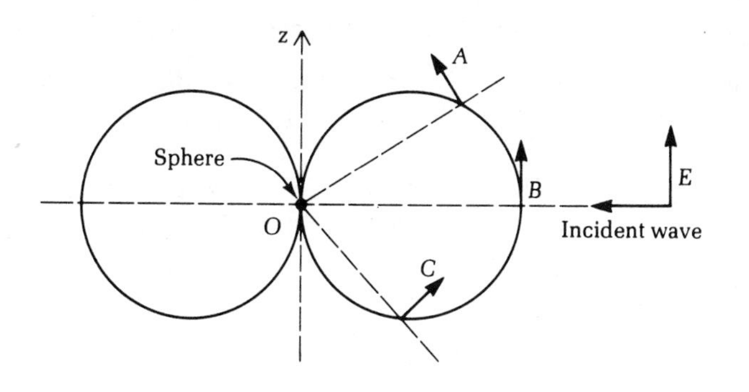

We can also see that, whereas the incident wave is polarized in the $\hat{z}$ direction, the polarization of the scattered field varies in space. We say that the scattering object "depolarizes" part of the incident wave.

Total Scattering Cross Section

To find the total power that the dielectric sphere reradiates, we may integrate the Poynting vector over a great sphere for which (8.6) is valid:

$$\langle \mathbf{S}^s \rangle = \frac{E_0^2(\epsilon - \epsilon_0)^2 k^4 a^6}{2\eta(\epsilon + 2\epsilon_0)^2 r^2} \sin^2\theta\, \hat{\mathbf{r}} \tag{8.7}$$

$$P^s = \int_0^\pi (\langle \mathbf{S}^s \rangle \cdot \hat{\mathbf{r}}) 2\pi r^2 \sin\theta\, d\theta$$

$$= \frac{8\pi}{3}\left(\frac{\epsilon - \epsilon_0}{\epsilon + 2\epsilon_0}\right)^2 k^4 a^6 \left(\frac{E_0^2}{2\eta}\right) \tag{8.8}$$

The **total scattering cross section,** σ_{total}, is defined as the ratio of the total scattered power, P^s, to the incident power density, $E_0^2/(2\eta)$:

$$\boxed{\sigma_{\text{total}} = \frac{8}{3}\left(\frac{\epsilon - \epsilon_0}{\epsilon + 2\epsilon_0}\right)^2 (k^4 a^4)(\pi a^2)} \quad \text{(total scattering cross section of a dielectric sphere)} \tag{8.9}$$

The above equation states that the scattered power is proportional to the *fourth power* of the frequency. This law is the celebrated Rayleigh law of scattering, discovered by J. W. S. Rayleigh (1842–1919), a British physicist.

Example 8.1 *This example gives an estimate of the power reradiated by a raindrop when it is in the path of a microwave. This power is described in terms of the total scattering cross section σ_{total}. Note that the latter is the ratio of the total reradiated power to the power density incident on the raindrop.*

Assume that a raindrop is a sphere of 3 mm diameter. Find its total scattering cross section at 10 GHz. Take the ϵ of the water to be $61\epsilon_0$ at 10 GHz.

Solution The wavelength in air is 0.030 m, which is twenty times greater than the radius of the raindrop. Thus, we can use the Rayleigh scattering formulas. Using (8.9), we find that

$$\frac{\sigma_{\text{total}}}{\pi a^2} = 0.024$$

Thus, the total scattering cross section is 0.024 times the geometrical cross section of the sphere.

Back-Scattering Cross Section

The "total scattering cross section" is not very meaningful for measuring the detectability of a scattering object by radar. Radar usually measures the scattered field in the backward direction. For this reason, we need to define a scattering cross section that is a function of direction.

The scattering cross section of an object in a given orientation is defined as the ratio of $4\pi r^2$ times the radiation intensity of the scattered wave in a specified direction to the power per unit area of the incident plane wave of a specified polarization. For the dielectric sphere in a linearly polarized uniform plane wave, as depicted in Figure 8.1, the scattering cross section $\sigma(\theta)$ is given as follows:

$$\begin{aligned}\sigma(\theta) &= \frac{4\pi r^2 \langle \mathbf{S}^s \rangle \cdot \hat{\mathbf{r}}}{\frac{1}{2}(E_0^2/\eta)} \\ &= 4\pi\left(\frac{\epsilon - \epsilon_0}{\epsilon + 2\epsilon_0}\right)^2 k^4 a^6 \sin^2\theta\end{aligned} \qquad \text{(8.10)}$$

(scattering cross section as a function of direction)

Note that the scattered intensity is maximum on the equatorial plane, including the forward and the backward directions ($\theta = \pi/2$). The value of σ in the direction of the incoming wave is called the **back-scattering cross section** of the object. When a radar beam illuminates an object, the back-scattering cross section determines the amount of intensity the radar can detect.

Example 8.2 *The back-scattering cross section is a measure of how much of the reradiation is going back in the direction of the source of the incident wave. We need to understand the difference between the back-scattering cross section and the total scattering cross section.*

Find the back-scattering cross section of the raindrop discussed in Example 8.1.

Solution Substituting in (8.10), yields

$$\frac{\sigma(\pi/2)}{\pi a^2} = 0.035$$

Thus, the back-scattering cross section of the raindrop is 0.035 times its geometric cross section.

Example 8.3 ***The intensity of scattering of microwaves by particles is also dependent on the dielectric property of the scatterer.***

Find the back-scattering cross section of a spherical ice particle of 3 mm diameter at 10 GHz.

Solution The permittivity of ice at 10 GHz is $3.2\epsilon_0$. Thus,

$$\frac{\sigma(\pi/2)}{\pi a^2} = 0.007$$

The difference in the back-scattering cross section of a water particle and that of an ice particle of the same size is fivefold. Thus, at 10 GHz, small water particles are much more efficient scatterers than ice particles.

Scattering of Light

Light scatters when it encounters particles in an otherwise homogeneous medium. When the particles are small with respect to the wavelength of the light, the Rayleigh law of scattering applies.

Consider, for example, a transparent tube filled with water that has been made turbid by adding a drop of milk. A light illuminates the tube from the right-hand side, as shown in Figure 8.3. An observer can see the tube filled with light because of the scattering of particles in the solution.

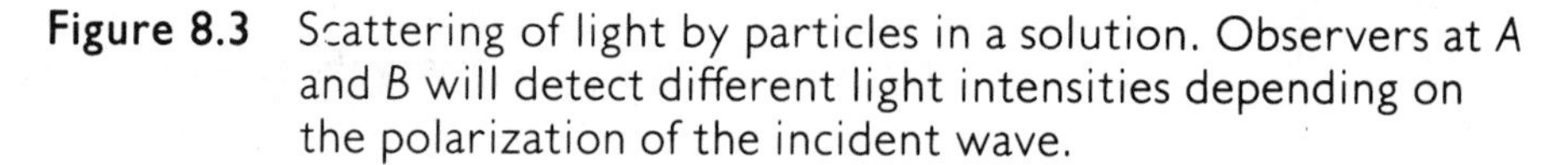

Figure 8.3 Scattering of light by particles in a solution. Observers at A and B will detect different light intensities depending on the polarization of the incident wave.

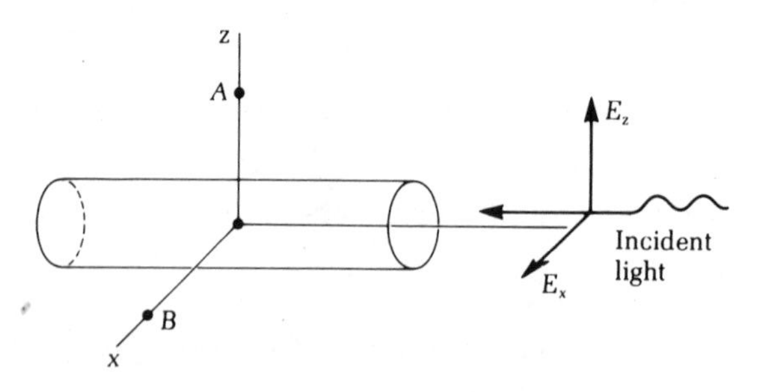

Let the incident light be polarized along the $\hat{z}$ direction. According to (8.6), an observer at point A, where $\theta = 0$, will see much less intense light than another observer at point B, where $\theta = 90°$. Conversely, if the incident light is polarized in the $\hat{x}$ direction, the intensity of the light observed at B will be less than that observed at A. This difference occurs because the quantity θ in (8.6) refers to the angle with respect to the polarization of the incident wave.

If the incident light is unpolarized and the observer at B is wearing a pair of polarized sunglasses, the observer will see the tube filled with light when the polarization of the glasses is parallel to the $\hat{z}$ direction. He will notice that the intensity of the light in the tube diminishes when the sunglasses are rotated by 90°.

Why the Sky Is Blue

Sunlight is scattered by small molecules as it enters the atmosphere. According to (8.9), a small particle scatters high-frequency waves much more efficiently than it scatters low-frequency waves. The frequency of violet light is approximately 6.9×10^4 Hz and that of red light is approximately 4.6×10^{14}. Thus,

$$\frac{\sigma_{\text{total}}\,(\text{violet})}{\sigma_{\text{total}}\,(\text{red})} = \left(\frac{6.9}{4.6}\right)^4 = 5.1$$

In other words, atmospheric particles scatter the violet-blue lights approximately five times more strongly than they scatter yellow-red lights. The color of the sky is therefore predominantly violet (to which our eyes are not very sensitive) and blue, mixed with green and yellow and very little red. These colors combine into a color that we call sky blue.*

8.2 FOURIER OPTICS AND HOLOGRAPHY

In analyses of circuit or linear systems, we know from the Fourier theorem that a time-varying signal—for example, a voltage or a current—may be thought of as the superposition of many time-harmonic signals of different frequencies. The Fourier transform or the Fourier series gives the amplitudes of these time-harmonic components of the signal.

Analogous to this example, a time-harmonic wave with nonuniform amplitude distribution as a function of space may be thought of as the superposition of many uniform plane waves of the same frequency and different wave numbers. Here, the Fourier transform is performed on the space coordinates rather than on the time variable. The phrase "Fourier optics" is used for theories and techniques that use the Fourier transform in spatial variables to analyze modern optics problems.

In this section we shall use a Fourier optics technique to show that a nonuniform beam of a given amplitude distribution will propagate in space in such a manner that it will become another nonuniform wave whose amplitude distribution is the Fourier transform of the original amplitude distribution. Let us consider the following example of diffraction: diffraction of light by a long slit. Figure 8.4 shows the geometry. A dark screen with a long horizontal slit is placed

* M. Minnaert, *Light and Colour*, New York: Dover, 1954, p. 238.

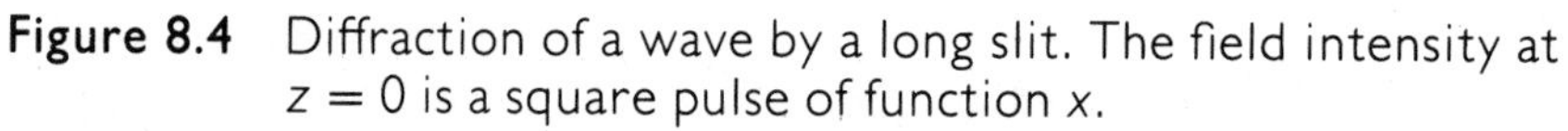

Figure 8.4 Diffraction of a wave by a long slit. The field intensity at $z = 0$ is a square pulse of function x.

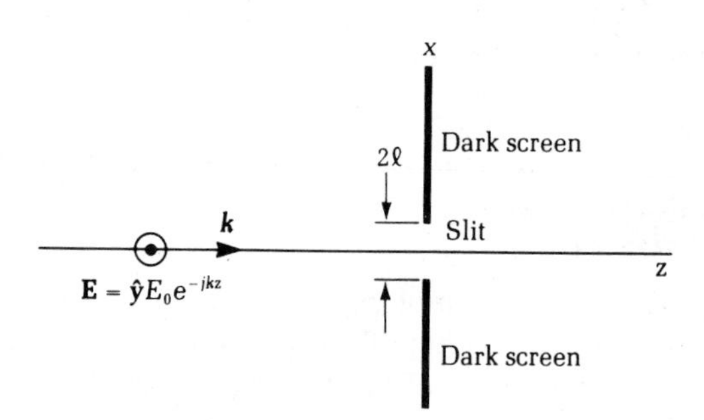

before a linearly polarized light source. Assume that the screen is sufficiently far from the light source that the slit may be considered to be illuminated by a uniform plane wave. Also assume that the field at the slit is approximately uniform, that it has the same intensity as the incident wave, and that it is zero everywhere else on the screen. Therefore, as Figure 8.5 shows, the intensity of the wave at $z = 0$ is a "square pulse" function.

We shall demonstrate that the square-pulse electromagnetic beam will diffract to become a beam whose spatial intensity is similar to a $(\sin x)/x$ distribution.

Referring to Figure 8.5, we know that

$$\mathbf{E}(x, 0) = \hat{y} E_0 \, U(\ell - |x|) \tag{8.11}$$

Figure 8.5 **(a)** Original distribution of the amplitude of a wave. **(b)** The amplitude of the wave after propagating a long distance: $x_0 = \pi z / k\ell$.

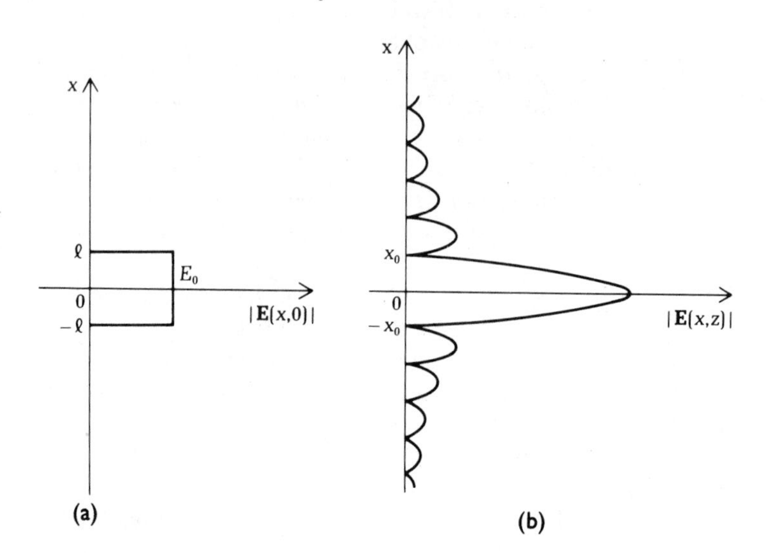

where $U(\alpha)$ is the step function that assumes the value zero for $\alpha < 0$ and the value unity when $\alpha > 0$. We may express the wave for $z > 0$ as the superposition of many plane waves:

$$\mathbf{E}(x, z) = \hat{y} \int_{-\infty}^{\infty} dk_x \,\mathrm{A}(k_x) \exp(-jk_x x - jk_z z) \tag{8.12}$$

Then, we equate (8.12) and (8.11) at $z = 0$ to find the spectrum $\mathrm{A}(k_x)$:

$$E_0 U(\ell - |x|) = \int_{-\infty}^{\infty} dk_x \,\mathrm{A}(k_x) e^{-jk_x x} \tag{8.13}$$

The above integral is the Fourier transform of $\mathrm{A}(k_x)$. Therefore, we can obtain $\mathrm{A}(k_x)$ by the inverse Fourier transform:

$$\begin{aligned} \mathrm{A}(k_x) &= \frac{E_0}{2\pi} \int_{-\infty}^{\infty} dx U(\ell - |x|) e^{jk_x x} \\ &= \frac{E_0 \ell \sin k_x \ell}{\pi k_x \ell} \end{aligned} \tag{8.14}$$

Substituting (8.14) into (8.12) yields the following:

$$\mathbf{E}(x, z) = \hat{y} \frac{E_0 \ell}{\pi} \int_{-\infty}^{\infty} dk_x \frac{\sin k_x \ell}{k_x \ell} \exp(-jk_x x - jk_z z) \tag{8.15}$$

Note that $k_z = \sqrt{k^2 - k_x^2}$. The integral in (8.15) may be evaluated either numerically using a computer or approximately using analytical methods.

To evaluate $\mathbf{E}(x, z)$ approximately, we transform the variable k_x to α, where

$$k_x = k\alpha$$

and $dk_x = k\, d\alpha$

$$\mathbf{E}(x, z) = \hat{y} \frac{E_0 k\ell}{\pi} \int_{-\infty}^{\infty} d\alpha \frac{\sin k\ell\alpha}{k\ell\alpha} \exp\left(-jkx\alpha - jkz\sqrt{1 - \alpha^2}\right) \tag{8.16}$$

Assume that $k\ell \gg 1$. This condition means that the width of the slit shown in Figure 8.4 is at least several wavelengths. Under this assumption, the first term in the integrand (8.16) has appreciable value only for small α. It follows that the square root in the exponent may be approximated as $(1 - \alpha^2/2)$. The result is

$$\mathbf{E}(x, z) = \hat{y} \frac{E_0 k\ell}{\pi} \int_{-\infty}^{\infty} d\alpha \frac{\sin k\ell\alpha}{k\ell\alpha} \exp\left[-jkx\alpha - jkz\left(1 - \frac{\alpha^2}{2}\right)\right]$$

Rearranging terms in the exponent yields the following:

$$\begin{aligned} \mathbf{E}(x, z) = \hat{y} \frac{E_0 k\ell}{\pi} \int_{-\infty}^{\infty} d\alpha \frac{\sin k\ell\alpha}{k\ell\alpha} \exp\left[\frac{jkz}{2}\left(\alpha - \frac{x}{z}\right)^2\right] \\ \times \exp\left[-jkz - j\frac{kz}{2}\left(\frac{x}{z}\right)^2\right] \end{aligned}$$

Note that the second exponent is independent of the integration variable α. At distances several wavelengths from the slit, $kz \gg 1$, and the first exponent is rapidly oscillatory except near $\alpha = x/z$. Thus, the integrand near $\alpha = x/z$ is the major contribution to the integral. A good approximation is to substitute $\sin[(k\ell x/z)]/(k\ell x/z)$ for $\sin k\ell\alpha/k\ell\alpha$ and take it out of the integral. Now we have

$$\mathbf{E}(x, z) = \hat{y}\frac{E_0 k\ell}{\pi}\frac{\sin(k\ell x/z)}{(k\ell x/z)}\exp\left[-jkz\left(1 + \frac{x^2}{2z^2}\right)\right] \times \int_{-\infty}^{\infty} d\alpha \exp\left[\frac{jkz}{2}\left(\alpha - \frac{x}{z}\right)^2\right]$$

This integral may be evaluated exactly (see Problem 8.6):

$$\int_{-\infty}^{\infty} d\alpha \exp\left[\frac{jkz}{2}\left(\alpha - \frac{x}{z}\right)^2\right] = \sqrt{\frac{2\pi}{-jkz}} = \sqrt{\frac{2\pi}{kz}}\exp\left(\frac{j\pi}{4}\right)$$

Finally, we have

$$\mathbf{E}(x, z) = \hat{y}\frac{E_0 k\ell}{\pi}\sqrt{\frac{2\pi}{kz}}\frac{\sin(k\ell x/z)}{(k\ell x/z)}\exp\left[-jkz\left(1 + \frac{x^2}{2z^2}\right) + \frac{j\pi}{4}\right] \tag{8.17}$$

The above approximate formula is obtained under the assumption that $k\ell \gg 1$ and that $kz \gg 1$. Figure 8.5 shows a plot of the amplitude of $\mathbf{E}(x, z)$ after the wave has propagated a long distance. We see that the original amplitude distribution and the diffracted distribution are the Fourier transform of each other. The first nulls occur at $x = \pm x_0$, with

$$x_0 = \frac{\pi z}{k\ell}$$

This place is where the $\sin A/A$ function in (8.17) vanishes.

Holography

In analyzing diffraction of a wave by a slit, we specify the distribution of light on a plane ($z = 0$). Based on that distribution, we can calculate the field everywhere else. In other words, if both the amplitude and the phase of a field are completely specified on a plane, we may determine the field everywhere else in space. Keeping this fact in mind, we shall discuss a method of image recording and display.

Figure 8.6 shows an object illuminated by a light source. The object scatters light, some of which is detected by an observer. Each of the observer's eyes views the object at a slightly different angle. The brain can detect the difference, and this difference gives the observer a "three-dimensional" picture of the object.

Now let us put a fictitious screen between the object and the observer, as shown in Figure 8.7. Suppose that in the recording mode the screen can faithfully

Figure 8.6 Light illuminates an object and is scattered by it. Some of the scattered light enters the pupil of the observer's eye and is focused on the retina to form an image.

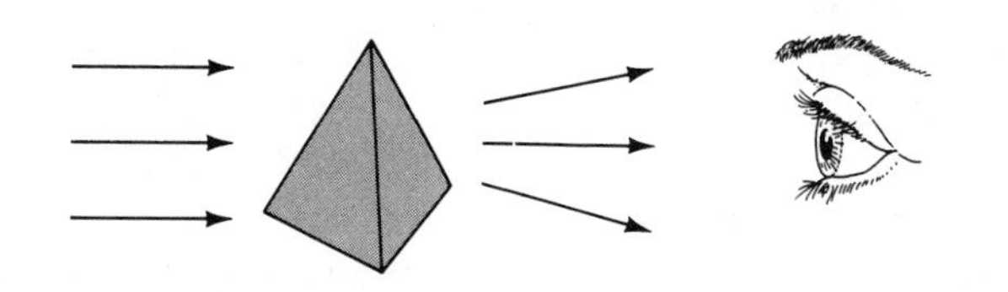

record the amplitude *and* the phase of the light incident on it. Also assume that the screen can regenerate the amplitude and phase of the light at all points on the screen when the screen is in the display mode. We first switch the screen to "recording mode." Then we remove the object, after which the screen is in the "display mode." We shall continue to see the three-dimensional object as long as the screen continues to reproduce the recorded amplitude and phase of the inci-

Figure 8.7 **(a)** A fictitious film records the amplitude and the phase of the light scattered from an object. **(b)** The film then regenerates the amplitude and the phase of the light. The observer will see a three-dimensional image, even though the object itself has been removed since the recording was made.

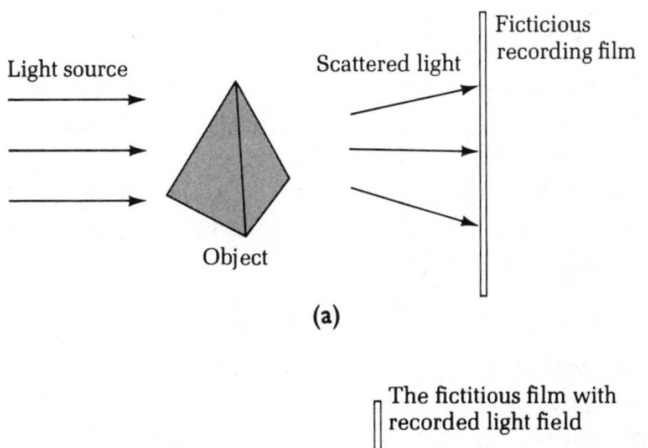

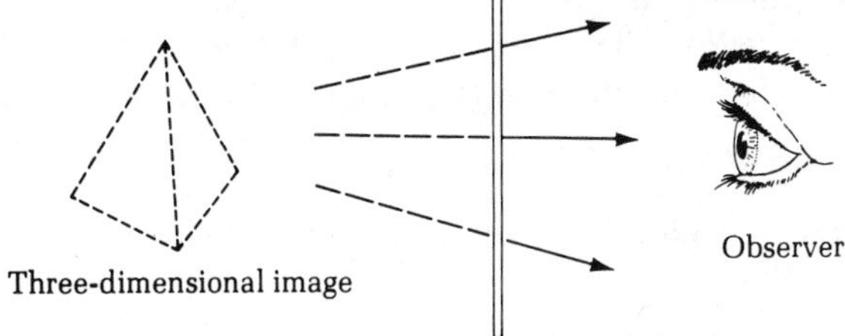

dent light that existed when the object was still there. Unfortunately, a screen that can record both the amplitude and the phase of an incident light has not yet been found. The photographic film available now can only record the amplitude but not the phase of the light. Thus, ordinarily we can only produce a two-dimensional image with such film. However, not long ago someone invented an ingenious way of recording both the amplitude and the phase of light waves using a film that can only record amplitude! This technique, which we shall now explain, is called holography.*

Denote the light scattered from the object and incident on the screen shown in Figure 8.8 (p. 271) as

$$S(x, y) = a_s(x, y) \exp[-j\phi_s(x, y)]$$

The function a_s is the amplitude distribution, and the function ϕ_s is the phase. In addition, the light from the source, which is called the reference wave, also illuminates the screen. Let reference wave be denoted as

$$R(x, y) = a_r(x, y) \exp[-j\phi_r(x, y)]$$

The total field on the screen is

$$T(x, y) = S(x, y) + R(x, y) \tag{8.18}$$

Now let us use an ordinary photographic film to record the total field. As mentioned earlier, this film records only the intensity of the total field—that is, $|T|^2$:

$$\begin{aligned} |T|^2 &= [a_s \exp(-j\phi_s) + a_r \exp(-j\phi_r)][a_s \exp(j\phi_s) + a_r \exp(j\phi_r)] \\ &= a_s^2 + a_r^2 + a_s a_r [\exp(-j\phi_s + j\phi_r) + \exp(j\phi_s - j\phi_r)] \\ &= SS^* + RR^* + SR^* + S^*R \end{aligned} \tag{8.19}$$

After the film records the intensity $|T|^2$, the object may be removed. The film with such a recording is called the **hologram**. Note that the hologram records the sum of the four terms on the right-hand side of (8.19).

In the display mode, the hologram is illuminated by a light that is equal to the reference wave that made the hologram. Thus, the field behind the film is proportional to $|T|^2 R(x, y)$. Or,

$$|T|^2 R = SS^*R + RR^*R + SR^*R + S^*RR \tag{8.20}$$

What an observer sees therefore, is the sum of four light waves. The third wave is $SR^*R = |R|^2 S$, which is the original field multiplied by a real constant. This is the three-dimensional image we want. However, this image is accompanied by three other fields. The first term in (8.20) can be made negligible if we make the reference light much more intense than the signal—that is, let

$$RR^* \gg SS^*$$

The second term is $|R|^2 R$, and it is simply a uniform background light.

---

* Invented by Dennis Gabor (1900–1979), a Hungarian-born electrical engineer, who won the 1971 Nobel prize in physics for this invention.

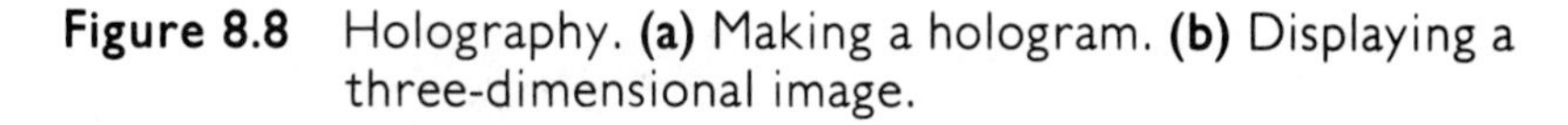

Figure 8.8 Holography. **(a)** Making a hologram. **(b)** Displaying a three-dimensional image.

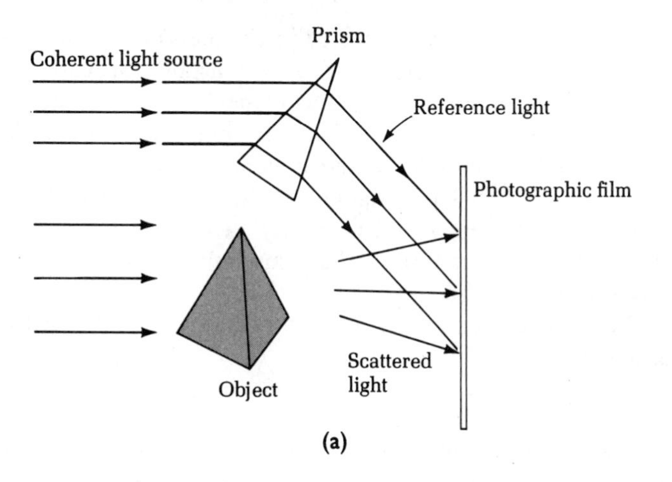

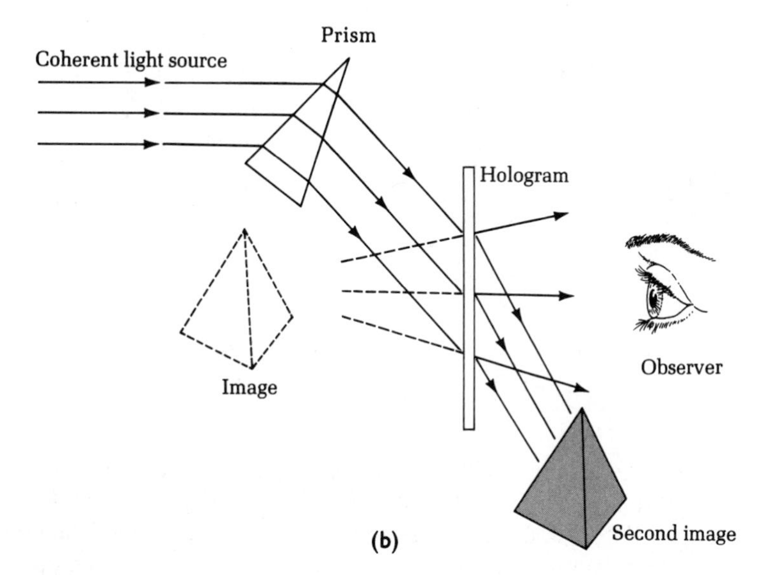

The third term in (8.20) is our image. If the hologram is illuminated by a normally incident uniform plane wave, then ϕ_r is a constant on the screen, and the fourth term is the conjugate of the image with a constant phase shift. Thus, two images will appear, as shown in Figure 8.8. The observer can select an observation angle so that only one image will be seen.*

---

* J. W. Goodman, *Introduction to Fourier Optics*, New York: McGraw-Hill, 1968, chap. 8.

8.3 GAUSSIAN BEAM

The uniform plane wave discussed in Chapter 3 is characterized by the following: (a) the amplitude of the wave is a constant throughout the space, and (b) constant-phase surfaces are parallel planes.

This section will discuss a more complex case: the characteristics of a wave with nonuniform amplitude and nonplanar phase fronts. An example is a light beam from a laser. The amplitude of the light beam at the output of the laser is strongest at the center and decreases away from the center. We wish to know whether the cross section of the beam will remain unchanged and whether the constant-phase surfaces are planar.

We assume that, at the reference plane $z = 0$, the electric field of a nonuniform wave is given by

$$\mathbf{E}(x, z = 0) = \hat{y} E_0 e^{-x^2/w^2} \tag{8.21}$$

In other words, the wave is polarized in the $\hat{y}$ direction, and its amplitude varies with x, as shown in Figure 8.9. To keep the mathematics as simple as possible, we assume that amplitude does not vary with y.

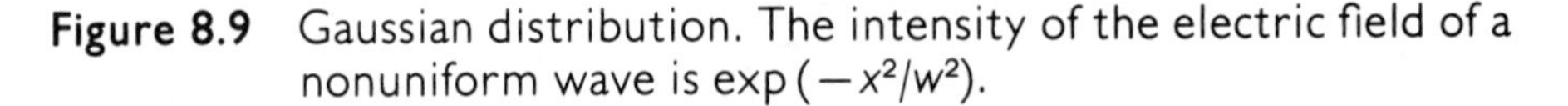

Figure 8.9 Gaussian distribution. The intensity of the electric field of a nonuniform wave is $\exp(-x^2/w^2)$.

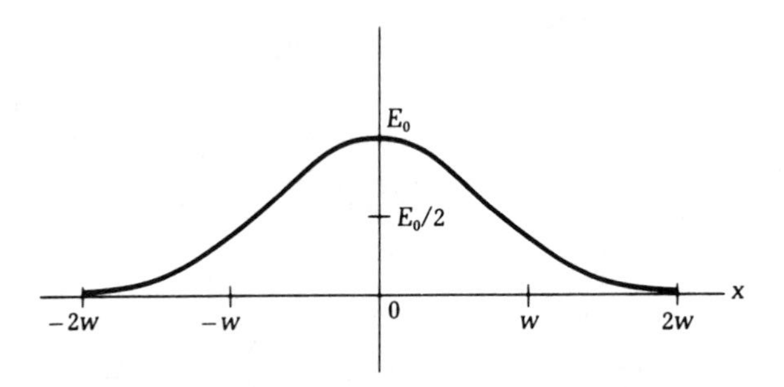

As Figure 8.9 illustrates, the amplitude of the wave is strongest at the center and decreases rapidly outward. The intensity at $x = w$ decreases to $1/e$, or 37%, or -8.7 dB, from the maximum value at the center. The quantity w is called the beam width. The choice of the function e^{-x^2/w^2} is based on the facts that, first, in actual cases it approximates the laser intensity quite well and, second, it is a mathematically well-behaved function. This particular distribution function is called the **Gaussian distribution**, in honor of the German mathematician and astronomer K. F. Gauss (1777–1855).

From Chapter 4 we know that a plane wave with $\hat{y}$ polarization may be expressed as

$$\mathbf{E} = \hat{y} A \exp(-jk_x x - jk_z z)$$

with $k_x^2 + k_z^2 = k^2$, where $k = \omega\sqrt{\mu\epsilon}$.

An ensemble of many plane waves may be expressed as

$$\mathbf{E}(x, z) = \hat{y} \int_{-\infty}^{\infty} dk_x \, \mathrm{A}(k_x) \exp(-jk_x x - jk_z z) \tag{8.22}$$

where $k_z = \sqrt{k^2 - k_x^2}$. The quantity $\mathrm{A}(k_x)$ can be thought of as the amplitude of that component of the plane wave whose wavenumber in the $\hat{x}$ direction is k_x. We may obtain $\mathrm{A}(k_x)$ by equating (8.21) with (8.22) at $\hat{z} = 0$:

$$\int_{-\infty}^{\infty} dk_x \, \mathrm{A}(k_x) e^{-jk_x x} = E_0 \, e^{-x^2/w^2} \tag{8.23}$$

We recognize that the integral is just the Fourier transform of $\mathrm{A}(k_x)$. Thus, $\mathrm{A}(k_x)$ may be obtained from the inverse Fourier transform:

$$\mathrm{A}(k_x) = \frac{1}{2\pi} \int_{-\infty}^{\infty} dx E_0 \, e^{-x^2/w^2} e^{jk_x x} \tag{8.24}$$

The above integration may be carried out analytically* to yield

$$\mathrm{A}(k_x) = \frac{E_0 \, w}{2\sqrt{\pi}} \exp\left(\frac{-w^2 k_x^2}{4}\right) \tag{8.25}$$

So far, we have shown that a Gaussian beam with amplitude distribution given by (8.21) may be thought of as the superposition of an infinite number of uniform plane waves going in all directions (or having a wide range of k_x values) whose amplitudes are corresponding values of $\mathrm{A}(k_x)$ given by (8.25).

Substituting (8.25) into (8.22), we obtain the amplitude of the Gaussian beam at an arbitrary plane z:

$$\mathbf{E}(x, z) = \hat{y} \frac{E_0 \, w}{2\sqrt{\pi}} \int_{-\infty}^{\infty} dk_x \exp\left(\frac{-w^2 k_x^2}{4}\right) \exp(-jk_x x - jk_z z) \tag{8.26}$$

with $k_z = \sqrt{k^2 - k_x^2}$. Note that, when $k_x > k$, $k_z = -j\sqrt{k_x^2 - k^2}$, so that the above expression remains bounded as z approaches infinity. Changing the variable from k_x to u, where $k_x = ku$ and $dk_x = k\,du$, yields the following result:

$$\mathbf{E}(x, z) = \hat{y} \frac{E_0 \, kw}{2\sqrt{\pi}} \int_{-\infty}^{\infty} du \, \exp\left(\frac{-k^2 w^2 u^2}{4}\right) \times \exp\left(-jkux - jk\sqrt{1 - u^2} z\right) \tag{8.27}$$

For a Gaussian beam with a cross section many wavelengths wide, as is usually the case for a laser beam, $kw \gg 1$, and the first exponential function is everywhere negligible except where u is very small compared with unity. Consequently, the

---

* $\int_{-\infty}^{\infty} \exp(-p^2 x^2 \pm qx) \, dx = \sqrt{\pi}/p \, \exp(q^2/4p^2)$. See I. S. Gradshteyn and I. M. Ryzhik, *Table of Integrals, Series, and Products*, New York: Academic Press, 1965, p. 307.

factor $(1 - u^2)^{1/2}$ in the second exponent may be approximated by $(1 - u^2/2)$, and (8.27) becomes

$$\mathbf{E}(x, z) = \hat{y}\frac{E_0 kw}{2\sqrt{\pi}}\int_{-\infty}^{\infty} du \exp(-\mathrm{p}^2u^2 + \mathrm{q}u + \mathrm{c}) \tag{8.28}$$

where $\mathrm{p}^2 = \frac{1}{4}k^2w^2 - \frac{1}{2}jkz$, $\mathrm{q} = -jkz$, and $\mathrm{c} = -jkz$. This integral is of the same form as (8.24); therefore, it can be integrated analytically:

$$\begin{aligned}\mathbf{E}(x, z) &= \hat{y}\frac{E_0 kw}{2\mathrm{p}} e^{\mathrm{q}^2/4\mathrm{p}^2} e^{-jkz} \\ &= \hat{y}E_0 \frac{1}{\sqrt{1 - j\dfrac{z}{z_f}}} e^{-jkz} \exp\left[\frac{-x^2}{w^2\left(1 + \dfrac{z^2}{z_f^2}\right)}\left(1 + j\frac{z}{z_f}\right)\right]\end{aligned} \tag{8.29}$$

where $z_f = kw^2/2$. The amplitude and the phase of the wave are predominantly determined by the exponential factors in (8.29) when $z \gg z_f$. Thus, after the beam has traveled a long distance, the width of the beam at z is

$$\text{width} = \frac{wz}{z_f} \tag{8.30}$$

and the phase is given by

$$\begin{aligned}\text{phase} &= -kz - \frac{x^2}{w^2}\frac{z_f}{z} + \frac{\pi}{4} \\ &= -k\left(z + \frac{x^2}{2z}\right) + \frac{\pi}{4} \\ &\approx -k\sqrt{z^2 + x^2} + \frac{\pi}{4}\end{aligned} \tag{8.31}$$

Figure 8.10 shows the beam width as a function of z and the constant-phase

Figure 8.10 Diffraction of a Gaussian beam. The beam width increases linearly with z as z/z_f approaching infinity.

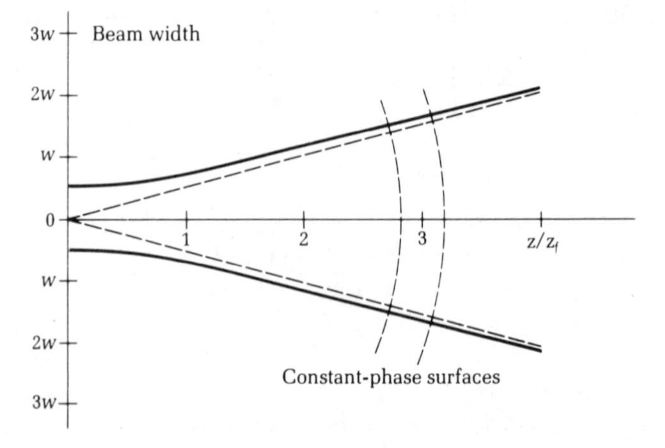

surfaces of a Gaussian beam. Thus, having traveled a long distance, a Gaussian beam will be wider, with its width linearly proportional to the distance it has traveled, and its phase front will become cylindrical. This phenomenon is called **Gaussian beam diffraction**.

Example 8.4 *This example gives an estimate of how much a laser beam can spread over a long distance.*

Assume that on the earth a laser beam of $\lambda = 6328$ Å is aimed at the moon. The initial beam width is 1 cm. Use (8.30) to estimate the width of the beam on the moon.

Solution $z_f = 496$ m. The distance between the earth and the moon is approximately 3.8×10^8 m. Thus, $w_z = 7.7$ km.

We see that the spot size of the laser beam is quite small compared with the diameter of the moon, which is approximately 3476 km.

In 1969 astronauts in the Apollo 11 space flight placed a laser reflector on the moon. A laser beam on earth illuminates the reflector, and the reflected light is received on earth. By measuring the time required for the laser beam to travel to the reflector and back, scientists precisely measured the distance between earth and the moon. This measurement is possible because the diffraction of the laser beam over this long distance is still mild enough to allow reflected energy from the moon reflector to be detected on earth.

Spherical Mirrors in Lasers

In a laser, atoms of a material are excited to higher energy levels. The laser action occurs when these atoms drop to lower energy levels and release the differential energy in electromagnetic waves at optical frequency. To synchronize the emission of light waves by atoms so that coherent light may be generated, a laser requires a resonant cavity. A cavity is formed by placing two reflecting mirrors facing each other, as shown in Figure 8.11a (p. 276). These two mirrors reflect light back and forth. If after two reflections a wave is superimposed upon itself, the condition for reinforcement, or resonance, is achieved. In a laser, the emission of light by atoms is synchronized with this resonance so as to emit sustained light.

The first laser cavities were formed with plane-parallel mirrors, as shown in Figure 8.11a. It was soon learned that the light in the laser is more accurately represented by a Gaussian beam. To accommodate this knowledge, spherical mirrors are used to match the wave front better, as shown in Figure 8.11b.*

* H. Kogelnik and T. Li, "Laser beams and resonators," *Applied Optics*, October 1966, pp. 1550–1567.

Figure 8.11(a) Light is generated at A, is reflected at B by the mirror, and is reflected again by the second mirror. Resonance occurs when the wave at C superimposes that at A.

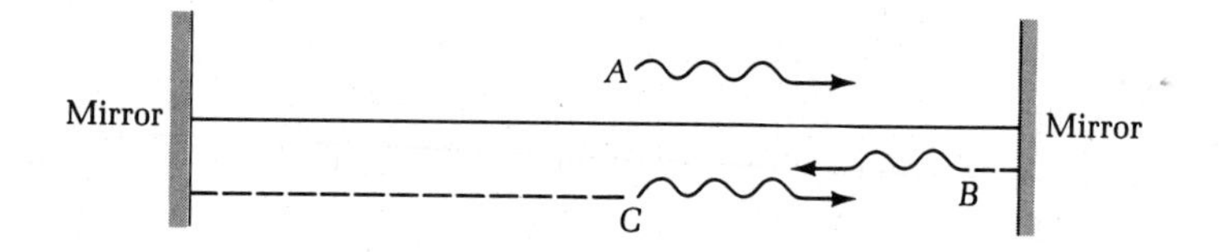

Figure 8.11(b) Spherical mirrors are used to form the laser cavity because they match the Gaussian beam wave front better.

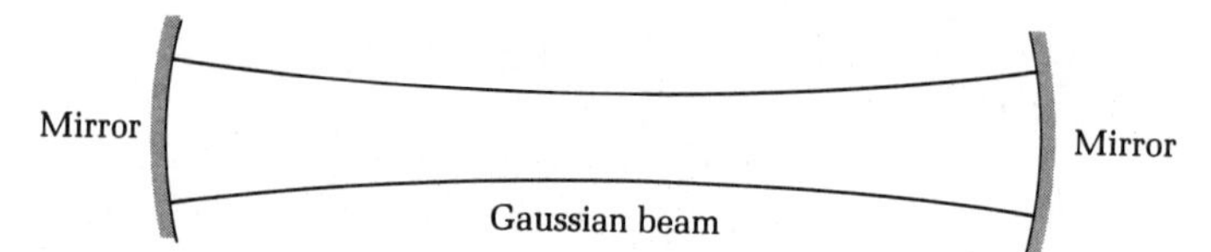

8.4 DOPPLER EFFECT

When the source of a time-harmonic wave is stationary with respect to an observer, the frequency of the observed signal is the same as that of the source. But when the source and the receiver are in relative motion, the apparent frequency detected by the receiver is usually different from the actual frequency. The difference in the transmitted and the received frequencies depends on the relative velocity of the source and the receiver. This discrepancy occurs in sound and electromagnetic waves and is known as the **Doppler effect**, named after the German physicist C. Doppler (1803–1853), who described the effect in 1842.

Consider the situation shown in Figure 8.12. We assume that at $t = t_0$ the transmitter is moving at a constant speed v in the θ direction with reference to the line SR. The wave the transmitter emits at time t_0 propagates outward and arrives at the receiver at time t', where

$$t' = t_0 + \frac{r_0}{c}$$

A period later, at time $t_0 + T$, the transmitter has moved to a new position S', as shown in Figure 8.12b. The distance SS' is equal to (Tv) meters. The wave emitted at time $t_0 + T$ propagates outward and arrives at the receiver at time t'', where

$$t'' = t_0 + T + \frac{1}{c}(S'R)$$

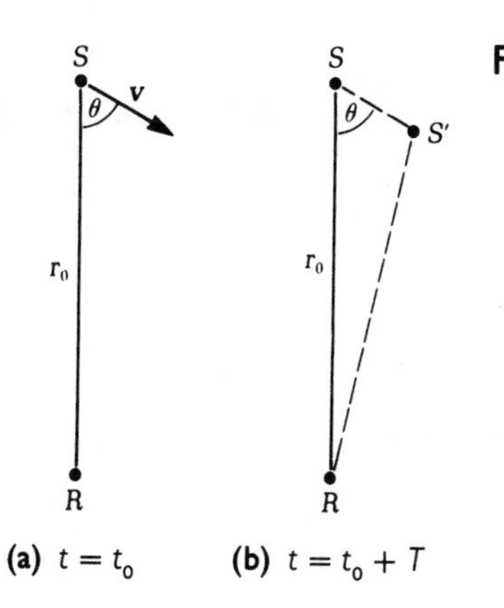

(a) $t = t_0$ (b) $t = t_0 + T$

Figure 8.12 A source is moving relative to a stationary receiver. **(a)** The distance between the source and the receiver is SR at $t = t_0$. **(b)** It is changed to $S'R$ at $t = t_0 + T$.

From trigonometry, we have

$$S'R = (r_0^2 - 2r_0 Tv \cos\theta + T^2v^2)^{1/2}$$

Thus, the apparent period of the wave detected by the receiver is T', where

$$T' = t'' - t'$$

or

$$T' = T + \frac{1}{c}(r_0^2 - 2r_0 Tv \cos\theta + T^2v^2)^{1/2} - \frac{1}{c}r_0$$

Assume that the velocity v of the transmitter and the period T are small so that $Tv \ll r_0$; then, taking the first two terms of the square root and using binomial expansion yields*

$$T' \approx T\left(1 - \frac{v}{c}\cos\theta\right) \tag{8.32}$$

The apparent frequency f' detected by the receiver is the inverse of T', or

$$\boxed{f' \approx f\left(1 + \frac{v}{c}\cos\theta\right)} \quad \textbf{(apparent frequency of a transmitter moving with a velocity } v\textbf{)} \tag{8.33}$$

if $v \ll c$. Note that θ is equal to zero when the transmitter is moving directly toward the receiver and when the apparent frequency detected by the latter is higher than the actual frequency. Conversely, the apparent frequency is lower when the transmitter moves away from the receiver.

* Choosing T as the second reference is unnecessary, and any reference time Δt may be used to replace T. Thus, $\Delta tv \ll r_0$ is always satisfied.

Example 8.5 *Sound waves also have the Doppler effect. This example explains a well-known phenomenon.*

Explain why a train whistle sounds different before and after the train has passed a listener.

Solution Let us assume that the frequency of the whistle is 1 kHz and that the speed of the train is 50 mi/h. The velocity of the sound is approximately 331 m/s. The train speed converted to MKS units is found to be 22.4 m/s.

When the train is approaching a listener,

$$f' = 1000\left(1 + \frac{v}{c}\right) = 1068 \text{ Hz}$$

As the train moves away, the frequency heard by the listener changes to

$$f'' = 1000\left(1 - \frac{v}{c}\right) = 932 \text{ Hz}$$

Thus, the pitch of the whistle is higher when the train is approaching than when it is leaving a listener. The difference in frequency is 136 Hz for a basic frequency of 1 kHz when the train speed is 50 mi/h.

Example 8.6 *A simple rule about the Doppler frequency shift of a moving transmitter is that it is equal to 1.1 Hz/Mach/1 MHz. This example shows where this number comes from.*

A jet airplane is flying at a speed of Mach 1 (the speed of sound in air). The electromagnetic wave radiated by an antenna in the aircraft is detected by a receiver located on the ground ahead of the airplane. Estimate the Doppler frequency shift detected by the gound receiver.

Solution The Doppler frequency shift of a time-harmonic electromagnetic wave radiated by a moving transmitter is approximately equal to 1.1 Hz per Mach per 1 MHz. This result can be obtained directly from (8.33). Again, the Doppler frequency shift is upward when the transmitter and the receiver are approaching each other, and downward when they are moving apart.

Red Shift

When a distant radio-source is moving away from the earth, the light it emits shifts to the lower frequency, or red, end of the spectrum. Thus, red shift in the light spectrum from a galactic body may suggest a velocity of recession. In fact, the observed red shift increases steadily and perhaps linearly with distance. This red shift gives us a picture of an expanding universe.

On June 14, 1973, E. J. Wampler and E. M. Burbidge, astronomers at the University of California, reported the discovery of a quasar that is believed to be the most distant recorded object in the universe. The object, designated 0Q172, may be ten billion light-years from earth.

Quasar is an acronym for quasi-stellar radio-source. A key characteristic of quasars is that they show large red shifts in the electromagnetic spectrum recorded from their light. Many astronomers interpret large red shifts as indicating great distance and rapid motion. Therefore, 0Q172, the object with the largest red shift ever recorded, is believed to be as far into the universe as any object yet observed.*

Doppler Frequency Shift of a Wave Reflected from a Moving Target

Equation (8.33) gives the Doppler-shifted frequency of a wave that comes from a moving transmitter and is detected by a stationary observer. In many applications, such as police radar, electromagnetic waves are radiated toward a moving target and are reflected or scattered back toward the radar. The radar receiver will detect that the reflected wave has a different frequency! This difference is also due to the Doppler effect.

For mathematical simplicity, we assume that the wave is incident on the surface of a plate of a perfect conductor that is moving with a velocity of $\boldsymbol{u}$ normal to its own surface. Figure 8.13 shows this situation.

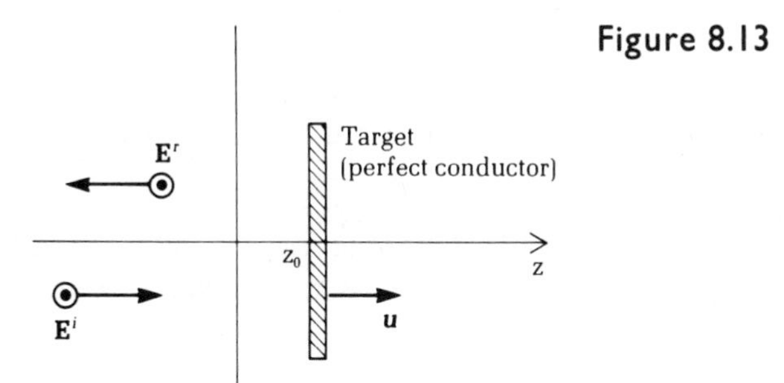

Figure 8.13 A wave strikes a moving target, and the frequency of the reflected wave shows a Doppler shift.

The incident wave is expressed by

$$\mathbf{E}^i = \hat{y} E_0 \exp(-jkz + j\omega t) \tag{8.34}$$

* T. P. Gill, *The Doppler Effect*, New York: Academic Press, 1965.
W. D. Metz, "Cosmology: The first large radio redshift," *Science* 181, no. 4105 (September 1973): 115.

The reflected wave is expressed by

$$\mathbf{E}^r = \hat{y} R E_0 \exp(jk'z + j\omega' t) \tag{8.35}$$

Because both the incident and the reflected waves are propagating in the same medium, the values of k and k' are given by

$$\begin{aligned} k &= \omega\sqrt{\mu\epsilon} \\ k' &= \omega'\sqrt{\mu\epsilon} \end{aligned} \tag{8.36}$$

The boundary condition requires that

$$\exp(-jkz_0 + j\omega t) + \mathrm{R}\exp(jk'z_0 + j\omega' t) = 0$$

Thus,

$$\mathrm{R} = -\exp[-j(k' + k)z_0 + j(\omega - \omega')t]$$

The reflected wave is then given by

$$\mathbf{E}^r = -\hat{y}E_0 \exp[-j(k + k')ut + j\omega t + jk'z]$$

where the relation $z_0 = ut$ has been used. Comparing this equation with (8.35) reveals that

$$\omega' = \omega - (k + k')u$$

or

$$\frac{\omega'}{\omega} = 1 - \frac{k}{\omega}\left(1 + \frac{k'}{k}\right)u$$

or

$$\frac{\omega'}{\omega} = 1 - \frac{u}{c}\left(1 + \frac{\omega'}{\omega}\right) \tag{8.37}$$

Equation (8.37) is obtained from the facts that $\omega/k = c$ and $k'/k = \omega'/\omega$, where c is the velocity of the wave in the medium. Solving (8.37) for ω'/ω yields

$$\boxed{\frac{\omega'}{\omega} = \frac{f'}{f} = \frac{1 - (u/c)}{1 + (u/c)}} \quad \textbf{(apparent frequency of the wave reflected by a moving reflector)} \tag{8.38}$$

or

$$\boxed{\frac{f'}{f} \approx 1 - \frac{2u}{c} \quad \text{if} \quad \frac{u}{c} \ll 1} \quad \textbf{(Doppler shift)} \tag{8.39}$$

Notice first that the Doppler shift is downward when the reflecting plate is moving away from the radar, as shown in Figure 8.13. If the target is moving toward the source of the electromagnetic wave, u must change sign, and the Doppler shift in frequency is then upward. These effects are similar to those in the case studied earlier in which the wave is radiated from a moving source and detected by a stationary receiver. But, quantitatively, the Doppler shift is twice as much in the present case shown in Figure 8.13 than in the previous case shown in Figure 8.12.

Example 8.7 *The Doppler frequency shift of an EM wave bounced off a moving object is twice of that of an EM wave radiated directly by the moving object. Note the difference between this example and Example 8.6.*

Assume that a police Doppler-radar* radiates at 10.525 GHz. What is the range of Doppler shift when the radar is used to detect the speeds of vehicles traveling at from zero to 100 mi/h?

Solution

$$u = \pm 100 \text{ mi/h} = \pm 44.7 \text{ m/s}$$

$$\Delta f = \frac{2u}{c} f$$

$$= \pm 3137 \text{ Hz}$$

Doppler Radar

Figure 8.14 (p. 282) shows a simple block diagram that illustrates the basic configuration of a Doppler radar. The oscillator generates a time-harmonic signal at frequency f_0, and the transmitting antenna radiates the signal as an electromagnetic wave. Part of the signal is coupled to the mixer in the detector circuit. The target echoes part of the electromagnetic energy back to the radar. The frequency of the wave returning in this way is different from the frequency of the wave originally transmitted. The returning wave is fed to the mixer via the receiving antenna. The mixer mixes the original f_0 signal with the echo, which is at $f_0 + \Delta f$, and generates a signal that has a frequency equal to the difference of the two, that is, Δf. This frequency is counted by a counter. According to (8.39), Δf is proportional to u by a constant factor. The measured Δf is multiplied by this constant factor, and the result is displayed.

Because the velocity of the electromagnetic wave is large, the above process can be accomplished in milliseconds or microseconds. Usually, the radar has a switch and uses only one antenna. In the transmitting mode, the switch connects

* P. D. Fisher, "Shortcomings of radar speed measurement," *IEEE Spectrum*, December 1980, p. 28.

Figure 8.14 Block diagram of a Doppler radar.

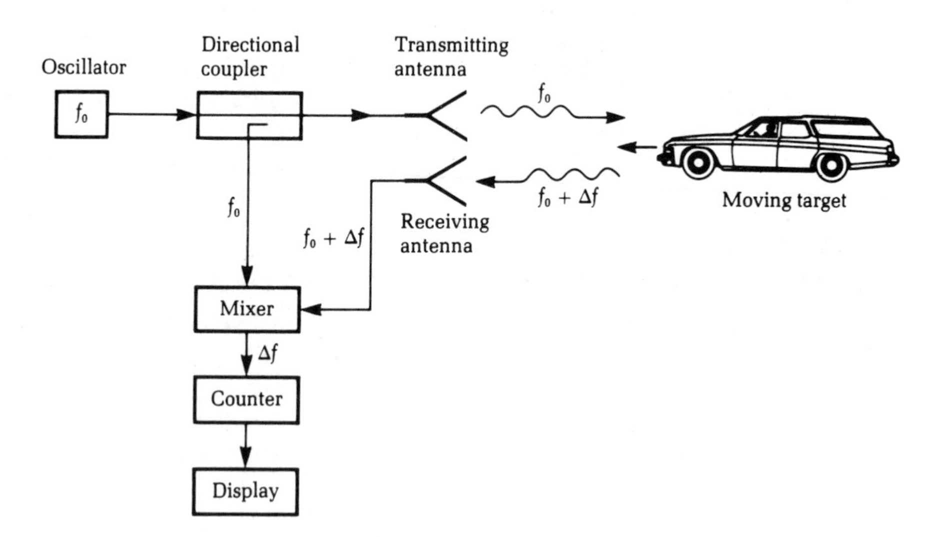

the oscillator output to the antenna, which serves as a radiator. The switch then cuts off the power and connects the antenna to the receiver circuit. Finally, the switch goes back to the transmitting mode for the next cycle after the frequency of the received echo is measured.

Frequency-Modulated Continuous-Wave Doppler Radar

A pulsed radar can measure the distance between the radar and the target. It sends out a pulse at t, and, if the echo from the target is received at $t + t_0$, the target must be r meters away, where $r = ct_0/2$. The pulsed radar cannot directly measure the velocity of the target.

A Doppler radar is able to measure the velocity of a moving target. This radar sends out continuous waves of frequency f_0. If the frequency of the echo reflected from the target is $f_0 + \Delta f$, the longitudinal velocity of the target must be equal to v where $v = (\Delta f/f_0)(c/2)$.

An FM-CW Doppler radar uses the combination of the two techniques. The radar radiates a continuous wave whose frequency changes over time, as shown in Figure 8.15. The frequency of the echo from the target is also a function of time. Thus, the waveform of the echo contains information about both the distance and the velocity of the target.

Referring to Figure 8.15, we see that the returned signal's frequency shift is proportional to the speed of the target. The waveform of the echo's spectrum also shows a definite shift in time scale. The time delay is proportional to the distance between the target and the radar.

Figure 8.15 An FM-CW Doppler radar sends out a frequency-modulated signal with the waveform shown in solid lines. The dotted lines are the waveform of the reflected signal. Comparing the original and the reflected signal reveals the distance and the velocity of the target.

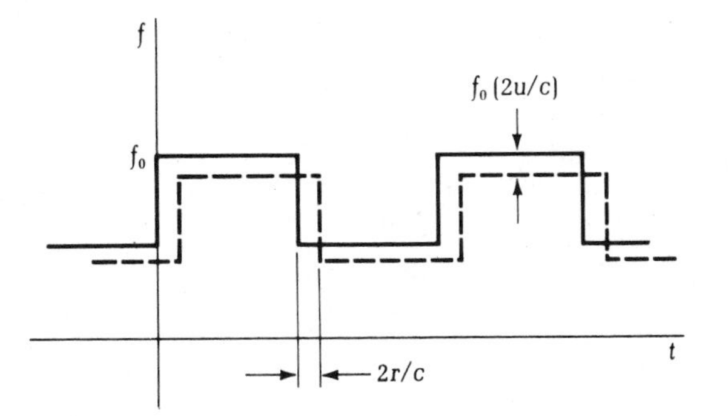

8.5 PLANE WAVES IN ANISOTROPIC MEDIA

A uniaxial medium is an anisotropic medium whose constitutive relations take the following simple forms:

$$\mathbf{B} = \mu\mathbf{H} \quad \text{and} \quad D_x = \epsilon_x E_x$$
$$D_y = \epsilon E_y$$
$$D_z = \epsilon E_z$$

Writing more compactly by using the permittivity tensor $\bar{\bar{\epsilon}}$, we have

$$\mathbf{D} = \bar{\bar{\epsilon}} \cdot \mathbf{E} \quad \text{where} \quad \bar{\bar{\epsilon}} = \begin{bmatrix} \epsilon_x & 0 & 0 \\ 0 & \epsilon & 0 \\ 0 & 0 & \epsilon \end{bmatrix} \tag{8.40}$$

Transforming to other coordinates, we find that the off-diagonal elements of the matrix are no longer zero. The coordinate system in which $\bar{\bar{\epsilon}}$ has only diagonal elements and in which two of these elements are the same is referred to as the principal coordinate system of the uniaxial medium.

Let us find the associated wave equation for such a uniaxial medium. Taking the curl of equation (2.28a), substituting (2.28b), and noting that $\mathbf{J} = 0$, we obtain

$$\nabla(\nabla \cdot \mathbf{E}) - \nabla^2\mathbf{E} = \omega^2\mu\mathbf{D} \tag{8.41}$$

We assume that $\mathbf{E}(z) = \mathbf{E}_0 e^{-jkz}$ where $\mathbf{E}_0$ is a constant vector. From $\nabla \cdot \mathbf{D} = 0$, we have $\partial D_z/\partial z = 0 = \epsilon\, \partial E_z/\partial z = -jk\epsilon E_z$, which dictates that $E_z = 0$—that is, that no component of $\mathbf{E}$ exists in the direction of propagation (the $\hat{z}$ direction).

For the assumed field variation, therefore, $\nabla \cdot \mathbf{E} = 0$ and

$$-\nabla^2 \mathbf{E} = \omega^2 \mu \mathbf{D} \tag{8.42}$$

Consider each component of the above equation:

$$(\hat{x} \text{ component}) \quad \left(\frac{\partial^2}{\partial z^2} + \omega^2 \mu \epsilon_x\right) \mathrm{E}_x = 0 \tag{8.43a}$$

$$(\hat{y} \text{ component}) \quad \left(\frac{\partial^2}{\partial z^2} + \omega^2 \mu \epsilon\right) \mathrm{E}_y = 0 \tag{8.43b}$$

$$(\hat{z} \text{ component}) \quad \mathrm{E}_z = 0 \tag{8.43c}$$

The solution of (8.43a) is as follows:

$$\mathrm{E}_x = E_{0x} e^{-jk_e z} \tag{8.44a}$$

where

$$\boxed{k_e = \omega \sqrt{\mu \epsilon_x}} \quad \textbf{(wavenumber of the extraordinary wave)} \tag{8.44b}$$

Similarly, the solution of the $\hat{y}$ component of the electric field is

$$\mathrm{E}_y = E_{0y} e^{-jk_0 z} \tag{8.45a}$$

where

$$\boxed{k_0 = \omega \sqrt{\mu \epsilon}} \quad \textbf{(wavenumber of the ordinary wave)} \tag{8.45b}$$

We call the wavenumber associated with the single dissimilar dielectric constant, ϵ_x, the **extraordinary** wavenumber, and the corresponding polarized wave the extraordinary wave. Similarly, the wavenumber associated with the two identical dielectric components, ϵ, is the **ordinary** wavenumber, and its corresponding polarized wave is the ordinary wave. Each separate polarization propagates with a different velocity. This property is known as **birefringence**.

How to Produce a Circularly Polarized Light

Consider a plate made of an anisotropic material, as shown in Figure 8.16a. Let a linearly polarized wave incident on the medium from the left be

$$\mathbf{E} = E_0(\hat{x} + \hat{y}) e^{-jkz} \quad \text{for} \quad z < 0 \quad k = \omega \sqrt{\mu_0 \epsilon_0} \tag{8.46}$$

If we neglect the effects of reflection at the boundaries $z = 0$ and $z = d$, then, upon exit at $z = d$, the wave is

$$\mathbf{E} = E_0(\hat{x} e^{-jk_e d} + \hat{y} e^{-jk_0 d}) e^{-jk(z-d)} \quad \text{for} \quad z > d \tag{8.47}$$

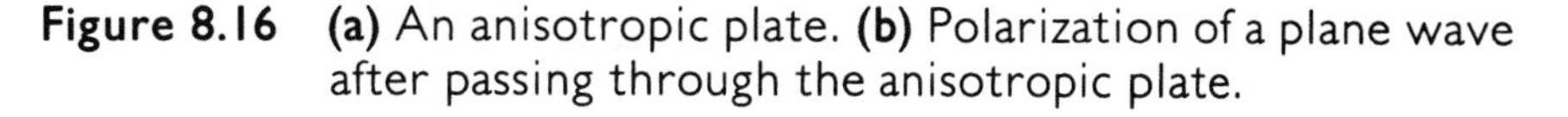

Figure 8.16 **(a)** An anisotropic plate. **(b)** Polarization of a plane wave after passing through the anisotropic plate.

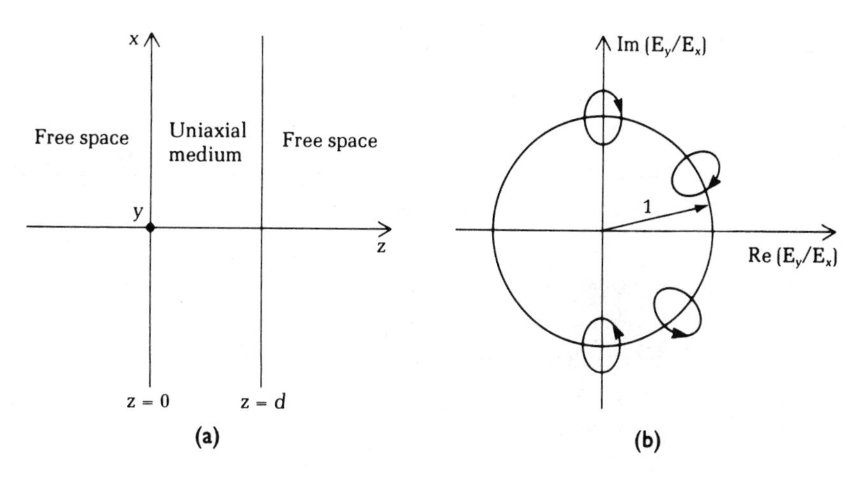

This form was obtained by breaking up the incident wave into components along the $\hat{x}$ and $\hat{y}$ and allowing each component to propagate through the medium (with its corresponding extraordinary or ordinary wavenumber) and then to recombine at $z = d$.

The form of the polarization upon exit (and for all $z > d$) can be determined from

$$\frac{E_y}{E_x} = exp(-jk_0 d + jk_e d) \tag{8.48}$$

Figure 8.16b indicates diagrammatically the possible types of polarizations.

What do we need to produce circular polarization? We require

$$-k_0 d + k_e d = \pm\frac{\pi}{2} + 2n\pi \tag{8.49}$$

or

$$d = \frac{\pm\frac{\pi}{2} + 2n\pi}{k_e - k_0}$$

The shortest distance d is

$$d = \frac{\frac{\pi}{2}}{k_e - k_0}$$

Define

$$k_e - k_0 = k_B$$

and

$$k_B \lambda_B = 2\pi$$

For circular polarization,

$$k_B d = \frac{\pi}{2} \quad \text{or} \quad d = \frac{\lambda_B}{4}$$

We therefore call a plate of uniaxial material with this thickness (plus an integer number of λ_B) a **quarter-wave plate.**

Dichroism and Polarized Sunglasses

Many crystals, such as natural crystals of calcite and quartz, exhibit anisotropic properties at optical frequencies. For calcite crystals the values of ϵ_x and ϵ are equal to 2.21 ϵ_0 and 2.76 ϵ_0, respectively. Note that these numbers are real. Therefore, depending on its polarization, light propagates in these crystals with different velocities, but it is not attenuated.

Some other anisotropic crystalline materials transmit light in one polarization and absorb it in another. Consequently, for one polarization, the permittivity is a real number, and for the other it is a complex number. This property is called **dichroism**. Tourmaline crystals are dichroic media.

In 1928, a young boy named Edwin H. Land developed a film through which a randomly polarized light became linearly polarized. This film was originally made of imbedded dichroic crystals and was later improved to contain long chains of iodine molecules. It is now commercially known as **Polaroid.*** Polaroid is an artificial dichroic material.

In Chapter 4 we learned that sunlight reflected from dielectric surfaces contains a longer horizontal component than vertical component (see Section 4.2, Figure 4.10). Therefore, if sunglasses are made of polarized film so oriented that the film absorbs horizontally polarized light and lets pass the vertically polarized light, then these glasses can better filter out glares than glasses made of tinted glass, which attenuates light equally regardless of its polarization.

Liquid Crystals

Liquid crystals are liquids whose molecules are arranged orderly. A liquid crystal can be activated by an electric field. In its unactivated state, the crystal is isotropic. When an electric field is applied, the molecules align themselves either parallel to the field or perpendicular to it, and the liquid crystal becomes a uniaxial medium.

---

* This Polaroid should not be confused with the instant photographic film of the same name, although the same inventor devised them both.

Figure 8.17 shows a typical arrangement of a liquid-crystal display.* It is operated in the so-called distortion-of-aligned-phases or DAP mode. Figure 8.17a shows the normal state of the crystal before activation. The light entering the crystal is polarized and then transmitted through the crystal with no alteration in polarization. The second polaroid absorbs all the light, and no light is transmitted. In its activated state, the crystal changes the polarization of the transmitted light, which propagates through the second polaroid and becomes visible.

Figure 8.17 Liquid crystal as a display device.

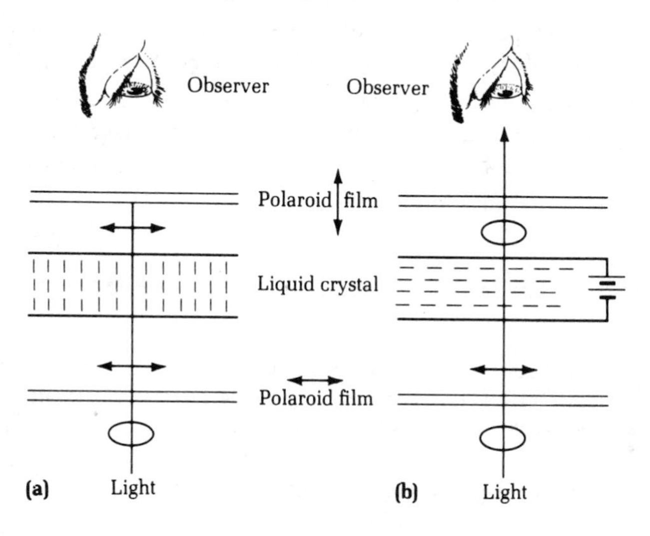

SUMMARY

1. Electromagnetic waves are scattered when there are particles in their paths. When the size of the particle is small compared with the wavelength, the scattering process is called Rayleigh scattering.
2. The scattering of an electromagnetic wave can be described in terms of the total or back-scattering cross section. For a spherical particle, the total Rayleigh scattering cross section is given by (8.9), and the back-scattering cross section is given by (8.10). Note that both are proportional to the fourth power of the frequency.
3. An electromagnetic wave passing through a slit will be diffracted to become a nonuniform wave whose space distribution function is the Fourier transform of the distribution function of the wave amplitude at the slit.
4. Holography is the technique to record the amplitude and the phase of the wave scattered by an object and then to regenerate the amplitude and the phase of the scattered wave to create a three-dimensional image of the object.

---

* See R. W. Gurtler and C. Maze, "Liquid crystal displays," *IEEE Spectrum*, November 1972, p. 25.

5. A Gaussian beam is an EM wave whose initial spatial amplitude distribution is like $\exp(-x^2/w^2)$, where x is the spatial coordinate and w a distance factor defined as the initial beam width. Such a beam will propagate in space and its amplitude distribution will spread out. The beam width will increase linearly with z, the distance the wave has traveled.
6. The apparent frequency of an EM wave radiated from a moving transmitter and detected by a stationary receiver is higher than the actual frequency when the transmitter is moving toward the receiver; and it is lower than the actual frequency if the transmitter is moving away from the receiver. The shift in the frequency is called the Doppler effect.
7. The Doppler shift is doubled if the EM wave is radiated by a stationary radar and reflected back to the radar receiver by a moving target.
8. EM waves with different polarizations will travel with different wave numbers in an anisotropic medium whose electromagnetic properties depend on the polarization of the wave. Polarized sunglasses and liquid crystal displays are two examples of anisotropic media.

Problems

8.1 The derivation of (8.5) considers only the electric field. Why is the magnetic field neglected? Hint: Compare the magnitude of E_s with ηH_s near the sphere, or the stored electric-energy density $(1/2)\epsilon|\mathbf{E}|^2$ with the stored magnetic-energy density $(1/2)\mu|\mathbf{H}|^2$.

8.2 Why is the rising or setting sun red?

8.3 The smoke emitted from engines of boats contains fine particles. Against a dark background the smoke looks blue, but against a bright background it looks yellow. Why?

8.4 Explain the appearance of shafts of sunlight through breaks in a cloud-covered sky.

8.5 Show that

$$I_1 = \int_{-\infty}^{\infty} e^{-x^2}\,dx = \sqrt{\pi}$$

Hint:

$$I_1^2 = \int_{-\infty}^{\infty} e^{-x^2}\,dx \cdot \int_{-\infty}^{\infty} e^{-y^2}\,dy$$

Then, transform x-y coordinates into cylindrical coordinates to perform the exact integration.

8.6 Show that

$$I_2 = \int_{-\infty}^{\infty} \exp(-p^2x^2 + qx)\,dx = \frac{\sqrt{\pi}}{p}\exp\left(\frac{q^2}{4p^2}\right)$$

Hint:

$$-p^2x^2 + qx = -\left(px - \frac{q}{2p}\right)^2 + \frac{q^2}{4p^2}$$

Then, use the result obtained in Problem 8.5 after transforming the integration variable from x to $px - q/2p$.

8.7 Assume that on earth a microwave beam of 10 GHz is radiated by a 20-meter-diameter disk antenna aimed at the moon. Estimate the size of the microwave beam on the moon.

8.8 A person leaving home by train mails a letter home every day. Suppose that the train travels 200 miles per day and that the mail moves at a speed of 200 miles per day. How frequently do the letters arrive home? Try to solve this problem by simple reasoning, not by substituting numbers in some formula.

8.9 A police Doppler radar is located on the roadside 20 meters from the center lane of a highway. A car in that lane is clocked by the radar as traveling at a speed of 120 km/hr. The car is approximately 60 meters away from the radar when the radar registers its reading. The radar has been calibrated to read the speed accurately when the target is traveling directly toward it. If you are the driver, would you argue that the radar does not record the actual speed of your car? What is the actual speed?

8.10 A Doppler radar sends a signal at 8.800 GHz, and the receiver displays a frequency spectrum of returned signals as shown in Figure P8.10. What can you say about the speed of the target(s)?

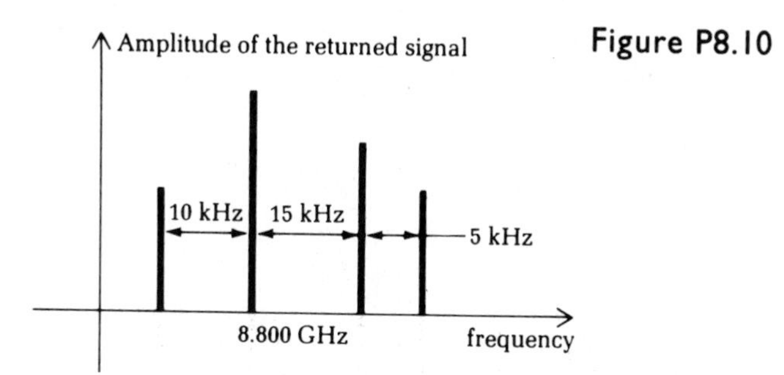

Figure P8.10

8.11 For the FW-CW Doppler radar discussed in Section 8.4, assume that f_0, the upper frequency of the radar, is 8.8 GHz. Suppose the radar is to measure target speeds ranging from 0 to 3 Mach and distances from 1 km to 10 km. Find the system's approximate frequency bandwidth and the time interval the system must be able to resolve.

8.12 If $d = \lambda_B/4$, as shown in Figure 8.16a, and if reflections at interfaces $z = 0$ and $z = d$ are negligible, a linearly polarized wave incident from the left will become a circularly polarized wave, as discussed in the text. What is the polarization of the exiting wave if the reflections at these interfaces are not negligible?

8.13 If $d = \lambda_B/2$, as shown in Figure 8.16a, what is the polarization of the exiting wave if the incident wave from the left is circularly polarized?

8.14 For a quartz crystal, $\epsilon_x = 2.41\epsilon_0$, and $\epsilon = 2.38\epsilon_0$. Find the minimum thickness of a quartz quarter-wave plate for a light having $\lambda = 6500$ Å.

8.15 In Figure P8.15 the Polaroid film at A is oriented such that it passes light polarized in the $\hat{x}$ direction and absorbs light polarized in the $\hat{y}$ direction. The film at B passes y-polarized light and absorbs x-polarized light. A randomly polarized light source, such as flashlight, sheds light from the left along z. Can an observer at C see the light? Explain.

Figure P8.15

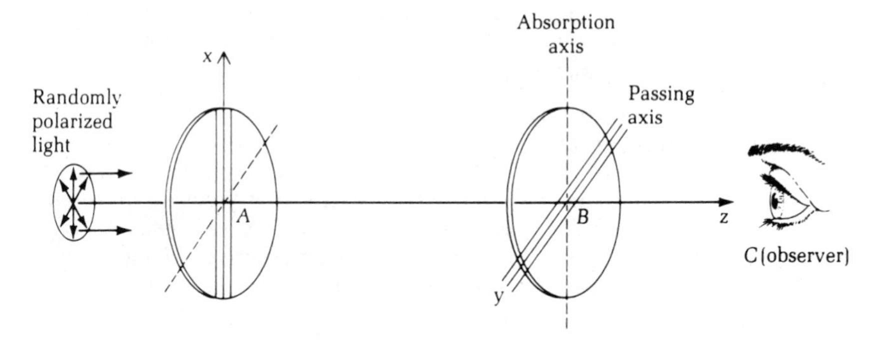

8.16 Consider the arrangement shown in Figure P8.16. This figure differs from Figure P8.15 only in the placement of a third Polaroid film at D between A and B. The absorption axis of the third film is 45° from either the x or the y axis. Now, can the observer at C see the light? (If you do not believe in your answer, do an experiment with three pairs of polarized sunglasses and see for yourself.)

Figure P8.16

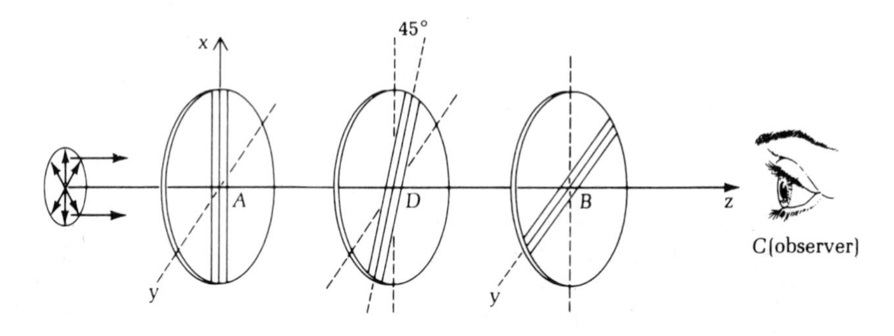

9 ELECTROSTATIC FIELDS

In the second half of this book, we investigate the behavior and the application of electromagnetic fields where ω approaches zero as a limit. Under the limit of $\omega = 0$, Maxwell's equations become decoupled; one set governs the electrostatic fields, whose sources are static charges, and the other set governs the magnetostatic fields, whose sources are steady currents.

Electric charges may be either uniformly distributed in a spherical volume, in which case they produce a **spherical charge**; uniformly distributed along a straight line, in which case they produce a **line charge**; or uniformly distributed over a plane, in which case they produce a **plane charge**. When the radius of a spherical charge approaches zero while the total charge in the volume remains unchanged, we call the charge a **point charge**. In earlier chapters we saw the usefulness of an idealized point-source like a Hertzian dipole antenna. The charge distributions mentioned above serve similar functions. In this chapter we develop techniques for calculating electric fields for a number of these charge distributions. We also develop the helpful concept of electrostatic potential.

9.1 ELECTROSTATIC POTENTIAL

In Chapter 7 we saw how oscillating currents and charges generate electromagnetic waves. When $\omega = 0$, we expect all fields to remain constant in time. That is, $\partial/\partial t$ is zero when operated on $\boldsymbol{B}$ or $\boldsymbol{D}$. Thus, the four Maxwell's equations become the following:

$$\nabla \times \boldsymbol{E} = 0 \tag{9.1}$$

$$\nabla \cdot \boldsymbol{D} = \rho_v \tag{9.2}$$

$$\nabla \times \boldsymbol{H} = \boldsymbol{J} \tag{9.3}$$

$$\nabla \cdot \boldsymbol{B} = 0 \tag{9.4}$$

Note that the first two equations do not contain magnetic fields and that the last two equations do not contain electric fields. Thus, when electromagnetic fields do not vary with time, the existence of the electric field does not depend on the existence of the magnetic field, and vice versa. In this chapter we shall study the electrostatic field produced by a stationary charge ρ_v. Mathematically, we will study the solution to the first two Maxwell's equations, (9.1) and (9.2). Chapter 13 studies the magnetostatic fields.

In simple media, $\boldsymbol{D}$ and $\boldsymbol{E}$ are related by a constant called permittivity:

$$\boxed{\boldsymbol{D} = \epsilon \boldsymbol{E}} \quad \textbf{(definition of permittivity)} \tag{9.5}$$

where ϵ is a real number.

The curl of a gradient of a scalar function is identically zero according to (2.11b). Thus, (9.1) implies that $\boldsymbol{E}$ may be represented as a gradient of some scalar function. Therefore, we write

$$\boxed{\boldsymbol{E} = -\nabla\Phi} \quad \textbf{(definition of scalar potential)} \tag{9.6}$$

which is exactly (7.3) with $\omega = 0$. Substituting (9.6) and (9.5) into (9.2) yields

$$\boxed{\nabla^2\Phi = \frac{-\rho_v}{\epsilon}} \quad \textbf{(Poisson equation)} \tag{9.7}$$

which is exactly (7.6) with $k = 0$. The solution to (9.7) is simply (7.8) with k equal to zero:

$$\boxed{\Phi(\boldsymbol{r}) = \frac{1}{4\pi\epsilon}\iiint_v \frac{\rho_v(\boldsymbol{r}')}{|\boldsymbol{r} - \boldsymbol{r}'|}\,dv'} \quad \textbf{(Poisson integral)} \tag{9.8}$$

Equation (9.7) is known as the **Poisson equation**, which is a partial differential equation. Equation (9.8) is known as the **Poisson integral**, which involves integration over all volumes containing the charge distribution $\rho_v(\boldsymbol{r}')$, where $\boldsymbol{r}$ is the coordinate of the observation point and $\boldsymbol{r}'$ is that of the charge distribution.

When the charge distribution is given, we can substitute it in (9.8), perform the integration, and obtain the potential at any point of interest. After the potential is calculated, the electric field follows directly from (9.6). In this section we shall find the potentials set up by a few typical charges—namely, a point charge, a line charge, and a plane charge. But before doing that, let us note that the unit of potential is **volt**. For example, for a 12-volt battery the potential at the positive terminal is 12 volts higher than the potential at the negative terminal.

Potential of a Point Charge

A charge distribution of particular interest is a point charge, which is an idealization of a charged particle, like an electron, a proton, or a nucleus. For a point

charge situated at the origin of a coordinate system and occupying a differential volume $\Delta V'$, the total charge q is given by

$$q = \rho_v \Delta V' \tag{9.9}$$

For example, an electron carries a charge of $q = -1.6 \times 10^{-19}$ C. It follows from the Poisson integral (9.8) that the potential of the point charge is given by

$$\boxed{\Phi(r) = \frac{q}{4\pi\epsilon r}} \quad \textbf{(scalar potential of a point charge in spherical coordinates)} \tag{9.10}$$

We see that the potential depends only on the coordinate r, and this fact is a direct consequence of the symmetry of the charge distribution. Thus, all points on a spherical surface centered at the origin have the same voltage. The surface on which all points have the same value of potential is called an **equipotential surface.**

Example 9.1

This example shows that the equipotential surfaces due to a point charge are concentric spherical surfaces centered at the charge.

The hydrogen atom, according to Bohr's model, has one electron revolving around the nucleus. The nucleus carries a charge of $q = 1.6 \times 10^{-19}$ C. The quantized orbits of the revolving electron are determined to be 0.52 Å, 2.08 Å, 4.68 Å, etc. The potentials experienced by the electron at these three equipotential surfaces are calculated from (9.10)— namely,

$$\Phi(r) = \frac{1.6 \times 10^{-19}}{4\pi\epsilon r}$$

which gives $\Phi(0.52\ \text{Å}) = 27.6$ V, $\Phi(2.07\ \text{Å}) = 6.92$ V, and $\Phi(4.68\ \text{Å}) = 3.07$ V. Figure 9.1 shows these potential surfaces.

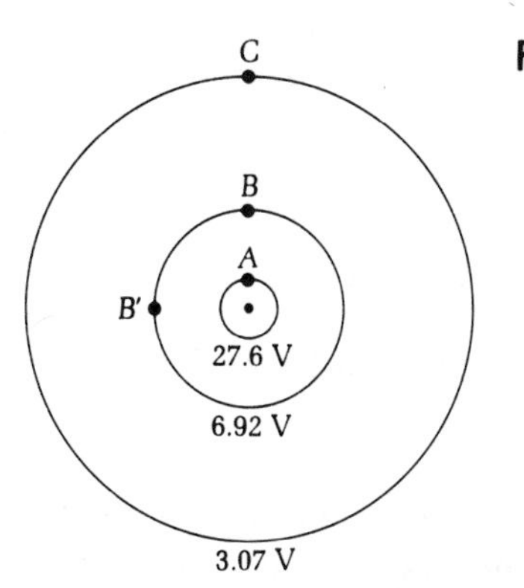

Figure 9.1 Equipotential surfaces near a hydrogen atom (Bohr's model).

Potential Difference

Potential difference V_{AB} between point A and point B in an electrostatic field is defined as the difference of the value of Φ at A and its value at B—that is,

$$V_{AB} = \Phi(\text{at } A) - \Phi(\text{at } B) \text{ volts}$$ **(potential difference between point A and point B)** **(9.11)**

Thus, in the field of the hydrogen atom shown in Figure 9.1, $V_{AB} = 20.68$ volts, $V_{AC} = 24.53$ volts, $V_{BB'} = 0$, $V_{BA} = -V_{AB} = -20.68$ volts, and $V_{AB'} = 20.68$ volts.

Note that, if we add a constant to the potential function (9.8) or (9.10), Equations (9.6) and (9.7) are not affected because differentiation of a constant always yields zero. Thus, it is permissible to add a constant to all values of potential at all points in space. This constant merely changes the reference point. The potential difference between any two points is not affected.

To take advantage of this flexibility in choosing the potential reference, the value of the potential of the earth is usually taken to be zero. Thus, in a circuit a point is called "grounded" if its voltage is equal to zero.

Potential Due to N Point Charges

Potential is a scalar function. From the Poisson equation (9.7) and the Poisson integral (9.8), we see that potential is a linear function of charge density. Thus, potential due to N point charges is the sum of N potentials due to each of the N charges.

$$\Phi = \sum_{n=1}^{N} \frac{q_n}{4\pi\epsilon r_n}$$ **(potential due to N point charges)** **(9.12)**

where r_n is the distance between the nth charge and the observation point.

Example 9.2 *To find the potential due to two point charges, we can use the general formula (9.12) and let $N = 2$.*

Two charges of q and $-q$ coulombs are located at (0, 0, 0.01) and (0, 0, −0.01), respectively, as shown in Figure 9.2a. Find Φ at (x, y, z).

Figure 9.2(a) Potential due to a two-charge system.

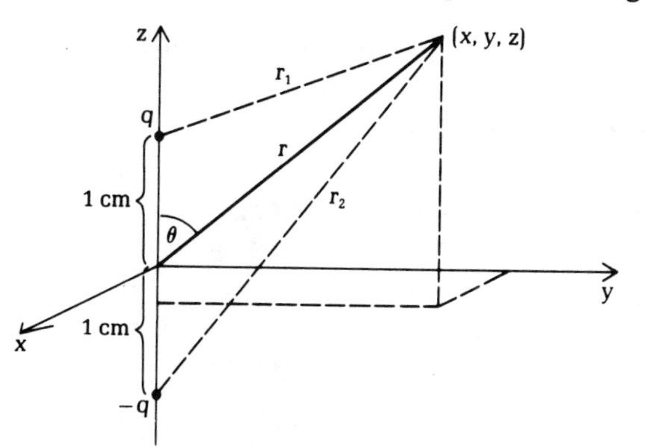

Solution

$$\Phi(x, y, z) = \frac{q}{4\pi\epsilon r_1} + \frac{-q}{4\pi\epsilon r_2} \tag{9.13}$$

Because

$$r_1 = \sqrt{x^2 + y^2 + (z - 0.01)^2}$$

and

$$r_2 = \sqrt{x^2 + y^2 + (z + 0.01)^2}$$

we have

$$\Phi(x, y, z) = \frac{q}{4\pi\epsilon}\left(\frac{1}{\sqrt{x^2 + y^2 + (z - 0.01)^2}} - \frac{1}{\sqrt{x^2 + y^2 + (z + 0.01)^2}}\right) \tag{9.14}$$

Note that the potential is zero on the x-y plane, where $z = 0$.

Potential Due to an Electrostatic Dipole

An electrostatic dipole is formed by two charges that have equal magnitude but opposite sign and are separated by an infinitesimal distance, as shown in Figure 9.2b. This charge distribution is similar to the one shown in Figure 9.2a except

Figure 9.2(b) An electrostatic dipole for which $d \to 0$.

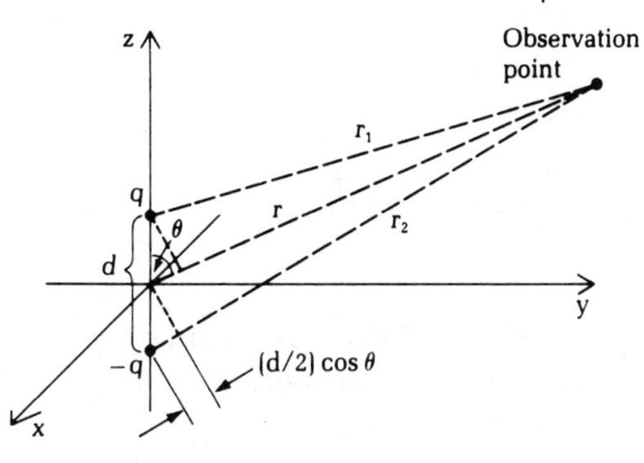

that in the latter case the separation of the positive and the negative charges is finite.

The potential of a dipole is very similar to that given by (9.13):

$$\Phi = \frac{q}{4\pi\epsilon}\left(\frac{1}{r_1} - \frac{1}{r_2}\right) \tag{9.15}$$

except that it is required that $d \ll r_1$, or r_2.

Using the spherical coordinate, we can approximate

$$r_1 = \left[x^2 + y^2 + \left(z - \frac{d}{2}\right)^2\right]^{1/2} = \left(r^2 - dz + \frac{d^2}{4}\right)^{1/2}$$

$$\approx r\left(1 - \frac{d}{r}\frac{z}{r}\right)^{1/2} \approx r\left(1 - \frac{1}{2}\frac{d}{r}\cos\theta\right) = r - \frac{1}{2}d\cos\theta \tag{9.16a}$$

Similarly,

$$r_2 \approx r + \frac{1}{2}d\cos\theta \tag{9.16b}$$

Thus,

$$\Phi = \frac{q}{4\pi\epsilon}\left(\frac{1}{r - \dfrac{d\cos\theta}{2}} - \frac{1}{r + \dfrac{d\cos\theta}{2}}\right) = \frac{q}{4\pi\epsilon}\frac{d\cos\theta}{r^2 - (d^2\cos^2\theta/4)}$$

Now, neglecting the smaller term containing d^2 and keeping the d term, we obtain

$$\boxed{\Phi(r, \theta) = \frac{p\cos\theta}{4\pi\epsilon r^2}} \quad \textbf{(potential due to an electric dipole)} \tag{9.17}$$

where $p = qd$, which is often referred to as the **dipole moment**. Comparing the above equation and (9.10), we notice that as the distance from the charge or charges is increased, the potential decreases like $1/r$ for a point charge but like $1/(r^2)$ for a dipole.

Example 9.3 *This example shows how to find the equipotential surfaces due to an electric dipole.*

Assume that the magnitude of the charges of an electrostatic dipole is $10^6 |q_e|$, where q_e is the charge of an electron and where $d = 10^{-6}$ m. Sketch the equipotential surfaces.

Solution Substituting the values of q and d in (9.17), we obtain the equation of the equipotential surface of voltage V,

$$r = 10^{-5}\left(\frac{14.4 \cos\theta}{V}\right)^{1/2} \quad \text{(meter)}$$

Figure 9.3 shows the surfaces.

Figure 9.3 Equipotential surfaces of a dipole with $p = 1.6 \times 10^{-19}$ C-m.

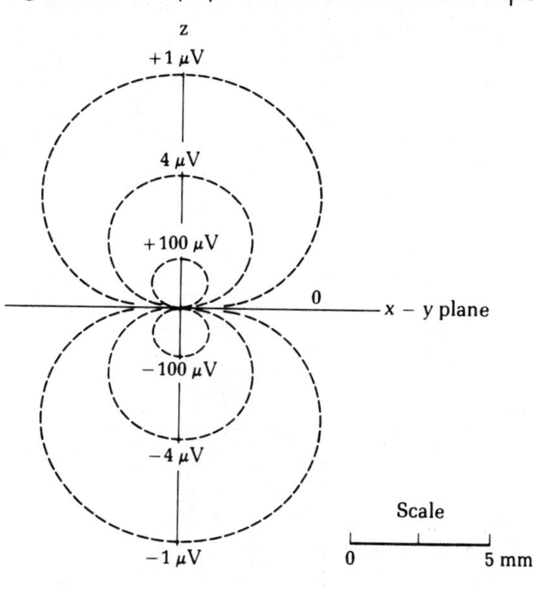

9.2 ELECTROSTATIC FIELDS

The electrostatic field $\boldsymbol{E}$ due to a point charge is found by substituting (9.10) into (9.6):

$$\boldsymbol{E} = -\nabla\Phi = -\nabla\frac{q}{4\pi\epsilon r}$$

Thus,

$$\boxed{\boldsymbol{E} = \frac{q}{4\pi\epsilon r^2}\hat{\boldsymbol{r}}} \quad \textbf{(electric field due to a point charge)} \qquad \textbf{(9.18)}$$

Example 9.4 *The electric field can be found by calculating the gradient of the potential function.*

Find the $\boldsymbol{E}$ field due to a point charge with $10^6\,|q_e|$ coulombs in air.

Solution

$$E = \frac{+1.44 \times 10^{-3}}{r^2}\hat{r}$$

The electric field is now pictured as lines with arrows pointing radially outward from the charge, as shown in Figure 9.4. Note that the arrows always point from the potential surface of higher voltage toward that of lower voltage.

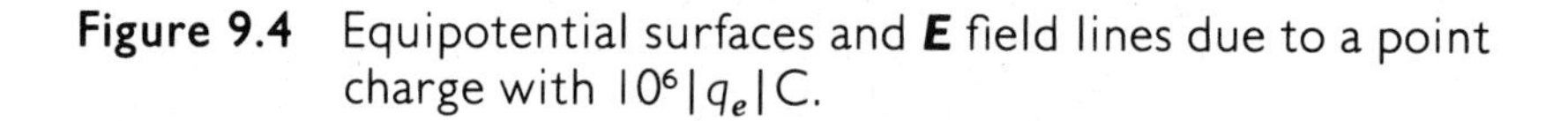

Figure 9.4 Equipotential surfaces and **E** field lines due to a point charge with $10^6|q_e|$C.

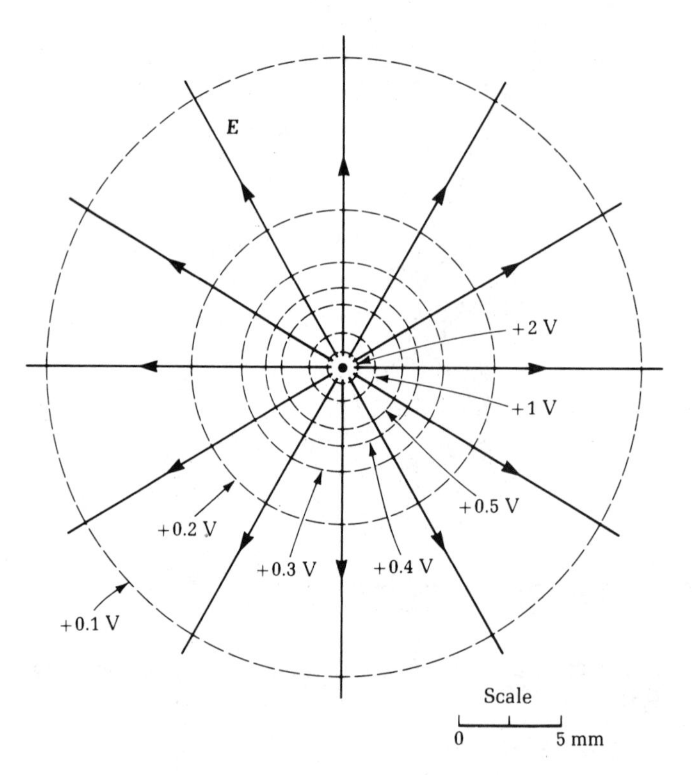

E Field Due to *N* Charges

We can obtain the electric field due to *N* charges by either of two methods. First, we can find the potential function of this charge system with the superposition principle expressed in (9.12). We then substitute the potential in (9.6) to yield ***E***.

$$\Phi(x, y, z) = \sum_{n=1}^{N} \frac{q_n}{4\pi\epsilon r_n} \qquad \textbf{(9.19a)}$$

$$E(x, y, z) = -\nabla \sum_{n=1}^{N} \frac{q_n}{4\pi\epsilon r_n} \qquad \textbf{(9.19b)}$$

The second method is the direct application of the superposition principle to field equation (9.18):

$$\boxed{E(x, y, z) = \sum_{n=1}^{N} \frac{q_n}{4\pi\epsilon r_n^2}\hat{r}_n} \quad \textbf{(electric field due to } N \textbf{ point charges)} \qquad \textbf{(9.19c)}$$

where r_n is the distance between the observation point and the nth charge and where the unit vector $\hat{r}_n$ is in the direction pointing from the nth charge toward the observation point.

Example 9.5 *To find the electric field due to N point charges, we may use (9.19b) or (9.19c), The latter is a vector equation. If the scalar potential is known, it will be easier to find the **E** field using the gradient method.*

Find the electric field $\boldsymbol{E}$ for the dipole shown in Figure 9.2b, and sketch the $\boldsymbol{E}$ field lines.

Solution The potential $\Phi(r, \theta)$ was found in (9.17). By using the spherical coordinate system, we find

$$\boldsymbol{E} = -\nabla\Phi(r, \theta) = -\hat{r}\frac{\partial}{\partial r}\Phi(r, \theta) - \hat{\theta}\frac{1}{r}\frac{\partial}{\partial \theta}\Phi(r, \theta)$$

$$= \frac{p}{4\pi\epsilon r^3}(\hat{r}\, 2\cos\theta + \hat{\theta}\sin\theta) \qquad \textbf{(9.20)}$$

The $\boldsymbol{E}$ field vectors are in the direction of decreasing Φ and are perpendicular to the equipotential surfaces shown in Figure 9.5. From Figure 9.5, we see that all $\boldsymbol{E}$ field lines originate from the positive charge and end at the negative charge.

The solution for the $\boldsymbol{E}$ field for an oscillating Hertzian dipole antenna is derivable from (7.14) by setting $\omega = 0$ because, in the static limit, the oscillating Hertzian dipole becomes the static dipole.

E Field Due to a Line Charge

Consider a thin line carrying electric charge ρ_ℓ coulombs per meter (note the unit of ρ_ℓ). We assume that the line extends from $z = -h$ to $z = h$ along the z axis, as shown in Figure 9.6a. Let us find the $\boldsymbol{E}$ field in the x-y plane. In cylindrical coordinates, because of azimuthal symmetry, we solve for the $\boldsymbol{E}$ field along the ρ axis.

Figure 9.5 Electric fields (solid lines) and equipotential surfaces (dotted lines) of an electrostatic dipole.

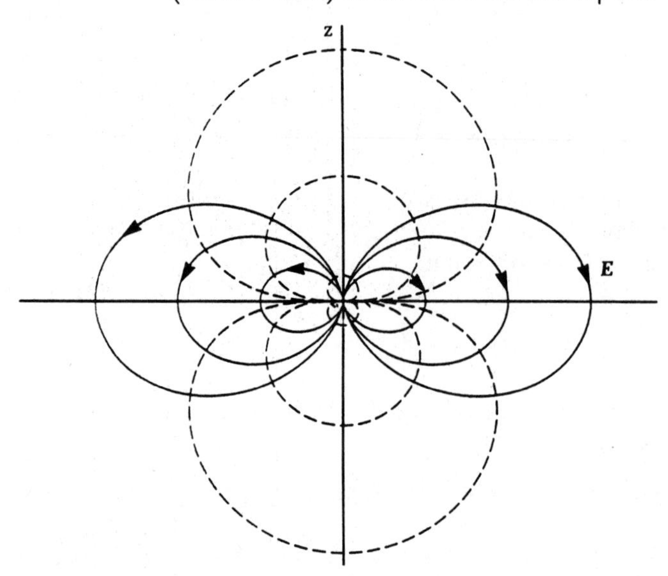

Figure 9.6(a) Field along ρ axis due to a line charge.

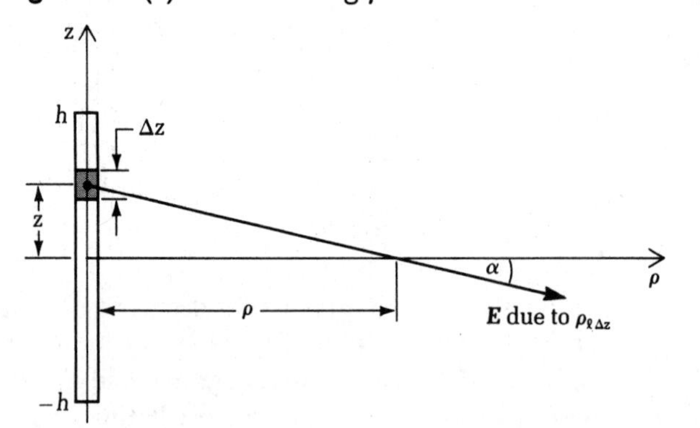

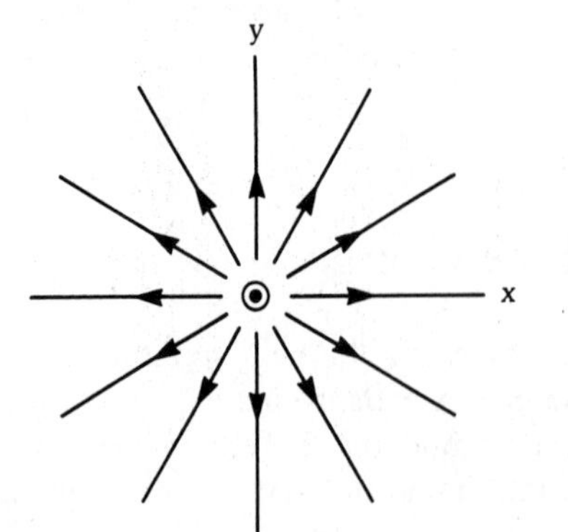

Figure 9.6(b) **E** field lines of a line charge.

To begin with, the line charge is divided into small segments of Δz in length. Each segment contains a charge of $\rho_\ell \Delta z$ coulombs, and each can be regarded as a point charge. The electric field due to a typical segment located at z is given by (9.18):

$$\Delta \boldsymbol{E} = \frac{\rho_\ell \Delta z}{4\pi\epsilon(\rho^2 + z^2)}(\hat{\rho} \cos \alpha - \hat{z} \sin \alpha)$$

Referring to Figure 9.6a,

$$\cos \alpha = \frac{\rho}{(\rho^2 + z^2)^{1/2}}$$

$$\sin \alpha = \frac{z}{(\rho^2 + z^2)^{1/2}}$$

Thus,

$$\Delta \boldsymbol{E} = \frac{\rho_\ell \Delta z(\hat{\rho}\rho - \hat{z}z)}{4\pi\epsilon(\rho^2 + z^2)^{3/2}}$$

Note that $\Delta \boldsymbol{E}$ is not a function of ϕ. The total $\boldsymbol{E}$ field is then the sum of all individual $\boldsymbol{E}$ fields due to all segments:

$$\boldsymbol{E} = \int_{-h}^{h} \frac{\rho_\ell dz(\hat{\rho}\rho - \hat{z}z)}{4\pi\epsilon(\rho^2 + z^2)^{3/2}}$$

The unit vectors $\hat{\rho}$ and $\hat{z}$ are pointing in fixed directions while z is varied from $-h$ to h in the integration. Therefore, these unit vectors can be taken out of the integral, and the integration then becomes a scalar integration:

$$\boldsymbol{E} = \frac{\rho_\ell}{4\pi\epsilon}\left[\hat{\rho}\rho \int_{-h}^{h} \frac{dz}{(\rho^2 + z^2)^{3/2}} - \hat{z} \int_{-h}^{h} \frac{z\, dz}{(\rho^2 + z^2)^{3/2}}\right]$$

Using the variable $z = \rho \tan \alpha$, $dz = \rho \sec^2 \alpha\, d\alpha$, we have

$$\boldsymbol{E} = \frac{\rho_\ell}{4\pi\epsilon}\left(\frac{\hat{\rho}}{\rho}\int_{-\alpha_0}^{\alpha_0} \cos \alpha\, d\alpha - \frac{\hat{z}}{\rho}\int_{-\alpha_0}^{\alpha_0} \sin \alpha\, d\alpha\right) = \hat{\rho}\frac{\rho_\ell}{2\pi\epsilon\rho} \sin \alpha_0 \tag{9.21}$$

where

$$\alpha_0 = \tan^{-1}\frac{h}{\rho}$$

We see that in the limit $h/\rho \ll 1$, which implies either that h is very small or that ρ is very large, $\sin \alpha_0 \approx h/\rho$, and we find

$$\boldsymbol{E} = \hat{\rho}\frac{\rho_\ell(2h)}{4\pi\epsilon\rho^2}$$

which is the result for a point charge with $\rho_\ell(2h)$ coulombs. In other words, a short line charge looks like a point charge from a large distance.

For the case of an infinitely long wire such that $h/\rho \gg 1$ and $\alpha_0 =$

$\tan^{-1}(h/\rho) \approx \pi/2$, we find from (9.21) that

$$\boxed{E = \hat{\rho}\frac{\rho_\ell}{2\pi\epsilon\rho}} \quad \textbf{(electric field due to a line charge)} \tag{9.22}$$

Notice that E is inversely proportional to the distance for a line charge, whereas for a point charge it is inversely proportional to the distance squared. Figure 9.6b sketches the E field lines in the x-y plane that are due to the line charge.

E Field Due to a Plane Charge

Consider an infinitely large plane deposited with a uniform charge density ρ_s coulombs per square meter (note the unit of ρ_s), as shown in Figure 9.7a. We are interested in finding the E field along the x axis produced by this plane charge.

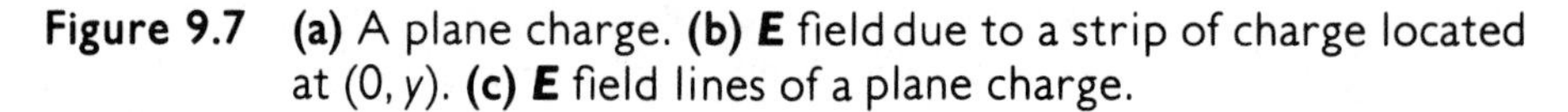

Figure 9.7 **(a)** A plane charge. **(b)** **E** field due to a strip of charge located at (0, *y*). **(c)** **E** field lines of a plane charge.

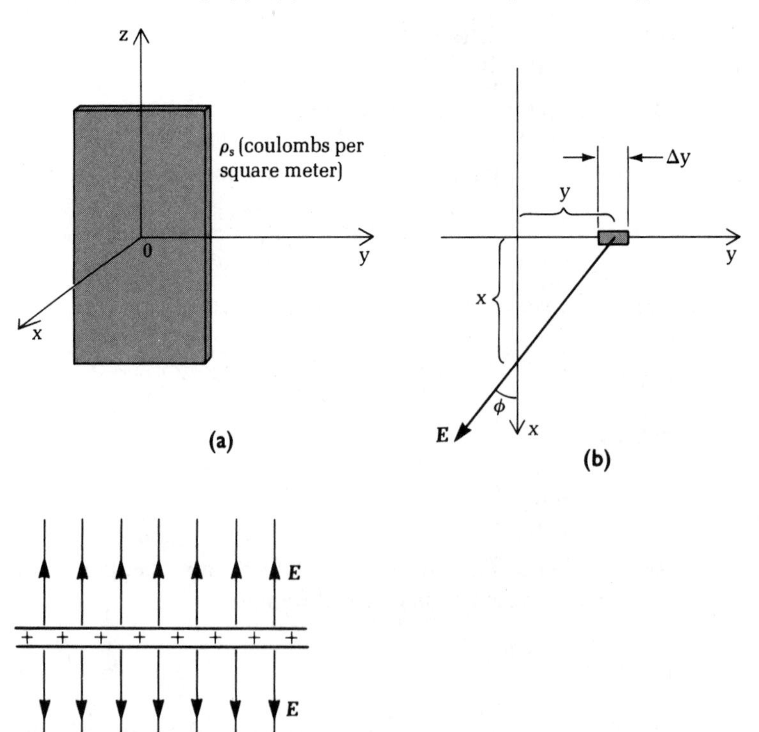

Because we know the solution for a line charge, we subdivide the plane charge into many thin strips each with width Δy. Figure 9.7b shows a typical strip. Each strip may be regarded as an infinitely long line charge of $\rho_\ell = \rho_s \Delta y$ coulombs per meter. The electric field at $(x, 0, 0)$ due to the typical line charge shown in Figure 9.7 is, referring to (9.22),

$$\Delta \boldsymbol{E} = \frac{\rho_s \Delta y}{2\pi\epsilon\sqrt{x^2 + y^2}}\left(\frac{\hat{\boldsymbol{x}}x - \hat{\boldsymbol{y}}y}{\sqrt{x^2 + y^2}}\right)$$

We then integrate y from $-\infty$ to $+\infty$ to obtain contributions from all strips:

$$\boldsymbol{E}(x) = \frac{\rho_s}{2\pi\epsilon}\left(\hat{\boldsymbol{x}}\int_{-\infty}^{\infty} dy \frac{x}{x^2 + y^2} - \hat{\boldsymbol{y}}\int_{-\infty}^{\infty} dy \frac{y}{x^2 + y^2}\right) \tag{9.23}$$

The second integral is zero because the integrand is an odd function. To evaluate the first integral, we let

$$y = x \tan\phi$$

$$dy = x \sec^2\phi \, d\phi \tag{9.24}$$

It follows that

$$\boldsymbol{E}(x) = \frac{\hat{\boldsymbol{x}}\rho_s}{2\pi\epsilon}\int_{-\pi/2}^{\pi/2} \frac{x^2 \sec^2\phi \, d\phi}{x^2(1 + \tan^2\phi)} = \frac{\rho_s}{2\epsilon}\hat{\boldsymbol{x}}$$

$$\boxed{\boldsymbol{E}(x) = \frac{\rho_s}{2\epsilon}\hat{\boldsymbol{x}}} \quad \textbf{(electric field due to a plane charge for } x > 0\textbf{)} \tag{9.25a}$$

Thus; the $\boldsymbol{E}$ field is a constant vector independent of the coordinates. For $x < 0$, we have

$$\boxed{\boldsymbol{E}(x) = \frac{\rho_s}{2\epsilon}(-\hat{\boldsymbol{x}})} \quad \textbf{(electric field due to a plane charge for } x < 0\textbf{)} \tag{9.25b}$$

This result holds because, according to (9.24), if $x < 0$, the upper limit of the integral is $-\pi/2$, and the lower limit is $\pi/2$, just the opposite of the case in which $x > 0$.

Because the $\boldsymbol{E}$ field is independent of z, we can sketch the $\boldsymbol{E}$ field lines due to an infinitely large plane charge, as shown in Figure 9.7c. The $\boldsymbol{E}$ field due to an infinitely large plane charge is a constant vector pointing away from the plane (assuming ρ_s is positive).

9.3 GAUSS' LAW AND ITS APPLICATIONS

Let us integrate Gauss' law (9.2) over a surface S enclosing volume V. We find

$$\oiint_S ds\,\hat{\boldsymbol{n}} \cdot \boldsymbol{D} = \iiint_v dv\,\rho_v \quad \textbf{(Gauss' law)} \tag{9.26}$$

where $\hat{\boldsymbol{n}}$ is the outward normal to the surface S. The integral on the left-hand side is a result of a vector calculus theorem called **divergence theorem** or **Gauss' theorem**, which is a mathematical theorem different from **Gauss' law**, a physical law. The divergence theorem states that for any vector $\boldsymbol{A}$,

$$\iiint_v dv\,\nabla \cdot \boldsymbol{A} = \oiint_S ds\,\hat{\boldsymbol{n}} \cdot \boldsymbol{A} \tag{9.27}$$

Equation (9.26) is called Gauss' law in integral form. It states that the total $\boldsymbol{D}$ flux leaving a closed surface is equal to the total charge *inside* the surface.

Figure 9.8 illustrates Gauss' law. Let the total $\boldsymbol{E}$ field due to all four point charges be $\boldsymbol{E}_t$. Then, without working out the exact formula for $\boldsymbol{E}_t$, we can obtain the following integrals through (9.26):

$$\oiint_{S_1} ds\,\hat{\boldsymbol{n}} \cdot \boldsymbol{E}_t = \frac{Q}{\epsilon}$$

$$\oiint_{S_2} ds\,\hat{\boldsymbol{n}} \cdot \boldsymbol{E}_t = 0$$

$$\oiint_{S_3} ds\,\hat{\boldsymbol{n}} \cdot \boldsymbol{E}_t = \frac{Q}{\epsilon}$$

Figure 9.8 Gauss' law applied to different surfaces.

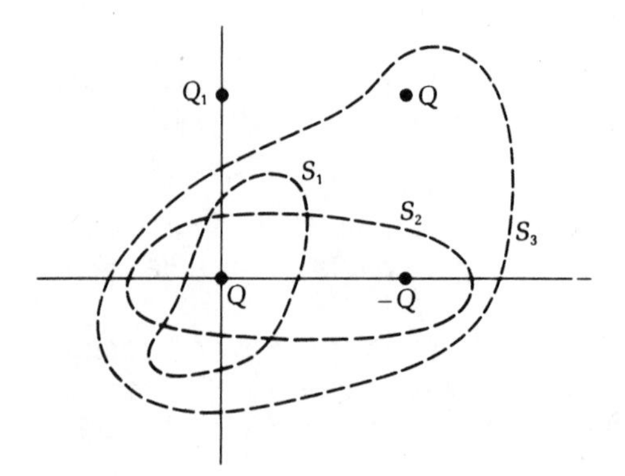

Note that integration of $\hat{n} \cdot E_t$ over S_1, S_2, or S_3 is not affected by Q_1 because, although Q_1 affects E_t, it does not affect the integration of $\hat{n} \cdot E_t$ over any surface that does not enclose Q_1.

Gauss' law in the form of (9.26) is very useful in solving a class of problems involving symmetrical charge distributions. The enclosed surface surrounding the charge distributions is called the **Gaussian surface**. In the following examples, we show that, by taking advantage of the symmetry of a given charge distribution and by judiciously choosing the Gaussian surface, we can determine the $\boldsymbol{E}$ field rather expeditiously by applying (9.26).

Example 9.6 *This example shows that we can reproduce the formula for the electric field due to a point charge by using Gauss' law.*

Find the $\boldsymbol{E}$ field due to a point charge with q coulombs located at the origin, as shown in Figure 9.9.

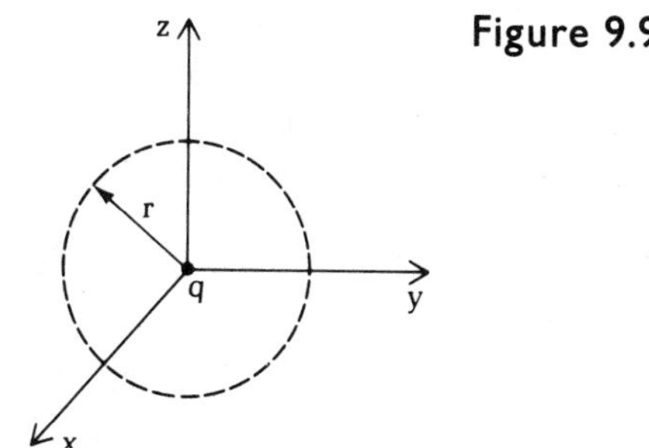

Figure 9.9 Sphere enclosing charge q.

Solution Let us enclose the charge with a spherical Gaussian surface of radius r centered at the charge. Now applying Gauss' law (9.26), we have

$$\oiint \hat{r} \cdot \boldsymbol{E}\, ds = \frac{1}{\epsilon} \iiint \rho_v\, dv = \frac{q}{\epsilon}$$

or

$$\oiint E_r\, ds = \frac{q}{\epsilon}$$

where E_r is the r component of $\boldsymbol{E}$ in spherical coordinates.

By symmetry, we note that E_r should be a constant on the spherical surface. Thus, E_r can be taken out of the integration sign:

$$E_r \oiint ds = \frac{q}{\epsilon}$$

The integral on the left-hand side is simply the total area of the surface of $4\pi r^2$; thus,

$$E_r 4\pi r^2 = \frac{q}{\epsilon}$$

or

$$E_r = \frac{q}{4\pi\epsilon r^2}$$

Also, from the symmetry we note that $\boldsymbol{E}$ can have only an $\hat{\boldsymbol{r}}$ component so that

$$\boldsymbol{E} = \frac{q}{4\pi\epsilon r^2}\hat{\boldsymbol{r}}$$

This result agrees with that obtained earlier in (9.18).

Example 9.7 *This example shows that Gauss' law can be used to obtain the electric field due to a line charge. The same formula has been derived by the integration method, as given by (9.22).*

Find the $\boldsymbol{E}$ field due to a line charge of ρ_ℓ coulombs per meter located at the origin, as shown in Figure 9.10.

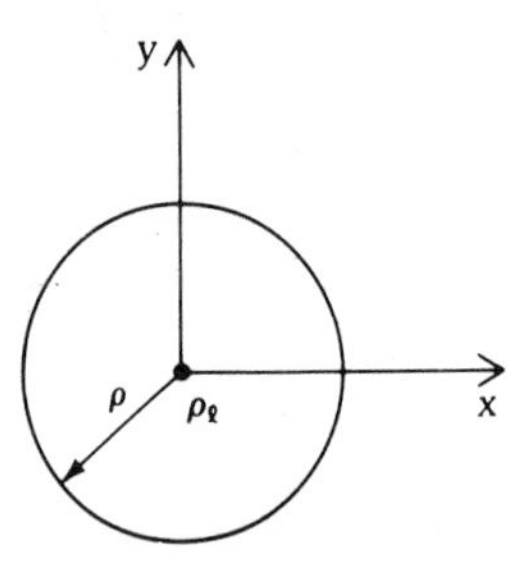

Figure 9.10 A cylindrical surface enclosing line charge ρ_ℓ (end view).

Solution We enclose the line charge with a cylindrical Gaussian surface of radius ρ and height $2h$ with $h \to \infty$. Applying Gauss' law, we find

$$\oiint \hat{\rho} \cdot \boldsymbol{E}\, ds = \frac{\rho_\ell(2h)}{\epsilon}$$

By symmetry, $\boldsymbol{E} = \hat{\rho} E_\rho$. Thus, we have

$$E_\rho(2\pi\rho)(2h) = \frac{\rho_\ell(2h)}{\epsilon}$$

It follows that

$$E_\rho = \frac{\rho_\ell}{2\pi\epsilon\rho}$$

which is exactly (9.22).

Example 9.8 *Again, this example shows that Gauss' law leads directly to the formula for the electric field due to a plane charge.*

Find the $\boldsymbol{E}$ field due to a plane charge located on the y-z plane, as shown in Figure 9.11.

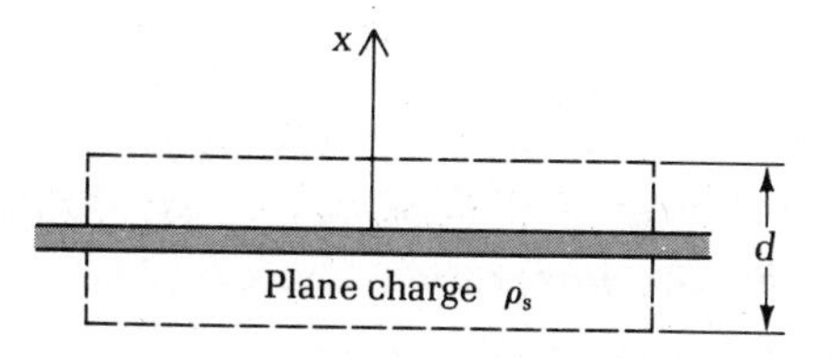

Figure 9.11 Rectangle enclosing plane charge ρ_s.

Solution We enclose the plane charge with a rectangular Gaussian surface with thickness d and area A where $A \to \infty$. By symmetry, $\boldsymbol{E}$ is in the $\hat{\boldsymbol{x}}$ direction for $x > 0$ and in the $-\hat{\boldsymbol{x}}$ direction for $x < 0$. Applying Gauss' law, we find

$$\hat{\boldsymbol{x}} \cdot (\hat{\boldsymbol{x}} E_x) A + (-\hat{\boldsymbol{x}}) \cdot (-\hat{\boldsymbol{x}} E_x) A = \frac{\rho_s A}{\epsilon}$$

Notice that the thickness d does not enter the left-hand side because $\boldsymbol{E}$ has no y or z component, i.e., $\hat{\boldsymbol{y}} \cdot \boldsymbol{E} = \hat{\boldsymbol{z}} \cdot \boldsymbol{E} = 0$. It follows from the above equation that

$$E_x = \frac{\rho_s}{2\epsilon}$$

which is exactly what we obtained for the plane charge ρ_s in (9.25).

Example 9.9 *A very important consequence of Gauss' law is concerned with the existence of electric charges in a conductor. This example shows that electric charges can exist only on the surface of a conductor, not inside it.*

Show that there can be no charge inside a perfectly conducting body.

Solution Assume that somewhere inside the perfect conductor is a charge q. Then, we draw a Gaussian surface S around that charge, as shown in Figure 9.12. From (9.26), we obtain

$$\oiint_S \hat{\boldsymbol{n}} \cdot \boldsymbol{E} = \frac{q}{\epsilon}$$

But by definition $\boldsymbol{E}$ is identically zero inside the perfect conductor. Thus, the integration yields zero, and it follows that $q = 0$ everywhere inside the perfect conductor.

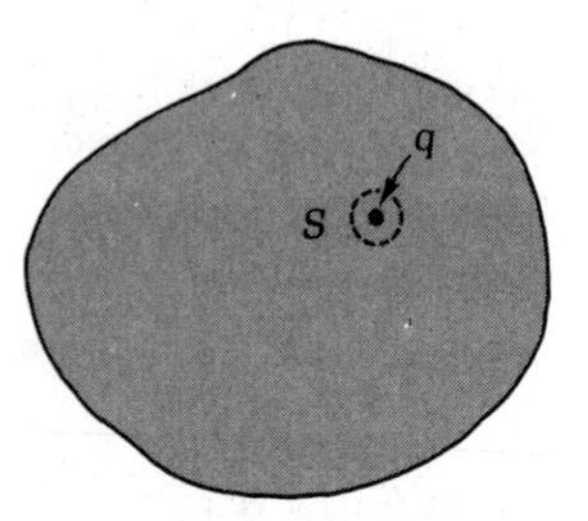

Figure 9.12 There cannot be any charge inside a perfectly conducting body—that is, q is always equal to zero.

Example 9.10 ***Earlier examples (9.6 through 9.8) show how to use Gauss' law to reproduce the results obtained by the integration method. This example shows that Gauss' law can be used to solve new problems.***

A solid conducting sphere of radius a carries a charge of q coulombs, as shown in Figure 9.13. Find $\boldsymbol{E}$ for $r > a$.

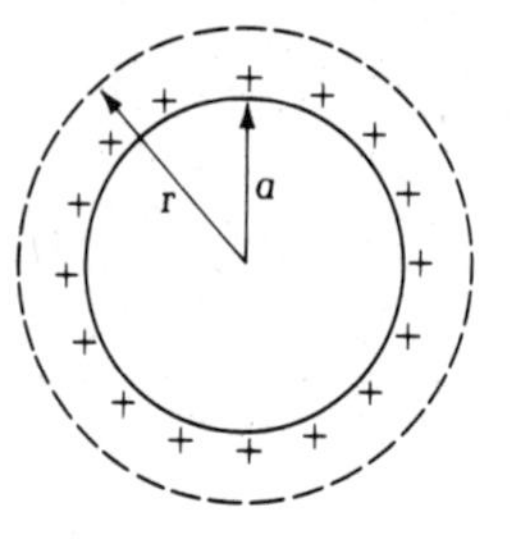

Figure 9.13 A conducting sphere carries q on its surface.

Solution Because the sphere is made of a perfect conductor, $\boldsymbol{E} = 0$, and there is no charge inside the sphere. The q coulombs of charge must then be distributed on the surface of the spherical conductor.

Draw a spherical surface S of radius r (with $r > a$) concentric with the sphere; then apply the integral Gauss' law (9.26):

$$\oiint_S \hat{\boldsymbol{r}} \cdot \boldsymbol{E}\, ds = \frac{q}{\epsilon}$$

or,

$$\oiint_S E_r\, ds = \frac{q}{\epsilon}$$

With the argument of symmetry, we again note that E_r is a constant on S, so that it can be taken out of the integral sign and the integration simply yields $4\pi r^2 E_r$. The final result is

$$\boldsymbol{E} = \frac{q}{4\pi\epsilon r^2}\hat{\boldsymbol{r}} \qquad \text{for} \quad r > a$$

Note that for $r > a$, the $\boldsymbol{E}$ field is as if the charge were concentrated at the center of the sphere like a point charge.

Example 9.11 *This example demonstrates the use of Gauss' law to obtain the electric field in a system consisting of a spherical conductor inside a conducting spherical shell. Although the geometry is more complex than the one discussed in Example 9.9, we can still take advantage of the symmetry of this system and solve for the electric field without too much difficulty.*

Consider a solid conducting sphere of radius a inside a concentric conducting spherical shell, as shown in Figure 9.14. The shell has an inner radius b and an outer radius c. A charge of q coulombs is placed on the inner solid conductor, and a charge of $-q'$ is on the outer conductor. They are uniformly distributed on the spherical surfaces. Find $\boldsymbol{E}$.

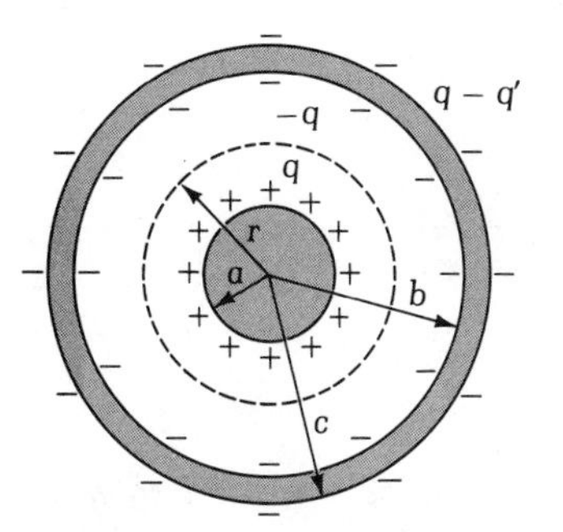

Figure 9.14 Charges in a conducting shell that is concentric with a conducting sphere.

Solution By symmetry, we see that the electric fields are obviously radially directed. From Gauss' law applied on a mathematical surface of radius r with $a < r < b$ concentric with the conducting spheres, we have

$$4\pi r^2 E_r = \frac{q}{\epsilon}$$

Thus,

$$\boldsymbol{E} = \hat{\boldsymbol{r}} \frac{q}{4\pi\epsilon r^2}$$

which is equivalent to that of a point charge of q placed at the origin.

Now, if the surface is drawn inside the outer conducting shell—that is, if $c > r > b$, we have

$$\int_S \boldsymbol{E} \cdot \hat{\boldsymbol{n}}\, ds = 0$$

This result obtains because $\boldsymbol{E}$ is zero in conductors. The above equation indicates that total charge is zero inside the spherical surface of radius r with $c > r > b$.

We know that there are q coulombs on the solid conductor. We therefore conclude that there must be $-q$ coulombs on the inner surface of the conducting shell. Outside of the outer sphere—that is, $r > c$—the electric field is found to be

$$\boldsymbol{E} = \hat{r}\frac{q - q'}{4\pi\epsilon r^2}$$

Because the total charge on the surface of the inner conductor with radius $r = a$ is q and the total charge on the inner surface of the outer spherical shell at radius $r = b$ is $-q$, the total charge on the outer surface with $r = c$ is equal to $q - q'$.

The above analysis shows that outside the spherical shell (where $r > c$) the electric field is as if there were a point charge of magnitude $(q - q')$ situated at the origin.

Figure 9.15 plots the electric field as a function of r.

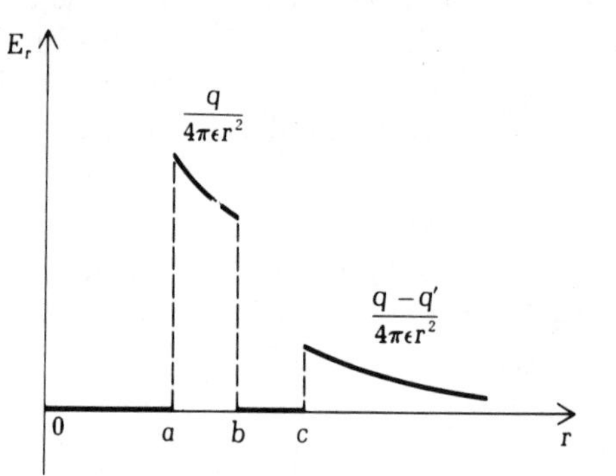

Figure 9.15 Electrostatic field for the concentric sphere-and-shell system shown in Figure 9.14.

Example 9.12 *This example shows how to use Gauss' law and solve for the electric field in a cylindrical system.*

Consider the coaxial line shown in Figure 9.16. The inner conducting cylinder has radius a, and the outer conductor has radius b. The inner conductor carries ρ_ℓ coulombs per meter of charge and the outer conductor carries $-\rho_\ell$ coulombs per meter. Find $\boldsymbol{E}$.

Solution Let the Gaussian surface be a cylindrical surface of radius ρ concentric with the conducting cylinders. When $a < \rho < b$ (as shown in Figure 9.16), Gauss' law applied to a section of the coaxial line of length h meters gives

$$\oiint_S \hat{n} \cdot \boldsymbol{E}\, ds = \frac{1}{\epsilon}\iiint_v \rho_v\, dv = \rho_\ell \frac{h}{\epsilon}$$

$$\underbrace{\iint \hat{\rho} \cdot \boldsymbol{E}\, ds}_{\text{cylindrical surface}} + \underbrace{\iint \hat{z} \cdot \boldsymbol{E}\, ds}_{\text{end surface at } z = h} + \underbrace{\iint -\hat{z} \cdot \boldsymbol{E}\, ds}_{\text{end surface at } z = 0} = \rho_\ell \frac{h}{\epsilon}$$

Figure 9.16 Charges and E_ρ in a coaxial line.

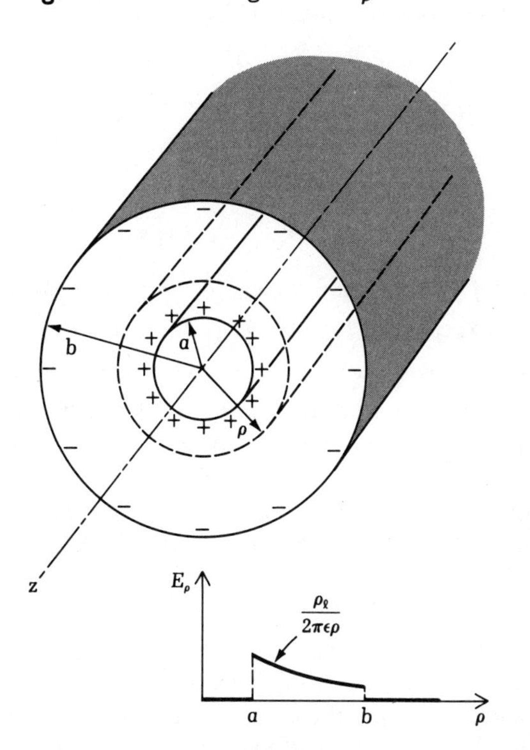

From symmetry, we know that $\boldsymbol{E}$ is radially directed. Hence, $\boldsymbol{E} \cdot \hat{\boldsymbol{n}}$ is zero on the two end surfaces of the cylindrical section. Furthermore, E_ρ is constant on the cylindrical surface of radius ρ. Thus, E_ρ may be taken out of the integral sign, and the remaining integration gives the total area of the cylindrical surface, which is equal to $2\pi\rho h$. It follows that

$$E_\rho 2\pi\rho h = \rho_\ell \frac{h}{\epsilon}$$

or

$$\boldsymbol{E} = \hat{\rho}\frac{\rho_\ell}{2\pi\epsilon\rho}$$

This result is exactly that of a line charge carrying a charge of ρ_ℓ coulombs per meter. Thus, for $a < \rho < b$, the $\boldsymbol{E}$ field is equal to that of an infinitely long line charge of ρ_ℓ coulombs per meter lying on the z axis [see (9.22)].

For $\rho < a$, the $\boldsymbol{E}$ field is zero because the medium is a perfect conductor. For $\rho > b$, we apply Gauss' law to a Gaussian surface with $\rho > b$ and obtain

$$\oiint_S \hat{\boldsymbol{n}} \cdot \boldsymbol{E}\, ds = \frac{1}{\epsilon} h[\rho_\ell + (-\rho_\ell)] = 0$$

The argument about symmetry still holds, but now the total charge inside the mathematical surface is zero. Thus,

$$\boldsymbol{E} = 0 \qquad \text{for} \quad \rho > b$$

Figure 9.16 plots E_ρ versus ρ.

Example 9.13

The geometry of the problem solved in this example is similar to that of Example 9.12, except that the charge is distributed in a volume instead of on a surface. The key point is to demonstrate how to obtain correctly the right-hand side of Gauss' law (9.26). It is the total charge enclosed by the Gaussian surface. The latter varies with ρ, and so does the total charge inside it.

A long cylindrical electron beam moving in air has a charge density of ρ_ℓ coulombs per meter. The radius of the beam is a, and the charge is distributed uniformly. Find $\boldsymbol{E}$.

Solution To calculate the field at ρ when ρ is much less than the length of beam, a good approximation is to assume that the beam is infinitely long. Now, draw a Gaussian surface concentric with the beam with $\rho > a$, as shown in Figure 9.17a. Applying Gauss' law to a section of the Gaussian surface of length h, we have

$$\oiint_S \hat{\boldsymbol{n}} \cdot \boldsymbol{E}\, ds = \frac{1}{\epsilon} \iiint_v \rho_v\, dv = \frac{1}{\epsilon} h\rho_\ell$$

From symmetry (the same arguments used in Example 9.12), we obtain

$$2\pi\rho h E_\rho = \frac{1}{\epsilon} h\rho_\ell$$

which yields

$$\boldsymbol{E} = \hat{\rho}\frac{\rho_\ell}{2\pi\epsilon\rho} \qquad \text{for} \quad \rho \geq a$$

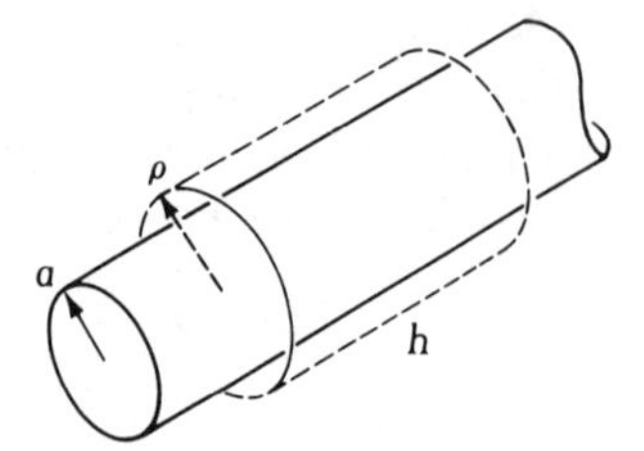

Figure 9.17(a) Gaussian surface outside the beam.

For $\rho < a$, the mathematical surface encloses only a portion of the charge in the beam, as illustrated in Figure 9.17b. The total charge enclosed in the mathematical surface of length h is equal to $h\rho_\ell(\pi\rho^2/\pi a^2)$. Gauss' law gives

$$\oiint_S \hat{\boldsymbol{n}} \cdot \boldsymbol{E}\, ds = \frac{h}{\epsilon}\rho_\ell \frac{\rho^2}{a^2}$$

or

$$2\pi\rho h E_\rho = \frac{h}{\epsilon}\rho_\ell \frac{\rho^2}{a^2}$$

so that

$$\boldsymbol{E} = \hat{\rho}\frac{\rho_\ell \rho}{2\pi\epsilon a^2} \qquad \text{for} \quad \rho \leq a$$

Figure 9.17c shows a plot of E_ρ versus ρ. Note that the field is zero at the center of the beam, increases with ρ, reaches maximum at $\rho = a$, and then decreases as $1/\rho$ for $\rho > a$.

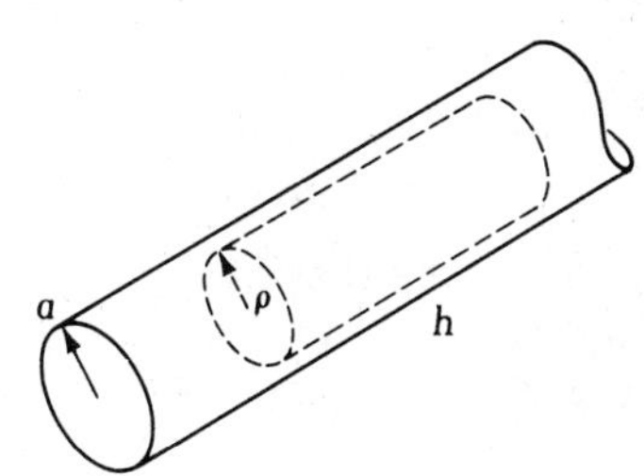

Figure 9.17(b) Gaussian surface inside the beam.

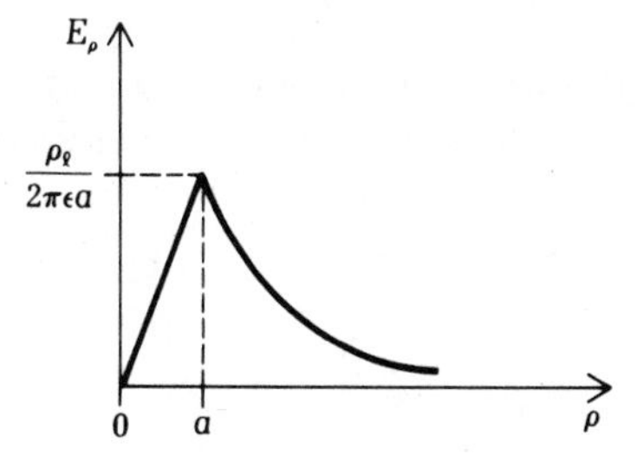

Figure 9.17(c) Variation of E_ρ versus ρ.

Example 9.14 *In this example Gauss' law is applied to a system in rectangular coordinates. Note the difference and similarity between this geometry and that of Example 9.8.*

Consider the parallel plate in Figure 9.18a. Equal amounts of positive and negative charges are placed on the lower and the upper plates and distributed uniformly on the inner surfaces of the two plates with densities equal to ρ_s coulombs per square meter and $-\rho_s$ coulombs per square meter, respectively. Find $\boldsymbol{E}$.

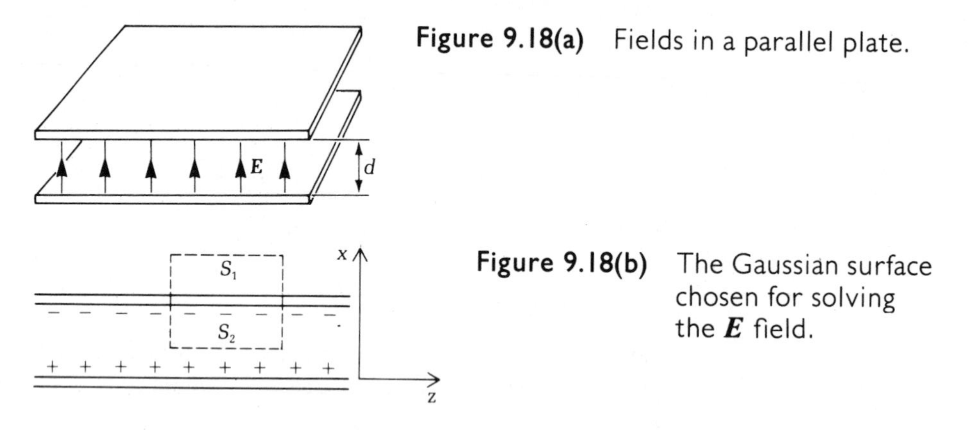

Figure 9.18(a) Fields in a parallel plate.

Figure 9.18(b) The Gaussian surface chosen for solving the $\boldsymbol{E}$ field.

Solution To apply Gauss' law, we draw a Gaussian surface in the form of a box with S_1 and S_2 as its top and bottom surfaces, as shown in Figure 9.18b. Gauss' law gives

$$\oiint_S \hat{n} \cdot \boldsymbol{E}\, ds = \frac{1}{\epsilon} \iiint_v \rho_v\, dv = \frac{-1}{\epsilon} \rho_s A$$

where A is the area of S_1 or S_2. Assuming that the plates are infinitely large, so that $\boldsymbol{E}$ is not a function of coordinates y and z, then, because $\boldsymbol{E}$ can only be directed in the $\hat{x}$ direction (because of the symmetry of the problem), we have

$$\iint_{S_1} E_x\, ds + \iint_{S_2} (-E_x)\, ds = -\frac{\rho_s A}{\epsilon}$$

Noting that there is no electric charge on the outer surfaces of the plates, we can set $\boldsymbol{E} = 0$ on S_1. E_x is a constant on S_2, so that the integral becomes $-E_x A$, or

$$-E_x A = -\frac{\rho_s A}{\epsilon}$$

Finally,

$$\boxed{\boldsymbol{E} = \frac{\rho_s}{\epsilon}\hat{x}} \quad \textbf{(electric field between parallel plates } 0 < x < d\textbf{)} \qquad \textbf{(9.28)}$$

The electric field is zero elsewhere.

Comparing the above result with (9.25), we note that the results differ by a factor of two. This difference exists because in the present case we have two plates, whereas (9.25) is the $\boldsymbol{E}$ field for a single sheet of charge.

9.4 CALCULATION OF POTENTIAL FROM E FIELD—LINE INTEGRAL

In Section 9.1 we defined the electrostatic potential difference V_{AB} between two points A and B in space as the difference of the potential function $\Phi(\mathbf{r})$ at these points: $V_{AB} = \Phi_A - \Phi_B$. When the electric field is known, the potential can be determined from (9.6):

$$\mathbf{E} = -\nabla\Phi = -\hat{\mathbf{x}}\frac{\partial\Phi}{\partial x} - \hat{\mathbf{y}}\frac{\partial\Phi}{\partial y} - \hat{\mathbf{z}}\frac{\partial\Phi}{\partial z} \tag{9.29}$$

Dot-multiplying this equation by the differential position vector $d\mathbf{r} = \hat{\mathbf{x}}\,dx + \hat{\mathbf{y}}\,dy + \hat{\mathbf{z}}\,dz$, we have

$$\mathbf{E}\cdot d\mathbf{r} = -\left(\frac{\partial\Phi}{\partial x}dx + \frac{\partial\Phi}{\partial y}dy + \frac{\partial\Phi}{\partial z}dz\right) = -d\Phi \tag{9.30}$$

Thus,

$$\boxed{V_{AB} = \Phi_A - \Phi_B = -\int_B^A \mathbf{E}\cdot d\mathbf{r}} \quad \textbf{(voltage difference is proportional to line integral of electric field)} \tag{9.31}$$

This integral is performed along the line defined by increments $d\mathbf{r}$ between the two points A and B (Figure 9.19). The result is zero if everywhere along the line the electric field $\mathbf{E}$ is perpendicular to the line such that $\mathbf{E}\cdot d\mathbf{r} = 0$. The result is maximum when the electric field is everywhere parallel to the line of integration.

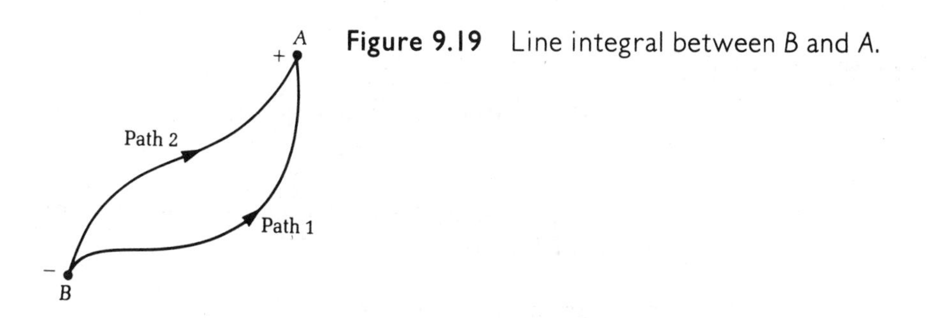

Figure 9.19 Line integral between B and A.

Consider the special case of a straight line connecting A and B. Remember that V_{AB}, the potential at A and B as a reference point, is negative if $\mathbf{E}$ points toward A and positive if $\mathbf{E}$ points away from A. For example, the point charge in Figure 9.4 has the electric field

$$\mathbf{E} = \hat{\mathbf{r}}\frac{q}{4\pi\epsilon r^2} \tag{9.32}$$

and

$$\Phi = \frac{q}{4\pi\epsilon r} \tag{9.33}$$

The electric field is in the $\hat{r}$ direction, and the potential is greater near the charge and decreases as r is increased.

Obviously, (9.33) is a direct consequence of (9.32) in light of the line integral (9.31).

$$\Phi(r) - \Phi(\infty) = -\int_{\infty}^{r} dr' \frac{q}{4\pi\epsilon r'^2} = \frac{q}{4\pi\epsilon r}$$

In (9.33), the point at infinity is taken as the reference point where $\Phi = 0$.

Potential of a Perfect Conductor

By definition, the electric field is zero inside a perfect conductor. From (9.31), we see that the potential difference between any two points in a perfect conductor is zero. Consequently, the potential Φ must be a constant throughout the space occupied by the conductor. Thus, each perfect conductor has a constant potential. However, this constant may differ from one conductor to the other when they are not connected. Conducting bodies connected by conducting wires have the same potential. If a conductor is connected by wire to the earth, we say that the conductor is grounded and that its potential is zero because at low frequencies and dc the earth is a good conductor and the potential everywhere on the earth is approximately constant. This constant is conventionally set equal to zero.

Example 9.15 *This example shows a direct way to obtain the potential from the electric field.*

Consider the parallel plate shown in Figure 9.18. Find the potential $\Phi(x)$ assuming $\Phi(0) = 0$.

Solution From Example 9.14, we know that

$$\boldsymbol{E} = \frac{\rho_s}{\epsilon} \hat{\boldsymbol{x}}$$

Thus,

$$\Phi(x) = -\int_0^x E_0 \, dx = -E_0 x \tag{9.34}$$

where $E_0 = \rho_s/\epsilon$.

Example 9.16 *This example shows how to obtain the potential from the electric field in cylindrical coordinates.*

Consider the coaxial line shown in Figure 9.16. Find $\Phi(\rho)$, assuming $\Phi(b) = 0$.

Solution According to the result obtained in Example 9.12, we have

$$E = \hat{\rho}\frac{\rho_\ell}{2\pi\epsilon\rho}$$

Thus,

$$\Phi(\rho) = -\int_b^\rho \frac{\rho_\ell}{2\pi\epsilon\rho'}\,d\rho' = \frac{\rho_\ell}{2\pi\epsilon}\ln\left(\frac{b}{\rho}\right) \tag{9.35}$$

Example 9.17 *To obtain the potential from the electric field, we must be careful to use the correct formula for the field. If the electric fields in different zones have different expressions, then the integration must be carried out from one zone to the next using the appropriate formulas.*

Consider the electron beam discussed in Example 9.13. Find the potential difference between $\rho = 0$ and $\rho = 2a$.

Solution

$$\Phi(2a) - \Phi(0) = -\int_{0\cdot}^{2a} E \cdot dr = -\int_0^a \frac{\rho_\ell \rho}{2\pi\epsilon a^2}\,d\rho - \int_a^{2a} \frac{\rho_\ell}{2\pi\epsilon\rho}\,d\rho$$

$$= -\frac{\rho_\ell}{4\pi\epsilon}(1 + 2\ln 2)$$

Example 9.18 *If a charge-carrying conducting body has a sharp point, then the electric field at that point is much stronger than the electric field over the smoother part of the body. This example explains why.*

Two conducting spheres of radii a and b, respectively, are connected by a long, thin conducting wire, as shown in Figure 9.20. A charge of q is placed on this structure. Find the charge on each sphere and the E field on the surface of each sphere.

Solution Because the two conducting spheres are far apart, we can assume that charges are uniformly distributed on both spheres as if they were infinitely far apart. Let q_a be

Figure 9.20 Two conducting spheres are connected by a long, thin conducting wire.

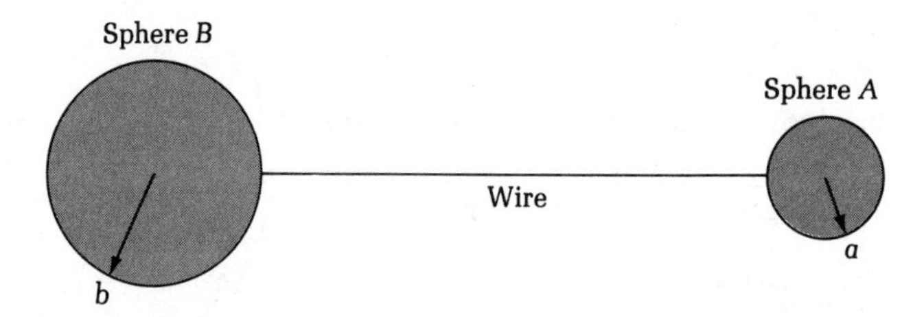

the charge carried on sphere A and q_b the charge on sphere B. Assume there is no charge on the conducting wire (because the wire is thin); then

$$q_a + q_b = q \tag{9.36}$$

The voltage on sphere A is V_a, where

$$V_a = \frac{q_a}{4\pi\epsilon a}$$

Similarly,

$$V_b = \frac{q_b}{4\pi\epsilon b}$$

Because these two conductors are connected by a conducting wire, their potential must be equal—that is, $V_a = V_b$. Or,

$$\frac{q_a}{4\pi\epsilon a} = \frac{q_b}{4\pi\epsilon b} \tag{9.37}$$

Solving (9.36) and (9.37), we obtain

$$q_a = q\left(\frac{a}{a+b}\right)$$

$$q_b = q\left(\frac{b}{a+b}\right)$$

The $\boldsymbol{E}$ fields on the surfaces of spheres A and B are, respectively,

$$E_a = \frac{q_a}{4\pi\epsilon a^2} = \frac{q}{4\pi\epsilon(a+b)a}$$

$$E_b = \frac{q_b}{4\pi\epsilon b^2} = \frac{q}{4\pi\epsilon(a+b)b}$$

Note that $E_a \gg E_b$ if $b \gg a$.

Lightning Rod

The results obtained in the previous example support the following qualitative statement: when a conducting body contains sharp points, the electric field on these points is much stronger than that on the smoother part of the conducting body. We can apply this result to explain how a lightning rod works.

A lightning rod is a sharp metal rod with one end placed on top of a tall structure and the other end connected to the ground, as shown in Figure 9.21a. When a cloud carrying electric charges approaches, the rod attracts opposite charges from the ground. The E field at the tip of the rod is much stronger than anywhere else. When the E field exceeds the breakdown strength, air near the tip is ionized, becomes conducting, and thereby provides a safe path for the electricity in the cloud to discharge to the ground. Figure 9.21b (p. 320) shows a lightning rod at work.

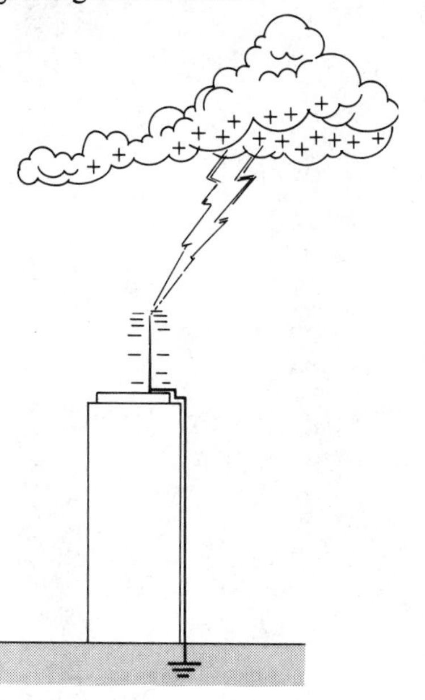

Figure 9.21(a) Lightning rod.

Line Integral for Electrostatic Potentials

We now show that by the nature of electrostatic fields, as long as the two end points are specified, the value V_{AB} as defined in (9.31) is independent of the path of integration. This property, so unique for electrostatic fields, follows directly from (9.1). **Stokes' theorem** in vector calculus states that for any vector field $\boldsymbol{A}$,

$$\iint d\boldsymbol{a} \cdot (\nabla \times \boldsymbol{A}) = \oint_C d\boldsymbol{r} \cdot \boldsymbol{A} \tag{9.38}$$

where the area integral is performed over an area enclosed by the closed contour C. Looking at Figure 9.19, we see that the closed contour C is made of path 1 and reversed path 2 with $d\boldsymbol{a}$ pointing out of the paper. Applying Stokes' theorem to (9.1), we obtain

$$\oint_C d\boldsymbol{r} \cdot \boldsymbol{E} = 0 \tag{9.39}$$

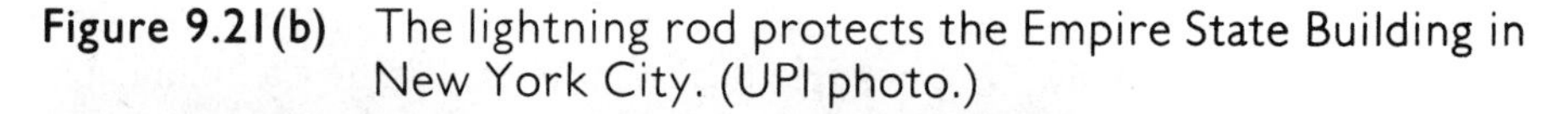
Figure 9.21(b) The lightning rod protects the Empire State Building in New York City. (UPI photo.)

The above equation expresses the conservative properties of the electrostatic fields. Applying (9.39) to the closed contour in Figure 9.19, we find that

$$\int_{\text{path 1}} d\boldsymbol{r} \cdot \boldsymbol{E} - \int_{\text{path 2}} d\boldsymbol{r} \cdot \boldsymbol{E} = 0$$

Thus,

$$\Phi_A - \Phi_B = -\int_{\text{path 1}} d\boldsymbol{r} \cdot \boldsymbol{E} = -\int_{\text{path 2}} d\boldsymbol{r} \cdot \boldsymbol{E}$$

Note that paths 1 and 2 are arbitrary paths connecting B and A. It follows that the value of V_{AB} is independent of the path linking A and B.

Kirchhoff's Voltage Law

In circuit theory, Equation (9.39) is known as Kirchhoff's voltage law. Note that (9.39) follows from (9.1), which is true only for static fields. If a time-varying magnetic field is present, then (9.39) is not valid. Thus, Kirchhoff's voltage law has its limitation: when a time-varying magnetic field links a circuit, Kirchhoff's voltage law will not apply unless the effect of the changing magnetic field is included.

Example 9.19 *This example shows that to obtain the potential from the electric field by integration method, the result is independent of the integration path.*

Let $\boldsymbol{E}$ be everywhere in the $\hat{z}$ direction and equal to 1 V/m. Find the potential V_{AB} with A at (3, 0, 5) and B at the origin, as shown in Figure 9.22.

Figure 9.22 Integration of the $\boldsymbol{E}$ field along different paths.

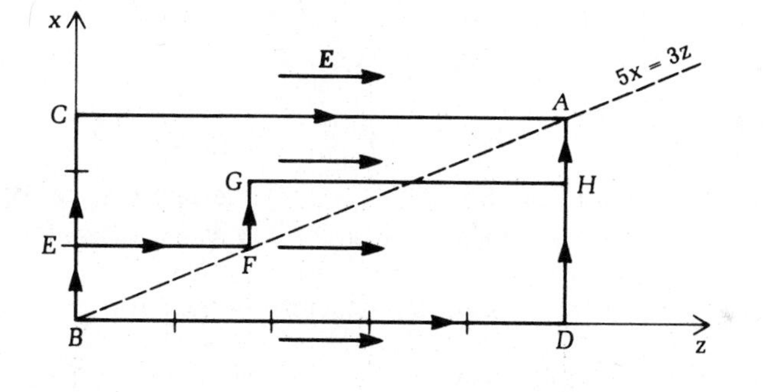

Solution We choose to integrate along three different paths. Along path $BECA$, $\boldsymbol{E}$ is perpendicular to BEC but parallel to CA. Thus,

$$V_{AB} = -3 \times 0 - 5 \times 1 = -5 \text{ volts}$$

Along path $BDHA$, $\boldsymbol{E}$ is parallel to BD but perpendicular to DHA; thus,

$$V_{AB} = -5 \times 1 - 3 \times 0 = -5 \text{ volts}$$

Along path $BEFGHA$, $\boldsymbol{E}$ is parallel to EF and GH but perpendicular to BE, FG, and HA. Thus,

$$V_{AB} = -1 \times 0 - 2 \times 1 - 1 \times 0 - 3 \times 1 = -5 \text{ volts}$$

The reader may try any other path and will find the same answer for V_{AB} because the given electric field is a static field satisfying $\nabla \times \boldsymbol{E} = 0$.

Example 9.20 *This example shows that if the potential depends on the integration path, then the electric field is not an electrostatic field.*

If the electric field is $\boldsymbol{E} = \hat{z}(x + 2)$ V/m, which points in the $\hat{z}$ direction but is not uniform, find the integration along $BECA$ shown in Figure 9.22. Compare it with that along $BDHA$.

Solution Along BC, the $\boldsymbol{E}$ is normal to the path, and hence the dot product is zero, and along $CA\, x = 3$, and therefore $\boldsymbol{E} = 5\hat{z}$. Thus,

$$V_{AB} = 0 - 5 \times 5 = -25 \text{ volts}$$

Along $BDHA$, we find

$$V_{AB} = -5 \times 2 - 0 = -10 \text{ volts}$$

The two values are different because $\nabla \times \boldsymbol{E}$ is not identically zero in the region where line integration is carried out. The given field is not an electrostatic field.

Example 9.21

This example shows how to integrate the electric field along a slanted straight line.

Given an electric field $\boldsymbol{E} = \hat{z}(x + 2)$, find the line integral of $\boldsymbol{E} \cdot d\boldsymbol{r}$ from B to A shown in Figure 9.22 along the straight line $5x = 3z$, $y = 0$.

Solution Since $\nabla \times \boldsymbol{E} \neq 0$, the integration depends on the path.

$$U = \int \boldsymbol{E} \cdot d\boldsymbol{r} = \int (x + 2)\, dz$$

But along the straight line, $5\, dx = 3\, dz$, therefore,

$$U = \frac{5}{3} \int_0^3 (x + 2)\, dx = 17.5$$

Example 9.22

Example 9.9 shows that there cannot be any electric charge in a solid conductor. This example shows that there cannot be any electric charge on the inner surface of a conductor having an empty cavity.

Consider an arbitrarily shaped conductor with an *empty* cavity in it, as shown in Figure 9.23. Show that the $\boldsymbol{E}$ field in the cavity is zero and that there is no charge on the surface of the cavity.

Solution From Example 9.9, we learn that $\boldsymbol{E} = 0$ and $\rho_v = 0$ in the conductor. Apply the integral Gauss' law (9.26) over the surface S', which lies just inside the conductor.

We obtain

$$\oint_{S'} \boldsymbol{E} \cdot \hat{\boldsymbol{n}}\, ds = q/\epsilon = 0$$

We conclude that the total charge inside S' is zero. Because the cavity is empty, we further deduce that the total charge on the inner surface of the conductor is equal to zero. At this point, we cannot conclude that the charge on the inner surface of the conductor is zero because there may be equal amounts of positive and negative charges on the inner surface, in which case the total charge would be zero. Let us investigate whether such a thing could happen. If we had a positive charge separated from a negative charge on the inner surface, then we would have an $\boldsymbol{E}$ field line C originating from the positive charge and ending at the negative charge. Thus, if we integrate $\boldsymbol{E} \cdot d\boldsymbol{r}$ *along the* $\boldsymbol{E}$ *field line* C, we obtain a positive number because $\boldsymbol{E}$ and $d\boldsymbol{r}$ are everywhere parallel:

$$\int_C \boldsymbol{E} \cdot d\boldsymbol{r} = V_{AB} > 0 \tag{9.40}$$

But we also know that the conductor is an equipotential body and the $V_{AB} = 0$ for any two points A and B on the conductor. This knowledge contradicts the above result. Thus, we conclude that charges on the surface of a cavity in a conductor must be of the same sign. This conclusion is true whether the cavity is empty or whether it contains some charges.

Now we put the two statements together: (a) according to Gauss' law the total charge inside S' is equal to zero because the cavity is empty, and (b) the sign of the charge on the inner surface of a hollow conductor is either all positive or all negative. We conclude that there is no charge at all on the inner surface of the hollow conductor. Furthermore, the $\boldsymbol{E}$ field is identically zero in the cavity because it must originate from a positive charge and end at a negative charge.

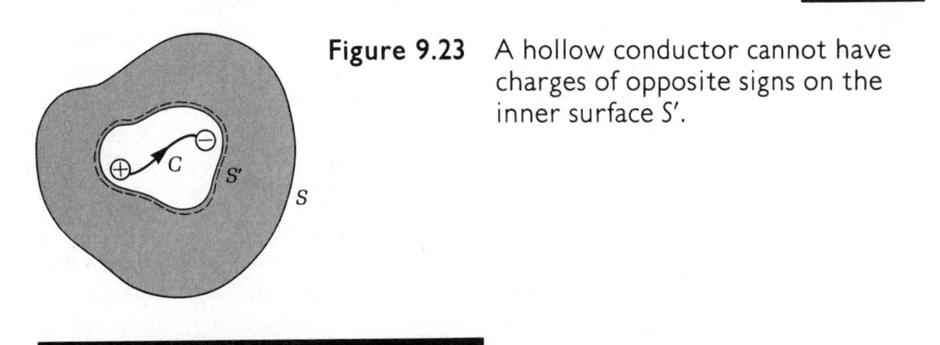

Figure 9.23 A hollow conductor cannot have charges of opposite signs on the inner surface S'.

The Van de Graaff Generator

The Van de Graaff generator is a device invented by Robert Van de Graaff (1901–1967). It can carry charges to a dome-shaped conductor in order to build up an electrostatic voltage as high as several million volts. Figure 9.24a illustrates its operating principle, and Figure 9.24b shows an earlier model.

Figure 9.24(a) Operating principle of a Van de Graaff generator.

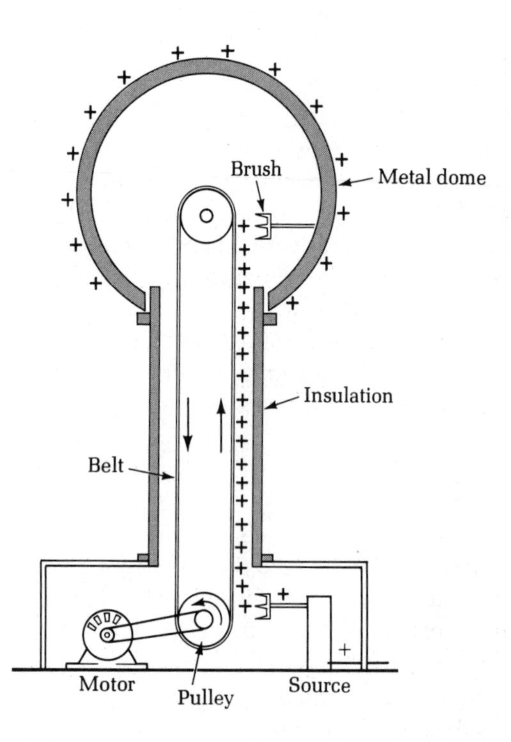

The insulator belt, made of silk or rubber, passes near a charge source of several kilovolts, and this source sprays charges on the belt. The belt is moved by two pulleys. As the charges are carried into a metal dome approximately one meter in diameter, a metal brush takes up these charges. This step is possible because, according to the previous example, the field in a cavity inside a conductor is zero. Although in the present case the metal is not entirely closed, the field inside is nevertheless very weak. Because charges cannot exist inside a conductor, those picked up by the brush quickly go to the outside of the metal dome. As the charge is being built up, the voltage also rises. A dome about one meter in diameter may achieve several million volts of voltage. The limit is the breakdown of air under the resulting intense $\boldsymbol{E}$ field.

Large Van de Graaff generators are used to accelerate charged particles for research in nuclear physics. Miniature Van de Graaff generators are also used in small-scale instruments to produce the high voltages required to generate neutrons. One such instrument is a "nuclear logging sonde."

In a neutron logging tool, a Van de Graaff generator produces a high voltage to accelerate charged deuterium particles. Neutrons are produced when these particles bombard a tritium target. When the logging tool is lowered into the oil well, these neutrons are absorbed by the surrounding rock formation. The absorption generates gamma rays, which are detected by a gamma-ray detector installed on the tool. The formation rock absorbs neutrons quickly when it con-

Figure 9.24(b) An earlier model of the Van de Graaff generator. (Courtesy of MIT Museum and Historical Collections.)

tains salt water. If the rock contains oil, the absorption is slower. Therefore, by measuring the decay rate of the intensity of the gamma rays, one can tell whether the tool is in an oil-bearing or in a water-bearing zone.*

Triboelectricity

Triboelectricity is electricity generated by friction. Static electric charges are generated when two materials in contact are suddenly separated or rubbed together. For example, by rubbing a glass rod with silk, a small amount of charge is transferred between them. The glass will become positively charged and the silk nega-

* D. A. Bronley, "The development of electrostatic accelerators," *Nuclear Instruments and Methods* 122 (1974): 1–34. D. W. Hilchie, "Neutron lifetime log." *Log Review* 1, Section 11 (Houston: Dresser-Atlas, 1974).

tively charged. Although the amount of charge generated this way is small, the voltage due to the charge may be quite high. Strolling across a vinyl floor can generate as much as 12,000 volts when humidity is low. Walking across a synthetic carpet can result in 1,500 to 35,000 volts of static voltage on a person. Discharge of triboelectricity through a person's fingers can sometimes cause a slight shock.

Electrostatic discharge of triboelectricity can damage electronic components when charged workers handle them. One source in the electronic industry estimated that as much as ten billion dollars (1983 value) were lost annually because of electrostatic discharge. A well-known electronics company reported that one of their computer divisions had an in-plant failure rate of 23 percent on a certain series of assemblies. The failure was attributed to electrostatic discharge. After an aggressive prevention program had been implemented, the failure rate dropped to less than 3 percent. Prevention of electrostatic discharge damage includes placing a conducting table mat and floor mat in the factory to conduct electricity to the ground.*

SUMMARY

1. When the frequency is zero, Maxwell's equations become decoupled; one set governs the electrostatic fields, and the other magnetostatic fields.
2. The electrostatic $\boldsymbol{E}$ field can be found from the scalar potential Φ by taking its gradient [Equation (9.6)].
3. The scalar potential Φ due to a point charge is given by (9.10). We can use summation or integration to obtain Φ due to multiple charges or a continuous distribution of charges.
4. The electrostatic $\boldsymbol{E}$ field due to a point charge is given by (9.18).
5. The electrostatic $\boldsymbol{E}$ field due to multiple charges or a continuous charge distribution of charges can be obtained by the superposition principle. The $\boldsymbol{E}$ field due to an infinitely long line charge is given by (9.22) in cylindrical coordinates. The $\boldsymbol{E}$ field due to a plane charge is given by (9.25) in rectangular coordinates.
6. Note that when we calculate the scalar potential Φ due to multiple charges, we can simply add potentials due to individual charges. But when we calculate the $\boldsymbol{E}$ field, we must add $\boldsymbol{E}$ fields due to individual charges using vector notations.
7. Gauss' law in integral form is given by (9.26). Note that the surface S in the left-hand side can be any mathematical closed surface, called the Gaussian surface, in space; and the right-hand side is the total charge inside that Gaussian surface.

---

* J. Bechtold, editor *Bench Briefs*, Hewlett-Packard Co., Mt. View, CA, March–May, 1983, pp. 1–7.

8. Gauss' law can be used to solve for the $\boldsymbol{E}$ field due to some symmetrical charge distributions such as a point charge, a uniform spherical distribution of charges, a charged beam, and so on.
9. The electrostatic $\boldsymbol{E}$ field can be found from the scalar potential Φ, and the reverse is also true: the difference of Φ between any two points can be found from $\boldsymbol{E}$ [Equation (9.31)].
10. We have proved that electric charges cannot exist inside a solid conductor and charges can only exist on its surface. Charges cannot exist inside a hollow conductor or on its inner surface if there is no charge in the cavity.
11. We have proved that an irregularly shaped charge-carrying conductor is an equipotential body; that is, the potentials at all points inside or on its surface are equal. However, the charge is more densely distributed at sharp edges and sharp points of the conductor than at its smoother part. Consequently, the electric field is also stronger at the edges and sharp points.

Problems

9.1 Consider the dipole arrangement shown in Figure 9.2a. Let $q = 1.6 \times 10^{-19}$ C. Find Φ at:

(a) $x = 0.1, y = 0.1, z = 0.1$
(b) $x = 1, y = 1, z = 1$

Use the exact formula (9.14) first. Then use the approximate formula (9.17), and find the accuracy of the latter. The medium is air.

9.2 Two point charges of $q_1 = +2q$ and $q_2 = -q$ are located at $(-1, 0, 0)$ and $(1, 0, 0)$, respectively, in rectangular coordinates. Show that the equipotential surface for $\Phi = 0$ is a spherical surface. Find the center and the radius of this surface.

9.3 Four point charges are equally spaced as shown in Figure P9.3. To simplify the algebra, let $Q = 4\pi\epsilon$.

(a) Find the explicit expression of $\Phi(x, y, z)$ due to these charges at an arbitrary point (x, y, z). Assume the potential at infinity to be zero.
(b) Find $\Phi(0, 0, 0)$.

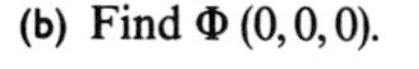

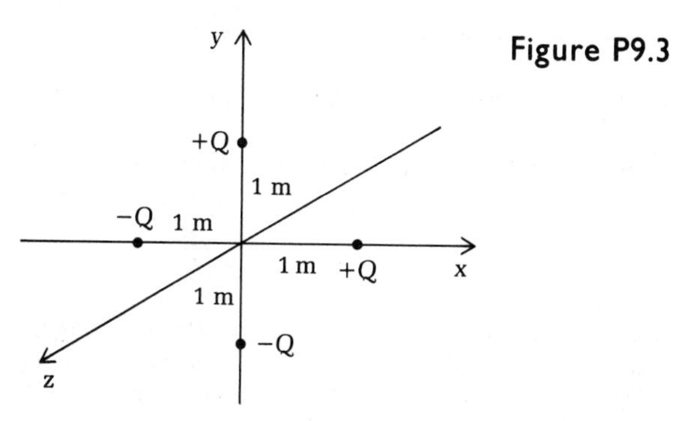

Figure P9.3

9.4 Two point charges are separated by a meters in air, as shown in Figure P9.4.

(a) Find the potential function $\Phi(x, y, z)$.
(b) Calculate Φ at $x = 100a$, $y = 100a$, $z = 0$.
(c) Show that, for distances much greater than a away from these charges, the potential is approximately given by

$$\Phi \approx \frac{-q}{4\pi\epsilon_0} \frac{1}{(x^2 + y^2 + z^2)^{1/2}}$$

where $(x^2 + y^2 + z^2)^{1/2} \gg a$. Use this approximate formula to calculate $\Phi(100a, 100a, 0)$, and compare it with the result obtained in (b).

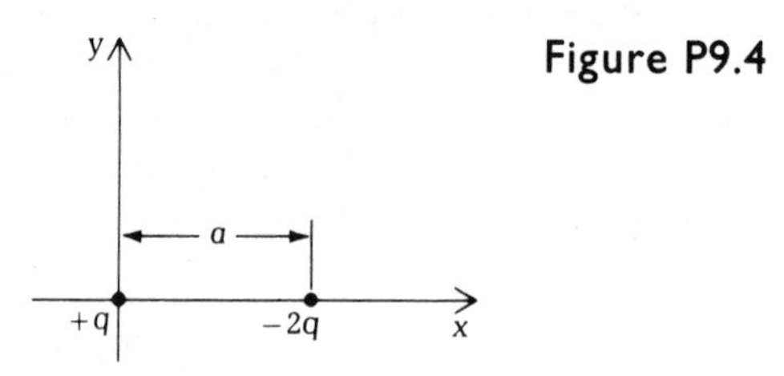

Figure P9.4

9.5 Find the $\boldsymbol{E}$ field in air due to a point of $10^6 q_e (q_e = -1.6 \times 10^{-19}$ C). Sketch a diagram similar to Figure 9.4.

9.6 For the charge distribution in Problem 9.2, find the $\boldsymbol{E}$ field at two points on the x axis where $\Phi = 0$.

9.7 For the charge distribution in Problem 9.3, first find $E_y\ (x, y, z)$ at an arbitrary point (x, y, z) and then find $E_y\ (0, 0, 0)$.

9.8 For the charge distribution shown in Figure 9.2a, show that the $\boldsymbol{E}$ field at the origin is $\boldsymbol{E} = -\hat{z}\ (5000q/\pi\epsilon)$.

9.9 Sketch the $\boldsymbol{E}$ field lines for the charge distribution given in Problem 9.2.

9.10 A line charge $2h$ meters long is located along the z axis, as shown in Figure 9.6a. The charge density is ρ_ℓ coulombs per meter.

(a) Calculate the electric field at $\rho = 0.1h$, $\phi = 0$, and $z = 0$ using the exact formula (9.21).
(b) Calculate the electric field at the same point using the assumption that the line is infinitely long.
(c) Find the percentage error of the value obtained in (b) as compared with the exact value.

9.11 For the same line charge described in Problem 9.10,

(a) Calculate the electric field at $\rho = 20h$, $\phi = 0$, $z = 0$ using the exact formula.
(b) Do the same using the assumption that the line is a point charge at the origin.
(c) Find the percentage error of value obtained in (b).

9.12 Find the electric field at (0, 0, z) in rectangular coordinates due to a charge distribution in the form of a thin circular ring defined by $x^2 + y^2 = 1$. The total charge on the ring is q and the charge is uniformly distributed on it.

9.13 A plane charge of ρ_s coulombs per square meter is located on the $x = 0$ plane, and another plane of $-\rho_s$ coulombs per square meter is located on the $x = 1$ plane. Find the total electric field in the region (a) $x > 1$, (b) $1 > x > 0$, and (c) $x < 0$.

9.14 A total of q coulombs of charge is uniformly distributed in a spherical region defined by $r \leq a$. Show that the electric field is given by

$$E_r = \begin{cases} \dfrac{qr}{4\pi\epsilon a^3} & \text{for} \quad r \leq a \\ \dfrac{q}{4\pi\epsilon r^2} & \text{for} \quad r \geq a \end{cases}$$

9.15 A charge distribution of the following form is set up in air:

$$\rho_v = 10^{-6} \cdot e^{-r} \text{ coulombs per cubic meter}$$

Use Gauss' law to find the $\boldsymbol{E}$ field everywhere. Hint: To find the total charge in a Gaussian surface, you must do the integration because the charge is not uniformly distributed. However, symmetry still exists with respect to ϕ and θ.

9.16 Electric charges are distributed uniformly in the region $-0.1 < x < +0.1$ with density $\rho_v = 10^{-6}\,\text{C/m}^3$. Elsewhere, the density is equal to zero. Find the $\boldsymbol{E}$ field everywhere. Plot E_x versus x. Find the potential difference $V_x - V_0$ for a point x with respect to the origin.

9.17 Find the potential difference $V_A - V_B$ for two points A and B located at $r = 0$ and $r = 1$ in the $\boldsymbol{E}$ field obtained in Problem 9.15.

9.18 The solution for the electric field of an oscillating Hertzian dipole with angular frequency ω is given in (7.14) as follows:

$$\boldsymbol{E} = \sqrt{\frac{\mu}{\epsilon}}\frac{jkI\,\Delta z\, e^{-jkr}}{4\pi r}\left\{\hat{r}\left[\frac{1}{jkr} + \frac{1}{(jkr)^2}\right]2\cos\theta + \hat{\theta}\left[1 + \frac{1}{jkr} + \frac{1}{(jkr)^2}\right]\sin\theta\right\}$$

Derive the solution (9.20) for a static dipole by setting $\omega = 0$. Notice that $k = \omega(\mu\epsilon)^{1/2}$ and $I\,\Delta z = \partial p/\partial t = j\omega p$.

9.19 In the electric field $\boldsymbol{E} = 3\hat{x} + 4\hat{y} - 5\hat{z}$, find $V_A - V_B$ if A is located at (1, 1, 2) and B is at the origin. Does the difference depend on the path of the integration?

9.20 Consider the spherical-shell problem shown in Figure 9.14. Find the potential $\Phi(r)$ at
(a) $r = c$;
(b) $b < r < c$;
(c) $b > r > a$, and
(d) $r = a$.
Assume $\Phi = 0$ at infinity. Plot $\Phi(r)$ versus r.

9.21 Repeat the preceding problem for the case in which the total charge on the conducting shell is equal to zero while all other conditions remain unchanged.

9.22 Consider the coaxial line shown in Figure P9.22. The inner conductor is a solid conducting cylinder with a radius equal to 0.1 m. The outer conductor has an inner radius equal to 0.4 m and an outer radius equal to 0.5 m. The medium between the inner and the outer conductor is air. The inner conductor carries a net charge of $-3\epsilon_0$ C/m and the outer conductor carries a net charge of $-15\epsilon_0$ C/m. The symbol ϵ_0 used here represents a constant equal to 8.854×10^{-12}.

(a) Find E_ρ in the region $0.1 \text{ m} < \rho < 0.4 \text{ m}$.
(b) Find E_ρ in the region $0.4 \text{ m} < \rho < 0.5 \text{ m}$.
(c) Find E_ρ in the region $\rho > 0.5$ m.
(d) Find Φ at $\rho = 0.2$ m, knowing that $\Phi = 0$ at $\rho = 1$ m.
(e) Sketch E_ρ as function of ρ for $0 < \rho < 1$ m. Mark the scale for E_ρ and ρ.

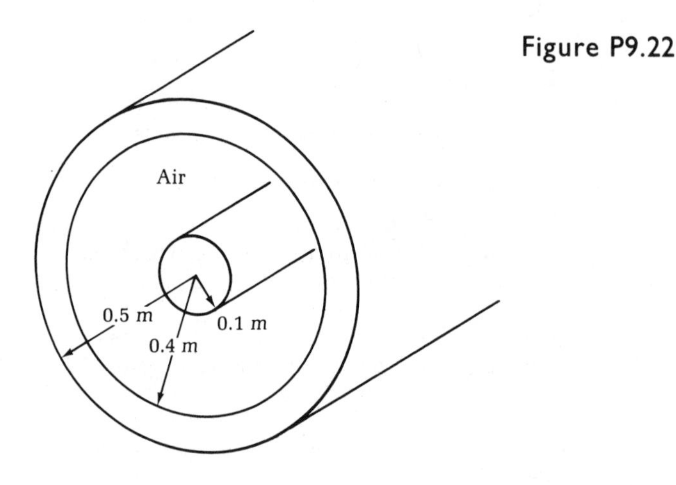

Figure P9.22

9.23 Model the dome of a Van de Graaff generator as a conducting sphere. The dome is charged to hold the maximum amount of electric charge Q_m before the air surrounding the dome breaks down. Use the following data:

radius of the dome = 0.11 m,
breakdown E of air = 3×10^6 V/m.

(a) Calculate the maximum Q_m accumulated on the dome just before the breakdown.
(b) Calculate the voltage of the dome in reference to the potential at infinity just before breakdown occurs.
(c) When the dome is charged with the maximum charge Q_m, a person uses a conducting rod to discharge the electricity. Assuming that the discharge takes 0.01 seconds to complete, how strong is the discharging current (on the average)?

10 ELECTRIC FORCE AND ENERGY

The previous chapter discussed several methods for calculating electrostatic fields from a variety of charge distributions. We want to know the exact electrostatic fields because in many applications charged particles can perform useful functions under the influence of electrostatic fields. For example, an electrostatic field directs an electron beam in a television or cathode ray tube to different spots on the screen to display pictures or other information. The previous chapter showed how to generate an electric field and how to calculate its value. We also need to quantify the interaction of the electric field and the charged particles, and we do so in this chapter. We first study the force exerted by the **E** field on charged particles and then examine the consequences of that interaction. Kinetic and stored energies are also discussed. Many interesting applications—for example, laser printing, CRT, and ion thruster—are presented.

10.1 ELECTRIC FORCE

The **Lorentz force law** states that an electromagnetic field exerts a force $\boldsymbol{F}$ on a charged particle moving with velocity $\boldsymbol{v}$ and carrying a charge q.

$$\boldsymbol{F} = q(\boldsymbol{E} + \boldsymbol{v} \times \boldsymbol{B}) \text{ Newtons} \tag{10.1}$$

This law holds both for static and for time-varying fields.

In this chapter we shall consider the electrostatic field in the absence of any magnetic field. Thus,

$$\boxed{\boldsymbol{F} = q\boldsymbol{E}} \quad \textbf{(force on } q \textbf{ in electrostatic field)} \tag{10.2}$$

Note that the above equation is a vector equation. The force is in the same direction as the $\boldsymbol{E}$ field if the charge is positive, and the force on a negative charge is in opposite direction to the applied $\boldsymbol{E}$ field.

Note that the $\boldsymbol{E}$ field in (10.1) and (10.2) refers to the external $\boldsymbol{E}$ field, or the field produced by all other charges except the one for which the force is being evaluated.

Coulomb's Law

Consider two particles separated by r meters and carrying q_1 and q_2 coulombs, respectively. The force on q_2 is given by

$$F_2 = q_2 E$$

where E is the electric field due to all charges except q_2. In the present case there is only one charge other than q_2. Let q_1 be located at the origin of the spherical coordinates. Using (9.18), we have

$$E = \hat{r}\frac{q_1}{4\pi\epsilon r^2}$$

Thus,

$$\boxed{F_2 = \hat{r}\frac{q_1 q_2}{4\pi\epsilon r^2}} \quad \textbf{(Coulomb's law)} \qquad \textbf{(10.3)}$$

We see that the electric force between two charged particles is proportional to the product of the two charges and inversely proportional to the square of the distance between them. The direction of the force is along the line joining the two particles. If q_1 and q_2 are opposite in sign, the force is attractive; otherwise, it is repulsive.

These facts were first experimentally obtained by Charles Coulomb in 1785. Equation (10.3) is known as **Coulomb's law**.

Example 10.1 *When considering forces between charged particles, the gravitational forces between them must also be considered. In this example, the gravitational force may be neglected because the electric force is much stronger.*

Two electrons are separated in air by 1 mm. Calculate the Lorentz force and the gravitational force between them. Compare the order of magnitude of these two forces.

Solution We substitute $q_1 = q_2 = -1.60 \times 10^{-19}$ C in (10.3) and obtain the repulsive force between these electrons:

$$F = 2.3 \times 10^{-22} \text{ N}$$

The gravitational force between two masses is attractive and the magnitude is given by

$$F_g = \frac{m_1 m_2}{r^2} \times 6.67 \times 10^{-11}$$

Substituting $m_1 = m_2 = 9.11 \times 10^{-31}$ kg into the above equation, we obtain

$$F_g = 5.5 \times 10^{-65}\ \text{N}$$

Thus, for all practical purposes, the gravitational force between electrons is negligible compared to the Lorentz force.

The Thomson Model of an Atom

At an early stage in the development of atomic theory, J. J. Thomson proposed that an atom consisted of a positive charge Ze, where Z is an integer, uniformly distributed over a sphere of radius a. He considered the electrons, each of charge $-e$, to be point charges imbedded in the positive charge.

We will show that with this model, contrary to experimental evidence, the electron would oscillate and radiate.

The positive charge density of the sphere is

$$\rho_v = \frac{Ze}{\frac{4\pi}{3} a^3} \tag{10.4}$$

The corresponding electric field inside the sphere is obtained from Gauss' law (see Problem 9.14),

$$E_r = \frac{Zer}{4\pi\epsilon_0 a^3} \tag{10.5}$$

Each electron experiences a force F_r in the radial direction.

$$F_r = -e(E_r) = -\frac{Ze^2 r}{4\pi\epsilon_0 a^3} \tag{10.6}$$

Let the electron mass be m, then, by Newton's law,

$$F_r = m\frac{d^2r}{dt^2} \tag{10.7}$$

Thus, we find the equation of motion for the electron to be

$$\frac{d^2r}{dt^2} = (-\omega_0^2)r \tag{10.8}$$

where

$$\omega_0^2 = \frac{Ze^2}{4\pi\epsilon_0 a^3 m} \tag{10.9}$$

Clearly, the solution for r is a time-harmonic function with angular frequency ω_0 or frequency f, where

$$f = \frac{\omega_0}{2\pi} = \frac{e}{2\pi}\sqrt{\frac{Z}{4\pi\epsilon_0 ma^3}}$$

For $Z = 1$, $a = 10^{-10}$ m, we have $f = 2.5 \times 10^{15}$ Hz, which would be in the ultraviolet range of the electromagnetic spectrum. In Chapter 7 we learned that oscillating currents radiate electromagnetic power. Thus, Thomson's atom model would lose power and would be unstable.

The Bohr Model of an Atom

According to the Bohr model of an atom, electrons revolve around a nucleus with a positive charge of Ze. The electron orbits are quantized according to the postulate

$$mrv = \frac{nh}{2\pi}$$

where m is the electron mass, v is the electron velocity, n is an integer number, and $h = 6.62 \times 10^{-34}$ J-s is the Planck constant. The centrifugal force mv^2/r as experienced by the electron is balanced by the Lorentz force $Ze^2/4\pi\epsilon r^2$. We have

$$\frac{mv^2}{r} = \frac{Ze^2}{4\pi\epsilon r^2}$$

Solving r from the above two equations, we obtain for the hydrogen atom with $Z = 1$,

$$r = \epsilon\frac{n^2h^2}{me^2\pi} = 0.52n^2 \times 10^{-10}m \tag{10.10}$$

The innermost orbital radius of the hydrogen electron is thus equal to 0.52 Å.

Motion of a Charged Particle in Uniform *E* Field

Consider the case in which a charged particle of mass m and charge q enters a uniform E field. Assume that at $t = 0$ the initial velocity of the charged particle is $\mathbf{v}_0 = v_{0x}\hat{x} + v_{0y}\hat{y} + v_{0z}\hat{z}$ and that the E field is $\mathbf{E} = E_0\hat{z}$. We want to calculate the particle's velocity and position at time t.

According to (10.2), the charged particle experiences a force $\mathbf{F}$, where

$$\mathbf{F} = qE_0\hat{z} \tag{10.11}$$

Because the force is in the $\hat{z}$ direction, only the velocity in that direction is changed. From Newton's law, we have

$$F_z = m\frac{dv_z}{dt} \tag{10.12}$$

Combining (10.11) and (10.12) yields

$$\frac{dv_z}{dt} = \frac{q}{m}E_0 \tag{10.13}$$

Integrating (10.13) once, we obtain

$$v_z(t) = \left(\frac{q}{m}E_0\right)t + C \tag{10.14}$$

Note that the initial condition is

$$v_z(0) = v_{0z}$$

The constant C in (10.14) may then be determined with this result:

$$v_z(t) = v_{0z} + \frac{qE_0}{m}t$$

The total velocity is then

$$\boldsymbol{v}(t) = \hat{\boldsymbol{x}}v_{0x} + \hat{\boldsymbol{y}}v_{0y} + \hat{\boldsymbol{z}}\left(v_{0z} + \frac{qE_0}{m}t\right) \tag{10.15}$$

The position of the particle at time t is

$$\boldsymbol{r} = \hat{\boldsymbol{x}}(v_{0x}t + x_0) + \hat{\boldsymbol{y}}(v_{0y}t + y_0) + \hat{\boldsymbol{z}}\left(v_{0z}t + z_0 + \frac{qE_0}{2m}t^2\right) \tag{10.16}$$

where (x_0, y_0, z_0) indicates the initial position of the particle at $t = 0$.

Example 10.2

In this example we use Newton's law and the Lorentz force law to predict the trajectory of charged particles moving in an electric field.

An electron with an initial velocity $\boldsymbol{v}_0 = 2 \times 10^7\hat{\boldsymbol{x}}$ m/s enters a parallel-plate capacitor charged with 100 V of voltage, as shown in Figure 10.1. The separation of the capacitor plates is 1 cm, and the plate measures 4 cm long in the $\hat{\boldsymbol{x}}$ direction. Find the trajectory of this electron.

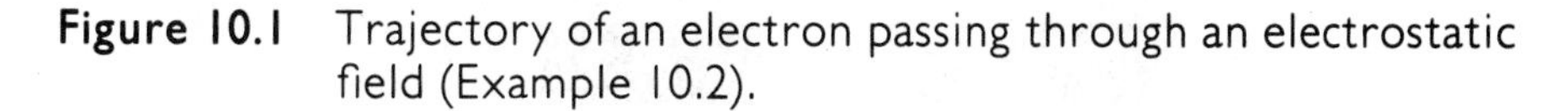

Figure 10.1 Trajectory of an electron passing through an electrostatic field (Example 10.2).

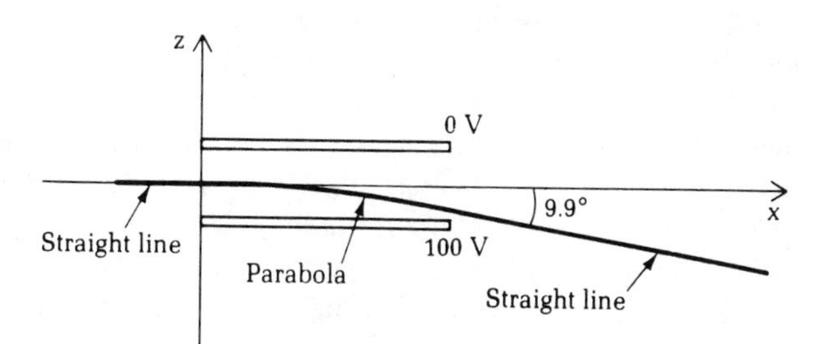

Solution The E field in the parallel-plate region is given by:

$$\mathbf{E} = \frac{V_0}{d}\hat{z} = 10^4 \text{ V/m } \hat{z}$$

Because the E field is along $\hat{z}$, the x component of the velocity of the electron will not be affected by the field. The initial position of the electron is $x_0 = y_0 = z_0 = 0$, and its initial velocity is $v_{0x} = 2 \times 10^7$ m/s and $v_{0y} = v_{0z} = 0$. Using (10.16), we find that at time t the position of the electron is

$$x = 2 \times 10^7 t$$

$$z = \frac{-1.6 \times 10^{-19} \times 10^4}{2 \times 9.1 \times 10^{-31}} t^2 = -8.8 \times 10^{14} t^2$$

Eliminating time t from the above two equations, we find the trajectory of the electron to be

$$z = -2.2x^2 \qquad \textbf{(10.17)}$$

which is a parabola. The electron will exit the plate region t_0 seconds after it enters the region with

$$t_0 = \frac{4 \times 10^{-2}}{2 \times 10^7} = 2 \times 10^{-9} \text{ s}$$

We find that the point of exit is at (x_e, z_e), where

$$x_e = 0.04 \text{ m}$$

$$z_e = -0.0035 \text{ m}$$

After the electron leaves the parallel-plate region, it follows a straight path that makes an angle of θ with respect to the x axis, with

$$\theta = \tan^{-1} \frac{v_z(t_0)}{v_x(t_0)} = -9.9^\circ$$

Figure 10.1 shows the entire trajectory.

Electrostatic Separation

Electrostatic forces can be used to separate two different granular materials. Such machines are widely used in industry for sorting minerals. Figure 10.2 illustrates their basic operating principle.

A mixture of two minerals, for example, phosphate ore composed of particles of quartz and phosphate rock, is placed in a vibrating feeder. The vibration causes friction between the quartz and the phosphate granules, and as a result the quartz grains become positively charged and the phosphate grains become negatively charged. This process of charging by friction is similar to what happens when one combs one's hair and finds out that the hair and the comb are charged with different polarities. In such a case, further combing would cause the two to neutralize each other by producing electric sparks, which are visible in the dark.

Figure 10.2 Electrostatic separator. Phosphate granules and quart granules are charged with different polarities when they pass through the vibrating feed, which produces friction among these granules.

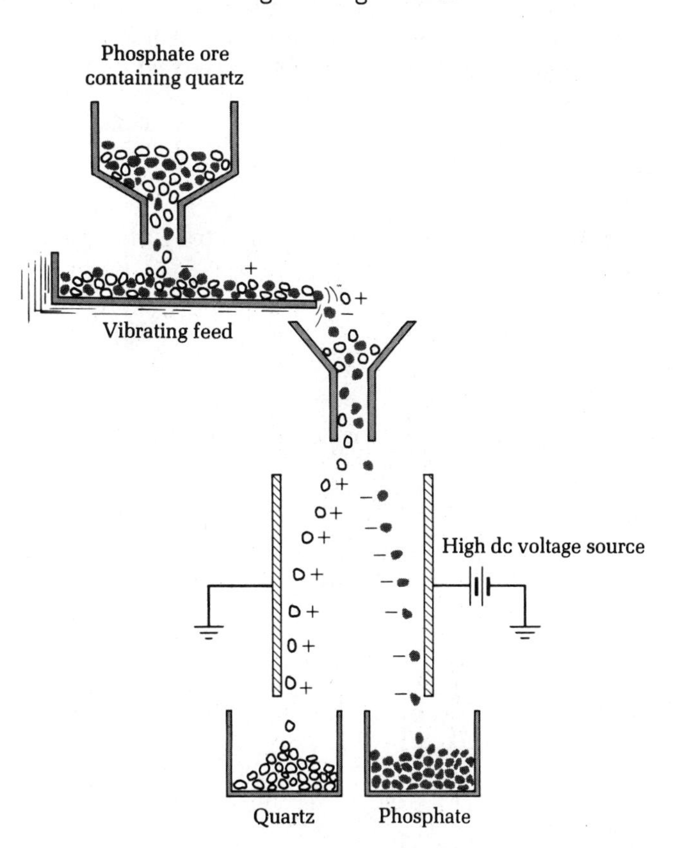

The vibrating feeder can produce approximately 10^{-5} C/kg of charge on the mineral granules,* which are then fed to a chute and dropped through a charged parallel plate. Because of the electrostatic forces, the quartz grains are attracted toward the negative plate, and the phosphate grains move toward the positive plate. Separate containers may be placed below the parallel plates to receive the quartz and phosphate separately, as shown in Figure 10.2.

* J. E. Lawver and W. P. Dyrenforth, "Electrostatic separation," in *Electrostatics and Its Applications*, ed. A. D. Moore, New York: Wiley, 1973, p. 229.

Example 10.3 *This is a numerical example of the electrostatic separation system.*

Assume in Figure 10.2 that the voltage between the parallel plates is 100 kV, that the separation of the plate is 0.5 m, and that the vertical dimension of the parallel plate is 1 m. The charge-to-mass ratio of a quartz particle is 9×10^{-6} C/kg. The particle starts to free fall at the middle of the top edge of the parallel plate. Find the position of the particle at the exit end of the plate.

Solution We set up the coordinate system as shown in Figure 10.3. The velocity in the $\hat{z}$ direction is governed by gravitational force—that is,

$$\frac{d^2z}{dt^2} = g = 9.8 \text{ m/s}^2$$

Figure 10.3 Calculation of the trajectory of a quartz granule (Example 10.3).

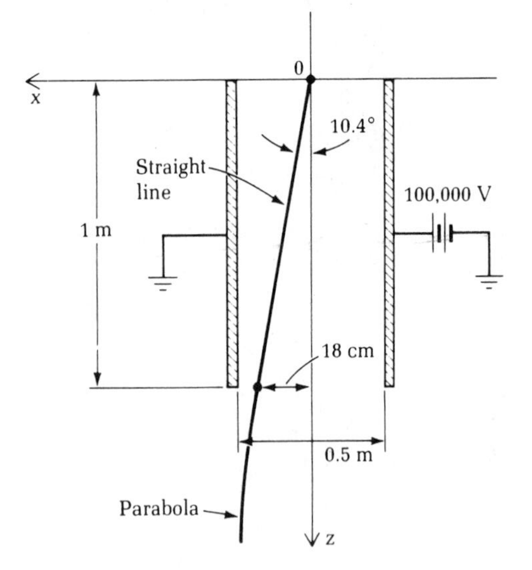

Integrating the above equation with respect to t and using the initial conditions $(dz/dt) = 0$ at $t = 0$, we obtain

$$\frac{dz}{dt} = 9.8t \text{ m/s}$$

Performing one more integration with the initial condition $z = 0$ at $t = 0$, we obtain

$$z = 4.9t^2 \text{ m}$$

Thus, the quartz granule will exit the plate at t_0, with

$$t_0 = \sqrt{\frac{1}{4.9}} = 0.45 \text{ s}$$

Now, the velocity of the quartz granule in the $\hat{x}$ direction is governed by the electrostatic force—that is,

$$m\frac{d^2x}{dt^2}\hat{x} = q\boldsymbol{E}$$

According to Example 9.15,

$$\boldsymbol{E} = \frac{V}{d}\hat{x} = 2 \times 10^5\hat{x} \text{ V/m}$$

Thus,

$$\begin{aligned}\frac{d^2x}{dt^2} &= \frac{q}{m} \times 2 \times 10^5\\ &= 9 \times 10^{-6} \times 2 \times 10^5\\ &= 1.8\end{aligned}$$

We integrate twice and substitute the initial conditions that $dx/dt = 0$ and $x = 0$ at $t = 0$, and we get

$$\frac{dx}{dt} = 1.8t$$

$$x = 0.9t^2$$

At $t = t_0 = 0.45$ s, the particle exits the plate at $x = 0.9 \times (0.45)^2 \approx 0.18$ m. Note that the variable t may be eliminated from the expressions for z and x to obtain an equation for z and x:

$$z = 5.44x$$

Thus, inside the parallel plate the trajectory of the quartz particle is a straight line. The $\boldsymbol{E}$ field is zero outside the plate, but gravitational force still exists; thus, the trajectory of the quartz after its exit from the parallel plate will be a parabola.

Electric Corona

In foggy or rainy weather, water droplets form on high-voltage power-transmission lines. These droplets represent sharp points on the otherwise smooth conductors. From Example 9.18 we learned that the $\boldsymbol{E}$ field is strongest at a conductor's sharp points. Thus, near the water droplets the electric field is higher than normal and often high enough to ionize the surrounding air. As a result, air is heated, and a faint glow is usually visible in the dark. This glow is called **corona discharge**. Corona is a plasma produced through gaseous discharge caused by very high electric fields.

These electric fields are usually nonuniform and exist only near the sharp points of a conductor or near a thin wire. Electric fields away from these regions are usually much weaker. Thus, the gaseous discharge is limited to a local region. The faint illumination is usually shaped like a ring around a thin wire or glow with streamers resembling a crown or a corona.*

A corona that occurs around power-transmission lines represents power loss. But generating a corona may be useful in some applications that we will discuss later. For the moment, let us consider the physics of corona generation.

Consider the coaxial line shown in Figure 10.4. Let us denote the voltage between the inner and the outer conductors as V_0. From Example 9.12, we have

$$\mathbf{E} = \frac{\rho_\ell}{2\pi\epsilon\rho}\hat{\boldsymbol{\rho}} \tag{10.18}$$

and

$$V(\rho) = \frac{\rho_\ell}{2\pi\epsilon}\ln\left(\frac{b}{\rho}\right) \tag{10.19}$$

where ρ_ℓ is the inner conductor's charge density, expressed in coulombs per meter. Noting that $V(a) = V_0$, we have

$$V_0 = \frac{\rho_\ell}{2\pi\epsilon}\ln\left(\frac{b}{a}\right) \tag{10.20}$$

Thus,

$$E_\rho = \frac{V_0}{\ln\left(\dfrac{b}{a}\right)}\frac{1}{\rho} \tag{10.21}$$

Figure 10.4 A corona tube. The central metal wire is very thin and is maintained at a voltage higher than that of the outer metal tube. A high electrostatic field near the thin wire produces a positive corona. If the polarity of the battery is reversed, a negative corona is excited near the thin wire.

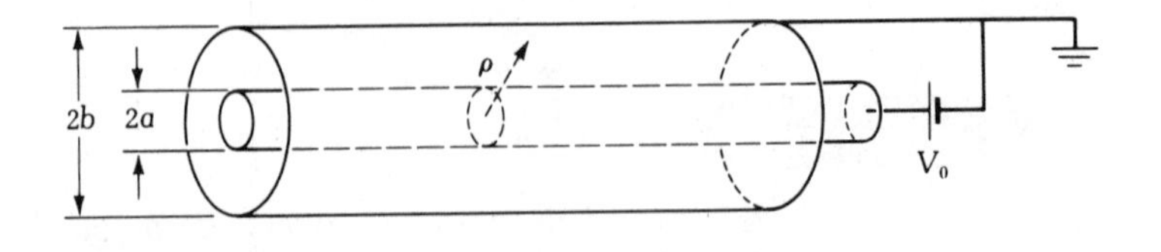

* A. D. Moore, *Electrostatics*, Garden City, N.Y.: Doubleday, 1968. L. B. Loeb, *Electrical Coronas*, Berkeley, Calif.: University of California Press, 1965, p. 171.

Obviously, the maximum electric-field intensity occurs at the surface of the inner conductor where $\rho = a$. If a is small, E_ρ could have a very large value.

An empirical formula tells us the strength of the electric field needed to produce a corona in air:*

$$E_c = 3 \times 10^6 m\left(\delta + 3 \times 10^{-2}\sqrt{\frac{\delta}{a}}\right) \text{ V/m} \tag{10.22}$$

where

m = wire-roughness factor, ranging from 0.5 to 0.9 in practice, $m = 1$ for a smooth wire

δ = the relative air density $= (T_0/T)(P/P_0)$

P = pressure of air (in units of atmospheric pressure)

T = air temperature, K

T_0 = 298K

P_0 = 1 atmospheric pressure

Using (10.22), we can calculate the minimum voltage V_0 that would produce a corona in the coaxial line shown in Figure 10.4. Let us assume that $b = 10$ cm, $a = 0.5$ cm, $m = 0.9$, at room temperature (20° C) and standard atmospheric conditions ($P = 1$ atm). With these data, (10.22) gives

$$E_c = 3.9 \times 10^6 \text{ V/m}$$

From (10.21), we obtain the corona voltage V_0

$$V_0 = E_c a \ln\left(\frac{b}{a}\right) = 58.4 \text{ kV}$$

Because of natural radiation, some of the air molecules are ionized. The percentage of such naturally ionized air is very low. However, in an intense electric field, these electrons are accelerated to high velocities. When they collide with an air molecule, the latter will usually be stripped of an electron, thus creating more fast-moving electrons in the air. This process continues, creates an avalanche in the region of high electric field, and thus forms the corona.

Electrostatic Filter

Electrostatic filters (or electrostatic precipitators) are widely used in industry to remove objectionable particles from gas emitted to the atmosphere. They are also used in homes to filter out pollen and dust for people with allergies.

Figure 10.5 illustrates the basic process of electrostatic filtering. The gas or air

---

* F. W. Peek, *Dielectric Phenomena in High-Voltage Engineering*, New York: McGraw-Hill, 1929, p. 53.

Figure 10.5 Electrostatic filter. Positive ions produced in the corona move toward the negative collector plate and become attached to particles in the gas. Both are collected on the negative collector wall.

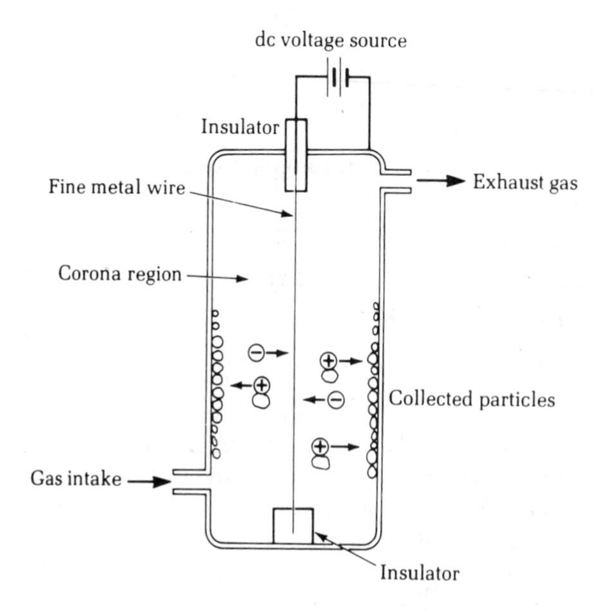

that is to be purified is forced into a coaxial metal-cylinder region. The central conductor is a thin wire, so that the electric field intensity near the wire is high enough to produce a corona. The positive ions produced in the corona migrate to the negative electrode and in so doing may collide with particles in the gas or air. The ions adhere to these particles, which are then attracted to the collection electrode by the electric field.*

Electrostatic Paint Spraying

Under the same operating principle as the electrostatic filter, paint sprayed from a spray gun may be charged by a corona and directed toward a target that is maintained at a lower potential than the corona electrode. Thus, the paint particles will be directed by electrostatic force to land on the target. This method can improve painting efficiency from 40% with the ordinary method to 70%.** Figure 10.6 illustrates one of several possible arrangements for electrostatic paint spraying.

---

* S. Oglesby, Jr., and G. B. Nichols, *Electrostatic Precipitation*, New York: Marcel Dekker, Inc., 1978, p. 4.

** E. P. Miller, "Electrostatic coating," in *Electrostatistics and Its Applications*, ed. A. D. Moore, New York: Wiley, 1973, p. 250.

Figure 10.6 Electrostatic paint spraying. Some of the paint drops that would otherwise miss the target are pulled in toward the target.

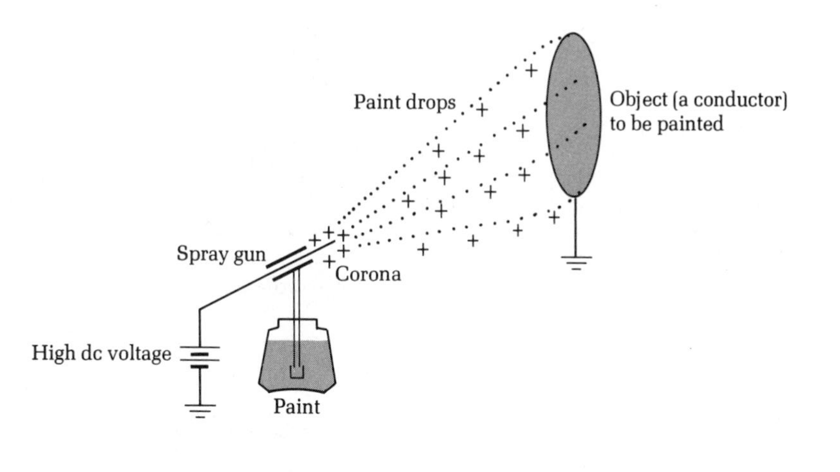

Xerography

Xerography is a technique that uses static electricity and photoconductive material to record images. It was invented in the 1930s and became widespread in the 1950s. Xerographic machines are now indispensable office equipment for copying, microfilm enlarging, facsimile printing, and computer-data printing. This electrostatic photographic process is different from the usual photographic process in that it does not require photosensitive chemical negatives. In fact, the negative is replaced by a photosensitive material, and the chemicals are replaced by electrons, and both may be used repeatedly, as we explain below.

In Figure 10.7a (p. 344), thin photoconductive material, such as amorphous selenium 20 to 100 μm thick, has been coated on a metal conductor. The photoconductor has a useful property: the spot that is exposed to light becomes a conductor, while the dark area remains an insulator. In the first step of the xerographic process, the photoconductor is charged by corona discharge. The procedure is done in darkness, so that opposite charges are uniformly distributed over the opposite surface of the photoconductor, as shown in Figure 10.7a.

The next step is to cast a light image on the photoconductor. The exposed area then becomes conductive, and the charges in that area leak away, while in the dark area the charges remain unchanged, as shown in Figure 10.7b. The number of charges that conduct away in a given area is proportional to the intensity of the light that is cast on the area and that thus forms an electrostatic latent image, as shown in Figure 10.7c. Then, negatively charged carbon particles are spread over the electrostatic image surface, as shown in Figure 10.7d. The areas with more charges attract more carbon particles. Finally, a positively charged sheet of paper is placed close to the image plane, so that carbon particles are attracted and attached to the paper, as shown in Figure 10.7e. Thus, the

Figure 10.7 Xerography.

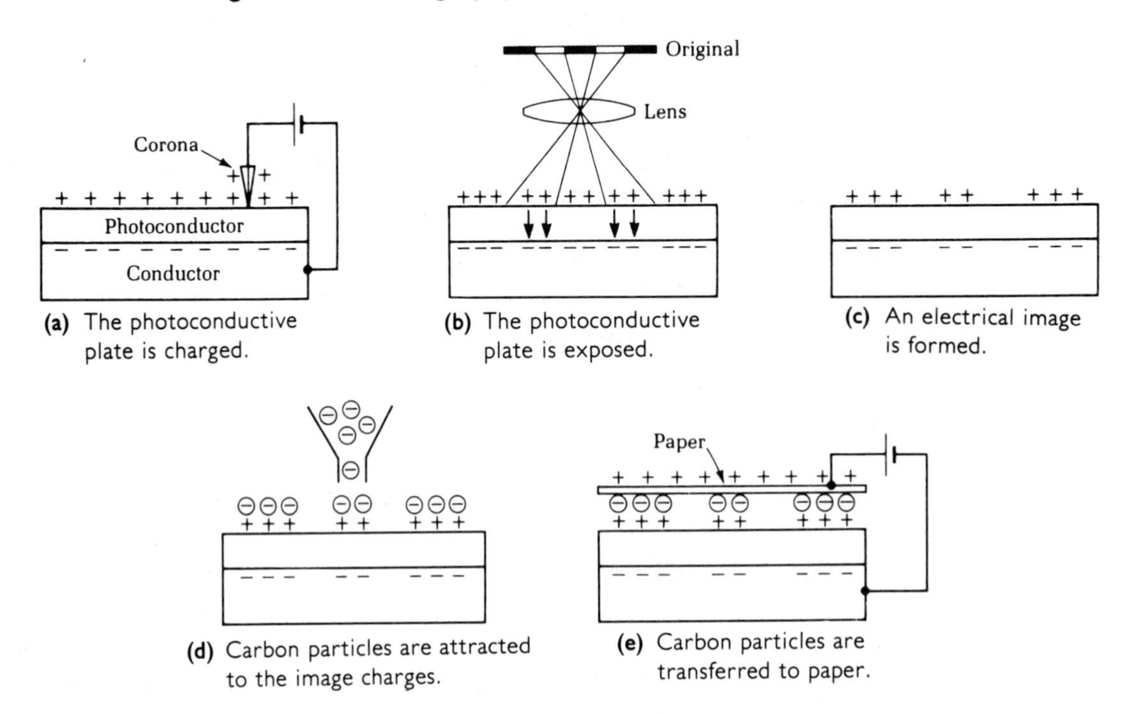

(a) The photoconductive plate is charged.

(b) The photoconductive plate is exposed.

(c) An electrical image is formed.

(d) Carbon particles are attracted to the image charges.

(e) Carbon particles are transferred to paper.

image on the photoconductive plate is transferred to the paper. The plate can then be cleaned and recharged for the next xerographic cycle.*

Laser Printer

Lasers are now being used as printers, producing faster and better-quality type than conventional dot-matrix and daisy-wheel printers. The operating principle of laser printers is similar to xerography. The major difference is in forming the latent image, a process illustrated in Figure 10.7b. In xerography, the image of the original copy is focused by a lens on a photoconductive plate. In laser printers, the image is formed on the photoconductive plate by a light beam emitted from a solid state laser. The photoconductive plate is usually rolled in cylinder form and is referred to as the drum. The laser is flashing on and off thousands of times per second as the light beam scans across the drum. Rather than moving the laser directly, a multifaceted mirror, spinning at approximately 100 revolutions per second, is used to reflect the light beam. Wherever the finely focused

* For advanced reading, refer to R. M. Schaffert, *Electro-Photography*, New York: The Focal Press, 1965, and J. H. Dessauer and H. E. Clark, eds, *Xerography and Related Processes*, New York: The Focal Press, 1965.

laser beam strikes the light-sensitive surface, the spot becomes conductive and charges in that area leak away. After a complete scan of a page, an image made of charges is then formed on the photosensitive drum. After this step, the remaining steps closely follow those used in xerography.

The laser printer combines the principle of a dot-matrix printer and that of a photocopying machine. Unlike the conventional dot-matrix printer, the laser printer is silent and prints faster. It can print an entire page in a few seconds and has a resolution of 14,000 dots per square centimeter.

10.2 WORK AND ENERGY

In an electrostatic field $\boldsymbol{E}$, a charged particle of q coulombs experiences a force given by (10.2)

$$\boldsymbol{F} = q\boldsymbol{E}$$

The work needed to move this particle from point A to point B is

$$\boxed{W = -\int_A^B q\boldsymbol{E} \cdot d\boldsymbol{r}} \qquad \textbf{(work needed to move } q \textbf{ from } A \textbf{ to } B\textbf{)} \qquad (10.23)$$

Note the minus sign in (10.23). If the particle moves in the direction against the force, the dot product of $\boldsymbol{E}$ and $d\boldsymbol{r}$ is negative, resulting in a positive W. Thus, external energy is really needed to move the charge in that direction. On the other hand, if $\boldsymbol{E}$ and $d\boldsymbol{r}$ are in the same general direction, their dot product is positive, resulting in a negative W. This negative work means that, instead of needing external work, the charge will provide work for the external system, or it will gain kinetic energy if it is not attached to any external system.

The equation (10.23) may be written as follows

$$W = -q\int_A^B \boldsymbol{E} \cdot d\boldsymbol{r}$$

From (9.31), we recognize that the above integral (with the minus sign) simply defines the difference in potential between B and A:

$$\boxed{W = q(\Phi_B - \Phi_A)} \qquad \textbf{(work needed to move } q \textbf{ from potential } \Phi_A \textbf{ to potential } \Phi_B\textbf{)} \qquad (10.24)$$

In other words, external work is needed to move a charge q from a lower potential point—for example, point A—to a higher potential point—for example, point B. This situation is analogous to potential in a gravitational field: external work is needed to move a rock from the ground to a certain height above the ground.

The work done to move a rock to a certain height is not wasted. It gives the rock a "stored energy" because, when released, the rock will fall and attain a kinetic energy equal to the work done to it. Similarly, the charge that is moved to a higher potential point has acquired a stored energy equal to the work done by the external force. This charge will fall back to the lower potential and at the same time gain kinetic energy when it is released.

Note that the work formula (10.23), just like the calculation of electrostatic potentials from the $\boldsymbol{E}$ field discussed in Section 9.4, is independent of the path of motion and is a function of the end points A and B only. This fact is a direct consequence of (9.39), which is valid for electrostatic fields. Needless to say, the above statements are incorrect for nonstatic electric fields.

Consider, for example, a charged particle of q coulombs at point A, where the potential is Φ_A. The particle moves from A to B, where the potential is Φ_B. Assume that the initial velocity of this particle at A is v_0, then the velocity at B is v, where v satisfies the following equation:

$$\frac{1}{2}mv^2 = \frac{1}{2}mv_0^2 + q(\Phi_A - \Phi_B) \quad \text{(kinetic energy at } B \text{ when } q \text{ goes from } A \text{ to } B) \tag{10.25}$$

Note that (10.25) is obtained by the principle of conservation of energy. The velocity v can be obtained from this equation without detailed knowledge of the force on this charged particle. In Section 10.1, we calculated the Lorentz force and then used Newton's law to obtain the trajectory of a charged particle. The discussion leading to (10.25) shows that, if the potential function is known instead of the $\boldsymbol{E}$ field and if the calculation of the trajectory is not required, the velocity of a charged particle can be predicted directly from (10.25).

Example 10.4

Charged particles are accelerated by the electric field. This example shows how to find the final velocity when a charged particle has traveled through a potential difference of V volts.

Assume that a charge of q coulombs initially at rest in an electric field is accelerated by the field and starts to move along the direction of the field. Find the velocity of the charge after it has traveled from B to A. V denotes the difference in potential between B and A.

Solution The kinetic energy of the charged particle at point A is equal to the potential energy qV that the particle gives up in order to acquire its final velocity v. Thus,

$$\frac{1}{2}mv^2 = qV$$

which gives

$$v = \sqrt{\frac{2qV}{m}} \tag{10.26}$$

For an electron with charge $q = -1.6 \times 10^{-19}$ C and mass $m = 9.11 \times 10^{-31}$ kg going through a difference in potential $V = -100$ volts, the velocity of the electron will become

$$v = 5.93 \times 10^6 \text{ m/s}$$

We must note however that, as v approaches the velocity of light, we must account for relativistic effects. The mass of the electron would then be $m(1 - v^2/c^2)^{-1/2}$, where c is the velocity of light.

Electron Volts

In SI units, energy is expressed in joules. This unit is too large in atomic physics. Instead, energy is sometimes measured in terms of electronvolts. An electronvolt is a unit of energy that an electron gains by going through a potential of one volt.

From (10.24), we know that

$$1 \text{ eV} = -1.6 \times 10^{-19} \times (-1) = 1.6 \times 10^{-19} \text{ J}$$

Thus, a particle of 1-BeV (a billion electronvolts) has a kinetic energy of 1.6×10^{-10} J.

Example 10.5 *This example gives the potential energy and the kinetic energy of an electron in the innermost orbit of a hydrogen atom.*

Find the potential energy, the kinetic energy, and the total energy for the electron of a hydrogen atom in its orbit with radius r.

Solution The potential due to the hydrogen nucleus at infinity is zero, and the potential at r is $q/4\pi\epsilon r$, where the charge of the nucleus $q = 1.6 \times 10^{-19}$ C. The potential energy (P.E.) of the electron is then

$$\text{P.E.} = -\frac{q}{4\pi\epsilon r}$$

Note that the negative potential energy indicates that negative work must be done to move the electron from infinity to r and that the electron is more stable with smaller r. From (10.23), we calculate the work done by noting the $\boldsymbol{E} = \hat{r}q/4\pi\epsilon r^2$ and that the electron charge is $-q$.

$$W = -\int_{\infty}^{r} (-q)\frac{q}{4\pi\epsilon r^2}\,dr = -\frac{q^2}{4\pi\epsilon r}$$

which is equal to the potential energy. Note also that the electron in its orbit experiences an electric force $-q^2/4\pi\epsilon r^2$, which is balanced by the centrifugal force mv^2/r.

$$m\frac{v^2}{r} = \frac{q^2}{4\pi\epsilon r^2}$$

The kinetic energy (K.E.) is given by

$$\text{K.E.} = \frac{1}{2}mv^2 = \frac{1}{2}\frac{q^2}{4\pi\epsilon r}$$

which is half the magnitude of the potential energy. The total energy of the hydrogen atom is therefore equal to P.E. + K.E. = $-(1/2)(q^2/4\pi\epsilon r)$, which is equal to 13.8 eV at the innermost orbit, with $r = 0.52$ Å. To ionize the electron from the atom, one must use a minimum energy of 13.8 eV.

Cathode Ray Tube

Cathode ray tubes, commonly called CRTs, are widely used as display devices for computer terminals and oscilloscopes. Figure 10.8 shows the basic operating principle of a CRT.

Figure 10.8 A cathode ray tube (CRT).

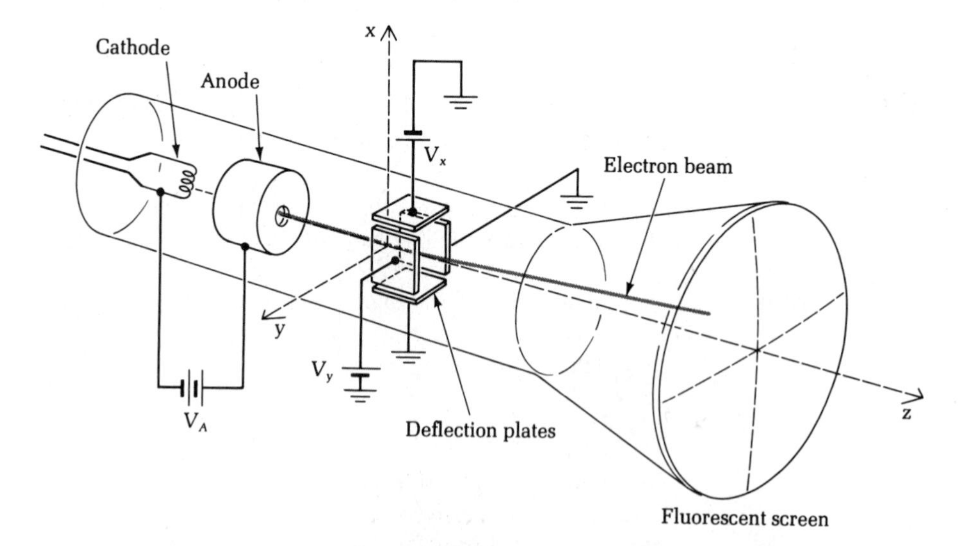

A beam of electrons is emitted from the cathode and accelerated by the anode. The beam passes through an opening in the anode and goes into a region of two orthogonal sets of parallel plates. The voltages on these two plates determine the deflection of the electron beam and the position where the beam hits the fluorescent screen to produce a bright spot. Changing the voltages V_x and V_y rapidly produces a pattern of spots on the fluorescent screen.

From (10.25), we can calculate the velocity of the electron at the anode:

$$v_z = \sqrt{\frac{2q_e V_A}{m_e}}$$

where $-q_e$ and m_e are the charge and the mass of the electron, respectively. For an anode voltage of 1 kV, we obtain

$$v_z = 1.87 \times 10^7 \text{ m/s}$$

the velocity with which an electron enters the region of the deflection plates. The electron is subject to two components of electrostatic force, f_x and f_y, where

$$f_x = m_e \frac{d^2x}{dt^2} = q_e \frac{V_x}{d_1}$$

$$f_y = m_e \frac{d^2y}{dt^2} = q_e \frac{V_y}{d_2}$$

$$z = v_z t$$

where x, y, z are the coordinates of the electron at time t ($t = 0$ corresponds to the moment the electron enters the parallel-plate region) and d_1 and d_2 are separations of x deflection plates and y deflection plates, respectively. If the deflection plates are ℓ meters long, the time when the electron leaves the region is t_0, where

$$t_0 = \frac{\ell}{v_z}$$

This time constant is important in that it sets an upper limit on how fast the deflection voltages V_x and V_y may be varied to direct the electron beam to different spots on the screen. In our present example, let $\ell = 3$ cm, and we have $t_0 = 1.60 \times 10^{-9}$ s.

At the exit, the velocity of the electron and its position may be expressed in terms of t_0:

$$v_x = \frac{q_e V_x}{m_e d_1} t_0$$

$$v_y = \frac{q_e V_y}{m_e d_2} t_0$$

$$v_z = v_z$$

$$x = \frac{q_e V_x}{2m_e d_1} t_0^2$$

$$y = \frac{q_e V_y}{2m_e d_2} t_0^2$$

$$z = v_z t_0$$

Note that the electron travels at a constant velocity from when it exits the region of the deflection plates until it hits the screen. In the present example, assuming that the distance between the fluorescent screen and the exit end of the deflection region is 15 cm, let us calculate the exact position where the electron beam will land on the screen.

Take $V_x = V_y = 100$ volts, and $d_1 = d_2 = 1$ cm. At $t = 1.6 \times 10^{-9}$ s (the time the electron is leaving the deflection region),

$$v_{x0} = v_{y0} = 2.81 \times 10^6 \text{ m/s}$$

$$x_0 = y_0 = 0.22 \text{ cm}$$

After that time,

$$x = x_0 + v_{x0}(t - t_0)$$

$$y = y_0 + v_{y0}(t - t_0)$$

$$z = z_0 + v_z(t - t_0)$$

The electron will hit the screen at $t = t_1$ and $x = x_1$, $y = y_1$, where

$$t_1 = t_0 + \frac{0.15}{v_z} = t_0 + 8.02 \times 10^{-9} \text{ s}$$

$$x_1 = y_1 = 2.47 \text{ cm}$$

Ion Thruster

An ordinary rocket engine propels the rocket forward by ejecting fuel backward. The thrust $\boldsymbol{F}$ (force) obtained is given by

$$\boldsymbol{F} = \frac{d}{dt}(m\boldsymbol{v}) = \boldsymbol{v}\frac{dm}{dt}$$

where dm/dt is the rate of ejection of the propellant and $\boldsymbol{v}$ is the velocity of the ejection, which is a constant. For chemical fuel, the ejection velocity may be supersonic, of the order of 10^3 m/s. It is difficult to achieve much higher velocities using chemical fuels.

Ion thrusters use ionized particles as propellant. The particles are accelerated by electrostatic force to a much higher velocity. Thus, for the same thrust, the ion thruster consumes less mass than a chemical engine does.

Figure 10.9a illustrates the basic operating principle of an ion thruster. Propellant enters the ionization chamber, where it is ionized either by heat or by bombardment by electrons.* Positive ions are accelerated and ejected through the screen. Electrons are absorbed by the anode. These electrons must also be ejected with the positive ions at the same velocity. Otherwise, the rocket will be

* D. B. Langmuir, E. Stuhlinger, and J. M. Sellen, Jr., *Electrostatic Propulsion*, New York: Academic Press, 1961, pp. 3–20.

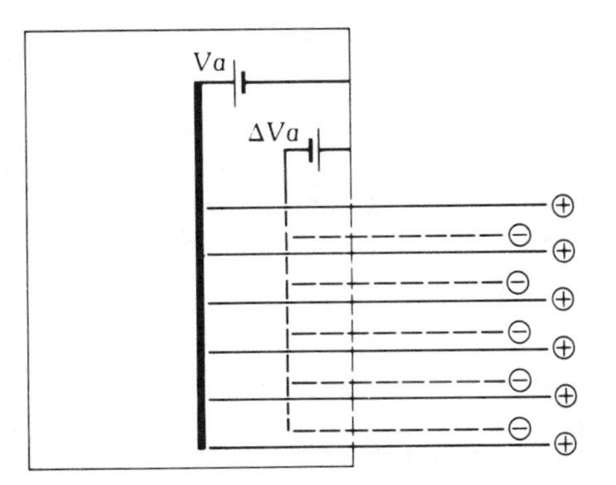

Figure 10.9(a) An ion thruster.

Figure 10.9(b) Some satellite ion thrusters. The largest one shown has a diameter equal to 30 cm capable of producing 129 millinewton of thrust. The smallest one produces about 5 millinewton of thrust. (Courtesy of Hughes Aircraft Company.)

increasingly negatively charged, and these ejected positive ions will consequently be attracted back to the thruster. Ejection of the electrons is accomplished by the cathode and a heated filament serving as the electron source.

The velocity of the positive ion v_i is given by

$$v_i = \left(\frac{2q}{m_i} V_a\right)^{1/2}$$

where q is the charge on the positive ion and m_i is its mass.

The velocity of the electron v_e is given by

$$v_e = \left(\frac{2q_e}{m_e}\Delta V_a\right)^{1/2}$$

where q_e is equal to 1.60×10^{-19} C, and $m_e = 9.11 \times 10^{-31}$ kg. To neutralize the ejected ion, v_e must be equal to v_i. Thus, the voltages V_a and ΔV_a must be related by the following equation:

$$\frac{V_a}{m_i} = \frac{\Delta V_a}{m_e}$$

Note that we have assumed that these ions carry a single positive charge. Because $m_i \gg m_e$, we have $V_a \gg \Delta V_a$.

To estimate the velocity of the ejected mixture of ions and electrons, let us use the following data:

$V_a = 2500$ volts

$m_i = 133 \times 1.67 \times 10^{-27}$ kg (corresponding to the mass of cesium which has an atomic weight of 133)

$q = 1.60 \times 10^{-19}$ C

We obtain

$$v_i = 0.6 \times 10^5 \text{ m/s}$$

A chemically fueled rocket engine cannot achieve this velocity.

To estimate the magnitude of the thrust produced by an ion thruster, let us assume that the ion current is 0.1 A. Assuming further that each cesium ion carries one positive charge, we can calculate the number of ions per volume needed to produce 0.1 A current:

$$I = nv_i qA$$

where A is the cross section of the ion beam. The rate of mass ejection is

$$\frac{dm}{dt} = nv_i m_i A$$

Thus,

$$F = v_i\frac{dm}{dt} = \frac{v_i m_i I}{q}$$

Using the data given above, we obtain

$$F = 8.3 \times 10^{-3}\text{ N}$$

a small number only equal to approximately 1.9×10^{-3} lb of force. Ion thrusters are used on synchronous satellites to adjust their position and to adjust antenna orientations. They are potentially useful for exploring deep space because of their large thrust-to-fuel-mass ratio as compared with chemical engines. Figure 10.9b shows some satellite ion thrusters.

Ink-Jet Printer

Most typewriters and computer line printers use so-called impact printing—that is, characters are cast on metal type, which make an impact on an ink ribbon to produce prints on paper. New requirements for performance and speed exceed the capabilities of most impact-printing technologies. The ink-jet printer is an alternative method. Figure 10.10a shows a schematic diagram of the system.

First, ink drops are sprayed from a nozzle. The nozzle vibrates at an ultrasonic frequency to provide pressure on the ink so that when the ink leaves the nozzle it forms drops of uniform size and spacing. These drops then pass a charging electrode, which charges them with variable amounts of electricity. Next they go through a pair of parallel plates charged to a fixed voltage. The ink drop is deflected vertically, with a displacement proportional to the amount of electricity it carries. For example, letting drops go without a charge produces a blank spot. These drops are not deflected and instead go directly to an ink collector. Otherwise, they are charged, and they land on the paper at predetermined spots. The print head moves horizontally at a constant speed and thus produces a pattern (a character or symbol) on the paper.

Figure 10.10(a) An ink-jet printer.

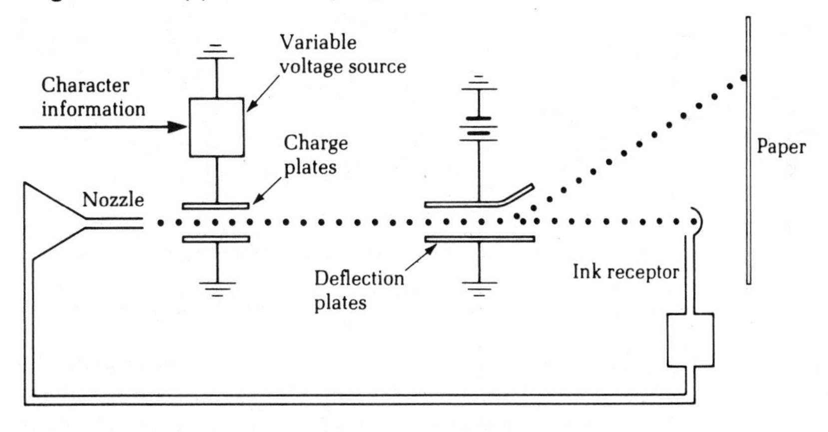

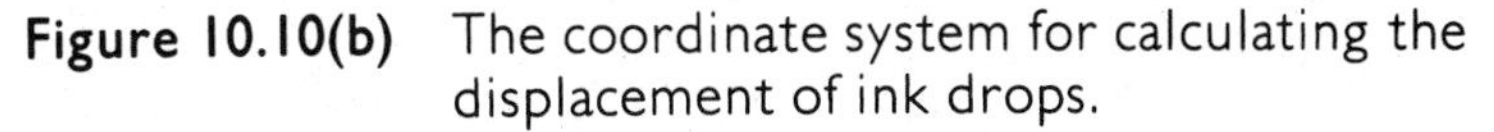

Figure 10.10(b) The coordinate system for calculating the displacement of ink drops.

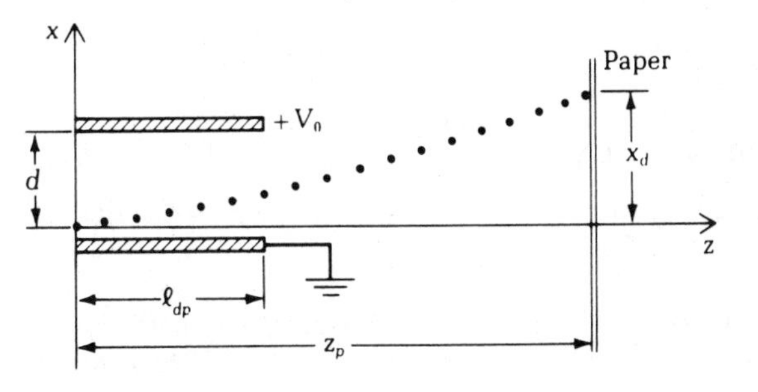

In ink-jet printing, ink drops may be formed at rates of approximately 10^5 drops per second.* For high-quality printing, 10^3 drops are needed to print a character. Thus, characters may be produced at a rate of 100 characters per second! Ink-jet printers have fewer moving parts than impact printers and are therefore more reliable. Above all, they are certainly much quieter.

Example 10.6

This is a numerical example of the ink-jet printer system.

Assume that an ink drop of diameter 0.033 mm is charged with -100×10^{-15} C of electricity. The density of the ink is 2 g/cc. The deflection plate is charged to 3300 V with a spacing of 1.6 mm. The length of the deflection plate is 1.3 cm, and the distance from the exit end of the deflection plate to the paper is 1.0 cm. The velocity of the ink drop is assumed to be 20 m/s. Find the vertical displacement of the drop.

Solution Let the origin of the coordinates be set at the entrance end of the deflection plates, as shown in Figure 10.10b. Inside the deflection plates, the following equations may be obtained:

$$v_x = \frac{dx}{dt} = \frac{qV_0}{md}t$$

$$x = \frac{qV_0}{2md}t^2$$

$$z = v_z t$$

For $z > 1.3$ cm outside the deflection plate region,

$$z = v_z t$$

$$x = x_0 + v_{x0}(t - t_0)$$

where t_0 is the time when the ink drop completes its journey through the deflection plates, and x_0 and v_{x0} are its position and velocity at the exit end of the plates.

To obtain a numerical answer, let us first calculate the mass of the drop.

$$m = \frac{4\pi}{3}\left(\frac{1}{2} \times 0.033 \times 10^{-1}\,\text{cm}\right)^3 \times (2 \times 10^{-3}\,\text{kg/cm}^3)$$

$$= 3.76 \times 10^{-11}\,\text{kg}$$

The velocity in the z direction is not affected by the deflection voltage. Therefore,

* W. L. Buehner, J. D. Hill, T. H. Williams, and J. W. Woods, "Application of ink jet technology to a word processing output printer," *IBM J. Res. Develop.*, January 1977, pp. 2–9.

$$t_0 = \frac{1.3 \times 10^{-2}}{20} = 6.5 \times 10^{-4} \text{ s}$$

$$v_{x0} = \frac{1 \times 10^{-13} \times 3300}{3.76 \times 10^{-11} \times 1.6 \times 10^{-3}} \times 6.5 \times 10^{-4} = 3.6 \text{ m/s}$$

$$x_0 = 1.2 \text{ mm}$$

$$t_1 = \frac{2.3 \times 10^{-2}}{20} = 1.15 \times 10^{-3} \text{ s}$$

$$x_1 = 1.2 \times 10^{-3} + 3.6 \times (11.5 - 6.5) \times 10^{-4} = 3 \text{ mm}$$

Note that because $x_0 = 1.2$ mm, which is greater than the spacing of the deflection plates, the plates must be curved outward at the exit end so that they will not intercept the ink drop.

Stored Electric Energy in a Two-Charge System

Consider a charge q_1 located at the origin, and bring in another charge of q_2 coulombs from infinity to r meters from q_1. Suppose that q_1 and q_2 are of the same sign; then there is a repulsive force against moving q_2 closer to q_1. This force is given by (10.3):

$$\boldsymbol{f}_{21} = \frac{q_1 q_2}{4\pi\epsilon r^2}\hat{\boldsymbol{r}}$$

Thus, the total work required to bring q_2 to r against this force, according to (10.23), is given as follows:

$$W = -\int_{\infty}^{r} \frac{q_1 q_2}{4\pi\epsilon r^2}\, dr = \frac{q_1 q_2}{4\pi\epsilon r}$$

As we mentioned earlier in this section, this work is not wasted. It becomes the stored electric energy in this two-charge system. Denote as U_E the stored electric energy in this system. We then see that

$$\boxed{U_E = \frac{q_1 q_2}{4\pi\epsilon r}} \quad \textbf{(stored electric energy in a two-charge system)} \qquad \textbf{(10.27)}$$

Stored Electric Energy in an *N*-Charge System

Now let us consider a system consisting of N charges: $q_1, q_2, \ldots, q_N$. What is the stored electric energy in this system?

To answer the above question, we shall try to see how much work has been required to put these N charges in their present locations. This work is equal to the stored electric energy. To start, we put q_1 at its position, with all other charges at infinity. This initial system does not store any electric energy. Now, we bring q_2 from infinity to its location, and the work done for this move is given by (10.27):

$$W_{12} = \frac{q_1 q_2}{4\pi\epsilon r_{12}} \tag{10.28}$$

Next, we bring q_3 from infinity. To do so, we must overcome the Lorentz force due to both q_1 and q_2. In other words, q_1 and q_2 have set up a potential field Φ_{12}, which, according to (9.12), is given by:

$$\Phi_{12} = \frac{q_1}{4\pi\epsilon r_1} + \frac{q_2}{4\pi\epsilon r_2}$$

where r_1 and r_2 are the distances to q_1 and q_2, respectively. To put q_3 where it is requires work equal to $\Phi_{12} q_3$, as we can see from (10.24). Thus, the total work needed to put in a three-charge system is given below:

$$W_{123} = \left(\frac{q_1 q_3}{4\pi\epsilon r_{13}} + \frac{q_2 q_3}{4\pi\epsilon r_{23}}\right) + \left(\frac{q_1 q_2}{4\pi\epsilon r_{12}}\right) \tag{10.29}$$

Rewriting the above equation in summation form yields the following:

$$W_{123} = \frac{1}{2} \sum_{m=1}^{3} \sum_{\substack{n=1 \\ m \neq n}}^{3} \frac{q_m q_n}{4\pi\epsilon r_{mn}} \tag{10.30}$$

In (10.30), the second summation sums all m from 1 to 3 except $n = m$. The factor 1/2 is due to the fact that the summation yields six terms because every term in (10.29) is counted twice.

Continuing this process for N charges yields the following final result:

$$\boxed{W = U_E = \frac{1}{2} \sum_{m=1}^{N} \sum_{\substack{n=1 \\ m \neq n}}^{N} \frac{q_m q_n}{4\pi\epsilon r_{mn}}} \quad \textbf{(stored electric energy in an } N\textbf{-charge system)} \tag{10.31}$$

Stored Electric Energy in a System with Continuous Charge Distribution

When the charge distribution is given in terms of charge density ρ_v (coulombs per cubic meter) instead of in terms of discrete charges, the stored electric energy may be found by dividing into infinitesimal volumes the volume containing ρ_v and using (10.31).

We recognize that the first summation over m in (10.31) is simply the potential function of the $N - 1$ charges excluding the nth charge. In the present case, we can replace the first summation with Φ, the scalar potential due to ρ_v:

$$U_E = \frac{1}{2}\sum_{n=1}^{\infty} \Phi q_n \tag{10.32}$$

Since q_n is infinitesimally small, whether Φ includes it is immaterial in this limiting case.* The summation in (10.32) can then be replaced by an integral:

$$U_E = \frac{1}{2}\int_V \Phi \rho_V \, dv \tag{10.33}$$

where the integration is over the volume V containing the charge. The above equation may be put in another form. Note that

$$\nabla \cdot \boldsymbol{D} = \rho_v$$

Thus,

$$U_E = \frac{1}{2}\int_{\text{all space}} \Phi \nabla \cdot \boldsymbol{D} \, dv \tag{10.34}$$

For mathematical convenience, the integration is extended to all space. This extension is permissible because $\nabla \cdot \boldsymbol{D}$ is zero anyway outside V. Now let us use the following vector identity:

$$\nabla \cdot (\Phi \boldsymbol{D}) = \nabla \Phi \cdot \boldsymbol{D} + \Phi \nabla \cdot \boldsymbol{D} \tag{10.35}$$

Substituting into (10.35) yields

$$U_E = \frac{1}{2}\int_{\text{all space}} \nabla \cdot (\Phi \boldsymbol{D}) \, dv + \frac{1}{2}\int_{\text{all space}} \boldsymbol{E} \cdot \boldsymbol{D} \, dv \tag{10.36}$$

In obtaining (10.36) we used the fact that $\boldsymbol{E} = -\nabla \Phi$. The first integral may be transformed into a surface integral using the divergence theorem (9.27):

$$\frac{1}{2}\int_{\text{all space}} \nabla \cdot (\Phi \boldsymbol{D}) \, dv = \frac{1}{2}\int_{\text{surface at } \infty} \Phi \boldsymbol{D} \cdot d\boldsymbol{S} \tag{10.37}$$

Note that a finite charge distribution will be like a point charge to an observer at infinity. Therefore,

$$\Phi \to \frac{Q}{4\pi\epsilon}\frac{1}{r} \qquad (\text{as } r \to \infty)$$

$$D_r \to \frac{Q}{4\pi r^2} \qquad (\text{as } r \to \infty)$$

where Q is the total charge in the volume.

Thus, as r approaches infinity, the product $\Phi \boldsymbol{D}$ in the integrand in (10.37) decreases like $1/r^3$. The surface at infinity increases like r^2, which is not rapid enough to offset the rapid decrease of the integrand. Consequently, the integral in

* It is immaterial as long as ρ_v is a smooth function of space, not like the delta function.

(10.37) is equal to zero. Recognizing this fact, we can now express the stored electric energy in terms of the second integral in (10.36) only—that is,

$$U_E = \frac{1}{2}\int_{\text{all space}} \mathbf{E}\cdot\mathbf{D}\,dv = \frac{\epsilon}{2}\int_{\text{all space}} (\mathbf{E}\cdot\mathbf{E})\,dv \qquad \text{(stored electric energy in space)} \qquad (10.38)$$

where we use the constitutive relation $\mathbf{D} = \epsilon\mathbf{E}$.

Example 10.7 *A system consisting of electric charges of the same sign contains stored electric energy. This example shows the conversion of kinetic energy into potential energy when two charges of the same sign are moving toward each other. The stored energy is calculated from (10.31).*

An electron is moving in air with a velocity equal to 10^5 m/s and is aiming directly at a stationary electron far away. How close can this moving electron get to the stationary electron before the repulsive force turns it back?

Solution The closer the two electrons get, the higher the stored energy of the system. The repulsive force slows the moving electron. As its velocity decreases, its kinetic energy is converted into stored electric energy. At the distance of the closest encounter, the kinetic energy is entirely converted to stored energy.

Let the distance of the closest encounter be r_0. According to (10.28), we obtain

$$\tfrac{1}{2}mv_0^2 = \frac{q_1 q_2}{4\pi\epsilon r_0} \qquad (10.39)$$

Substituting $m = 9.11 \times 10^{-31}$ kg, $v_0 = 10^5$ m/s, $q_1 = q_2 = -1.60 \times 10^{-19}$ C, $\epsilon_0 = 1/36\pi \times 10^{-9}$ F/m, we obtain $r_0 = 5.1 \times 10^{-8}$ m.

Example 10.8 *This example shows that the parallel-plate capacitor may be considered as a device for storing electric energy because it contains electric fields. The stored energy is calculated from (10.38).*

A parallel-plate capacitor is charged to 100 V. Assume that the separation between the two plates is 1 cm, the area of each plate is 100 cm^2, and the medium is air. Find the total electric energy stored in this capacitor.

Solution We can use (10.38) to calculate the stored electric energy. The electric field in the parallel plate is given by

$$E_0 = \frac{V_0}{d} = 10^4 \text{ volts/m}$$

$$U_E = \frac{1}{2}\epsilon_0 \int_{\text{parallel plate}} (10^8)\, dv = 4.42 \times 10^{-8}\,\text{J} \qquad \textbf{(10.40)}$$

10.3 Capacitance

As we learned in Section 9.4, the potential of a perfect conductor is a constant. Between two perfect conductors A and B, the two constants may be different and may give rise to a difference in potential, $V = \Phi_A - \Phi_B$. This difference in potential indicates that electric fields exist between the two conductors, as shown in Figure 10.11. All the electric-field lines are perpendicular to the conducting

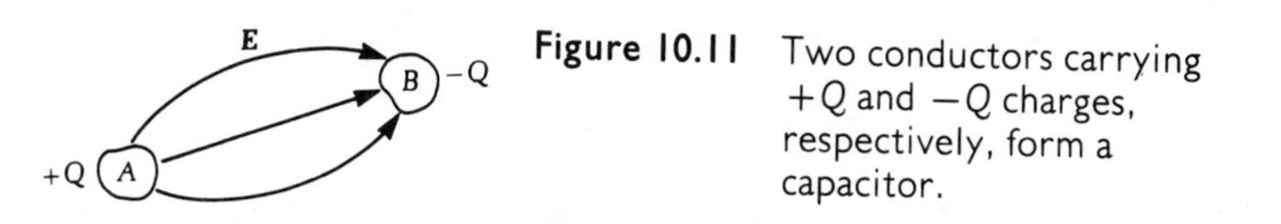

Figure 10.11 Two conductors carrying $+Q$ and $-Q$ charges, respectively, form a capacitor.

surface because the boundary condition (4.4) requires that the tangential electric field be zero there. Assume that the two conductors carry equal and opposite total charges. Let the charge on each conductor be Q. The capacitance is defined by

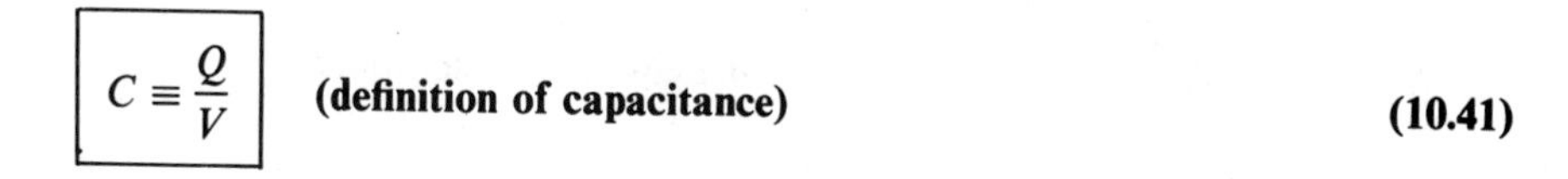

$$\boxed{C \equiv \frac{Q}{V}} \qquad \textbf{(definition of capacitance)} \qquad \textbf{(10.41)}$$

Because the voltage V is proportional to Q (Maxwell's equations are linear equations with respect to the source), the capacitance defined in (10.41) is a constant determined by the geometric configuration of the two conductors.

Parallel-Plate Capacitor

A parallel-plate capacitor consists of two flat conductors of area A separated by a distance a, as shown in Figure 10.12. Neglecting fringing fields, the electric field between the plates is a constant vector in the $\hat{x}$ direction

$$\mathbf{E} = \hat{x}E_0$$

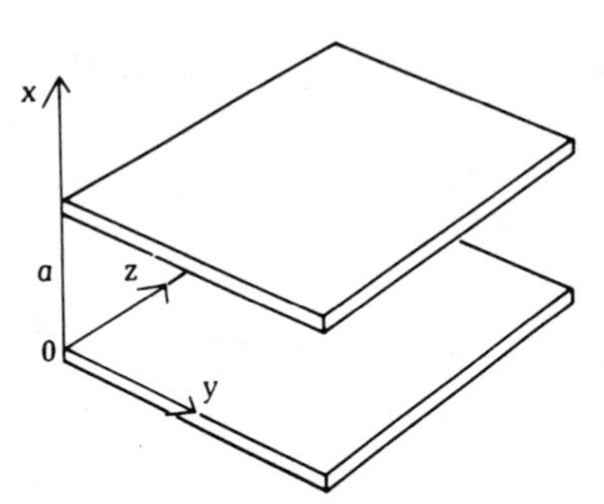

Figure 10.12 A parallel-plate capacitor.

This equation is in fact (5.13) with $k = 0$ and $\mathbf{H} = 0$. The difference in potential between the plates is

$$V = \int_0^a dx E_0 = E_0 a$$

The total charge on the plate at $x = 0$, according to the boundary condition (4.5), is

$$Q = \epsilon E_0 A$$

Thus, the capacitance of the parallel-plate capacitor is

$$C = \frac{Q}{V} = \frac{\epsilon A}{a} \tag{10.42}$$

Note that the capacitance is dependent on the geometry of the capacitor and on the dielectric between the two conductors. It is independent of the field strength or of the amount of charge on the conductors.

Example 10.9 *This is a numerical example to compute the capacitance of a parallel-plate capacitor.*

Find the capacitance of a parallel-plate capacitor with $A = 1\ \text{m}^2$, $a = 1$ mm, and $\epsilon = 9\epsilon_0$.

Solution $C = 7.96 \times 10^{-8}$ F

Note that the capacitance is a small number in terms of farads.

Spherical Capacitor

A spherical capacitor is made of two concentric conductors of radii a and b, respectively, as shown in Figure 10.13. Let the total charge on the inner sphere

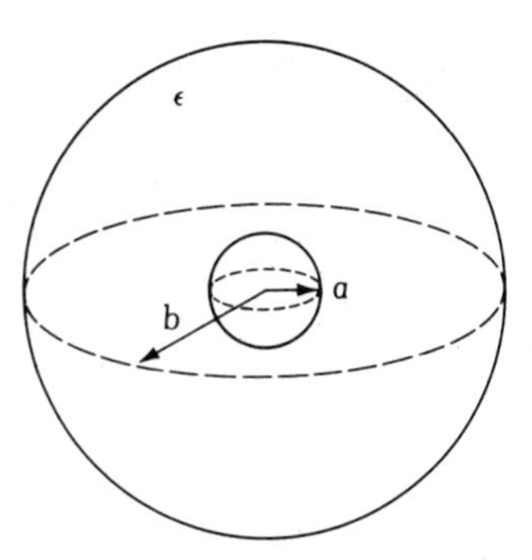

Figure 10.13 A concentric-sphere capacitor.

with radius a be Q. The electric-field lines are radial

$$\boldsymbol{E} = \hat{\boldsymbol{r}}\frac{Q}{4\pi\epsilon r^2}$$

The difference in potential is

$$V = \int_a^b dr \frac{Q}{4\pi\epsilon r^2} = \frac{Q}{4\pi\epsilon}\left(\frac{1}{a} - \frac{1}{b}\right)$$

The capacitance is easily determined from the above two equations

$$C = \frac{Q}{V} = \frac{4\pi\epsilon}{\dfrac{1}{a} - \dfrac{1}{b}} \tag{10.43}$$

We see that as the outer sphere recedes to infinity,

$$C = 4\pi\epsilon a \tag{10.44}$$

The capacitance of the dome of a Van de Graaff generator (see Section 9.4) with $a = 0.5$ m may be approximately calculated using (10.44), which yields $C = 55.6 \times 10^{-12}$ F. Note that in this case the outer conductor is actually the structure of the room or the earth's surface.

Example 10.10 *This example shows that the earth can be considered as a capacitor and that its capacitance is a small number in terms of farads despite its physical size.*

What is the capacitance of the spherical "conductor" called the earth?

Solution We use (10.44). In this case, ϵ is the medium surrounding the earth which is, of course, equal to ϵ_0. The "outer conductor" is at infinity. For the earth, $a = 6.5 \times 10^6$ m; thus,

$$C = 7.2 \times 10^{-4} \text{ F}$$

Coaxial-Cylinder Capacitor

A coaxial-cylinder capacitor consists of two concentric conducting cylinders with inner radius a and outer radius b, as shown in Figure 10.14. We assume that the

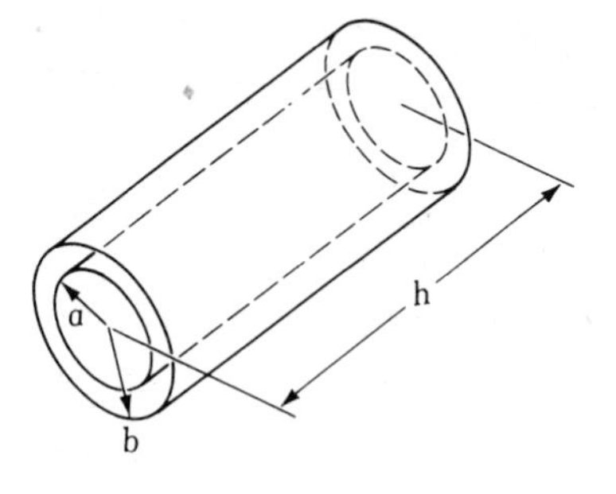

Figure 10.14 A coaxial-cylinder capacitor.

length is h and that the dielectric has permittivity ϵ. The electric field is given by (9.22):

$$\boldsymbol{E} = \hat{\boldsymbol{\rho}} \frac{\rho_\ell}{2\pi\epsilon\rho} \quad \text{(cylindrical coordinates)}$$

The difference in potential is

$$V = \int_a^b d\rho \frac{\rho_\ell}{2\pi\epsilon\rho} = \frac{\rho_\ell}{2\pi\epsilon} \ln\left(\frac{b}{a}\right)$$

The total charge on the inner conductor is

$$Q = \int_0^h dz \int_0^{2\pi} a\, d\phi \frac{\rho_\ell}{2\pi a} = \rho_\ell h$$

Thus, the capacitance is

$$C = \frac{Q}{V} = \frac{2\pi\epsilon h}{\ln\left(\frac{b}{a}\right)} \quad \textbf{(10.45)}$$

The coaxial line discussed in Chapter 5 has a capacitance of $2\pi\epsilon/\ln(b/a)$ farads per meter.

Energy Stored in a Capacitor

A capacitor may be thought of as a device to store electric charge. This view is sometimes misleading because the total charge in a capacitor is actually zero. A better definition of the capacitor is that it is a device to store electric energy. From (10.33) and Figure 10.11, we see that the stored electric energy in a capacitor is

$$U_E = \frac{1}{2} \int \Phi \rho_v \, dv$$

Referring to Figure 10.11 and remembering that Φ is equal to Φ_A on A and equal to Φ_B on B, we see that

$$U_E = \frac{1}{2}\left(\Phi_A \int_A \rho_v \, dv + \Phi_B \int_B \rho_v \, dv\right) = \frac{1}{2}(\Phi_A Q_A + \Phi_B Q_B)$$

For a capacitor, $Q_A = Q$, $Q_B = -Q$; therefore,

$$U_E = \tfrac{1}{2}(\Phi_A - \Phi_B)Q = \tfrac{1}{2} Q V_{AB}$$

Note that

$$C = \frac{Q}{V_{AB}}$$

We have

$$\boxed{C = \frac{2U_E}{(V_{AB})^2}} \quad \textbf{(equivalent definition of capacitance)} \qquad \textbf{(10.46)}$$

Sometimes, (10.46) is used to define the capacitance C.

Example 10.11 *This example shows that there are two ways to compute the stored electric energy in a capacitor. One way is to integrate the electric field—that is, to use (10.38). The other is to compute it in terms of the capacitance and electric charge using (10.46).*

Assume that the coaxial-cylinder capacitor shown in Figure 10.14 is charged to 100 V, with $a = 0.5$ cm, $b = 0.6$ cm, $h = 2$ cm, and $\epsilon = \epsilon_0$. Find the stored electric energy in the capacitor.

Solution We will use two methods to solve this problem.

Method I: Let us first calculate the capacitance C using (10.45):

$$C = 6.09 \times 10^{-12} \text{ F}$$

Then, we use (10.46) to calculate U_E:

$$U_E = 3.05 \times 10^{-8} \text{ J}$$

Method II: We shall solve this problem the hard way. Instead of using (10.46), we use (10.38). Our concept is that a capacitor is a device to store electric energy and that the latter is in the form of an electric field distributed over a volume.

The electric field in the coaxial-cylinder region is given by

$$\boldsymbol{E} = \frac{V_0}{\rho}\hat{\boldsymbol{\rho}}$$

The difference in potential between the inner and outer conductor is

$$100 = \int_a^b \boldsymbol{E}\, d\rho \cdot \hat{\boldsymbol{\rho}} = V_0 \ln\left(\frac{b}{a}\right)$$

Thus,

$$V_0 = \frac{100}{\ln 1.2} = 548$$

Substituting $\boldsymbol{E}$ into (10.38), we obtain

$$U_E = \frac{1}{2}\epsilon \int \frac{V_0^2}{\rho^2}\, dv$$

In cylindrical coordinates, $dv = (d\rho)(\rho\, d\phi)(dz)$; therefore,

$$U_E = \frac{1}{2}\epsilon \int_0^h dz \int_0^{2\pi} d\phi \int_a^b \rho\, d\rho \frac{(V_0)^2}{\rho^2}$$

$$= \epsilon h \pi V_0^2 \ln\left(\frac{b}{a}\right)$$

$$= 3.05 \times 10^{-8}\ \text{J}$$

Dielectric Breakdown

The specifications of a capacitor include not only the capacitance value but also the maximum voltage to which the capacitor can be charged. What happens if the capacitor is overcharged? The answer is that the dielectric that separates the two conductors will break down and the capacitor will be short-circuited or burned out.

The breakdown of a dielectric is similar to the ionization of air discussed in Section 10.1. In the dielectric insulator of a capacitor, although the number of free electrons is very small, impurities and other defects do contain some electrons. These electrons are accelerated by the electric field and collide with the lattice structure of the dielectric material. If the electric field is high enough to produce a hole-electron pair that is also accelerated by the electric field, more collisions result, and complete breakdown finally occurs.

Table 10.1 lists some common insulating materials with their breakdown electric-field strength.

Table 10.1 Breakdown Electric-Field Strength of Insulators.

Material	Relative permittivity	Breakdown **E** field (10^6 V/m)
Air	1.0	Approximately 3 [see (10.22)]
Oil	2.3	15
Paper	1.5–4.0	15
Polystyrene	2.7	20
Glass	6.0	30
Mica	6.0	200

Example 10.12 *This is a numerical example showing the relationship between the breakdown electric field and the maximum voltage that a Van de Graaff generator can maintain.*

For a Van de Graaff generator, with radius $a = 0.5$ m, what is the maximum voltage V and the corresponding charge storage before the breakdown occurs?

Solution The total charge is calculated from $E = Q/4\pi\epsilon a^2$ with $E = 3 \times 10^6$ V/m. We thus have

$$Q = 3\pi\epsilon \times 10^6 = 8.3 \times 10^{-5} \text{ C}$$

The corresponding maximum voltage is

$$V = \frac{Q}{4\pi\epsilon a} = Ea = 1.5 \times 10^6 \text{ volts}$$

Coaxial Cable with Multiple Layers of Insulation

The electric field in the coaxial cable shown in Figure 10.15a is given by

$$\mathbf{E} = \frac{V_0}{\rho}\hat{\boldsymbol{\rho}}$$

where V_0 is a constant. To ensure that the cable will not break down, we must keep the maximum electric field inside the cable below the breakdown field-strength of the material. The maximum field in a cable is located at the surface of the inner conductor—that is, at $\rho = a$. Thus, the coaxial cable must be filled with an insulator whose breakdown field-strength is greater than V_0/a.

A material of high breakdown field-strength is usually more expensive. As shown in Figure 10.15b, the electric field decreases as $1/\rho$, so that, away from the inner conductor, it is not necessary to use high-strength material. An alternative way to insulate high-voltage cable is to use two layers of insulation, as shown in Figure 10.15c. To find the electric field in this coaxial region, we use the Gauss' theorem presented in Section 9.3.

Figure 10.15 Electric fields in coaxial lines. **(a)** A coaxial line with one layer of insulation. **(b)** Electric field distribution of (a). **(c)** A coaxial line with two layers of insulation, with Gaussian surface drawn at ρ. **(d)** Corresponding electric field distribution of the line shown in (c).

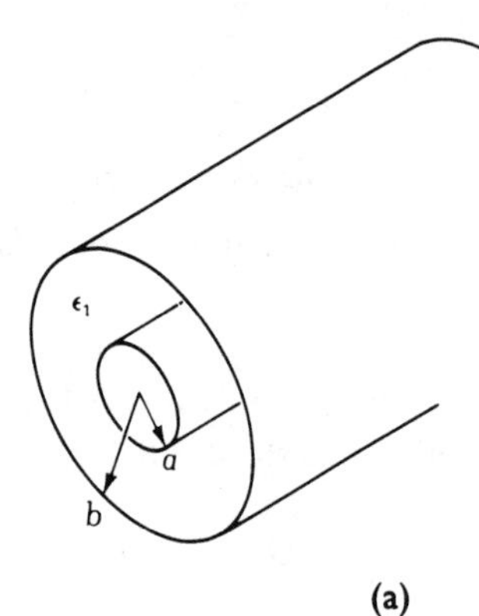

(a)

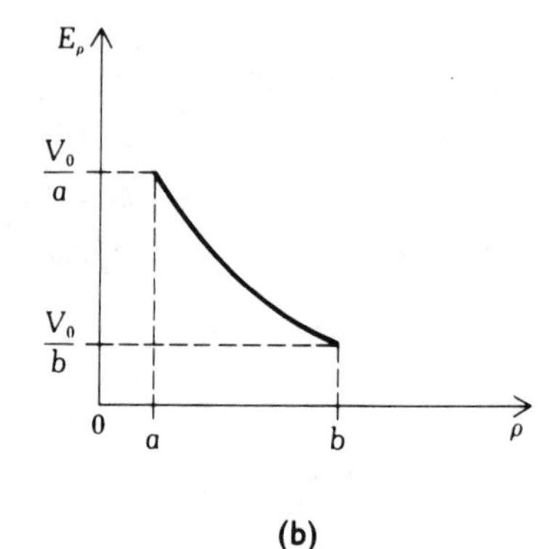

(b)

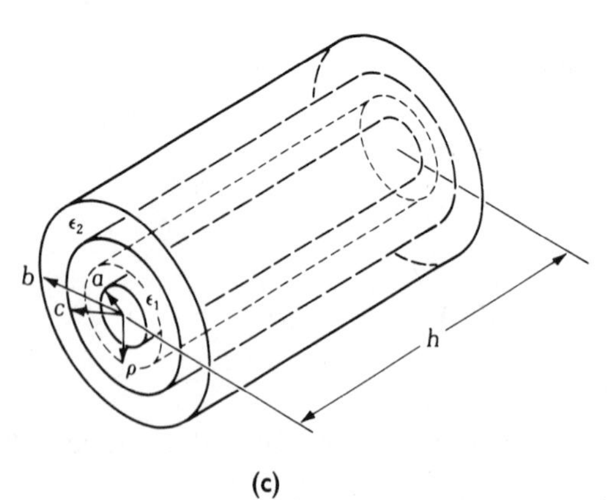

(c)

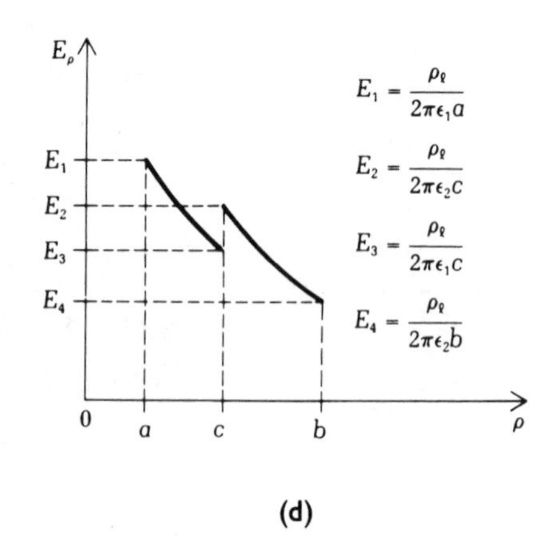

(d)

Let the charge on the inner conductor be ρ_ℓ coulombs per meter. Draw a Gaussian surface of radius ρ and length h as shown in Figure 10.15c. Applying Gauss' law (9.26), we obtain:

$$\int \mathbf{D} \cdot \hat{\mathbf{n}}\, ds = \rho_\ell \cdot h$$

Note that because of the symmetry of this problem, we can assume that $\boldsymbol{D}$ is along the $\hat{\rho}$ direction and is uniform on the cylindrical Gaussian surface. The product $\boldsymbol{D} \cdot \hat{\boldsymbol{n}}$ is zero on the two circular end surfaces. Thus,

$$2\pi\rho h D_\rho = \rho_\ell h$$

$$D_\rho = \frac{\rho_\ell}{2\pi\rho} \tag{10.47}$$

This result is independent of ϵ_1 and ϵ_2. Thus,

$$\boldsymbol{E} = \begin{cases} \dfrac{\rho_\ell}{2\pi\epsilon_1\rho}\hat{\rho} & \text{for} \quad c \geq \rho \geq a \\ \dfrac{\rho_\ell}{2\pi\epsilon_2\rho}\hat{\rho} & \text{for} \quad b \geq \rho \geq c \end{cases} \tag{10.48}$$

The maximum electric field in the first layer is

$$E_{1\,\text{Max}} = \frac{\rho_\ell}{2\pi\epsilon_1 a} \tag{10.49}$$

and that in the second medium is

$$E_{2\,\text{Max}} = \frac{\rho_\ell}{2\pi\epsilon_2 c} \tag{10.50}$$

Figure 10.15d shows the variation of E as a function of ρ for the case $\epsilon_2 < \epsilon_1$.

Example 10.13 *This example shows how to answer a practical question: What is the maximum voltage rating on a coaxial capacitor?*

A coaxial line similar to the one shown in Figure 10.15a has $a = 5$ mm and $b = 15$ mm. The insulation is polystyrene. To what maximum voltage can the coaxial line be charged?

Solution From (10.47), which is valid for a coaxial line with one or with multiple layers, we obtain:

$$E_\rho = \frac{\rho_\ell}{2\pi\epsilon_1\rho} \qquad b \geq \rho \geq a \tag{10.51}$$

E_ρ is maximum at $\rho = a$, which must be kept smaller than 20×10^6 V/m (Table 10.1)— that is,

$$2 \times 10^7 = \frac{\rho_\ell}{2\pi\epsilon_1 a} \tag{10.52}$$

The voltage is obtained by integrating E_ρ from a to b:

$$V = \frac{\rho_\ell}{2\pi\epsilon_1}\ln\left(\frac{b}{a}\right) \tag{10.53}$$

Substituting (10.52) into (10.53), we obtain

$$V = a \times \ln\left(\frac{b}{a}\right) \times 2 \times 10^7 = 1.10 \times 10^5 \text{ volts}$$

Example 10.14 *This example shows how to solve the problem of coaxial capacitors with multiple insulating layers.*

The same coaxial line discussed in the previous example is to be insulated by polystyrene and paper ($\epsilon_r = 3.0$), in a manner similar to the situation shown in Figure 10.15c. With what value of c is the maximum electric field in polystyrene just equal to the breakdown strength in polystyrene and the maximum electric field in paper also equal to the breakdown strength in paper? What is the voltage between the two conductors?

Solution According to (10.49) and (10.50),

$$\frac{\rho_\ell}{2\pi\epsilon_1 a} = 20 \times 10^6 \tag{10.54}$$

$$\frac{\rho_\ell}{2\pi\epsilon_2 c} = 15 \times 10^6 \tag{10.55}$$

Taking the ratio of these two equations, we obtain:

$$\frac{\epsilon_2 c}{\epsilon_1 a} = \frac{20}{15}$$

$$c = 6 \text{ mm}$$

The voltage can be calculated from integration of (10.48):

$$V = \int_a^c \frac{\rho_\ell}{2\pi\epsilon_1\rho}\,d\rho + \int_c^b \frac{\rho_\ell}{2\pi\epsilon_2\rho}\,d\rho$$

$$= \frac{\rho_\ell}{2\pi\epsilon_1}\ln\left(\frac{c}{a}\right) + \frac{\rho_\ell}{2\pi\epsilon_2}\ln\left(\frac{b}{c}\right)$$

From (10.54) and (10.55), we obtain

$$V = 2 \times 10^7 a \ln\left(\frac{c}{a}\right) + 1.5 \times 10^7 c \ln\left(\frac{b}{c}\right)$$

$$= 1.01 \times 10^5 \text{ volts}$$

Force on a Dielectric Material

Consider the parallel-plate capacitor shown in Figure 10.16. A dielectric slab is partially inside the parallel-plate region. The capacitor is isolated from other systems, so that the charge on each plate remains at a constant value, Q. Will the dielectric slab be pulled in, pushed out, or remain stationary?

Figure 10.16 A parallel-plate capacitor carries $+Q$ and $-Q$ on its lower and upper plates, respectively. The dielectric slab is subject to a force which tends to pull it in.

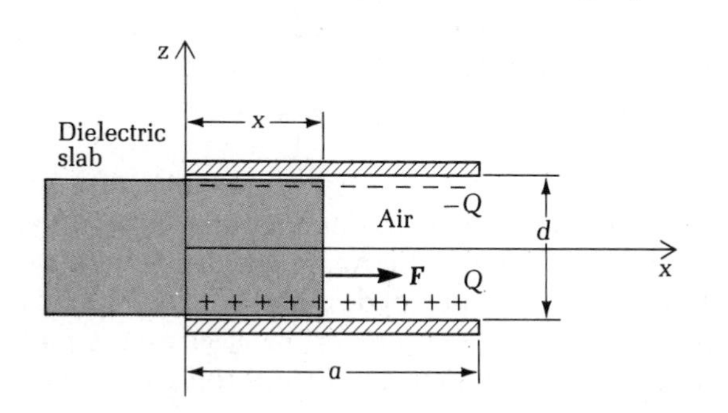

We assume that the electric field is constant inside the parallel-plate region and that the fringe fields near the edges may be neglected. The tangential electric field is continuous across the boundary, so the electric field in air and the electric field in the dielectric slab are equal.

If $x = 0$—that is, if the capacitor is filled with air—the electric field is given by

$$E_0 = \frac{Q}{\epsilon_0 ah}$$

where h is the length of the capacitor and a is its width. The voltage V_0 is given by

$$V_0 = E_0 d = \frac{Qd}{\epsilon_0 ah}$$

The total stored electric energy is

$$U_{E0} = \frac{1}{2}\epsilon_0(E_0)^2(adh) = \frac{Q^2 d}{2\epsilon_0 ah}$$

If the capacitor is entirely filled with the dielectric, the corresponding E_1, V_1, and

U_{E1} are as follows:

$$E_1 = \frac{Q}{\epsilon a h}$$

$$V_1 = \frac{Qd}{\epsilon a h}$$

$$U_{E1} = \frac{Q^2 d}{2\epsilon a h}$$

Notice that the electric-field strength, the voltage, and the stored energy are all smaller than their corresponding values when the capacitor is filled with air. Because a system always tends to go to the lower energy state, we conclude that, if the dielectric slab is partially filling the capacitor as shown in Figure 10.16, the electrostatic force will pull it into the capacitor.

The electrostatic force $\boldsymbol{F}$ can be calculated from

$$\boldsymbol{F} = -\nabla U_E$$

The physical interpretation is similar to that of $\boldsymbol{E} = -\nabla\Phi$, where the electric field is pointing in the direction of decreasing potential. Likewise, the electrostatic force is pointing in the direction of decreasing electric stored energy U_E.

We can calculate the magnitude of the electrostatic force with the following approach. For the situation shown in Figure 10.16, the stored energy is given by

$$U_E(x) = \tfrac{1}{2} E^2[\epsilon_0(a - x) + \epsilon x]hd \tag{10.56}$$

Notice that E in the $\hat{z}$ direction is the same in both regions because of the boundary conditions of continuity of tangential electric fields.

But what is E? The value E may be found as follows:

$$D = \begin{cases} \epsilon E & \text{(in dielectric)} \\ \epsilon_0 E & \text{(in air)} \end{cases}$$

Thus,

$$\rho_s = \begin{cases} \epsilon E & \text{(in dielectric)} \\ \epsilon_0 E & \text{(in air)} \end{cases}$$

Therefore,

$$(\epsilon E)xh + (\epsilon_0 E)(a - x)h = Q$$

or

$$E = \frac{Q}{[\epsilon x + \epsilon_0(a - x)]h}$$

Substituting the above equation into (10.56), we obtain

$$U_E(x) = \frac{Q^2 d}{2[\epsilon x + \epsilon_0(a - x)]h}$$

The force on the slab is then given by

$$F = \frac{-\partial U_E(x)}{\partial x} = \frac{Q^2 d}{2[\epsilon x + \epsilon_0(a - x)]^2 h}(\epsilon - \epsilon_0) \tag{10.57}$$

The electrostatic system is trying to lower its stored energy by exercising a force F to pull the dielectric into the spacing between the two parallel plates. Notice that in this problem we are considering the force in the $\hat{x}$ direction pulling the dielectric slab into the capacitor region in order to rest in a smaller energy state. The electric force between the two plates carrying the opposite charge Q in the $\hat{y}$ direction is not the issue here because we shall assume that the two plates are somehow held apart by external forces and are not free to move toward each other. It should be noted here that (10.56) is valid only for $x < a$. Consequently, the force given by (10.57) is valid only under the same condition.

Now consider a different case shown in Figure 10.17. Notice that in Figure 10.16 the charge Q is a constant because the capacitor is itself an isolated system. However, in Figure 10.17, the capacitor is connected to a constant voltage source, and the voltage V is a constant. The capacitor and the battery now form a complete system. Again we ask the question: will the dielectric slab be pulled in, pushed out, or remain stationary?

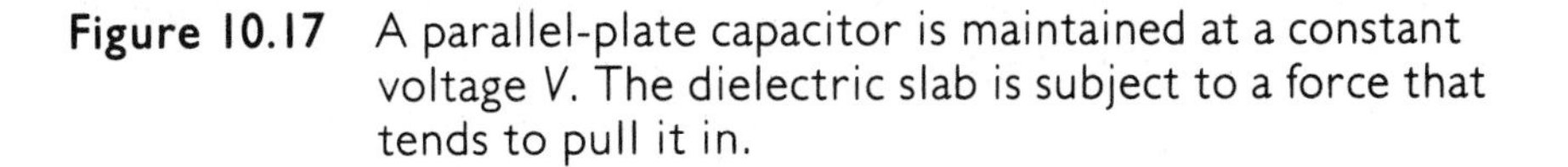
Figure 10.17 A parallel-plate capacitor is maintained at a constant voltage V. The dielectric slab is subject to a force that tends to pull it in.

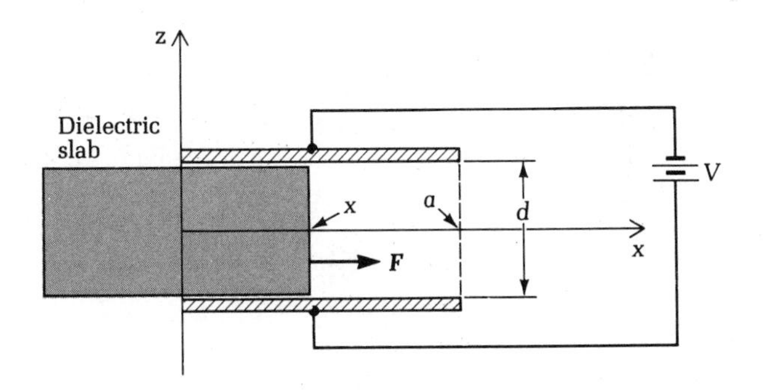

Let us first set $x = 0$, so that the dielectric slab is completely outside the parallel-plate region. Because V is a constant, we have

$$E_0 = \frac{V}{d}$$

$$Q_0 = \epsilon_0 E_0 ah = \epsilon_0 Va\frac{h}{d}$$

$$U_{E0} = \frac{1}{2}\epsilon_0 E_0^2 ahd = \frac{V^2 \epsilon_0 ah}{2d}$$

Then, we calculate E_1, Q_1, and U_{E1}, when the dielectric slab fills the capacitor:

$$E_1 = \frac{V}{d} = E_0$$

$$Q_1 = \epsilon E_1 ah = \epsilon Va\frac{h}{d} > Q_0$$

$$U_{E1} = \frac{1}{2}\epsilon E_1^2 ahd = \frac{V^2\epsilon ah}{2d} > U_{E0}$$

Before we conclude that, because $U_{E1} > U_{E0}$, the tendency is for the electrostatic force to push the dielectric out in order to lower the stored energy, we must consider what happened to the other component of the entire system—namely, the battery.

If the dielectric slab completely occupies the parallel-plate region, the charge increases from Q_0 to Q_1. These charges are delivered by the constant-voltage battery. The battery thus spends its stored energy in an amount equal to $V(Q_1 - Q_0)$ in order to deliver the extra amount of charge to the plates. Thus, the total stored energy of the entire system is represented by $U_{E1} + U_{EB}$, where,

$$\begin{aligned} U_{EB} &= -V(Q_1 - Q_0) \\ &= -V^2(\epsilon - \epsilon_0)a\frac{h}{d} \end{aligned}$$

Therefore,

$$\begin{aligned} U_{E1} + U_{EB} &= \frac{V^2\epsilon ah}{2d} - \frac{V^2(\epsilon - \epsilon_0)ah}{d} \\ &= U_{E0} - \frac{V^2a(\epsilon - \epsilon_0)h}{2d} \end{aligned}$$

In other words, the total stored energy after the dielectric slab has occupied the capacitor is less than that when it is filled with air. Thus, the slab shown in Figure 10.17 is pulled into the parallel-plate region in the same way as the slab shown in Figure 10.16.

We can show that, when the dielectric slab is in the situation shown in Figure 10.17, the force on it is given as follows:

$$F = -\frac{\partial U_E}{\partial x} = \frac{V^2(\epsilon - \epsilon_0)h}{2d} \tag{10.58}$$

From the foregoing discussion, we may conclude that electrostatic forces always try to pull a dielectric material into a region of high electric field.

Electrostatic Adhesive Surface

Figure 10.18 shows fine conductors embedded in a dielectric surface. Every other conductor is charged to positive voltages. This array of conductors produces an electric field in the air just above the surface of the dielectric. Now, if a paper is

Figure 10.18 Electrostatic adhesive surface. The paper is being pulled toward the electrostatic fields that are generated by conducting wires embedded in an insulator.

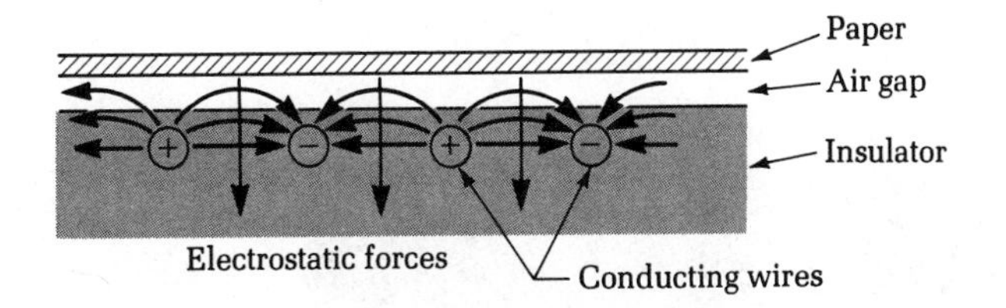

placed on top of the dielectric, the electrostatic force pulls it closer to the surface, and in this way the paper adheres to the surface. When the conductors are discharged, the adhesive force disappears, of course.

This electrostatic adhesive surface is widely used in desktop calculator-driven curve tracers. A typical voltage used to charge the embedded conductors is 300 volts, and typical spacings between them are approximately 2 mm.*

Electrophoresis

Electrophoresis is a method for separating charged molecules, mainly proteins and nucleic acids, by allowing them to migrate in an electric field. It is usually carried out in gels formed in tubes, slabs, or on a flat bed. The method is widely used in biochemistry to separate macromolecules for analytical or preparative purposes.

Figure 10.19 (p. 374) shows the result of electrophoresis for the separation of different topological forms of a closed circular DNA (deoxyribonucleic acid). DNA samples were prepared by incubating the native form of DNA (the supercoiled form shown) with a DNA-relaxation enzyme and an increasing amount (from left to right) of a drug capable of breaking the DNA. The reaction resulted in a mixture of the unreacted supercoiled DNA, the relaxed DNA products, the linear DNA, and intermediates with different degrees of supercoiling. The DNA mixture was then put into a loading well. A constant voltage (about 80 volts) was applied across a flat-bed agarose gel immersed in an electrolyte. The current density across the width of the gel was approximately 1 mA/cm. After 16 hours, the gel was stained with a fluorescent DNA dye and photographed when exposed to ultraviolet light. The DNA molecules, carrying negative charges at neutral pH, migrated from cathode to anode. The results showed that the supercoiled DNA is most mobile in gel because of its compactness. Consequently, it is most easily pushed by the electric force, whereas the relaxed DNA is the least mobile. Although these kinds of DNA are chemically identical, the electrophoresis process can be employed to separate them.

---

* P. Lorrain and D. R. Corson, *Electromagnetism*, San Francisco: W. H. Freeman and Co., 1978, p. 189.

Figure 10.19 DNA molecules of different geometrical shapes separated by electrophoresis. (Courtesy of Dr. Linus Shen, Abbott Laboratories, Abbott Park, IL.)

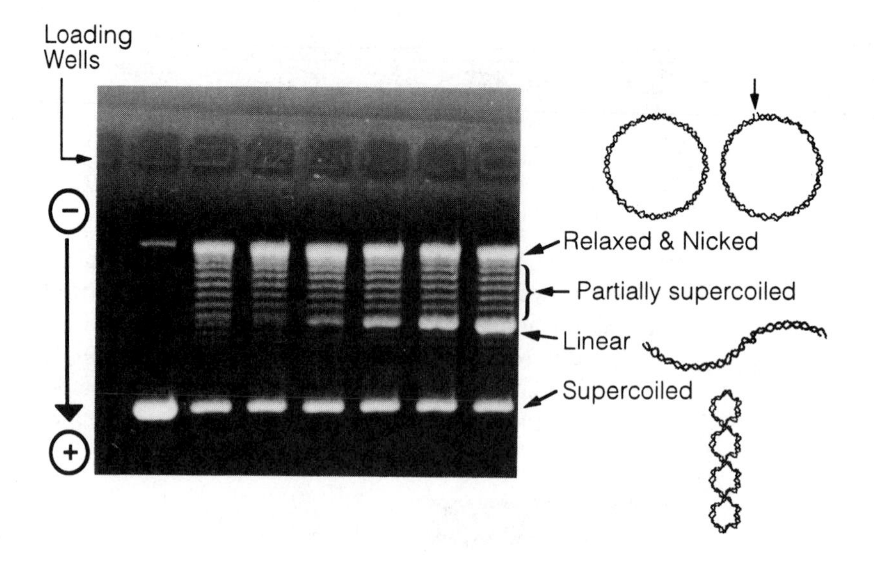

SUMMARY

1. A charged particle is subject to a force in an $\boldsymbol{E}$ field [Equation (10.2)]. The particle is accelerated in accordance to Newton's law.
2. The work done to carry a charge q from point A to point B in an electric field is the integration of electric force times the distance as given by (10.23). It is also equal to the potential difference between B and at A times the charge, as given by (10.24).
3. A moving charged particle has traveled from point A to point B, then its velocity at point B is determined by the equation of conservation of energy: the sum of the kinetic energy and the potential energy is a constant, as given by (10.25).
4. There is stored electric energy in an assembly of N charges. It is given by (10.31). We can also say that the electric energy is stored wherever an $\boldsymbol{E}$ field exists. It is given by (10.38).
5. A capacitor is a device that stores electric energy. As such, it is a device where a concentrated $\boldsymbol{E}$ field can exist.
6. Capacitance is a measure of the capacity of a capacitor. It is defined as the ratio of the total electric charge on the positive plate to the potential difference between the positive plate and the negative or ground plate [Equation (10.41)].
7. Capacitance can also be defined in terms of a device's capacity to store electric energy: the ratio of the total stored electric energy and half the square of the potential difference between the two plates [Equation (10.46)].

Problems

10.1 A point charge of q coulombs is located at the origin (0, 0, 0), and a second point charge of q' coulombs is at (1, 0, 0). A small test charge is placed at (3, 0, 0), and it is found that the total force on the test charge is equal to zero. Find q' in terms of q.

10.2 Two identical small balls are attached to weightless strings 15 cm long. Each ball carries 10^{-9} C of charge, and each has a mass of 1 g. They achieve an equilibrium state under the influence of electrostatic force and gravitational force, as shown in Figure P10.2. Find the angle α. Hint: α is small.

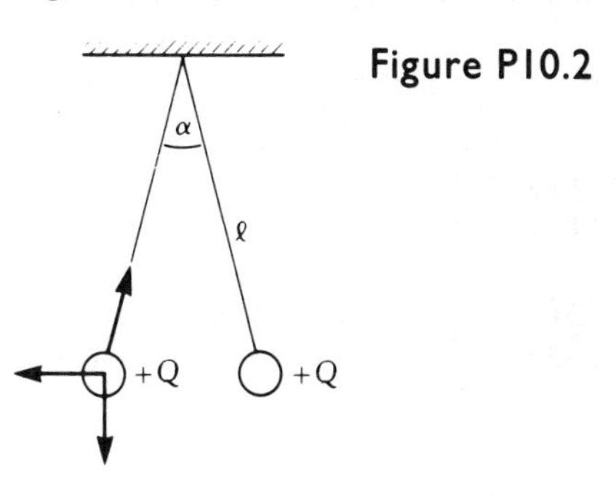

Figure P10.2

10.3 Consider a long line charge with $\rho_\ell = 10^{-6}$ C/m. Find the force acting on a dust particle carrying -10^{-9} C, 1 m away from the line charge.

10.4 A line charge with $\rho_\ell = 10^{-6}$ C/m is located in air at $x = 1$, $y = 0$. A plane charge with $\rho_s = 10^{-6}$ C/m^2 is located at $x = 0$. A positive point charge of 10^{-9} C is at $(\frac{1}{2}, 0, 0)$ in rectangular coordinates. What is the total force acting on this point charge?

10.5 What would happen to the electron if the plate in Example 10.2 were 6 cm wide in the $\hat{x}$ direction?

10.6 In a seed sorting machine, undesirable seeds are deposited with an electrostatic charge while they pass an automatic color-sensitive or size-sensitive monitor. The good seeds are

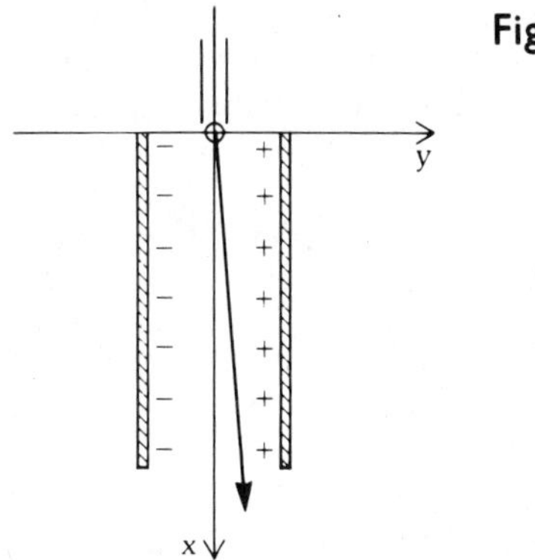

Figure P10.6

passed uncharged. All seeds are dropped between a high-voltage parallel-plate region to sort out the undesirable seeds. Let the charge on the undesirable seed be q, its mass be m, the voltage between the parallel plates be V, and the plate separation be d. Assume that the seeds enter the parallel-plate region at velocity v_0, and find the displacement y of the bad seed as a function of x. Figure P10.6 illustrates this situation. Consider only the trajectory inside the parallel plate.

10.7 At room temperature (20° C) and standard atmosphere, what should be the size of the corona wire if $b = 3$ cm, $V_0 = 10$ kV, and the roughness factor of the wire is equal to 0.8? (Refer to Figure 10.4.) There are two solutions. The first satisfies the condition that $a \ll b$; the second satisfies $(b - a) \ll b$.

10.8 How much potential difference is needed to accelerate an electron from zero speed to 1/10th the speed of light?

(a) Obtain your answer by neglecting the relativistic effect.
(b) Use the relativistic mass formula given in Example 10.4 to obtain your answer.

10.9 Refer to Figure 10.1. If the voltage applied to the parallel plate is the sawtooth signal shown in Figure P10.9, find the locus of the electron on the fluorescent screen located at $x = 20$ cm.

Figure P10.9

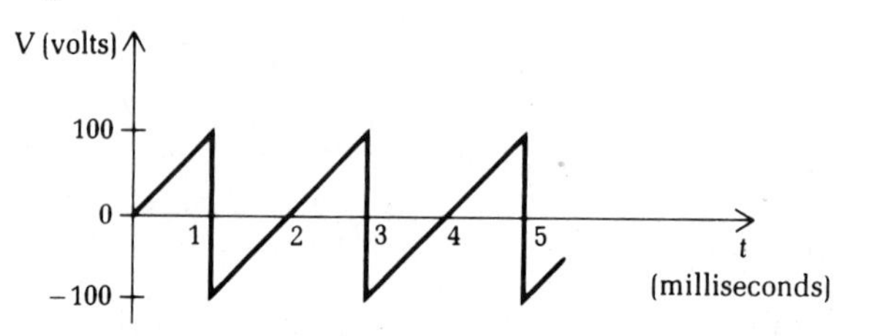

10.10 In an ion thruster, a voltage V_1 is used to eject cesium ions and a voltage V_2 is used to eject electrons. It is necessary to eject both at the same speed. Find the ratio V_1/V_2. Assume that both particles are initially at rest.

10.11 An electron is accelerated by a difference in potential of 1 kV between the anode and the cathode. It enters the parallel-plate region with this kinetic energy. Its velocity makes a 5° angle with the plane of the parallel plate at the entrance end, as shown in Figure P10.11.

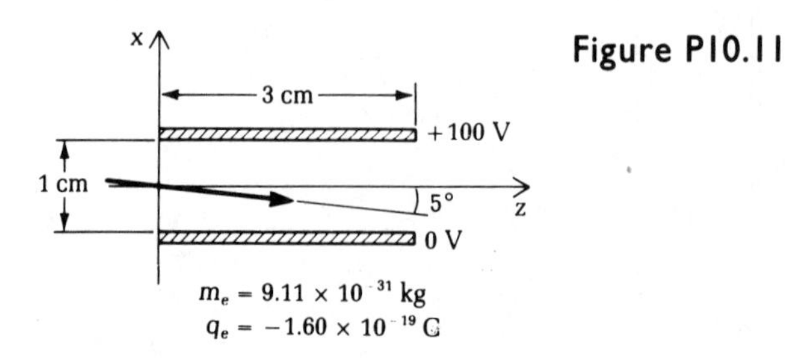

Figure P10.11

(a) Find v_0, v_{0z}, and v_{0x} at $t = 0$.
(b) Obtain two equations for the coordinates of the electron (x, z) as functions of t. Note that $x = 0$ and $z = 0$ at $t = 0$.
(c) Find the position of the electron at the exit end of the parallel plate.

10.12 An ink particle of 2×10^{-11} kg is charged with $+10^{-13}$ C of electricity. The initial velocity of the particle is 25 m/s in the $\hat{x}$ direction when it enters the parallel region. The voltage between the plates is 2000 volts; plate separation is 2 mm; and its length in the $\hat{x}$ direction is 1.5 cm. The distance between the end of the plate and the paper is 2 cm (see Figure P10.12). Find the location on the paper where the ink will impact.

Figure P10.12

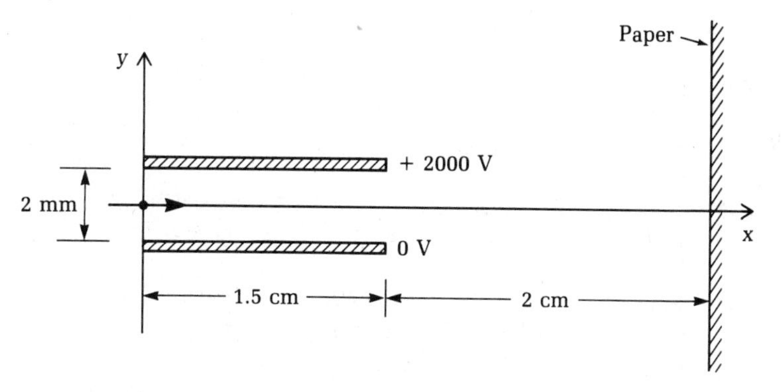

10.13 Find the capacitance of the spherical capacitor shown in Figure 10.13 by using (10.38) and (10.46). Start from

$$\boldsymbol{E} = \frac{Q}{4\pi\epsilon r^2}\hat{r} \qquad \text{for} \quad b > r > a$$

and show that your result agrees with (10.45).

10.14 Find the capacitance of the cylindrical capacitor shown in Figure 10.14 by using (10.38) and (10.46). Start from

$$\boldsymbol{E} = \frac{\rho_\ell}{2\pi\epsilon\rho}\hat{\rho} \qquad \text{for} \quad b > \rho > a$$

and show that your result agrees with (10.49).

10.15 Consider the parallel-plate capacitor shown in Figure 10.12. What is the maximum capacitance one can obtain by using mica as the insulator? Let the area of the plate be 10 cm^2 and the voltage rating of the capacitor be 2 kV, with a safety factor of 10. Use Table 10.1 for the value of ϵ for mica.

10.16 Consider the cylindrical capacitor shown in Figure 10.14. What is the maximum capacitance one can obtain by using oil as the insulator? Take $a = 1$ cm, $h = 2$ cm, and the voltage rating $= 2$ kV, with a safety factor of 5, using Table 10.1 for the value of ϵ for oil.

10.17 A parallel-plate capacitor is filled with two dielectric materials in a configuration shown in Figure P10.17. The total area of the plate is A.

(a) Find the capacitance C in terms of A, d, ϵ_1, and ϵ_2.
(b) Suppose that the positive plate carries Q coulombs of charge, and find Q_1 and Q_2 in terms of Q, where Q_1 and Q_2 are charges on the left- and on the right-hand sides of the plate, respectively. Neglect fringing fields.

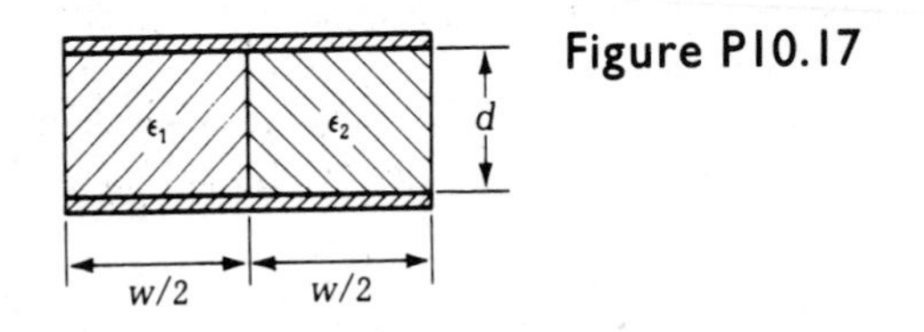

Figure P10.17

10.18 Consider the capacitor shown in Figure P10.17. Let $\epsilon_1 = 3\epsilon_0$, $\epsilon_2 = 5\epsilon_0$, $d = 0.6$ mm, and $A = 20$ cm^2. The potential between the plates is 300 V. Find the total stored electric energy in this capacitor.

10.19 Find the capacitance per unit length of a coaxial capacitor with two layers of insulating materials, as shown in Figure 10.15c. Express C/h in terms of a, b, c, ϵ_1, and ϵ_2.

10.20 Find the capacitance C of a parallel-plate capacitor with two layers of insulating materials, as shown in Figure P10.20. Express C in terms of A (the area of the plate), d_1, d_2, ϵ_1, and ϵ_2.

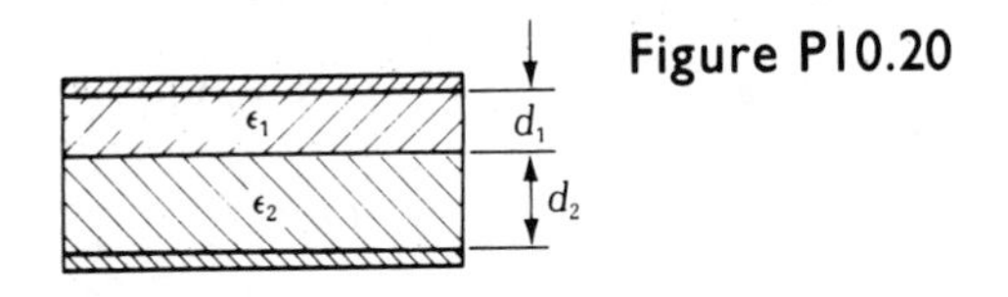

Figure P10.20

10.21 Refer to the capacitor shown in Figure P10.20. Let $\epsilon_1 = 3\epsilon_0$, $\epsilon_2 = 5\epsilon_0$, $d_1 = 0.3$ mm, $d_2 = 0.3$ mm, and $A = 20$ cm^2. The voltage across the capacitor is 300 V. Find the total stored electric energy in this capacitor.

10.22 Derive (10.58).

10.23 A parallel-plate capacitor carries $+Q$ on one plate and $-Q$ on the other plate. The area of each plate is A and the separation between the plates is S. The medium is air.

(a) Find the total stored energy U_E in this capacitor in terms of Q, A, S, and ϵ_0.
(b) What is the electrostatic force acting on the plates? Is it attractive or repulsive? Hint: Find the change in U_E with respect to S.

10.24 A cylindrical capacitor as shown carries a total charge of $+Q$ and $-Q$ on the inner and the outer conductors, respectively. The capacitor is not connected to the ground or to any voltage source. A dielectric washer, with inner and outer diameters matching the dimensions of the cylindrical capacitor, is placed x meters inside the capacitor as shown in Figure P10.24.

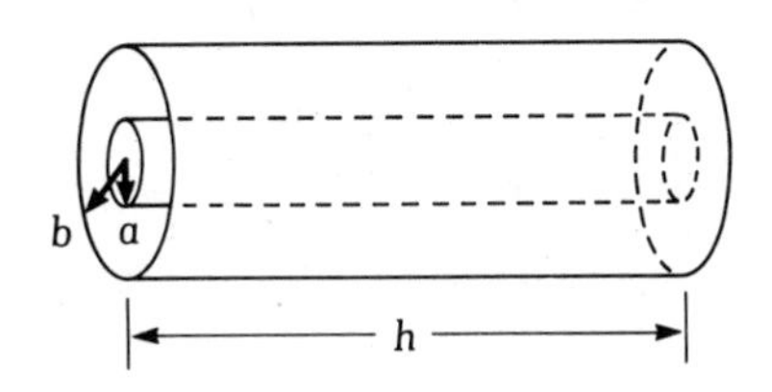

Figure P10.24

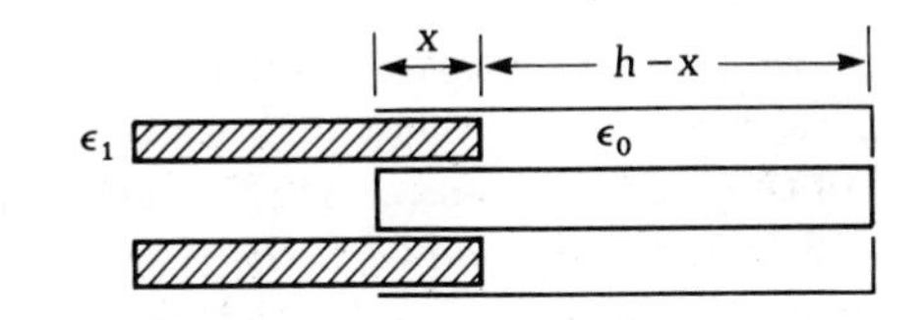

(a) Find the electric field in the capacitor in terms of Q. Hint: $\rho_1 x + \rho_0(h - x) = Q$, where ρ_1 and ρ_0 are line-charge densities in the dielectric and air regions, respectively. Also,

$$E = \frac{\rho_1}{2\pi\epsilon_1\rho} = \frac{\rho_0}{2\pi\epsilon_0\,\rho}$$

(b) Find the voltage between the two conductors.

(c) Find the force acting on the dielectric washer.

11 SOLUTION TECHNIQUES

In the previous chapter we discussed many applications of electrostatic fields. The reader will appreciate that in designing electrostatic devices one must know the distribution of the electric fields or of the electrostatic potential. For example, in designing CRTs or ink-jet printers, the trajectory of the electron beam or the ink drop must be controlled within a very stringent tolerance. This tight control is possible only if one knows the electric-field distribution precisely.

So far, we have learned how to obtain the electric fields for a few special cases, such as parallel-plate capacitors, spherical charge distributions, or cylindrical conductors. These cases are all characterized by symmetrical charge distributions. For these cases, we apply Gauss' law in integral form to find the solutions. In this chapter we will study other techniques for solving electrostatic-field problems. Specifically, we will study in detail the method of images and the method of separation of variables.

11.1 POISSON AND LAPLACE EQUATIONS

Chapter 9 introduced the Poisson equation, which is given in differential form by (9.7):

$$\boxed{\nabla^2\Phi = \frac{-\rho_v}{\epsilon}} \quad \textbf{(Poisson equation)} \tag{11.1}$$

In a region where ρ_v is equal to zero, the equation is simplified and is called the Laplace equation:

$$\boxed{\nabla^2\Phi = 0} \quad \textbf{(Laplace equation)} \tag{11.2}$$

The above equations are partial-differential equations. Usually, the values of the potential function Φ are known on some boundaries. The task is to solve (11.1) or (11.2) subject to those boundary conditions.

Example 11.1 *Some electrostatic problems can be solved in different ways. This example shows how to solve the parallel-plate problem using the differential equation approach.*

The upper plate of a parallel-plate capacitor is maintained at -100 V, and the lower plate is maintained at 0 V. The plates are 10 cm apart, and they are infinitely large. Find Φ in the parallel-plate region.

Solution Let us set up a coordinate system shown in Figure 11.1. Because these plates are infinitely large, we can conclude by symmetry that Φ is a function of x only and that it is not a function of y or z. There is no charge in the parallel-plate region. Thus, the present problem is to solve the Laplace equation (11.2) subject to the boundary conditions that $\Phi = 0$ at $x = 0$ and $\Phi = -100$ V at $x = 0.1$ m.

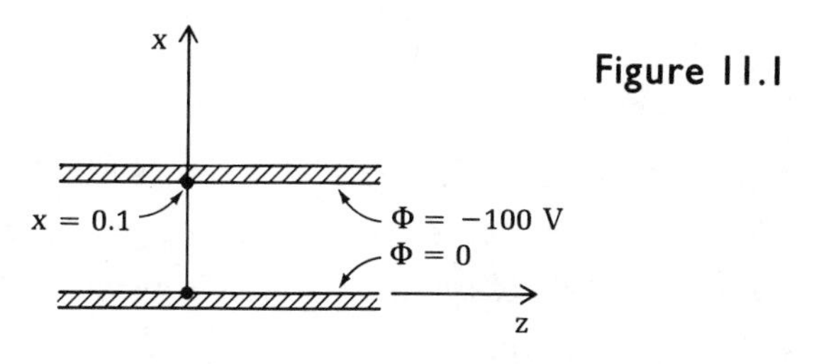

Figure 11.1 Solving Laplace's equation in the parallel-plate region subject to boundary conditions.

Using the expression of the Laplacian operator, we may write as follows (11.2) in the rectangular coordinate system:

$$\frac{\partial^2 \Phi}{\partial x^2} + \frac{\partial^2 \Phi}{\partial y^2} + \frac{\partial^2 \Phi}{\partial z^2} = 0$$

Because Φ is a function of x only, we obtain

$$\frac{\partial^2 \Phi}{\partial x^2} = 0 \tag{11.3}$$

Integrating (11.3) twice with respect to x yields

$$\Phi(x) = c_1 x + c_2 \tag{11.4}$$

where c_1 and c_2 are integration constants. Now we apply the boundary conditions:

$$\Phi(0) = 0 = c_2$$

$$\Phi(0.1) = -100 = 0.1c_1 + c_2$$

Solutions of c_1 and c_2 are easily found to be

$$c_2 = 0$$

$$c_1 = -1000$$

The final answer is

$$\Phi(x) = -1000x \tag{11.5}$$

Note that the electric field may be found by taking the gradient of (11.5). According to (9.6),

$$\boldsymbol{E} = -\nabla\Phi = -\frac{\partial\Phi}{\partial x}\hat{\boldsymbol{x}} = 1000\hat{\boldsymbol{x}}$$

We see that the electric field is pointing in the $\hat{\boldsymbol{x}}$ direction perpendicular to the surfaces of the plates.

Note that in Examples 9.14 and 9.15 the same problem is solved using Gauss' law in terms of charges on the plates. Here the solution is in terms of voltage difference between the plates. Using (9.34) and substituting $x = 0.1$ and $\Phi = -100$ yields $E_0 = \rho_s/\epsilon = 1000$. Thus, the answers obtained here and in Examples 9.14 and 9.15 are the same.

Example 11.2

The electrostatic problem in Example 11.1 can be solved by using Gauss' law or the Laplace equation. Some other problems, which cannot be solved by using Gauss' law for lack of symmetry in their geometry, can be solved using the differential equation approach.

Two semi-infinite conducting plates are arranged at an angle ϕ_0, as shown in Figure 11.2. One plate is charged to 0 volts and the other to V_0 volts. A gap at the tip insulates one plate from the other. Find the potential Φ in the region $0 < \phi < \phi_0$.

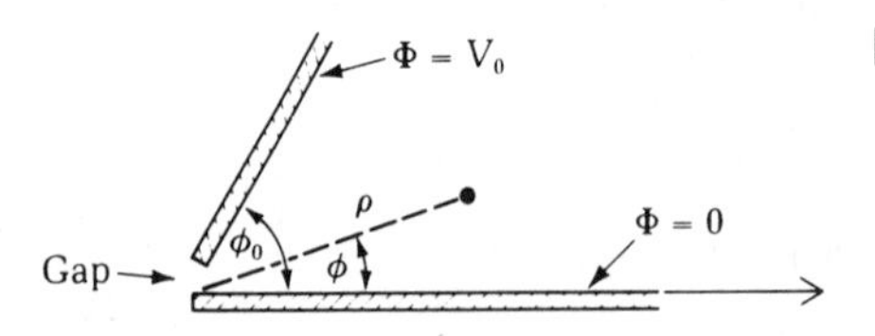

Figure 11.2 Solution of Laplace's equation in the nonparallel-plate region subject to boundary conditions.

Solution Obviously, the cylindrical coordinate system should be used, as shown in Figure 11.2. There is no charge in the region $0 < \phi < \phi_0$; thus, the Laplace equation (11.2) should be solved subject to the given boundary conditions.

Because of the symmetry of this problem and the fact that the boundary conditions for Φ involve the ϕ coordinate only, we can assume that Φ is a function of ϕ only and that it is independent of ρ and z. Consequently, we may write (11.2) as follows:

$$\frac{1}{\rho^2}\frac{\partial^2\Phi}{\partial\phi^2} = 0 \tag{11.6}$$

Excluding the small air gap, where $\rho = 0$, (11.6) may be expressed as follows:

$$\frac{\partial^2 \Phi}{\partial \phi^2} = 0$$

Integrating the above equation twice with respect to ϕ yields

$$\Phi = c_1 \phi + c_2 \tag{11.7}$$

Substituting the boundary conditions that $\Phi = 0$ at $\phi = 0$ and $\Phi = V_0$ at $\phi = \phi_0$, we obtain the final solution:

$$\Phi = \left(\frac{V_0}{\phi_0}\right)\phi \tag{11.8}$$

Note that we obtain the electric field by taking the gradient of (11.8):

$$\boldsymbol{E} = -\nabla\Phi = -\frac{1}{\rho}\frac{V_0}{\phi_0}\hat{\boldsymbol{\phi}} \tag{11.9}$$

The electric fields are in the negative $\hat{\boldsymbol{\phi}}$ direction originating from the higher potential plate and ending at the lower potential plate.

Child-Langmuir Law

Let us consider a vacuum-tube diode consisting of a cathode and an anode. The cathode is heated to emit electrons, which are attracted toward the anode and thus constitute a current flow. If the anode is at the wrong polarity—that is, if the anode potential is lower than that of the cathode—convection of current will stop. To simplify the analysis of the diode, let us consider the cathode and the anode as two parallel plates, as shown in Figure 11.3. The anode is charged to V volts, and electrons are flowing toward it. If the amplitude of the current is small, the presence of electrons may be neglected, and the potential field distribution is similar to that for Example 11.1—that is,

$$\Phi(y) = \frac{V}{d}y$$

and

$$\boldsymbol{E} = -\hat{\boldsymbol{y}}\frac{V}{d}$$

Figure 11.3 A diode with space charge in the parallel-plate region.

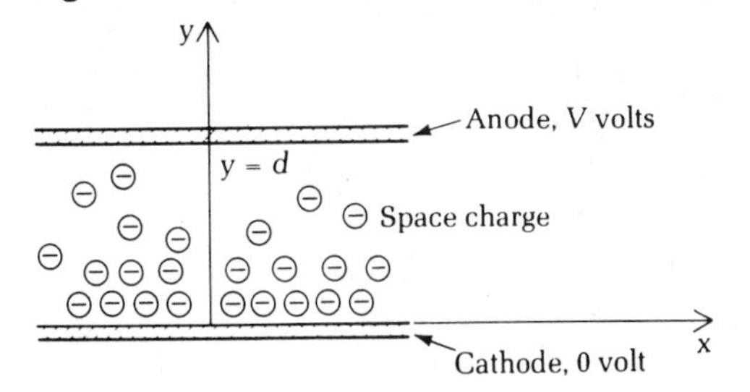

If the diode is used as a rectifier and the flow of the current is strong, then the electron density in the parallel-plate region is not negligible and must be taken into account. These negative charges are called the **space charges**. The space-charge density tends to increase because a positive electric field tends to pull more electrons out of the cathode. However, as the negative charges build up, repulsive forces between them and the electrons slow down the emission of electrons from the cathode. Finally, an equilibrium is reached, at which the total electric field at the surface of the cathode is equal to zero. The mathematical version of the above statement is as follows:

$$E_y(0) = -\left.\frac{d\Phi}{dy}\right|_{y=0} = 0 \tag{11.10a}$$

$$\Phi(0) = 0 \tag{11.10b}$$

$$\Phi(d) = V \tag{11.10c}$$

At equilibrium, the stream of electrons from the cathode to the anode constitutes a steady current $J = -\rho_v v$, where ρ_v is the space-charge density and v is the velocity of the electron. Note that J and v are positive and ρ_v is negative. The velocity of the electron at potential Φ at an arbitrary position y is given by [see (10.26)]

$$v = \sqrt{\frac{2e\Phi(y)}{m}}$$

where e is the magnitude of the electron charge (1.60×10^{-19} C) and m is its mass (9.11×10^{-31} kg). Using the above equation, the space-charge density ρ_v may now be expressed in terms of J and Φ:

$$\rho_v = \frac{-J}{v} = -J\sqrt{\frac{m}{2e\Phi}} \tag{11.11}$$

Now the problem is reduced to solving Poisson's equation (11.1) with the charge density given by (11.11) and subject to the boundary conditions (11.10). Because of symmetry, we can assume that Φ is a function of y only. Therefore,

$$\frac{d^2\Phi}{dy^2} = \frac{J}{\epsilon}\sqrt{\frac{m}{2e\Phi}} \tag{11.12}$$

Note that Φ appears on the right-hand side of the equation, too. To solve (11.12), we multiply it on both sides by $2d\Phi/dy$ to yield

$$2\frac{d\Phi}{dy}\frac{d^2\Phi}{dy^2} = a\Phi^{-1/2}\frac{d\Phi}{dy}$$

where

$$a = \frac{2J}{\epsilon}\sqrt{\frac{m}{2e}} \tag{11.13}$$

Both sides of the above equations can now be integrated separately to give

$$\left(\frac{d\Phi}{dy}\right)^2 = 2a\Phi^{1/2} + c_1$$

Using the conditions (11.10), we have $c_1 = 0$. The above equation may be written as follows:

$$\Phi^{-1/4}\frac{d\Phi}{dy} = \sqrt{2a}$$

Integrating the above equation yields

$$\frac{4}{3}\Phi^{3/4} = \sqrt{2a}y + c_2$$

The constant c_2 is found to be equal to zero because of the boundary (11.10b). Finally, we use (11.10c) to obtain

$$\frac{4}{3}V^{3/4} = \sqrt{2a}d \tag{11.14}$$

Eliminating a from (11.14) and (11.13), we find that J is given as follows:

$$J = \frac{4\epsilon}{9d^2}\sqrt{\frac{2e}{m}}V^{3/2} \tag{11.15}$$

Note that the current J and the applied voltage V are not linearly related. The relation of J and V expressed by (11.15) is known as the **Child-Langmuir law.***

The potential function is thus determined to be

$$\Phi(y) = \left(\frac{3}{4}\sqrt{2a}y\right)^{4/3} = \left(\frac{V^{3/4}}{d}y\right)^{4/3} = V\left(\frac{y}{d}\right)^{4/3}$$

The electric field is given by

$$\boldsymbol{E} = -\hat{y}\frac{4}{3}\frac{V}{d}\left(\frac{y}{d}\right)^{1/3}$$

which is equal to zero at the cathode.

We observe that in solving the second-order differential equation (11.12), the boundary conditions (11.10b) and (11.10c) specified the potential values at $y = 0$ and $y = d$. The added boundary condition (11.10a) enables us to find the steady current J, which is a constant related to V, as governed by the Child-Langmuir law in (11.15).

* C. D. Child, "Discharge from hot CaO," *Phys. Rev.* 32 (1911): 492. I. Langmuir, "The effect of space charge and residual gases on thermionic current in high vacuum," *Phys. Rev.* 2 (1913): 450.

Uniqueness Theorem

Before we continue solving more Poisson or Laplace equations, we shall ask an important question: If a solution satisfies the Poisson or Laplace equation and the boundary conditions, are there other solutions? In other words, is the solution unique? The answer is that there is only one solution.

To prove it, let us assume that there are two solutions Φ_1 and Φ_2, both satisfying Poisson's equation and the boundary conditions—that is, in a region R,

$$\nabla^2\Phi_1 = \frac{-\rho_v}{\epsilon} \quad \text{in } R \tag{11.16a}$$

$$\nabla^2\Phi_2 = \frac{-\rho_v}{\epsilon} \quad \text{in } R \tag{11.16b}$$

$$\Phi_1 = V \quad \text{on the boundary of } R \tag{11.17a}$$

$$\Phi_2 = V \quad \text{on the boundary of } R \tag{11.17b}$$

Now let us define Φ_3 as the difference between Φ_1 and Φ_2—that is,

$$\Phi_3 = \Phi_1 - \Phi_2 \quad \text{in } R \tag{11.18}$$

Applying the Laplacian operator on both sides of (11.18), we obtain

$$\nabla^2\Phi_3 = \nabla^2(\Phi_1 - \Phi_2) = \nabla^2\Phi_1 - \nabla^2\Phi_2 \tag{11.19}$$

Substituting (11.16a) and (11.16b) into (11.19) yields

$$\nabla^2\Phi_3 = 0 \quad \text{in } R \tag{11.20}$$

Also, because of (11.17a) and (11.17b), we have

$$\Phi_3 = 0 \quad \text{on the boundary of } R \tag{11.21}$$

Now let us consider the following vector identity:

$$\nabla \cdot (\Phi_3 \nabla\Phi_3) \equiv \nabla\Phi_3 \cdot \nabla\Phi_3 + \Phi_3 \nabla^2\Phi_3 \tag{11.22}$$

We integrate both sides of (11.22) over the volume R:

$$\iiint_R \nabla \cdot (\Phi_3 \nabla\Phi_3)\, dv = \iiint_R |\nabla\Phi_3|^2\, dv + \iiint_R \Phi_3 \nabla^2\Phi_3\, dv \tag{11.23}$$

Through the divergence theorem, the integral on the left side of (11.23) can be expressed in terms of the integral over the boundary surface of R. The second integral on the right of (11.23) is zero because of (11.20). Therefore, we have

$$\oiint_{\text{boundary of } R} \Phi_3 \frac{\partial\Phi_3}{\partial n}\, ds = \iiint_R |\nabla\Phi_3|^2\, dv \tag{11.24}$$

But because of (11.21), the integral at the left is zero. We are thus left with the following equation:

$$\iiint_R |\nabla\Phi_3|^2\, dv = 0$$

In the above equation, the integrand is always nonnegative, and the integration is zero. The result can occur only if $\nabla\Phi_3$ is zero **everywhere** in R. In other words,

$$\Phi_3 = C \qquad \text{in } R$$

But (11.21) says that Φ_3 is zero on the boundary of R. Therefore, $C = 0$—that is,

$$\Phi_3 = 0 \qquad \text{in } R$$

We have now shown that Φ_3 is identically zero in R. Remember that $\Phi_3 = \Phi_1 - \Phi_2$. We conclude that Φ_1 is identical to Φ_2. This conclusion proves the uniqueness theorem: there is only one unique solution that satisfies both the Poisson (or the Laplace) equation in R and the boundary condition on the boundary of R.

Let us now review Examples 11.1 and 11.2. In the former we say that because of symmetry Φ may be assumed to be independent of y and z. This assumption leads to the simple differential equation (11.3). Similarly, in Example 11.2, we assume that Φ is a function of ϕ only. Are these assumptions justified? Are there possibly other solutions in which Φ may be more complicated and not just a function of one variable? From the uniqueness theorem, we know that the solutions obtained in Examples 11.1 and 11.2 are unique solutions to these problems and that there can be no other solutions.

The solution of Poisson or Laplace equations with boundary conditions may sometimes look very complicated but is actually rather simple. In fact, in certain cases solutions may be obtained by intuition or even by guesswork. A simple way to see if a solution thus obtained is correct is to substitute it back and see if it satisfies both the Poisson (or Laplace) equation and the given boundary conditions. If it does, then we are assured that we have found *the* solution.

Example 11.3 *This example shows that a seemingly complex problem may have a surprisingly simple solution. It does not matter how simple a solution is. As long as it satisfies both the Poisson (or Laplace) equation and the boundary condition, it is the one and only solution.*

An irregularly shaped, hollow conductor is maintained at V_0 volts, as shown in Figure 11.4. There is no charge in the cavity R. What is Φ in the cavity?

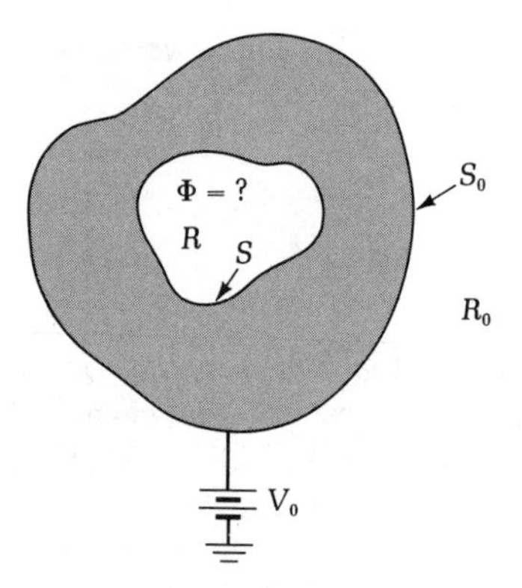

Figure 11.4 An irregularly shaped hollow conductor is charged to V_0 volts. The shaded volume is the conductor. The unshaded region is a cavity designated as R in the conductor.

Solution This boundary-value problem requires the solution of the Laplace equation (11.2) in R, subject to the boundary condition that $\Phi = V_0$ on S. The solution seems to be difficult because neither R nor S has any symmetry. However, the uniqueness theorem allows us to guess the solution. We notice that Φ is equal to V_0 on S (the boundary of R) because the conductor is an equipotential body. Let us then propose that Φ be equal to V_0 *everywhere* in R:

$$\Phi = V_0 \qquad \text{in } R \tag{11.25}$$

The above equation satisfies (11.2) because the Laplacian of a constant is equal to zero. It also satisfies the boundary condition. From the uniqueness theorem, we known that (11.25) is *the* solution and that there are no other solutions! We can now make the following general statement: every point on or in a conductor, which may or may not contain hollow cavities, is at equal potential, as long as there are no charges in any of the cavities.

Example 11.4 ***This example is similar to Example 11.3 but the result is very different. Remember that a solution to the electrostatic problem must satisfy the Poisson (or Laplace) equation and all boundary conditions.***

Find Φ in the outside region R_0 shown in Figure 11.4.

Solution No simple solution for this problem can be found. It may occur to the reader that (11.25) may also be the solution in R_0. Isn't it true that (11.25) satisfies the Laplace equation (11.2) in R_0 and the boundary condition $\Phi = V_0$ on S_0? Then why is it not a solution?

The answer is that, unlike the region R, R_0 has two boundaries. One is S_0, the other is the surface at infinity. At infinity, we know that Φ must approach zero. Therefore, the boundary condition for Φ in the exterior region is

$$\Phi = \begin{cases} V_0 & \text{on } S_0 \\ 0 & \text{at infinity} \end{cases}$$

Obviously (11.25) cannot satisfy the boundary condition at infinity. Consequently, it is not a solution in R_0. In fact, because of the region's irregular shape, no analytical solution exists. Approximate numerical solutions may be found with a digital computer.

11.2 IMAGE METHOD

In many practical electrostatic problems, charges are located near a conductor. An electron just emitted from the cathode and a power transmission line hanging over the conducting earth are typical examples. Let us consider the case of a point charge q near an infinite planar conductor, as shown in Figure 11.5. The potential of the conductor is set equal to zero. To find the potential Φ in the upper half-space, we need to solve for the Poisson equation in the half-space $z > 0$, subject to the boundary condition that $\Phi = 0$ at $z = 0$ and at infinity.

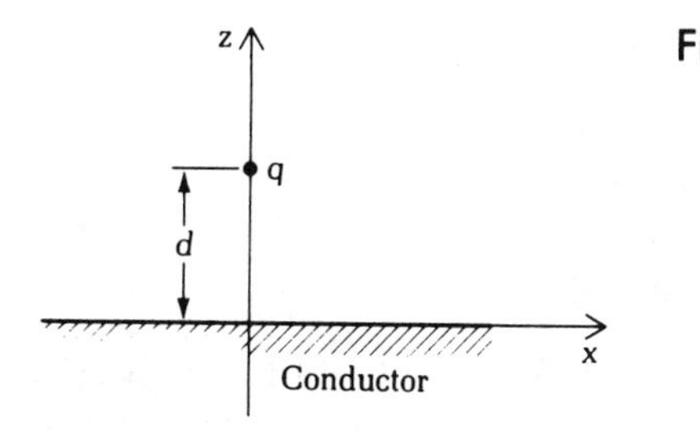

Figure 11.5 A point charge above a conducting plane.

In the absence of the conductor, the solution for a point charge in free space is

$$\Phi(x, y, z) = \frac{q}{4\pi\epsilon} \cdot \frac{1}{\sqrt{x^2 + y^2 + (z - d)^2}} \tag{11.26}$$

The potential function Φ in (11.26) satisfies the Poisson equation for $z > 0$ and the boundary conditions $\Phi = 0$ at infinity. However, the potential is not zero at $z = 0$. Therefore, it is not a solution to the problem shown in Figure 11.5.

Instead of trying to solve the Poisson equation directly, let us recall the problem we solved earlier in Section 9.1—namely, the electrostatic potential of a dipole. Figure 11.6 (p. 390) shows the dipole configuration. The potential of the dipole is found to be

$$\Phi(x, y, z) = \frac{q}{4\pi\epsilon}\left(\frac{1}{\sqrt{x^2 + y^2 + (z - d)^2}} - \frac{1}{\sqrt{x^2 + y^2(z + d)^2}}\right) \tag{11.27}$$

Because (11.27) is a solution of the dipole problem, it satisfies the Poisson equation with the charge distribution being the two point charges shown in Figure 11.6. Notice that $\Phi = 0$ at infinity and at $z = 0$.

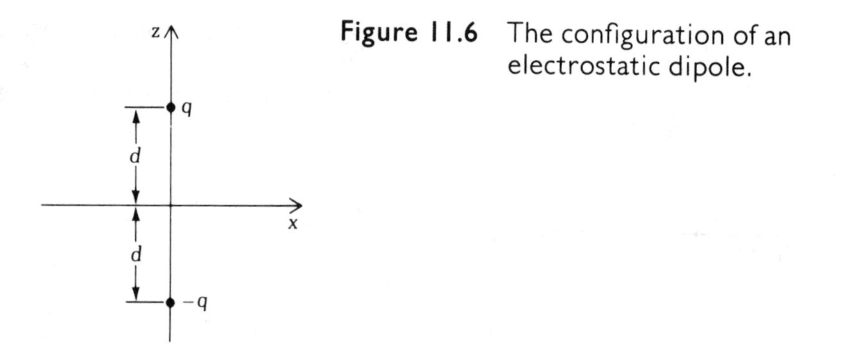

Figure 11.6 The configuration of an electrostatic dipole.

Now let us compare Figure 11.5 and Figure 11.6. We notice that in the upper region ($z > 0$) the charge distribution shown in Figure 11.5 is identical to that shown in Figure 11.6. Both only have a point charge located at $(0, 0, d)$. Furthermore, in both cases, Φ is equal to zero at $z = 0$ and at infinity. These are sufficient reasons for us to believe that (11.27) must also be the solution of the problem shown in Figure 11.5. In other words, we claim that the solution to the problem in Figure 11.5 is as follows:

$$\Phi(x, y, z) = \begin{cases} \dfrac{q}{4\pi\epsilon}\left(\dfrac{1}{\sqrt{x^2 + y^2 + (z - d)^2}} - \dfrac{1}{\sqrt{x^2 + y^2 + (z + d)^2}}\right) \\ \quad \text{for } z \geq 0 \\ 0 \quad \text{for } z \leq 0 \end{cases} \tag{11.28}$$

The potential Φ given in (11.28) is identical to that given in (11.27) in the region $z > 0$. We know that (11.27) satisfies the Poisson equation in that region—a region containing no other charges except a point charge q located at $(0, 0, d)$. Thus, (11.28) also satisfies the Poisson equation in the region $z > 0$ shown in Figure 11.5. The Φ given by (11.28) also satisfies the boundary conditions in that it is zero at $z = 0$ and at infinity. According to the uniqueness theorem, (11.28) is *the* solution of the boundary-value problem shown in Figure 11.5. Note that the electrostatic potential in the region $z > 0$ shown in Figure 11.5 is the superposition of the potential due to the point charge q and the potential due to the "image" of the point charge. The image charge is located at the image point as if the conductor were a mirror. The value of the image is equal to $-q$. To an observer in the region $z > 0$, it appears as if there were two charges.

Example 11.5 *Once the potential function is obtained, the electric field can be computed directly from the potential. This example shows the electric-field lines from a point charge near a conducting plane.*

Find the electric field in the region $z \geq 0$ shown in Figure 11.5.

Solution To obtain the electric field, we can take the gradient of (11.28):

$$E = -\nabla\Phi = \frac{q}{4\pi\epsilon}\left\{\frac{\hat{x}x + \hat{y}y + (z-d)\hat{z}}{[x^2 + y^2 + (z-d)^2]^{3/2}} - \frac{\hat{x}x + \hat{y}y + (z+d)\hat{z}}{[x^2 + y^2 + (z+d)^2]^{3/2}}\right\} \quad \mathbf{(11.29)}$$

Figure 11.7 shows the electric-field lines.

Figure 11.7 Electric-field lines from a positive point charge near a conducting plate.

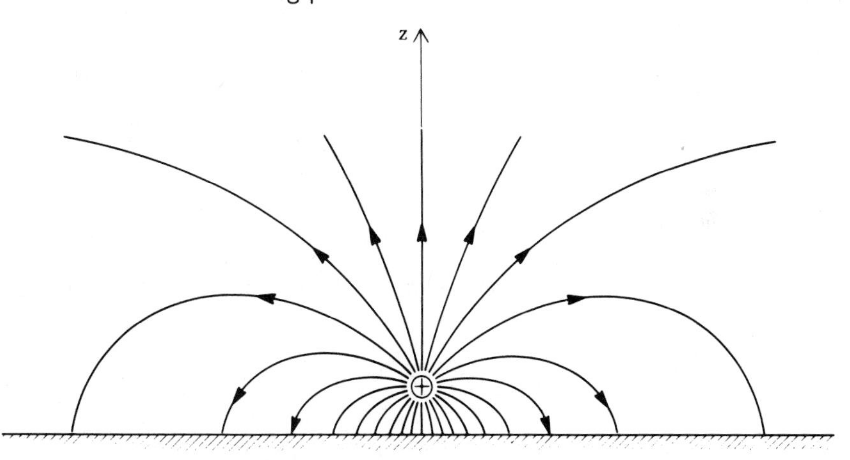

Example 11.6

Again, once the potential function is obtained, the electric field everywhere and the induced charge on the conducting plane may be found directly from the potential. The electric force on the charge can also be computed.

Find the charge distribution on the surface of the conductor shown in Figure 11.5, and then calculate the total charge induced on it. Also find the electric force on q.

Solution At the surface, where $z = 0$, we have

$$E(x, y, 0) = \frac{-qd\hat{z}}{2\pi\epsilon(x^2 + y^2 + d^2)^{3/2}} \quad \mathbf{(11.30)}$$

Note that E is normal to the conductor, as it should be.

The surface charge is obtained from the boundary condition

$$\hat{n} \cdot (D_1 - D_2) = \rho_s$$

Here, $\hat{n} = \hat{z}$, $D_2 = 0$, and D_1 is ϵ times the E given in (11.30). Therefore, we obtain

$$\rho_s = \frac{-qd}{2\pi(x^2 + y^2 + d^2)^{3/2}} \quad \mathbf{(11.31)}$$

Figure 11.8 shows the distribution of ρ_s. Note that the induced surface charges are concentrated in the region directly below the charge q and that the density decreases rapidly outward.

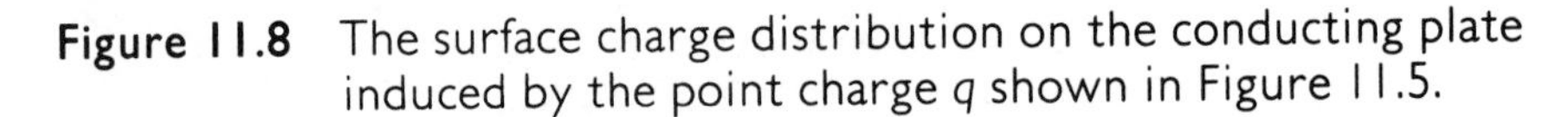

Figure 11.8 The surface charge distribution on the conducting plate induced by the point charge q shown in Figure 11.5.

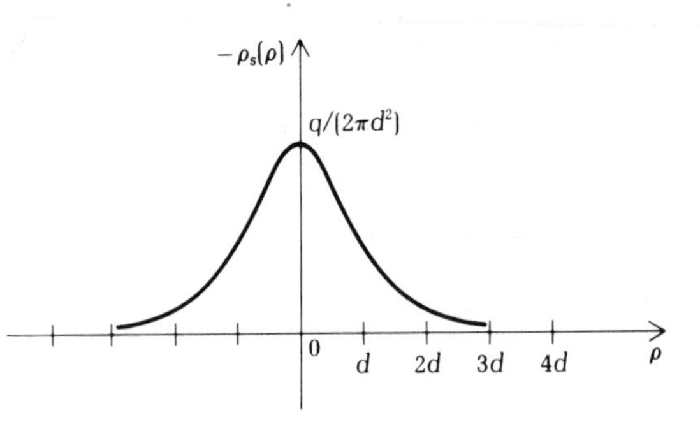

We may calculate the total amount of the induced charge by integration, which gives

$$
\begin{aligned}
q_{\text{induced}} &= \int \rho_s\, ds = \int_{-\infty}^{\infty} dx \int_{-\infty}^{\infty} dy \frac{-qd}{2\pi(x^2 + y^2 + d^2)^{3/2}} \\
&= \frac{-qd}{2\pi}\int_{-\infty}^{\infty} dx \int_{-\pi/2}^{\pi/2} d\theta \frac{\cos\theta}{x^2 + d^2} = \frac{-qd}{\pi}\int_{-\infty}^{\infty} dx \frac{1}{x^2 + d^2} \\
&= \frac{-qd}{\pi}\int_{-\pi/2}^{\pi/2} d\theta' \frac{1}{d} = -q
\end{aligned}
$$

In the above calculation, we made the change of variables $y = (x^2 + d^2)^{1/2} \tan\theta$ and $x = d \tan\theta'$. Note that the total surface charge induced on the surface of the conductor is exactly equal to the negative of the charge located above the conductor.

The force on q due to all induced charges on the conductivity plate is equal to the force on q due to its image. Thus,

$$
\boldsymbol{F} = \frac{q^2}{16\pi\epsilon d^2}(-\hat{z})
$$

Example 11.7 *This example shows how to generalize the image theory involving a point charge near a conducting plate to other similar cases. Sometimes you need an infinite number of images to satisfy the boundary conditions.*

A point charge is located between two conducting plates separated by a distance d (Figure 11.9a). Use the image method to find the electrostatic potential Φ between the plates.

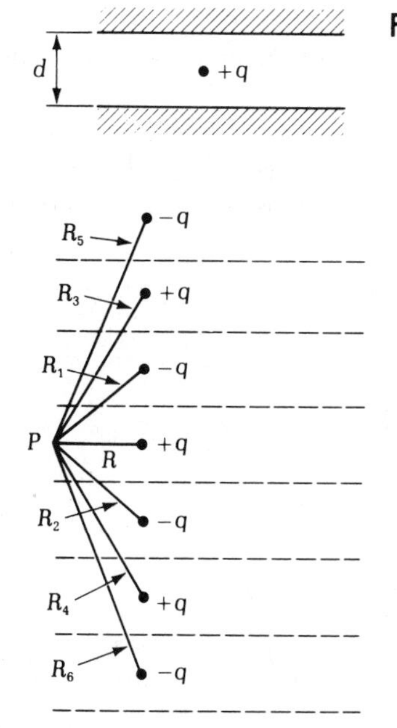

Figure 11.9(a) A point charge located between two conducting plates.

Figure 11.9(b) There are infinite number of image charges.

Solution In using the image method, we must obey two rules:

1. The image charge or charges must be placed in the region containing the conductor. They cannot be located in the "real" space where Φ is to be found.
2. The image charge or charges must be placed so that the potential at the conductor's surfaces is equal to zero or a constant.

If we cannot satisfy both rules, then the problem cannot be solved by the image method. The first rule ensures that the Poisson equation will be satisfied with the correct charge distribution, and the second rule guarantees that the boundary conditions will be met.

We now place image charges in the regions occupied by conducting plates. To satisfy the boundary condition at the top plate, we place a negative charge at R_1 from the observation point P. To satisfy the boundary condition at the bottom boundary, we have to put the pair of images at R_2 and R_4. These new images violate the boundary condition at the top plate, and consequently we have to place images at R_3 and R_5, and so on and so forth. The potential at P is as follows:

$$\Phi = \frac{q}{4\pi\epsilon}\left(\frac{1}{R} - \frac{1}{R_1} - \frac{1}{R_2} + \frac{1}{R_3} + \frac{1}{R_4} - \frac{1}{R_5} - \frac{1}{R_6} + \cdots\right) \quad \textbf{(11.32)}$$

Example 11.8 *This is another example of finding the image charges (magnitudes and locations) for a point charge near a conducting structure.*

A point charge is located near the corner of a horizontal and a vertical conducting plate, as shown in Figure 11.10. Find the electrostatic potential Φ in the region $x > 0$ and $y > 0$.

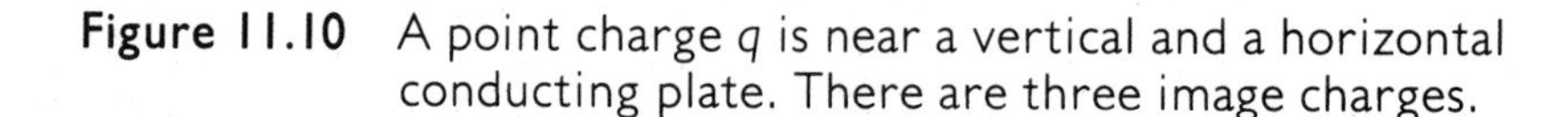

Figure 11.10 A point charge q is near a vertical and a horizontal conducting plate. There are three image charges.

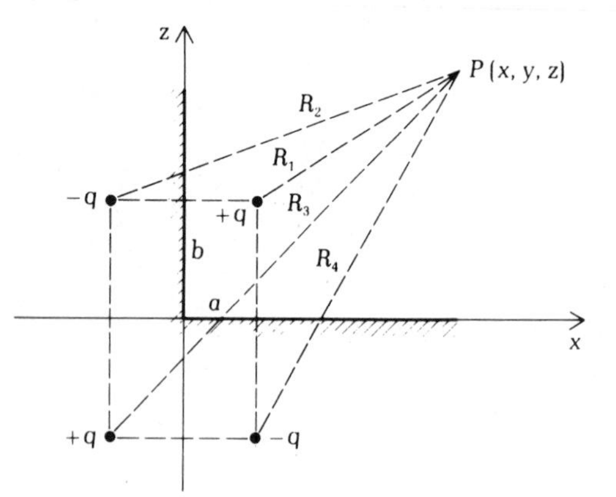

Solution Applying the rules of the image method to this problem, we obtain three image charges, as shown in Figure 11.10. Notice that two of the images carry $-q$ charges and that one carries a $+q$ charge. By inspection, we can see that Φ is equal to zero on the vertical wall as well as on the horizontal wall. Thus, the potential in the real space ($x > 0$, $y > 0$) is the superposition of the potentials due to four point charges—namely,

$$\Phi(x, y, z) = \frac{q}{4\pi\epsilon}\left(\frac{1}{R_1} - \frac{1}{R_2} + \frac{1}{R_3} - \frac{1}{R_4}\right) \tag{11.33}$$

where

$$R_1 = [(x - a)^2 + y^2 + (z - b)^2]^{1/2}$$
$$R_2 = [(x + a)^2 + y^2 + (z - b)^2]^{1/2}$$
$$R_3 = [(x + a)^2 + y^2 + (z + b)^2]^{1/2}$$
$$R_4 = [(x - a)^2 + y^2 + (z + b)^2]^{1/2}$$

A Point Charge Near a Corner of Conducting Walls

For a point charge near a corner of conducting walls, as shown in Figure 11.11, we can use the image method to place appropriate image charges in the region occupied by the conductor. Remember the two rules:

1. The image charges can only be placed in the region occupied by the conductor. They cannot be located in the real space.
2. The strategy of placing the image charges is to make the potential equal to zero or a constant on the conducting walls.

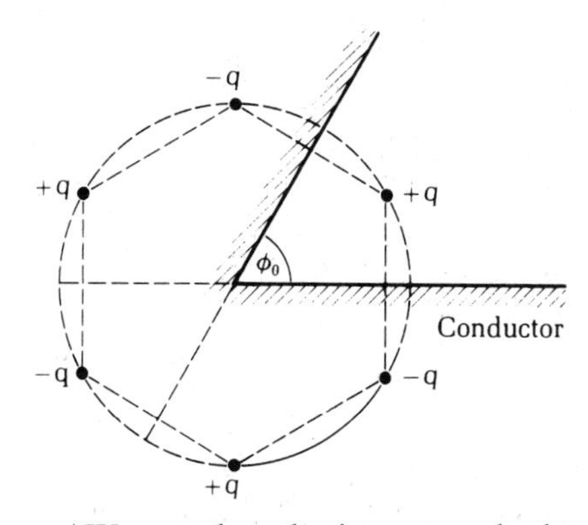

Figure 11.11 A point charge near a corner of a conducting wall, $\phi_0 = 60°$. There are five image charges.

We see that the image method can only be used if the angle ϕ_0 is equal to $180°/n$, where n is an integer and there should be $(2n - 1)$ images. For $n = 1$, there is one image charge, as shown in Figure 11.5. For $n = 2$, there are three image charges, as shown in Figure 11.10. For $n = 3$—that is, $\phi_0 = 60°$—we expect to have five image charges, and we indeed do, as illustrated in Figure 11.11.

A Line Charge Near a Conducting Cylinder

We can also apply the image method to the case of an infinitely long charge near an infinitely long conducting cylinder, as shown in Figure 11.12. The potential function Φ outside the conducting cylinder is the superposition of the line charge

Figure 11.12 **(a)** A line charge near a conducting cylinder. **(b)** The image is a line charge of $-\rho_\ell$ located at ρ_0, where $\rho_0 = a^2/d$.

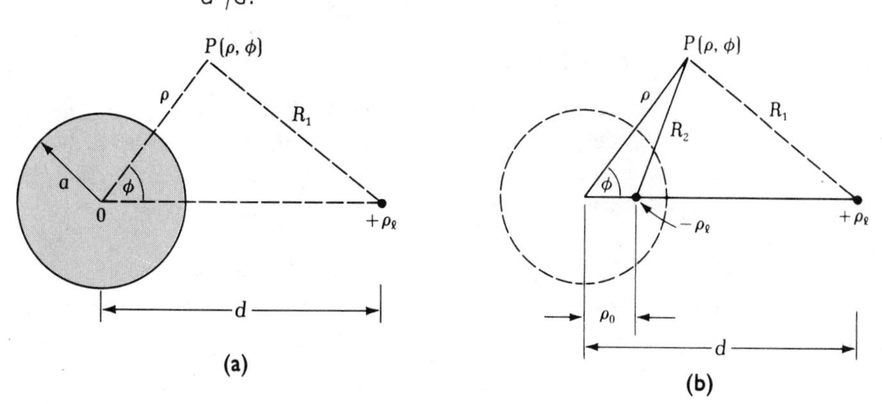

ρ_ℓ and its image, which has a magnitude of $-\rho_\ell$ and is located at $\rho = \rho_0$. We shall find the value ρ_0 in terms of the parameters a and d.

Let us first find the potential due to a line charge ρ_ℓ located on the z axis. From (9.22), we know that the electric field of this line charge is

$$E(\rho) = \hat{\rho}\frac{\rho_\ell}{2\pi\epsilon\rho}$$

The potential may be found by the line integral, the method discussed in Section 9.4—namely,

$$\Phi(\rho) = -\int_c^\rho E(\rho)\cdot\hat{\rho}\,d\rho$$
$$= \frac{\rho_\ell}{2\pi\epsilon}\ln\left(\frac{c}{\rho}\right)$$

The constant c is the coordinate of the reference point where the potential is zero. The above formula is valid for ρ_ℓ located on the z axis. If ρ_ℓ is parallel to but not exactly on the z axis, such as in the case shown in Figure 11.12b, the potential function can be modified to yield the following:

$$\Phi(\rho, \phi) = \frac{\rho_\ell}{2\pi\epsilon}\ln\left(\frac{R_0}{R_1}\right)$$

where R_1 is the distance between ρ_ℓ and the point (ρ, ϕ) and R_0 is the distance between the reference point and ρ_ℓ. The total potential due to the line charges ρ_ℓ and $-\rho_\ell$ shown in Figure 11.12b is the superposition of the respective potentials due to ρ_ℓ and $-\rho_\ell$; that is,

$$\Phi(\rho, \phi) = \frac{\rho_\ell}{2\pi\epsilon}\ln\left(\frac{R_0}{R_1}\right) - \frac{\rho_\ell}{2\pi\epsilon}\ln\left(\frac{R_0}{R_2}\right) = \frac{\rho_\ell}{2\pi\epsilon}\ln\left(\frac{R_2}{R_1}\right) \tag{11.34}$$

Note that R_0 cancels out in (11.34). This cancellation implies that the reference point is chosen at equal distances from ρ_ℓ and $-\rho_\ell$. In fact, according to (11.34), the potential is equal to zero at any point on the plane that bisects the line joining ρ_ℓ and $-\rho_\ell$. In Figure 11.12b, R_1 and R_2 are given as follows:

$$R_1 = (\rho^2 + d^2 - 2\rho d\cos\phi)^{1/2}$$
$$R_2 = (\rho^2 + \rho_0^2 - 2\rho_0\rho\cos\phi)^{1/2}$$

Notice that the solution (11.34) holds only in regions outside the conducting cylinder. On the surface of the conducting cylinder at $\rho = a$, the potential should be a constant, and it can be a constant only if R_2/R_1 is independent of ϕ at $\rho = a$. More specifically,

$$\Phi(\rho = a, \phi) = \frac{\rho_\ell}{4\pi\epsilon}\ln\left(\frac{\rho_0}{d}\frac{\dfrac{a^2+\rho_0^2}{2a\rho_0} - \cos\phi}{\dfrac{a^2+d^2}{2ad} - \cos\phi}\right) \tag{11.35}$$

must be a constant independent of ϕ. The above condition can be met if the following equation holds:

$$\frac{a^2 + d^2}{2ad} = \frac{a^2 + \rho_0^2}{2a\rho_0}$$

From the above equation, we obtain

$$\boxed{\rho_0 = \frac{a^2}{d}} \quad \textbf{(position of image in a conducting cylinder)} \qquad \textbf{(11.36)}$$

Substituting (11.36) into (11.35) yields

$$\Phi(a, \phi) = \frac{\rho_\ell}{2\pi\epsilon} \ln\left(\frac{a}{d}\right) \qquad \textbf{(11.37)}$$

The above equation gives the potential of the cylindrical conductor. If the cylinder is grounded so that its potential is equal to zero, then (11.34) must be modified with the constant given in (11.37) subtracted from it—that is,

$$\Phi(\rho, \phi) = \frac{\rho_\ell}{2\pi\epsilon}\left[\ln\left(\frac{R_2}{R_1}\right) - \ln\left(\frac{a}{d}\right)\right]$$

Example 11.9 *The image of a line charge near a conducting cylinder has been obtained. This example shows how to use the result to solve other related problems.*

A long cylindrical conductor is charged to V_0 volts and is h meters above the ground, as shown in Figure 11.13a (p. 398). The potential of the ground is taken to be zero. Find this system's capacitance per unit length.

Solution We solve this problen applying the result just obtained. First, we redraw Figure 11.13a as Figure 11.13b, in which we have a conducting cylinder charged to V_0 and a line charge ρ_ℓ at a distance d away. The quantity V_0 is related to ρ_ℓ according to (11.37)—that is,

$$V_0 = \frac{\rho_\ell}{2\pi\epsilon} \ln\left(\frac{a}{d}\right) \qquad \textbf{(11.38)}$$

Note that ρ_ℓ and V_0 are opposite in sign since $a/d < 1$. The potential outside the cylinder is given by (11.34). Obviously, the potential is equal to zero when R_1 is equal to R_2. In other words, the potential is zero on the surface AB shown in Figure 11.13b. Now we compare the region above the earth shown in Figure 11.13a with the region above the plane AB shown in Figure 11.13b. These regions have identical boundary conditions, and there are no charges in them except for the surface charges at the boundary. According to the uniqueness theorem, the

Figure 11.13(a) A conducting cylinder over a flat conducting plane.

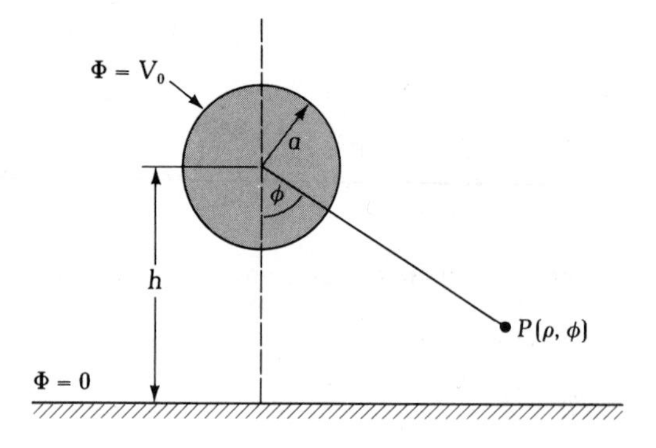

Figure 11.13(b) A line charge near a conducting cylinder. The boundary condition of the upper region is identical to that of the region shown in Figure 11.13(a).

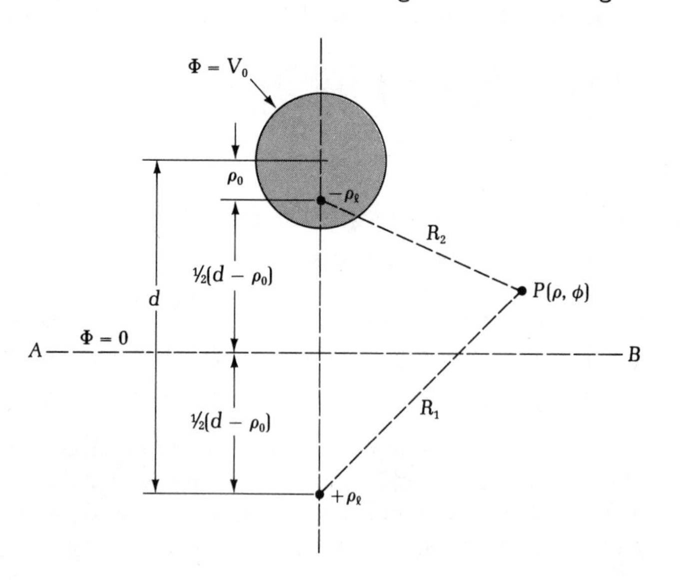

potentials in these regions are identical. However, the potential shown in Figure 11.13b has been solved and is given by (11.34). Consequently, the potential in the region shown in Figure 11.13a is obtained simply by substituting h for $\frac{1}{2}(d + \rho_0)$ in (11.34):

$$h = \frac{1}{2}(d + \rho_0)$$

Using (11.36), we then obtain

$$d = h + \sqrt{h^2 - a^2} \tag{11.39}$$

This system's capacitance per unit length is equal to the charge per unit length divided by the difference in potential. The total charge per unit length carried on the cylinder is equal to $-\rho_\ell$. The capacitance is thus given by

$$C = \frac{-\rho_\ell}{V_0} = \frac{2\pi\epsilon}{\ln\frac{d}{a}} = \frac{2\pi\epsilon}{\ln\left[\frac{h}{a} + \sqrt{\left(\frac{h}{a}\right)^2 - 1}\right]}$$

or

$$C = \frac{2\pi\epsilon}{\cosh^{-1}\left(\frac{h}{a}\right)} \left(\frac{\text{F}}{\text{m}}\right) \tag{11.40}$$

This result is the rigorous solution of the capacitance per unit length of a cylindrical conductor over a perfect conducting plate. It is useful in calculating the capacitance of a power transmission line hanging above the ground.

Example 11.10 *This example shows that once the potential is obtained, the electric field can be computed directly from the potential.*

Calculate the maximum electric-field strength on the surface of the conducting cylinder shown in Figure 11.13a.

Solution The maximum electric field is located at the surface of the conductor at the point nearest the ground because of the fact that at this point the electric fields due to $+\rho_\ell$ and $-\rho_\ell$ charges, respectively, add up constructively. From another point of view, we know that the charge on the conductor tends to be attracted toward the ground, so that the greatest charge concentration on the conductor must be at the point nearest the ground. According to the boundary condition, a large charge concentration also means a large electric field.

Let us now proceed to calculate the maximum electric field. We obtain the electric field by taking the gradient of Φ given in (11.34). Referring to the cylindrical coordinates defined in Figure 11.12b, we obtain:

$$\boldsymbol{E} = -\nabla\Phi$$

$$= \frac{-\rho_\ell}{4\pi\epsilon}\left[\hat{\rho}\left(\frac{2\rho - 2\rho_0\cos\phi}{R_2^2} - \frac{2\rho - 2d\cos\phi}{R_1^2}\right) + \hat{\phi}\left(\frac{2\rho_0\sin\phi}{R_2^2} - \frac{2d\sin\phi}{R_1^2}\right)\right]$$

At the point on the cylinder nearest to the ground, $\rho = a$, and $\phi = 0$. We find

$$\boldsymbol{E} = \hat{\rho}\frac{V_0}{a\ln\left(\frac{d}{a}\right)}\left(\frac{d+a}{d-a}\right) \tag{11.41}$$

A Point Charge Near a Conducting Sphere

The image method can also solve the electrostatic problem of a point charge near a grounded conducting sphere, as illustrated in Figure 11.14. The image of the real charge q is a point charge of q' located at r_0 inside the conducting sphere. We shall find the values of q' and r_0 in terms of a, d, and q. According to (9.12), the potential Φ due to q and q' is given as follows:

$$\Phi(r, \theta) = \frac{1}{4\pi\epsilon}\left(\frac{q}{R_1} + \frac{q'}{R_2}\right) \tag{11.42}$$

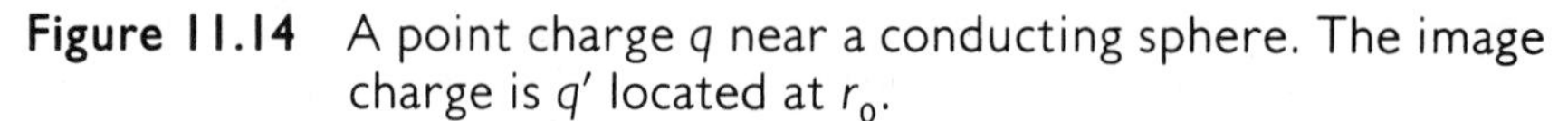
Figure 11.14 A point charge q near a conducting sphere. The image charge is q' located at r_0.

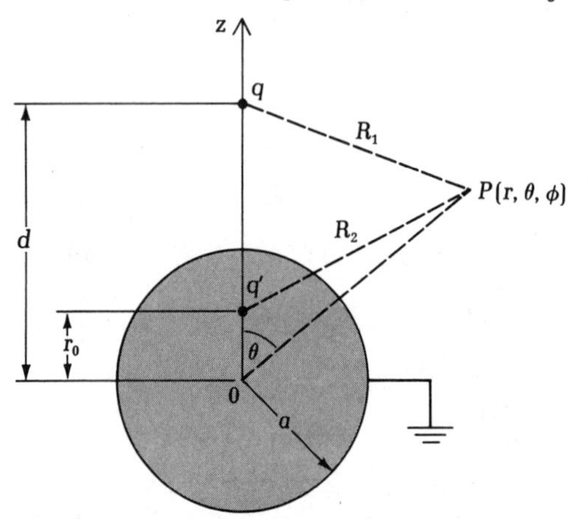

where

$$R_1 = (r^2 + d^2 - 2rd\cos\theta)^{1/2}$$

$$R_2 = (r^2 + r_0^2 - 2r_0 r\cos\theta)^{1/2}$$

We use spherical coordinates in the above equation.

At $r = a$, we require that $\Phi = 0$—that is,

$$\Phi(a, \theta) = \frac{1}{4\pi\epsilon}\left[\frac{q}{(a^2 + d^2 - 2ad\cos\theta)^{1/2}} + \frac{q'}{(a^2 + r_0^2 - 2r_0 a\cos\theta)^{1/2}}\right]$$

$$= \frac{1}{4\pi\epsilon}\left[\frac{q}{\sqrt{2ad}\left(\dfrac{a^2 + d^2}{2ad} - \cos\theta\right)^{1/2}} + \frac{q'}{\sqrt{2ar_0}\left(\dfrac{a^2 + r_0^2}{2ar_0} - \cos\theta\right)^{1/2}}\right] = 0$$

Because the above equation must hold for all θ, we obtain the following two conditions:

$$\frac{a^2 + d^2}{2ad} = \frac{a^2 + r_0^2}{2r_0 a}$$

$$\frac{q}{\sqrt{2ad}} = \frac{-q'}{\sqrt{2ar_0}}$$

Solving the above equations, we obtain

$$r_0 = \frac{a^2}{d} \quad \textbf{(position of image in a conducting sphere)} \quad \textbf{(11.43a)}$$

$$q' = -q\left(\frac{a}{d}\right) \quad \textbf{(value of the image charge)} \quad \textbf{(11.43b)}$$

In conclusion, the potential Φ is due to the point charge q and its image q', which is located in the conducting sphere at r_0 distance off the center of the sphere toward the direction of the real charge. The total charge induced at the surface of the conducting sphere is equal to $-aq/d$.

When the conducting sphere is insulated from the ground and carries no net charge, then (11.42) is no longer valid and must be modified to give the correct result. Because the net charge on the sphere is zero and the induced surface charge due to the external charge q is $-qa/d$, $+qa/d$ of charge must be distributed on the surface of the conducting sphere. The distribution of this qa/d charge on the surface of the sphere is uniform because the surface of the sphere is an equipotential surface. From Example 9.10, we know that the potential due to charges uniformly distributed on a spherical conductor is found in the same way as when the charge is concentrated at the center of the sphere. Thus, we simply need to add this term to the solution (11.42):

$$\Phi(r, \theta) = \frac{q}{4\pi\epsilon}\left(\frac{1}{R_1} - \frac{a}{R_2 d} + \frac{a}{rd}\right) \quad \textbf{(11.44)}$$

This equation is the potential function of a point charge near a conducting sphere that is insulated from the ground and that carries no net charge. Note that the potential on this insulated sphere is given as follows:

$$\Phi(a, \theta) = \frac{q}{4\pi\epsilon d}$$

A Point Charge Near a Dielectric Interface

Consider a point charge q located in medium 1 h meters away from a flat boundary of medium 2, as illustrated in Figure 11.15. Media 1 and 2 are characterized by permittivities ϵ_1 and ϵ_2, respectively.

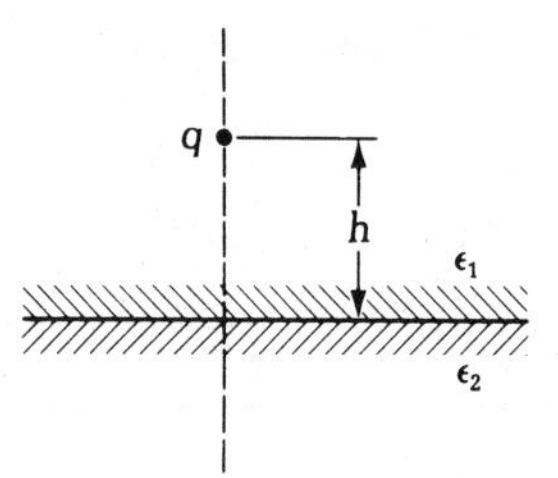

Figure 11.15 A point charge near an interface of two dielectric media.

In medium 1, the electrostatic potential is due to the point charge q and its image, which is located at the mirror-image point of the charge q. The potential in medium 1 may be found as if there were two charges q and q' located in a homogeneous medium characterized by ϵ_1. Figure 11.16a shows this situation, and Figure 11.16b shows the situation of medium 2. In other words, the potential in medium 2 is found as if there were a point charge q'' located at the position of the real charge q and as if the whole space were homogeneous and characterized by ϵ_2. We shall find the values of q' and q'' in terms of q, ϵ_1, and ϵ_2.

Figure 11.16 **(a)** In the situation shown in Figure 11.15, it would appear to an observer in medium 1 that the space was homogeneous and that there were two charges. **(b)** To an observer in medium 2, it would appear that there was only one charge located in a homogeneous medium characterized by ϵ_2.

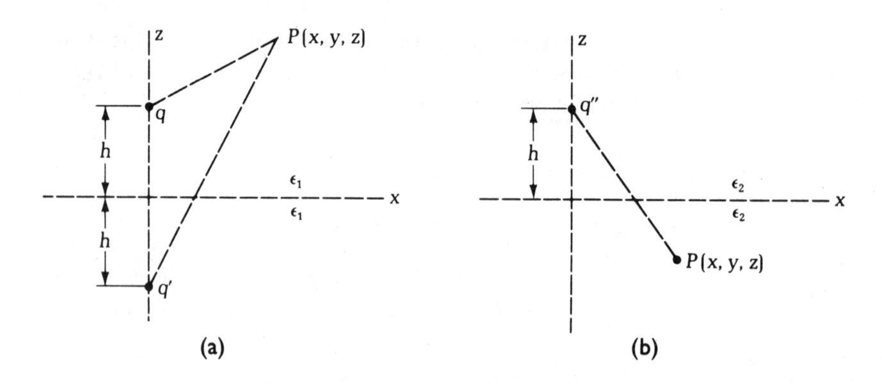

Let us denote $\Phi_1(x, y, z)$ and $\Phi_2(x, y, z)$ as the potential functions in media 1 and 2, respectively. According to the statements given above, we can set up the following equations for Φ_1 and Φ_2:

$$\Phi_1(x, y, z) = \frac{1}{4\pi\epsilon_1}\left[\frac{q}{\sqrt{x^2 + y^2 + (z - h)^2}} + \frac{q'}{\sqrt{x^2 + y^2 + (z + h)^2}}\right] \quad \textbf{(11.45a)}$$

$$\Phi_2(x, y, z) = \frac{1}{4\pi\epsilon_2}\frac{q''}{\sqrt{x^2 + y^2 + (z - h)^2}} \quad \textbf{(11.45b)}$$

We require that the following two conditions be satisfied:

$$\Phi_1(x, y, 0) = \Phi_2(x, y, 0) \tag{11.46a}$$

$$\epsilon_1 \frac{\partial \Phi_1}{\partial z}\bigg|_{z=0} = \epsilon_2 \frac{\partial \Phi_2}{\partial z}\bigg|_{z=0} \tag{11.46b}$$

The condition (11.46a) simply requires that the electrostatic potential be continuous across the boundary, and the condition (11.46b) requires the continuity of the normal component of the D field.

Applying (11.46) to (11.45), we obtain

$$\frac{1}{4\pi\epsilon_1}\left(\frac{q}{\sqrt{x^2 + y^2 + h^2}} + \frac{q'}{\sqrt{x^2 + y^2 + h^2}}\right) = \frac{q''}{4\pi\epsilon_2\sqrt{x^2 + y^2 + h^2}}$$

$$\frac{qh}{(x^2 + y^2 + h^2)^{3/2}} - \frac{q'h}{(x^2 + y^2 + h^2)^{3/2}} = \frac{q''h}{(x^2 + y^2 + h^2)^{3/2}}$$

The above equations may be simplified to yield the following:

$$q + q' = q''\left(\frac{\epsilon_1}{\epsilon_2}\right)$$

$$q - q' = q''$$

This set of equations can be solved for q' and q'':

$$q' = q\left(\frac{\epsilon_1 - \epsilon_2}{\epsilon_1 + \epsilon_2}\right) \quad \textbf{(image charge viewed from medium 1)} \tag{11.47a}$$

$$q'' = q\left(\frac{2\epsilon_2}{\epsilon_1 + \epsilon_2}\right) \quad \textbf{(image charge viewed from medium 2)} \tag{11.47b}$$

Interestingly, in the limit that ϵ_2 approaches to infinity, $q' = -q'$ and $q'' = 2q$. This result is exactly the case of a point charge near a perfectly conducting plate. According to (11.45b), with $q'' = 2q$ and $\epsilon_2 \to \infty$, the potential in the perfect conductor is equal to zero, as it should be.

Example 11.11 *Once the image charge is found, the potential function and the electric field can be computed directly.*

For the case shown in Figure 11.15, sketch the electric-field lines. Assume $\epsilon_2 = 2\epsilon_1$.

Solution Using (11.47), we obtain

$$q' = \frac{-q}{3}$$

$$q'' = \frac{4q}{3}$$

Figure 11.17 shows the electric-field lines. Note that in medium 2 the field lines are straight radial lines that tend to converge at the point $(0, 0, h)$.

Figure 11.17 Electric-field lines from a point charge near a dielectric interface.

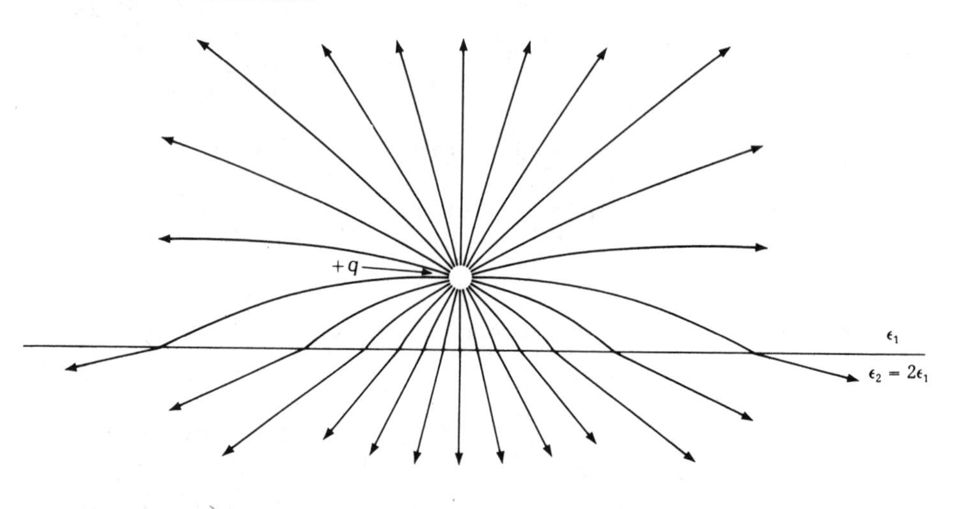

Electric Fields Due to Power Lines

Humans live in an environment full of electromagnetic fields. Some fields are generated by natural sources like the sun. Some are artificial, like radio and television broadcasts. We have presented the IEEE safety limits for human exposure to high-frequency electromagnetic fields (Figure 3.5). Notice that the safety guideline for EM fields below 3 kHz has not been established.

Power lines and electric appliances generate low frequency (60 Hz in the United States) electromagnetic fields. Are these fields safe? The question about the safety of low-frequency EM fields is still very controversial and experts' opinions are varied. Those who argue that EM fields generated by power lines pose a serious health hazard quote statistical evidence.* Some studies have found sta-

* P. Brodeur, *The Great Power-Line Cover-Up*, Boston: Little, Brown, & Co., 1994.

tistical links between routine exposure to strong electromagnetic fields and the likelihood of developing cancer. However, the link is considered "weak" because the absolute risk is small. For example,* a power substation operator's leukemia risk may be 2 in 10,000 instead of the average risk of 1 in 10,000.

Some scientists hold the opinion that electromagnetic fields generated by present power lines and appliances pose no threat to health.** Their arguments are based on the computation that power-line fields are weaker than those that exist naturally and that the energy penetrated into the human body is weaker than thermal noises.

Here we will compute the electric-field strength generated by a typical power transmission line. The line consists of three parallel wires located at R_0 meters above the ground. Each wire consists of a bundle of conductors, but for our purpose we model the bundle of conductors as a wire having an effective radius a. The wires are spaced s meters apart, as shown in Figure 11.18. The wires carry line charges, resulting in voltages equal to V_1, V_2, and V_3, respectively. We further assume that these voltages are three-phase voltages—that is, $V_1 = V_0 \exp(-j2\pi/3)$, $V_2 = V_0$, and $V_3 = V_0 \exp(+j2\pi/3)$.

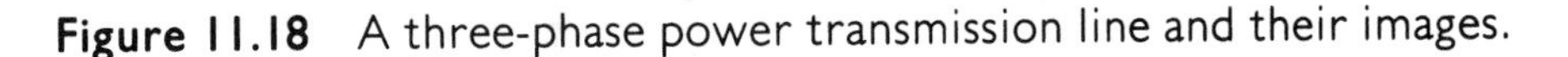
Figure 11.18 A three-phase power transmission line and their images.

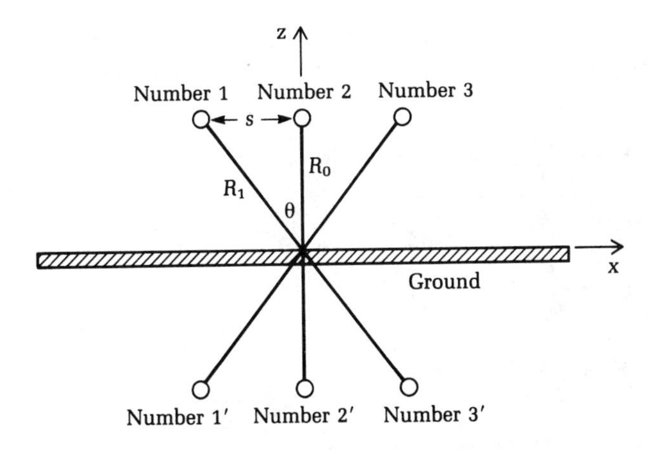

Because the voltages are given as three-phase voltages, it is implied that they vary cosinusoidally with time at a constant frequency. Rigorously speaking, the problem is a dynamic problem and the fields are not electrostatic. However, the frequency of the power line is low, and the wavelength in air is very large (5000 km at 60 Hz) as compared with the physical lengths R_0, s, and a, which are 10 m or less. We can therefore use the formulas derived for electrostatic fields to compute electric fields generated by power lines with negligible error.†

* P. Thomas, "Power struggle," *Harvard Health Letter*, vol. 18, July 1993, p. 1.

** W. R. Bennett, Jr., "Cancer and power lines," *Physics Today*, April 1994, p. 23.

† A more rigorous justification is given in Chapter 15.

Referring to Figure 11.18, the ground is modeled as a perfect conductor and, according to the image theory, three image wires with opposite voltages are found. The total electric field can then be calculated by summing the electric fields generated by individual wires and their images.

The electric field due to a wire carrying a line charge ρ_ℓ is given by (9.22):

$$E_\rho = \frac{\rho_\ell}{2\pi\epsilon\rho} \tag{11.48}$$

The voltage is obtained by integrating the $\boldsymbol{E}$ field along ρ:

$$V(\rho) = \frac{\rho_\ell}{2\pi\epsilon} \ln\left(\frac{R_0}{\rho}\right)$$

where R_0 is the distance from the center of the wire to a reference plane, which we take as the ground. The voltage at a wire of radius a is then

$$V(a) = \frac{\rho_\ell}{2\pi\epsilon} \ln\left(\frac{R_0}{a}\right) \tag{11.49}$$

Substituting (11.49) in (11.48) yields

$$E_\rho = \frac{V(a)}{\rho \ln(R_0/a)} \tag{11.50}$$

Using (11.50), we obtain the $\boldsymbol{E}$ field on the ground directly under the Number 2 wire due to all three wires and their images. The electric field due to the Number 1 wire is as follows:

$$\boldsymbol{E}_1 = \frac{V_1}{R_1 C}(\hat{\boldsymbol{x}} \sin\theta - \hat{\boldsymbol{z}} \cos\theta)$$

where $R_1 = (s^2 + R_0^2)^{1/2}$, and $C = \ln(R_0/a)$. Similarly,

$$\boldsymbol{E}_2 = \frac{V_2}{R_0 C}(-\hat{\boldsymbol{z}})$$

$$\boldsymbol{E}_3 = \frac{V_3}{R_1 C}(-\hat{\boldsymbol{x}} \sin\theta - \hat{\boldsymbol{z}} \cos\theta)$$

$$\boldsymbol{E}_{1'} = \frac{-V_1}{R_1 C}(\hat{\boldsymbol{x}} \sin\theta + \hat{\boldsymbol{z}} \cos\theta)$$

$$\boldsymbol{E}_{2'} = \frac{-V_2}{R_0 C}\hat{\boldsymbol{z}}$$

$$\boldsymbol{E}_{3'} = \frac{-V_3}{R_1 C}(-\hat{\boldsymbol{x}} \sin\theta + \hat{\boldsymbol{z}} \cos\theta)$$

Summing all six equations shows that the x component of the total electric field is zero, and the z component is given as follows:

$$E_{tz} = \frac{-2\cos\theta}{R_1 C}(V_1 + V_3) - \frac{2V_2}{R_0 C}$$

Substituting $V_1 = V_0 \exp(-j2\pi/3)$, $V_2 = V_0$, and $V_3 = V_0 \exp(+j2\pi/3)$ into the above equation yields

$$E_{tz} = \frac{-2V_0 s^2}{R_1^2 R_0 C} \tag{11.51}$$

For a high-voltage transmission line, the following data may be used:*

$$V_0 = 525 \text{ kV}$$

$$R_0 = 10.6 \text{ m}$$

$$s = 10 \text{ m}$$

and

$$a = 0.15 \text{ m}$$

Substituting these data in (11.51) gives

$$E_{tz} = -11 \text{ kV/m}$$

In comparison, the natural electric field on earth is approximately 120 V/m. It should be noted if a person is exposed to an electric field with the above-computed strength, the field inside the body is much smaller (see footnote reference to Bennetti, 1994, op. cit.).

We will compute the magnetic fields generated by power lines in Chapter 13. Now let us come back to the question of safety. Inasmuch as the controversy is unsettled, a team at the Carnegie Mellon University advocates "prudent avoidance"—that is, to avoid unnecessary exposure to any artificial EM fields. From an engineering point of view, it also means to design electric wiring in houses and in consumer products such that the resulting electromagnetic field leakage is minimized.

11.3 METHOD OF SEPARATION OF VARIABLES

In this section we shall introduce a standard mathematical method for solving the Laplace equation, subject to certain boundary conditions. First we will study a two-dimensional boundary-value problem. Then we will generalize the procedure to the three-dimensional case.

Consider the rectangular metal pipe of dimensions a and b shown in Figure 11.19. The z dimension is assumed to be infinitely long, so that the electrostatic potential and the electric field are independent of z. There are only two variables, x and y. For this reason, we call this problem a two-dimensional boundary-value

* J. J. LaForest, ed., *Transmission Line Reference Book*, Palo Alto, CA: Electric Power Institute, 1982, p. 332.

problem. As shown in Figure 11.19, three sides of the pipe are maintained at zero potential, while the fourth side (the top plate) is maintained at a constant voltage V_0. Insulating gaps on both sides of the top plate prevent short-circuiting.

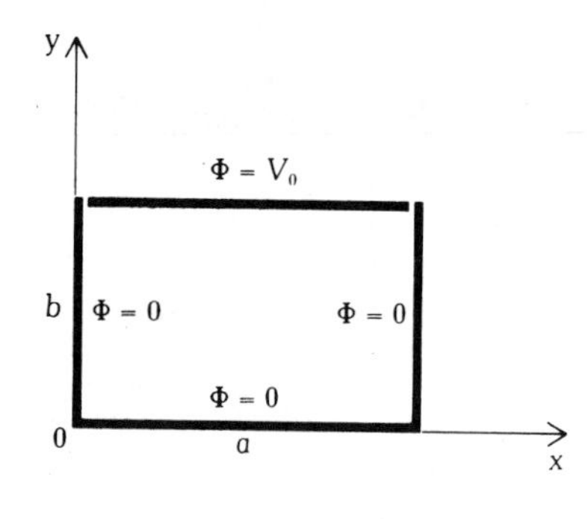

Figure 11.19 A rectangular metal pipe is grounded on three sides and is maintained at V_0 at the top plate. There is no volume charge inside the pipe. The top plate is insulated from the other walls at the two gaps. The potential inside the pipe may be found by using the method of separation of variables.

Assuming that the volume charge density inside the pipe is equal to zero, we wish to find the potential $\Phi(x, y)$ inside the pipe. Mathematically, the present problem can be stated as follows: find $\Phi(x, y)$ such that

$$\nabla^2\Phi(x, y) = 0 \quad \text{for} \quad 0 < x < a \quad \text{and} \quad 0 < y < b \tag{11.52}$$

and

$$\Phi(0, y) = 0 \quad \text{for} \quad 0 < y < b \tag{11.53a}$$

$$\Phi(a, y) = 0 \quad \text{for} \quad 0 < y < b \tag{11.53b}$$

$$\Phi(x, 0) = 0 \quad \text{for} \quad 0 < x < a \tag{11.53c}$$

$$\Phi(x, b) = V_0 \quad \text{for} \quad 0 < x < a \tag{11.53d}$$

Note that while ρ_v is zero inside the pipe, there are surface charges on its walls. However, these surface charges are not uniformly distributed on the interior walls. We will know the distribution function when we find the potential function.

Because Φ is a function of x and y, (11.52) can be rewritten as follows:

$$\frac{\partial^2\Phi}{\partial x^2} + \frac{\partial^2\Phi}{\partial y^2} = 0 \tag{11.54}$$

We now apply the procedure of separation of variables by letting

$$\boxed{\Phi(x, y) = X(x)Y(y)} \quad \textbf{(separation of variables)} \tag{11.55}$$

In other words, we assume that the potential function Φ can be expressed as a product of $X(x)$, which is a function of the variable x only and of $Y(y)$, which is a function of the variable y only. Note that not all functions of x and y may be expressed as a product of two functions of x and y, respectively. For example, the function $f(x) = x + y$ cannot be expressed in the form given by (11.55). The reader may then question the justification of making the assumption (11.55). The justification is from the uniqueness theorem discussed in Section 11.1. According to the uniqueness theorem, it does not matter what assumption one makes as long as the final solution satisfies (11.52) and (11.53). If it does satisfy (11.52) and (11.53), then we have found the unique solution.

Substituting (11.55) into (11.54), we obtain

$$X''Y + XY'' = 0$$

where X'' and Y'' are the second derivatives of $X(x)$ and $Y(y)$, respectively. Rearranging the above equation yields the following new form:

$$\frac{X''}{X} = -\frac{Y''}{Y} \tag{11.56}$$

Note that X is a function of x only, as is X''. Therefore, the left-hand side of (11.56) is either a function of x only or is a constant. Using a similar argument, we conclude that the right-hand side of (11.56) is either a function of y only or is a constant. Because (11.56) must hold for all x and y is the region $0 < x < a$ and $0 < y < b$, it can only hold if both sides are equal to a constant—that is,

$$\frac{X''}{X} = C \tag{11.57a}$$

$$\frac{Y''}{Y} = -C \tag{11.57b}$$

Note that the original second-order partial-differential equation (11.54) has now been transformed into two second-order ordinary differential equations (11.57).

The constant C that appears in (11.57) may be positive, negative, or simply zero. Let us consider these three possibilities in detail.

Case A: $C > 0$

If we assume that C is positive, then C may be expressed as follows: $C = k^2$. Consequently, $X(x)$ is the combination of e^{kx} and e^{-kx} or $\sinh(kx)$ and $\cosh(kx)$, and $Y(y)$ is the combination of $\sin(ky)$ and $\cos(ky)$.

Case B: $C < 0$

In this case, $C = -k^2$ and $X(x)$ is the combination of $\sin(kx)$ and $\cos(kx)$, and $Y(y)$ is the combination of $\sinh(ky)$ and $\cosh(ky)$.

Case C: $C = 0$

In this case, $X(x) = A_1 x + B_1$, and $Y(y) = A_2 y + B_2$.

Which case shall we choose? We could try them out one at a time until we satisfy all four boundary conditions given in (11.53). However, a better alternative is to examine the boundary condition for clues. From (11.53a) and (11.53b), we notice that Φ must vanish at $x = 0$ and again at $x = a$. Therefore, we know that $X(x)$ must have at least two zeros in the range $0 \leq x \leq a$. We then go back to examine the three cases listed above. For Case A, $X(x)$ is expressed as $\sinh(kx)$, which has only one zero and $\cosh(kx)$, which has none. Similarly, for Case C, $X(x)$ can have at most one zero. For Case B, however, $X(x)$ is expressed as $\sin(kx)$ and $\cos(kx)$, both of which have an infinite number of zeros. The choice is now obvious. To satisfy the boundary conditions, we must choose Case B.

Consequently, we solve (11.57a) and (11.57b) to yield the following results:

$$X(x) = A_1 \sin(kx) + B_1 \cos(kx) \tag{11.58a}$$

$$Y(y) = A_2 \sinh(ky) + B_2 \cosh(ky) \tag{11.58b}$$

$$\Phi(x, y) = [A_1 \sin(kx) + B_1 \cos(kx)][A_2 \sinh(ky) + B_2 \cosh(ky)] \tag{11.58c}$$

We now apply the boundary condition (11.53a) and require that

$$\Phi(0, y) = 0 = B_1 Y(y)$$

The above condition may be satisfied by letting $B_1 = 0$. Similarly, from (11.53c), we obtain $B_2 = 0$. Now the potential function takes the following form:

$$\Phi(x, y) = A \sin(kx) \sinh(ky) \tag{11.59}$$

where A and k are arbitrary constants. To satisfy (11.53b), we substitute $x = a$ into (11.59) and obtain the following:

$$0 = A \sin(ka) \sinh(ky) \qquad \text{for} \quad 0 < y < b$$

This result is possible only if $ka = m\pi$, where m is an integer. Thus, the potential function takes the following form:

$$\Phi(x, y) = A \sin\left(\frac{m\pi x}{a}\right) \sinh\left(\frac{m\pi y}{a}\right) \tag{11.60}$$

where $m = 1, 2, 3, \ldots$. Note that (11.60) satisfies the Laplace equation (11.54) and the first three of the four boundary conditions given in (11.53). If we have a series consisting of terms corresponding to different values of m, then the series also will satisfy the Laplace equation (11.54). In other words, let

$$\Phi(x, y) = \sum_{m=1}^{\infty} A_m \sin\left(\frac{m\pi x}{a}\right) \sinh\left(\frac{m\pi y}{a}\right) \tag{11.61}$$

The reader should verify that (11.61) also satisfies the first three of the four boundary conditions given in (11.53). We now impose the fourth boundary condition (11.53d), which yields

$$V_0 = \sum_{m=1}^{\infty} A_m \sin\left(\frac{m\pi x}{a}\right) \sinh\left(\frac{m\pi b}{a}\right) \qquad \text{for} \quad 0 < x < a \tag{11.62}$$

We recognize that (11.62) is simply the Fourier-series expansion of the constant V_0 in the interval $0 < x < a$. Using the standard technique of calculating the

Fourier coefficients, we can find the values of A_m in (11.62) for all m. The procedure is shown below.

Multiplying both sides of (11.62) by $\sin(n\pi x/a)$ and then integrating from 0 to a, we obtain

$$\int_0^a V_0 \sin\left(\frac{n\pi x}{a}\right) dx = A_n \sinh\left(\frac{n\pi b}{a}\right)\left(\frac{a}{2}\right) \tag{11.63}$$

because of the orthogonality property of the functions $\sin(n\pi x/a)$ and $\sin(m\pi x/a)$ in the interval $0 < x < a$:

$$\int_0^a \sin\left(\frac{n\pi x}{a}\right) \sin\left(\frac{m\pi x}{a}\right) dx = \begin{cases} 0 & \text{if } m \neq n \\ a/2 & \text{if } m = n \end{cases}$$

The left-hand side of (11.63) may be integrated analytically to yield the following result:

$$A_n = \frac{2V_0[1 - \cos(n\pi)]}{n\pi \sinh\left(\dfrac{n\pi b}{a}\right)}$$

which is zero when n is an even integer. The final solution of Φ is given below:

$$\Phi(x, y) = \frac{4V_0}{\pi} \sum_{n=\text{odd}}^{\infty} \frac{\sin\left(\dfrac{n\pi x}{a}\right) \sinh\left(\dfrac{n\pi y}{a}\right)}{n \sinh\left(\dfrac{n\pi b}{a}\right)} \tag{11.64}$$

where "n = odd" means $n = 1, 3, 5, \ldots$. This solution is the unique solution to the given problem because it satisfies the Laplace equation and all the boundary conditions. At $y = b$, in particular, $\Phi(x, b) = V_0$ for $0 \leq x \leq a$. Equation (11.64) yields

$$\sum_{n\text{-odd}}^{\infty} \frac{4}{n\pi} \sin\frac{n\pi x}{a} = 1$$

We can see that the series does add up to be equal to unity for $0 \leq x \leq a$. Figure 11.20 illustrates this result with the sum of the first three terms.

Figure 11.20 $\Phi(x, b)$ given by (11.64) and its Fourier components.

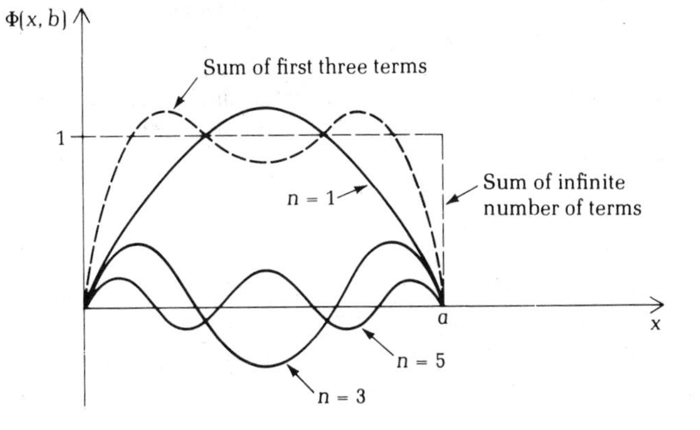

Although we have formally solved the boundary-value problem with the solution in the form of an infinite series, a practical question still needs to be answered. Is the infinite series (11.64) converging fast enough so that in practice Φ may be calculated numerically in a reasonable amount of computer time? To answer this question, we must examine how terms in the series behave as the index n becomes large. Notice that

$$\sinh(nu) = \frac{e^{nu} - e^{-nu}}{2}$$

For large nu, the second term is negligible, and we have the following asymptotic behavior for $\sinh(nu)$:

$$\sinh(nu) \rightarrow \frac{e^{nu}}{2}$$

Therefore, the nth term in (11.64) may be approximated as follows:

$$\frac{1}{n}\sin\left(\frac{n\pi x}{a}\right)\exp\left[\frac{-n\pi(b-y)}{a}\right]$$

This result is exponentially small except near $y = b$. But at $y = b$ the potential is known to be V_0. Consequently, the series (11.64) converges very rapidly at any point in the interior of the pipe.

Figure 11.21a shows the equipotential lines inside a rectangular pipe, with $V_0 = 100$ volts, $a = 20$ units, and $b = 15$ units. These lines are obtained first by computing the potential given by (11.64) at various points (e.g., 20 by 15 points) in the pipe, and then using a graphics software to draw the contour lines. The 3-D potential distribution is shown in Figure 11.21b. The infinite series given in (11.64) is truncated at $n = 31$.

Example 11.12 *The separation of variables technique used to solve the two-dimensional electrostatic problem can be generalized to the three-dimensional case.*

Find the electrostatic potential inside the rectangular metal box shown in Figure 11.22 (p. 414). All five sides of the box are maintained at zero potential, while the sixth side (the top plate) is maintained at V_0 volts. The volume charge density inside the box is equal to zero.

Solution We shall follow the procedure of separation of variables as outlined in this section and generalize it to the present case, which is a three-dimensional problem. The mathematical statement of the problem is as follows:

$$\frac{\partial^2\Phi}{\partial x^2} + \frac{\partial^2\Phi}{\partial y^2} + \frac{\partial^2\Phi}{\partial z^2} = 0 \quad \text{for} \begin{cases} 0 < x < a \\ 0 < y < b \\ 0 < z < c \end{cases} \tag{11.65}$$

$$\Phi(0, y, z) = 0 \quad \text{for} \quad 0 < y < b \tag{11.66a}$$

$$\Phi(a, y, z) = 0 \qquad\qquad 0 < z < c \tag{11.66b}$$

Figure 11.21 **(a)** The equipotential lines inside the metal pipe shown in Figure 11.19, with $a = 20$ units, $b = 15$ units, and $V_0 =$ 100 volts, using Equation (11.64). **(b)** The 3-D plot of the potential distribution.

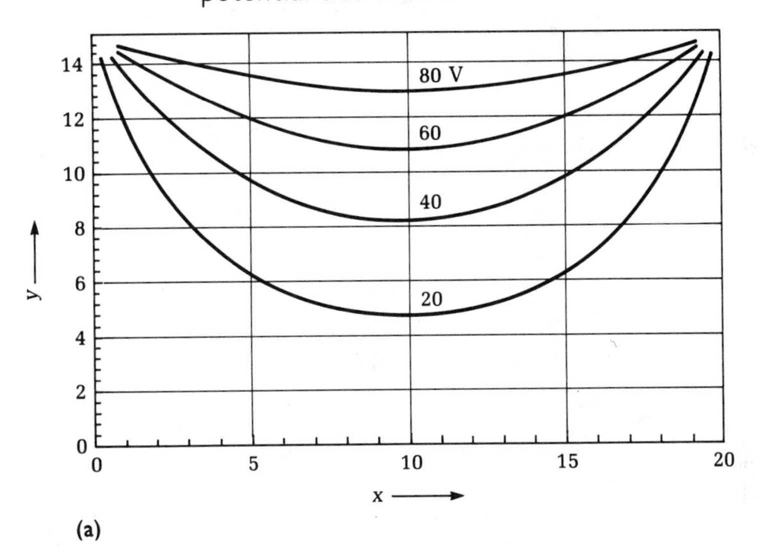

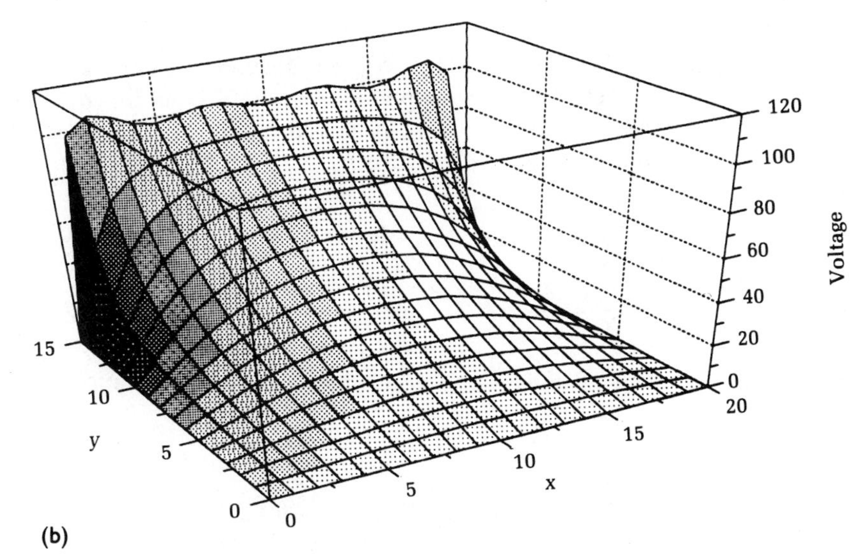

Figure 11.22 The top plate of a rectangular metal box is maintained at V_0 volts while the rest of the box is grounded. There is no charge inside the box. The electrostatic potential may be found by using the method of separation of variables.

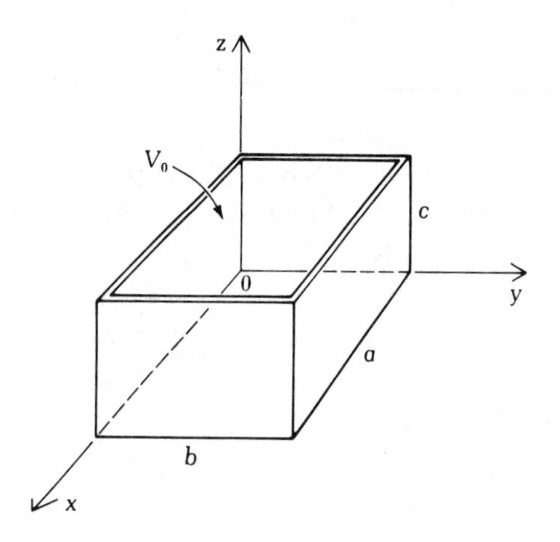

$$\left.\begin{aligned}\Phi(x,0,z)&=0\\ \Phi(x,b,z)&=0\end{aligned}\right\} \text{ for } \begin{aligned}&0<x<a\\ &0<z<c\end{aligned} \qquad \begin{aligned}&(11.66c)\\ &(11.66d)\end{aligned}$$

$$\left.\begin{aligned}\Phi(x,y,0)&=0\\ \Phi(x,y,c)&=V_0\end{aligned}\right\} \text{ for } \begin{aligned}&0<x<a\\ &0<y<b\end{aligned} \qquad \begin{aligned}&(11.66e)\\ &(11.66f)\end{aligned}$$

Let $\Phi(x, y, z) = X(x)Y(y)Z(z)$, and substitute this expression into (11.65). We obtain

$$\frac{X''}{X} = -\left(\frac{Y''}{Y} + \frac{Z''}{Z}\right) \tag{11.67}$$

The left-hand side of (11.67) is a function of x only, whereas the right-hand side is a function of y and z. Nonetheless, both sides are equal for all x, y, and z inside the box. This equality is possible only if both sides are equal to the same constant—that is,

$$\frac{X''}{X} = C_1 \tag{11.68a}$$

$$\frac{Y''}{Y} + \frac{Z''}{Z} = -C_1 \tag{11.68b}$$

Rearranging (11.68b) we obtain the following equation:

$$\frac{Y''}{Y} = -C_1 - \frac{Z''}{Z} \tag{11.68c}$$

The left-hand side of (11.68c) is a function of y only, whereas the other side is a function of z only. Nonetheless, both sides are equal for all y and z inside the box.

This equality is possible only if both sides are equal to the same constant—that is,

$$\frac{Y''}{Y} = C_2 \tag{11.69a}$$

$$\frac{Z''}{Z} = -(C_1 + C_2) \tag{11.69b}$$

We have now transformed the original second-order partial-differential equation (11.65) into a set of three second-order ordinary differential equations (11.68a), (11.69a), and (11.69b). The arbitrary constants C_1 and C_2 may be positive, negative, or simply zero. Therefore, we have nine possibilities. However, a closer examination of the boundary conditions (11.66a–11.66f) gives us some clue. Notice that Φ must vanish at $x = 0$ and at $x = a$. Therefore, $X(x)$ must be a sinusoidal function. Consequently, C_1 must be a negative constant. For a similar reason, we conclude that C_2 is also negative, so that $Y(y)$ is a sinusoidal function of y. With these conclusions, we let $C_1 = -k_1^2$ and $C_2 = -k_2^2$, and we obtain the following results:

$$X(x) = A_1 \sin(k_1 x) + B_1 \cos(k_1 x) \tag{11.70a}$$

$$Y(y) = A_2 \sin(k_2 y) + B_2 \cos(k_2 y) \tag{11.70b}$$

$$Z(z) = A_3 \sinh(\sqrt{k_1^2 + k_2^2}\, z) + B_3 \cosh(\sqrt{k_1^2 + k_2^2}\, z) \tag{11.70c}$$

$$\Phi(x, y, z) = X(x)Y(y)Z(z) \tag{11.70d}$$

Applying the boundary condition (11.66a) to (11.70a) or (11.70d), we find the following result:

$$B_1 = 0$$

Similarly, applying (11.66c) and (11.66e) to (11.70b), (11.70c), and (11.70d) yields the following results:

$$B_2 = 0$$

$$B_3 = 0$$

Furthermore, to satisfy (11.66b) and (11.66d), we must impose the following conditions on k_1 and k_2:

$$k_1 = \frac{m\pi}{a}$$

$$k_2 = \frac{n\pi}{b}$$

where m and n are arbitrary integers. At this point, the potential function Φ is given in the following form:

$$\Phi(x, y, z) = A \sin\left(\frac{m\pi x}{a}\right) \sin\left(\frac{n\pi y}{b}\right) \sinh\left[\sqrt{\left(\frac{m\pi}{a}\right)^2 + \left(\frac{n\pi}{b}\right)^2}\, z\right] \tag{11.71}$$

Note that (11.71) satisfies the Laplace equation (11.65) and the first five boundary conditions given in (11.66). By inspection, we know that (11.71) in its present form cannot satisfy the sixth boundary condition (11.66f). Only a series of terms similar to (11.71) can satisfy the boundary condition. For this reason, we try the following expression for $\Phi(x, y, z)$:

$$\Phi(x, y, z) = \sum_{m=1}^{\infty} \sum_{n=1}^{\infty} A_{mn} \sin\left(\frac{m\pi x}{a}\right) \sin\left(\frac{n\pi y}{b}\right) \sinh\left[\sqrt{\left(\frac{m}{a}\right)^2 + \left(\frac{n}{b}\right)^2} \pi z\right] \tag{11.72}$$

Note that (11.72) still satisfies the Laplace equation (11.65) and the first five boundary conditions given in (11.66). The sixth boundary condition (11.66f) may be satisfied if the following equality holds:

$$V_0 = \sum_{m=1}^{\infty} \sum_{n=1}^{\infty} A_{mn} \sin\left(\frac{m\pi x}{a}\right) \sin\left(\frac{n\pi y}{b}\right) \sinh\left[\sqrt{\left(\frac{m}{a}\right)^2 + \left(\frac{n}{b}\right)^2} \pi c\right] \tag{11.73}$$

for $0 < x < a$ and $0 < y < b$. We recognize that (11.73) is the two-dimensional Fourier-series expansion of the constant V_0 in the rectangular area defined by $0 < x < a$ and $0 < y < b$. We may use the standard procedure for calculating the Fourier coefficients to obtain A_{mn} that appears in (11.73). To this end, we multiply both sides of (11.73) by the function $\sin(p\pi x/a)\sin(q\pi y/b)$, where p and q are integers. We then integrate both sides over the area $0 < x < a$ and $0 < y < b$. Because of the orthogonality of $\sin(p\pi x/a)$ versus $\sin(m\pi x/a)$ and of $\sin(q\pi y/b)$ versus $\sin(n\pi y/b)$, we obtain the following result:

$$\int_0^a dx \int_0^b dy V_0 \sin\left(\frac{p\pi x}{a}\right) \sin\left(\frac{q\pi y}{b}\right) = A_{pq} \frac{ab}{4} \sinh\left[\sqrt{\left(\frac{p}{a}\right)^2 + \left(\frac{q}{b}\right)^2} \pi c\right] \tag{11.74}$$

The integration on the left-hand side of the above equation may be carried out analytically with the following result:

$$V_0 \frac{a[1 - \cos(p\pi)]b[1 - \cos(q\pi)]}{p\pi q\pi} = A_{pq} \frac{ab}{4} \sinh\left[\sqrt{\left(\frac{p}{a}\right)^2 + \left(\frac{q}{b}\right)^2} \pi c\right]$$

Solving from A_{pq}, we obtain:

$$A_{mn} = \begin{cases} \dfrac{16V_0}{mn\pi^2 \sinh[\sqrt{(m/a)^2 + (n/b)^2}\pi c]} & \text{for } m = \text{odd},\ n = \text{odd} \\ 0 & \text{otherwise} \end{cases}$$

Here we have replaced the indices p and q with m and n, respectively. The final expression of $\Phi(x, y, z)$ is given as follows:

$$\Phi(x, y, z) = \frac{16V_0}{\pi^2} \sum_{m=\text{odd}}^{\infty} \sum_{n=\text{odd}}^{\infty} \frac{\sin\left(\frac{m\pi x}{a}\right) \sin\left(\frac{n\pi y}{b}\right) \sinh[\sqrt{(m/a)^2 + (n/b)^2}\pi z]}{mn \sinh[\sqrt{(m/a)^2 + (n/b)^2}\pi c]} \tag{11.75}$$

The series (11.75) converges very rapidly except at points that are close to the top plate. Using the asymptotic form of sinh (u) for large u, we have, for large values of m and n, the mnth term as:

$$\frac{1}{mn}\exp\left[-\sqrt{\left(\frac{m}{a}\right)^2+\left(\frac{n}{b}\right)^2}\,\pi(c-z)\right]$$

The above term shows how fast the mnth term in (11.75) decreases as m and n increase. It does not include the sine terms because they just oscillate, with maximum values equal to unity.

SUMMARY

1. Electrostatic potential can be obtained by solving the Poisson equation (when charge density ρ_v is not zero) or the Laplace equation (when ρ_v is zero). The solution must also satisfy the boundary conditions.
2. The uniqueness theorem assures us that there is only one solution that can satisfy the Poisson (or Laplace) equation and the boundary conditions.
3. For a point charge q near a perfect conducting plane, its image is a charge $-q$ located at the same distance on the other side of the plane.
4. For a point charge near a corner of conducting walls, there are $2n-1$ images if the angle of the corner is equal to $180°/n$ and n is an integer. If n is not an integer, the electrostatic potential cannot be obtained by simple image method because the image will be a complicated distribution of charges.
5. For a line charge ρ_ℓ near and parallel to a conducting cylinder, the image is a line charge $-\rho_\ell$ inside the cylinder. Its location is derived (see Equation (11.36)].
6. For a point charge q near a conducting sphere, the image is a point charge in the sphere. The magnitude and location of the image are given [see Equation (11.43)].
7. For a point charge q located in medium 1 near a planar boundary between media 1 and 2, and for an observer in medium 1, the image is q' located at the same distance on the other side. For an observer in medium 2, it would seem to have only one charge of q'' located at the position where the actual charge is. Values of q' and q'' are derived [see Equation (11.47)].
8. The potential function in the space where the actual charge is located is the sum of contributions from the actual charge and its image or images as if all charges were located in a homogeneous space.
9. Electrostatic potential can be obtained by solving the Laplace equation using the method of separation of variables when the potential is a function of two or three coordinate variables. The solution is usually in the form of an infinite series.

Problems

11.1 Consider the three boundary-value problems shown in Figure P11.1. The solution of Case I is Φ_1, and the solution of Case II is Φ_2. In Case III, the charges q_1 and q_2 are the same charges that appear in Cases I and II, and they appear in exactly corresponding positions. Express Φ_3 in terms of Φ_1 and Φ_2.

Figure P11.1

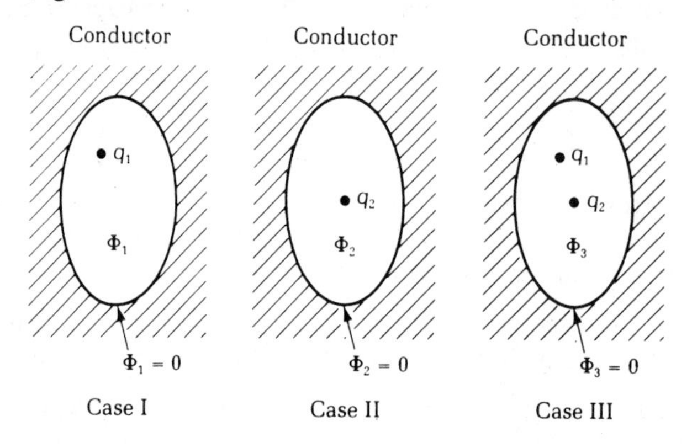

11.2 Consider the three boundary-value problems shown in Figure P11.2. The solution of Case I is Φ_1, and the solution of Case II is Φ_2. In Case III, the charges q_1 and q_2 are the same charges that appear in Cases I and II, and they appear in exactly corresponding positions. Note the differences in the boundary conditions for the three cases. Can Φ_3 be expressed in terms of Φ_1 and Φ_2? If so, obtain the expression. If not, explain why not.

Figure P11.2

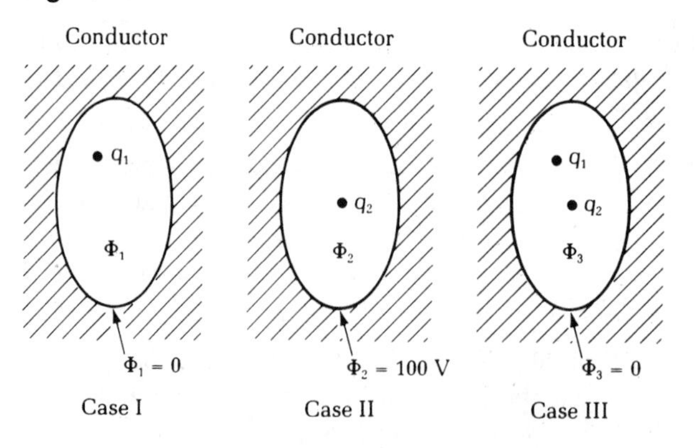

11.3 The radius of the inner conductor of a coaxial line is a and that of the outer conductor is b. The potential of the inner conductor is V_a and that of the outer conductor is V_b. There is no volume charge density between b and a. Start from the Laplace equation to obtain the potential in the coaxial line.

11.4 Two concentric conducting spheres have radii a and b, respectively ($b > a$). The outer sphere is at V_b, and the inner sphere is maintained at V_a volts. There is no space charge between the conductors. Start from the Laplace equation to obtain the potential $\Phi(r)$ for $b > r > a$.

11.5 In Figure P11.5 a conducting cone is at a potential V_0, and a small gap separates its vertex from a conducting plane. The axis of the cone is perpendicular to the conducting plane, which is maintained at zero potential. The angle of the cone is θ_1. Because of the symmetry of this problem and the fact that the boundary conditions on the potential Φ involve θ only, Φ is independent of r and ϕ when spherical coordinates are used. Find the potential $\Phi(\theta)$ in the region $\theta_1 \leq \theta \leq 90°$. Hint: $\int (1/\sin\theta)\,d\theta = \ln(\tan\theta/2)$. Find the surface-charge density on the cone.

Figure P11.5

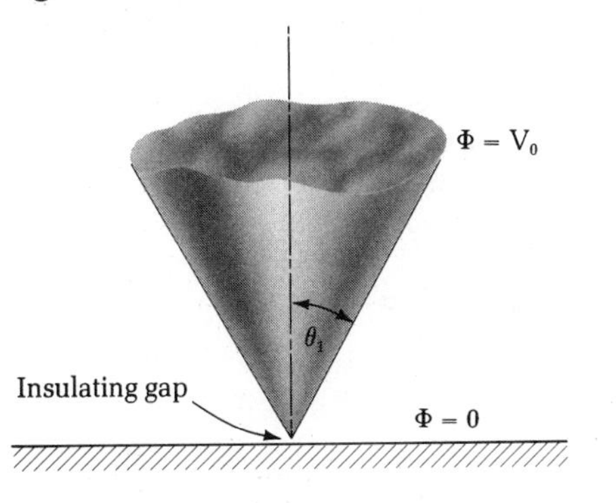

11.6 The upper plate of the parallel-plate capacitor discussed in Example 11.1 is maintained at 100 V, and the lower plate is at 80 V. All other conditions remain unchanged. Find Φ.

11.7 Model a dc vacuum-tube rectifier as two parallel plates with a space charge in between, as shown in Figure 11.3. Let the separation be 1 cm. Find the voltage needed to produce 1 A/m^2 current.

11.8 Find the surface-charge distribution on the vertical and the horizontal conducting walls for the case discussed in Example 11.8. Plot ρ_s for $z > 0$ and $x = y = 0$. Let $q = 10^{-6}$ C, and $a = b = 1$.

11.9 Find the images of a point charge near a corner of a conductor similar to the one shown in Figure 11.11 except that $\phi_0 = 45°$.

11.10 Find the electrostatic force that acts on the point charge q shown in Figure 11.10.

11.11 Calculate the capacitance per meter of a 12-inch (0.3048 m)-diameter steel pipe located 6 ft (1.83 m) above and parallel to the ground.

11.12 Example 11.10 states that the maximum electric field on the surface of the conducting cylinder is located at the point nearest the ground. Show the validity of this statement by plotting out E_ρ on the surface as a function of ϕ. Use the following data: $V_0 = 100$ V, $h = 2$ m, and $a = 1$ m.

11.13 A conductor in an electric power transmission line is charged to 500 kV. What is the minimum radius the conductor should have in order to avoid corona? Assume that the conductor is 5 meters above the ground in the configuration shown in Figure 11.13a. Take the breakdown electric field of air to be 3×10^6 V/m.

11.14 For the point charge q located d meters from the grounded conducting sphere shown in Figure 11.14, find the surface charge distribution as a function of θ.

11.15 Equation (11.42) gives the potential due to a point charge in the presence of a *grounded* conducting sphere. Equation (11.44) gives the potential due to a point charge in the presence of an *isolated* sphere carrying *no* net charge. From these results, find the potential due to a point charge q, d meters from an isolated conducting sphere carrying a net charge of q_0.

11.16 A line charge ρ_ℓ is inside a conducting tunnel of radius a, as shown in Figure P11.16. Notice that the line charge is b meters off center. Find the potential function in the tunnel. Hint: This is a complementary problem of the one shown in Figure 11.12.

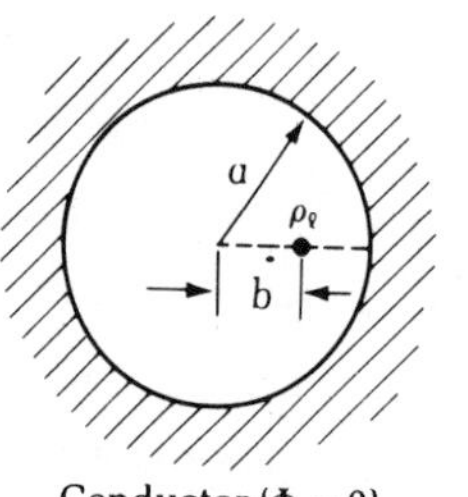

Figure P11.16

11.17 Calculate the force per meter acting on the line charge in the tunnel shown in Figure P11.16.

11.18 A point charge q is inside a spherical cavity of a conductor, as shown in Figure P11.18. The radius of the cavity is a and the cavity is filled with air.

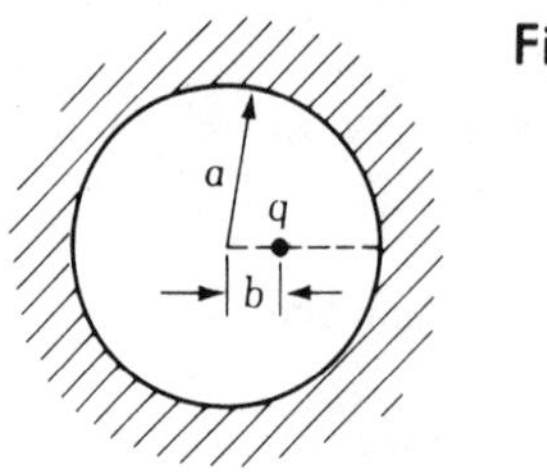

Figure P11.18

(a) Find the potential Φ in the cavity when $b = 0$.
(b) Find the surface-charge density of the cavity wall when $b = 0$.

(c) Find the potential Φ in the cavity when $b = a/2$.
(d) Find the surface-charge density of the cavity wall when $b = a/2$.

11.19 Calculate the electrostatic force acting on the point charge in the cavity shown in Figure P11.18.

11.20 Sketch the E lines due to a point charge near the interface of two dielectric media. The situation is similar to the one shown in Figure 11.17, except that $\epsilon_2 = 0.5\epsilon_1$.

11.21 A rectangular conducting trough of width a and height b is maintained at zero potential, as shown in Figure P11.21. The potential on the top plate, which covers the trough, is known to be $\Phi(x, b) = 200 \sin(2\pi x/a)$ volts. Find the potential Φ in the trough. There is no volume charge in the trough.

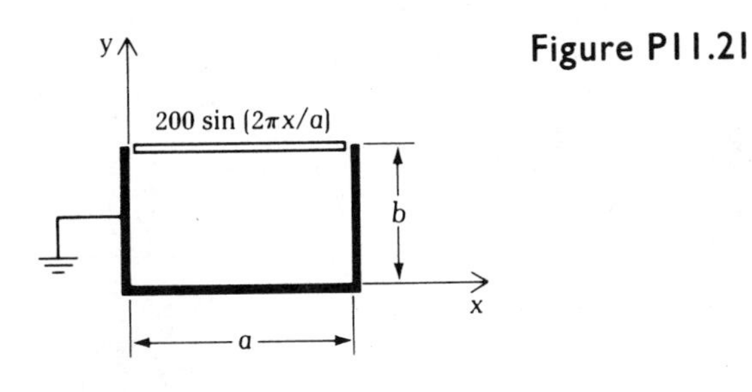

Figure P11.21

11.22 Three sides of a rectangular conducting pipe are grounded, while the fourth side is maintained at 100 V, as shown in Figure P11.22. Find the potential in the pipe. There is no volume charge in the pipe.

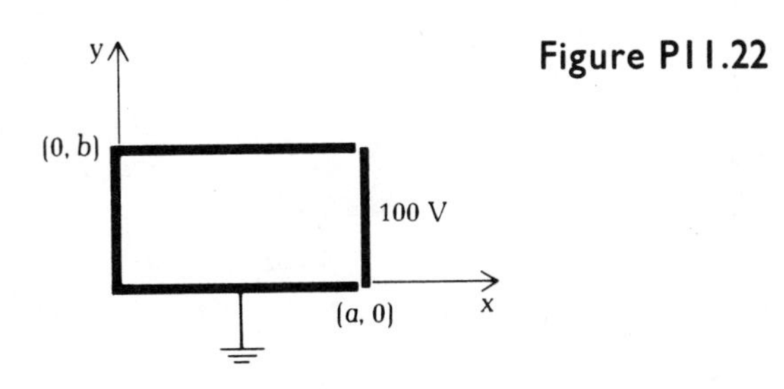

Figure P11.22

11.23 The boundary potentials of a rectangular conducting pipe are shown in Figure P11.23. Find the potential in the pipe. There is no volume charge in the pipe.

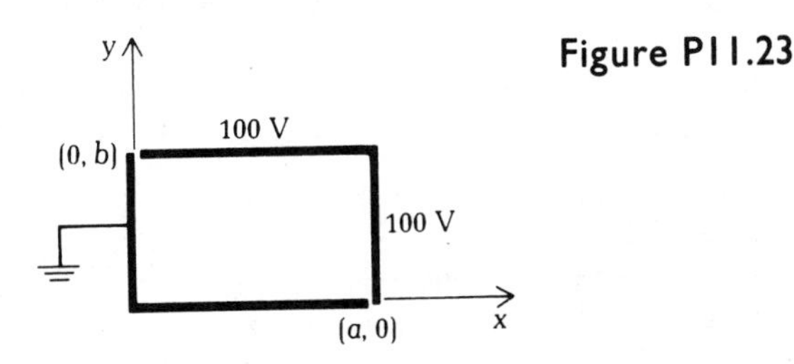

Figure P11.23

11.24 Consider the four boundary-value problems shown in Figure P11.24. One of them has the following solution:

$$\Phi(x, y) = \sum_{n=1}^{\infty} C_n \exp\left(\frac{n\pi x}{a}\right) \sin\left(\frac{n\pi y}{a}\right)$$

Identify the case that has the above solution and find the constants C_n. Write out explicitly the first two nonzero terms of $\Phi(x, y)$.

Figure P11.24

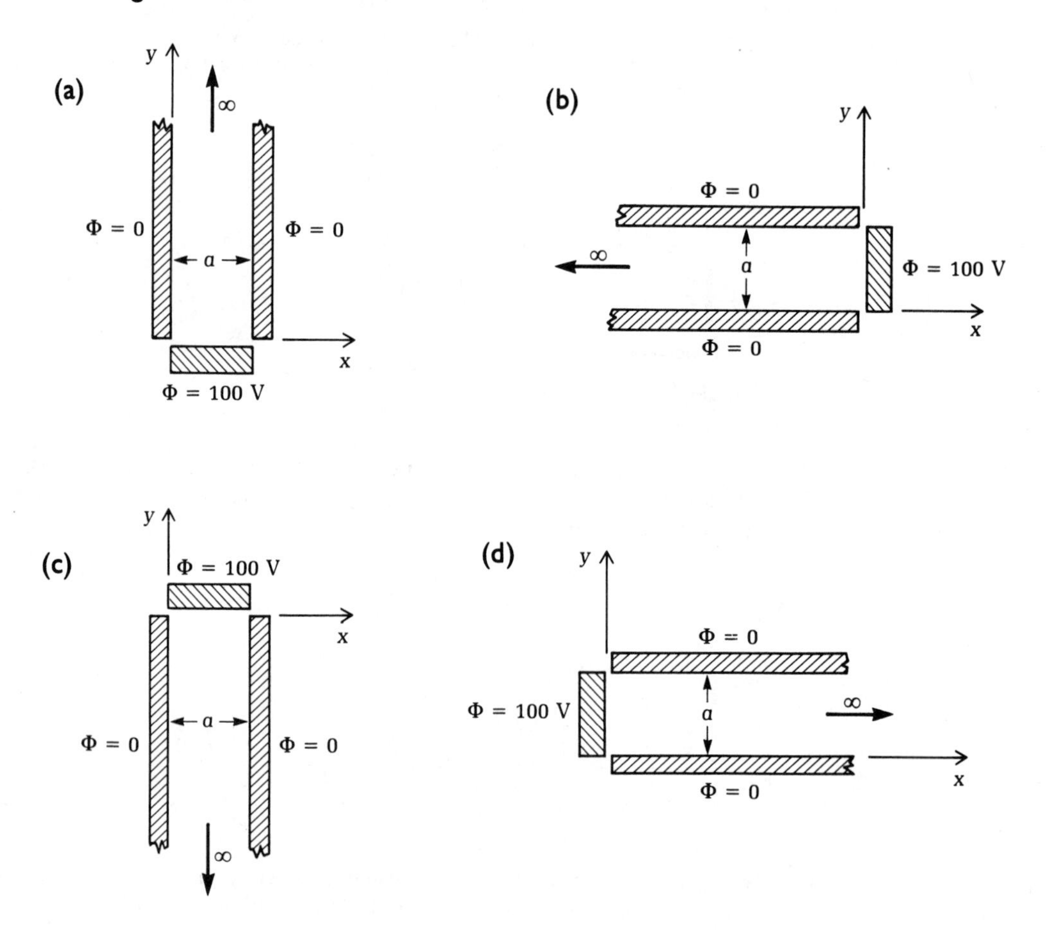

11.25 A spherical capacitor is filled with a dielectric material of ϵ_1 in half of the space and with another material of ϵ_2 in the remaining space, as shown in Figure P11.25.

(a) Find the potential function $\Phi(r)$ in the region $a < r < b$. The potential at $r = a$ is V_0 and it is zero at $r = b$. Hint: The potential satisfies the Laplace equation (11.2) and it may be assumed that it is a function of r only.

(b) Find the electric field in the region $a < r < b$.

(c) Find the $\boldsymbol{D}$ field in the region $a < r < b$. Hint: The $\boldsymbol{D}$ field in medium 1 is different from that in medium 2. Note: the boundary conditions on the tangential $\boldsymbol{E}$ and on the normal $\boldsymbol{D}$ fields are satisfied using the suggested approach.

(d) Find the total charge on the inner conductor and the capacitance of this capacitor.

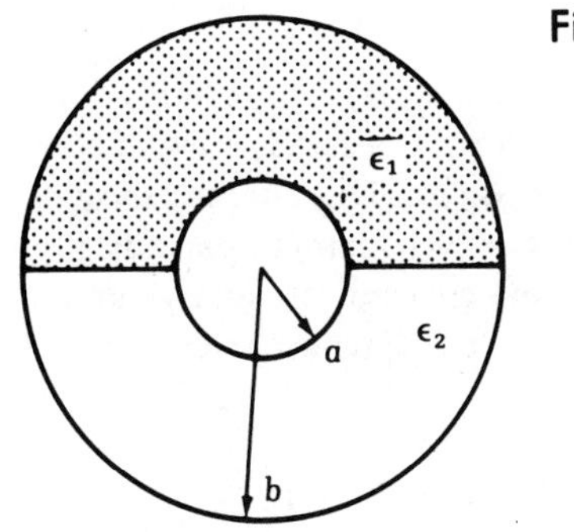

Figure P11.25

12 DIRECT CURRENTS

In Chapters 10 and 11 we studied electrostatic fields due to stationary electric charges. In this chapter we discuss the steady flow of electric charges that constitutes a **direct current**, or **dc**. This current should be distinguished from **ac**, the **alternating current**, which changes periodically with time. Some other textbooks also call direct current stationary current or steady current.

Ohm's law relates the current $\mathbf{J}$, whether dc or ac, to the electric field $\mathbf{E}$.

$$\mathbf{J} = \sigma \mathbf{E} \tag{12.1}$$

The continuity equation in the static limit is

$$\nabla \cdot \mathbf{J} = 0 \tag{12.2}$$

and the electrostatic equation is

$$\nabla \times \mathbf{E} = 0 \tag{12.3}$$

We shall study the origins of Ohm's law, explore the analogy between direct currents and electrostatic fields, and revisit the concepts of conductance and resistance learned in circuit theory.

12.1 OHM'S LAW

There are two types of electric currents: convection currents and conduction currents. Convection currents arise from the transport of charged particles, like electron beams in a CRT. Conduction currents represent the drift motion of charge carriers through a medium. For instance, in a metal, the heavier positive ions stay fixed in the lattice structure, while the free electrons, drifting from one atom to another under the influence of an applied electric field, constitute an electric current.

In a linear medium, the conduction current density $\boldsymbol{J}$ and the electric field $\boldsymbol{E}$ obey Ohm's law:

$$\boxed{\boldsymbol{J} = \sigma \boldsymbol{E}} \quad \textbf{(Ohm's law)}$$

where σ is the **conductivity** of the medium in mhos per meter (or siemens per meter). The inverse of the conductivity is called the **resistivity** $\rho_r = 1/\sigma$, which is given in ohms-meter.

Table 12.1 lists the resistivity and conductivity of some materials at room temperature. The materials listed near the top of the table, such as aluminum, copper, gold, nickel, and silver, are **good conductors**. The materials listed near the bottom of the table, such as amber, glass, petroleum oil, and quartz, are **good insulators**. Those listed in the middle, such as germanium and silicon, are neither good conductors nor good insulators. They are referred to as **semiconductors**.

Table 12.1 Resistivity and Conductivity of Materials.*

Material	$\rho_r(\Omega\text{-m})$	$\sigma(\text{mho/m})$
Aluminum	2.83×10^{-8}	3.53×10^{7}
Copper	1.69×10^{-8}	5.92×10^{7}
Gold	2.44×10^{-8}	4.10×10^{7}
Nickel	7.24×10^{-8}	1.38×10^{7}
Silver	1.62×10^{-8}	6.17×10^{7}
Germanium	0.45	2.20
Silicon	640.00	1.56×10^{-3}
Seawater (average)	0.25	4.00
Amber	5.00×10^{14}	2.00×10^{-15}
Glass	10^{10}–10^{14}	10^{-10}–10^{-14}
Petroleum oil	10^{14}	10^{-14}
Quartz	7.50×10^{17}	1.30×10^{-18}

* Data from American Institute of Physics Handbook (New York: McGraw-Hill, 1957).

As we saw from the Bohr model of the atom, the total energy of an electron is the sum of its kinetic energy and its potential energy. A material becomes conductive when the energy of some of its electrons is so high that it is not attached to any particular atom. We say that the energy of these electrons is in the material's **conduction band**. Electrons are bound to the atom when their energy is so low that conduction cannot occur. We say that the energy of these electrons is in the material's **valence band**. As shown in Figure 12.1, the energy gap between

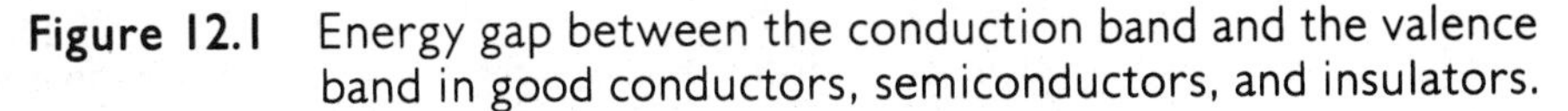

Figure 12.1 Energy gap between the conduction band and the valence band in good conductors, semiconductors, and insulators.

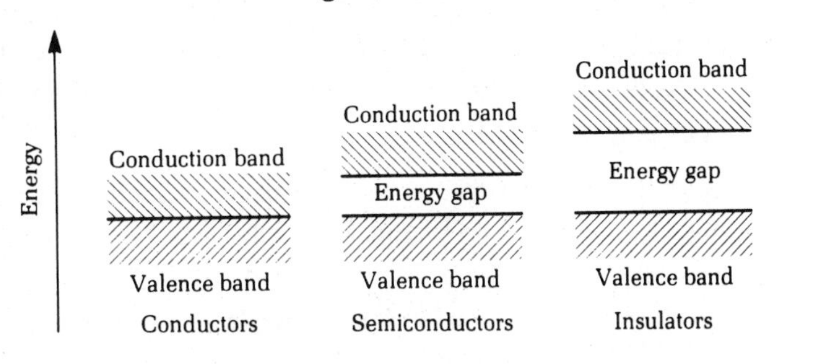

the valence band and the conduction band is narrow for good conductors and wide for good insulators, whereas the gap for semiconductors is between these extremes.

The conduction process of metallic conductors can be stated quantitatively as

$$\boldsymbol{J} = -\rho_e \mu_e \boldsymbol{E} = \sigma \boldsymbol{E}$$

where ρ_e is the free-electron charge density and μ_e is the mobility of the electron in the metal. For copper, with $\rho_e \sim -1.8 \times 10^{10}$ C/m^3 and $\mu_e = 3.2 \times 10^{-3}$ m^2/V/s, the conductivity $\sigma = -\rho_e \mu_e = 5.8 \times 10^7$ mho/m. For semiconductors, the conductivity is a function of both hole and electron concentrations and mobilities. We have

$$\sigma = -\rho_e \mu_e + \rho_h \mu_h \tag{12.4}$$

where ρ_h and μ_h are, respectively, the hole concentrations and mobilities. For silicon, $\mu_e \sim 0.12$, $\mu_h = 0.025$, and $\rho_h = -\rho_e = 0.011$ C/m^3, which yields $\sigma \sim 1.6 \times 10^{-3}$ mho/m.

Near room temperature, the resistivity ρ_r increases almost linearly with temperature. At very low temperatures, the resistivity of some metals, such as aluminum below 1.14 K, drops abruptly to zero, and the metals become superconductors. Some ceramic materials have been found to become superconductors at 90 K or higher temperatures (see Section 3.4). Although the conductivity of metallic conductors decreases with temperature, the conductivity of semiconductors increases with temperature because, as temperature increases, the mobilities decrease but the charge densities of semiconductors increase very rapidly.

Joule's Law

The conduction electrons inside a conductor move under the influence of the electric field in accordance with the Lorentz force law. Let the charge density be ρ_e. Then the force per unit volume is $\boldsymbol{f} = \rho_e \boldsymbol{E}$ newtons per cubic meter. Under this force the electrons should accelerate and increase their velocity. However, the electrons will be impeded by constant collision with the atoms of the conductor and will attain an average constant drifting velocity $\boldsymbol{v}$. The work performed moving all charges inside a small volume dV during a time interval dt is given by

$$dW_e = \rho_e \boldsymbol{E} \cdot \boldsymbol{v} \, dV \, dt$$

This work is transformed into heat. The corresponding dissipated power per unit volume is $P_d = dW_e/(dt\,dV) = \rho_e \boldsymbol{E} \cdot \boldsymbol{v}$. Because $\boldsymbol{J} = \rho_e \boldsymbol{v}$, we obtain

$$\boxed{P_d = \boldsymbol{J} \cdot \boldsymbol{E}} \quad \textbf{(Joule's law; ohmic loss per volume)} \tag{12.5}$$

This equation is known as the point form of **Joule's law**. The integral form of Joule's law is obtained by integrating P_d over the volume V of a current-carrying conductor.

Kirchhoff's Current Law

Integrating (12.2) over an arbitrary closed surface S and using the divergence theorem, we obtain

$$\oiint_S \boldsymbol{J} \cdot \hat{\boldsymbol{n}}\, ds = 0 \tag{12.6}$$

which states that the total current flowing from a volume or a junction is always equal to zero.

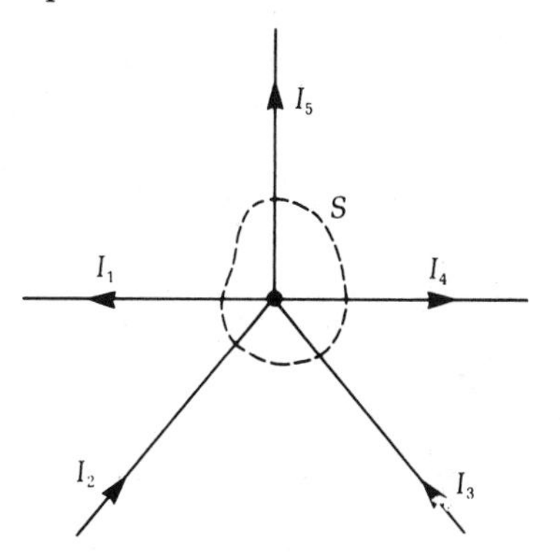

Figure 12.2 Kirchhoff's current law (KCL) is derived from the equations for the direct currents. KCL states that the sum of all currents flowing out of a junction is equal to zero.

Applying (12.6) to a node in the electric circuit shown in Figure 12.2, we find

$$I_1 - I_2 - I_3 + I_4 + I_5 = 0$$

which can be written in the following form:

$$\sum_n I_n = 0$$

This formula represents one of the basic postulates in circuit theory—namely, Kirchhoff's current law (KCL).

Conductance

Consider a conductive cylinder of length ℓ and cross-sectional area A, as shown in Figure 12.3. Let an applied voltage V generate a uniform electric field along the axial direction. The magnitude of the electric field is

$$E = \frac{V}{\ell}$$

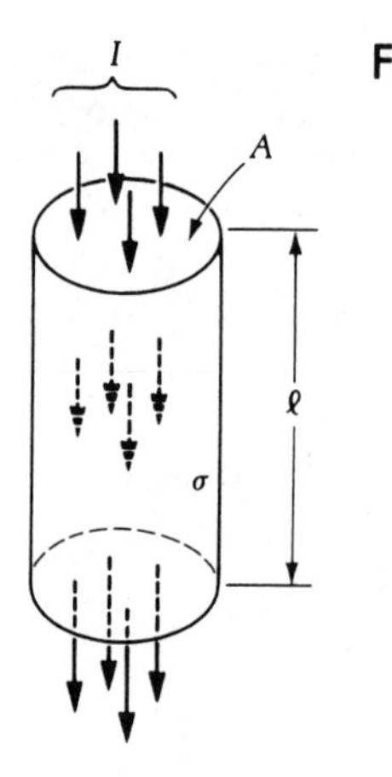

Figure 12.3 A conducting cylinder of cross-sectional area A and length ℓ has a conductance equal to $\sigma A/\ell$ mhos, where σ is the conductivity of the material of the cylinder.

The total current flowing from one end to the other is

$$I = AJ = \sigma AE$$

The last equality follows from Ohm's law (12.1).

The conductance G is defined as the ratio of the current to the voltage—that is,

$$G = \frac{I}{V} \tag{12.7}$$

For the cylinder shown in Figure 12.3,

$$G = \frac{I}{V} = \frac{\sigma A}{\ell}$$

The unit for G is mho (or siemens).

Resistivity and Resistance

The resistivity ρ_r is the reciprocal of the conductivity σ. Similarly, the resistance R is the reciprocal of the conductance G. The units for ρ_r and R are ohms-meter and ohms, respectively.

$$\rho_r = \frac{1}{\sigma} \text{ ohms-meter}$$

$$R = \frac{1}{G} \text{ ohms}$$

For the cylinder of finite length shown in Figure 12.3,

$$R = \frac{\rho_r \ell}{A}$$

12.2 ANALOGY BETWEEN DIRECT CURRENT AND ELECTROSTATIC FIELD

Consider the two sets of identical parallel plates shown in Figures 12.4a and 12.4b. The area of each plate is denoted as A, and the plate separation is denoted as d. In Figure 12.4a the parallel-plate region is filled with a dielectric material of permittivity ϵ. The voltage difference between the two plates is V. Let Q be the total charge on each plate and C be the capacitance of the parallel plates. The potential function $\Phi(y)$ is

$$\Phi(y) = V\frac{y}{d} \tag{12.8}$$

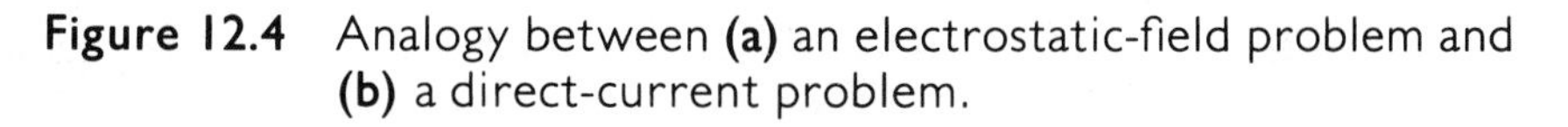

Figure 12.4 Analogy between **(a)** an electrostatic-field problem and **(b)** a direct-current problem.

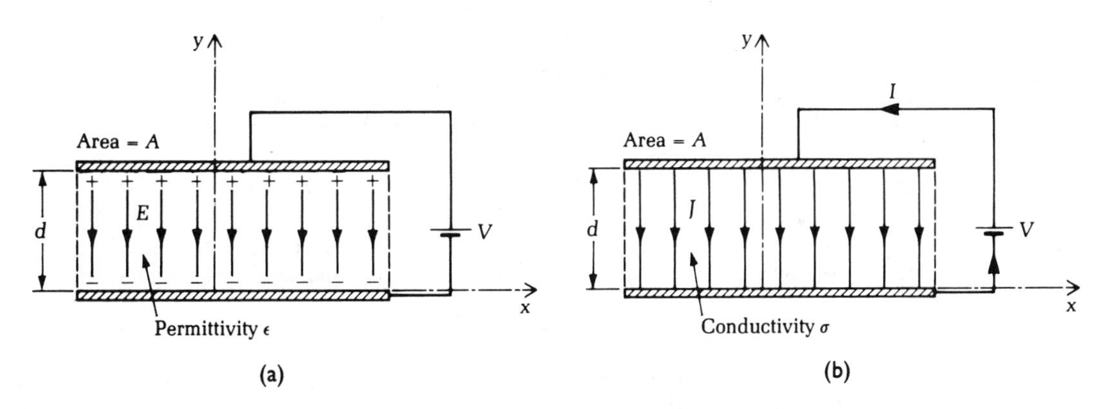

It follows that

$$\boldsymbol{E} = -\hat{y}\frac{V}{d} \tag{12.9}$$

$$\rho_s = \epsilon E = \frac{\epsilon V}{d} \tag{12.10}$$

$$Q = \rho_s A = \frac{\epsilon V A}{d} \tag{12.11}$$

$$C = \frac{Q}{V} = \frac{\epsilon A}{d} \tag{12.12}$$

The parallel plate shown in Figure 12.4b is filled with a conducting material of conductivity σ. V denotes the voltage difference between the two plates. Let I

be the total current and G be the conductance. We have the following equations:

$$\Phi(y) = V\frac{y}{d} \tag{12.13}$$

$$\boldsymbol{E} = -\hat{y}\frac{V}{d} \tag{12.14}$$

$$\boldsymbol{J} = \sigma\boldsymbol{E} = -\hat{y}\frac{\sigma V}{d} \tag{12.15}$$

$$I = JA = \frac{\sigma V A}{d} \tag{12.16}$$

$$G = \frac{I}{V} = \frac{\sigma A}{d} \tag{12.17}$$

Careful examinination of Equations (12.8)–(12.17) reveals the analogy between the electrostatic-field problem and the direct-current problem shown in Table 12.2.

Table 12.2 The Analogy Principle.

Electrostatic-field problem	Direct-current problem
Φ	Φ
$\boldsymbol{E} = -\nabla\Phi$	$\boldsymbol{E} = -\nabla\Phi$
ϵ	σ
Q (on the conductor)	I (leaving the conductor)
ρ_s (on the surface of a conductor)	J_n (leaving the surface of a conductor)
C	G

We see that if the geometry of an electrostatic-field problem is the same as that of a direct-current problem, as it is, for example, in the analogous problems shown in Figures 12.4a and 12.4b, then the expressions of the potential function and of the related electric field of the direct-current problem are identical to those of the electrostatic problem when ϵ is replaced by σ and Q by I. In addition, the capacitance in the electrostatic-field problem becomes the conductance in the direct-current problem.

As with electrostatics, the direct-current problems can be cast in the form of the Laplace equation. From (12.3), we have

$$\boldsymbol{E} = -\nabla\Phi \tag{12.18}$$

where Φ is the potential function. Substituting in (12.2) and using (12.1), we obtain the Laplace equation

$$\nabla^2\Phi = 0 \tag{12.19}$$

Once we solve (12.19) for Φ subject to the appropriate boundary conditions, we can obtain the solutions for $\boldsymbol{E}$ from (12.18) and the current $\boldsymbol{J} = \sigma\boldsymbol{E}$ from (12.1).

However, in electrostatics, we have the Poisson equation, which does not have an analogy in direct-current problems because the analogy of $\boldsymbol{J} = \sigma \boldsymbol{E}$ is $\boldsymbol{D} = \epsilon \boldsymbol{E}$. In electrostatics, we have

$$\nabla \cdot \boldsymbol{D} = \rho_v$$

whereas in direct-current problems, we have

$$\nabla \cdot \boldsymbol{J} = 0$$

A point current source similar to a point charge in electrostatics does not exist. This fact needs not impede the powerful tool of analogy in solving direct-current problems if we assume that for practical considerations point currents do exist. We shall illustrate this assertion with the following examples and applications.

Example 12.1

This example shows the circuit model of a capacitor that has an imperfect insulator; that is, the dielectric material in the capacitor has a nonzero conductivity.

Find the conductance and the capacitance of a parallel-plate capacitor filled with material characterized by permittivity ϵ and conductivity σ, as shown in Figure 12.5a.

Figure 12.5 **(a)** The parallel plate filled with a material characterized by σ and ϵ. **(b)** The equivalent circuit.

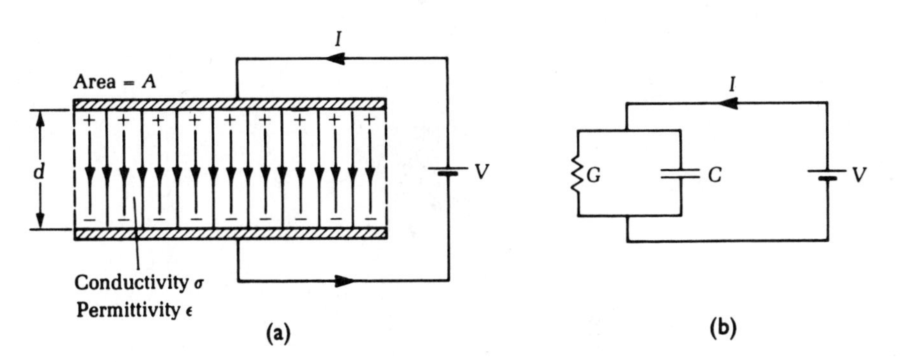

Solution The capacitance C is given by (12.12) and the conductance by (12.17).

$$C = \frac{\epsilon A}{d}$$

$$G = \frac{\sigma A}{d}$$

Figure 12.5b shows the equivalent circuit of this system. Note that the system shown in Figure 12.5a has two electric properties: (a) it stores electrostatic energy, and (b) it dissipates electric power. The simplest model of this system is the equivalent circuit of a capacitor is parallel with a conductor, with the capacitance and resistance given by the above equations. The capacitor stores energy, and the conductor dissipates power. They are in parallel instead of in series because, if we disconnect the battery V from the system, the charges in the plates tend to neutralize each other. Thus, as the equivalent circuit predicts, a transient current flows through the material after the battery is disconnected. Chapter 15 will discuss a more rigorous derivation of the equivalent circuit using electroquasistatics (see Example 15.1).

Example 12.2 *This example shows that the conductance or the resistance between two electrodes are related to the capacitance between the same.*

Find the resistance between two concentric spherical electrodes of radii a and b, respectively, when the material between the electrodes is characterized by permittivity ϵ and conductivity σ. Find the equivalent circuit of this system, shown in Figure 12.6.

Figure 12.6 Current flow from the inner conducting sphere to the outer conducting shell. A thin, insulated wire feeds the current through the shell.

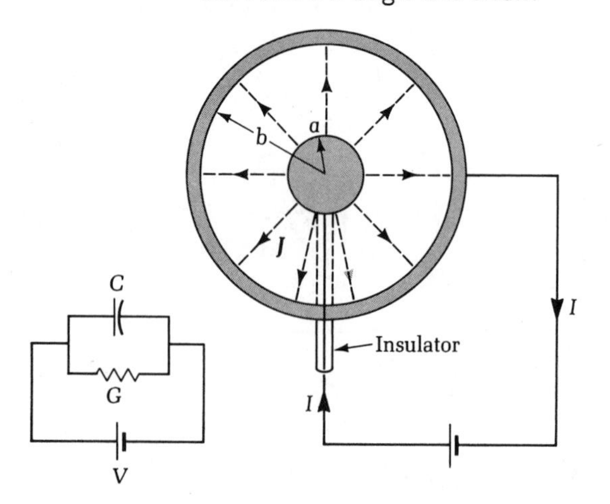

Solution We can consider the inner sphere as a source with current flowing radially to the outer electrode. A thin, insulated cable supplies current to the inner electrode, as shown in Figure 12.6. This system satisfies the continuity equation (12.2). In practice, however, the effect of the thin, insulated return path may be neglected because it contributes little either to the resistance or to the capacitance of the system.

The capacitance of this system has already been solved. According to the result obtained in (10.43), Section 10.3,

$$C = \frac{4\pi\epsilon}{\dfrac{1}{a} - \dfrac{1}{b}}$$

From the analogy principle (Table 12.2), we obtain G from the above equation by simple substitution:

$$G = \frac{4\pi\sigma}{\dfrac{1}{a} - \dfrac{1}{b}}$$

Consequently,

$$R = \frac{1}{4\pi\sigma}\left(\frac{1}{a} - \frac{1}{b}\right)$$

Example 12.3 *The circuit model for the parallel-plate capacitor with an imperfect insulator discussed in Example 12.1 can also be used for a coaxial capacitor, as shown in this example.*

A coaxial line of inner radius a and outer radius b is filled with a material characterized by a permittivity ϵ and a conductivity σ, as shown in Figure 12.7. Find the

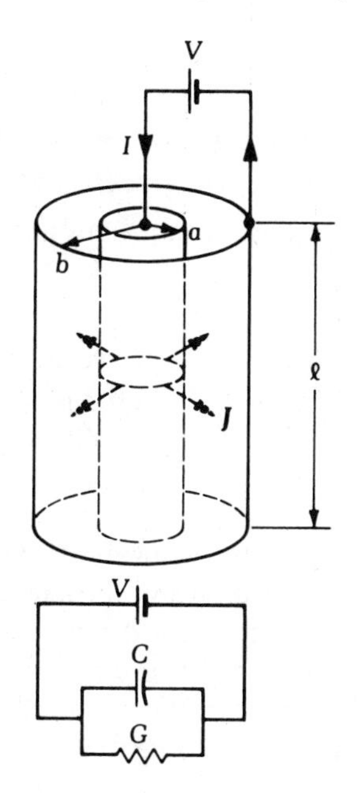

Figure 12.7 Conductance and resistance for a section of coaxial line may be calculated by using the analogy principle.

equivalent circuit of this system when it is connected to a dc battery. Assume $\ell \gg b > a$.

Solution Because ℓ is much larger than the cross-sectional dimension of the line, the electrostatic field is approximately the same as an infinitely long line. Both the electric field and the current are radially directed. From (10.45), the capacitance per unit length of an infinitely long line with a similar cross-sectional area is given as follows:

$$\frac{C}{\ell} = \frac{2\pi\epsilon}{\ln(b/a)}$$

Thus, the total capacitance C is given below

$$C = \frac{2\pi\epsilon\ell}{\ln(b/a)} \text{ farads}$$

Using the analogy principle, we obtain

$$G = \frac{2\pi\sigma\ell}{\ln(b/a)} \text{ mhos}$$

and

$$R = \frac{\ln(b/a)}{2\pi\sigma\ell} \text{ ohms}$$

Figure 12.7 shows the equivalent circuit.

Example 12.4

The capacitance between a conducting pipe near a ground is obtained in Chapter 11 using the image theorem. The result can be used to find the conductance between two buried pipes in a conducting medium, as in this example.

Two identical steel pipes, each 1 km long, are buried in the ground. The diameters of these steel pipes are 12 in. (0.3048 m), and the pipes are 100 m apart. The ground conductivity is 0.01 mho/m. What is the resistance between these two pipes?

Solution We shall use the analogy principle to solve this problem. The analogous problem for this case is the electrostatic problem discussed in Example 11.9, in which we calculated the capacitance between a pipe and a perfectly conducting ground. That situation can be modified slightly to suit the present case. Refer to Figure 12.8, in which two cylindrical conductors are separated $2h$ apart and charged to V_0 and $-V_0$, respectively. It is clear that the potential is zero on the plane in the middle of these pipes. Therefore, the potential function in the upper half-space shown in Figure 12.8 is identical to that in Example 11.9 and Figure 11.13.

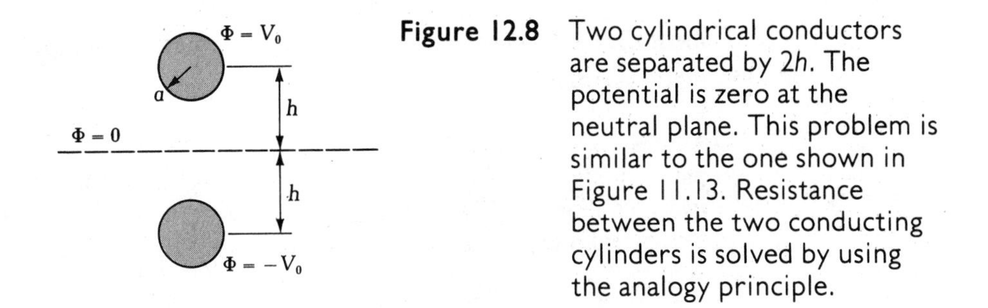

Figure 12.8 Two cylindrical conductors are separated by $2h$. The potential is zero at the neutral plane. This problem is similar to the one shown in Figure 11.13. Resistance between the two conducting cylinders is solved by using the analogy principle.

In the system shown in Figure 12.8, the difference in potential between the two conductors is $2V_0$, whereas in Figure 11.13 it is just V_0. The charge carried on each conductor is assumed to be the same. Consequently, the capacitance of the system shown in Figure 12.8 should be one-half of that shown in Figure 11.13 because of the definition of the capacitance:

$$C = \frac{Q}{V}$$

Thus, the capacitance for the present case is half of that given in (11.40):

$$C = \frac{\pi\epsilon}{\cosh^{-1}(h/a)} \text{(F/m)} \tag{12.20}$$

Consequently, the total capacitance for a pair of steel pipes ℓ meters long is given as follows:

$$C_t = C\ell = \frac{\pi\epsilon\ell}{\cosh^{-1}(h/a)} \text{(F)}$$

Using the analogy principle, the total conductance between the steel pipe buried in a medium of conductivity σ is given as follows:

$$G_t = \frac{\pi\sigma\ell}{\cosh^{-1}(h/a)} \text{(mho)}$$

The total resistance is found to be

$$R_t = \frac{\cosh^{-1}(h/a)}{\pi\sigma\ell} \text{(ohm)} \tag{12.21}$$

For $h = 50$ m, $a = 0.1524$ m.

$$\cosh^{-1}\left(\frac{h}{a}\right) \cong \ln\left(\frac{2h}{a}\right) = 6.49$$

Therefore,

$$R_t = \frac{6.49}{\pi \times 0.01 \times 1000} = 0.21\,\Omega$$

Example 12.5 *As this example shows, we can also use the image theorem to find the potential function due to a current source near the boundary of two conducting media. The theorem is analogous to the theorem for point charges in two dielectric media.*

An electrode puts out a constant current of I amperes. The electrode is near a planar boundary of two media with conductivities equal to σ_1 and σ_2, respectively, as shown in Figure 12.9. Find the potential function in both media. Sketch the current lines assuming $\sigma_2 = 2\sigma_1$.

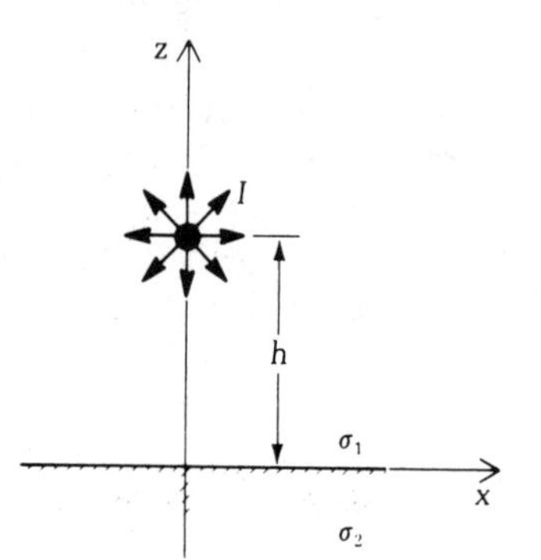

Figure 12.9 A current source puts out I amperes near a planar boundary between two media that have conductivities σ_1 and σ_2, respectively.

Solution The present situation is analogous to the one shown in Figure 11.15. By the principle of analogy, we obtain the potential function from (11.45):

$$\Phi_1(x, y, z) = \frac{1}{4\pi\sigma_1}\left[\frac{I}{\sqrt{x^2 + y^2 + (z - h)^2}} + \frac{I'}{\sqrt{x^2 + y^2 + (z + h)^2}}\right] \quad \text{for} \quad z \geq 0 \tag{12.22a}$$

$$\Phi_2(x, y, z) = \frac{1}{4\pi\sigma_2}\frac{I''}{\sqrt{x^2 + y^2 + (z - h)^2}} \quad \text{for} \quad z \leq 0 \tag{12.22b}$$

where

$$I' = I\left(\frac{\sigma_1 - \sigma_2}{\sigma_1 + \sigma_2}\right) \tag{12.23a}$$

$$I'' = I\left(\frac{2\sigma_2}{\sigma_1 + \sigma_2}\right) \tag{12.23b}$$

The flow of current follows the electric-field lines, shown in Figure 11.17. There is no need to repeat the sketch here because it would look identical to Figure 11.17 except that q, ϵ_1, and ϵ_2 would be replaced by I, σ_1, and σ_2, respectively.

Well Logging

After a well is drilled to a certain depth in search for petroleum deposits, drilling is temporarily halted to allow the well to be "logged." Well logging is the technique of lowering down the well instruments (called "sondes") to detect whether the well has traversed an oil-bearing layer of the earth. If such a layer is detected, then a well-log analyst can estimate from the reading of the responses of several sondes how much petroleum is in the reservoir.

A commonly used logging tool is a sonde that measures the resistivity of the formation surrounding the well. Information about resistivity interests the explorer because the geological structure suspected of having petroleum deposits usually consists of porous rocklike sandstone. The fluid that fills these porous spaces is usually either brine, which is conductive, or oil, which is very resistive (see Table 12.1). Thus, an oil-bearing bed sandwiched between two beds saturated with saline water manifests itself by a much higher resistivity reading as a resistivity sonde passes through the oil deposit.

Here we discuss the basic operating principle of a well-logging resistivity sonde. Figure 12.10 shows a constant direct current I being delivered from the surface of the ground down to the electrode A by way of an insulated cable. The

Figure 12.10 Resistivity logging. A tool to measure the resistivity of a geological formation to see if the formation contains petroleum.

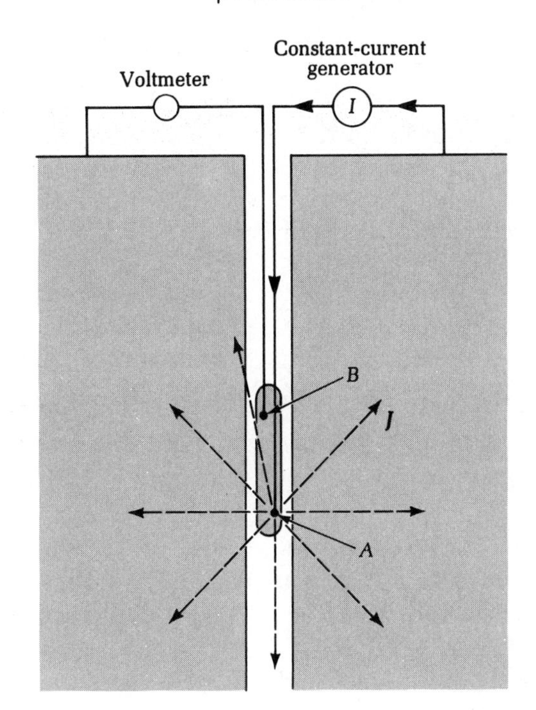

entire ground serves as the current's return electrode. To simplify the problem, we assume that the conductivity of the fluid found in the well is equal to that of the surrounding formation. Thus, we may consider the current electrode to be located in an infinite, homogeneous media. The potential function everywhere in the formation is then given below:

$$\Phi(r) = \frac{I}{4\pi\sigma r} \tag{12.24}$$

where r is in the spherical coordinate system and σ is the conductivity of the formation. Note that (12.24) is obtained with the analogy principle. The analogous case is the potential function of a point charge Q in a homogeneous dielectric medium of permittivity ϵ.

A voltage electrode B is located d meters from A to measure the potential at that point. The potential is directly proportional to the flow of current in the formation. Because B is separated from A by d

$$\Phi_B = \frac{I}{4\pi\sigma d} \tag{12.25}$$

This voltage is measured and recorded at the surface. Because Φ_B, I, and d are now all known quantities, the conductivity may be inferred:

$$\sigma = \frac{I}{4\pi d\Phi_B}$$

or, equivalently, the resistivity:

$$\rho_r = (4\pi d)\frac{\Phi_B}{I} \tag{12.26}$$

The resistivity is usually calculated from (12.26) and recorded as a function of the depth of the well. The chart in which the resistivity is recorded is called a resistivity "log," and it looks like the one shown in Figure 12.11. From this log, we notice strong evidence that the layer between 5040 ft and 5060 ft contains either oil or gas.*

Example 12.6 *This example demonstrates the principle of a simple well-logging tool. We can predict what the tool will read as it is near the boundary of two different conducting media.*

The well-logging resistivity sonde shown in Figure 12.10 is located near a boundary of two beds whose resistivities are $\rho_{r1} = 10$ Ω-m and $\rho_{r2} = 100$ Ω-*m*, respec-

* For further reading, see *Log Review 1* (Houston: Dresser-Atlas, 1974).

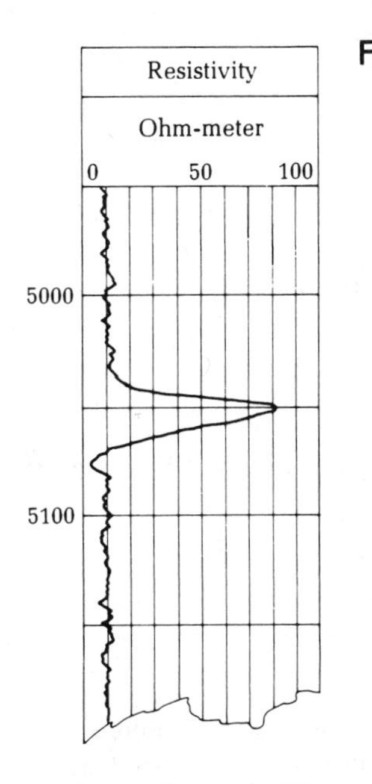

Figure 12.11 A resistivity log. Strong evidence of an oil-bearing layer appears at a depth of 5050 feet.

tively, as shown in Figure 12.12. Usually the reading of ρ_r is calibrated so that it reads the exact value only if the medium is infinite and homogeneous. Calculate the apparent resistivity detected by this sonde. Take $d = 16$ in.

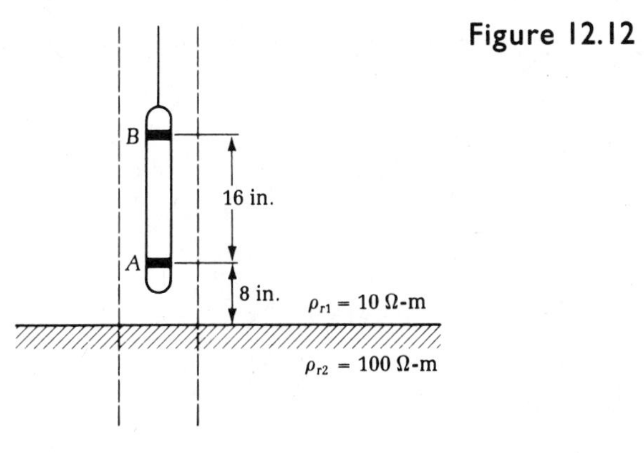

Figure 12.12 A well-logging resistivity tool is near a boundary of two thick beds of different resistivities. The presence of the boundary causes the tool to read an incorrect resistivity. See Example 12.6 for a quantitative analysis of the problem.

Solution The resistivity read by the sonde will be influenced by medium 2. Thus, the sonde will not read $\rho_r = 10$ Ω-m, although it is located entirely in medium 1. To find the expected reading, we must first calculate the potential detected at B. We solved the potential problem in Example 12.5. In the present case, we have $\sigma_1 = 0.1$, $x = 0$, $y = 0$, $z - h = 16\,\text{in.} \times (2.54/100)\,\text{m/in.} = 0.406\,\text{m}$, $z + h = 32 \times 2.54/100\,\text{m} = 0.813\,\text{m}$, and

$$I' = I\left(\frac{0.1 - 0.01}{0.1 + 0.01}\right) = 0.818I$$

Therefore, according to (12.22a),

$$\Phi_B = \frac{I}{4\pi \times 0.1}\left[\frac{1}{0.406} + \frac{0.818}{0.813}\right] = \frac{I}{4\pi}(34.7)$$

Substituting the above value into (12.26), we obtain

$$\rho_r = 4\pi \times 16 \times \frac{2.54}{100} \times \frac{1}{4\pi} \times 34.7 = 14.1\ \Omega\text{-m}$$

SUMMARY

1. Ohm's law states that $\boldsymbol{J} = \sigma \boldsymbol{E}$, where σ is defined as the conductivity of the medium.
2. Joule's law states that $\boldsymbol{J} \cdot \boldsymbol{E}$ is the power lost by the electric field to heating.
3. Conductance is defined as the ratio of current flowing between two electrodes to the voltage difference between them. The resistance is the inverse of the conductance.
4. The electrostatic-field problem involving potential, dielectric constant, and charges is analogous to the direct-current problem involving the potential, conductivity, and current sources. The analogy principle is summarized in Table 12.2.

Problems

12.1 A parallel plate is filled with two materials in a configuration shown in Figure P12.1. The total area of the plate is A. The conductivity of one material is σ_1 and that of the other material is σ_2. Find the resistance of this parallel plate, and express it in terms of A, d, σ_1, and σ_2.

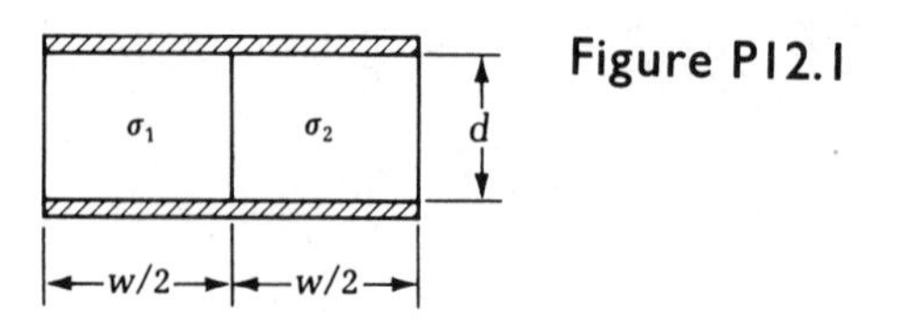

Figure P12.1

12.2 A parallel plate is filled with two materials in a configuration shown in Figure P12.2. Find its resistance and express it in terms of A, the area of the plate, and d_1, d_2 σ_1, and σ_2, which are defined in the figure.

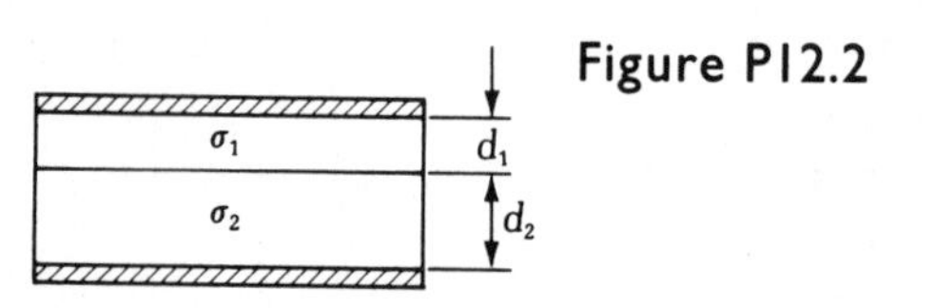

Figure P12.2

12.3 A coaxial line has two layers of insulation. Figure P12.3 shows the geometry. Find

(a) the potential Φ_1 for $a < \rho < b$

(b) the potential Φ_2 for $b < \rho < c$

(c) the resistance of a section of such a line ℓ meters long

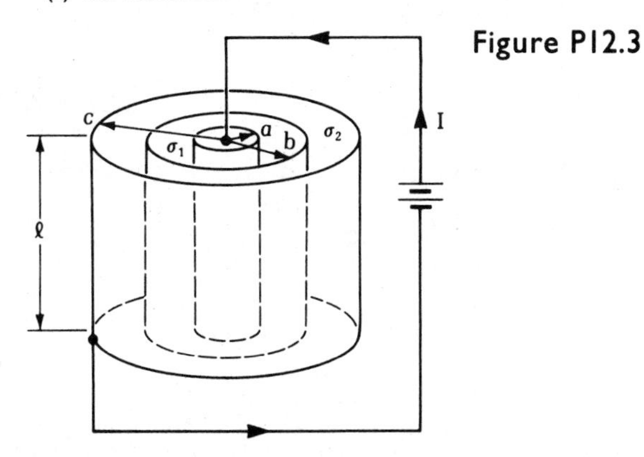

Figure P12.3

12.4 A spherical conductor of radius a is inside a spherical conducting shell of radius c. Two materials are used to fill the space between these conductors. The conductivities of these materials are σ_1 and σ_2, respectively. Figure P12.4 shows the configuration. Find the resistance of this device and express it in terms of a, b, c, σ_1, and σ_2.

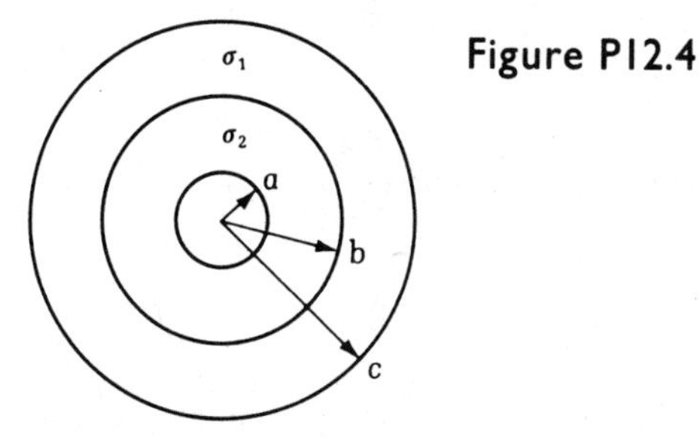

Figure P12.4

12.5 Two oil wells are 1 km apart. The resistance between two steel pipes in these wells is measured at 3.41 Ω. What is the conductivity of the ground near these wells? Use the following data: the length of both pipes = 1 km, and the diameter of both pipes = 10 cm.

12.6 A current electrode is near a perfectly conducting plate that is bent to form a 90° corner, as shown in Figure P12.6. The output from the electrode is I amperes, and the material filling the space has a conductivity equal to σ. Find the potential function $\Phi(x, y, z)$.

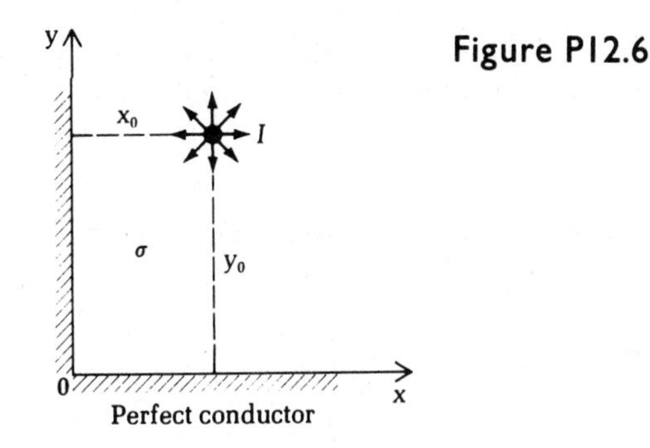

Figure P12.6

12.7 A current electrode is near a perfectly conducting plate that is bent to form a 60° corner, as shown in Figure P12.7. The electrode produces 10 A of current, and the material filling the region defined by $0 < \phi < 60°$ is water with conductivity equal to 0.01 mho/m. Find the potential at point B shown in the figure.

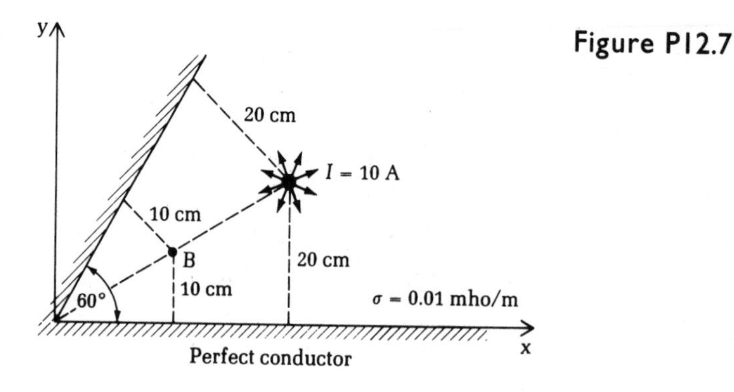

Figure P12.7

12.8 A point electrode puts out I amperes of current above a conducting plane, as shown in Figure P12.8.

(a) Find $\Phi(x, y, z)$ for $z > 0$.
(b) Find the current density $J_s(x, y)$ at the surface of the conductor.
(c) Sketch the paths of the current flow.

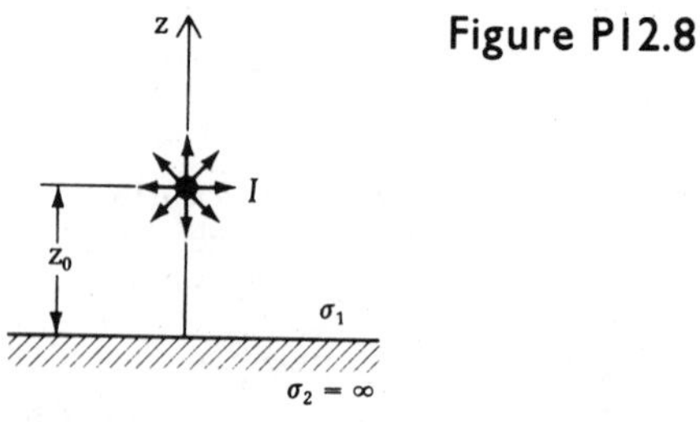

Figure P12.8

12.9 For the case shown in Figure 12.9, find the percentage of the current emitted from the electrode that crosses the boundary and enters in medium 2.

12.10 A source 4 meters below an interface of two conducting media emits 2 A of direct current, as shown in Figure P12.10.

(a) Calculate the potential at point B.
(b) Calculate the potential at point C.

Figure P12.10

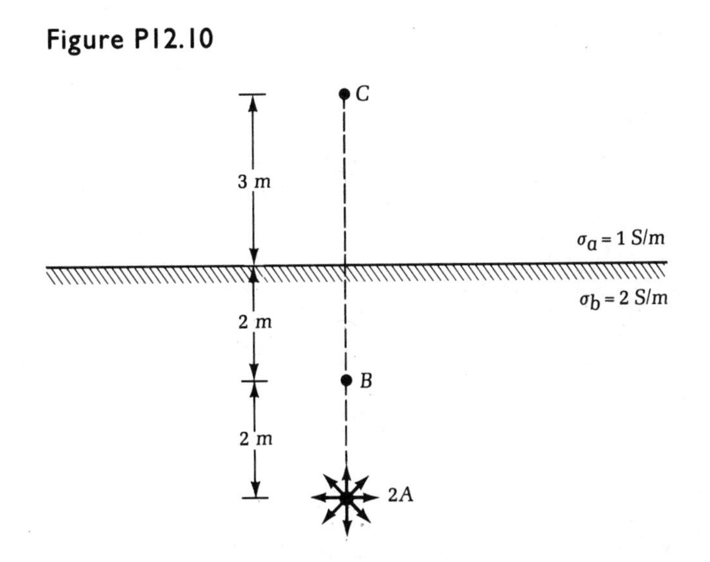

12.11 A well-logging resistivity tool similar to the one shown in Figure 12.12 is near a boundary between two beds, as shown in Figure P12.11. The boundary is making a 60° angle with the well. Find the apparent resistivity measured by this tool at the position shown.

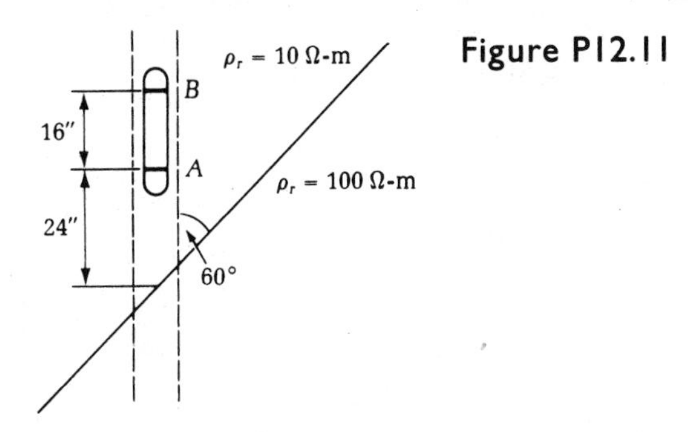

Figure P12.11

12.12 Refer to Example 12.6. Obtain ρ_a (the apparent resistivity measured by the tool) as a function of tool position for $z_0 = +160$ in. to $z_0 = -160$ in., where z_0 is the position of the center of the tool (the midpoint between electrodes A and B) relative to the boundary. Calculate ρ_a for at least 21 points, and plot ρ_a versus z_0.

12.13 Repeat Problem 12.12 for the situation shown in Figure P12.11.

12.14 A point electrode is located at $(0, y_1, 0)$, and a perfectly conducting sphere of radius a is located at $(-\ell, 0, 0)$ as shown in Figure P12.14. The electrode gives I amperes of current. The conductivity of the medium is σ. Find the potential Φ on the y axis. Hint: Use (11.44).

Figure P12.14

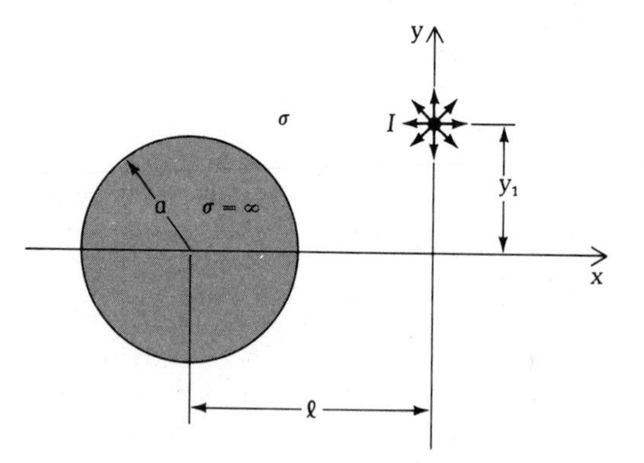

12.15 Consider a well-logging resistivity tool similar to the one shown in Figure 12.10. Let the spacing between the current electrode A and the potential electrode B be 6 m. The tool measures the conductivity of the earth formation as it travels in a well. Assume that the well passes near a mineral deposit modeled by a perfectly conducting sphere, as shown in Figure P12.15. Find the apparent resistivity measured by the tool as a function of y. Use the following data: $\sigma = 0.01$ mho/m for the ground; the radius of the mineral deposit = 50 m; and the distance between the center of the sphere and the well = 70 m. Plot σ_{apparent} versus y for $-70 < y < 70$. Hint: Use the result obtained in the preceding problem.

Figure P12.15

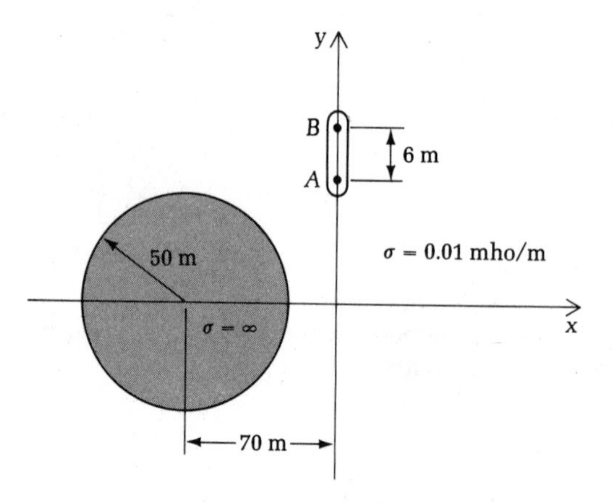

13 MAGNETOSTATIC FIELDS

We studied electrostatics in Chapters 9–11, and in Chapter 12 we discussed the origin and applications of the direct current $\boldsymbol{J}$. The current is the source of the magnetostatic field, as we can see from Maxwell's equations in the static limit. The magnetic fields are governed by the following two equations:

$$\nabla \cdot \boldsymbol{H} = 0 \tag{13.1}$$

$$\nabla \times \boldsymbol{H} = \boldsymbol{J} \tag{13.2}$$

The first equation is the magnetic Gauss' law. We have used the constitutive relation $\boldsymbol{B} = \mu \boldsymbol{H}$, with μ a constant, to derive (13.1). We shall study the solutions of these equations. The magnetic field $\boldsymbol{H}$ manifests itself in the Lorentz force law

$$\boldsymbol{F} = q\boldsymbol{v} \times (\mu \boldsymbol{H}) \tag{13.3}$$

Therefore, in this chapter we will discuss forces and torques acting on charge particles and current-carrying wires, magnetic energies, and the concept of inductance.

13.1 MAGNETOSTATIC FIELDS

When a direct current flows along a wire, it produces steady magnetic fields in its surroundings. The relationship between the current density $\boldsymbol{J}$ and the magnetic field $\boldsymbol{H}$ is given by (13.2). In its present form, (13.2) is not very useful for finding $\boldsymbol{H}$ from a given $\boldsymbol{J}$. However, (13.2) can be transformed so that the resulting new form is useful in calculating $\boldsymbol{H}$ from $\boldsymbol{J}$. To accomplish this change, we integrate both sides of (13.2) over an area A (see Figure 13.1):

$$\iint_A \nabla \times \boldsymbol{H} \cdot \hat{\boldsymbol{n}}\, da = \iint_A \boldsymbol{J} \cdot \hat{\boldsymbol{n}}\, da \tag{13.4}$$

From the Stokes theorem discussed in Section 9.4, the left-hand side of (13.4) may be expressed as an integral along a closed path C, with the following result:

$$\boxed{\oint_C \boldsymbol{H} \cdot d\boldsymbol{\ell} = I} \quad \textbf{(Ampère's law, integral form)} \tag{13.5}$$

Figure 13.1 Integration of $\boldsymbol{H}$ along the contour C is equal to the total current I that flows through the area A bounded by C. The direction of C and the normal $\hat{\boldsymbol{n}}$ of A are related by the right-hand rule. I is contributed by J_2, J_3, and J_4 but not by J_1 or J_5.

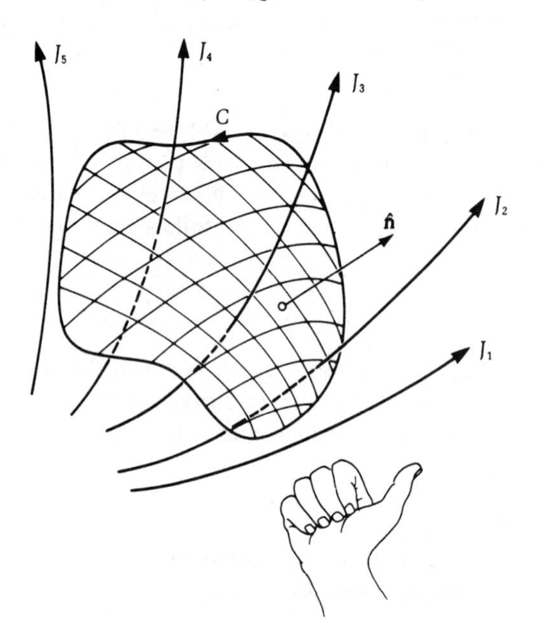

where C is the path that forms the boundary of A and I is the total current that flows through the area A. The direction of the contour C and the normal vector $\hat{\boldsymbol{n}}$ of A are related by the right-hand rule, shown in Figure 13.1. Equation (13.5) is called Ampère's law in integral form, and (13.2) is Ampère's law in differential form.

Magnetostatic Field Due to a Line Current

Consider an infinitely long wire carrying a direct current I, as shown in Figure 13.2a. We shall use Ampère's law (13.5) to find the magnetostatic field $\boldsymbol{H}$.

Let us set up the cylindrical coordinate systems with the z axis coinciding with the wire, as shown in Figure 13.2a. We apply Ampère's law (13.5) over a circle centered at a point along the z axis with radius ρ, and we obtain the following equation:

$$\int_0^{2\pi} H_\phi \rho \, d\phi = I$$

By symmetry, we see that H_ϕ is independent of ϕ and that, as such, it can be taken out of the integral sign. The result is as follows:

$$H_\phi 2\pi\rho = I$$

that is,

$$\boxed{\boldsymbol{H} = \hat{\boldsymbol{\phi}}\frac{I}{2\pi\rho}} \qquad \textbf{(}\boldsymbol{H}\textbf{ field due to current on a long wire)} \qquad \textbf{(13.6)}$$

Figure 13.2b shows a plot of H_ϕ versus ρ. Thus, the magnetostatic field due to the current in an infinitely long wire is in the $\hat{\boldsymbol{\phi}}$ direction and decreases like $1/\rho$ in the radial direction.

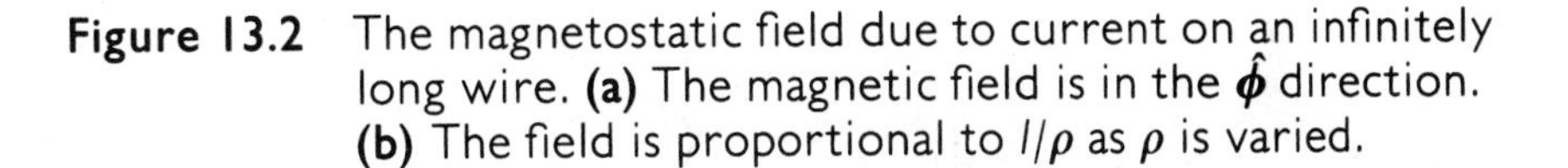

Figure 13.2 The magnetostatic field due to current on an infinitely long wire. **(a)** The magnetic field is in the $\hat{\boldsymbol{\phi}}$ direction. **(b)** The field is proportional to I/ρ as ρ is varied.

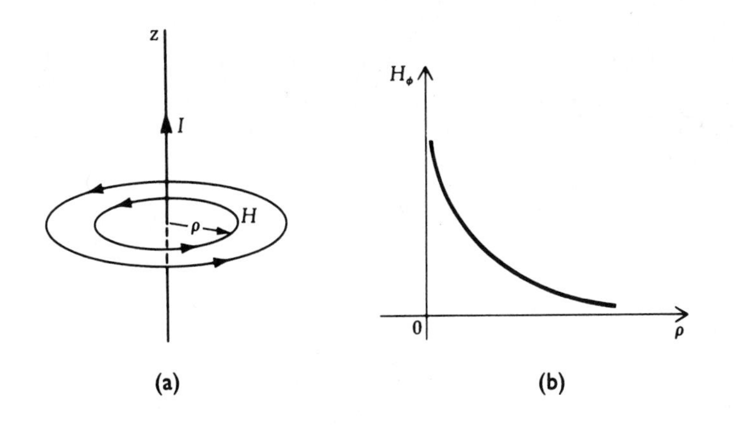

Magnetostatic Field Due to Current in a Coaxial Line

We see that the magnetostatic field due to a long wire of zero radius increases without limit as ρ approaches zero. Figure 13.3a (p. 448) shows a more realistic model, a coaxial line consisting of a solid inner conductor of radius a and a cylindrical shell of inner radius b and thickness $(c - b)$. The inner conductor carries I amperes of current in the $\hat{z}$ direction, and the outer shell carries the return current in the negative $\hat{z}$ direction. Because the currents are dc, the skin depth is infinite, and the currents are uniformly distributed in these conductors. We shall solve the magnetostatic field associated with these currents.

We apply Ampère's law (13.5) on a circular area A centered at a point on the z axis with radius ρ. Let us assume that $\rho < a$, so that the area is inside the inner conductor, as shown in Figure 13.3b. Note that the total current flowing through area A is less than I because the area is inside the inner conductor. To be exact, (13.5) takes the following form under the conditions shown in Figure 13.3b:

$$\int_0^{2\pi} H_\phi \rho \, d\phi = \left(\frac{\pi\rho^2}{\pi a^2}\right) I \tag{13.7a}$$

Figure 13.3 **(a)** A coaxial line carries I and $-I$ amperes of current on its inner and outer conductors, respectively. **(b)** Ampère's law applied to a circle with $\rho < a$. **(c)** $a < \rho < b$. **(d)** $b < \rho < c$. **(e)** $\rho > c$. **(f)** Plot of H_ϕ versus ρ, $H_1 = I/2\pi a$, and $H_2 = I/2\pi b$.

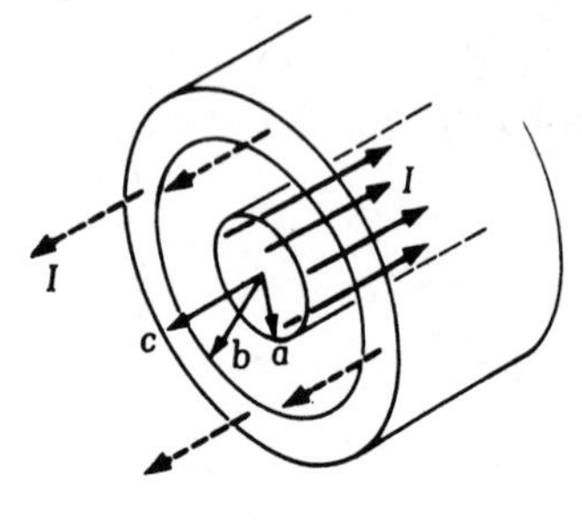

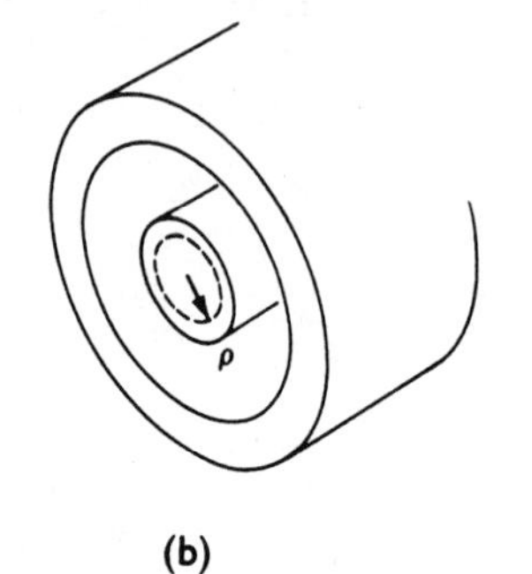

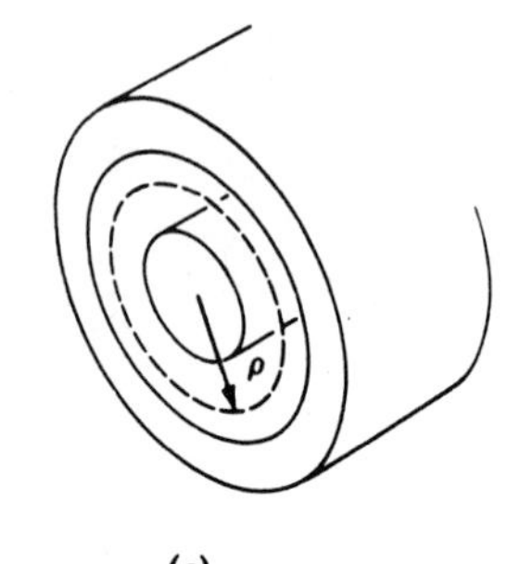

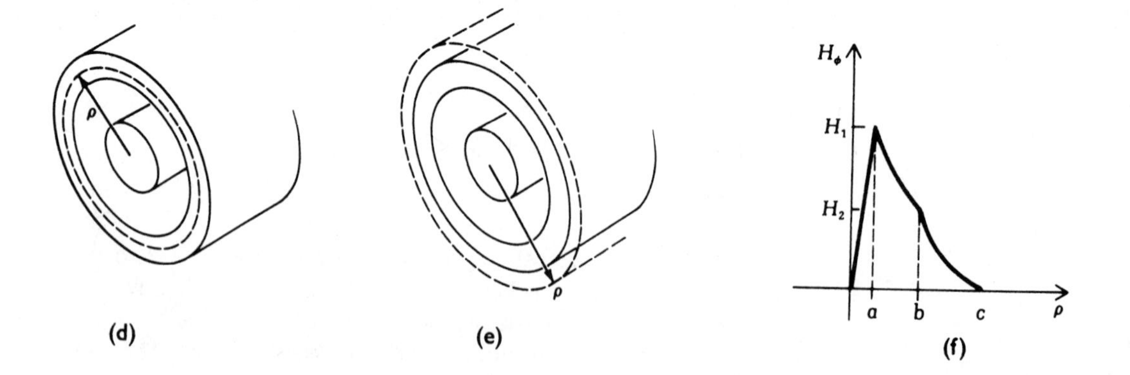

Again, H_ϕ may be taken out of the integral sign on account of the symmetry of this problem. The result is

$$H_\phi = \frac{I\rho}{2\pi a^2} \qquad 0 < \rho < a \tag{13.7b}$$

Let us consider the situation in which $a < \rho < b$, as shown in Figure 13.3c. The current flowing through area A is now exactly I. Hence, (13.5) yields the following result:

$$2\pi\rho H_\phi = I$$

or

$$H_\phi = \frac{I}{2\pi\rho} \qquad a < \rho < b \tag{13.7c}$$

Figure 13.3d shows the situation for $b < \rho < c$. The total current flowing through area A is I minus that portion of the return current that is passing through the ring between ρ and b in the outer conductor. Consequently, (13.5) yields

$$2\pi\rho H_\phi = I - I\left(\frac{\pi\rho^2 - \pi b^2}{\pi c^2 - \pi b^2}\right)$$

or, equivalently,

$$H_\phi = \frac{I}{2\pi\rho}\left(\frac{c^2 - \rho^2}{c^2 - b^2}\right) \qquad b < \rho < c \tag{13.7d}$$

Finally, for $\rho > c$, the total current flowing through area A is equal to zero, as shown in Figure 13.3c, and we obtain

$$H_\phi = 0 \qquad \rho > c \tag{13.7e}$$

Figure 13.3f sketches a plot of H_ϕ versus ρ.

Magnetostatic Field Due to a Current Sheet

Consider a sheet of metal carrying current flowing in the $\hat{z}$ direction, as shown in Figure 13.4a. Figure 13.4b shows an end view of the plate. The density of the current is J_s amperes per meter. The sheet may be thought of as consisting of many current-carrying wires arranged in parallel, as illustrated in Figure 13.4c. From (13.6), we know that each wire generates $\hat{\phi}$-directed $\boldsymbol{H}$ fields. As we combine the wires to form a sheet, all fields above the sheet add up in the $-\hat{x}$ direction and line up in the $+\hat{x}$ direction below the sheet. The fields in the $\hat{y}$ direction cancel each other out. With this qualitative picture of the $\boldsymbol{H}$ field distribution in mind, we can now proceed to obtain the $\boldsymbol{H}$ field quantitatively.

We choose the contour C shown in Figure 13.4b and use Ampère's law (13.5) to obtain the following:

$$\int_{AB+BC+CD+DA} \boldsymbol{H} \cdot d\boldsymbol{\ell} = J_s w$$

Figure 13.4 **(a)** A sheet of current flowing in the $\hat{\mathbf{z}}$ direction. The current density is J_s amperes per meter. **(b)** End view of (a). Current flows out of the paper. **(c)** The current sheet may be thought of as a combination of parallel wires each of which produces $\hat{\boldsymbol{\phi}}$-directed $\boldsymbol{H}$ fields. These $\boldsymbol{H}$ fields combine into $-\hat{\mathbf{x}}$- and $\hat{\mathbf{x}}$-directed total $\boldsymbol{H}$ fields above and below the current sheet, respectively.

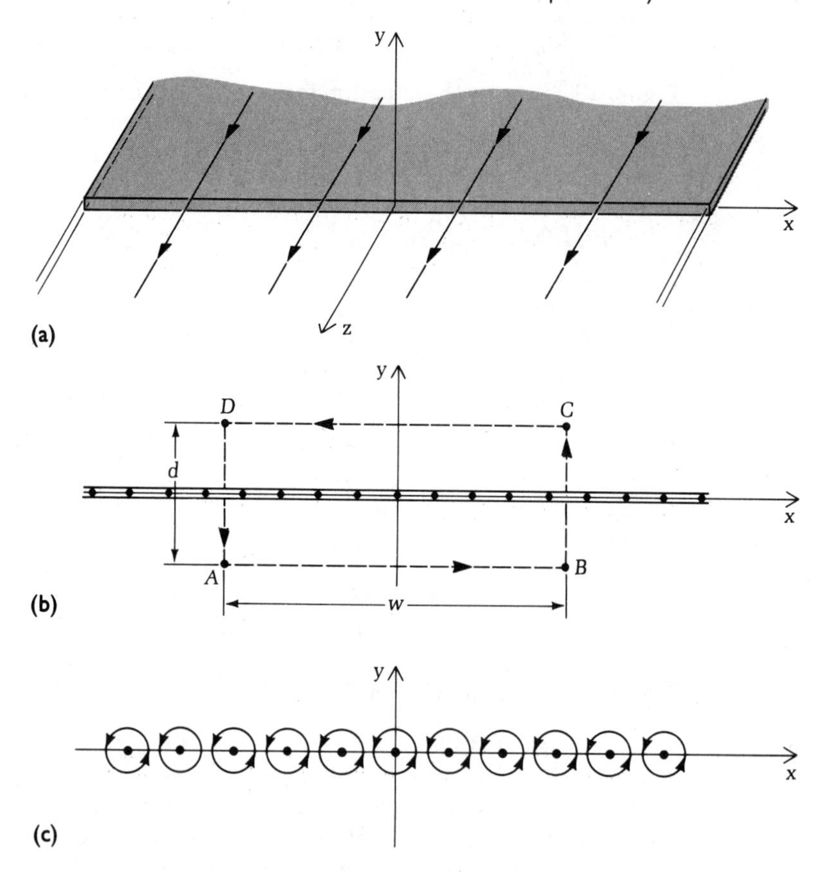

But

$$\boldsymbol{H} \cdot d\boldsymbol{\ell} = \begin{cases} H_x\,dx & \text{on } AB \\ 0 & \text{on } BC \\ (-H_x)(-dx) & \text{on } CD \\ 0 & \text{on } DA \end{cases}$$

Furthermore, because of this problem's symmetry, H_x may be considered to be independent of x. We therefore obtain the following result:

$$2H_x w = J_s w$$

or

$$\boldsymbol{H} = \begin{cases} \dfrac{-\hat{x}J_s}{2} & y > 0 \\ \dfrac{\hat{x}J_s}{2} & y < 0 \end{cases}$$ **(*H* field due to a current on a sheet)** **(13.8)**

Example 13.1

The superposition principle can be used to obtain the magnetic field due to several current sources. The total field is the vector sum of fields due to individual sources.

A parallel-plate waveguide carries I amperes of direct current flowing in the $\hat{z}$ direction in the upperplate, and the current returns via the lower plate, as shown in Figure 13.5. Assume that $w \gg a$. Find the corresponding magnetostatic field.

Figure 13.5 The magnetostatic field due to the opposite currents on the parallel-plate conductors may be found by superposing the fields due to individual sheets.

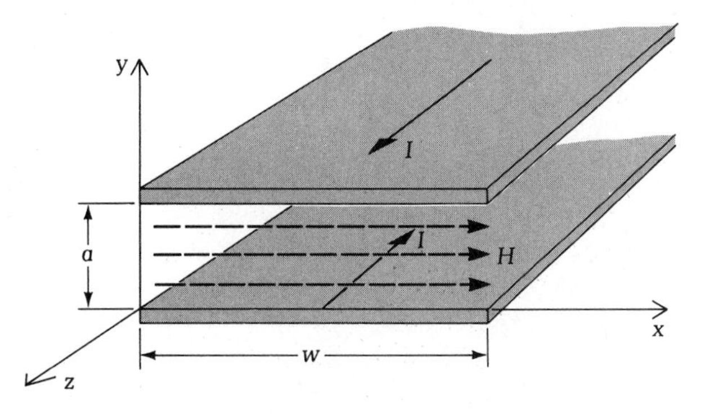

Solution Because w is much greater than a, we may use (13.8) for each plate and superimpose the fields.

For $y > a$, H_x due to the upper plate is negative, and that due to the lower plate is positive because the current on the lower plate is in the opposite direction. Thus, the two fields cancel each other—that is,

$$\boldsymbol{H} = 0 \quad \text{for} \quad y > a \qquad \textbf{(13.9a)}$$

For $0 < y < a$, H_x due to the upper plate is positive, and H_x due to the lower plate is also positive. Therefore, they reinforce each other to yield

$$\boldsymbol{H} = \hat{x}J_s = \frac{\hat{x}I}{w} \quad \text{for} \quad 0 < y < a \tag{13.9b}$$

$$\boldsymbol{H} = 0 \qquad\qquad \text{for} \quad y < 0 \tag{13.9c}$$

Equation (13.9c) is obtained because of the cancellation of H_x in that region. Note that the result obtained here is identical to that obtained in Chapter 5 for a parallel-plate waveguide if we let $k = 0$ in (5.13) and interchange the x and y coordinates.

Magnetic Field in a Solenoid

A solenoid consists of many turns of coil wound to form a cylindrical structure, as shown in Figure 13.6. We assume that the long solenoid has n turns per meter. Because of the symmetry of this configuration, we assume that the magnetostatic field is independent of z (because the solenoid is very long) and of ϕ (because the coils are circular). Let $\boldsymbol{H} = \hat{\rho}H_\rho(\rho) + \hat{\phi}H_\phi(\rho) + \hat{z}H_z(\rho)$, where H_ρ, H_ϕ, and H_z are components of $\boldsymbol{H}$ in the $\hat{\rho}$, $\hat{\phi}$, and $\hat{z}$ directions, respectively. Note that they are functions of ρ only. We shall prove that H_ρ and H_ϕ are equal to zero.

Figure 13.6 A long solenoid, formed by coils wound into a cylindrical structure. Closeness of the coil winding is measured by the quantity n, number of coils per meter.

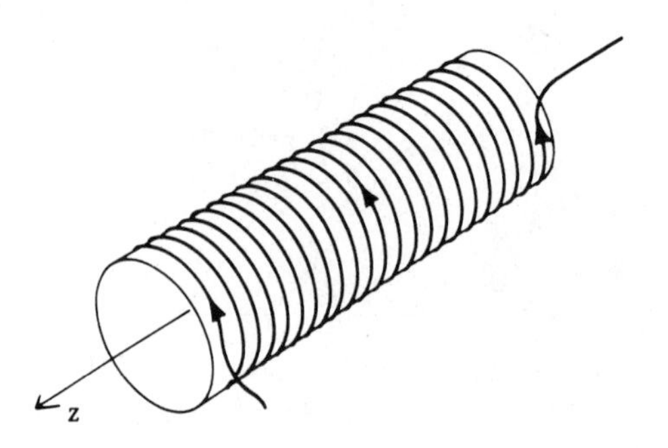

Let us integrate $\boldsymbol{H}$ throughout a cylindrical volume V at the center of the solenoid with radius ρ and length ℓ, as shown in Figure 13.7a. According to (13.1) and $\boldsymbol{B} = \mu\boldsymbol{H}$, we have

$$\int_v (\nabla \cdot \boldsymbol{H})\, dv = 0$$

From the divergence theorem, the above equation is transformed into an equation involving the surface integral:

$$\oint_S \boldsymbol{H} \cdot \hat{\boldsymbol{n}}\, ds = 0$$

This surface integral includes integration over the two circular end pieces and over the cylindrical surface. The contributions to the integral cancel each other out over the two end pieces, and integration over the cylindrical surface yields the following:

$$2\pi\rho\ell H_\rho = 0$$

This result proves that H_ρ is equal to zero everywhere.

Figure 13.7 **(a)** Integration of $\nabla \cdot \boldsymbol{H}$ over a volume V bounded by S. **(b)** Integration of $\boldsymbol{H} \cdot \boldsymbol{d\ell}$ along path S_1 or S_2. **(c)** Integration of $\boldsymbol{H} \cdot \boldsymbol{d\ell}$ along path A-B-C-D or C-C'-D'-D.

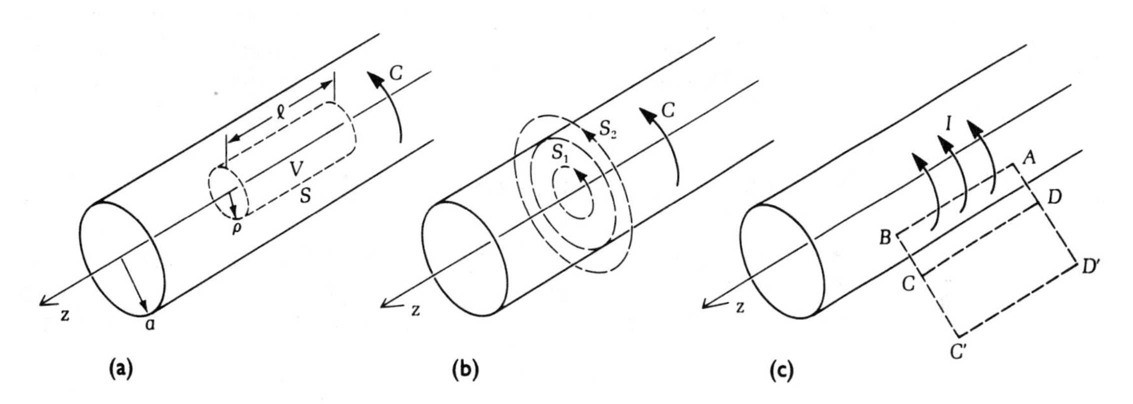

Let the contour S_1 or S_2 shown in Figure 13.7b be far from either end of the solenoid. From Ampère's law (13.5), we obtain the following equation:

$$\oint_{S_1 \text{ or } S_2} H_\phi \rho\, d\phi = 0$$

because no current is flowing through either S_1 or S_2. We concluded earlier that H_ϕ is independent of ϕ; therefore, it may be taken out of the above integral sign to yield the following result:

$$2\pi\rho H_\phi = 0$$

Thus, H_ϕ is equal to zero everywhere, and consequently H_z is the only component that may exist near a solenoid.

We now apply (13.5) to the contour $DCC'D'$ outside the solenoid shown in Figure 13.7c. We obtain

$$\ell(H_z \text{ on } DC - H_z \text{ on } C'D') = 0$$

Because the positions of CD and $C'D'$ are arbitrary, the above equation leads to the following statement

$$H_z = \text{constant} \qquad \text{for} \quad \rho > a$$

Physically we know that H_z must vanish as ρ approaches infinity. Thus, the constant must be zero or

$$H_z = 0 \qquad \text{for} \quad \rho > a \tag{13.10a}$$

Similarly, we apply (13.5) to the contour $ABCD$ and obtain the following result:

$$\ell(H_z) = nI\ell$$

Therefore,

$$\boxed{H_z = nI} \qquad \textbf{for } \rho < a \textbf{ (}H\textbf{ field in a long solenoid)} \tag{13.10b}$$

We conclude that the magnetostatic field associated with an infinitely long solenoid that has n turns per meter and that carries a direct current I is equal to (nI) in the $\hat{z}$ direction inside the solenoid and is equal to zero outside the solenoid.

Magnetic Field in a Toroidal Coil

As shown in Figure 13.8, a toroidal coil is a structure like a ring with coils wound around it. A toroid is essentially a solenoid (Figure 13.6) bent to form a circular ring. Thus, in a toroidal coil, we expect the magnetostatic field to be in the $\hat{\phi}$ direction. There will be no magnetic field outside a toroid.

Applying Ampère's law (13.5) to a circular path of radius ρ inside the toroid, we obtain

$$2\pi\rho H_\phi = NI \tag{13.11}$$

where N is the total number of windings on the toroidal coil. Therefore,

$$H_\phi = \frac{NI}{2\pi\rho} \qquad \text{for} \quad (b - a) < \rho < (b + a) \tag{13.12}$$

If $b \gg a$, then H_ϕ is approximately uniform inside. The average value of H inside is given as follows:

$$\boxed{H_\phi = \frac{NI}{2\pi b}} \qquad \textbf{if } b \gg a \textbf{ (}H\textbf{ field in a toroid)} \tag{13.13}$$

Note that $N/(2\pi b)$ is the number of windings per meter. Thus, (13.13) is equivalent to the field in a solenoid given by (13.10b) when $b \gg a$. Also, because of the

Figure 13.8(a) Schematic diagram of a toroid. The magnetic field in the toroid goes counterclockwise when the current in the coil is flowing in the indicated direction.

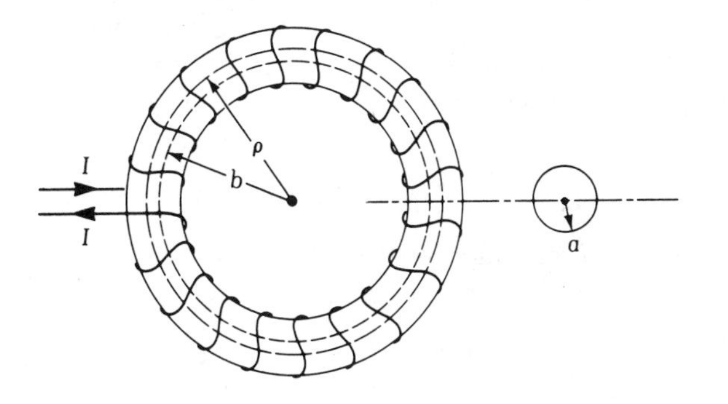

Figure 13.8(b) Actual toroidal coils used in push-button dialing telephones. (Courtesy of Western Electric Co.)

way the coil is wound on the toroid shown in Figure 13.8, the magnetic field is in the counterclockwise direction. Should the current be reversed, the magnetic field would then be clockwise.

Biot-Savart Law

So far we have studied several cases in which magnetostatic fields are found from given current distributions. These cases have one thing in common: they all possess certain kinds of symmetry, so that Ampère's law expressed as (13.5) may be used to solve for the magnetic fields.

For cases that are not highly symmetrical, (13.5) may not be useful for obtaining magnetic fields. To solve for such fields, we have to go back to the basic set of equations (13.1) and (13.2). Recall that Chapter 7 introduced a vector potential $\boldsymbol{A}$, which then leads to the solution of electromagnetic fields due to a short dipole antenna. We found in Chapter 7 that, for a short wire that has a length Δz, is located at the origin, and is carrying a current $I\, e^{j\omega t}$ in the $\hat{z}$ direction, the associated magnetic field is given by (7.13), which is repeated below:

$$\Delta \boldsymbol{H} = \hat{\phi}\frac{jkI\,\Delta z e^{-jkr}}{4\pi r}\left(1 + \frac{1}{jkr}\right)\sin\theta \qquad \textbf{(13.14)}$$

The above equation employs the spherical coordinates. The magnetostatic field due to a current segment $I\,\Delta z\hat{z}$ may be obtained from (13.14) by letting k approach zero. The result is

$$\Delta \boldsymbol{H} = \hat{\phi}\frac{I\,\Delta z\,\sin\theta}{4\pi r^2} \qquad \textbf{(13.15a)}$$

Note that the above equation may be rewritten as follows:

$$\Delta \boldsymbol{H} = \frac{I\,\Delta z\hat{z} \times \hat{r}}{4\pi r^2} \qquad \textbf{(13.15b)}$$

If the current element is not located at the origin and is not oriented in the $\hat{z}$ direction, then we must modify (13.15b); and we obtain the following equation:

$$d\boldsymbol{H} = \frac{I\,d\boldsymbol{\ell}_1 \times \hat{\boldsymbol{r}}_{12}}{4\pi r_{12}^2} \qquad \textbf{(Biot-Savart law)} \qquad \textbf{(13.16)}$$

Figure 13.9a shows the situation in which (13.15) is applicable, and Figure 13.9b illustrates the case in which (13.16) is applicable.

Figure 13.9 Biot-Savart law. **(a)** A current segment is located at the origin and oriented in the $\hat{\mathbf{z}}$ direction. **(b)** A current segment is located at an arbitrary point and oriented in an arbitrary direction.

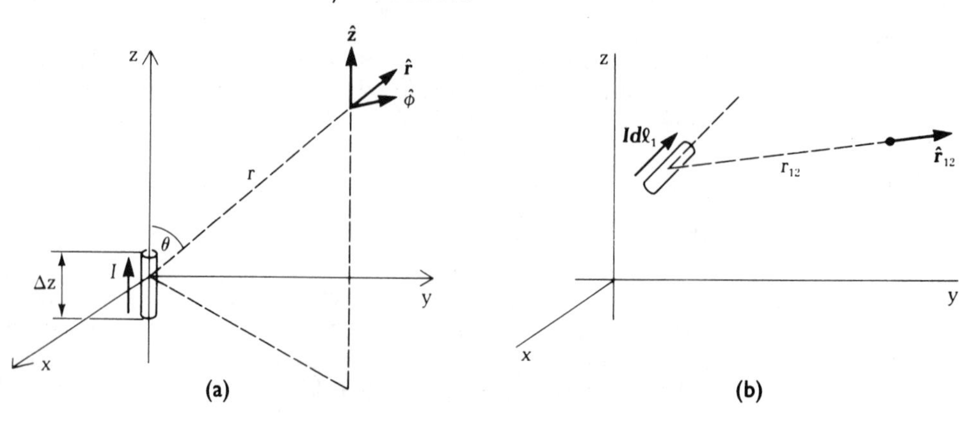

Equation (13.16) enables us to calculate what each individual segment of a current-carrying circuit contributes to the magnetic field. By summing up all these contributions, we can find the total magnetic field due to a complete circuit. Equation (13.16) is known as the **Biot-Savart law**.

Example 13.2

This example is a demonstration of the Biot-Savart law and the superposition principle. The total field is the vector sum of fields due to individual sources.

Using the Biot-Savart law, find the magnetic field $\boldsymbol{H}$ due to an infinitely long wire carrying current I, as shown in Figure 13.10.

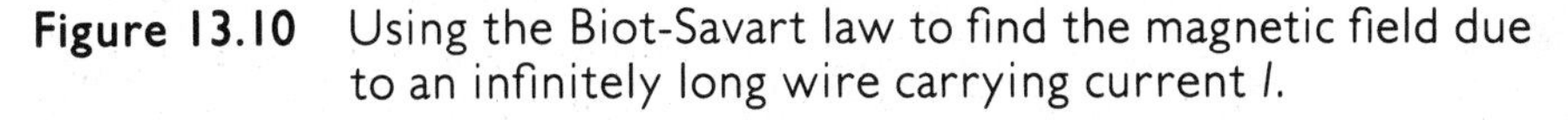

Figure 13.10 Using the Biot-Savart law to find the magnetic field due to an infinitely long wire carrying current I.

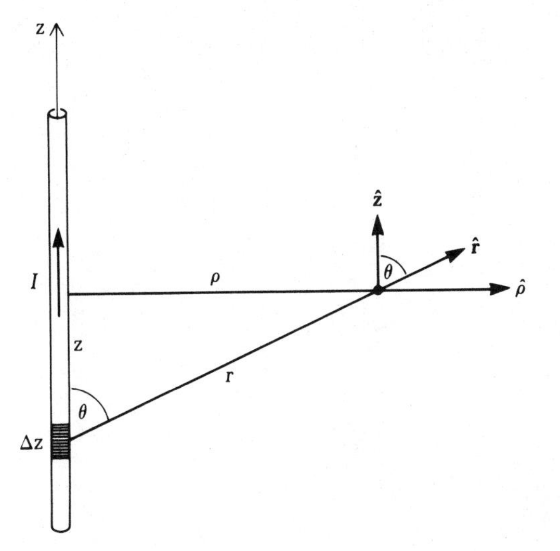

Solution From the Biot-Savart law, the magnetic field $d\boldsymbol{H}$ due to an infinitesimal section dz of the wire is given by (13.15a). Writing in cylindrical coordinates, we have

$$d\boldsymbol{H} = \hat{\boldsymbol{\phi}}\frac{I\,dz\,\sin\theta}{4\pi r^2} = \hat{\boldsymbol{\phi}}\frac{I\,dz}{4\pi(\rho^2 + z^2)}\frac{\rho}{\sqrt{\rho^2 + z^2}}$$

The total magnetic field $\boldsymbol{H}$ is obtained by integrating z from $-\infty$ to $+\infty$.

$$\boldsymbol{H} = \hat{\boldsymbol{\phi}}\frac{I}{4\pi}\int_{-\infty}^{\infty} dz\frac{\rho}{(\rho^2 + z^2)^{3/2}}$$

Letting $z = \rho\tan\alpha$, $dz = \rho\sec^2\alpha\,d\alpha$, we find

$$\boldsymbol{H} = \hat{\boldsymbol{\phi}}\frac{I}{4\pi}\int_{-\pi/2}^{\pi/2} \rho\sec^2\alpha\,d\alpha\frac{\rho}{\rho^3\sec^3\alpha} = \hat{\boldsymbol{\phi}}\frac{I}{2\pi\rho}$$

This answer is identical to (13.6), which was obtained by directly applying Ampère's law.

Example 13.3 *This example is another example of computing the magnetic field using the Biot-Savart law. Current-carrying loops are commonly used to generate magnetic fields (see the Helmholtz coils described in Problem 13.10).*

A circular loop of radius a carries a direct current I, as shown in Figure 13.11. Calculate the magnetostatic field $\boldsymbol{H}$ on the z axis.

Figure 13.11 A circular loop carries a current counterclockwise (viewed from the top). The magnetic field on the z axis is calculated in Example 13.3.

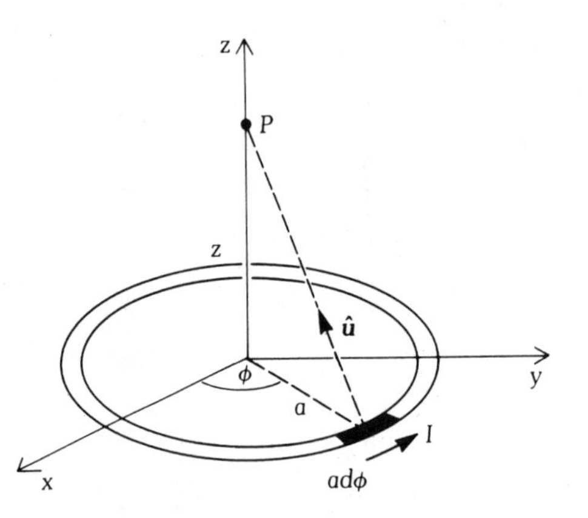

Solution We shall use (13.16) to solve this problem. Let us take a typical segment of the circular loop that is located at ϕ, which has an infinitesimal length $a\,d\phi$, and is oriented in the $\hat{\boldsymbol{\phi}}$ direction. At point P the magnetic field due to this segment of current is

$$d\boldsymbol{H} = \frac{Ia\,d\phi\hat{\boldsymbol{\phi}} \times \hat{\boldsymbol{u}}}{4\pi(a^2 + z^2)} \tag{13.17}$$

where the unit vector $\hat{\boldsymbol{u}}$ is pointing in the direction from the current segment toward the point P, as shown in Figure 13.11. Expressing $\hat{\boldsymbol{u}}$ in terms of $\hat{\boldsymbol{\rho}}$ and $\hat{\boldsymbol{z}}$ in cylindrical coordinates, we obtain:

$$\hat{\boldsymbol{u}} = \frac{-a\hat{\boldsymbol{\rho}} + z\hat{\boldsymbol{z}}}{\sqrt{a^2 + z^2}} \tag{13.18}$$

Substituting (13.18) into (13.17), we obtain

$$dH = \frac{Ia\,d\phi(a\hat{z} + z\hat{\rho})}{4\pi(a^2 + z^2)^{3/2}} \tag{13.19}$$

This equation can be integrated from $\phi = 0$ to 2π to obtain the total $\boldsymbol{H}$ field at P. Note that z and $\hat{z}$ remain unchanged as ϕ varies. However, $\hat{\rho}$ varies with ϕ. Integration of $\hat{\rho}$ over a circular path is equal to zero. We see this result by letting $\hat{\rho} = \hat{x}\cos\phi + \hat{y}\sin\phi$. The integral can then be performed along the constant directions $\hat{x}$ and $\hat{y}$. The result of the integration gives zero. Consequently, we obtain the following:

$$\boldsymbol{H} = \int_0^{2\pi} d\phi \frac{Iaa\hat{z}}{4\pi(a^2 + z^2)^{3/2}} = \hat{z}\frac{Ia^2}{2(a^2 + z^2)^{3/2}} \tag{13.20}$$

The above expression is for the $\boldsymbol{H}$ field at any point on the z axis. The expression of the $\boldsymbol{H}$ field at an arbitrary point in space is more complicated, but with a similar procedure it can be found in principle.

Magnetic Fields due to Power Lines

We computed the electric fields due to power lines in Section 11.2 and discussed the controversy about the safety issue. Here we show how to compute the magnetic fields generated by a typical power transmission line. The line consists of three parallel wires located at R_0 meters above the ground. Each wire consists of a bundle of conductors, but for our purpose we model the bundle of conductors as a single wire having an effective radius a. The wires are spaced s meters apart, as shown in Figure 13.12. The wires carry currents designated as I_1, I_2, and I_3,

Figure 13.12 A three-phase power transmission line and their images. The $\hat{\boldsymbol{y}}$ direction (pointing into the paper) is the positive direction for currents.

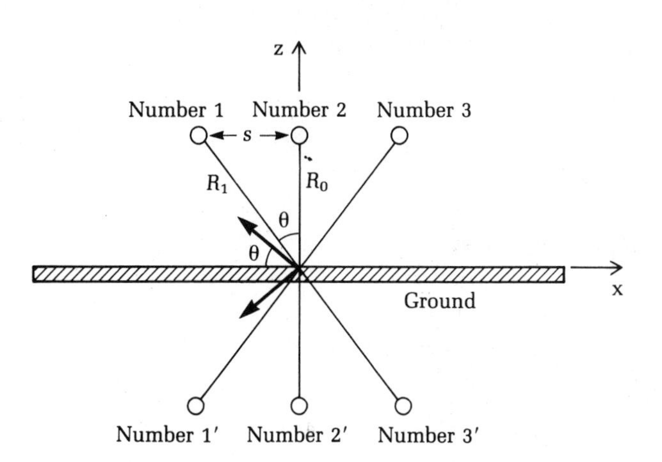

respectively. We further assume that these currents are three-phase currents; that is, $I_1 = I_0 \exp(-j2\pi/3)$, $I_2 = I_0$, and $I_3 = I_0 \exp(+j2\pi/3)$.

As mentioned in Section 11.2, rigorously speaking, the present problem is a dynamic problem and the fields are not magnetostatic. However, the frequency of the power line is low, and the wavelength in air is very large (5000 km at 60 Hz) as compared with the physical lengths R_0, s, and a, which are 10 m or less. We can therefore use the formulas derived for magnetostatic fields to compute magnetic fields generated by power lines with negligible error.*

Referring to Figure 13.12, the effect of the ground can be taken into account using the image theory. A current flowing in the $\hat{y}$ direction on a wire parallel to the ground can be thought to be produced by positive charges moving in the $\hat{y}$ direction. According to the image theory, there correspond negative charges moving in the $\hat{y}$ direction, or equivalently, a current of the same amount in the $-\hat{y}$ direction. Therefore, the Number 1′ wire carries $-I_1$, and Number 2′ and Number 3′ wires carry $-I_2$ and $-I_3$, respectively. The total magnetic field can then be calculated by summing the magnetic fields generated by individual wires and their images.

The magnetic field due to a wire carrying a line current I is given by (13.6):

$$H_\phi = \frac{I}{2\pi\rho}$$

Therefore, the $\boldsymbol{H}$ field on the ground directly under the Number 2 wire due to the Number 1 wire is as follows:

$$\boldsymbol{H}_1 = \frac{I_1}{2\pi R_1}(-\hat{x}\cos\theta - \hat{z}\sin\theta)$$

where $R_1 = (s^2 + R_0^2)^{1/2}$. Similarly, at the same location,

$$\boldsymbol{H}_2 = \frac{I_2}{2\pi R_0}(-\hat{x})$$

$$\boldsymbol{H}_3 = \frac{I_3}{2\pi R_1}(-\hat{x}\cos\theta + \hat{z}\sin\theta)$$

$$\boldsymbol{H}_{1'} = \frac{I_1}{2\pi R_1}(-\hat{x}\cos\theta + \hat{z}\sin\theta)$$

$$\boldsymbol{H}_{2'} = \frac{I_2}{2\pi R_0}(-\hat{x})$$

$$\boldsymbol{H}_{3'} = \frac{I_3}{2\pi R_1}(-\hat{x}\cos\theta - \hat{z}\sin\theta)$$

* A more rigorous justification is given in Chapter 15.

Summing all six equations shows that the $\hat{z}$ component of the total magnetic field is zero, and the $\hat{x}$ component is given as follows:

$$H_{tx} = \frac{-2 \cos \theta}{2\pi R_1}(I_1 + I_3) - \frac{2I_2}{2\pi R_0}$$

Substituting $I_1 = I_0 \exp(-j2\pi/3)$, $I_2 = I_0$, and $I_3 = I_0 \exp(+j3\pi/3)$ into the above equation yields

$$H_{tx} = \frac{-2I_0 s^2}{2\pi R_1^2 R_0} \tag{13.21}$$

For a high-voltage transmission line, the following data may be used:*

$$I_0 = 1 \text{ kA}$$

$$R_0 = 10.6 \text{ m}$$

$$s = 10 \text{ m}$$

and

$$a = 0.15 \text{ m}$$

Substituting these data in (13.21) and noting $B = \mu_0 H$, we have

$$B_{tx} = -0.18 \times 10^{-4} \text{ Tesla}$$

This is the magnetic flux on the ground directly under the Number 2 wire. At other locations on the ground, the flux takes different values with the same order of magnitudes. In comparison, the earth's magnetic field is approximately 0.5×10^{-4} Tesla.

Figure 13.13 shows the range of magnetic flux density generated by electric power lines and appliances at normal distances from these sources.** In comparison, the earth's magnetic field is also shown. It should be noted, however, that the earth's magnetic field is basically a steady field, whereas the fields produced by electric power sources vary sinusoidally with time. Note also that by careful redesign of the wiring in electric blankets, the average magnetic flux exposure is reduced by an order of magnitude.

The question of safety of the electromagnetic fields emanating from power lines and appliances is discussed in Section 11.2. Because experts' opinions vary widely and the controversy is by no means settled, we advocate (1) "prudent avoidance"—that is, to avoid unnecessary exposure to any artificial EM fields; and (2) to design electrical wiring in houses and in consumer products such that the resulting electromagnetic field leakage is minimized.

---

* J. J. LaForest, ed., *Transmission Line Reference Book*, Palo Alto, CA: Electric Power Institute, 1982, p. 332.

** P. Thomas, "Power Struggle," *Harvard Health Letter*, vol. 18, July 1993, p. 1. W. R. Bennett, Jr., "Cancer and power lines," *Physics Today*, April 1994, p. 23.

Figure 13.13 Typical human exposure to magnetic flux density generated by electric power sources.

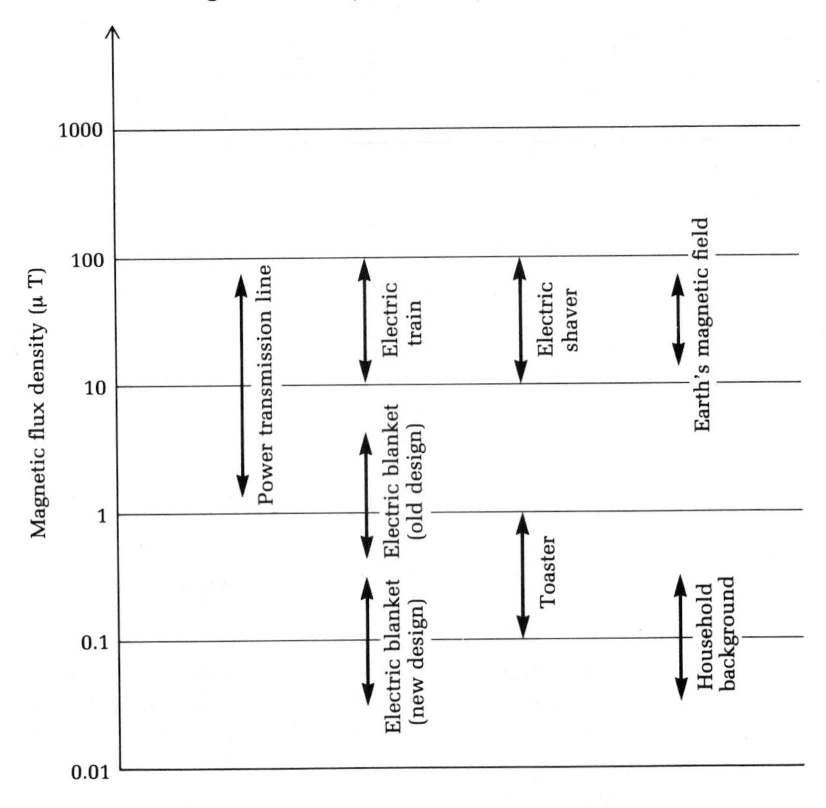

13.2 MAGNETIC FORCE AND TORQUE

In Chapter 10 we introduced the force on a charged particle q placed in an electrostatic field. The Lorentz force law states that

$$\boldsymbol{F} = q(\boldsymbol{E} + \boldsymbol{v} \times \boldsymbol{B}) \tag{13.22}$$

The second term in the above equation is ineffective when we are concerned with electrostatic fields only. In this chapter, however, we are solely concerned with magnetostatic fields. Consequently, we will study the specialized case in which $\boldsymbol{E}$ is equal to zero and the Lorentz force becomes

$$\boxed{\boldsymbol{F} = q\boldsymbol{v} \times \boldsymbol{B}} \quad \textbf{(magnetic force on a moving charge)} \tag{13.23}$$

The above equation describes the force due to a magnetic field on a charge of q coulombs moving with velocity $\boldsymbol{v}$. If we have many charges moving at a common velocity $\boldsymbol{v}$, they constitute a current. Referring in Figure 13.14, we assume that there are n charges per unit volume in the space, that all move at a constant velocity $\boldsymbol{v}$, and that each carries q coulombs of charge. The current I passing through an area A is defined to be the total charge passing through A in one second. Therefore,

$$I = nqAv \quad \textbf{(13.24)}$$

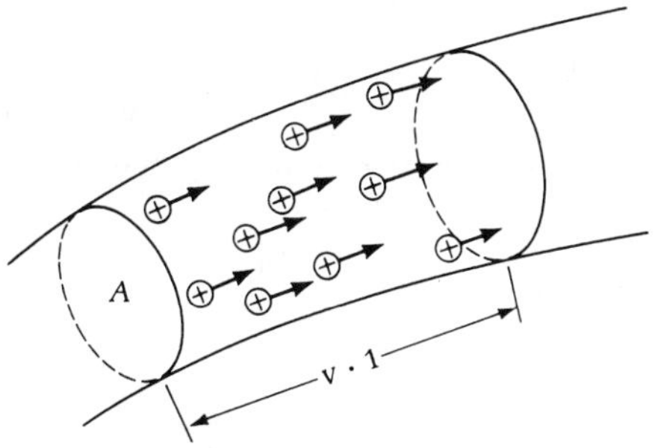

Figure 13.14 There are n charges per cubic meter moving with $\mathbf{v}$ in space. Each charge carries q coulombs. The current I in the cylinder with cross section A is defined to be $I = nqAv$.

The total magnetic force on an assembly of charges in a volume dV, according to (13.23) is

$$d\boldsymbol{F} = nq\boldsymbol{v}\,dV \times \boldsymbol{B} \quad \textbf{(13.25)}$$

Using (13.24), and the relation $dV = A\,d\ell$, we find

$$\boxed{d\boldsymbol{F} = I\,d\boldsymbol{\ell} \times \boldsymbol{B}} \quad \textbf{(magnetic force on a current-carrying element)} \quad \textbf{(13.26)}$$

The above equation expresses the magnetic force on a conductor that has length $d\ell$ and carries a current I in the direction of $d\boldsymbol{\ell}$.

Example 13.4

A charged particle moving through a uniform magnetic field experiences a force that makes it take a circular trajectory. This phenomenon has led to many important applications.

A charged particle of mass m and charge q is initially moving with a velocity $\boldsymbol{v}$ in a uniform magnetic field $\boldsymbol{B}_0$ that is pointing into the paper or in the $-\hat{z}$ direction, as shown in Figure 13.15. Find the particle's trajectory.

Solution Assume that the charged particle is initially at position A as shown. Then $\boldsymbol{v} = v_0\,\hat{x}$. Using (13.23), we obtain the force $\boldsymbol{F}$:

$$\boldsymbol{F} = qv_0\,\hat{x} \times B_0(-\hat{z}) = qv_0\,B_0\,\hat{y}$$

Figure 13.15 A charge is moving circularly under the influence of a magnetostatic field.

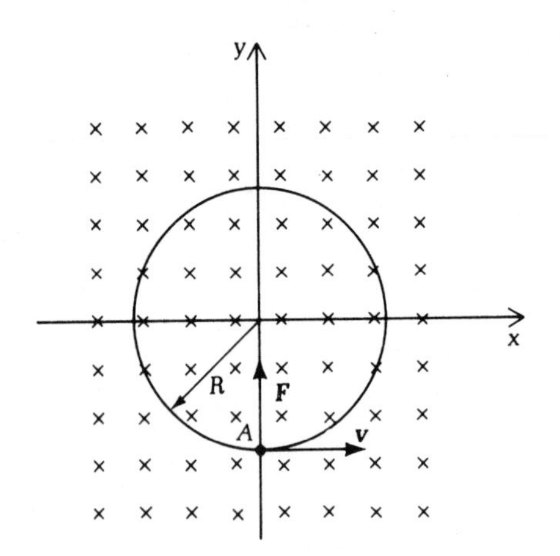

Note that $\boldsymbol{F}$ is perpendicular to $\boldsymbol{v}$. Because of the cross product, $\boldsymbol{F}$ is always perpendicular to $\boldsymbol{v}$. This perpendicularity suggests a circular motion. To maintain a circular motion, the force must be equal to the centrifugal force—that is,

$$qv_0 B_0 = m\frac{v_0^2}{R}$$

Thus, the charged particle will move along a circle with radius R, where

$$R = \frac{mv_0}{qB_0} \tag{13.27a}$$

Notice that the angular frequency of the particle ω_c is independent of the velocity v_0.

$$\omega_c = \frac{v_0}{R} = \frac{qB_0}{m} \tag{13.27b}$$

The quantity $f_c = \omega_c/2\pi$ is called the **cyclotron frequency** of the particle in the magnetic field of strength B_0. If we inject several identical charged particles of different velocities into this magnetic field, they will move in different circles but with the same frequency. According to (13.27), the faster particles will move along larger circles and the slower ones along smaller circles.

Remember that, because $\boldsymbol{F}$ is always perpendicular to $\boldsymbol{v}$, the magnetic force never does work to the charge. In other words, the magnetic field may change the direction of motion of the moving charge, but it does not increase or decrease its speed.

Isotope Separation

Isotopes are elements occupying the same place in the periodic table but having different atomic weights. Hundreds of different isotopes occur in nature. For example, the well-known element of uranium has three natural isotopes: Uranium 238 (99.28%), Uranium 235 (0.715%), and Uranium 234 (0.005%). Because these isotopes occupy the same place in the periodic table, their chemical properties are similar. Consequently, something other than chemical methods must be used to separate isotopes from their natural mixtures.

Figure 13.16 illustrates a method of isotope separation. The compound of isotopes is first vaporized, then ionized by electric discharge. Next, these ions are passed through slits in a parallel-plate region, where they are accelerated by the electrostatic field. The kinetic energy of these accelerated ions is thousands of times greater than their average thermal energy. Thus, their kinetic energy is approximately equal to qV, where q is the charge of the ionized isotope and V is the potential difference between the plates. Note that kinetic energy is not dependent on mass. Thus, all isotopes have the same kinetic energy. However, because of their different masses, these isotopes have different velocities. The ratio of the velocities of ^{235}U and ^{238}U is obtained as follows:

$$\frac{v_5}{v_8} = \left(\frac{m_8}{m_5}\right)^{1/2}$$

Here, v_5 and m_5 are the velocity and mass of ^{235}U, respectively, and v_8 and m_8 are the velocity and mass of ^{238}U.

Figure 13.16 An electromagnetic isotope separator. The electric field in the parallel-plate region accelerates the ionized isotopes, which are then passed through a magnetic field. Heavier isotopes travel along a slightly larger circular path than do lighter ones.

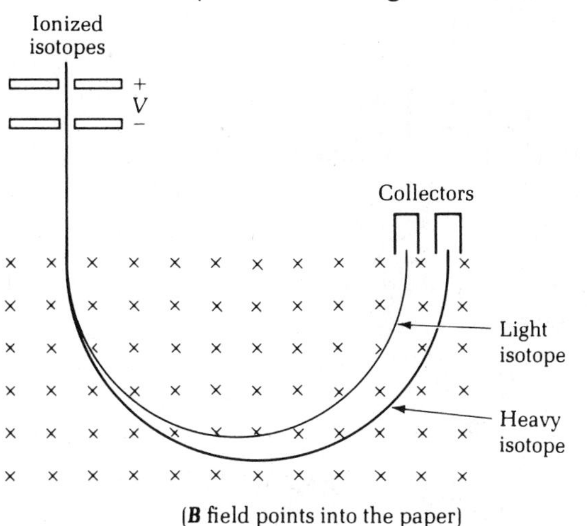

These particles then enter a strong uniform magnetic field. According to the result obtained in Example 13.4, they follow circular paths of different radii. From Example 13.4, we obtain

$$\frac{R_5}{R_8} = \frac{m_5 v_5}{m_8 v_8} = \left(\frac{m_5}{m_8}\right)^{1/2} = \left(\frac{235}{238}\right)^{1/2} = 0.9937$$

where R_5 and R_8 are respectively the radii of the circular paths along which the ^{235}U and ^{238}U particles travel. Thus, to obtain the separated isotopes, collectors may be positioned accordng to this ratio. The first appreciably sized samples of pure ^{235}U were produced in 1942 by an electromagnetic separator based on the principle illustrated in Figure 13.16.*

Cyclotron

A cyclotron can be used to produce high-energy charged particles (Figure 13.17a). Between two large pieces of an electromagnet, two hollow, D-shaped

Figure 13.17(a) A schematic diagram of a cyclotron.

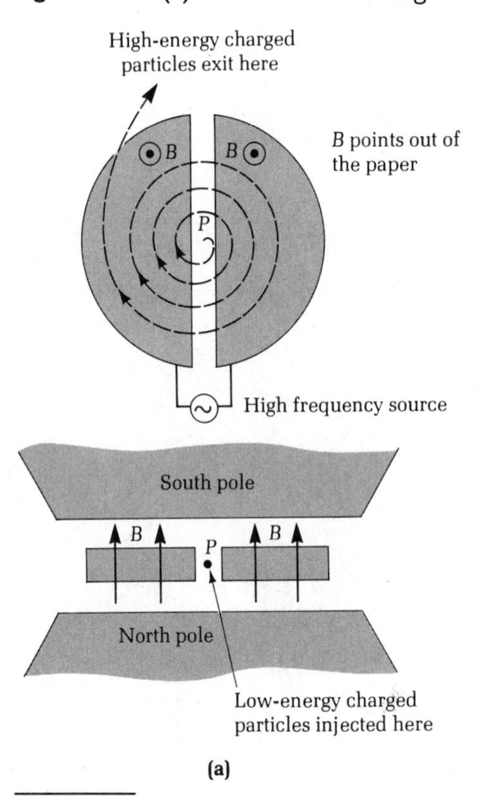

(a)

* H. D. Smyth, *Atomic Energy for Military Purposes*, Princeton, N.J.: Princeton University Press, 1945, p. 188.

copper electrodes are connected to a high-frequency source of alternating voltage. Starting at a point P, a positively charged particle is attracted to the D-shaped region on the right-hand side. The charge moves in a semicircular path inside the D-shaped region, where the electric field is absent but the magnetic field still exists. It comes back to the electric field and gets accelerated to the left D-shaped region because the alternating electric field has changed its own polarity. As we know, to produce this action, the alternating electric field should be applied at the same frequency as that of the circular motion of the particle given by (13.27):

$$f_c = \frac{qB_0}{2\pi m}$$

Notice that f_c is independent of the linear velocity of the particle and that it remains unchanged as the particle is accelerated repeatedly when passing through the gap. Thus, a constant-frequency source may be used to generate the alternating voltage synchronous with the angular motion of the particle. After many such revolutions, the particle exits with high velocity and high energy. Figure 13.17b shows an early model of a cyclotron.

Figure 13.17(b) A cyclotron built at Harvard University in 1931. (Courtesy of Harvard University, photographed by P. H. Donaldson of Harvard University, 1939.)

Example 13.5 *An important application of the interaction between the magnetic field and charged particles is the cyclotron. This example shows the relevant parameters of such a device.*

A deuterium particle with mass $m = 3.34 \times 10^{-27}$ kg and electric charge $q = 1.6 \times 10^{-19}$ C is produced and accelerated in a cyclotron with a radius of 1 m and a magnetic field of 20,000 G (2 Wb/m^2). Calculate the energy of the particle at the exit.

Solution The angular velocity of the deuterium particle depends only on q, B_0, and m. From (13.27b), we have

$$\omega_c = \frac{qB_0}{m} = 9.6 \times 10^7 \text{ rad/s}$$

The linear velocity at the exit is $R\omega_c$, or

$$v = R\omega_c = 9.6 \times 10^7 \text{ m/s}$$

The kinetic energy of the particle at the exit is W, where

$$W = \frac{1}{2}mv^2 = 1.53 \times 10^{-11} \text{ J} = 96 \text{ MeV}$$

Notice that, in the above calculation, a more accurate relativistic approach should be used when the velocity of the particle is approaching the velocity of light, which the particle velocity can never exceed. From the above calculations, one can also see that the cyclotron is not suitable for accelerating electrons because the electron's mass is so small that the electron reaches the limit of the velocity of light at energies of only a few MeV. The corresponding relativistic increase of the electron mass affects the cyclotron frequency, and the electrons soon get out of "sync" with the applied alternating voltage.

Hall Effect

Consider a current of positive charge flowing inside a conducting slab, as shown in Figure 13.18a. When the current is placed in a magnetic field perpendicular to the slab, the magnetostatic force ($\boldsymbol{J} \times \boldsymbol{B}$) acts on these charges and displaces them in the $\hat{x}$ direction, as shown. The charges give rise to a voltage V_h across the slab's two sides, which, in the absence of the magnetic field, have no difference in potential. This phenomenon is called the **Hall effect**, and the voltage V_h is called the **Hall voltage.**

When the current is due to moving negative charges, we see from Figure 13.18b that the Hall voltage is opposite to that illustrated in Figure 13.18a. By measuring the Hall voltage, one can distinguish a p-type semiconductor (in which positive holes are drifting to constitute the current flow) from an n-type semiconductor (in which negative electrons are flowing).

Figure 13.18 The Hall effect. The magnetic force pushes charges flowing in a conductor along the $\hat{x}$ direction. The result is that a voltage in the transverse direction is created. **(a)** The current is due to drifting positive charges or holes. **(b)** The current is due to negative charges.

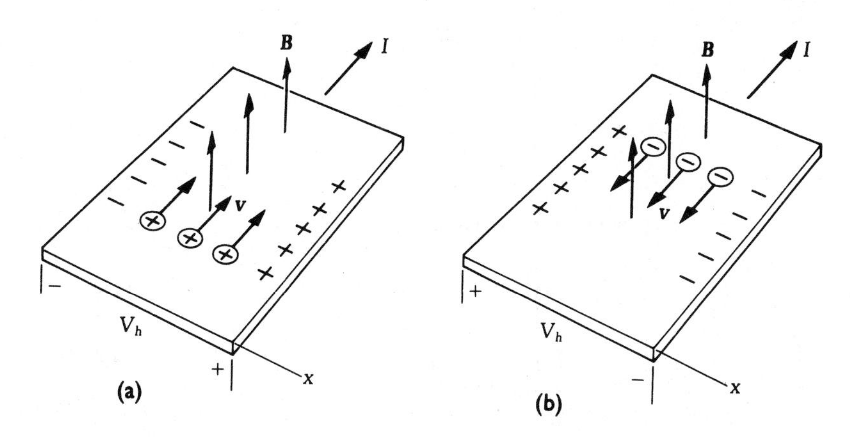

Magnetic Deflection of the Electron Beam

In a cathode ray tube (CRT) or in the picture tube of a television receiver, an electron beam is deflected to predetermined spots on the display screen either by an electrostatic field or by a magnetostatic field. Section 10.2 described the operating principle of a CRT. Figure 10.8 showed a sketch of the CRT.

The deflecting plates shown in Figure 10.8 generate an electrostatic field to deflect the electron beam. They can be replaced by coils to produce magnetic fields to accomplish the same task, as illustrated in the following example.

Assume that an electron with a velocity of 2×10^7 m/s enters a magnetic field of $\boldsymbol{B}_0 = 5 \times 10^{-4}\ \hat{y}$ webers per square meter, as shown in Figure 13.19. The magnetic field is assumed to be restricted to a 4-cm-long region. Let us find the trajectory of this electron.

From the analysis carried out in Example 13.4, we know that the electron follows a circular path once it enters the magnetic field. The radius of the circle is R, where

$$R = \frac{mv_0}{qB_0} = 0.228 \text{ m}$$

According to the coordinates shown in Figure 13.19, the velocity of the electron is given by

$$v_x = v_0 \cos(\omega_c t) \tag{13.28a}$$

$$v_z = -v_0 \sin(\omega_c t) \tag{13.28b}$$

Figure 13.19 An electron is deflected by a magnetic field pointing into the paper. The trajectory in the magnetic-field region is a circular arc.

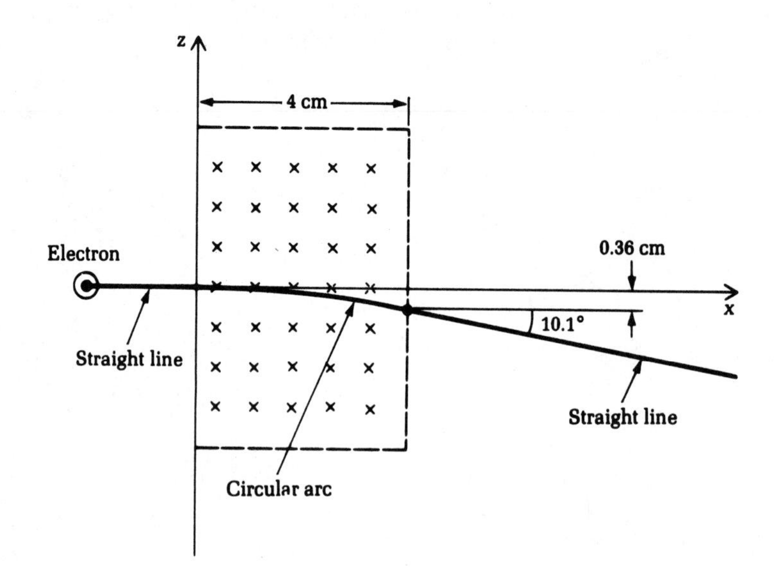

where $v_0 = 2 \times 10^7$ m/s and ω_c is given as follows:

$$\omega_c = \frac{qB_0}{m} = 8.78 \times 10^7 \text{ rad/s}$$

We have set t equal to zero at the time the electron enters the magnetic field. We obtain the position of the electron in the region of the magnetic field from (13.28) by integration:

$$x = \frac{v_0}{\omega_c} \sin(\omega_c t)$$

$$z = -\frac{v_0}{\omega_c}[1 - \cos(\omega_c t)]$$

At the point of exit from the magnetic field, $x_1 = 0.04$. Therefore,

$$\sin \omega_c t_1 = \frac{0.04\omega_c}{v_0} = 0.176$$

It follows that

$$\cos \omega_c t_1 = 0.984$$

The point of exit is therefore at $x_1 = 0.04$ m, and

$$z_1 = -0.0036 \text{ m}$$

The electron will follow a straight line after the exit. The straight line makes an angle θ with the x axis, where

$$\tan\theta = \frac{-v_z}{v_x} = \frac{0.176v_0}{0.984v_0}$$

$$\theta = 10.1^\circ$$

Figure 13.19 shows the entire trajectory. It is interesting to compare this result with the one Figure 10.1 shows for Example 10.2. Notice that we have chosen the magnitude of the force acting on the electron to be the same as that in Example 10.2. The trajectory in the electric field is a parabola, whereas it is a circular path in the magnetic field. The exit positions and the angles of the straight-line trajectories are almost identical in both cases.

Effect of Earth's Magnetic Field on CRT Display*

The natural magnetic field on earth is of the order of 0.5×10^{-4} Tesla. The horizontal component of the magnetic field points approximately in the north-south direction and the vertical component points downward in the northern hemisphere and upward in the southern hemisphere. [A sketch of the earth's magnetic field lines is shown in Figure 14.8(a), p. 500.]

From the numerical results given in the preceding topic, it is seen that the strength of the earth's natural magnetic field is strong enough to affect the trajectory of the electron beam in the CRT. Manufacturers of CRTs always adjust the deflecting magnetic yoke in an environment that simulates the natural magnetic field at the product's destination. Using a CRT compensated for the northern hemisphere in the southern hemisphere, and vice versa, could result in a shift of 2 to 3 mm in the display's center and a 1 to 2 mm tilt from the corner to the diagonally opposite corner. Displayed colors may also lose their sharpness. In addition, CRTs are usually fine-tuned at the factory while facing east. Even when a display has been adjusted for the correct hemisphere operation, it may be slightly distorted if oriented in the north-south direction. Televisions are similarly affected. However, because pictures on the television are usually changed rapidly, the effect of the earth's magnetic field is less noticeable.

Example 13.6 *A current-carrying element in a magnetic field produced by other sources experiences a magnetic force. This example shows how to compute that force.*

A rectangular loop $ABCD$, as shown in Figure 13.20, is near an infinitely long wire carrying 10 A of current. Find the total magnetic force on the loop.

---

* K. Fitzgerald, "Don't move that CRT," *IEEE Spectrum*, December, 1989, p. 16.

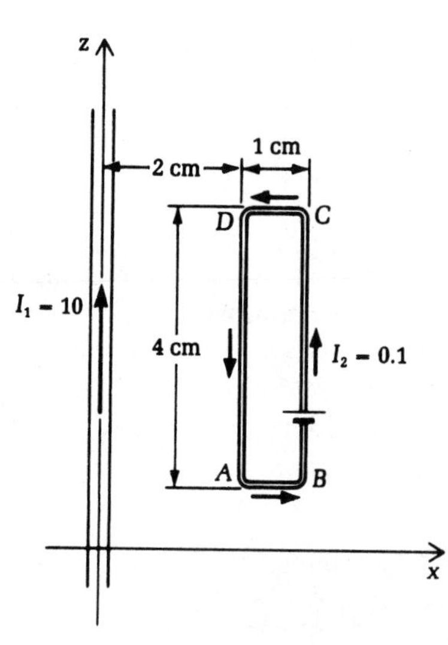

Figure 13.20 A rectangular loop is near an infinitely long wire. The total force on the loop from the magnetic field of the wire is calculated in Example 13.6.

Solution The magnetic field $\boldsymbol{H}$ due to the infinitely long wire is given by (13.6):

$$\boldsymbol{H} = \frac{I_1}{2\pi x}\hat{y}$$

According to (13.26), the force on the segment AB is given below

$$\boldsymbol{F}_{AB} = \int_{0.02}^{0.03} I_2\, dx\hat{x} \times \frac{I_1\mu_0}{2\pi x}\hat{y} = \frac{\mu_0}{2\pi}\ln\left(\frac{3}{2}\right)\hat{z} = 0.81 \times 10^{-7}\hat{z} \text{ newtons}$$

Forces acting on segments BC, CD, and DA are obtained similarly:

$$\boldsymbol{F}_{BC} = I_2 \times 0.04\hat{z} \times \frac{\mu_0 I_1}{2\pi(0.03)}\hat{y} = 2.67 \times 10^{-7}(-\hat{x}) \text{ newtons}$$

$$\boldsymbol{F}_{CD} = -\boldsymbol{F}_{AB} = 0.81 \times 10^{-7}(-\hat{z}) \text{ newtons}$$

$$\boldsymbol{F}_{DA} = I_2 \times 0.04(-\hat{z}) \times \frac{\mu_0 I_1}{2\pi(0.02)}\hat{y} = 4.00 \times 10^{-7}\hat{x} \text{ newtons}$$

The total force on the loop is the vector sum of the four forces given above. Therefore,

$$\boldsymbol{F} = \boldsymbol{F}_{AB} + \boldsymbol{F}_{BC} + \boldsymbol{F}_{CD} + \boldsymbol{F}_{DA} = 1.33 \times 10^{-7}\hat{x} \text{ newtons}$$

Magnetic Torque

Consider a rectangular loop carrying a direct current I. The loop is placed in a uniform magnetostatic field with $\boldsymbol{B} = B_x\hat{x}$, as shown in Figure 13.21a. The magnetic forces on each of the four sides of the loop may be obtained by using

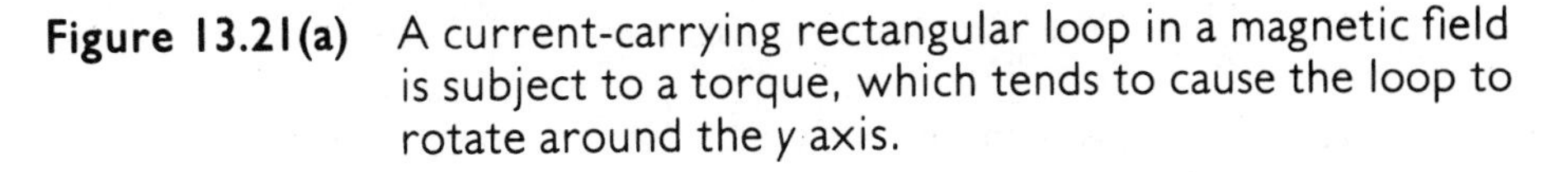
Figure 13.21(a) A current-carrying rectangular loop in a magnetic field is subject to a torque, which tends to cause the loop to rotate around the y axis.

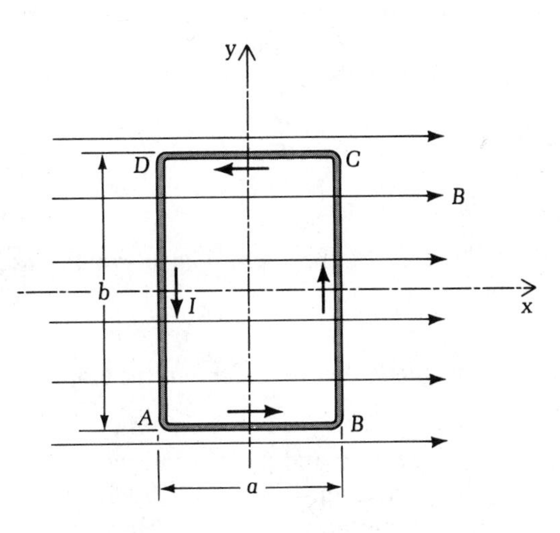

Figure 13.21(b) An irregularly shaped current-carrying loop in a magnetic field is equivalent to many rectangularly shaped subloops each carrying the same current I.

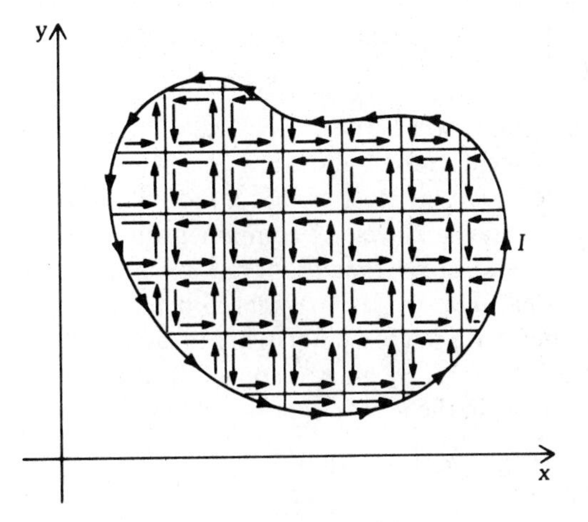

(13.26). The results are as follows:

$$\boldsymbol{F}_{AB} = 0$$

$$\boldsymbol{F}_{BC} = -\hat{\boldsymbol{z}} I b B_x$$

$$\boldsymbol{F}_{CD} = 0$$

$$\boldsymbol{F}_{DA} = \hat{\boldsymbol{z}} I b B_x$$

The total force on this loop is equal to zero. Therefore, under these conditions the loop does not have a linear motion. However, the loop tends to rotate around the y axis in the direction described by the right-hand rule: the thumb of the right hand points in the $\hat{y}$ direction, while the fingers follow the direction of rotation of the loop. The torque produced as the result of interaction between the current-carrying loop and the magnetostatic field is given below:

$$\begin{aligned} \boldsymbol{T} &= \sum_i \boldsymbol{r}_i \times \boldsymbol{F}_i = \frac{1}{2}a\hat{x} \times \boldsymbol{F}_{BC} - \frac{1}{2}a\hat{x} \times \boldsymbol{F}_{DA} = IabB_x\hat{y} \\ &= IAB_x\hat{y} \qquad \text{for} \quad \boldsymbol{B} = B_x\hat{x} \end{aligned} \tag{13.29}$$

where $A = ab$ is the area of the loop. The **magnetic moment** $\boldsymbol{m}$ of a current-carrying loop is defined to be

$$\boxed{\boldsymbol{m} = AI\hat{\boldsymbol{u}}} \qquad \textbf{(magnetic moment of a current loop)} \tag{13.30}$$

where A is the area of the loop and $\hat{\boldsymbol{u}}$ is the unit vector indicating the orientation of the loop when using the right-hand rule; the fingers of the right hand follow the current, while the thumb points in the $\hat{\boldsymbol{u}}$ direction, which is perpendicular to the loop. In the above example, $\hat{\boldsymbol{u}} = \hat{z}$. If we use the definition of $\boldsymbol{m}$ given in (13.30), we may rewrite the torque on the rectangular loop given by (13.29) as follows:

$$\boxed{\boldsymbol{T} = \boldsymbol{m} \times \boldsymbol{B}} \qquad \textbf{(torque on a current loop)} \tag{13.31}$$

The above equation is derived specifically for the case of a rectangular loop with a uniform $\boldsymbol{B}$ field in the $\hat{x}$ direction. We shall prove that it also holds for an arbitrarily shaped loop (as long as it is a flat or planar loop) with $\boldsymbol{B}$ oriented in an arbitrary direction. If $\boldsymbol{B}$ has a $\hat{z}$ component in addition to its $\hat{x}$ component, it is not too difficult to prove that the forces on opposite sides of the loop cancel each other and that they are also parallel to the x-y plane. Thus, the $\hat{z}$ component of $\boldsymbol{B}$ produces neither a net force nor a net torque. Following a procedure similar to the one that led to (13.29), we can show that the effect of a $\hat{y}$-directed $\boldsymbol{B}$ field is to produce a torque in the $-\hat{x}$ direction—that is

$$\boldsymbol{T} = 0 \qquad \text{for} \quad \boldsymbol{B} = B_z\hat{z} \tag{13.32}$$

$$\boldsymbol{T} = IAB_y(-\hat{x}) \qquad \text{for} \quad \boldsymbol{B} = B_y\hat{y} \tag{13.33}$$

Combining (13.29), (13.32), and (13.33), we obtain (13.31).

Now let us show that, for (13.31) to hold, the shape of the loop is immaterial. Consider the irregularly shaped loop shown in Figure 13.21b. The loop is divided into many small rectangularly shaped subloops. Viewed from the top, each subloop carries a counterclockwise current. The torque for a typical ith subloop is given by (13.31):

$$\boldsymbol{T}_i = \boldsymbol{m}_i \times \boldsymbol{B} = IA_i\hat{z} \times \boldsymbol{B}$$

Thus, the total $\boldsymbol{T}$ is the sum of all subloops:

$$\boldsymbol{T} = \sum_i IA_i \hat{z} \times \boldsymbol{B} = I\hat{z} \times \boldsymbol{B} \sum_i A_i = I\hat{z} \times \boldsymbol{B}A$$

The above equation is identical to (13.31). Notice that there are opposing currents on the common boundary of two neighboring subloops and that they effectively cancel out. The only exceptions are the boundaries on the perimeter of the loop. Thus, one may justifiably consider an irregularly shaped current loop to be equivalent to many small rectangular loops each carrying the same current I. This conclusion completes our proof for (13.31).

DC motor

A direct-current motor, or dc motor, is a device that takes advantage of (13.31) to produce a torque by passing a direct current through a loop that is immersed in a magnetostatic field. Figure 13.22 shows a simplified sketch of a dc motor. A rectangular loop of N turns (called the **armature**) is placed in a magnetic field produced by another set of coils (called the **field magnet**). A dc current is passed through the loop via two semiloops (called the **commutator**), so that the current always flows in one direction and the magnetic moment always has an upward component. The torque on the loop is then given by (13.31)

$$\boldsymbol{T} = NIAB \sin \alpha(-\hat{z})$$

Figure 13.22 Direct-current (dc) motor. **(a)** The two semicircular rings ensure that current always flows in the same direction. **(b)** The magnetic moment $\boldsymbol{m}$ always points toward the upper half of the x-y plane.

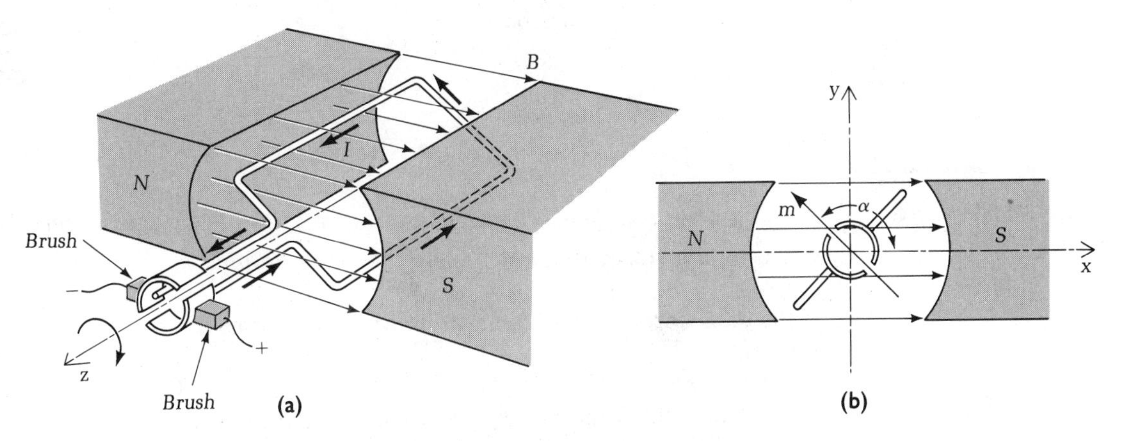

where α is the angle between the direction of the loop and the magnetic field. Notice that the angle α varies between 0 and π so that $\sin \alpha$ is always positive. Therefore, $\boldsymbol{T}$ is always in the $-\hat{z}$ direction, and the dc motor always rotates in the same direction.

13.3 STORED MAGNETIC ENERGY AND INDUCTANCE

Just as an electric field represents stored electric energy, a magnetic field represents stored magnetic energy. The discussion of Poynting's theorem in Chapter 2 identified the following term as the stored magnetic energy U_H:

$$U_H = \frac{1}{2}\int \mu \boldsymbol{H} \cdot \boldsymbol{H}\, dV = \frac{1}{2}\int \boldsymbol{B} \cdot \boldsymbol{H}\, dV \quad \textbf{(stored magnetic energy)} \qquad \textbf{(13.34)}$$

Example 13.7 *A device that holds magnetic fields is also holding magnetic energy. This example shows how to compute the stored magnetic energy in a solenoid.*

Find the stored magnetic energy per unit length in a solenoid of radius a with n turns per meter of coil and I amperes of current in the coil.

Solution According to (13.10), we find that the magnetic field in a solenoid is given as follows:

$$\boldsymbol{H} = \hat{z} n I$$

Substituting into (13.34), we find the stored magnetic energy per unit length:

$$U_H = \frac{1}{2}\mu n^2 I^2 (\pi a^2) \text{ joules per meter} \qquad \textbf{(13.35)}$$

Example 13.8 *A common device to hold magnetic energy is a toroid. This example shows how to compute stored magnetic energy in a thin toroid.*

Find the total stored magnetic energy in the toroid shown in Figure 13.8. Assume that $b \gg a$.

Solution The magnetic field in a toroid is given by (13.13) for $b \gg a$. Therefore, using (13.34), we obtain

$$U_H = \frac{1}{2}\mu\left(\frac{NI}{2\pi b}\right)^2 (\pi a^2)(2\pi b) = \frac{1}{4}\mu (NI)^2 \left(\frac{a^2}{b}\right) \text{ joules} \qquad \textbf{(13.36)}$$

Inductance

Let a direct current I flow in the closed contour, as shown in Figure 13.23. The current produces a magnetic field linking the loop. The total flux of B field going through the loop, called the **flux linkage** ψ, may be found by the following integral:

$$\psi = \int d\mathbf{a} \cdot \mathbf{B} \qquad \textbf{(definition of flux linkage)} \qquad \textbf{(13.37)}$$

Figure 13.23 A current flowing in a loop generates a magnetic field that goes through the loop. The self-inductance is defined to be the total magnetic flux linkage divided by I.

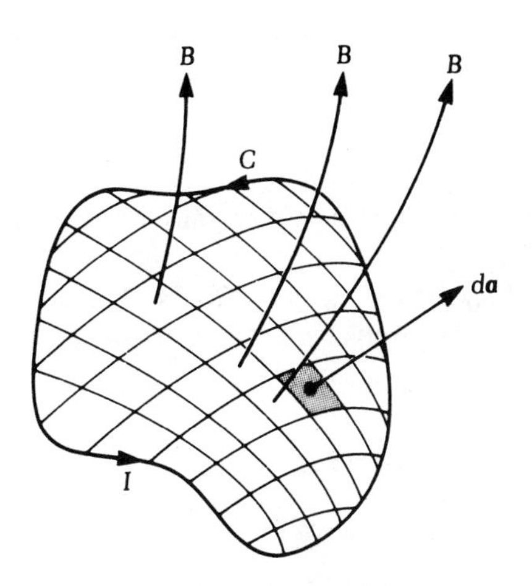

where $d\mathbf{a}$ is a differential area oriented in the direction shown. When the closed loop consists of N turns and carries the same current I, the total flux linkage is $\Lambda = N\psi$. The inductance L of the loop is defined to be the ratio of the total magnetic-flux linkage Λ to the current I in the loop.

$$L = \frac{\Lambda}{I} = \frac{N\psi}{I} \qquad \textbf{(definition of inductance)} \qquad \textbf{(13.38)}$$

The unit of inductance is webers per ampere, which is also designated as henrys.

The flux produced by I may also link other loops near it. Let there be two loops with currents I_1 and I_2. The flux produced by I_1, which links the loop carrying I_2, is denoted by ψ_{21}. The mutual inductance is defined as

$$\boxed{M_{21} = \frac{N_2 \psi_{21}}{I_1}} \quad \textbf{(definition of mutual inductance } \boldsymbol{M_{21}}\textbf{)} \tag{13.39a}$$

By symmetry, one can define a mutual inductance M_{12}, which is related to the flux linkage on loop 1 due to current I_2:

$$\boxed{M_{12} = \frac{N_1 \psi_{12}}{I_2}} \quad \textbf{(definition of mutual inductance } \boldsymbol{M_{12}}\textbf{)} \tag{13.39b}$$

We shall prove that

$$\boxed{M_{12} = M_{21}} \quad \textbf{(an important result about mutual inductances)} \tag{13.40}$$

by using energy concepts.

Recall that a capacitor may be regarded as a system holding electric charges. Or, as pointed out in Section 10.3, a capacitor may be equivalently defined as a system storing electric energy. In this section we define inductance using (13.38). We shall show that an inductor may equivalently be defined as a device to store magnetic energy. Specifically, an equivalent definition for the inductance L is that

$$\boxed{L = \frac{U_H}{\frac{1}{2}I^2}} \quad \textbf{(alternate definition of inductance)} \tag{13.41}$$

In terms of vector potential $\boldsymbol{A}$ defined in Section 7.1, we use (13.34) to write

$$\begin{aligned} U_H &= \frac{1}{2} \iiint_V dV (\nabla \times \boldsymbol{A}) \cdot \boldsymbol{H} \\ &= \frac{1}{2} \left[\iiint_V dV \nabla \cdot (\boldsymbol{A} \times \boldsymbol{H}) + \iiint_V dV \boldsymbol{A} \cdot (\nabla \times \boldsymbol{H}) \right] \end{aligned} \tag{13.42}$$

where we use the following vector identity:

$$\nabla \cdot (\boldsymbol{A} \times \boldsymbol{H}) = \boldsymbol{H} \cdot (\nabla \times \boldsymbol{A}) - \boldsymbol{A} \cdot (\nabla \times \boldsymbol{H}) \tag{13.43}$$

The divergence theorem can change the first term in (13.42) to a surface integral. The surface integral is zero because the surface must enclose all the magnetic energy. Therefore, the surface must be at infinity where $\boldsymbol{A}$ and $\boldsymbol{H}$ are zero. In the second term of (13.42), Ampère's law (13.2) gives $\nabla \times \boldsymbol{H} = \boldsymbol{J}$, which results in an integrand of the form $\boldsymbol{A} \cdot \boldsymbol{J} dV$. We may write $\boldsymbol{J} dV = I\, d\boldsymbol{\ell}$ (amperes per meter), where $d\boldsymbol{\ell}$ denotes the differential filament and the direction of current flow for the loop shown in Figure 13.23. The volume integral now becomes a line integral around the loop. We thus have

$$U_H = \frac{1}{2} \oint \boldsymbol{A} \cdot d\boldsymbol{\ell} \tag{13.44}$$

Using Stoke's theorem and substituting in (13.41), we obtain

$$L = \frac{1}{I} \iint \nabla \times \boldsymbol{A} \cdot d\boldsymbol{a} = \frac{1}{I} \iint \boldsymbol{B} \cdot d\boldsymbol{a}$$

which is seen to be identical to (13.38) for a single turn ($N = 1$) of the loop.

The mutual inductances M_{12} and M_{21} may be defined similarly to the alternate definition for self-inductance as follows:

$$M_{12} = \frac{1}{I_1 I_2} \iiint_V \mu \boldsymbol{H}_1 \cdot \boldsymbol{H}_2 \, dV \tag{13.45a}$$

$$M_{21} = \frac{1}{I_2 I_1} \iiint_V \mu \boldsymbol{H}_2 \cdot \boldsymbol{H}_1 \, dV \tag{13.45b}$$

which can be shown to be equivalent to (13.39). Clearly, (13.40) follows immediately from (13.45).

Example 13.9 *A device that can store magnetic energy is called an inductor. This example shows how to compute the inductance of a solenoidal inductor.*

Calculate the inductance per unit length of a solenoid of radius a with n turns per unit length.

Solution The magnetic energy stored in a solenoid is found in Example 13.7. Using the definition of L given by (13.41), we obtain the inductance per unit length of the solenoid as follows:

$$L = \mu n^2 \pi a^2 \text{ henrys per meter} \tag{13.46}$$

Example 13.10 *This example shows how to compute the inductance of a toroidal inductor.*

Find the inductance of the toroid shown in Figure 13.8. Assume that $b \gg a$.

Solution The total magnetic energy stored in the toroid is found in Example 13.8. Using the definition (13.41) for L, we obtain

$$L = \frac{\mu N^2 a^2}{2b} \tag{13.47}$$

Toroids are commonly used as inductances for applications such as band-pass filters. Telephone companies make hundreds of toroid coils every year for use in filters that recognize the frequencies in telephones with push-button dialing.*

Example 13.11

When two inductors are close to each other, the magnetic field produced by one inductor may link through the other, and vice versa. The degree of mutual linkage (or coupling) is measured by the mutual inductance between them. This example shows how to compute the mutual inductance between two solenoidal inductors.

Consider two concentric solenoids with radii equal to R_1 and R_2, respectively. Figure 13.24 shows the configuration of these two solenoids. Assume that these solenoids are of the same length ℓ. The length ℓ is long compared with the radii R_1 and R_2, so that the end effect may be neglected. Find the mutual inductance of these solenoids. The outer solenoid has N_1 turns, and the inner one has N_2 turns.

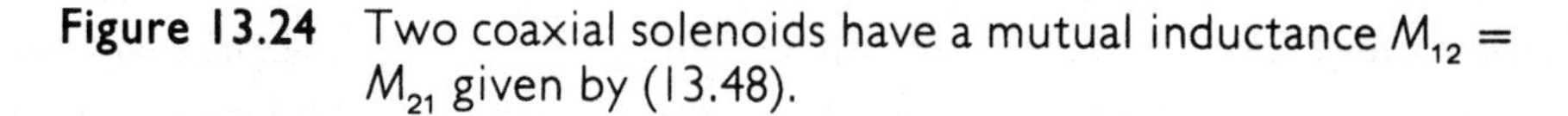

Figure 13.24 Two coaxial solenoids have a mutual inductance $M_{12} = M_{21}$ given by (13.48).

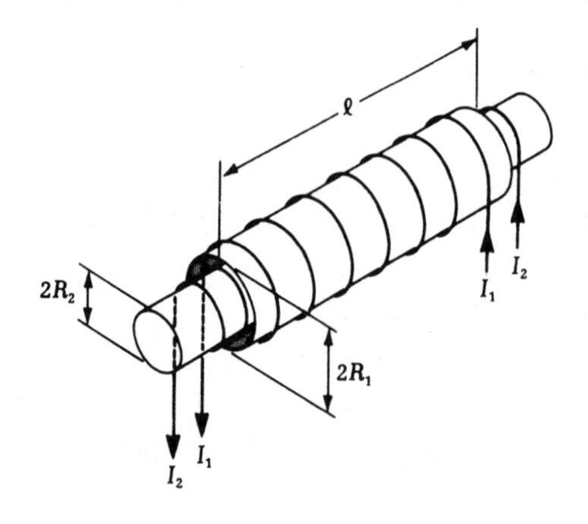

* G. W. Ciszak, "More efficient winding of toroidal coils," *The Western Electric Engineer* 25, no. 3 (1981): 11.

Solution The flux produced by current I_1 in the outer solenoid is found from (13.10b):

$$B_1 = \frac{\mu_0 N_1 I_1}{\ell}$$

Thus, the flux linkage through each turn of the coils of the inner solenoid is given below:

$$\psi_{21} = \frac{\mu_0 N_1 I_1 \pi R_2^2}{\ell}$$

Using (13.39a), we obtain the mutual inductance M_{21}:

$$M_{21} = \frac{\mu_0 N_1 N_2 \pi R_2^2}{\ell} \qquad \textbf{(13.48a)}$$

We can also solve the problem the other way—that is, we can first calculate the field due to I_2:

$$B_2 = \frac{\mu_0 N_2 I_2}{\ell}$$

The flux linkage through each turn of the coil of the outer solenoid is B_2 times the cross-sectional area of the *inner* solenoid, not the outer solenoid because the magnetic field due to I_2 outside the inner solenoid is equal to zero [see (13.10a)]. Consequently, we have

$$\psi_{12} = \frac{\mu_0 N_2 I_2 \pi R_2^2}{\ell}$$

Using (13.39b), we obtain the mutual inductance M_{12}:

$$M_{12} = \frac{\mu_0 N_2 N_1 \pi R_2^2}{\ell} \qquad \textbf{(13.48b)}$$

Note that $M_{12} = M_{21}$.

Coupling Coefficient

Consider a coil 1 through which a current I_1 produces a flux ψ_{11}. The coil 1 couples to a coil 2 with a fraction of ψ_{11}. We write

$$\psi_{21} = \kappa_1 \psi_{11}$$

Obviously $|\kappa_1|$ is less than or equal to unity. The self-inductance of coil 1 is

$$L_1 = \frac{N_1 \psi_{11}}{I_1}$$

The mutual inductance between the two coils is

$$M_{21} = \frac{N_2 \kappa_1 \psi_{11}}{I_1} = \kappa_1 \frac{N_2}{N_1} L_1$$

Likewise,

$$M_{12} = \kappa_2 \frac{N_1}{N_2} L_2$$

Because $M_{12} = M_{21} = M_0$, we find

$$M_0^2 = \kappa_1 \kappa_2 L_1 L_2$$

or

$$M_0 = \pm \kappa \sqrt{L_1 L_2}$$

where

$$\kappa = \pm \sqrt{\kappa_1 \kappa_2}$$

is called the coupling coefficient between the two coils. It has values between -1 and 1. The maximum M_0 between the two coils is thus $(L_1 L_2)^{1/2}$.

Inductance of a Coaxial Line

A coaxial line like that shown in Figure 13.3a may be regarded as a closed-loop circuit, except that the loop is connected at infinity. The current on the inner conductor flows in one direction and returns through the outer conductor by a connection at infinity. Because the circuit is a bona fide closed-loop circuit, we should be able to find the inductance of the coaxial line. To simplify the algebra, let us assume that the current flows only in thin layers at $r = a$ and $r = b$, respectively, so that the model of the coaxial line is that shown in Figure 13.25.

Because a magnetic field is present in the coaxial line, we know that magnetic energy is stored there. The magnetic field is given by (13.7):

$$H_\phi = \begin{cases} \dfrac{I}{2\pi\rho} & b > \rho > a \\ 0 & \text{elsewhere} \end{cases}$$

Substituting the above expression in (13.34), we obtain

$$U_H = \frac{1}{2}\mu \int_0^{2\pi} d\phi \int_a^b \rho \, d\rho \frac{I^2}{4\pi^2 \rho^2} = \frac{\mu I^2}{4\pi} \ln\left(\frac{b}{a}\right)$$

This result is the stored magnetic energy per unit length of the coaxial line. Consequently, we can calculate the inductance per unit length of the line from (13.41):

$$L = \frac{\mu}{2\pi} \ln\left(\frac{b}{a}\right) \tag{13.49}$$

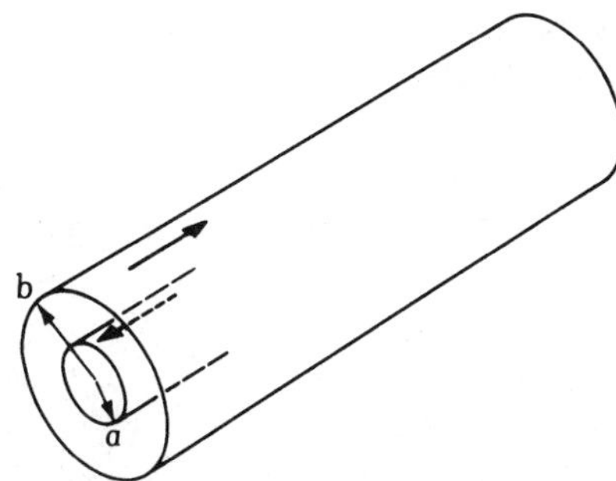

Figure 13.25 A coaxial line made of two thin cylindrical shells has an inductance given by (13.49).

This inductance per unit length also appears in the transmission-line representation of the coaxial line in (6.19) of Chapter 6.

SUMMARY

1. The magnetostatic field is produced by direct currents. Ampère's law relates the $\boldsymbol{H}$ field and the current I.
2. For symmetrical current-carrying structures, such as wires, coaxial lines, sheets, solenoids, and toroidal coils, the resulting $\boldsymbol{H}$ fields can be found by using Ampère's law in integral form—that is, Equation (13.5).
3. For general current-carrying elements, the resulting $\boldsymbol{H}$ field can be found by using the Biot-Savart law—that is, Equation (13.16).
4. A charged particle moving in a magnetic field experiences a force given by Equation (13.23). This magnetic force is always perpendicular to the velocity of the particle. Consequently, it changes the direction of the particle motion but it does not change the particle's speed.
5. A current-carrying element experiences a magnetic force when it is placed in an external magnetic field (the field not generated by its own current).
6. A current-carrying coil experiences a magnetic force and a magnetic torque when it is placed in an external magnetic field.
7. Presence of magnetic field in a region means that energy is stored in that region. A device that holds the magnetic energy is called an inductor.
8. The inductance of an inductor can be defined as the ratio of stored magnetic energy to the current that produces the magnetic field in the device. See Equation (13.41).

Problems

13.1 Find the magnetic field $\boldsymbol{H}$ at the center of a square loop carrying a current I. The side of the square loop is b meters long.

13.2 A circular loop that has radius a and that carries a current I produces the same magnetic-field strength at its center as that at the center of a square loop that has side b and that carries the same current I. Find the ratio of b to a.

13.3 Consider a large conducting plate of thickness d located at $-d/2 \le y \le d/2$, as shown in Figure P13.3. Uniform current of density J is flowing in the $\hat{z}$ direction. Find $\boldsymbol{H}$ in all regions.

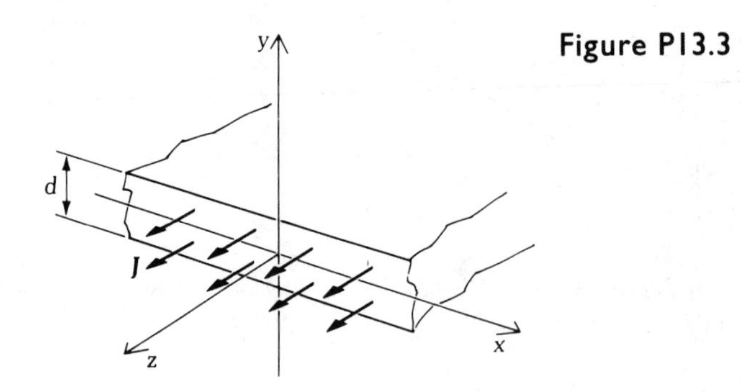

Figure P13.3

13.4 The earth's magnetic field at the north magnetic pole is approximately 0.62 G (1 G = 10^{-4} Wb/m^2). Assume that this magnetic field is produced by a loop of current flowing along the equator. Estimate the magnitude of this current. The radius of the earth is approximately 6500 km.

13.5 In a cylindrical region, the current density is given as follows:

$$J_z = \begin{cases} 2\ \text{A/m}^2 & \text{for } 0 < \rho < a \\ 4\ \text{A/m}^2 & \text{for } a < \rho < b \\ 0 & \text{for } b < \rho \end{cases}$$

Find the $\boldsymbol{H}$ field in all three regions (see Figure P13.5).

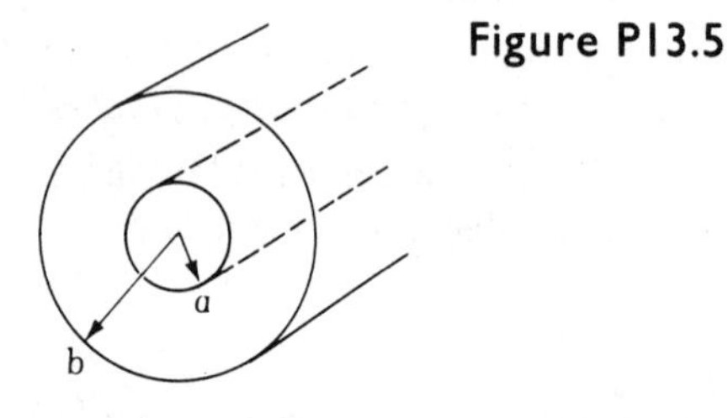

Figure P13.5

13.6 An infinitely long tubular conductor has outer radius b and inner radius a offset by a distance c from the axis of the outer cylinder, as shown in Figure P13.6. This eccentric tubular conductor carries a direct current of I amperes. Find the $\boldsymbol{H}$ field at point A shown in the figure. Hint: Consider the tube to be a superposition of two solid cylinders that have radii b and a and that carry uniform current density J in opposite directions.

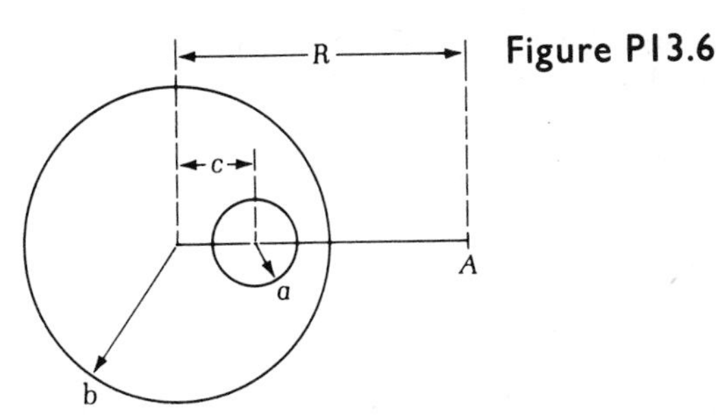

Figure P13.6

13.7 An infinitely long wire is bent to form a 90° corner, as shown in Figure P13.7. A direct current I flows in the wire. At point A find the $\boldsymbol{H}$ field due to this current. Follow the steps given below.

(a) Use the Biot-Savart law to express the $\boldsymbol{H}$ field at A due to a typical segment of wire dy on the wire axis. Express the field in rectangular coordinates.
(b) Integrate the result obtained in (a) to the $\boldsymbol{H}$ field due to the semi-infinite wire $\boldsymbol{OC}$. Note: to facilitate integration, let $y = a \tan \theta$, so that $dy = a \sec^2 \theta \, d\theta$.
(c) Find the $\boldsymbol{H}$ field at A due to the semi-infinite wire $\boldsymbol{BO}$.
(d) Add the results obtained in (b) and (c) to yield the total field at A due to the current in the wire $\boldsymbol{BOC}$.

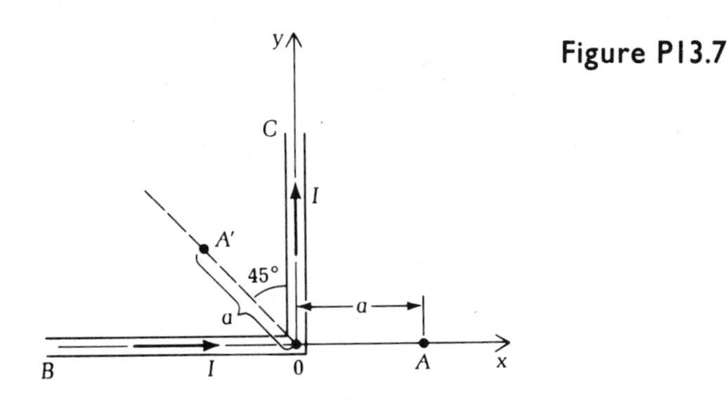

Figure P13.7

13.8 Follow a similar procedure to the one described in Problem 13.7 to find the $\boldsymbol{H}$ field at point A', as shown in Figure P13.7.

13.9 Consider a circular loop carrying a current I counterclockwise, as shown in Figure 13.11. Plot the magnetic field H_z on the z axis for $-a/2 < z < a/2$. Find the value z_0 in terms of a, such that, if $|z| < z_0$, then H_z is uniform within 10% of the value of H_z at the center of the loop.

13.10 To improve the uniformity of the magnetic field along the axis of a circular loop (see Problem 13.9), one may use two identical loops separated by a distance equal to their radii, as shown in Figure P13.10. Such a pair of current-carrying loops is called **Helmholtz coils**. Find H_z as a function of z on the axis of the Helmholtz coils. Plot H_z for $-a < z < a$. Find, in terms of a, the value z_0 such that, within the range $|z| < z_0$, H_z is uniform within 10% of the magnetic field at the middle of the two coils. Compare your result with that obtained in Problem 13.9 for a single loop.

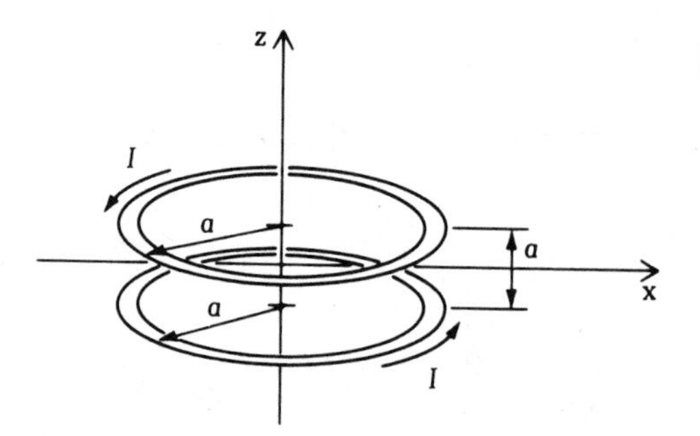

Figure 13.10 Helmholtz coils.

13.11 A direct current I flows in an equilateral triangular circuit, as shown in Figure P13.11. The length at each side is $2a$. Find the $\boldsymbol{H}$ field at the center of the triangle. Hint:

$$\int_{-a}^{a} \frac{dx}{(x^2 + b^2)^{3/2}} = \frac{2a}{b^2(a^2 + b^2)^{1/2}}$$

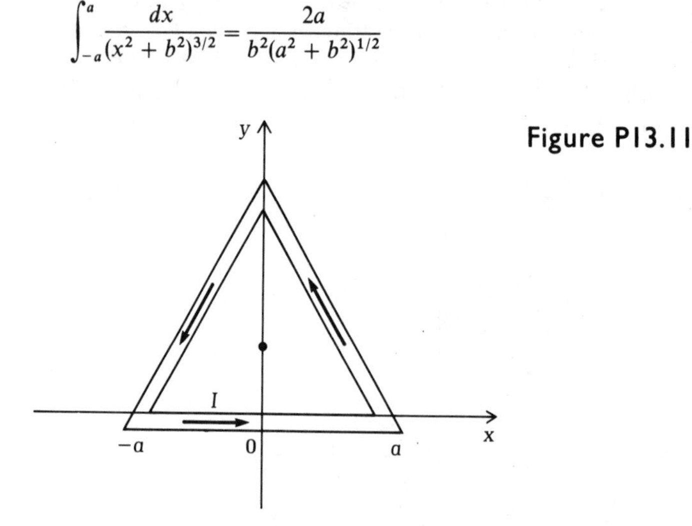

Figure P13.11

13.12 A surface charge of ρ_s C/m^2 is uniformly distributed on a record disk. The inner radius of the disk is a and the outer radius is b. The record disk is turning at a constant angular velocity ω rad/s in the clockwise direction. Find the magnetic field at the center of the disk due to the surface charge on the turning disk. Ignore the presence of the metal post on the turntable.

13.13 The earth's magnetic field at the equator is approximately $B = 10^{-4}$ Wb/m^2. Calculate the cyclotron frequency of the electron in the ionosphere.

13.14 Because natural uranium contains a slight amount of Uranium 234, the electromagnetic isotope separator can also yield ^{234}U. If the radius of the circular path for ^{238}U particles (see Figure 13.16) is equal to 10 m, where should one place collectors for ^{235}U and ^{234}U particles? Express spacings in meters.

13.15 Refer to Figure 13.19. The magnetic field is changed from 5×10^{-4} to 10^{-3} Wb/m^2. All other parameters remain unchanged. Find the following:

(a) the position of the electron at the exit side of the magnetic-field region
(b) the exit angle (the angle between the trajectory and the x axis after the electron has passed through the magnetic field).

13.16 A positive charge q with mass m is at the origin at $t = 0$ in a uniform magnetic field $\boldsymbol{B} = B_0 \hat{z}$, as shown in Figure P13.16.

(a) What are the coordinates of the position of the particle as functions of t if $\boldsymbol{v} = v_0 \hat{y}$ at $t = 0$? Express them in terms of the given parameters. What is the geometrical shape of the trajectory?
(b) Repeat (a) if $\boldsymbol{v} = v_0 \hat{y} + v_z \hat{z}$ at $t = 0$.

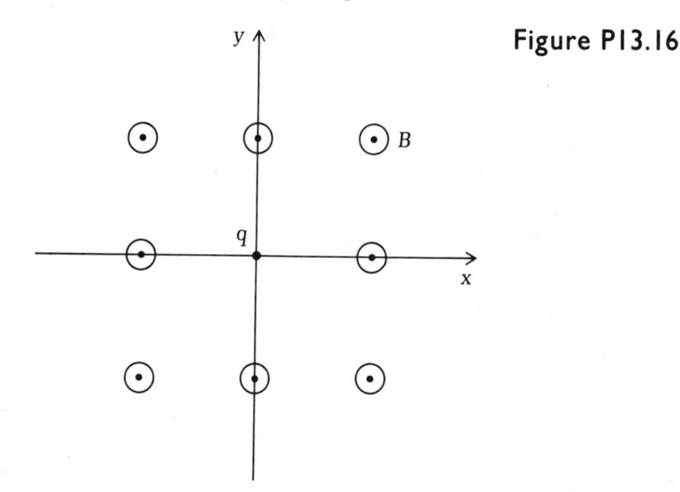

Figure P13.16

13.17 Two parallel wires are carrying 100 A of current in opposite directions. On each wire find the force per unit length due to the magnetic field produced by the other wire. Is the force repulsive or attractive? Assume that the lines are 1.5 m apart.

13.18 Two identical circular loops of radii a are separated by a distance d, where $d \ll a$. One of the coils carries I amperes of current clockwise, and the other carries I amperes counterclockwise. Find the force between these coils. Hint: Because these coils are close together, you can approximate the magnetic field that is at one coil and is produced by the current on the other as $H_1 = I_2/(2\pi d)$, the field due to an infinitely long wire. Let $a = 1$ m and $d = 0.05$ m. How much current is needed to produce a force of 9.8 N?

13.19 A circular loop of radius 0.5 m and 100 turns is excited by a 2 A direct current. This loop is placed in the earth's magnetic field, which is approximately equal to 5×10^{-5} Wb/m^2 pointing north. How do you orient this loop to produce a maximum torque? What is the value of this torque? Find that orientation of the loop in which it experiences no torque.

13.20 The triangular circuit described in Problem 13.11 is placed in a uniform external field $\boldsymbol{B} = B_0 \hat{x}$. Find the force on each of the three sides and the torque on the whole circuit. Give both the amplitude and the direction of the forces and the torque. Describe how the circuit will react if it is free to move or rotate.

13.21 An infinitely long conductor of radius a carries a direct current I as shown in Figure P13.21.

(a) Find the $\boldsymbol{H}$ field in the region $0 < \rho < a$.
(b) Calculate the stored magnetic energy per unit length in the region $0 < \rho < a$.
(c) Find the inductance per unit length of the conductor. Consider only the magnetic energy in the region $0 < \rho < a$.

Figure P13.21

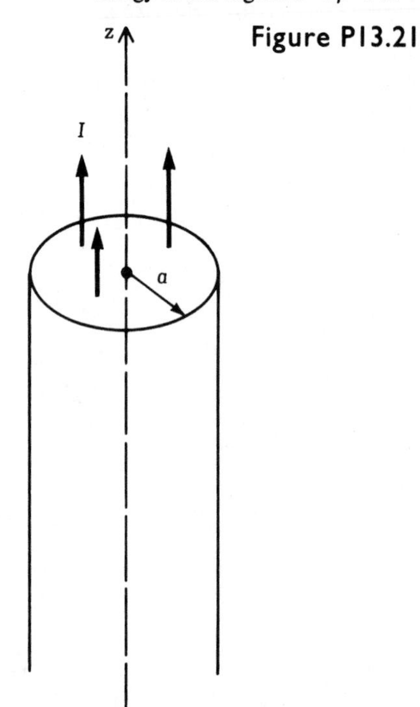

13.22 Three infinitely long parallel wires each carry 10 A of current in the $\hat{z}$ direction, as shown in Figure P13.22. Find the force per unit length acting on the Number 3 wire due to the magnetic fields produced by the other two wires. Give the numerical value of the force, its direction, and its unit.

Figure P13.22

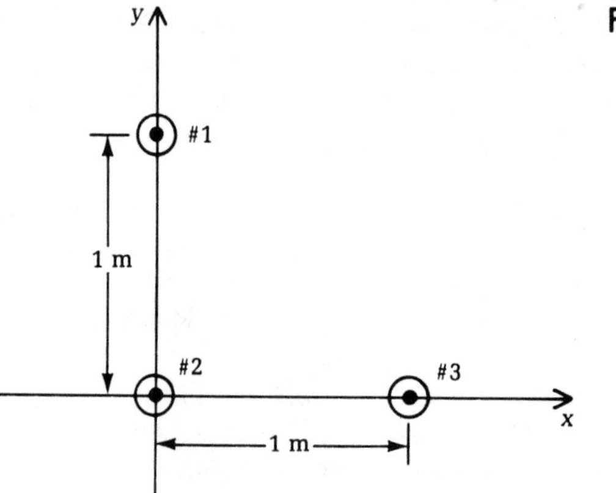

13.23 The magnetic field in a coaxial line is given by

$$H_\phi = \begin{cases} 1/\rho & \text{for } 0.1\,\text{m} < \rho < 0.2\,\text{m} \\ 0 & \text{elsewhere} \end{cases}$$

The medium is air. What is the total stored magnetic energy per unit length in the line? Give the numerical value and indicate its unit.

13.24 (a) Calculate the stored magnetic energy per unit length of the parallel-plate conductors shown in Figure 13.5.
(b) If the parallel plate is used as a capacitor to store electric energy, find the voltage V_0 for which the stored electric energy is equal to the stored magnetic energy found in (a). Let $I = 1\,\text{A}$, $w = 10\,\text{cm}$, and $a = 1\,\text{cm}$. Express V_0 in volts. The medium is air.

13.25 Calculate the inductance per unit length of the coaxial line shown in Figure 13.3a.

13.26 Calculate the inductance per unit length of the parallel-plate conductors shown in Figure 13.5.

14 MAGNETIC MATERIALS AND MAGNETIC CIRCUITS

Magnetic materials are widely used in traditional and modern devices. Permanent magnets have been used by humankind for almost 5000 years, and various ferrites are used as recording media in computers and audio and video equipment. We concentrate in this chapter the discussion of these ferromagnetic materials and their applications.

In the design of a magnetic device, such as a recording head on a computer disk drive or a transformer in a power network, we must know how much excitation is needed to obtain a desired strength of the magnetic field. This is the reason we also want to study the magnetic circuit in this chapter. We will show that the magnetic circuit problem is analogous to the electric circuit problem.

14.1 MAGNETIC MATERIALS

In Chapter 13 we discussed many examples of how to find the magnetic field from known current distributions. The key formula relating the magnetic field and the current is Ampère's law:

$$\nabla \times \boldsymbol{H} = \boldsymbol{J} \qquad \textbf{(14.1a)}$$

or

$$\boxed{\oint \boldsymbol{H} \cdot d\boldsymbol{\ell} = I} \qquad \textbf{(Ampère's law)} \qquad \textbf{(14.1b)}$$

Note that current is related to $\boldsymbol{H}$, the magnetic-field strength. When we study the interaction between a magnetic field and charge particles or current-carrying wires, the Lorentz force is expressed not by $\boldsymbol{H}$ but by $\boldsymbol{B}$, the magnetic-flux density:

$$\boldsymbol{F} = q\boldsymbol{v} \times \boldsymbol{B} \qquad \textbf{(14.2a)}$$

or

$$F = I\, d\ell \times B \tag{14.2b}$$

The field strength H and the flux density B are related by the constitutive relation:

$$\boxed{B = \mu H} \quad \textbf{(relation between } B \textbf{ and } H \textbf{ fields)} \tag{14.3}$$

Different materials have different values of μ, which characterizes the magnetic property of the medium. It is called the **magnetic permeability**, or simply **permeability**, of the medium. In vacuum,

$$\boxed{\mu = \mu_0 = 4\pi \times 10^{-7}\ \text{H/m}} \quad \textbf{(permeability of vacuum)}$$

Other media may have permeabilities greater or smaller than that of vacuum.

In many practical applications, we need to design a special coil configuration to produce a magnetic field for a special purpose. Usually we want the magnetic field B to be strong, and we want to be able to produce it with the smallest possible current in the coil. Because the current is proportional to H and B is proportional to μH, these considerations often call for materials with high permeabilities. Many materials have high permeabilities compared with μ_0. These materials are called **ferromagnetic**. Let us briefly study why materials have different μ compared with that of vacuum.

Basically, we can say that, because of the interaction between the external magnetic field and moving charges in the atoms of the material, the magnetic flux B either increases or decreases from its free-space value, so that μ is greater or smaller than μ_0, respectively. Internal magnetic fields are mainly produced by electrons orbiting around nuclei or by electrons spinning around themselves. Both cases are analogous to current loops that produce B fields without external current excitation. Figure 14.1 illustrates this analogy.

The equivalent current loop constitutes a magnetic moment $\boldsymbol{m}$, as defined in (13.30):

$$\boxed{\boldsymbol{m} = I\pi a^2 \hat{z}} \quad \textbf{(magnetic moment of a current loop)} \tag{14.4}$$

where πa^2 is the area of the loop. The size of the loop is of atomic or microscopic scale. For a macroscopic volume that contains many such loops, we can visualize the situation to be similar to the one shown in Figure 14.2. We can approximate each column as a solenoid. Let us assume that there are N such loops in the column; according to (13.10b), the magnetic field H is given by

$$H_z = \frac{N}{\ell} I \tag{14.5}$$

Figure 14.1 (a) Orbital electron or (b) spinning electron is equivalent to (c) a current loop that has a magnetic moment pointing in the $\hat{z}$ direction.

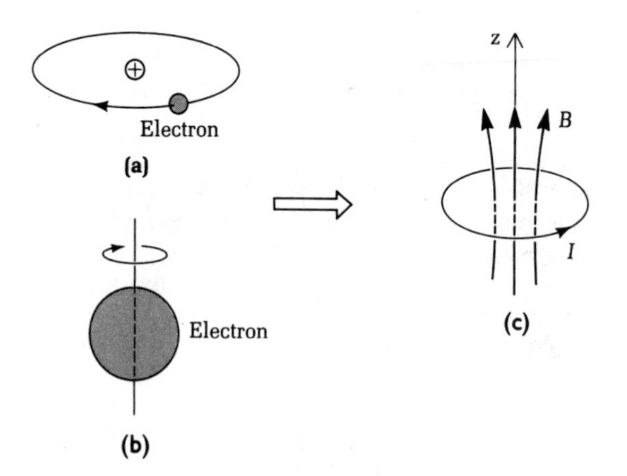

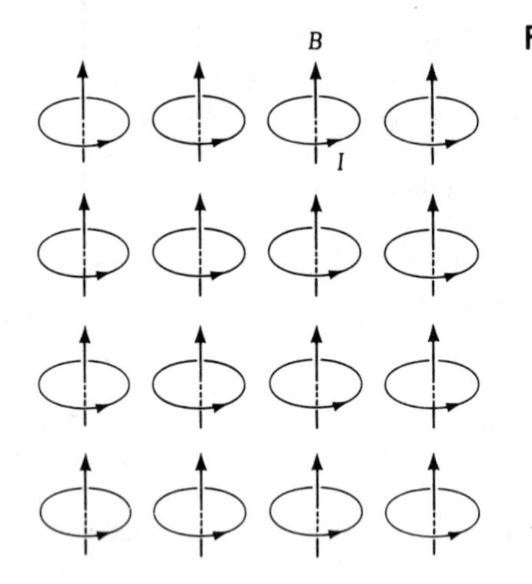

Figure 14.2 In a volume containing many magnetic moments, each column may be thought of as a solenoid for which the magnetic field is known.

where ℓ is the length of the column. Rewriting the above equation, we have the following:

$$\boldsymbol{H}_{\text{internal}} = \frac{NI(\pi a^2)\hat{z}}{\ell(\pi a^2)} = \frac{N\boldsymbol{m}}{\text{volume}} = \boldsymbol{M} \tag{14.6}$$

Note that $\boldsymbol{M}$ is the magnetic-moment density, which is equal to the number of magnetic moments $\boldsymbol{m}$ per unit volume. We see that the magnetic-field intensity produced by the spinning electrons is equal to the magnetic-moment density. The magnetic flux due to this internal $\boldsymbol{H}$ is then given as follows:

$$\boldsymbol{B}_{\text{internal}} = \mu_0 \boldsymbol{M} \tag{14.7}$$

Note that (14.6) and (14.7) are expressions of internal fields due to spinning electrons in the material.

In a class of materials called **diamagnetic** materials, the magnetic fields produced by orbital and spinning electrons cancel out completely. However, under an external magnetic field, a combined interaction of the external magnetic field, the electrons, and quantum mechanical forces slightly distort the balance, so that the magnetic fields produced by the orbital and spinning electrons do not cancel out completely. As a result, a magnetic-moment density is induced in the medium **opposite** the applied external field. Thus, in this case, the total flux density (external plus induced) may be expressed as follows:

$$\boldsymbol{B}_{\text{total}} = \mu_0(\boldsymbol{H}_{\text{external}} + \boldsymbol{M}) \tag{14.8}$$

where $\boldsymbol{M}$ is opposite to $\boldsymbol{H}_{\text{external}}$ in diamagnetic materials. The magnitude of $\boldsymbol{M}$ is proportional to the external applied $\boldsymbol{H}$:

$$\boldsymbol{M} = \chi_m \boldsymbol{H}_{\text{external}} \tag{14.9}$$

Therefore, we have

$$\boldsymbol{B}_{\text{total}} = \mu_0(1 + \chi_m)\boldsymbol{H}_{\text{external}} \tag{14.10}$$

From here on, we will write

$$\boxed{\begin{aligned} \boldsymbol{B} &= \mu \boldsymbol{H} \\ \mu &= \mu_0(1 + \chi_m) \end{aligned}} \quad \textbf{(magnetic susceptibility)} \qquad \begin{aligned} &\textbf{(14.11a)} \\ &\textbf{(14.11b)} \end{aligned}$$

with the understanding that $\boldsymbol{H}$ is the external applied H field and $\boldsymbol{B}$ is the total magnetic flux due to the external source and to the internal spinning electrons. The quantity χ_m is called the **magnetic susceptibility** of the material. For diamagnetic materials, χ_m is negative and is of the order of -10^{-5}. Table 14.1 lists some of the diamagnetic materials.

Table 14.1 Diamagnetic and paramagnetic materials.

Substance	χ_m
Copper	-0.94×10^{-5}
Lead	-1.70×10^{-5}
Water	-0.88×10^{-5}
Vacuum	0.00
Air	$+3.60 \times 10^{-7}$
Platinum	$+2.90 \times 10^{-4}$
Aluminum	$+2.10 \times 10^{-5}$
Liquid Oxygen	$+3.50 \times 10^{-3}$

In **paramagnetic** materials, the magnetic field due to the orbital and spinning electrons do not cancel completely. However, because of thermal agitation, the magnetic moments point in random directions and produce no net magnetic-moment density. Under the influence of an external field, they align themselves because of the torque produced [see (13.3)]. Note that the torque is equal to zero only if $\boldsymbol{m} \times \boldsymbol{B}$ is zero—a condition achieved when $\boldsymbol{m}$ is parallel to $\boldsymbol{B}$. Clearly then, in paramagnetic materials, the resulting B field is greater than the corresponding value in vacuum. The magnetic susceptibility χ_m of a paramagnetic material is positive and is of the order of 10^{-5} to 10^{-3}. Table 14.1 also lists some common paramagnetic materials.

The third kind of material mentioned earlier in this section exhibits very high susceptibility. This kind of material is useful because a strong B field may be produced in it with relatively small current in the coil. The model of this **ferromagnetic** material is as follows.

The material has many small **domains** that are approximately 10^{-6} m in linear dimension. Each domain contains many magnetic dipoles produced by spinning electrons, which are aligned in parallel by the strong force between neighboring dipoles. The direction of the alignment of the magnetic dipole differs from one domain to the other, so that there is usually no spontaneous magnetic field in the material as a whole. This absence of magnetic field corresponds to point O in the magnetization curve shown in Figure 14.3a. Figure 14.4a shows the corresponding situation for the domain. As an external H field is applied in the $\hat{x}$ direction, for example, the domains in which the magnetic dipoles are parallel to the applied field grow in size, and the sizes of other domains diminish. The magnetic field produced by the spinning electrons and the original external field combine to result in a strong total B field. This result is represented by the curve between point 0 and point P_1 shown in Figure 14.3a, and Figure 14.4b sketches the corresponding situation for the domain. As the external H field is

Figure 14.3 **(a)** Magnetization curve. **(b)** Permeability versus H. Maximum μ occurs at H_1 where the line OP_3 is tangent to the magnetization curve.

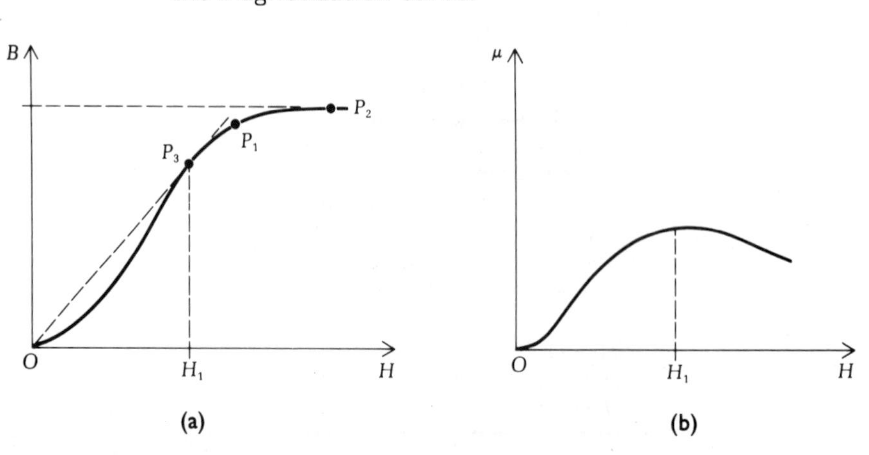

Figure 14.4 Domains in a ferromagnetic material. **(a)** In the absence of an external field, magnetic moments in each domain are parallel, but domains are oriented randomly. **(b)** Domains are distorted under the influence of an external field. Domains with magnetic moments parallel to the external field expand while others shrink. **(c)** All domains align with the external field at the saturation point.

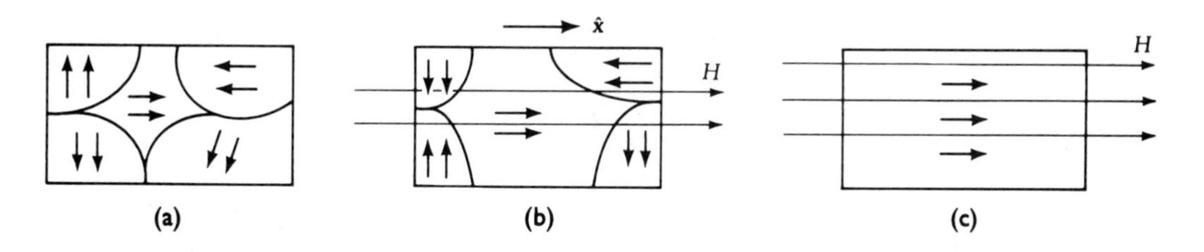

increased further, the entire body is magnetized to become a single domain, and all the spinning electrons are aligned. The B field therefore reaches the saturation value and cannot be significantly increased further. This saturation point is represented by point P_2 shown in Figure 14.3a, and Figure 14.4c shows the corresponding situation for the domain. (To be precise, the magnetic flux B increases at a rate equal to μ_0 as H is increased beyond P_2. But this increase is practically negligible as compared with the rate near point P_3 in Figure 14.3). The magnetic permeability μ is a nonlinear function of H. Notice that μ is simply the slope of the line joining the origin of the B-H curve to a point on the B-H curve. Figure 14.3b shows a plot of μ versus H. The maximum μ is achieved at the point where the line joining the origin to a point on the magnetization curve is tangent to the curve. This line is the line OP_3 shown in Figure 14.3a.

Hysteresis

Let us consider a toroid similar to the one shown in Figure 13.8. The material inside the toroid is ferromagnetic. The H field in the toroid is given by (13.13):

$$H_\phi = \frac{NI}{2\pi b} \tag{14.12}$$

Let us vary the current and measure the resulting B field in the toroid. We assume that in the beginning the ferromagnetic toroid has no spontaneous magnetic field. As I increases, H increases according to (14.12). The B field also starts to increase from zero. If we plot B versus H, the plot will be similar to the curve OA_1, shown in Figure 14.5. If, after initial magnetization to the point A_1, the H field is decreased by reducing the current in the coil, the corresponding B field is also reduced, but the B-H curve does not follow the original magnetization curve! Instead, it traces a curve similar to A_1B_1 shown in Figure 14.5. Note that

Figure 14.5 A typical hysteresis curve of iron. B_1 is the remanence of the iron, and H_1 is the coercive force.

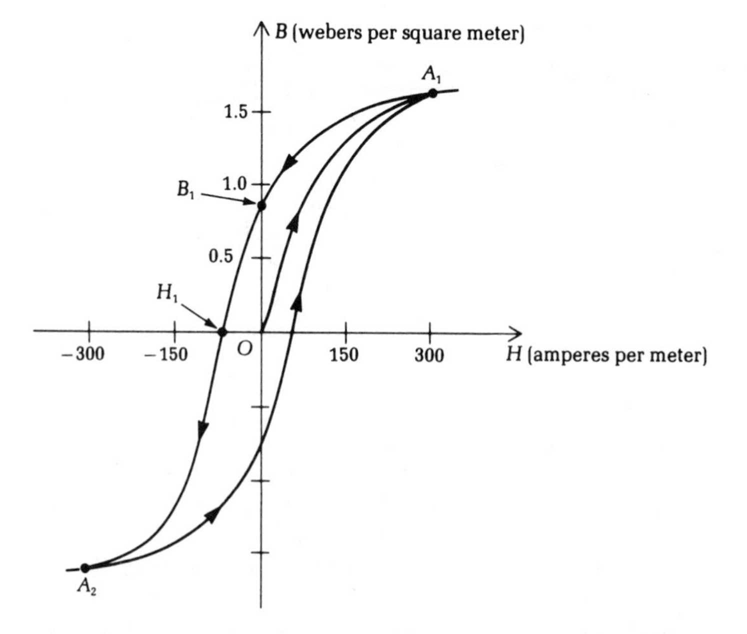

at point B_1, the current in the coil of a toroid is equal to zero, as is H, yet residual magnetic flux B is nonetheless present in the toroid. This residual flux is due to the fact that the magnetic moments of some of the domains in the ferromagnetic material are still aligned in the same direction. The magnitude of this residual B is called **remanence.**

If we now reverse the current to reverse H, the B-H curve will trace a curve B_1H_1, as shown in Figure 14.5. Note that it takes some negative value of H to null the B field. The value of H in the negative direction of initial magnetization necessary to nullify the B field is called the **coercive force.** If the reverse current is increased beyond this point, the B field begins to reverse, and the B-H curve follows the curve H_1A_2 shown in Figure 14.5. If the current is now reduced, the B-H curve traces a new path A_2A_1. The closed loop $A_1B_1H_1A_2A_1$ is called a **hysteresis loop.** If the current is varied in a smaller cycle, the corresponding hysteresis loop will be smaller. If the material is brought to saturation in both ends of the magnetization curve, the remanence B is called the **retentivity** of the ferromagnetic material, and the coercive force H is called the **coercivity** of the material.

Permanent Magnet

The portion of the hysteresis loop that is in the second quadrant is known as the **demagnetization curve**. Figure 14.6 shows the demagnetization curve of an alloy

Figure 14.6 Demagnetization curve of alnico V alloy.

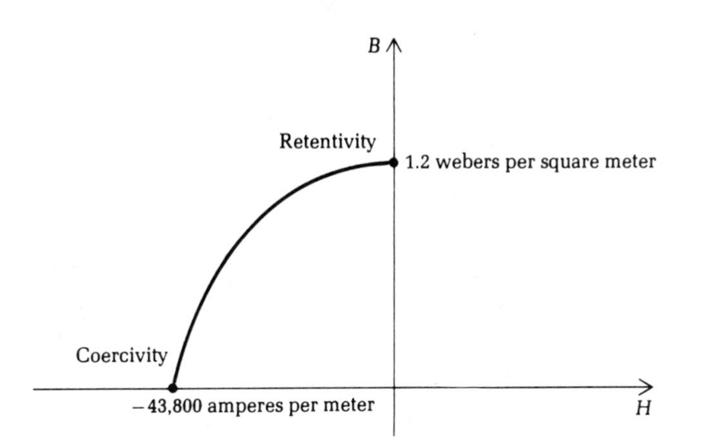

called alnico V, which has a high retentivity, so that it can be used as a strong permanent magnet. Notice that its coercivity is also high. It takes 43,800 A/m of H to demagnetize the material. High retentivity and coercivity are desirable properties of a permanent magnet. Note that permanent magnets lose their permanent magnetism when the temperature is higher than a critical value called the **Curie temperature**. For most ferromagnetic materials, the Curie temperatures are approximately 500° C.

Now assume that a permanent magnet is made of alnico V and that the retentivity is

$$B_r = 1.2 \text{ Wb/m}^2 \tag{14.13}$$

In an equivalent solenoid with air as its core, let us calculate the current that would produce the same magnetic flux. For a solenoid of n turns per meter, the H field is given by (13.10b):

$$H_z = nI$$

Therefore,

$$B_z = \mu_0 nI$$

Let B_z be equal to the remanent flux density given in (14.13). We have

$$nI = \frac{1.2}{\mu_0} = 0.95 \times 10^6 \text{ ampere turns per meter}$$

This large number of ampere turns may be surprising at first. It is a result of the spinning electrons in the permanent magnet. Remember that there are on the order of 10^{30} or more electrons per cubic meter in the magnet!

If we cut a piece of the alnico V permanent magnet into a disk form, as shown in Figure 14.7, we can estimate that the magnetic field will be smaller than the retentivity of 1.2 Wb/m^2 because it is a thin disk. The disk may be modeled as a

Figure 14.7 **(a)** A thin disk of a permanent magnet. **(b)** Its equivalent current loop. **(c)** Permanent magnetic disk subject to a torque exerted by an external magnetic field B_0.

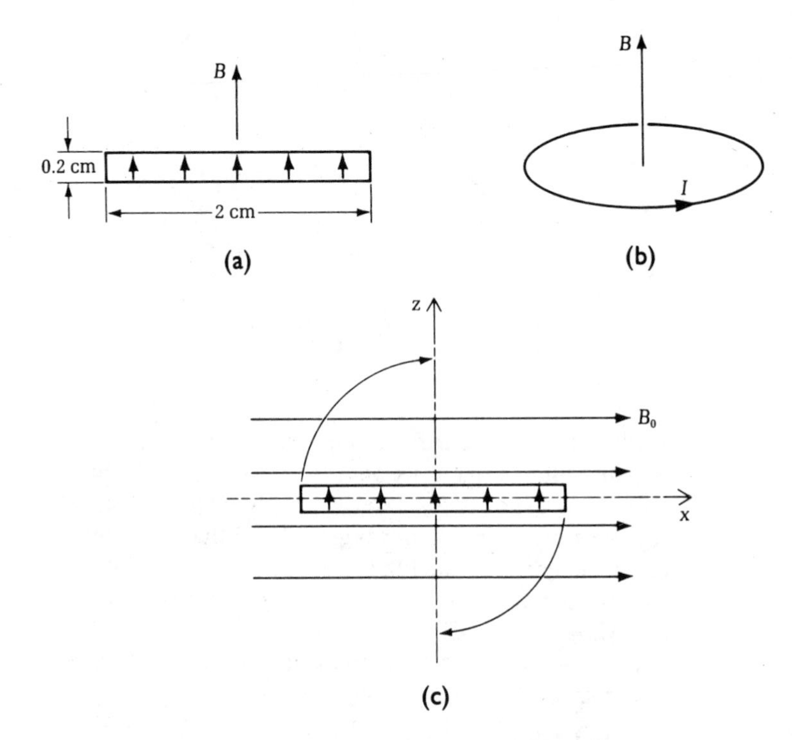

circular loop current with a current I. We have found that a solenoid of 0.95×10^6 A-t/m could produce the same remanent field of 1.2 Wb/m^2. Thus, a disk 2 mm thick is equivalent to a circular current loop with current I, where

$$I = 0.95 \times 10^6 \times 0.002 = 1.9 \times 10^3 \text{ A}$$

Now, to find B, we use the formula of the magnetic field on the axis of a circular loop given by (13.20):

$$\boldsymbol{B} = \frac{Ia^2\mu_0}{2(a^2 + z^2)^{3/2}}\hat{\boldsymbol{z}}$$

for any point on the axis of the disk z meters away. In particular, at the surface of the disk $z = 0$, and we have

$$\boldsymbol{B} = \frac{I\mu_0}{2a}\hat{\boldsymbol{z}} = 0.12\hat{\boldsymbol{z}} \text{ webers per square meter}$$

If the permanent magnetic disk is placed in the earth's magnetic field given below;

$$\boldsymbol{B}_e = 0.6 \times 10^{-4}\hat{\boldsymbol{x}} \text{ webers per square meter}$$

it is subject to a torque that tends to turn the disk so as to align it with the earth's magnetic field. To calculate the torque, we first calculate the magnetic moment $\boldsymbol{m}$:

$$\boldsymbol{m} = \pi a^2 I \hat{z} = 0.60\hat{z}$$

Substituting into (13.31), we have

$$\boldsymbol{T} = \boldsymbol{m} \times \boldsymbol{B}_e = 0.36 \times 10^{-4}\hat{y} \text{ newtons per meter}$$

High-Permeability and Permanent-Magnet Materials

Table 14.2 lists some materials of high permeability and some materials for permanent magnets.

Table 14.2 High-permeability and permanent-magnet materials.*

High-permeability material	Relative permeability	Saturation flux density (Wb/m²)	Permanent-magnet material	Retentivity (Wb/m²)	Coercivity (A/m)
Cast iron	100–600	2.0	Carbon steel	0.86	3800
Permendure	800–4500	2.4	Alnico V	1.20	43,800
Supermalloy	10^5–10^6	0.8	Cunico	0.34	52,500

* R. M. Bozorth, *Ferromagnetism*, New York: Van Nostrand Reinhold, 1951; or E. C. Jordan *Reference Data for Engineers*, 7th ed., Indianapolis, Ind.: Howard W. Sams & Co., 1985, pp. 4–25.

Earth's Magnetic Field

It is believed that electric currents flow in the fluid iron core of the earth and produce a magnetic field as does the loop of current shown in Figure 13.11. The magnetic-flux density is about 0.62×10^{-4} Wb/m² at the magnetic north pole and is about 0.5×10^{-4} Wb/m² at 40° N latitude.* The axis of the effective magnetic-current loop is inclined 11° from the axis of the rotation of the earth, as shown in Figure 14.8a (p. 500). Note that the magnetic **north** pole of the earth is the **south** pole of the magnet if you consider the earth as a giant permanent magnet—that is, the B field points down into the ground at the magnetic north pole. A permanent magnetic bar would tend to rotate so that its magnetic moment would point to the magnetic north pole of the earth. According to a

* A magnetic-field strength map of the earth is presented in D. N. Lapedes, ed., *McGraw-Hill Encyclopedia of the Geological Sciences*, New York: McGraw-Hill, 1977, p. 299.

Figure 14.8(a) Earth's magnetic field. The magnetic north is 11° off the axis of rotation of the earth. The magnetic field points down at the northern pole.

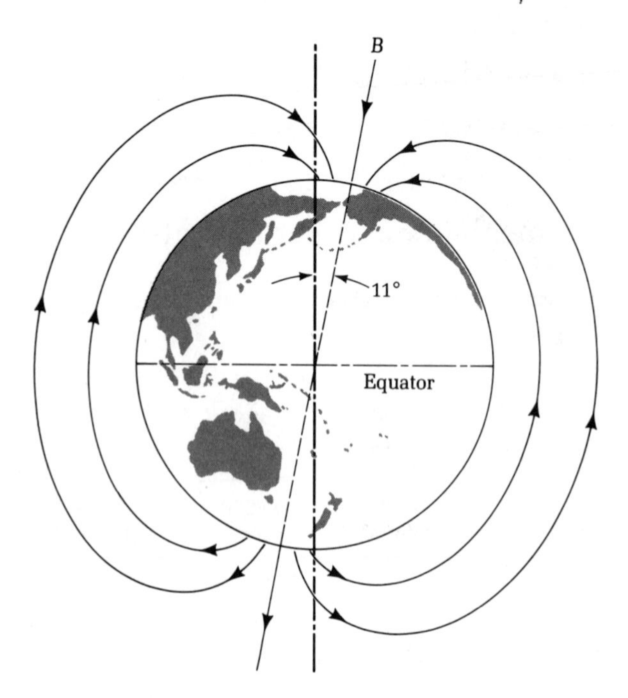

Chinese legend recorded in an ancient history book, a tribal chieftain named Huang Di (Figure 14.8b) discovered an iron bar that tended to point to the north all the time. He mounted it on a chariot, and with its help he was able to lead his men through thick fog to defeat an enemy troop. That battle was possibly the first useful application of electromagnetism because it supposedly happened in approximately 2700 B.C.

Earlier, we mentioned that a permanent magnet would lose its permanent magnetism if the temperature was raised above a critical point called the Curie temperature. The reverse is also true. For example, a ferromagnetic rock in lava is not magnetized until the lava is cooled below the Curie temperature, at which point the rock is magnetized by the earth's magnetic field. The amazing fact discovered by geologists is that the lava layers have alternating directions of magnetization. These alternating layers suggest that the earth's magnetic field has reversed its direction many times in the last 6 million years. The most recent reversal occurred only approximately 30,000 years ago. In fact, an Australian student not long ago discovered an ancient fireplace in which early humans cooked their food. He carefully measured the magnetization of some of the rocks in the fireplace and found that they were magnetized opposite to the present geomagnetic field. The student boldly proposed to his professor that some 30,000 years ago the geomagnetic field was pointing the opposite way!

Figure 14.8(b) Huang Di, as depicted in a stone rubbing of later centuries, is believed to have used a permanent magnet as a compass in 2700 B.C. (By courtesy of *China Reconstructs*.)

More amazing things have been discovered in the new field of geology called **paleomagnetism**, the examination of magnetized rocks at different places and different depths to study the history of the geomagnetic field and the history of the earth's surface. For example, the magnetic north pole has been wandering during the past 500 million years. Some 250 million years ago, it was located in Europe. Some geophysicists explain that the pole did not actually move around but that the surface of the earth, the lithosphere, did.* The lithosphere slid over the core of the earth relative to the magnetic pole, which was stationary.

Magnetic-Core Memory

A class of magnetic materials called **ferrites** has almost rectangular hysteresis loops. Figure 14.9 (p. 502) shows a sketch of such a loop. We see that the material has two possible remanent states, represented respectively by points A and C in Figure 14.9. These two states can be used to represent "one" and "zero" in a binary-number memory system.

Consider a toroid made of ferrite material. A wire carrying a current I is placed coaxial to the toroid, as shown in Figure 14.10. This toroid is somewhat

* F. Press and R. Siever, *Earth*, San Francisco: W. H. Freeman, 1978, chap. 18.

Figure 14.9 A typical hysteresis loop of a "square-loop" ferrite.

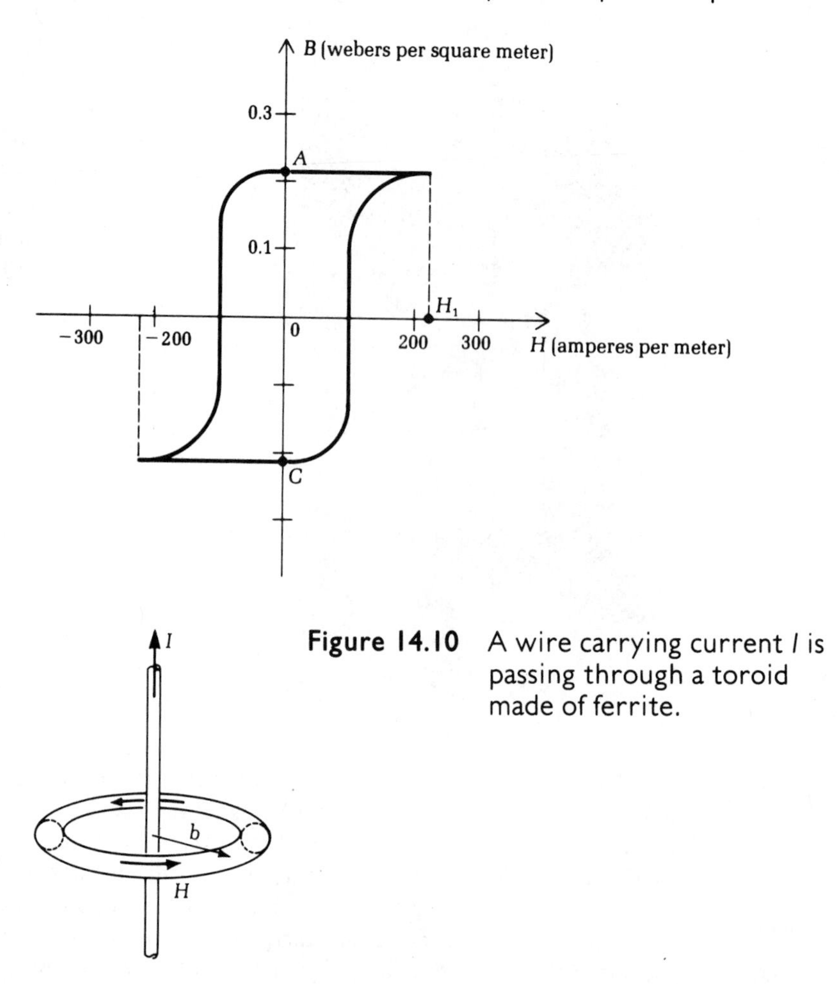

Figure 14.10 A wire carrying current I is passing through a toroid made of ferrite.

different from the one shown in Figure 13.8, which has many turns of coil wound on its surface. In the present case, we can say that the toroid has only one turn. Using the results obtained in (13.13), we have

$$H_\phi = \frac{I}{2\pi b} \tag{14.14}$$

Notice that this result is also the magnetic-field strength due to an infinitely long wire as given by (13.6).

Let us assume that the toroid is initially at the state C shown in Figure 14.9. If the current I is positive and H_ϕ given by (14.14) is greater than H_1, then the toroid's flux will be reversed and will stay reversed even after I is stopped. In other words, if we send a pulse with sufficiently high amplitude, the pulse will change the state of the toroid from C to A. On the other hand, if the toroid is initially at state A, the positive-current pulse will not change the toroid's state. The width of the pulse is determined by how fast the toroid can change its flux. For ferrites, the switching time is of the order of a microsecond.*

Magnetic-core memory in digital computers consists of tens of thousands of tiny ferrite toroids, called cores, arranged in arrays similar to the one shown in Figure 14.11. Each ferrite core is passed by two orthogonal wires. Neighboring cores are arranged orthogonally, so that mutual coupling between them is kept at a minimum.

Figure 14.11 Arrangement of magnetic-memory cores using the coincident-current scheme. The dotted line represents the sensing wire for reading the stored data.

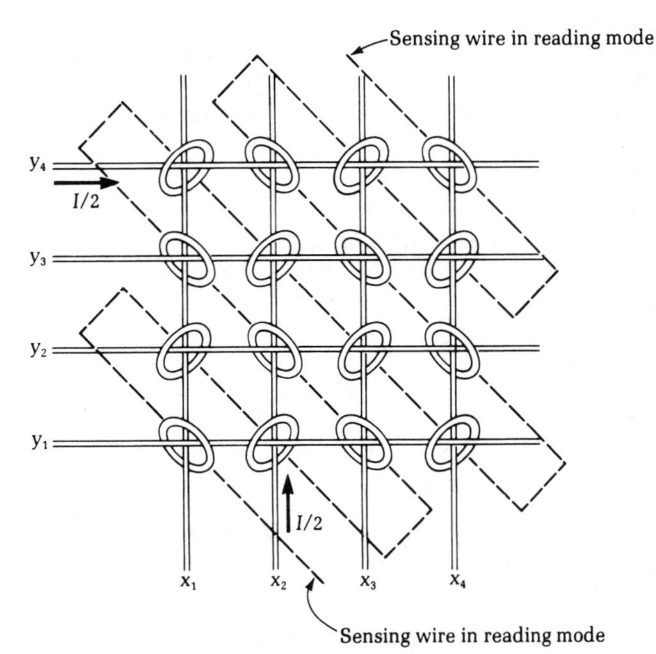

As shown in Figure 14.9, assume that all cores are in state C, which represents "zero." Suppose that we now want to register "one" in the memory core $x_2 y_4$. We can send a current pulse of $I/2$ amplitude on wire x_2 and also on wire y_4. The amplitude $I/2$ is designed to correspond to an H field of $H_1/2$. This H field alone is not strong enough to switch the core from point-C state to point-A state. We draw this conclusion from Figure 14.9. But the particular core $x_2 y_4$ is located at the intersection of x_2 wire and y_4 wire. Thus, it receives a total current

* C. J. Quartly, *Square-Loop Ferrite Circuitry*, Englewood Cliffs, NJ.: Prentice-Hall, 1962, p. 14.

excitation equal to I, which corresponds to an H field of H_1. Thus, this particular core changes its state from C to A (or the "one" state), while all others remain at "zero" or C state. This process exactly accomplishes our goal of storing a bit of information in that particular core.

To read the information we have stored on the magnetic-core memory, we put a third wire through the cores. Suppose we want to read the value stored at core $x_2 y_4$. We send two current pulses of $-I/2$ down wire x_2 and wire y_4. These pulses will not disturb the status of any core except possibly the $x_2 y_4$ core. If that core has a value of "one," it is an "A" state. The two negative currents combine to generate an H field, which is just strong enough to reverse the magnetic flux. When the magnetic flux is reversed, the sensing (or reading) wire detects an induced voltage, according to the Faraday induction law, which we will discuss in detail in Chapter 16. On the other hand, if the state of core $x_2 y_4$ is "zero" or C state, the magnetic flux changes very little because of saturation. Thus, in this case, the induced voltage on the reading wire is small. This is why the cores are made of ferrites that have almost rectangular hysteresis loops. If the hysteresis loop is not rectangular, the flux change in "zero" or in C may not be negligible, and the reading wire may detect some voltage instead of a negligible voltage.

Note that because we know exactly which core in the memory we are reading, we need only one sensing wire to go through all cores. Also note that the reading scheme just described is destructive, in other words, the memory is erased after each reading. To save the memory, one must use a register to save the information read and then restore the core to its proper state by writing the information back.*

Example 14.1 *This example shows the amount of current needed to reverse the magnetic flux in a ferrite memory core.*

A typical ferrite memory core has an effective radius $b = 0.6$ mm. Assume that the H field needed to switch the flux is $H_1 = 220$ A/m; what should be the amplitude of the current pulse on each wire of the coincident-current magnetic-core memory device shown in Figure. 14.11?

Solution Let the magnitude of the current pulse be I; the corresponding magnetic field H should be $H_1/2$, or 110 A/m. Since the current is not perpendicular to the area of the core, we must take its perpendicular component to find H using (14.14). That is,

$$110 = \frac{I \cos 45^\circ}{2\pi \times 0.0006}$$

$$I = 0.6 \text{ A}$$

* M. M. Mano, *Digital Logic and Computer Design*, Englewood Cliffs, N.J.: Prentice-Hall, 1979, p. 297.

Magnetic-Disk Memory

Many computer systems use magnetic-disk memory devices as secondary memory. Compared to the primary memory, which uses either integrated-circuit memory or magnetic cores, the secondary memory is slower in retrieval time but usually costs less in storage capacity.

There are two kinds of magnetic disks. A flexible disk or floppy disk is a thin disk made of Mylar plastic coated with ferrite material. High performance disks use a rigid aluminum disk with ferrite coating. The floppy disk is either 9 cm (3.5 in.) or 13 cm ($5\frac{1}{4}$ in.) in diameter, and a regular disk is either 20 cm (8 in.) or 36 cm (14 in.) in diameter. Because of advancing technology, the data density of the disk has been increasing rapidly. At present, it is of the order of 200 Mbit/cm^2.

When in operation, the disk is spinning and is coupled to a magnetic head. The magnetic head is a ring-shaped device made of a ferrite core wound with coil. A small gap cut on the ring allows the magnetic field to extend into the surrounding air near the gap when the head is writing data on the disk or to pick up the magnetic field in the disk when the head is reading data from the disk. Figure 14.12 (p. 506) shows sketches of the disk and the head. The head flies aerodynamically over the disk surface. High density recording requires a small clearance between the disk surface and the head. At present, the smallest clearance is 0.1 to 1 μm.*

The disk is divided into many circular tracks from on the order of 0.5 mm wide in floppy disks to on the order of 0.025 mm wide in rigid disks. Each track is divided into small segments on the order of 0.005 mm (5 μm) long. A bit of information is stored in such a small cell.

In the writing mode, current is passed through the coil of the head to produce magnetic field in the ferrite core. The magnetic flux stays in the core region, except in the air gap where it spreads out. The magnetic field in the air is coupled to the disk and magnetizes the cell. The current may be reversed if the next cell is to be magnetized in the opposite direction. For a code called **double-frequency modulation**, a "one" is represented by a reversal of magnetization in a cell and a "zero" is represented by the absence of a reversal.** An additional reversal is inserted between each cell to provide a timing signal. Figure 14.13 shows a pattern of magnetization on a strip of a track on a disk.

In the reading mode, the head passes by the cells and picks up the remanent magnetic fields in the cells. The reversal of the flux induces voltages on the coil because of the Faraday induction law (see Chapter 16). Thus, a series of pulses marking both the cell boundaries and the "zero" or "one" information contained in the cells appears as the output of the coil voltage (see Figure 14.13b).

Note two important points here. First, the magnetization, or the alignment of magnetic moments in the ferrite coating on the disk, is done parallel to the

---

* C. Tsang, M. M. Chen, and T. Yogi, "Gigabit-density magnetic recording," *IEEE Proceedings*, Vol. 81, September 1993, pp. 1344–1359.

** R. M. White, "Disk-storage technology," *Scientific American*, August 1980, p. 138.

Figure 14.12 **(a)** A spinning magnetic disk and the read/write head. **(b)** In the writing mode, current is passed through the coil to produce a magnetic field. The magnetic flux in the head stays in the core except in the air gap where the flux spreads out.

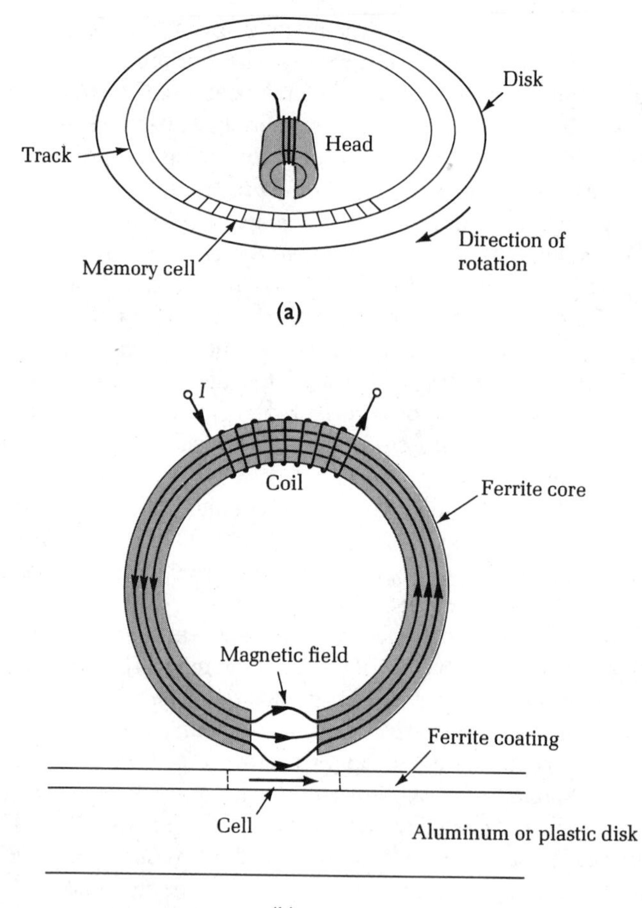

surface of the disk. The size of the cell containing a bit is determined mainly by the consideration that neighboring dipole moments tend to align themselves. Consequently, if the magnetic-flux reversal occurs too closely, a permanent reversal will not be achieved. Thus, the main volumes of magnetization in adjacent cells must be kept at a distance from each other to avoid a strong coupling between the two.

Second, because the magnetization by the head relies on the fringe field in the gap, most of the magnetization in the disk occurs near the surface. If the magnetization is made vertically—that is, perpendicular to the surface of the disk—it could increase the density of the remanent flux and therefore reduce the size of

Figure 14.13 Double-frequency-modulation binary coding on a magnetic disk. The value *1* is represented by a magnetization reversal in a memory cell, and the value *0* is represented by absence of the reversal. When reading, the magnetic flux is reversed from cell to cell to generate boundary markers. **(a)** Magnetization pattern on the disk. **(b)** The output voltage of the reading head.

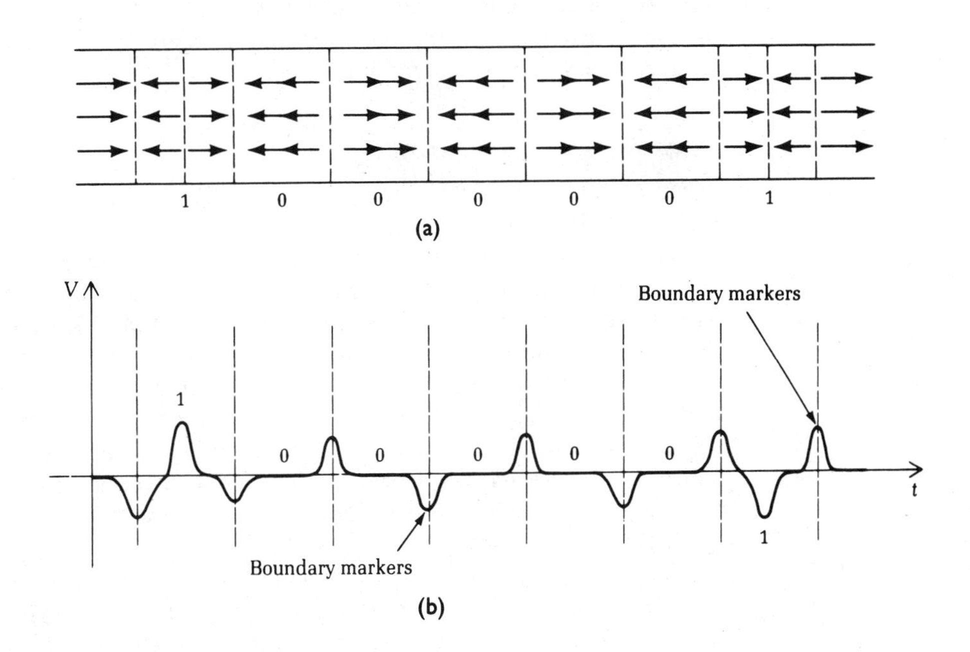

the cell and increase the density of the stored data. The data-packing density of the rigid disk is on the order of 10^5 bits/cm^2. It is estimated that vertical recording could increase the density 100 times.

The Sea Floor as a Magnetic Memory

After World War II, scientists used sensitive magnetic-field detectors, originally developed to detect submarines, to measure the local magnetic fields of rocks on the sea floor. Near Iceland, they discovered parallel magnetized stripes hundreds of miles long on the sea floor. Interestingly, the stripes were magnetized alternately in the direction of the present geomagnetic field and opposite to that direction. In our earlier discussion of geomagnetic fields, we noted that the geomagnetic field has reversed itself many times in the past six million years. So scientists theorized that the ocean floor near Iceland had been splitting apart during the same period. Each time it cracked, molten rock surged up from the bottom and was magnetized by the geomagnetic field when it cooled down below the Curie temperature. The direction of the magnetization, of course, was along the prevailing geomagnetic field at that time.

It is intriguing to notice the striking similarity between the pattern of the man-made binary recording device shown in Figure 14.13a and what nature has recorded on the sea floor. The difference is that one magnetization reversal occurred on a magnetic-disk memory several micrometers (10^{-6} m) long, the other on the sea floor several kilometers long. Writing or reading the record on the former takes a microsecond, whereas the recording of data on the sea floor spanned a period of several million years!*

Thermomagnetic Copying

Earlier in this section, we learned that a ferromagnetic material will lose its permanent magnetism when it is heated above the Curie temperature T_c. When the material is cooled below T_c, the material will be magnetized in the direction of whatever external magnetic field is present at that time. Consider two materials that have different Curie temperatures T_{c1} and T_{c2}, respectively, with $T_{c1} > T_{c2}$. These two materials are put in close contact and are then heated to a temperature below T_{c1} but higher than T_{c2}. After cooling down to room temperature, the magnetization of the first material will not change. The second material will lose its magnetization first and then be magnetized by the first material. This method is called thermomagnetization and can be used for high-speed videotape duplication.

The conventional method of making video copies uses home-type video recorders connected in parallel to a single program source. Typically, over 1000 such recorders are used for mass production. The duplication is done in real-time speed so that the process is rather slow and labor-intensive. The thermomagnetization method can copy videotapes at more than 50 times the real-time speed without compromising quality.

Videotapes are made of polyester film-base coated with magnetic particles. When chromium dioxide (CrO_2) is used as the recording medium for the copy tape, then its Curie temperature is approximately 100° C. The master tape is coated with cobalt-modified iron oxide (Co Fe Oxide) whose T_c is approximately 400° C. A mirror master tape is first prepared on which the magnetization is the mirror image of that on the master tape. The magnetic surface of the mirror master tape and that of the copy tape are then pressed between two rollers, as shown in Figure 14.14. A laser light is directed at the tapes. The upper roller is made of glass so that the laser light can penetrate through the roller and heat the tape locally. The intensity of the laser light is designed to heat the tape to about 150° C as the tapes roll through at a speed of 2.5 m/s. The CrO_2 particles lose

* E. Orowan, "The origin of the ocean ridges," *Scientific American*, November 1969, p. 103 and P. M. Hurley, "The confirmation of continental drift," *Scientific American*, April 1968, p. 53.

Figure 14.14 Thermomagnetic copying of video tapes.

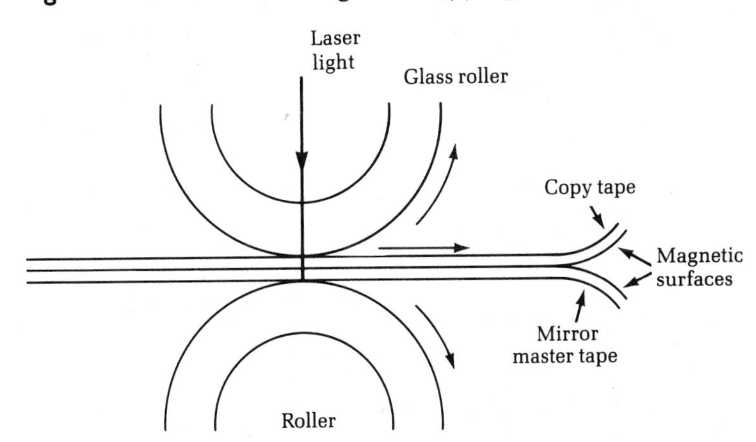

their magnetization when the temperature rises above their Curie temperature. The magnetization in the mirror master tape is not affected because the temperature is still far below its T_c. After the heated spot leaves the laser beam, the local temperature drops below 100° C in about 0.4 milliseconds. The CrO_2 particles are then magnetized with the same pattern as those on the mirror master tape.*

14.2 MAGNETIC CIRCUITS

In electric-circuit theory, we are interested in calculating currents and voltages at various branches or nodes of a network that is excited by external voltage or current sources. All calculations are based on two equations, Kirchhoff's voltage law (KVL) (Section 9.4) and Kirchhoff's current law (KCL) (Section 12.1):

$$\sum_m V_m = 0 \quad \text{(KVL)} \tag{14.15}$$

$$\sum_m I_m = 0 \quad \text{(KCL)} \tag{14.16}$$

KVL applies to any closed path in a network, and V_m are the voltage drops encountered along the closed path. The voltage rise of an external voltage source is considered to be a negative voltage drop. KCL is applicable to any node in an electric network.

As is an electric circuit, a magnetic circuit is encountered in many applications. A simple example is the magnetic core shown in Figure 14.10. We wish to know how much current is necessary to magnetize or demagnetize this core. A

* M. P. Chouinard and J. E. Gantzhorn, Jr., "High-speed videotape duplication," *Dupont Magazine*, Vol. 79, No. 2, April 1985, p. 14a.

slightly more complicated example is the magnet used in the dc motor shown in Figure 13.22 or the ferrite head of the magnetic-disk memory device shown in Figure 14.12. We see that in both cases the magnetic cores are cut to contain air gaps. It is of practical interest to know how much current is needed to generate magnetic flux in such cores. We consider this problem a magnetic-circuit problem.

Analysis of a magnetic circuit is based on the following equations:

$$\sum_m H_m \ell_m = nI \qquad (14.17)$$

$$\sum_m B_m A_m = 0 \qquad (14.18)$$

(magnetic-circuit equations)

The above equations are simply the second and third of Maxwell's equations, specialized for the magnetostatic case:

$$\nabla \times \boldsymbol{H} = \boldsymbol{J} \quad \text{or} \quad \oint \boldsymbol{H} \cdot d\boldsymbol{\ell} = I \qquad (14.19)$$

$$\nabla \cdot \boldsymbol{B} = 0 \quad \text{or} \quad \oint \boldsymbol{B} \cdot \hat{\boldsymbol{n}}\, ds = 0 \qquad (14.20)$$

In (14.17) H_m is the magnetic field intensity in the mth branch of a magnetic circuit and ℓ_m is its length. The equation is applicable to any closed path. I is the current flowing across the area bounded by the closed path, and n is the number of turns of the coil. In (14.18) B_m is the magnetic-flux density in the mth branch of a junction in a magnetic circuit, and A_m is the cross section of the branch. The analogy between the electric circuit and the magnetic circuit is obvious from comparing (14.15), (14.16), (14.17), and (14.18).

Electric Circuit	Magnetic Circuit
voltage drop V	$H\ell$ magnetovoltage drop
voltage source V	nI magnetomotive force
current I	$\Psi = BA$ magnetic flux

Example 14.2 *In practical cases, we are often required to design a circuit that produces a certain amount of magnetic flux in an air gap. This example demonstrates how to solve this kind of magnetic-circuit problem.*

Consider the magnetic circuit shown in Figure 14.15. Assume that the material of the core is iron. Figure 14.5 shows the magnetization curve of the iron. To produce a magnetic-flux density of 1.6 Wb/m^2 (T) in the core as well as in the air gap (neglecting fringing fields), what should the current in the coil be if the coil has 1000 turns?

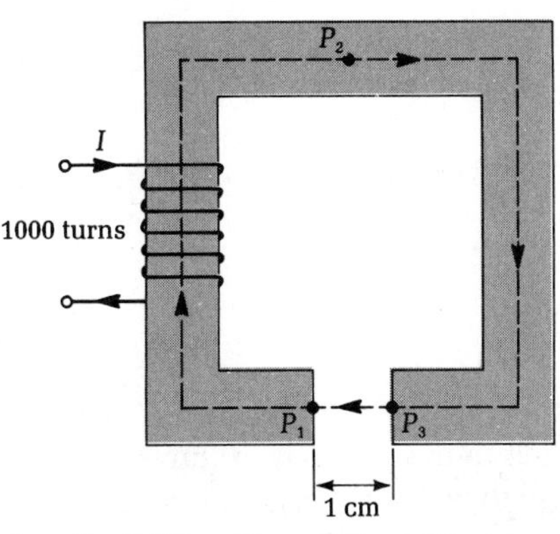

Lengths: $P_1P_2P_3 = 30$ cm, $P_3P_1 = 1$ cm
Core material: iron (Figure 14.5)

Figure 14.15 Calculating the current needed to produce the desired B field in the air gap in a magnetic circuit. (See Example 14.2.)

Solution The following equation can be set up according to (14.17):

$$H_c \ell_c + H_a \ell_a = nI \tag{14.21}$$

where H_c is the field in the core and H_a is the field in the air gap. From Figure 14.15, we have $\ell_c = 0.3$, $\ell_a = 0.01$, and $n = 1000$. In the core and in the air gap, $B = 1.6$. From Figure 14.5, we obtain

$$H_c = 260 \text{ A/m}$$

Using the equation $B = \mu_0 H$ in air, we obtain

$$H_a = \frac{1.6}{4\pi \times 10^{-7}} = 1.3 \times 10^6 \text{ A/m}$$

Substituting into (14.21), we have

$$nI = 260 \times 0.3 + 1.3 \times 10^6 \times 0.01 = 78 + 13{,}000 = 13{,}078 \text{ A-t}$$

With $n = 1000$, I is found to be

$$I = 13.08 \text{ A}$$

Note that most of the magnetovoltage drop occurs in the air gap. We further demonstrate this point in the next example.

Example 14.3

Note that in the previous example most of the magnetovoltage drop (H times the length) occurs in the air gap. This example further demonstrates this point.

Assume that the air gap is plugged by an iron block, so that the iron core shown in Figure 14.15 is a continuous structure. What is the current needed to produce the same magnetic flux—that is, $B = 1.6$ Wb/m^2—in the core?

Solution The equation (14.21) is modified and becomes as follows:

$$H_c \ell = nI$$

where $\ell = 0.30 + 0.01 = 0.31$ m. Using the value $H_c = 260$ A/m read from Figure 14.5, we have

$$nI = 260 \times 0.31 = 80.6$$

$$I = 0.08 \text{ A}$$

Comparing this result with that of the previous example, we realize that the large amount of current needed in the previous example was due to the presence of the air gap. Thus, the magnetic-core material is analogous to the conductor in an electric circuit, and the air gap is analogous to a highly resistive material.

Example 14.4 *This example is a more complicated magnetic circuit problem. It shows that the solution procedure is very similar to that used in solving electric circuit problems.*

Figure 14.16 shows a magnetic circuit. The magnetic field in the air gap is $B = 1.2$ Wb/m^2. Assuming that the permeability of the magnetic material is $\mu = 1000\mu_0$, find the current I in each of the two coils.

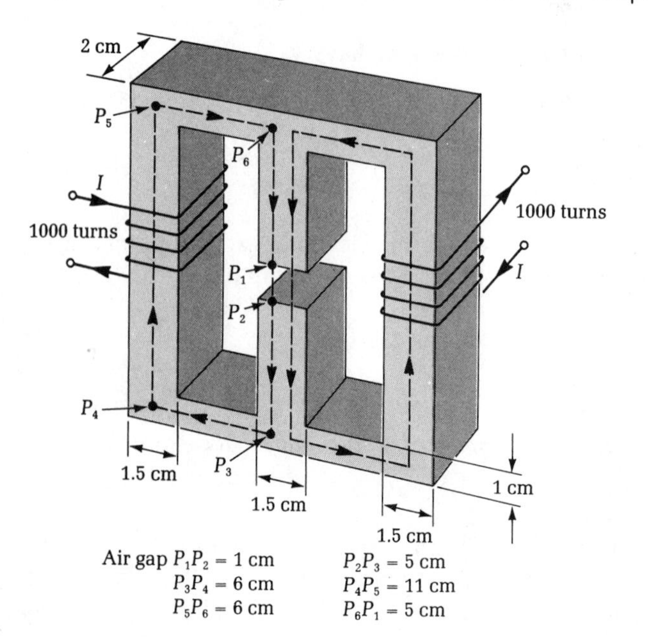

Figure 14.16 A magnetic circuit solved in Example 14.4.

electric circuit, and the air gap is analogous to a highly resistive material.

Solution Because of the symmetry of this magnetic circuit, we can consider only the loop $P_1P_2P_3P_4P_5P_6P_1$. The total flux Ψ in the air gap is given as follows:

$$\Psi = B_a A_a = 1.2 \times 0.02 \times 0.015 = 3.6 \times 10^{-4} \text{ Wb}$$

Half of this flux flows in the left half of the circuit. Thus, the B field in the branches P_3P_4, P_4P_5, and P_5P_6 are as follows

$$B_{34} = B_{56} = \frac{1.8 \times 10^{-4}}{0.01 \times 0.02} = 0.9 \text{ Wb/m}^2$$

$$B_{45} = \frac{1.8 \times 10^{-4}}{0.02 \times 0.015} = 0.6 \text{ Wb/m}^2$$

The corresponding H fields are found accordingly:

$$H_{\text{air}} = \frac{B_a}{\mu_0} = \frac{1}{\mu_0}(1.2)$$

$$H_{23} = H_{61} = \frac{1.2}{1000\mu_0} = \frac{1}{\mu_0}(1.2 \times 10^{-3}) \text{ A/m}$$

$$H_{34} = H_{56} = \frac{1}{\mu_0}(0.9 \times 10^{-3}) \text{ A/m}$$

$$H_{45} = \frac{1}{\mu_0}(0.6 \times 10^{-3}) \text{ A/m}$$

Multiplying H's with the corresponding lengths and using (14.17), we obtain:

$$\begin{aligned} 1000I &= \frac{1}{\mu_0}(1.2 \times 0.01 + 1.2 \times 10^{-3} \times 0.1 \\ &\quad + 0.9 \times 10^{-3} \times 0.12 + 0.6 \times 10^{-3} \times 0.11) \\ &= 9783 \\ I &= 9.783 \text{ A} \end{aligned}$$

Although (14.18) does not seem to have been used in previous examples, in fact it has been. The statement that half of the flux flows to the left branch of the circuit and the calculation of B_{34}, B_{45}, and B_{56} assume implicitly that flux is conserved in the magnetic circuit just as current is in an electric circuit. This statement is exactly what (14.18) implies.

Example 14.5 *The B-H curve of a magnetic material is usually nonlinear. Consequently, to solve the magnetic-circuit problem containing a nonlinear magnetic material may require using a graphical or numerical method, as this example demonstrates.*

Consider the magnetic circuit shown in Figure 14.17a. The current in the coil is 10 A, and the coil has 100 turns. The core is made of steel, which has the initial magnetization curve shown in Figure 14.17b. Find the magnetic flux *B* in the air gap.

Figure 14.17(a) An iterative procedure is used to find the magnetic flux in the air gap. (See Example 14.5.)

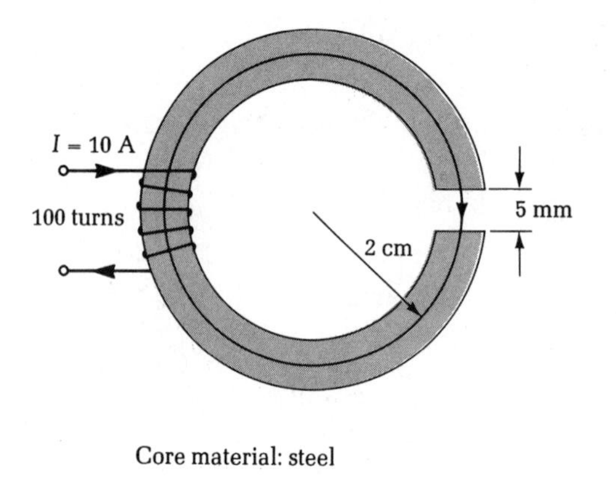

Figure 14.17(b) The initial magnetization curve of steel and the straight line corresponding to Equation (14.22).

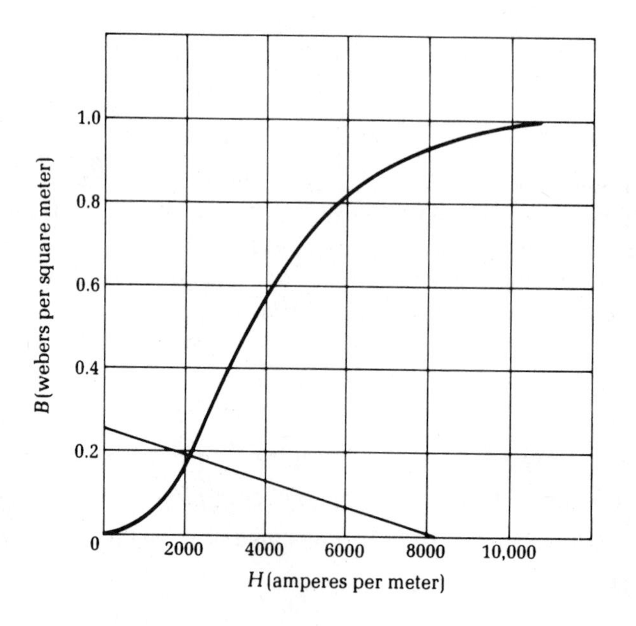

Solution Let us assume that the magnetic flux in the core and in the air gap is B. Using (14.17), we obtain the following equation:

$$\frac{B}{\mu_0}\ell_a + H_c\ell_c = 1000 \tag{14.22}$$

where $\ell_a = 0.005$ m and $\ell_c = (2\pi \times 0.02) - 0.005 = 0.121$ m. The above equation cannot be solved analytically because the relation between B and H_c is shown in a graphical form (Figure 14.17b), which is a nonlinear function. Instead, we may use the following iterative procedure.

Let (14.22) be expressed in a different form as follows:

$$B = \frac{(1000 - H_c\ell_c)\mu_0}{\ell_a} \tag{14.23}$$

Our experience in solving the previous examples shows that the magnetovoltage drop is mainly due to that in air. Thus, our first approximation is to neglect the $H_c\ell_c$ term in (14.23) to obtain the "zeroth-order" approximation for B, which is denoted as $B^{(0)}$:

$$B^{(0)} = \frac{1000\mu_0}{\ell_a} = 0.251 \text{ Wb/m}^2$$

From $B^{(0)}$, we find the corresponding $H^{(0)}$ (Figure 14.17b):

$$H^{(0)} = 2400 \text{ A/m}$$

The iterative method calls for substituting the above value into (14.23) to obtain the "first-order" approximation of B, which is denoted as $B^{(1)}$.

$$B^{(1)} = \frac{(1000 - 2400 \times 0.121)\mu_0}{0.005} = 0.178 \text{ Wb/m}^2$$

The corresponding $H^{(1)}$ may be read from Figure 14.17b:

$$H^{(1)} = 2100 \text{ A/m}$$

We obtain the "second-order" approximation of B by substituting the above $H^{(1)}$ value for H in (14.23):

$$B^{(2)} = \frac{(1000 - 2100 \times 0.121)\mu_0}{0.005} = 0.187 \text{ Wb/m}^2$$

This procedure can be repeated to find the nth iterative result of $B^{(n)}$. When a digital computer is available, the magnetization curve can be approximated by a standard polynomial-curve fitting and stored in the computer memory. A simple program may be written to carry out the iterative procedure, which requires very little computer time (see Problem 14.14).

The problem at hand can also be solved by a graphical method. Note that (14.22) or, equivalently, (14.23) is an equation of a straight line on the B-H plane. As shown in Figure 14.17b this line intersects the B axis at 0.251 Wb/m^2 and the H axis at 8264 A/m. It also intersects the nonlinear magnetization curve at $B = 0.19$ Wb/m^2. This result agrees fairly well with the result obtained by the iterative method.

SUMMARY

1. The B field (magnetic-flux density) is equal to the magnetic permeability μ times the H field (magnetic-field strength).
2. When μ is less than μ_0, the permeability of the free space, then the material is called a diamagnetic material. A material is paramagnetic if its μ is greater than μ_0. A ferromagnetic material has a very high μ/μ_0 ratio.
3. Many ferromagnetic materials exhibit hysteresis characteristics. The relation between B and H traces loops on the B-H plane. It is possible to have magnetic flux B when H is zero. This residual flux is called remanence. The H field needed to nullify the residual flux is called the coercive force—that is, the H field that will bring B down to zero.
4. Permanent magnets are characterized by retentivity and coercivity. Retentivity is the remanence flux density after the magnet has been brought to saturation and then the H field has been withdrawn. The coercivity is the coercive force that is required to bring the B field down to zero in the same saturation condition.
5. In practical cases, we are often required to design a circuit that produces a certain amount of magnetic flux in an air gap. This is called a magnetic-circuit problem. It is solved by using Ampère's law and Gauss' law for magnetic fields. Equations (14.17) and (14.18) are those laws applied to magnetic circuits.
6. The procedure to solve magnetic circuit problems is very similar to that in solving electric circuit problems. The magnetic voltage drop is the H field times the distance along a closed path in the circuit. The voltage source is nI, a number of turns of coils times the electric current. The magnetic flux Ψ (B times the area) is analogous to current in an electric circuit.

Problems

14.1 Refer to the magnetization curve shown in Figure 14.3. The material is a nonlinear medium because μ depends on the magnitude of H. For magnetostatic fields, μ is equal to the slope of the line joining the origin to the (H, B) point on the magnetization curve. In this way, Figure 14.3b is obtained from Figure 14.3a. Now, if the material is placed in a time-harmonic field, the effective μ will be different from the μ for the magnetostatic fields. Consider a field $H = H_0 + H_1 \cos(\omega t + \phi)$, where H_0 is the bias magnetostatic field and H_1 is the amplitude of the time-harmonic component of the total field. Let $H_1 \ll H_0$; then the effective permeability of a material is the slope of the tangent of the magnetization curve at H_0. Sketch the effective μ versus H_0 for the curve shown in Figure 14.3a. Compare it with the magnetostatic μ shown in Figure 14.3b, and show that the μ's in these two cases are equal to each other at P_3.

14.2 Point out the differences between the following pairs of terms: (a) diamagnetic versus paramagnetic, (b) remanence versus retentivity, and (c) coercive force versus coercivity.

14.3 What are approximate values of the retentivity and the coercivity of the ferrite shown in Figure 14.9?

14.4 Consider the carbon steel, alnico V, and cunico materials listed in Table 14.2. Which has the highest permanent magnetic-field strength? Which has the most difficulty in losing its permanent magnetism once it is magnetized?

14.5 A permanent magnet of radius 1.5 cm and thickness 0.3 cm is put in a magnetic field that is parallel to the disk, as in the situation depicted in Figure 14.7. The torque on the disk is equal to 1.2×10^{-3} N-m, and the magnetic field is equal to 10^{-3} Wb/m^2. What is the remanence of the permanent magnet?

14.6 To write "one" in the memory core $x_2 y_3$ shown in Figure 14.11, how should the current pulses be sent along the wires? Specify the polarity of these pulses.

14.7 Consider the magnetic-core memory sketched in Figure 14.11 and the corresponding hysteresis curve for the cores shown in Figure 14.9. Now suppose that, because of malfunction in the circuitry, a positive pulse of amplitude I, which alone is capable of producing the switching magnetic field strength H_1, is sent down the line y_2 and that simultaneously an identical pulse is sent down the line x_3. Assume that all cores are initially in the "zero" state, which corresponds to having the magnetic flux circulation point either toward the upper left or the lower left (using the right-hand rule). What are the states of all of the cores after these pulses have passed through?

14.8 Compare the hysteresis loops of two ferrites shown in Figure P14.8. The curve labeled Number 1 is "thinner" than that labeled Number 2. Which ferrite core requires less switching current? Which ferrite has a better ability to withstand magnetic interferences?

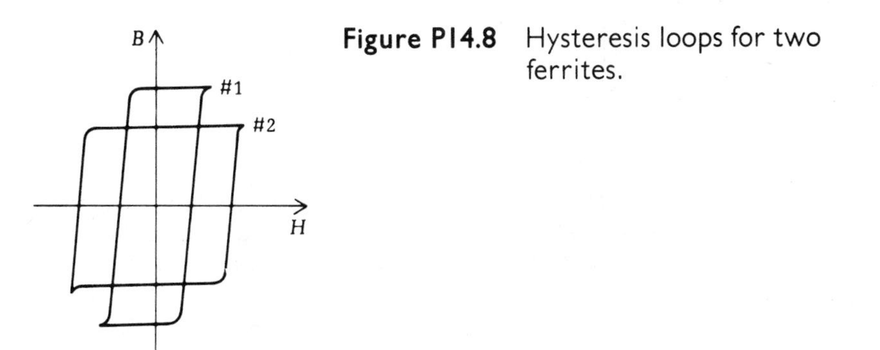

Figure P14.8 Hysteresis loops for two ferrites.

14.9 What is the minimum amount of the current I needed to magnetize the ferrite film of a magnetic tape? The hysteresis loop of the ferrite material and the magnetic circuit of the recording head are shown in Figure P14.9.

(a) Assume that a cross-sectional area of the air gap of the recording head is the same as that of the magnetic core; that is, the fringing magnetic field in the air gap is neglected. Find I.

(b) To take into account the spread of the magnetic field in the air gap of the recording head, assume that the effective cross-sectional area of the air gap is 20% larger than that of the magnetic core. Find I.

Figure P14.9

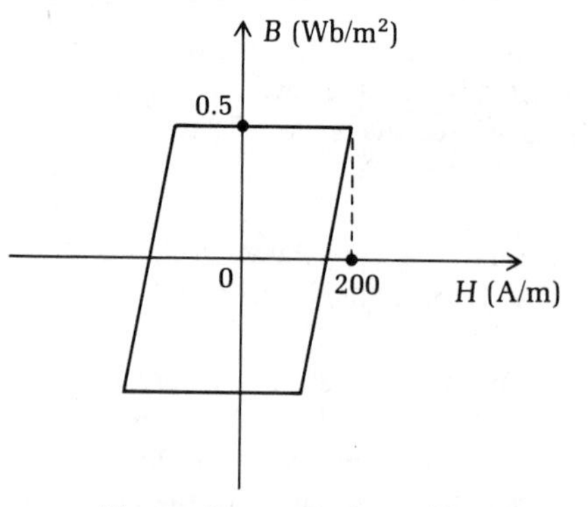

Hysteresis curve for ferrite film

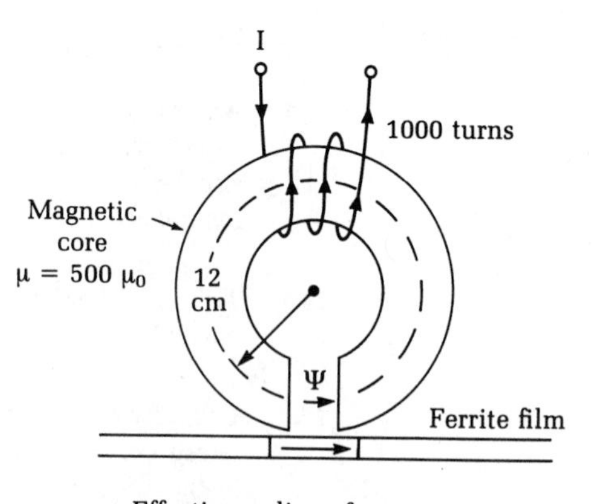

Effective radius of core = 12 cm
Air gap = 0.4 cm

14.10 The magnetic circuit shown in Figure P14.10 is made of a material with $\mu = 600\mu_0$. Find the flux densities B_1 and B_2, and indicate their directions.

Figure P14.10

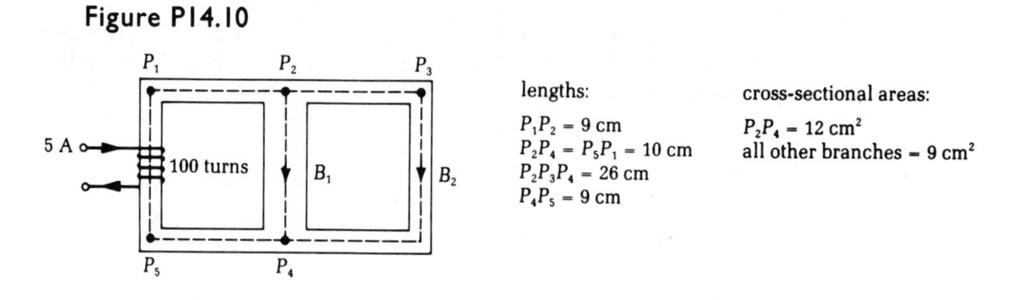

14.11 To produce a magnetic flux of 0.5 Wb/m² in the air gap of the magnetic circuit shown in Figure P14.11, what should be the magnitude of the current in the coil? Take $\mu = 200\mu_0$. The cross-sectional area of all branches is equal to 4 cm².

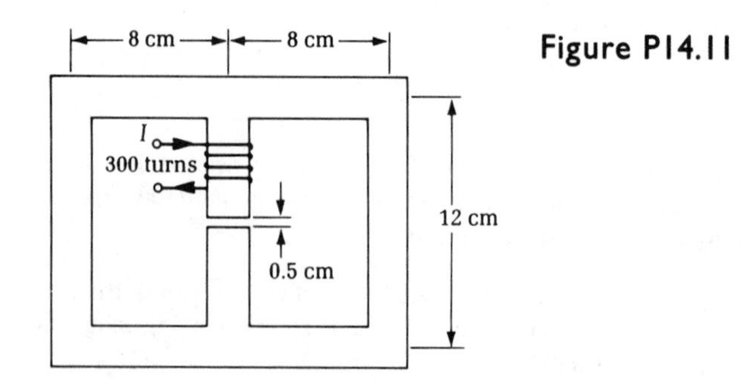

Figure P14.11

14.12 Consider the magnetic circuit shown in Figure P14.12. The magnetic flux density B_{air} is equal to 0.1 T in the air gap. Assuming no flux leakage, find the current I. Hint: find the following quantities in order: sum of $H\ell$ in the branch 2-7-8-5, then H and B of branch 2-3-4-5, then Ψ of these branches, and then Ψ of 5-6-1-2.

Figure P14.12

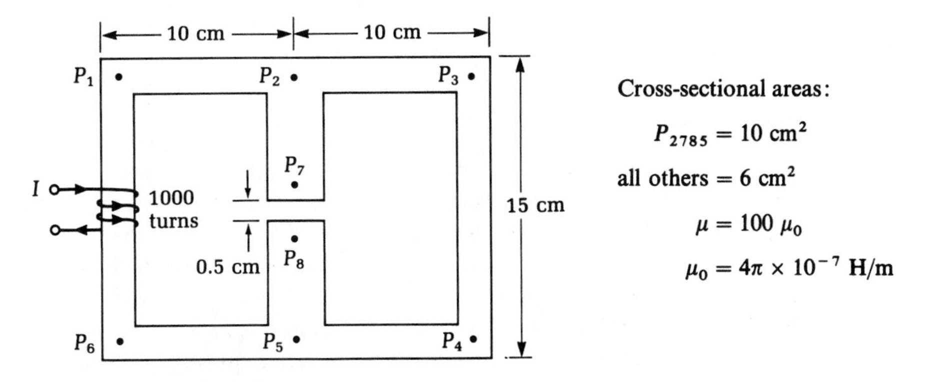

14.13 Find the approximate value of B in the magnetic circuit shown in Figure 14.17a for excitation current $I = 15$ A instead of 10 A. All other conditions given remain unchanged. Carry out the iteration a sufficient number of times to obtain an accuracy to the third digit.

14.14 Write a computer program to carry out the iteration procedure outlined in Example 14.5. First, approximate the nonlinear curve in Figure 14.17b by a polynomial of fifth order. Then carry out the iteration five times to obtain the fourth-order approximation for B.

15 ELECTROQUASISTATIC FIELDS

In Chapters 2 through 8 we studied time-harmonic fields, which are periodic functions of time. In Chapters 9 through 14 we studied static fields, which do not vary with time. In this and the following chapter, we shall consider fields that are not static but that vary very slowly with time. These slowly varying electromagnetic fields are called quasistatic fields. Electromagnetic fields produced by power transmission lines and in generators, motors, transformers, and electric appliances are quasistatic fields.

After we define the quasistatic fields mathematically, we shall separate them into two categories known as electroquasistatics and magnetoquasistatics. This chapter gives examples of electroquasistatic fields, and Chapter 16 discusses magnetoquasistatic fields.

15.1 QUASISTATIC APPROXIMATION

When electromagnetic fields vary slowly with time, the terms involving time derivatives are small quantities. When all time derivatives are neglected in Maxwell's equation, the following set of equations, defined as the **zeroth-order equations**, is obtained:

$$\nabla \times \boldsymbol{E}^{(0)} = 0 \tag{15.1a}$$

$$\nabla \times \boldsymbol{H}^{(0)} = \boldsymbol{J}^{(0)} \tag{15.1b}$$

$$\nabla \cdot \boldsymbol{D}^{(0)} = \rho_v^{(0)} \tag{15.1c}$$

$$\nabla \cdot \boldsymbol{B}^{(0)} = 0 \tag{15.1d}$$

Note that the above equations are similar to the electrostatic and magnetostatic field equations.

Very often the zeroth-order fields are accurate enough for many applications. However, if it turns out that more accuracy is needed, **first-order terms** are calculated according to the following equations:

$$\nabla \times \boldsymbol{E}^{(1)} = -\frac{\partial \boldsymbol{B}^{(0)}}{\partial t} \tag{15.2a}$$

$$\nabla \times \boldsymbol{H}^{(1)} = \boldsymbol{J}^{(1)} + \frac{\partial \boldsymbol{D}^{(0)}}{\partial t} \tag{15.2b}$$

$$\nabla \cdot \boldsymbol{D}^{(1)} = \rho_v^{(1)} \tag{15.2c}$$

$$\nabla \cdot \boldsymbol{B}^{(1)} = 0 \tag{15.2d}$$

We see that those terms involving time derivatives are put back into the Maxwell's equations. However, these terms are approximate quantities because they are time derivatives of the zeroth-order fields, not of the exact fields. It has been said that when fields vary slowly with time, time derivative terms are small quantities. Thus the error will likely be small when approximate terms instead of exact terms are used.

When even more accurate results are needed, then the **second-order** terms may be calculated using the following equations:

$$\nabla \times \boldsymbol{E}^{(2)} = -\frac{\partial \boldsymbol{B}^{(1)}}{\partial t} \tag{15.3a}$$

$$\nabla \times \boldsymbol{H}^{(2)} = \boldsymbol{J}^{(2)} + \frac{\partial \boldsymbol{D}^{(1)}}{\partial t} \tag{15.3b}$$

$$\nabla \cdot \boldsymbol{D}^{(2)} = \rho_v^{(2)} \tag{15.3c}$$

$$\nabla \cdot \boldsymbol{B}^{(2)} = 0 \tag{15.3d}$$

The process can continue indefinitely. In general, for the ***n*th-order** terms, we find

$$\nabla \times \boldsymbol{E}^{(n)} = -\frac{\partial \boldsymbol{B}^{(n-1)}}{\partial t} \tag{15.4a}$$

$$\nabla \times \boldsymbol{H}^{(n)} = \boldsymbol{J}^{(n)} + \frac{\partial \boldsymbol{D}^{(n-1)}}{\partial t} \tag{15.4b}$$

$$\nabla \cdot \boldsymbol{D}^{(n)} = \rho_v^{(n)} \tag{15.4c}$$

$$\nabla \cdot \boldsymbol{B}^{(n)} = 0 \tag{15.4d}$$

We can prove that the exact fields are equal to the sums of their corresponding partial fields. That is, let

$$\boldsymbol{E} = \boldsymbol{E}^{(0)} + \boldsymbol{E}^{(1)} + \boldsymbol{E}^{(2)} + \cdots + \boldsymbol{E}^{(n)} + \cdots \tag{15.5a}$$

$$\boldsymbol{H} = \boldsymbol{H}^{(0)} + \boldsymbol{H}^{(1)} + \boldsymbol{H}^{(2)} + \cdots + \boldsymbol{H}^{(n)} + \cdots \tag{15.5b}$$

$$\boldsymbol{B} = \boldsymbol{B}^{(0)} + \boldsymbol{B}^{(1)} + \boldsymbol{B}^{(2)} + \cdots + \boldsymbol{B}^{(n)} + \cdots \tag{15.5c}$$

$$\boldsymbol{D} = \boldsymbol{D}^{(0)} + \boldsymbol{D}^{(1)} + \boldsymbol{D}^{(2)} + \cdots + \boldsymbol{D}^{(n)} + \cdots \tag{15.5d}$$

$$\boldsymbol{J} = \boldsymbol{J}^{(0)} + \boldsymbol{J}^{(1)} + \boldsymbol{J}^{(2)} + \cdots + \boldsymbol{J}^{(n)} + \cdots \tag{15.5e}$$

$$\rho_v = \rho_v^{(0)} + \rho_v^{(1)} + \rho_v^{(2)} + \cdots + \rho_v^{(n)} + \cdots \tag{15.5f}$$

then $\boldsymbol{E}$, $\boldsymbol{H}$, $\boldsymbol{B}$, $\boldsymbol{D}$, $\boldsymbol{J}$, ρ_v are the exact electromagnetic fields, current, and charge that satisfy the Maxwell's equations. The preceding statement can be proved by adding equations (15.1a), (15.2a), (15.3a), and so on to obtain the following result:

$$\nabla \times [\boldsymbol{E}^{(0)} + \boldsymbol{E}^{(1)} + \boldsymbol{E}^{(2)} + \cdots + \boldsymbol{E}^{(n)} + \cdots] = 0 - \left(\frac{\partial}{\partial t}\right)[\boldsymbol{B}^{(0)} + \boldsymbol{B}^{(1)} + \boldsymbol{B}^{(2)} + \cdots + \boldsymbol{B}^{(n)} + \cdots]$$

Substituting (15.5a) and (15.5c) in the above equation yields

$$\nabla \times \boldsymbol{E} = -\frac{\partial \boldsymbol{B}}{\partial t}$$

which is the first Maxwell's equation. Other equations may be obtained in a similar manner.*

In summary, for quasistatic fields, the zeroth-order fields are solved first from the zeroth-order Equations (15.1). The first-order fields may be solved from (15.2) and the solution already obtained for the zeroth-order fields. The first-order fields may be added to the zeroth-order fields to improve the accuracy of the solution. Second-order and higher-order fields may be obtained in a similar manner when even more accurate results are needed.

Quasistatic Fields in a Parallel-Plate Region

Consider the parallel-plate region shown in Figure 15.1. A time-varying voltage source $V(t)$ is placed at $z = 0$ across the plates, where

$$V(t) = V_0 \cos \omega t \tag{15.6}$$

where V_0 is assumed to be a constant independent of ω.

Figure 15.1 A parallel plate connected to a time-harmonic voltage source.

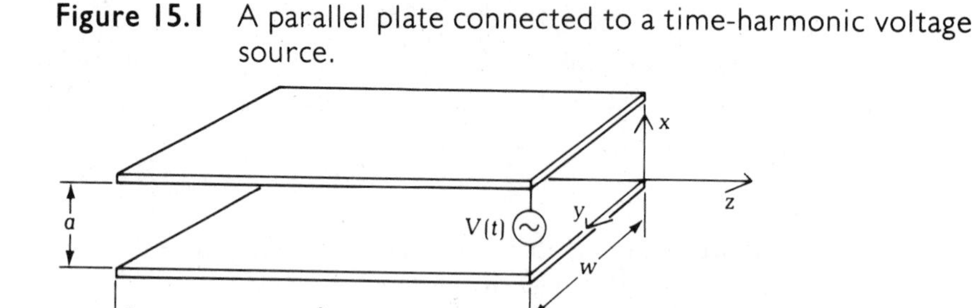

The zeroth-order electric field is just the static electric field. For a given parallel plate with voltage V and separation a, the electrostatic field is given as follows:

$$\boldsymbol{E}^{(0)} = -\hat{\boldsymbol{x}}\frac{V(t)}{a} = -\hat{\boldsymbol{x}}\frac{V_0}{a}\cos \omega t \tag{15.7}$$

* A different derivation of these quasistatic equations is given in: R. M. Fano, L. J. Chu, and R. B. Adler, *Electromagnetic Fields, Energy, and Forces*, New York: Wiley 1960, chap. 6.

There is no current density $\boldsymbol{J}$ anywhere in the parallel-plate region. Therefore, the zeroth-order $\boldsymbol{H}$ is zero:

$$\boldsymbol{H}^{(0)} = 0$$

The first-order $\boldsymbol{E}^{(1)}$ is zero according to (15.2a) because $\boldsymbol{H}^{(0)}$ is zero. The first-order $\boldsymbol{H}^{(1)}$ can be found from (15.2b). There is no current density anywhere in the parallel-plate region so that $\boldsymbol{J}$ is zero for all orders. Equation (15.2) gives the following relation:

$$\nabla \times \boldsymbol{H}^{(1)} = \hat{\boldsymbol{x}} \frac{V_0 \epsilon \omega}{a} \sin(\omega t) \tag{15.8}$$

Assuming that the width of the plates is much greater than their separation—that is, that $w \gg a$—we can consider the plates infinitely wide and the fields independent of y. Consequently, Equation (15.8) becomes

$$-\hat{\boldsymbol{x}} \frac{\partial H_y^{(1)}}{\partial z} = \hat{\boldsymbol{x}} \frac{V_0 \epsilon \omega}{a} \sin(\omega t)$$

Integrating both sides of the above equation yields

$$H_y^{(1)} = -\frac{V_0 \epsilon \omega}{a} z \sin(\omega t) + C$$

The constant C is an integration constant to be determined by the boundary condition. The surface current on the upper plate is

$$\boldsymbol{J}_s^{(1)} = \hat{\boldsymbol{n}} \times \boldsymbol{H}^{(1)} = -\hat{\boldsymbol{x}} \times \boldsymbol{H}^{(1)}$$

Because the parallel plate ends at $z = 0$, the surface current must be zero there. Consequently,

$$C = 0$$

In summary, the first-order fields are as follows:

$$\boldsymbol{H}^{(1)} = -\hat{\boldsymbol{y}} \omega z \frac{\epsilon V_0}{a} \sin \omega t \tag{15.9a}$$

$$\boldsymbol{E}^{(1)} = 0 \tag{15.9b}$$

We have now obtained the zeroth- and the first-order electric and magnetic fields and can continue the process to obtain higher-order terms. The general Equations (15.4) may be used repeatedly to yield higher-order terms. Proceeding, we notice that

$$\nabla \times \boldsymbol{H}^{(2)} = \frac{\partial}{\partial t} \epsilon \boldsymbol{E}^{(1)}$$

Because $\boldsymbol{E}^{(1)} = 0$, we have

$$\boldsymbol{H}^{(2)} = 0$$

Again, from (15.3a),

$$\nabla \times \boldsymbol{E}^{(2)} = -\mu \frac{\partial}{\partial t} \boldsymbol{H}^{(1)} = \hat{y} \mu \omega^2 z \frac{V_0 \epsilon}{a} \cos \omega t$$

The electric field is a function of z, and t points in the $\hat{x}$ direction. From

$$\frac{\partial}{\partial z} E_x^{(2)} = \mu \omega^2 z \frac{\epsilon V_0}{a} \cos \omega t$$

we find

$$\boldsymbol{E}^{(2)} = \hat{x} \frac{V_0}{a} k^2 \cos \omega t \left(\frac{1}{2} z^2 + C_1 \right) = \hat{x} \frac{V_0}{2a} k^2 z^2 \cos \omega t \tag{15.10}$$

where $k^2 = \omega^2 \mu \epsilon$. The integration constant C_1 is equal to zero because we want $\boldsymbol{E}(z = 0)$ to be equal to $-\hat{x}(V_0/a) \cos \omega t$, which is determined by the fact that the voltage source is attached at $z = 0$.

Continuing the process, we find the third-order magnetic field

$$\boldsymbol{H}^{(3)} = \hat{y} \frac{\mu \epsilon^2 \omega^3 V_0}{3 \cdot 2a} z^3 \sin \omega t = \hat{y} \frac{V_0}{3 \cdot 2a\eta} k^3 z^3 \sin \omega t \tag{15.11}$$

where $\eta = (\mu/\epsilon)^{1/2}$. The third-order electric field $\boldsymbol{E}^{(3)}$ is equal to zero, and the fourth-order electric field

$$\boldsymbol{E}^{(4)} = -\hat{x} \frac{V_0}{4 \cdot 3 \cdot 2a} k^4 z^4 \cos \omega t \tag{15.12}$$

and so on. The general form of the nth-order field is seen to be

$$\boldsymbol{E}^{(n)} = -\hat{x} \begin{cases} (-1)^{n/2} \dfrac{V_0}{n!a} (kz)^n \cos \omega t & n = 0, 2, 4, \ldots \\ 0 & n = 1, 3, 5, \ldots \end{cases}$$

$$\boldsymbol{H}^{(n)} = -\hat{y} \begin{cases} (-1)^{(n-1)/2} \dfrac{V_0}{n!a\eta} (kz)^n \sin \omega t & n = 1, 3, 5, \ldots \\ 0 & n = 0, 2, 4, \ldots \end{cases}$$

The infinite sum of the above terms yields the full-wave solution

$$\boldsymbol{E} = \sum_n \boldsymbol{E}^{(n)} = -\hat{x} \frac{V_0}{a} \cos kz \cos \omega t \tag{15.13a}$$

$$\boldsymbol{H} = \sum_n \boldsymbol{H}^{(n)} = -\hat{y} \frac{V_0}{a\eta} \sin kz \sin \omega t \tag{15.13b}$$

In terms of the complex notation, we find

$$\mathbf{E} = -\hat{x} \frac{V_0}{a} \cos kz \tag{15.14a}$$

$$\mathbf{H} = \hat{y} \frac{jV_0}{a\eta} \sin kz \tag{15.14b}$$

such that Re$\{\mathbf{E}e^{j\omega t}\}$ yields (15.13a) and Re$\{\mathbf{H}e^{j\omega t}\}$ yields (15.13b). The solution in (15.14a) and (15.14b) represents the superposition of a wave in the positive $\hat{z}$ direction and a wave in the negative $\hat{z}$ direction.

$$\mathbf{E} = -\hat{x}\frac{V_0}{2a}(e^{-jkz} + e^{jkz})$$

$$\mathbf{H} = -\hat{y}\frac{V_0}{2a\eta}(e^{-jkz} - e^{jkz})$$

which are the TEM waves in the parallel-plate waveguide, as discussed in Chapter 5. The total **E** and **H** in (15.14a) and (15.14b) are standing waves satisfying the open-circuit boundary condition at $z = 0$, where $\mathbf{J}(z = 0) = \hat{n} \times H(z = 0) = 0$.

15.2 CIRCUIT THEORY AND ELECTROQUASISTATICS

In circuit theory, the capacitance C, the inductance L, and the resistance R are characterized by two variables—namely, the voltage $V(t)$ and the current $I(t)$. For an element with capacitance C, $V(t)$ and $I(t)$ are related by

$$I(t) = C\frac{dV(t)}{dt} \tag{15.15}$$

For an element with inductance L, we have

$$V(t) = L\frac{dI(t)}{dt} \tag{15.16}$$

The resistance R provides a linear relationship between V and I

$$V(t) = RI(t) \tag{15.17}$$

The inverse of R is the conductance G, and we have

$$I(t) = GV(t) \tag{15.18}$$

In the previous chapters on static fields, we defined the capacitance, the inductance, and the resistance from the point of view of the static field.

From the electroquasistatic point of view, we have zeroth-order electric fields that give rise to first-order currents associated with the first-order magnetic field. Recall that our static definition for capacitance is

$$Q = CV$$

Taking the time derivative on both sides and noting that $I = dQ/dt$, we obtain

$$I = C\frac{dV}{dt}$$

which is the same as (15.15). In general, the current is composed of two terms: one is associated with the potential of the zeroth order, and the other is (15.18). We thus have

$$I = GV + C\frac{dV}{dt} \tag{15.19}$$

This formula represents a parallel combination of a conductance and a capacitance (Figure 15.2).

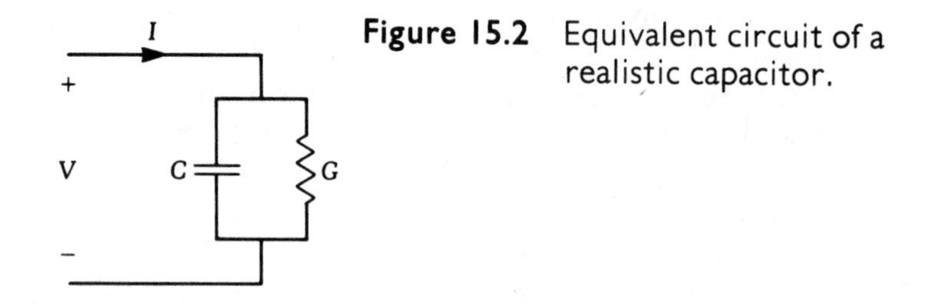

Figure 15.2 Equivalent circuit of a realistic capacitor.

From the magnetoquasistatic point of view, we have zeroth-order magnetic fields that give rise to first-order electric fields. Recall that our static definition for inductance is

$$\Psi = LI$$

Chapter 16 will show that the time derivative of the magnetic flux yields a voltage V. Thus, taking the time derivative on both sides of the above equation gives

$$V = L\frac{dI}{dt}$$

which is the same as (15.16). In general, the zeroth-order current produces zeroth-order magnetic fields, which in turn generate first-order electric fields. The voltage associated with the first-order electric field is given in (15.16). The zeroth-order current also causes a voltage drop in a resistance. We thus have

$$V = RI + L\frac{dI}{dt} \tag{15.20}$$

This formula represents a series combination of a resistance R and an inductance L (Figure 15.3).

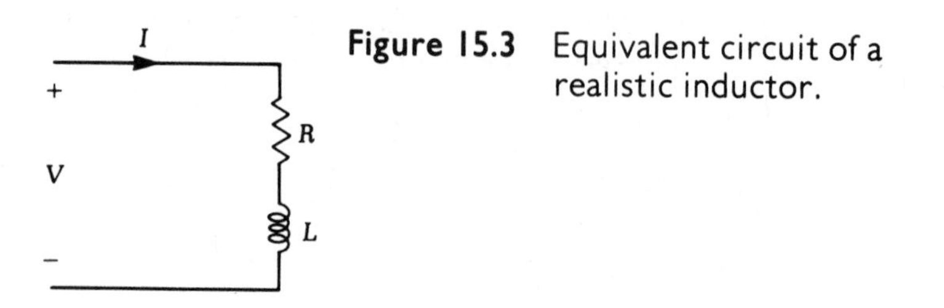

Figure 15.3 Equivalent circuit of a realistic inductor.

Magnetoquasistatic Limit of a Short-Circuited Transmission Line

As we learned in Chapter 6, the voltage and current distributions on a short-circuited transmission line (Figure 15.4) take the following forms:

$$V = V_0 \sin kz \tag{15.21a}$$

$$I = j\frac{V_0}{Z_0}\cos kz \tag{15.21b}$$

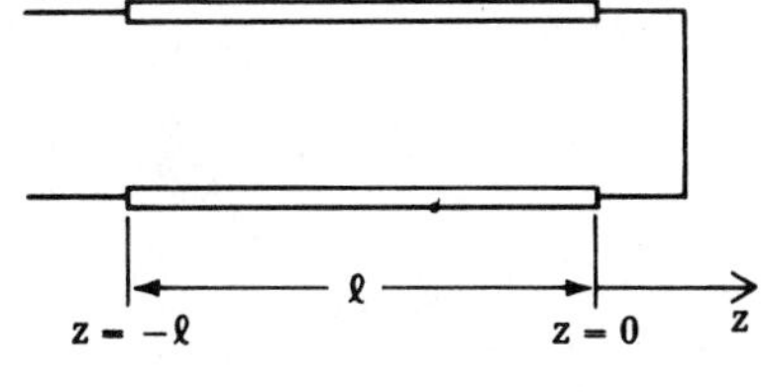

Figure 15.4 A short-circuited transmission line.

where $k = \omega\sqrt{LC}$ and $Z_0 = \sqrt{L/C}$. We see that $V = 0$ at the short-circuit end, where $z = 0$. In accordance with the magnetoquasistatic assumptions, the zeroth-order current is, by approximating $\cos kz \approx 1$,

$$I = \frac{jV_0}{Z_0}$$

Approximating $\sin kz \approx kz$ in (15.21), we see that the zeroth-order voltage is zero and that the first-order term is

$$V \approx V_0 kz$$

At the input end, where $z = -\ell$, the input impedance is found to be

$$Z = \frac{V}{I} = jk\ell Z_0 = j\omega\sqrt{LC}\ell\sqrt{\frac{L}{C}} = j\omega(L\ell)$$

This impedance is the impedance of an inductor with inductance $L\ell$, where L is the characteristic inductance per unit length of the transmission line and ℓ is its length.

The next chapter contains more examples for the magnetoquasistatic fields. In this chapter we concentrate on the electroquasistatic fields.

Electroquasistatic Limit of an Open-Circuited Transmission Line

As we learned in Chapter 6, the voltage and current on an open-circuited transmission line (Figure 15.5) is

$$V = V_0 \cos kz \tag{15.22a}$$

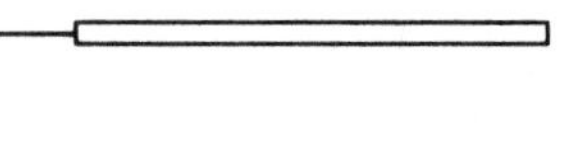

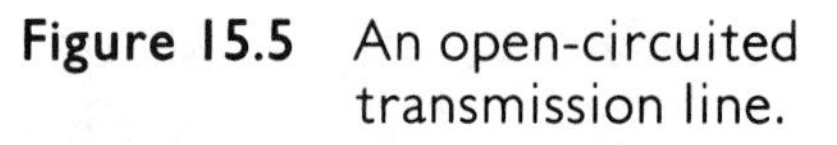

Figure 15.5 An open-circuited transmission line.

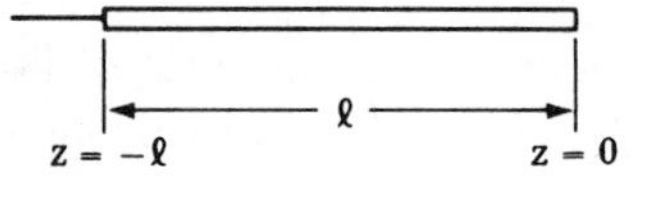

$$I = \frac{V_0}{jZ_0} \sin kz \tag{15.22b}$$

We see that $I = 0$ at $z = 0$. In accordance with the electroquasistatic approximation, the zeroth-order voltage is

$$V \approx V_0$$

and the first-order current is

$$I \approx \frac{V_0}{jZ_0} kz$$

Thus, in the electroquasistatic limit, the voltage is a constant across the transmission line, and the current varies linearly with z and is zero at the open-circuited end, where $z = 0$. At the input end, where $z = -\ell$, the input impedance is found to be

$$Z = \frac{V}{I} = -j\frac{Z_0}{k\ell} = \frac{1}{j\omega C\ell}$$

which is the impedance of a capacitor with capacitance $C\ell$, with C the characteristic capacitance per unit length for the transmission line and ℓ the length of the transmission line.

Note that the foregoing discussion shows that the quasistatic limit is valid when $kz \ll 1$, so that either the frequency must be very low or the length of the transmission line $z = \ell$ must be much smaller than a wavelength.

Electroquasistatic Fields of a Dipole

Chapter 7 gives the exact solution for the electroquasistatic limit of a dipole field. The electric and magnetic fields due to a current element with dipole moment $I\ell$ are

$$\mathbf{E} = \frac{\eta jkI\ell e^{-jkr}}{4\pi r}\left\{\hat{r}\left[\frac{1}{jkr} + \left(\frac{1}{jkr}\right)^2\right]2\cos\theta + \hat{\theta}\left[1 + \frac{1}{jkr} + \left(\frac{1}{jkr}\right)^2\right]\sin\theta\right\} \tag{15.23a}$$

$$\mathbf{H} = \hat{\phi}\frac{jkI\ell e^{-jkr}}{4\pi r}\left(1 + \frac{1}{jkr}\right)\sin\theta \tag{15.23b}$$

In the electroquasistatic limit, we look for the zeroth-order electric field, which is independent of ω. In (15.23a), we let $\omega = 0$ by noting that $k = \omega\sqrt{\mu\epsilon} = 0$. Also note that the source for the electroquasistatic field is charge and not current. Thus, we write $I\ell = j\omega q\ell$, where $q\ell$ is the electric dipole moment whose time derivative gives rise to the current dipole moment $I\ell$. From (15.23a), we find

$$\mathbf{E} \approx \sqrt{\frac{\mu}{\epsilon}}\frac{j\omega\sqrt{\mu\epsilon}(j\omega q\ell)}{4\pi r}\left[\hat{\mathbf{r}}\left(\frac{1}{j\omega\sqrt{\mu\epsilon}r}\right)^2 2\cos\theta + \hat{\boldsymbol{\theta}}\left(\frac{1}{j\omega\sqrt{\mu\epsilon}r}\right)^2 \sin\theta\right]$$

$$= \frac{q\ell}{4\pi\epsilon r^3}(\hat{\mathbf{r}}2\cos\theta + \hat{\boldsymbol{\theta}}\sin\theta)$$

The first-order magnetic field is derived from the current moment $I\ell$, and as we see from (15.23b), we find

$$\mathbf{H} = \hat{\boldsymbol{\phi}}\frac{I\ell}{4\pi r^2}\sin\theta$$

This result is identical to the magnetic field given by the Biot-Savart law for an infinitesimal current element pointing in the $\hat{z}$ direction.

Notice that in the above approximations we have demanded $kr \ll 1$, so that the frequency must be very low or the observation distance r must be much shorter than the wavelength.

Example 15.1

This example shows how to use the quasistatic theory to obtain an equivalent circuit of a practical capacitor filled with an imperfect insulator.

Find the equivalent circuit for a parallel plate filled with a material characterized by conductivity σ and permittivity ϵ (Figure 15.6a).

Figure 15.6 **(a)** The parallel plate is filled with a material characterized by σ and ϵ. **(b)** The equivalent circuit of the parallel plate shown in (a) is an ideal capacitor C in parallel with an ideal resistor G.

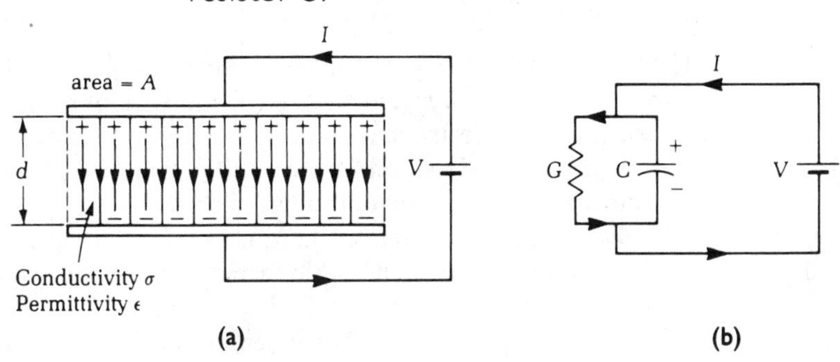

Solution The zeroth-order electric field is

$$E^{(0)} = \frac{V}{d}$$

The zeroth-order current associated with $E^{(0)}$ is

$$I^{(0)} = \iint_A dS\sigma E^{(0)} = \sigma \frac{A}{d} V = GV$$

where A is the area of the plate surface and $G = \sigma A/d$ is the conductance. The zeroth-order total surface charge is

$$Q_s^{(0)} = \iint_A dS\rho_s^{(0)} = \epsilon \frac{A}{d} V = CV$$

where $C = \epsilon A/d$ is the capacitance. The first-order current derived from $Q_s^{(0)}$ is

$$I^{(1)} = \frac{dQ_s^{(0)}}{dt} = C\frac{dV}{dt}$$

The total quasistatic current is

$$I = I^{(0)} + I^{(1)} = GV + C\frac{dV}{dt}$$

Figure 15.6b shows the equivalent circuit, which was also derived in Chapter 12.

Condenser Microphones

In the electroquasistatic field, the first-order current is obtained from the time derivative of the zeroth-order charge distributions. For a capacitor carrying charge $Q = CV$, the first-order current is

$$I = \frac{dQ}{dt} = \frac{d}{dt}(CV) = C\frac{dV}{dt} + V\frac{dC}{dt}$$

The second term is zero when the capacitance is not a function of time. If the capacitor changes its shape with time, the variation of C with time creates an electric current. Electroquasistatic transducers are devices that use the second term by mechanically vibrating a capacitor to generate electric current. Examples of transducers are condenser microphones, vibrating reed electrometers, and piezoelectric driving systems, like quartz and Rochelle salt used in supersonic generation, crystal oscillators, loudspeakers, phonograph pickups, and sonic depth-finders in the ocean.

The operating principle of a condenser microphone is based on the quasistatic fields induced by the movement of one side of a condenser (better known as a capacitor). In a condenser microphone, a metal-plate diaphragm is tightly stretched but still capable of movement in response to sound transients. It is often made of a plastic, such as polyester film, coated with an extremely fine, thin covering of gold to make it conductive. The diaphragm forms one plate of a capacitor whose plastic backing is a dielectric facing the fixed back-plate electrode (Figure 15.7). The capacitor is connected to a dc voltage source of between 50 and 200 V. Under the pressure of sound, the diaphragm vibrates and causes the capacitance to change, thereby generating an electric signal.*

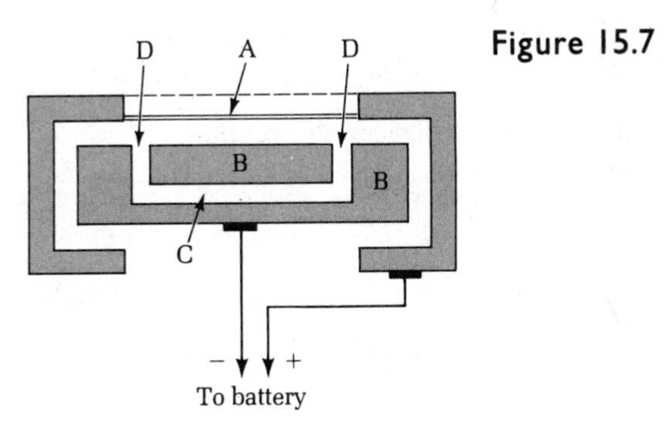

Figure 15.7 A simplified cross-sectional view of a condenser microphone. *A* is the diaphragm, *B* is the hollow base plate with cavity *C*, which has openings *D*. *A* and *B* form a capacitor (condenser), and the cavity regulates the damping of the vibrating diaphragm to produce a uniform frequency response.

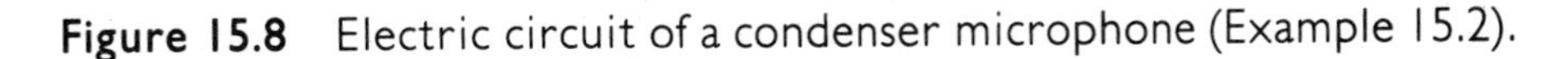

Example 15.2 *This is a quantitative example showing how a condenser microphone works.*

Figure 15.8 shows the circuit of a condenser microphone. For a time-harmonic sound pressure creating a diaphragm movement of $s(t) = S_1 \cos \omega t$, find the output voltage $v(t)$ connected to a transistor amplifier. Let $V_0 = 50$ volts, $R = 10^7$ Ω, and the area of the diaphragm $A = 8\ \text{cm}^2$; in the absence of sound pressure, $S_0 = 25\ \mu\text{m}$, $S_1 = 1\ \mu\text{m}$, and frequency $f = 1$ kHz.

Figure 15.8 Electric circuit of a condenser microphone (Example 15.2).

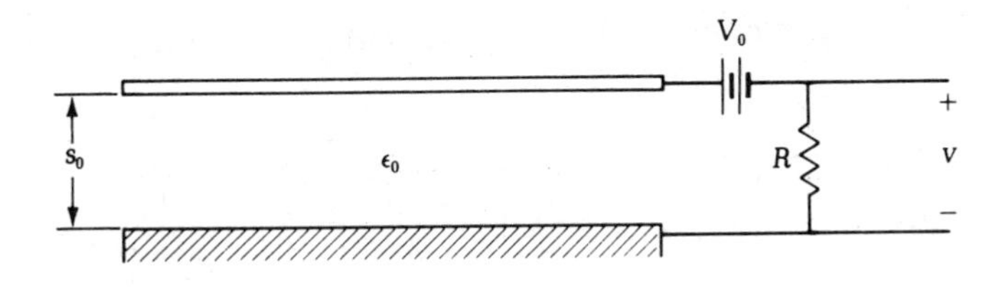

* L. Burroughs, *Microphones: Design and Application*, Plainview, NY, Sagamore, Publishing Co., 1974, p. 42.

Solution The capacitance of the capacitor in the presence of sound pressure is found to be

$$C(t) = \frac{\epsilon_0 A}{S(t)}$$

where $S(t) = S_0 + s(t)$ and $s(t)$ is due to the sound pressure. The zeroth-order total charge on the capacitor is

$$Q(t) = C(t)[V_0 + v(t)]$$

The first-order quasistatic current is then

$$I(t) = -\frac{dQ(t)}{dt}$$

Because the voltage across R is $v = RI$, we find

$$v(t) = -R\frac{dQ(t)}{dt} = -R\frac{dC(t)}{dt}[V_0 + v(t)] - RC(t)\frac{dv(t)}{dt}$$

which can be rearranged to give the differential equation for $v(t)$:

$$RC\frac{dv(t)}{dt} + \left[1 - R\frac{\epsilon_0 A}{S^2(t)}\frac{ds(t)}{dt}\right]v(t) = R\frac{\epsilon_0 A}{S^2(t)}\frac{ds(t)}{dt}V_0$$

Notice that both $v(t)$ and $ds(t)/dt$ are first-order quantities. Neglecting terms of orders higher than the first, we obtain

$$RC_0\frac{dv(t)}{dt} + v(t) = RC_0 V_0 \frac{d}{dt}\left(\frac{s(t)}{S_0}\right)$$

where

$$C_0 = \frac{\epsilon_0 A}{S_0}$$

Using the complex notation with $v(t) = \text{Re}\{Ve^{j\omega t}\}$ and $s(t) = \text{Re}\{S_1 e^{j\omega t}\}$, we find

$$V = \frac{S_1/S_0}{1 + \dfrac{1}{j\omega RC_0}} V_0$$

From the given numerical values, we find the magnitude of the output voltage to be

$$|V| = \frac{2}{\sqrt{1 + \left(\dfrac{25 \times 10^{-6}}{2\pi \times 10^3 \times 10^7 \times 8.85 \times 10^{-12} \times 8 \times 10^{-4}}\right)^2}} = 2.0 \text{ volts}$$

Notice that the induced voltage $|V|$ is directly proportional to the diaphragm vibration amplitude $|S_1|$ when $\omega RC_0 \gg 1$.

Poynting Power in Electroquasistatics

In electroquasistatics, the electric field $\boldsymbol{E}$ is related to the potential Φ by

$$\boldsymbol{E} = -\nabla\Phi$$

Consider a volume V containing interconnected circuit elements enclosed by the surface A (Figure 15.9). The Poynting power enters the surface and is given by

$$\begin{aligned}
P &= -\oint\!\!\oint_A d\boldsymbol{a} \cdot (\boldsymbol{E} \times \boldsymbol{H}) = \oint\!\!\oint_A d\boldsymbol{a} \cdot [(\nabla\Phi) \times \boldsymbol{H}] \\
&= \iiint_V dV \nabla \cdot [(\nabla\Phi) \times \boldsymbol{H}] = \iiint_V dV \nabla \cdot [\nabla \times (\Phi\boldsymbol{H}) - \Phi\nabla \times \boldsymbol{H}] \\
&= -\iiint_V dV \nabla \cdot (\Phi\nabla \times \boldsymbol{H}) = -\oint\!\!\oint_A d\boldsymbol{a} \cdot \Phi\nabla \times \boldsymbol{H} \\
&= -\oint\!\!\oint d\boldsymbol{a} \cdot \Phi\left(\boldsymbol{J} + \frac{\partial \boldsymbol{D}}{\partial t}\right)
\end{aligned}$$

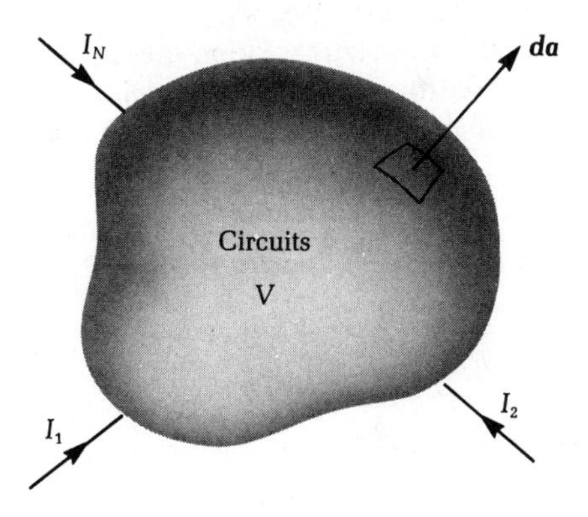

Figure 15.9 Application of Poynting's theorem for quasistatic fields to a circuit in the volume V.

Ignoring the displacement current $\partial\boldsymbol{D}/\partial t$, we find

$$P = -\oint\!\!\oint d\boldsymbol{a} \cdot \Phi\boldsymbol{J} = \sum_{n=1}^{N} V_n\left(-\iint_{A_n} d\boldsymbol{a} \cdot \boldsymbol{J}\right) = \sum_{n=1}^{N} V_n I_n$$

where I_n represents the current flowing into the volume through the surface A_n, whose surface normal is pointing outward. Thus, we have established that the circuit theory's concept of power input is valid only when the displacement current is negligible.

SUMMARY

1. When the frequency is so low that the wavelength is much greater than the physical size of the region of interest, Maxwell's equations can be decomposed into subsets of equations. The first set of equations is obtained by setting the frequency equal to zero. It is called the zeroth-order equations. It is exactly the electrostatic and magnetostatic equations.
2. The first-order equations are obtained by replacing the terms involving time derivatives in Maxwell's equations with the zeroth-order terms, and terms involving spatial derivatives (divergence and curl) are denoted first-order fields.
3. Similarly, the *m*th-order equations are Maxwell's equations formed by using the time-derivatives of $(m-1)$th-order fields and spatial derivatives of *m*th-order fields.
4. The total fields are the sums of the zeroth-order, first-order, etc. fields. Usually, only zeroth-order and maybe the first-order terms are needed to achieve the required accuracy in practical low-frequency situations.

Problems

15.1 Consider the short-circuited parallel plate shown in Figure P15.1, with width w, length ℓ, and separation a. Find the zeroth-, first-, second-, and third-order electric and magnetic fields. Show that the sum of the quasistatic solution is equal to the full-wave solution as presented in Chapter 6 for a short-circuit transmission line. Assume that the current at $z = 0$ is $I(t) = I_0 \cos(\omega t)$ and that all fields are functions of t and z only.

Figure P15.1 A parallel plate with short-circuit current $I = I_0 \cos(\omega t)$ at $z = 0$.

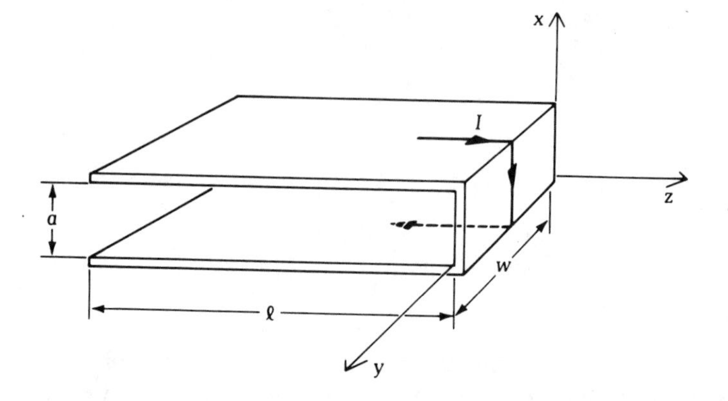

15.2 Calculate the total zeroth-order stored electric energy in the parallel-plate region shown in Figure 15.1. What is the zeroth-order stored magnetic energy in the same region?

15.3 Calculate the total first-order stored magnetic energy in the parallel-plate region shown in Figure 15.1. What is the first-order stored electric energy in the same volume? Denote $U_{Hm}^{(1)}$ the maximum total first-order stored magnetic energy in the region and $U_{Em}^{(0)}$ the maximum total zeroth-order stored electric energy in the same volume. Show that

$$\frac{U_{Hm}^{(1)}}{U_{Em}^{(0)}} = \frac{1}{3}(k\ell)^2$$

15.4 Calculate the total second-order stored electric energy in the parallel-plate region shown in Figure 15.1. Compare it with the zeroth-order stored electric energy. If ℓ is 0.1λ long, what can you say about the relative magnitudes of the zeroth-order stored electric energy, the first-order stored magnetic energy, and the second-order stored electric energy?

15.5 Find the total zeroth-order stored electric and magnetic energies in a parallel plate with a short-circuit current $I = I_0 \cos(\omega t)$ at $z = 0$ (refer to Problem 15.1).

15.6 Find the higher-order stored electric and magnetic energies in the parallel-plate region shown in Figure P15.1 up to the third order. If $\ell = 0.1\lambda$, compare the relative magnitudes of these stored energies. Use the total zeroth-order stored magnetic energy found in Problem 15.5 for comparison.

15.7 A coaxial line ℓ meters long is filled with a material characterized by ϵ and σ. The radii of the inner and the outer conductors are a and b, respectively. The voltage between the coaxial conductors is $V_0 \cos(\omega t)$. Find the zeroth-order electroquasistatic field $E^{(0)}$, the current $I^{(0)}$, the charge $Q^{(0)}$, and the first-order current $I^{(1)}$. Express these in terms of the parameters V_0, a, b, ℓ, ϵ, σ, and ωt.

15.8 Two concentric spherical electrodes of radii a and b, respectively, are filled with a material characterized by ϵ and σ. The voltage between the electrodes is $V_0 \cos(\omega t)$. Find the zeroth-order electroquasistatic field $E^{(0)}$, the current $I^{(0)}$, the charge $Q^{(0)}$, and the first-order current $I^{(1)}$. Express these in terms of V_0, a, b, ϵ, σ, and ωt.

15.9 Show that the time needed to charge a Van de Graaff generator (shown in Figure 9.24a) with radius R to a maximum voltage of V_{max} by applying a charging current I is equal to $4\pi\epsilon_0 R V_{max}/I$. Calculate the charging time t if $R = 1$ m, $V_{max} = 10^6$ V and $I = 10^{-5}$ A.

15.10 A parallel plate is filled with two materials in a configuration shown in Figure P15.10. The total area of the plate is A. The dielectric constant and the conductivity of one material are ϵ_1 and σ_1, respectively. Those of the other material are ϵ_2 and σ_2. The upper and the lower plates are connected to the two terminals of a low-frequency source. Find the equivalent circuit for this parallel plate, and express the circuit parameters in terms of A, d, ϵ_1, σ_1, ϵ_2, and σ_2.

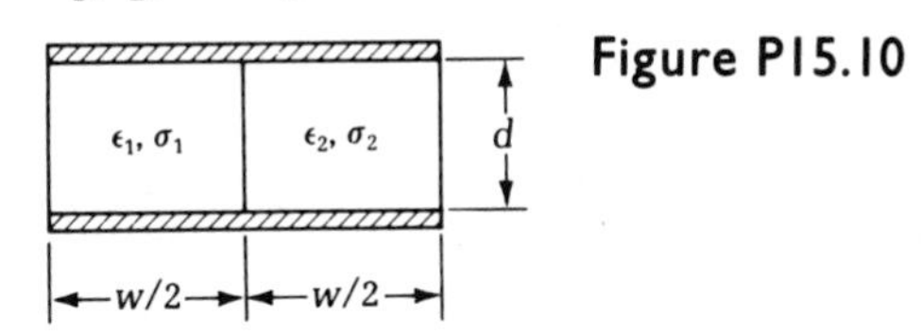

Figure P15.10

15.11 A parallel plate is filled with two materials in a configuration shown in Figure P15.11. The upper and the lower plates are connected to the two terminals of a low-frequency source. Find its equivalent circuit, and express the circuit parameters in terms of A, the area of the plate, and d_1, d_2, ϵ_1, ϵ_2, σ_1, and σ_2, which are defined in the figure.

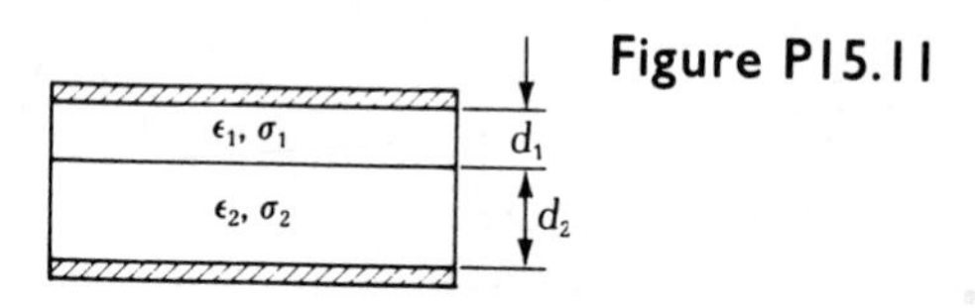

Figure P15.11

15.12 A spherical conductor of radius a is inside a spherical conducting shell of radius c. Two materials are used to fill the space between these conductors. The dielectric constants and the conductivities of these materials are ϵ_1, σ_1, ϵ_2, σ_2, respectively. Figure P15.12 shows the configuration. The inner conductor and the outer conductor are connected to the two terminals of a low-frequency source. Find the equivalent circuit of this system, and express the circuit parameters in terms of a, b, c, ϵ_1, ϵ_2, σ_1, and σ_2.

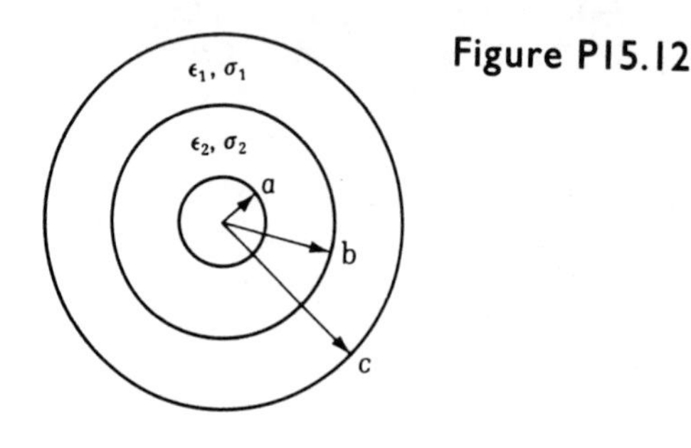

Figure P15.12

16 MAGNETOQUASISTATIC FIELDS

In the preceding chapter we introduced the concept of quasistatic fields and studied the electroquasistatic fields. Bear in mind that quasistatic approximations are valid when the physical dimensions of the system are much smaller than a wavelength. At 60 Hz, for example, the wavelength in air is 5000 km. Therefore, almost all physical 60-Hz systems in air are quasistatic. Under the quasistatic limit, we derived the equations governing the magnetoquasistatic field in Chapter 15. These equations are the subject of this chapter. They are listed below:

$$\nabla \times \boldsymbol{H} = \boldsymbol{J} \qquad \text{(zeroth-order magnetic fields)} \tag{16.1a}$$

$$\nabla \cdot \boldsymbol{B} = 0 \tag{16.1b}$$

$$\nabla \times \boldsymbol{E} = -\frac{\partial \boldsymbol{B}}{\partial t} \qquad \text{(first-order electric field)} \tag{16.1c}$$

where $\boldsymbol{H}$, $\boldsymbol{B}$, and $\boldsymbol{J}$ are of the zeroth order and $\boldsymbol{E}$ is of the first order. If necessary, we can readily calculate higher-order fields by following the procedures described in the preceding chapter.

In this chapter we will first study Faraday's law given by (16.1c). We will use higher-order corrections if the situation warrants. We will study traditional devices, such as transformers, generators, and motors from the magnetoquasistatic viewpoint, and we will also discuss rail guns, induction heating, nondestructive evaluation, and the magneplane.

16.1 FARADAY'S LAW

Consider a rectangular loop in the vicinity of an infinitely long wire carrying a current $I(t)$, as shown in Figure 16.1. To calculate the magnetic field, we can use the quasistatic approach, which gives

$$\boldsymbol{H} = \hat{y}\frac{I(t)}{2\pi x} \tag{16.2}$$

Figure 16.1 Voltage is induced on the rectangular loop by the time-varying magnetic field produced by the time-varying current in the infinitely long conductor.

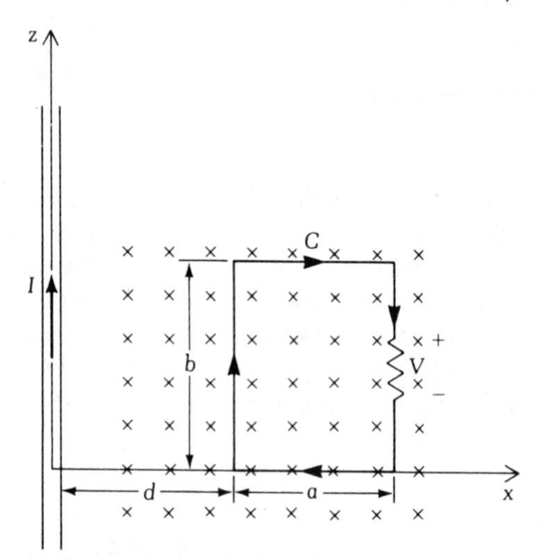

We then obtain the electric field $\boldsymbol{E}$ from Faraday's law

$$\nabla \times \boldsymbol{E} = -\frac{\partial}{\partial t}\boldsymbol{B} \tag{16.3}$$

Integrating both sides of (16.3) over the area of the square loop shown in Figure 16.1, we obtain the following integral form of (16.3):

$$\iint_A (\nabla \times \boldsymbol{E}) \cdot \boldsymbol{da} = -\iint_A \boldsymbol{da} \cdot \frac{\partial}{\partial t}\boldsymbol{B} \tag{16.4}$$

where, as shown in Figure 16.1, the surface normal is pointing into the paper in the same direction as the magnetic field, and the contour C bounding the area of the loop is in the clockwise direction. By Stokes' theorem, the left-hand side of (16.4) is equated to a contour integral over the contour C. Because the loop is stationary, we may take the time derivative outside the integral. The closed-contour integral is defined as the **electromotive force (EMF)**. Consequently, we have

$$\text{EMF} = \oint_C \mathbf{E} \cdot \boldsymbol{d\ell} = -\frac{\partial}{\partial t}\iint_A \boldsymbol{B} \cdot \boldsymbol{da} = -\frac{\partial \Psi}{\partial t} \quad \textbf{(induced EMF in a closed loop)} \tag{16.5}$$

where

$$\Psi = \iint_A \boldsymbol{da} \cdot \boldsymbol{B} \quad \textbf{(magnetic-flux linkage)} \tag{16.6}$$

is the magnetic flux linking the loop. The integral of $\boldsymbol{E} \cdot d\boldsymbol{\ell}$ over the closed loop, that is, the EMF, equals the total voltage drops in the loop, as indicated by V across the resistor in Figure 16.1. We see from (16.5) that Faraday's law states that the EMF is equal to the rate of decrease of the magnetic-flux linkage through the loop. When the magnetic field produced by the infinitely long line is increasing with time, the changing magnetic field induces an EMF in the nearby loop, and this EMF tends to produce a counterclockwise current so as to create a magnetic field in the direction opposite to that due to the infinitely long line. On the other hand, when the magnetic field is decreasing with time, the induced EMF in the loop creates a clockwise current so as to generate a magnetic field to compensate for the decreasing magnetic field. In other words, the induced EMF always tends to resist change in the status quo of the circuit. This assertion is known as the **Lenz law**.

Example 16.1 *This example demonstrates how to compute the induced voltage in a loop due to the magnetic field from another source.*

Calculate the voltage across the resistor shown in Figure 16.1. Assume $I = I_0 t/T$.

Solution Using (16.5) and (16.6), we first calculate the total flux linkage through the loop.

$$\Psi = \int_0^b dz \int_d^{d+a} dx \frac{\mu I}{2\pi x}$$

$$\Psi = \frac{\mu I_0 t}{2\pi T} b \ln\left(\frac{d+a}{d}\right) \tag{16.7}$$

According to (16.5), we have

$$V = -\frac{\mu I_0}{2\pi T} b \ln\left(\frac{d+a}{d}\right)$$

The polarity of this voltage is opposite to that shown in Figure 16.1.

Example 16.2 *This example shows that Faraday's induction works whether or not the field varies like a sinusoidal function of time or any other function of time.*

If the current shown in Figure 16.1 is time-harmonic with amplitude I and frequency ω, what is V?

Solution By virtue of (16.5) and (16.7), we find

$$V = -j\omega \frac{\mu I}{2\pi} b \ln\left(\frac{d+a}{d}\right)$$

Example 16.3 *An important consequence of Faraday's induction is demonstrated in this example. It shows that placement of the lead wires of a voltmeter may affect the reading of the voltmeter even when the tips of the leads are connected to the same points in a circuit.*

As shown in Figure 16.2, a loop consisting of two series resistors is linked by a magnetic flux that is increasing at a rate of one weber per second. The flux is pointing into the paper and is limited to the region $A_1A_2A_3A_4$. Find the readings of the voltmeters V_1, V_2, and V_3, which are deployed as shown. The input impedances of these voltmeters are assumed to be infinite.

Figure 16.2 A circuit in a time-varying magnetic field V_1, V_2, and V_3 are readings of the voltmeters.

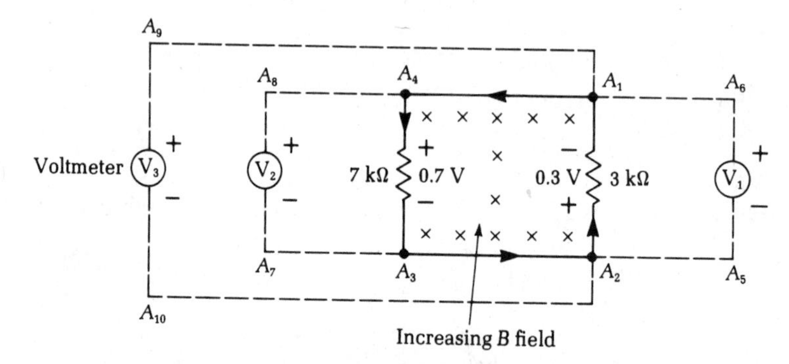

Solution The induced EMF in the loop is equal to 1 V, and the induced current flowing in the loop $A_1A_2A_3A_4$ is 0.1 mA in the counterclockwise direction. Assume that the magnetic flux generated by the induced current is negligible compared with the external flux. For the loop $A_1A_6A_5A_2$, we have $\Psi = 0$ and $d\Psi/dt = 0$. Therefore, EMF = 0, and the sum of all voltage drops is equal to zero—that is,

$$+0.3 + V_1 = 0$$

which gives

$$V_1 = -0.3 \text{ volt}$$

To find V_1, we may also use the loop $A_6A_5A_2A_3A_4A_1$; the loop has an EMF $= -d\Psi/dt = -1$ volt. The sum of the voltage drops is in the clockwise direction and is equal to EMF. We have

$$-1 = V_1 - 0.7$$

which again gives $V_1 = -0.3$ volt. For the loop $A_4A_3A_7A_8$, EMF = 0, and we have

$$-V_2 + 0.7 = 0$$

which yields

$$V_2 = 0.7 \text{ volt}$$

Again, we may use another loop such as $A_7A_8A_6A_5$ to find V_2. Because it is linked by EMF $= -1$ volt, we have

$$-1 = V_1 - V_2$$

which gives $V_2 = 0.7$ volt.

We can easily obtain the same result by considering the loop $A_7A_8A_1A_2$. The magnetic flux links through the loop $A_1A_2A_{10}A_9$, and EMF $= -d\Psi/dt = -1$ volt. We have

$$-1 = -0.3 - V_3$$

and it follows that

$$V_3 = 0.7 \text{ volt}$$

Notice that, although V_1 and V_3 are connected to the same nodes A_1 and A_2, one reads -0.3 volt, and the other reads $+0.7$ volt. Hopefully, with this example, we will appreciate that, in the presence of a time-varying magnetic field, a voltage measurement across the same nodes in a circuit depends on how the wires are laid out.

Example 16.4

This example presents a more complicated situation than the one given in Example 16.3. One of the lead wires has an extra turn and the voltmeter reading is again affected as a result.

Consider the circuit shown in Figure 16.3. It is different from the one shown in Figure 16.2 in that an extra turn is present in the loop $A_1A_2A_3A_4$. Assume that

Figure 16.3 A circuit with two turns in a time-varying magnetic field.

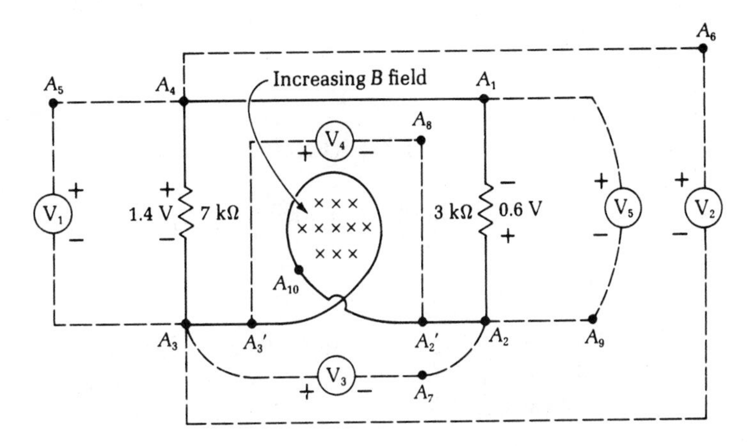

the flux is concentrated in the region of the inner loop and that all other conditions remain the same as in Figure 16.2. Find the readings of voltmeters V_1, V_2, V_3, V_4, and V_5.

Solution Because the loop $A_1A_2A_3A_4$ has two turns, the EMF induced along the loop is equal to -2 volts. The current in the circuit is 0.2 mA in the reverse direction—that is, along $A_1A_4A_3A_{10}A_2$. To find the reading of V_1, we first notice that no flux links the loop $A_4A_3A_5$. Therefore, EMF = 0, and

$$1.4 - V_1 = 0$$

which gives

$$V_1 = 1.4 \text{ volts}$$

Now let us consider the loop $A_4A_6A_3A_4$. The changing flux in this single loop produces an EMF equal to -1 volt, which gives

$$-1 = V_2 - 1.4$$

$$V_2 = 0.4 \text{ volt}$$

To find V_3, we consider the loop $A_7A_3A_4A_1A_2$ and obtain

$$-1 = -V_3 - 1.4 - 0.6$$

$$V_3 = -1 \text{ volt}$$

For V_4, we examine the loop $A_8A_2'A_{10}A_3'$. The EMF is -2 volts. Therefore, we have

$$-2 = V_4$$

Finally, for V_5, we consider the loop $A_9A_2A_1$, which has no flux linkage, and it follows that

$$0 = V_5 + 0.6$$

$$V_5 = -0.6 \text{ volt}$$

Voltage Across an Inductor

In Section 13.3, we defined self-inductance L as

$$L = \frac{N\Psi}{I}$$

where Ψ is the flux linkage through the cross-sectional area of the inductor, N is the number of twins in the inductor, and I is the current flowing in the inductor. Figure 16.4 shows an inductor with N turns of coils. Now, if the current is time-

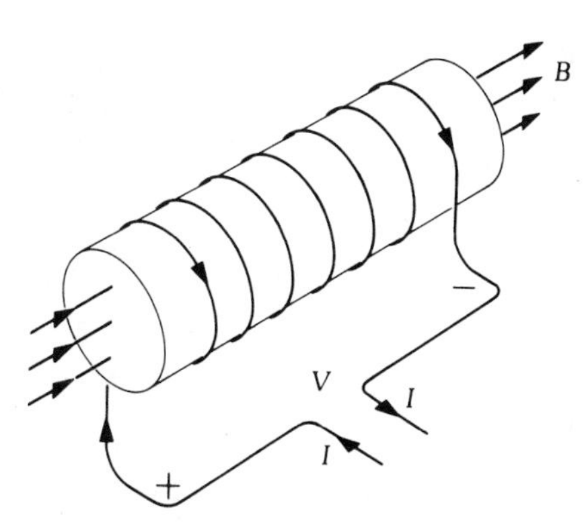

Figure 16.4 Voltage across an inductor.

varying, then the EMF induced on each turn of the inductor is

$$\text{EMF} = -\frac{d\Psi}{dt}$$

The total EMF induced in the N turns is then

$$\text{EMF} = -N\frac{d\Psi}{dt} = -\frac{d}{dt}(LI)$$

The voltage drop across an inductor should be the negative of the induced EMF, which gives

$$V = \frac{d}{dt}(LI) = L\frac{dI}{dt} \tag{16.8}$$

Figure 16.4 shows the polarity of the voltage V. Notice that V is positive, so that it tends to resist the increase of the current. This fact is again a consequence of the Lenz law.

Induced EMF on a Moving Wire

Whereas (16.5) and (16.6) give the EMF induced in a stationary circuit, EMF is also induced on a moving wire that is cutting across a magnetic field.

Consider a conducting bar moving with a velocity $\boldsymbol{v}$ along a pair of stationary tracks made of perfect conductors, as shown in Figure 16.5. A uniform and steady magnetic field $\boldsymbol{B} = \hat{y}B$ is applied. A positive charge q on the moving bar is exerted by the Lorentz force

$$\boldsymbol{F} = q\boldsymbol{v} \times \boldsymbol{B}$$

The charge experiences an equivalent electric field $\boldsymbol{E}$ with

$$\boldsymbol{E} = \boldsymbol{v} \times \boldsymbol{B} \tag{16.9}$$

which is in the $\hat{z}$ direction. This electric field moves positive charges to the upper

Figure 16.5 An electromotive force is induced on the moving bar A_2A_3 as it cuts through the magnetic field.

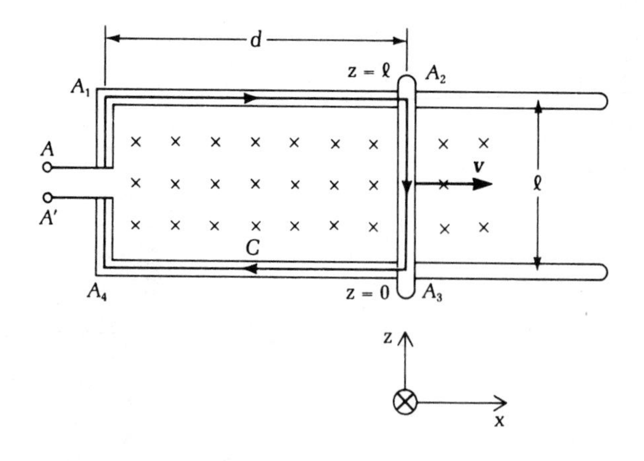

track and negative charges to the lower one. The net effect is a voltage across the two tracks.

Let the separation of the two tracks be ℓ; then

$$\boxed{\text{EMF} = \int_0^{\ell} d\boldsymbol{\ell} \cdot (\boldsymbol{v} \times \boldsymbol{B})} \quad \textbf{(induced voltage on a moving wire)} \tag{16.10}$$

where $d\boldsymbol{\ell} = -\hat{z}\,dz$. This motion-induced voltage is another form of the electromotive force (EMF). Note that no electric field is generated if $\boldsymbol{v}$ is parallel to $\boldsymbol{B}$. The electric field is maximum when $\boldsymbol{v}$ is perpendicular to $\boldsymbol{B}$. The EMF is a result of the moving conductor "cutting through" the magnetic field lines.

Now let us consider the combined effect of a time-varying magnetic field and a moving conductor in a circuit. For the same configuration as that shown in Figure 16.5, assume that the applied magnetic field $\boldsymbol{B}(t)$ is time-varying. Using (16.6), we integrate $\boldsymbol{B}$ over the area bounded by the contour C. The area includes the moving conducting bar. The result is

$$\begin{aligned} \text{EMF} &= -\iint \frac{\partial \boldsymbol{B}}{\partial t} \cdot d\boldsymbol{a} + \int_0^{\ell} d\boldsymbol{\ell} \cdot (\boldsymbol{v} \times \boldsymbol{B}) \\ &= -\iint \frac{\partial \boldsymbol{B}}{\partial t} \cdot d\boldsymbol{a} - \int_0^{\ell} \boldsymbol{B} \cdot (\boldsymbol{v} \times d\boldsymbol{\ell}) \end{aligned} \tag{16.11}$$

where the EMF is in the direction C, as shown in Figure 16.5. Notice that both $d\boldsymbol{a}$ and $\boldsymbol{v} \times d\boldsymbol{\ell}$ are in the $\hat{y}$ direction. The EMF on the closed contour C yields two terms on the right-hand side of (16.11). The second term represents the increase of magnetic flux due to the increase in the area generated by the motion of the conducting bar. Thus, the sum of the first and the second integrals at the right side of (16.11) is equal to the total time derivative of the magnetic flux linking the area A. We find

$$\text{EMF} = -\frac{d}{dt}\int_A \boldsymbol{B}\cdot d\boldsymbol{a} = -\frac{d\Psi}{dt} \qquad \textbf{(induced voltage in a loop with time-varying boundaries in a time-varying } \boldsymbol{B} \textbf{ field)} \qquad \textbf{(16.12)}$$

Compare with (16.5), and note that (16.12) uses the total derivative d/dt instead of the partial derivative $\partial/\partial t$. The above result is generally true when the whole loop is moving and deforming in arbitrary directions.

In summary, the induced EMF is negative when Ψ is increasing and positive when it is decreasing. When Ψ is increasing and the EMF is negative, closing the terminal at AA' generates a current flowing in a direction opposite to the direction of path C. This current, according to Ampère's law, produces a magnetic field opposing that of the original increasing magnetic field. Conversely, when Ψ is decreasing and EMF is positive, the current produces a magnetic field in the same direction as the original magnetic field. This fact confirms once again the Lenz law: the circuit tends to resist any change in the status quo.

Example 16.5 *This example shows that, if one branch of a circuit is moving in a magnetic field that is time-varying, then the voltage induced in the circuit has two terms, one due to the change in the magnetic flux and the other due to the moving branch.*

Consider the sliding conducting bar in Figure 16.5. Assume a uniform time-varying magnetic field $\boldsymbol{B}(t) = \hat{y}B_0 \cos \omega t$. The bar is moving at a velocity $\boldsymbol{v} = \hat{x}v$. Determine the induced EMF.

Solution Using (16.11), we obtain

$$\text{EMF} = \omega \sin \omega t \iint B_0\, da - \int_0^{\ell} dz\, vB_0 \cos \omega t = \ell B_0(\omega d \sin \omega t - v \cos \omega t)$$

Example 16.6 *This example shows an application of Faraday's induction in a magnetic memory device.*

Assume that a core in a core memory is switched from state A to state C in a microsecond (see Figure 14.9). Estimate the voltage induced on the sensing wire.

Assume that the switching is done linearly with time and that the cross-sectional area of the core is 3×10^{-7} m^2.

Solution The total flux of a single core linking the sensing wire shown in Figure 14.11 is the remanent flux density times the cross section of the core. According to Figure 14.9, the remanent flux is 0.21 Wb/m^2. Therefore,

$$\Psi = 0.21 \times 3 \times 10^{-7} \text{ Wb} = 0.63 \times 10^{-7} \text{ Wb}$$

$$\frac{\Delta\Psi}{\Delta t} = \frac{2 \times 0.63 \times 10^{-7}}{10^{-6}} = 0.13 \text{ V}$$

The factor 2 in the above equation is due to the fact that the flux reverses its direction during the switching.

Example 16.7

A conducting disk rotating in a magnetic field will have an EMF induced on the disk, as this example shows. Note that the total induced voltage along the path from the center to the edge must be obtained by integration.

A metal disk is rotating in a uniform magnetic field, as shown in Figure 16.6. What is the voltage between the two contact brushes A and A'? Assume that the disk is rotating at a speed of 6,000 r/min, $B = 1.1$ Wb/m^2, and $a = 10$ cm.

Figure 16.6 **(a)** An electromotive force induced between the edge and the center of a metal disk rotating in a magnetic field. A and A' are connected to two contact brushes. **(b)** End view of the disk looking in the $\hat{\mathbf{z}}$ direction.

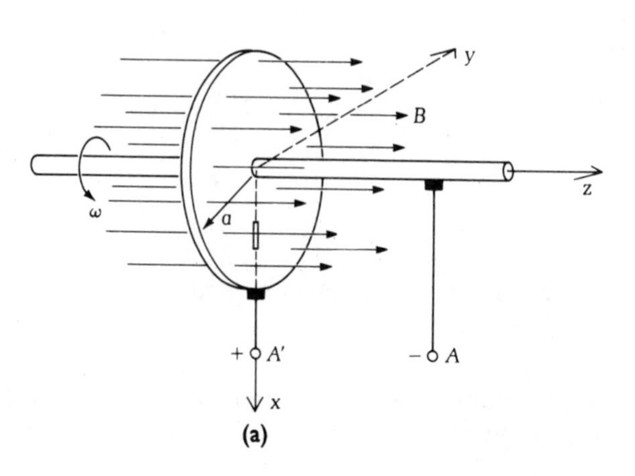

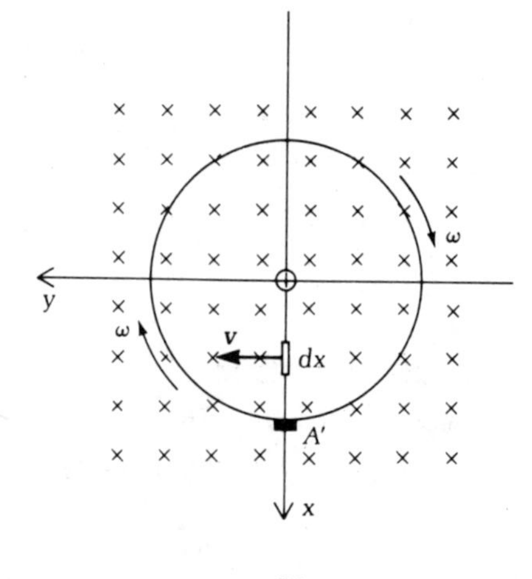

Solution Consider a short piece of conducting strip at x between the center of the disk and the contact brush A' shown in Figure 16.6b. The velocity of this strip is

$$\boldsymbol{v} = \hat{y}x\omega$$

The electromotive force along this strip, according to (16.9), is

$$d\boldsymbol{E} = \boldsymbol{v} \times \hat{z}B = \hat{x}x\omega B \tag{16.13}$$

Therefore,

$$\text{EMF} = \int_0^a x\omega B\hat{x} \cdot \hat{x}\, dx = \frac{1}{2}\omega B a^2 \tag{16.14}$$

Substituting $\omega = 2\pi \times 6000/60$ rad/s, $B = 1.1$ Wb/m^2, and $a = 0.1$ m into (16.14), we have

$$\text{EMF} = 3.46 \text{ volts} = V_{A'A}$$

The polarity of the EMF is indicated in Figure 16.6a. Such a metal disk rotating in a magnetic field perpendicular to its axis is known as the **Faraday disk**, which we discuss further in the next section, along with the **homopolar generator**.

Force on a Coil

Consider the solenoid shown in Figure 16.7. A steady current of I amperes supplied by a constant-current generator flows through its coils. The magnetic field in the solenoid is given by (13.10), which is repeated below:

$$H_z = \begin{cases} nI & \text{for } \rho < a \\ 0 & \text{for } \rho > a \end{cases} \tag{16.15}$$

where n is the number of turns per meter of the coil. As we see from the Lorentz law $\boldsymbol{F} = \boldsymbol{J} \times \boldsymbol{B}$, the $\hat{z}$-directed magnetic field will exert a $\hat{\rho}$-directed force on the

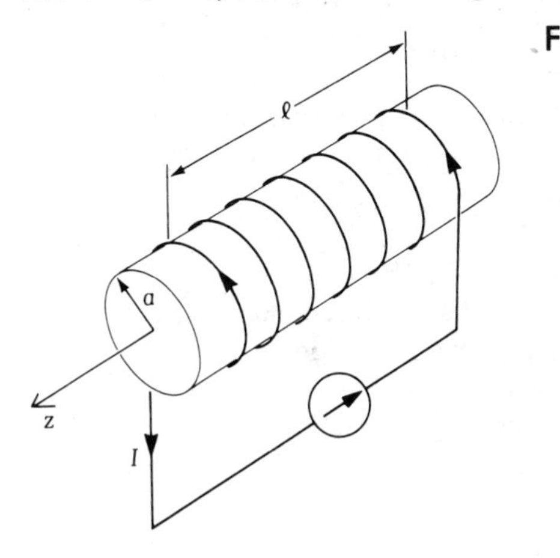

Figure 16.7 A system consisting of a solenoid and a constant-current generator.

wire carrying a $\hat{\phi}$-directed current. We can also obtain this radially expanding force on the coil from energy considerations as follows:

Example 13.7 found the stored magnetic energy in the solenoid. Assume that the solenoid is ℓ meters long; then

$$U_H = \frac{1}{2}\mu n^2 I^2(\pi a^2)\ell \text{ joules}$$

or

$$U_H = \frac{1}{2}LI^2 \tag{16.16}$$

The last equation is obtained from (13.46). Let us examine what happens when the radius of the solenoid is increased from a to $a + \Delta a$. The formula (16.15) indicates that there is no change in the magnetic-field strength because the current is supplied by a constant-current generator. However, the cross-sectional area is increased; consequently, the inductance and the stored magnetic energy are also increased:

$$L' = \frac{L(a + \Delta a)^2}{a^2} \tag{16.17a}$$

$$U'_H = \frac{1}{2}(L')I^2 = \frac{U_H(a + \Delta a)^2}{a^2} \tag{16.17b}$$

The amount of the increase in the stored magnetic energy is

$$\Delta U_H = \frac{1}{2}(L' - L)I^2 \tag{16.18}$$

The change in the inductance induces a voltage V_L across the solenoid

$$V_L = \frac{d}{dt}(LI) = I\frac{dL}{dt} \tag{16.19}$$

because the current I is a constant in time. The current generator has to overcome this counter-voltage, and, considering (16.19), the total work that the generator has to supply is

$$\Delta W = V_L I \,\Delta t = (L' - L)I^2 \tag{16.20}$$

This amount represents a decrease in the energy stored in the generator. The total energy change in the generator and in the inductor is thus

$$\Delta U_S = -\Delta W + \Delta U_H = -\frac{1}{2}(L' - L)I^2 \tag{16.21}$$

which decreases when the cross-sectional area of the solenoid is increased. Because all physical systems tend to go to a lower energy state, we expect the coil that confines the magnetic flux to be subject to an expansion force. The following equation determines the magnitude of this force:

$$F = -\frac{\Delta U_S}{\Delta a} \tag{16.22}$$

From (16.21) and (16.17), we find

$$-\Delta U_S = \frac{1}{2}(L' - L)I^2 = \frac{1}{2}LI^2\frac{2a\,\Delta a + (\Delta a)^2}{a^2}$$

Therefore, the force acting on the coil is, in the limit $\Delta a \to 0$,

$$F = \frac{1}{2}LI^2\frac{2a + \Delta a}{a^2} = \frac{1}{a}LI^2 \tag{16.23}$$

This result is the expansion force acting on the coils of the solenoid. The inductance of the solenoid shown in Figure 16.7 is given in Example 13.7:

$$L = \mu n^2 \pi a^2 \ell$$

Therefore,

$$\frac{dL}{da} = \mu n^2 \pi 2a\ell = \frac{2L}{a}$$

Substituting the above equation in (16.23) yields

$$F = \frac{1}{2}\frac{d}{da}(LI^2) = \frac{I^2}{2}\frac{dL}{da} \tag{16.24}$$

We see that the expansion force is equal to the derivative of the stored magnetic energy with respect to the physical dimension of the system holding the stored energy.

Rail Guns

From the preceding discussion, we learned that the coil of a solenoid is subject to an expansion force. Generally, conductors of any circuit confining a magnetic field are subject to forces that tend to expand the circuit. This phenomenon is analogous to the pressure force on a pipe or to a container that holds water.

Consider a pair of rails consisting of two parallel conducting plates, as shown in Figure 16.8. A current I is flowing through the rails across an end plate which short-circuits the rails. This circuit is similar to a loop in a solenoid, and we expect the rails as well as the end plate to be subject to magnetic forces that tend to expand the circuit. In the following we find the force acting on the end plate.

The first step is to find the inductance of the rail with the short-circuiting end. Let us assume that the rails are closely spaced so that the magnetic field between the plates may be approximated by (13.9). Notice that the configuration shown in Figure 16.8 is similar to that shown in Figure 13.5. Therefore, we have

$$H_x = \frac{I}{w} \tag{16.25}$$

Figure 16.8 A rail gun. Current flows in the rails across the end plate, which is also the projectile. The latter is pushed in the $\hat{\mathbf{z}}$ direction by the magnetic force.

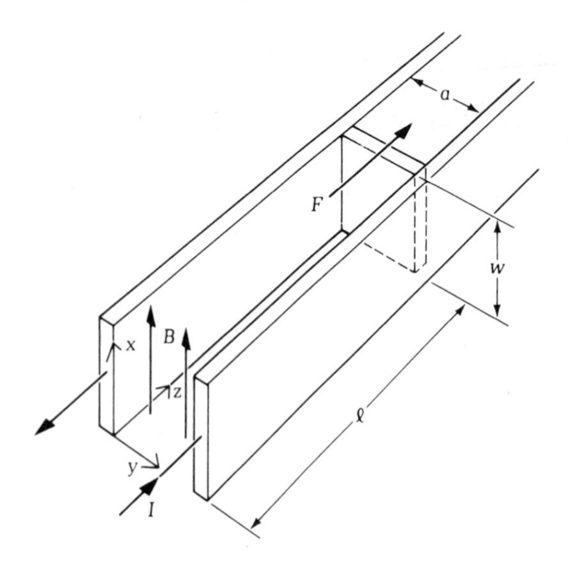

The total flux through the area bounded by the loop is given as follows:

$$\Psi = \frac{\mu \ell a I}{w} \tag{16.26}$$

We obtain the inductance L from the definition given in Section 13.3:

$$L = \frac{\Psi}{I} = \mu \frac{\ell a}{w} \tag{16.27}$$

Considering (16.24) and recognizing that ℓ is now the changing physical dimension, we find

$$F = \frac{I^2}{2}\frac{dL}{d\ell} = \frac{\mu a I^2}{2w} \tag{16.28}$$

Equation (16.28) gives the magnitude of the force acting on the short-circuited end plate shown in Figure 16.8. If a very strong current is passed through these rails, the end piece may be accelerated to a very high vlelocity. This is the operating principle of a **rail gun**.

Conventional guns make use of chemical energy stored in explosive materials to accelerate the projectiles. These guns are noisy and generate smoke. In practice, the speed of sound in the exploding gases limits the projectiles' speed to 1 to 2 km/s. On the other hand, rail guns could accelerate projectiles at least twice as fast as conventional guns, and the would be silent and more accurate. Theoretically, the speed of the projectiles is limited only by the velocity of light. A rail gun 5 m long has been built to accelerate 2-g projectiles to a speed of 11 km/s.* Rail guns are only one of many possible applications of electromagnetic launchers. Other examples are (1) fuel pellets accelerated to 150 km/s so as to trigger thermonuclear fusion, (2) electromagnetic artillery, and (3) space vehicles or aircraft. An electromagnetic launcher was proposed** to shoot satellites directly into earth orbit. The barrel would be 500 m long and the exit speed would be 5.5 km/s. It could put a 1264-kg satellite into a 500-km orbit.

Example 16.8 *This example gives the order of magnitude of various parameters in a rail gun. Note that a very large current is required in the rail gun.*

Assume that the rail gun shown in Figure 16.8 has the following data: $a = 0.1$ m, $w = 0.3$ m, $I = 300$ kA. What is the force acting on the projectile (the end plate)? The projectile weighs 3 g. Express the force in terms of gee (earth's gravitational force on the projectile). If the projectile is subject to the force along a 3-m long barrel, what is its exit speed?

Solution We use (16.28) and obtain

$$F = 1.9 \times 10^4 \text{ N}$$

$$mg = 3 \times 10^{-3} \times 9.8 \text{ N}$$

$$\frac{F}{mg} = 0.64 \times 10^6 \text{ gee}$$

To find the exit speed, we need to solve

$$F = \frac{d(mv)}{dt} = m\frac{d^2z}{dt^2}$$

which yields

$$t = \sqrt{\frac{2mz}{F}} = \sqrt{\frac{2 \times 3 \times 3 \times 10^{-3}}{1.9 \times 10^4}} = 0.97 \text{ ms}$$

and

$$v = \frac{F}{m}t = 6.2 \text{ km/s}$$

* H. Kolm and P. Mongeau, "An alternative launching medium," *IEEE Spectrum*, Vol. 19, No. 4, April 1982, pp. 30–36.

** J. L. Brown et al., "Earth-to-orbit railgun launcher," *IEEE Trans. on Magnetics*, Vol. 29, January 1993, p. 373.

Note that the rails are also subject to the magnetic force, which tends to push them apart. Thus, they must be fastened securely so that the only moving part is the projectile serving as the end plate. Furthermore, because of the large amount of current needed for electromagnetic launchers, the "end plate" is actually a volume of plasma because ordinary conductors would melt. The real projectile is placed ahead of the plasma column and pushed by the latter during acceleration. The high-current source for the rail gun may be a homopolar generator, which we will discuss in the next section.

Induction Heating

Consider a long circular shell of length ℓ, radius a, and conductivity σ placed in a uniform time-varying magnetic field along the $\hat{z}$ axis, as shown in Figure 16.9. The magnetic field due to the external source is

$$\mathbf{H}^e = \hat{z} H_0 e^{j\omega t} \tag{16.29}$$

Applying Faraday's law, we obtain

$$\mathrm{E}_\phi 2\pi a = -j\omega\mu_0 H_0 \pi a^2 \tag{16.30}$$

Figure 16.9 A thin, cylindrical shell is placed in an alternating magnetic field. Eddy currents are induced.

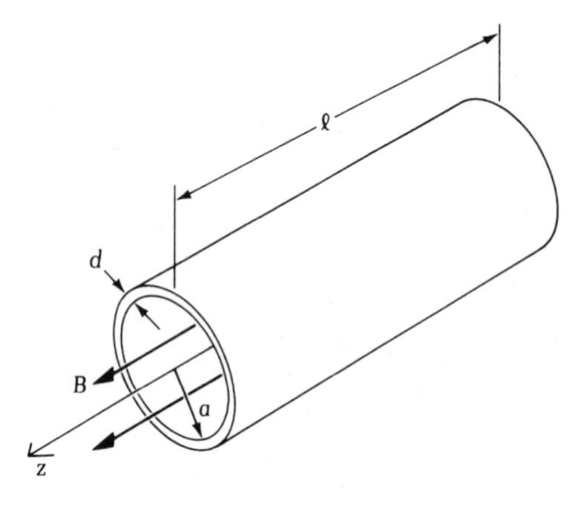

The current density following in the shell is

$$\mathrm{J}_\phi = \sigma \mathrm{E}_\phi = -j\frac{\omega\mu_0 \sigma a}{2} H_0 \tag{16.31}$$

Note that this current goes to infinity if the shell is made of a perfect conductor. As a result, if it were the correct answer to the problem at hand, the power dissipation in the shell would become infinite because the ohmic loss is equal to

$\boldsymbol{J} \cdot \boldsymbol{E}$. But we know that correct power dissipation in a perfect conductor should be zero. Therefore, we conclude that (16.31) does not give the total current in a conducting shell placed in a time-varying magnetic field.

To find the correct answer, we note that J_ϕ also produces a magnetic field in the direction opposite to that of the applied external field and that we must therefore take it into account.* Thus, to correct the above results, we need to consider the first-order magnetic field, which is generated by the first-order J_ϕ, as given by (16.31).

The $\hat{\phi}$-directed current in the shell resembles the current in the solenoid shown in Figure 13.6. In fact, the current in the solenoid expressed as nI amperes per meter is equivalent to the current in the cylindrical shell expressed as $J_\phi d$ amperes per meter. The magnetic field in a solenoid is given by (13.10). For the present case, we simply substitute $J_\phi d$ for nI and obtain the induced magnetic field as follows:

$$H_z^{(i)} = J_\phi d = \sigma E_\phi d$$

where the superscript (i) denotes the induced H field. The total magnetic field is the sum of the externally applied field and $H_z^{(i)}$:

$$\mathbf{H} = \hat{z}(H_0 + \sigma E_\phi d) \tag{16.32}$$

Using (16.32), we obtain from Faraday's law:

$$E_\phi 2\pi a = -j\omega\mu_0(H_0 + \sigma E_\phi d)\pi a^2$$

Solving for E_ϕ gives

$$E_\phi = \frac{-j\dfrac{\omega a}{2}\mu_0 H_0}{1 + j\dfrac{\omega\mu_0 \sigma a d}{2}}$$

The conduction current in the shell is

$$J_\phi = \frac{-j\sigma\omega a\mu_0 H_0/2}{1 + j\omega\mu_0 \sigma a d/2} = \frac{-jH_0 a/\delta^2}{1 + jad/\delta^2} \tag{16.33}$$

where

$$\delta = \sqrt{\frac{2}{\omega\mu_0 \sigma}}$$

is the skin depth. The total dissipated power in the shell is found to be

$$\langle P_d \rangle = \frac{1}{2}\text{Re}\int dV J_\phi^* E_\phi = \frac{1}{2}(2\pi a \ell d)\sigma |E_\phi|^2$$

$$= (\pi a \ell d\sigma)\frac{(\omega a\mu_0 H_0/2)^2}{1 + (\omega\mu_0 \sigma a d/2)^2} \tag{16.34}$$

* H. A. Haus and J. R. Melcher, *Electromagnetic Fields and Energy*, Englewood Cliffs, N.J.: Prentice-Hall, 1989, p. 481.

We see that $\langle P_d \rangle$ is equal to zero if the shell either is made of a perfect insulator (for which $\sigma = 0$) or is itself a perfect conductor (for which $\sigma = \infty$). This result is what we expect of a correct answer. The correct induced current is given by (16.33). This $\hat{\phi}$-directed current whirling around the magnetic field is commonly referred to as the **eddy current**. By placing a finite conductor in a strong time-varying magnetic field, the eddy current produces ohmic loss in the conductor, and the dissipated power may be high enough to melt it. Using magnetic induction for heating or melting metal is called induction heating. The metal industry commonly uses induction furnaces because they can produce higher temperatures (up to 2800° C) than ordinary furnaces that burn coal or oil.

Example 16.9 *This example gives an estimate of the time required to melt a metal pipe by induction heating.*

An aluminum cylinder is placed in a 60-Hz magnetic field of $B = 1$ Wb/m^2. Use the following data to determine how long it will take to melt this aluminum cylinder starting from room temperature (20° C).

For aluminum:

$$\ell = 10 \text{ cm}$$

$$a = 0.8 \text{ cm}$$

$$d = 0.1 \text{ cm}$$

$$\sigma = 3.54 \times 10^7 \text{ mho/m}$$

$$\text{specific heat} = 0.214 \text{ cal/g}^\circ \text{C}$$

$$\text{density} = 2.7 \text{ g/cm}^3$$

$$\text{melting point} = 660^\circ \text{C}$$

$$1 \text{ calorie} = 4.18 \text{ J}$$

Solution The volume of the cylinder is $\pi[(a+d)^2 - a^2]\ell = 5.34$ cm^3. Its mass is

$$m = 2.7 \times 5.34 = 14.4 \text{ g}$$

To raise this cylinder from 20° to 660° C requires an energy U where

$$U = 14.4 \times (660 - 20) \times 0.214 \times 4.18 = 8.26 \times 10^3 \text{ J}$$

According to (16.34), we calculate the total dissipated power as

$$\langle P_d \rangle = 202 \text{ W}$$

Thus, it takes 41 s to melt this cylindrical aluminum shell.

Nondestructive Evaluation of Pipes

One of several methods for inspecting metal pipes for possible cracks is the so-called eddy-current method. Figure 16.10a shows a schematic diagram of the system. The pipe under test is passed through a coil, which excites eddy current in the pipe. The input impedance of the coil consists of two parts. The imaginary part is due to the inductance of the coil. The real part is due to the power lost in the pipe because of eddy currents. If the pipe has a longitudinal crack, like that shown in Figure 16.10b, the crack will impede the eddy current and the induction loss in that section of the pipe will be greatly reduced. Thus, by monitoring the real part of the input impedance of the coil, we may detect longitudinal cracks in a pipe. Because this testing method does not damage the pipe, it is called a method of nondestructive evaluation, or NDE.

Figure 16.10 Nondestructive evaluation: inspecting a pipe for cracks. **(a)** The system. **(b)** The longitudinal crack. It impedes the eddy current and consequently produces an abnormal reading in the impedance meter. **(c)** A circumferential crack. The scheme shown in (a) will not detect it.

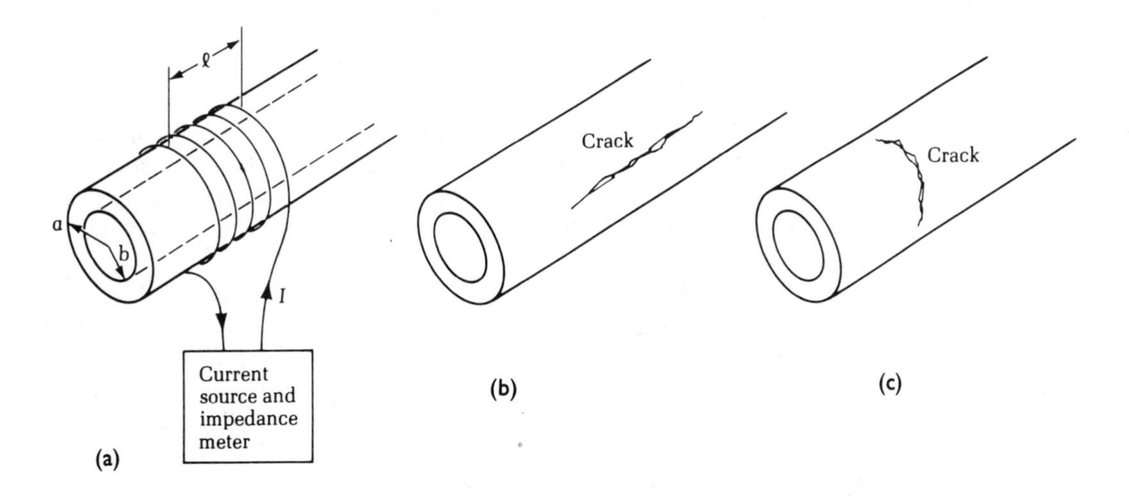

However, note that the scheme shown in Figure 16.10a cannot detect a circumferential crack, such as that shown in Figure 16.10c. Neither will this arrangement be able to detect any crack that is deep inside the pipe—that is, deep with respect to the skin depth.

In the following, we shall find the approximate value of the impedance of a coil used for NDE. Let us denote the impedance Z; then

$$Z = R + j\omega L$$

The inductance L of a solenoid is given by (13.46):

$$L = \mu_0 n^2 \pi a^2 \ell \text{ henrys}$$

where n is the number of turns per meter of the solenoid, a the radius of the solenoid, and ℓ its length.

The resistance R is related to the induction loss $\langle P_d \rangle_{\text{total}}$ in the pipe.

$$\langle P_d \rangle_{\text{total}} = \frac{1}{2}\text{Re}\{\mathrm{VI}^*\} = \frac{1}{2}\text{Re}\{\mathrm{ZII}^*\} = \frac{1}{2}R|\mathrm{I}|^2$$

Therefore, we have

$$R = \frac{2\langle P_d \rangle_{\text{total}}}{|\mathrm{I}|^2} \tag{16.35}$$

Assuming that the radius of the pipe is much greater than the skin depth and that the thickness of the pipe is approximately equal to the skin depth—namely, $\delta \approx d = a - b \ll a$—we may use (16.34) and neglect 1 and replace d with δ in the denominator to obtain

$$\langle P_d \rangle \approx (\pi a \ell d \sigma)\left(\frac{\omega a \mu_0 H_0}{2}\right)^2 \frac{\delta^2}{a^2} = \frac{\pi a \ell d}{\sigma \delta^2} H_0^2$$

Because $H_0 = nI$, we find from (16.35)

$$R = \frac{2\pi a \ell d}{\sigma \delta^2} n^2 = n^2 \pi \omega \mu_0 a \ell d$$

The input impedance is then determined to be

$$\mathrm{Z} = R + j\omega L = \omega L\left(\frac{d}{a} + j\right)$$

We see that the real part of the input impedance is small compared with the imaginary part. Consequently, in NDE that uses the eddy-current method to detect longitudinal cracks in pipes, the challenge is to design a detector that can accurately measure the real part of the input impedance, which is predominantly inductive.

Note that for the case in which the radius of the pipe is not smaller than the skin depth, the solution for $\langle P_d \rangle_{\text{total}}$ is quite complicated. Interested readers may refer to Popovic for a more rigorous approach to that problem.*

16.2 TRANSFORMERS, GENERATORS, AND MOTORS

A transformer is a device in which two parts of a circuit are coupled by magnetic flux in order to change the voltage or the impedance of the circuit. Two parts of a circuit are coupled by winding the primary and the secondary coils on a

---

* B. D. Popovic, *Introductory Engineering Electromagnetics*, Reading, Mass.: Addison-Wesley, 1971, p. 507.

common ferromagnetic core. Figure 16.11 shows a schematic diagram of a transformer and its windings. One of the reasons for using ferromagnetic material for the core is that ferromagnetic material has a larger permeability, so that most of the magnetic flux stays in the core and very little leaks. Thus, the flux linking through the primary coil is also the flux linking through the secondary coil. Coupling between the transformer and the nearby circuit is minimized.

Figure 16.11 A transformer. Note the polarity of V and the direction of the current.

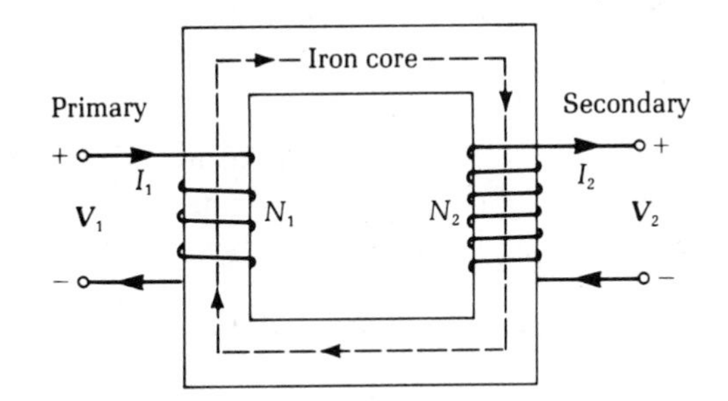

Let the effective length of the core be ℓ and its cross-sectional area be A. Recalling the technique of analyzing magnetic circuits discussed in Section 14.2, we obtain the following equation for the transformer shown in Figure 16.11.

$$\boxed{N_1 I_1 - N_2 I_2 = H\ell} \quad \textbf{(equation for the transformer shown in Figure 16.11)} \tag{16.36}$$

We obtain the above equation directly from (14.17). However, we must be careful to get the correct signs for the currents in respect to the way coils are wound. Because $\Psi = BA = \mu HA$, (16.36) may be written as follows:

$$N_1 I_1 - N_2 I_2 = \frac{\Psi\ell}{\mu A} \tag{16.37}$$

For an ideal transformer, μ is infinity; hence we have the following relation:

$$\boxed{\frac{I_1}{I_2} = \frac{N_2}{N_1}} \quad \textbf{(current ratio for an ideal transformer)} \tag{16.38}$$

With ferromagnetic material as the core, ordinary transformers have μ so high that (16.38) is always a good approximation.

From Faraday's law, we have

$$V_1 = N_1 \frac{d\Psi}{dt}$$

$$V_2 = N_2 \frac{d\Psi}{dt}$$

Consequently,

$$\boxed{\frac{V_1}{V_2} = \frac{N_1}{N_2}} \quad \textbf{(voltage ratio for an ideal transformer)} \qquad (16.39)$$

Equations (16.38) and (16.39) are valid for voltages and currents that are either time-harmonic or are of other kinds of time variation. For time-harmonic voltages and currents, we can relate the impedance at the primary port to that of the secondary port.

Because $Z_1 \equiv V_1/I_1$ and $Z_2 \equiv V_2/I_2$, we obtain the following equation from (16.38) and (16.39):

$$\boxed{\frac{Z_1}{Z_2} = \left(\frac{N_1}{N_2}\right)^2} \quad \textbf{(impedance ratio for an ideal transformer)} \qquad (16.40)$$

An ideal transformer does not dissipate any power. We draw this conclusion from the product of (16.39) and the conjugate of (16.38). It follows that

$$\frac{V_1 I_1^*}{V_2 I_2^*} = 1$$

Therefore, the input power is equal to the output power.

In reality, however, some power is always lost in transformers. There are two major components of power loss in a transformer: eddy-current loss and hysteresis loss. We now turn to a discussion of these losses.

Eddy-Current Loss in a Transformer

A transformer core contains time-varying magnetic flux. In the preceding section we learned that time-varying magnetic flux induces eddy-current loss in the conductor. There are two ways to reduce the eddy-current loss in a transformer. First, we can use materials that have high magnetic permeability but low electric conductivity. Ferrites are examples of such materials. For cast iron, σ is approximately 300 mho/m, but for ferrites the typical value of σ is 10^{-2} mho/m. Second, we can reduce eddy-current loss by assembling the core from laminations. In other words, thin steel plates in the shape of hollow squares like those shown in Figure 16.11 are coated with oxide or insulating varnish and then stacked together to form the magnetic core. Such structures greatly reduce the eddy currents, which always flow in the direction perpendicular to the magnetic flux.

Hysteresis Loss

In Section 14.1 we discussed magnetization curves for ferromagnetic materials. These curves are called hysteresis loops. Figure 14.5 shows a typical curve for iron, and we repeat it here as Figure 16.12. Let us assume that a material with such a hysteresis loop is used as the core of a transformer. The transformer is used in an alternating-current circuit. Consequently, the material is going through the hysteresis loop the same number of times in a second as the frequency of the current. To simplify the problem, we leave the second coil open.

Figure 16.12 A hysteresis loop of a ferromagnetic material. The area bounded by the loop represents the energy lost going through a complete cycle of magnetization and demagnetization.

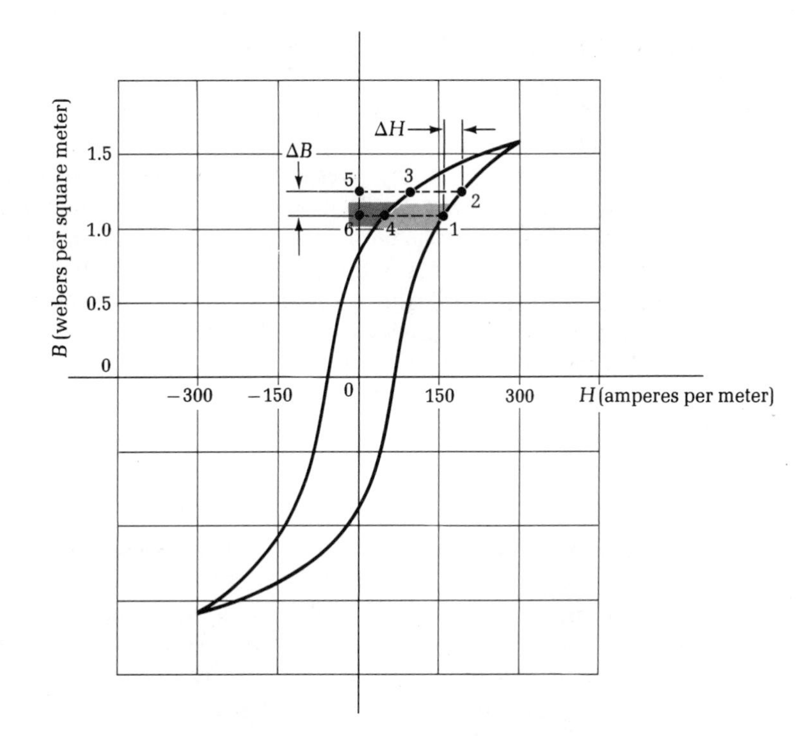

Let us consider the situation at time t when the value of H is H_1 at point 1, as indicated in Figure 16.12. The corresponding B value is B_1. In Δt seconds, the primary current is increased from I_1 to $I_1 + \Delta I$. The corresponding H is increased by ΔH and B by ΔB. In other words, the status of the core's magnetization is moved from point 1 to point 2 shown in Figure 16.12.

During this Δt period, a counter-EMF is induced, and, according to Faraday's law, we have,

$$\text{EMF} = -\frac{A\,\Delta B N_1}{\Delta t} \quad \textbf{(16.41)}$$

Where A is the cross-sectional area of the core and N_1 is the number of turns in the primary coil. The work done by the external source in changing status from 1 to 2 is given by

$$W = -(\text{EMF})I_1\,\Delta t = A\,\Delta B N_1 I_1 \quad \textbf{(16.42)}$$

According to (16.36) and the fact that $I_2 = 0$ (because the secondary coil is open-circuited), we have

$$W = A\,\Delta B H\ell = (\Delta B H)\,(\text{volume}) \quad \textbf{(16.43)}$$

Note that ΔBH is approximately the area of the horizontal strip marked 1-2-5-6.

Now let us investigate another time period when the magnetization status is changed from point 3 to point 4 as marked in the hysteresis loop. This time the magnetic flux is decreased by ΔB, resulting in an EMF equal in amplitude and opposite in sign to that given by (16.41). Thus, the magnetic core gives energy back to the external source. The energy released is equal to the area bounded by 3-5-6-4. On balance, however, the external circuit gives more energy to the material core than it receives from it because the area bounded by 1-2-5-6 is greater than that bounded by 3-5-6-4. The difference is the energy corresponding to the area 1-2-3-4 on the hysteresis loop. Thus, when the external source completes a cycle of the hysteresis loop, the energy lost is equal to the area bounded by the hysteresis loop. The entire energy lost per cycle in the core material may be calculated from the area of the hysteresis loop multiplied by $A\ell$, the volume of the core. Electric power lost as a result of hysteresis is converted into heat. Needless to say, a good transformer material will have a large permeability but a very "thin" hysteresis loop.

Example 16.10 *This example shows how to design a power transformer.*

A transformer is used to lower 2400 V to 240 V at 60 Hz. Its core has an effective cross-sectional area of 250 cm^2. The maximum flux density in the core should be as large as possible but not greater than 0.9 Wb/m^2. Find the number of turns in the primary and in the secondary coils. Neglect fringing fields.

Solution Denote as N_1 and N_2 the number of turns in the primary and secondary coils, respectively. From (16.39), we obtain

$$\frac{N_1}{N_2} = \frac{2400}{240} = 10$$

The approximate flux Ψ is found to be

$$\Psi \leq 0.9\ (\text{Wb/m}^2) \times 250 \times 10^{-4}\ (\text{m}^2) = 2.25 \times 10^{-2}\ \text{Wb}$$

According to Faraday's law, we have

$$V_1 = \left| N_1 \frac{d\Psi}{dt} \right| = |j\omega N_1 \Psi| = \omega N_1 \Psi$$

Substituting $V_1 = 2400$, $\omega = 120\pi$, and the value of Ψ just obtained, we have

$$N_1 \geq \frac{2400}{120\pi \times 2.25 \times 10^{-2}} = 282.94$$

N_1 and N_2 should both be integers. To satisfy the requirement that Ψ be limited to 0.9 Wb/m^2, we choose N_1 to be the smallest integer that is greater than 282.94 and that will also result in an integer value for N_2. The solution is therefore as follows:

$$N_1 = 290 \text{ turns}$$

$$N_2 = 29 \text{ turns}$$

$$\Psi = \frac{2400}{120\pi \times 290} = 2.20 \times 10^{-2}\ \text{Wb}$$

$$B = \frac{\Psi}{A} = 0.88\ \text{Wb/m}^2$$

Example 16.11 *This example shows how to estimate loss in a transformer due to hysteresis.*

A transformer is used in a 60-Hz network. The material of the core is iron with its hysteresis loop shown in Figure 16.12. Estimate the hysteresis loss in watts. Assume that the volume of the core is 0.03 m^3.

Solution Because no analytical expression of the hysteresis curve is given, we can estimate the area of the hysteresis loop by counting the number of squares bounded by the loop. We estimate that the area bounded by the upper curve (on which points 3 and 4 are located) and the $B = 0$ line is approximately 6 squares and that the area bounded by the lower curve (on which points 2 and 1 are located) is approximately 3.3 squares. Thus, the total area of the entire loop is

$$A_s = 2 \times (6.0 - 3.3) = 5.4 \text{ squares}$$

Each square has an "area" equal to 75 Wb-A/m³ or 75 J/m³. Thus, the hysteresis loss per loop is

$$U = 75 \times 5.4 = 405 \text{ J/m}^3$$

The power loss is obtained by multiplying U by the volume of the core and 60 because the transformer is operated by 60 Hz—that is,

$$P_{\text{hysteresis}} = 405 \times 60 \times 0.03 = 729 \text{ W}$$

Betatrons

The betatron, developed in 1940 by D. W. Kerst, is an electron accelerator. In the betatron, electrons circulate in a doughnut-shaped vacuum chamber and are accelerated by the induced EMF of a time-varying magnetic field (Figure 16.13). By Faraday's law, the induced EMF is

$$\text{EMF} = E_\phi 2\pi R = -\frac{d\Psi}{dt}$$

where R is the radius of the electron's circular orbit and E_ϕ is the induced electric field due to the time-harmonic magnetic flux Ψ.

Let the velocity of the electron be v_ϕ; then the force acting on the electron is

$$m\frac{dv_\phi}{dt} = -eE_\phi = \frac{e}{2\pi R}\frac{d\Psi}{dt}$$

which yields

$$v_\phi = \frac{e}{2\pi mR}\Psi \tag{16.44}$$

A stable orbit of constant radius R is achieved when the centrifugal force due to v_ϕ is balanced by the radial Lorentz force

$$m\frac{v_\phi^2}{R} = ev_\phi B(R) \tag{16.45}$$

where $B(R)$ is the magnitude of the magnetic field at radius R pointing in the $\hat{z}$ direction. Comparing the two equations (16.44) and (16.45), we find

$$\Psi = 2\pi R^2 B(R)$$

Because the flux Ψ is the result of the integration of the magnetic field over the area from $\rho = 0$ to $\rho = R$, it follows that

$$\Psi = \int_0^{2\pi} d\phi \int_0^R \rho\, d\rho B(\rho) = 2\pi R^2 B(R)$$

Carrying out the integration over ϕ yields

$$\int_0^R \rho\, d\rho B(\rho) = R^2 B(R)$$

Differentiating the above equation with respect to R, we obtain:

$$RB(R) = 2RB(R) + R^2 B'(R)$$

The above equation leads to the following differential equation:

$$\frac{B'}{B} = \frac{-1}{R}$$

The solution of B is then

$$B(\rho) = \frac{C}{\rho} \qquad \textbf{(16.46)}$$

where C is a constant. We see that the magnetic field must have larger values at smaller radii and must decrease in the outward direction. To produce such an inhomogeneous field, the pole magnet must be shaped properly, as shown in Figure 16.13.

Figure 16.13 The betatron: a device to accelerate electrons.

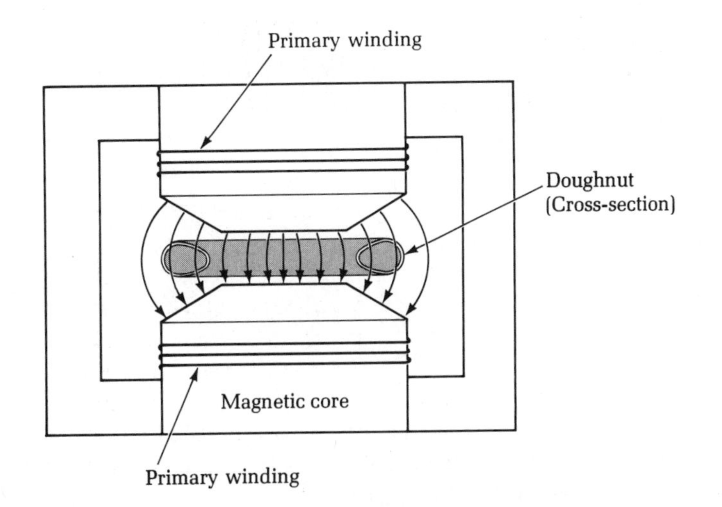

Note that, to accelerate an electron, the magnetic flux must increase with time. If the magnet is produced by an alternating current of frequency f with period T, then the total acceleration of the electron must be completed within $T/4$ when the flux increases with time. Therefore, only electrons that start at the right instant can attain maximum acceleration. A betatron produces accelerated electrons in groups that appear at a frequency equal to the frequency of the time-varying magnetic field.

As the electrons are held in an orbit of constant radius R, a further multiple acceleration to higher energies can be produced by a cyclotron-like alternating electric field. The time variation of the magnetic field and of the accelerating electric field must be synchronized with the acceleration of each electron group. This synchronously occurring betatron and cyclotron acceleration is the guiding principle for the operation of a **synchrotron**.

High-energy electrons produced by a betatron can be used for research in physics or for the production of X rays for medicine or industry.

Homopolar Generators

A homopolar generator consists of a Faraday disk that has conductivity σ and radius b and that rotates in a uniform magnetic field B_0, as shown in Figure 16.14a. Figure 16.14b shows an equivalent circuit. Let the z axis of the cylindrical

Figure 16.14 **(a)** A homopolar generator, consisting of a Faraday disk rotating in a magnetic field excited by an external dc current. **(b)** The equivalent circuit of the homopolar generator.

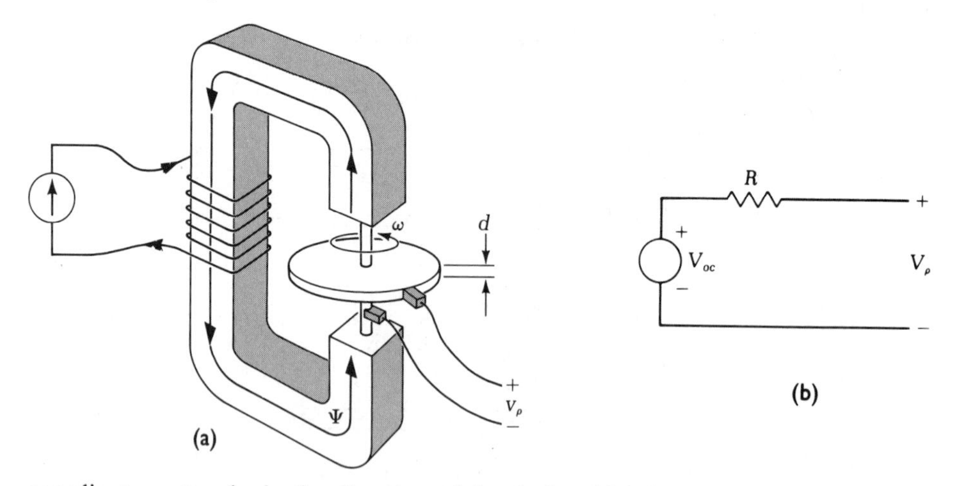

coordinate system be in the direction of the shaft, which has a radius a. Consider a small element located at ρ. The EMF induced on this element per unit length is given by (16.13):

$$\Delta E = \omega \rho B_0$$

Because the disk is made of a conductor of finite conductivity, the voltage drop per unit length due to the resistance of the disk is J_ρ/σ. Therefore, the following equation holds:

$$E_\rho = \omega \rho B_0 - \frac{J_\rho}{\sigma}$$

The total $\hat{\rho}$-directed current flowing in the disk is equal to $2\pi\rho d J_\rho$. Thus,

$$E_\rho = \omega\rho B_0 - \frac{i_\rho}{2\pi\rho\sigma d}$$

where ω is the angular speed of the disk, d is the thickness of the disk, J_ρ is the current density, and i_ρ is the total current in the $\hat{\rho}$ direction.

The induced voltage is determined to be

$$v_\rho = -\int_a^b d\rho\left(\frac{i_\rho}{2\pi\sigma\rho d} - \omega\rho B_0\right) = \frac{-i_\rho}{2\pi\sigma d}\ln\frac{b}{a} + \frac{\omega B_0}{2}(b^2 - a^2)$$

For the equivalent circuit shown in Figure 16.14b, we find that

$$R = \frac{1}{2\pi\sigma d}\ln\frac{b}{a}$$

is the equivalent resistance and

$$v_{oc} = \frac{\omega B_0}{2}(b^2 - a^2)$$

is the induced open-circuit voltage. As an example, consider the case in which

$$\sigma = 5.7 \times 10^7 \text{ mho/m}$$

$$B_0 = 1 \text{ Wb/m}^2$$

$$d = 5 \text{ mm}$$

$$b = 10 \text{ cm}$$

$$a = 1 \text{ cm}$$

$$\omega = 120\pi \text{ rad/s}$$

We find

$$v_{oc} = \frac{120\pi}{2}(0.1^2 - 0.01^2) = 1.87 \text{ volts}$$

The equivalent resistance R, as shown in Figure 16.14b, is

$$R = \frac{1}{2\pi \times 5.7 \times 10^7 \times 0.005}\ln\frac{0.1}{0.01} = 1.29 \times 10^{-6}\ \Omega$$

The short-circuit current will be

$$i_{sc} = \frac{v_{oc}}{R} = 1.45 \times 10^6 \text{ A}$$

Thus, homopolar generators are typically low-voltage and high-current devices.

AC Generators

Figure 16.15 illustrates the basic operating principle of an alternating-current (or ac) generator. A rectangular coil, known as the armature coil, is rotated by a mechanical device (a turbine, for example) in a dc magnetic field at an angular

Figure 16.15 An ac generator. **(a)** Full view and **(b)** end view, where ***m*** is the magnetic moment due to the current in the loop.

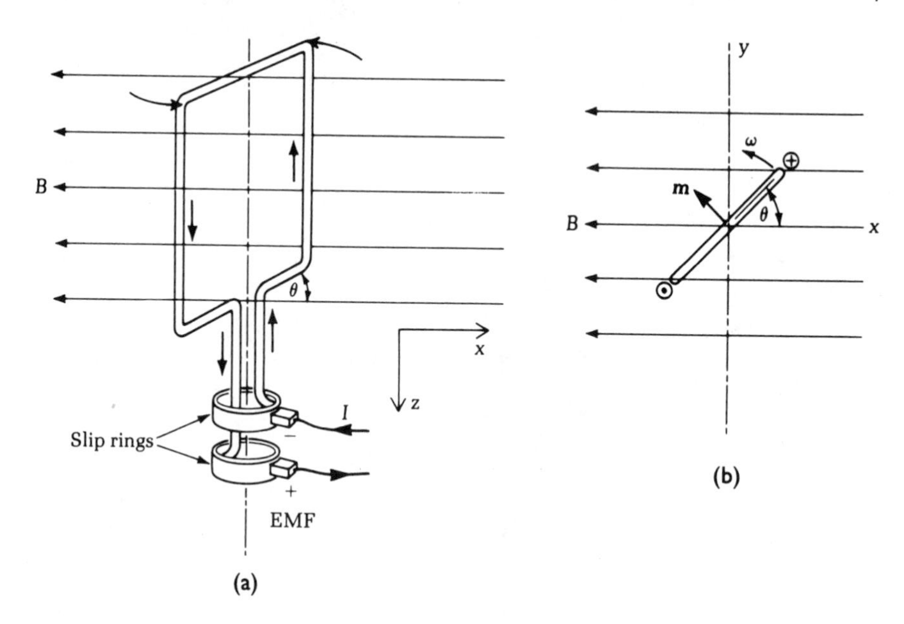

frequency ω. The loop has an area A. The flux linkage through the loop is given as follows:

$$\Psi = BA \sin \theta \tag{16.47}$$

Using Faraday's law, we have

$$\text{EMF} = -\frac{d\Psi}{dt} = -BA \cos \theta \frac{d\theta}{dt} = -BA\omega \cos (\omega t) \tag{16.48}$$

Because of the minus sign in (16.48), the instantaneous value of the EMF at $\theta = \omega t = \pi/4$ is a negative number. The leads of the coil are connected to the outside circuit through two slip rings. The voltage that appears at these rings is time-harmonic, as we can see from (16.48).

Suppose that the current which is flowing through the armature coil is $I \cos (\omega t + \alpha)$. The instantaneous electric power delivered to the outside circuit by the generator is

$$P_e(t) = V(t)I(t) = -[BA\omega \cos (\omega t)][I \cos (\omega t + \alpha)] \tag{16.49}$$

The torque on the armature coil, on the other hand, is given by (13.31):

$$\boldsymbol{T} = \boldsymbol{m} \times \boldsymbol{B}$$

In the present case, $m = AI \cos(\omega t + \alpha)$ with the direction shown in Figure 16.15. Therefore,

$$\boldsymbol{T} = \hat{z} AI \cos(\omega t + \alpha) B \cos(\omega t)$$

Note that the torque is against the rotation of the coil. The mechanical power to overcome this electromagnetic drag force is $-T\omega$, or

$$P_m(t) = -\omega T = -ABI\omega \cos(\omega t) \cos(\omega t + \alpha) \tag{16.50}$$

Comparing (16.49) and (16.50), we can prove that $P_e(t) = P_m(t)$. In other words, power is conserved at any given time. The generator can be thought of as a device that converts mechanical power into electric power.

Energy Conversion by a DC Motor

In Section 13.2, we discussed the operating principle of a dc motor. Let us assume that the loop of the armature is pointing in the direction that makes an angle α with the magnetic field, as shown in Figure 13.22b, which is repeated here in Figure 16.16. If the current is I and the external magnetic field is B_0, then the torque is $\boldsymbol{T}$, which is given as follows:

$$\boldsymbol{T} = NIAB_0 \sin \alpha(-\hat{z}) \tag{16.51}$$

where N is the total number of turns and A the area of the coil. When the armature turns a $[-\Delta\alpha]$ angle, the mechanical work done by the motor is $T(-\Delta\alpha)$, or

$$W_m = NIAB_0 \sin \alpha \, \Delta\alpha \tag{16.52}$$

Figure 16.16 A dc motor is an energy conversion device. Armature rotates in the magnetic field because of the magnetic force on the current-carrying coil. At the same time, Faraday induction produces counterelectromotive force.

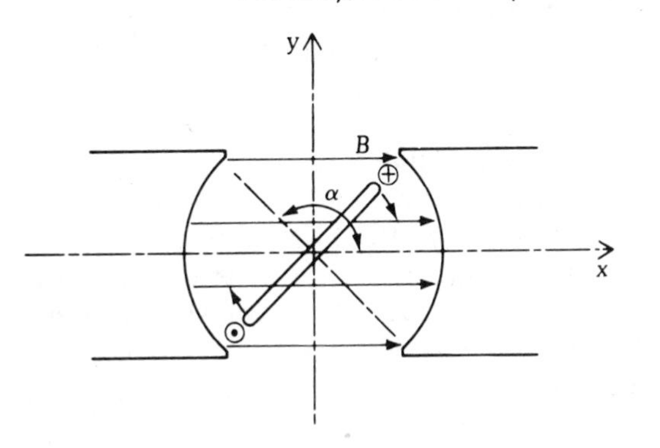

Note that the total flux through the coil is given as follows:

$$\Psi = NB_0 A \cos \alpha \tag{16.53}$$

Suppose that the armature rotates a $(-\Delta\alpha)$ angle in Δt seconds; then the induced voltage on the coil is given by Faraday's law:

$$\text{EMF} = -\frac{d\Psi}{dt} = NB_0 A \sin \alpha \frac{\Delta\alpha}{\Delta t} = -NB_0 A \sin \alpha \left(\frac{-\Delta\alpha}{\Delta t}\right) \tag{16.54}$$

The current I flows against this EMF according to the Lenz law, and the product of I and the EMF is equal to the power delivered to the motor by the external electric source. During the Δt period, the total electric energy delivered to the motor is W_E, where

$$W_E = I(\text{EMF})\Delta t = INB_0 A \sin \alpha \, \Delta\alpha \tag{16.55}$$

Comparing (16.52) with (16.55), we see that the mechanical work performed by the motor is equal to the electric power delivered to it by the external electric source. Therefore, energy is conserved, as it should be. Thus, a dc motor may be considered a device that converts electric energy to mechanical energy.

Rotating Magnetic Field

In our discussion of dc motors in Section 13.2 and of ac generators in this section, the **armature** of the motor or of the generator has been represented by a coil in air. This simple model serves well to demonstrate the basic operating principle of these machines. To discuss induction motors and synchronous motors, however, we need to improve our model for the armature. Figure 16.17a shows a more realistic model. The **rotor** is a ferromagnetic cylinder. The magnetic field is excited by coils wound on the **stator**, which is also made of a ferromagnetic material. The magnetic field excited by the current in the coil flows from the stator to the rotor and passes through the air gap between them. Depending on the geometric layout of the coils and on the current in them, different configurations of the magnetic field in a motor or in a generator may be produced. In this section we shall discuss the excitation of a rotating magnetic field produced by coils fed with three-phase alternating currents. Understanding how a rotating magnetic field can be excited by stationary coils in the stator is a prerequisite to understanding our subsequent discussion of induction motors and synchronous machines.

Three sets of coils are embedded in the stator to excite the magnetic field. These coils, labeled in Figure 16.17 *a-a'*, *b-b'*, and *c-c'*, are fed by *three-phase currents.* Figure 16.18 shows the waveforms of these currents. Current I_a represents the current flowing into the *a-a'* coils at the *a* terminal and exiting at the *a'* terminal. At time $t = 0$, current I_a is positive, so that it flows into the paper at the *a* terminal and out of the paper at the *a'* terminal, as shown in Figure 16.17a. The "+" symbol denotes current flowing in the into-the-paper direction, and the

Figure 16.17 A set of three coils arranged to produce a rotating magnetic field in the rotor-stator gap. Coils *a*-*a*′, *b*-*b*′, and *c*-*c*′ are connected to the three-phase currents I_a, I_b, and I_c as shown in Figure 16.18.

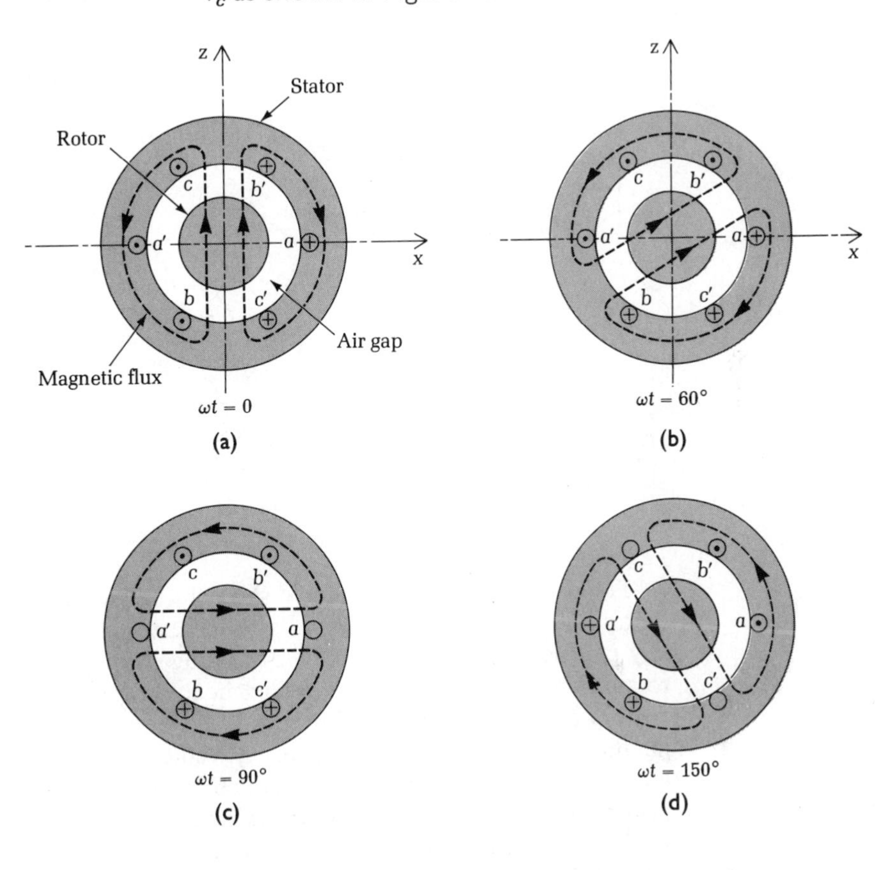

"•" denotes current flowing in the out-of-the-paper direction. At the same time ($t = 0$), currents I_b and I_c are both negative; thus, currents in coils b and c flow from b' to b and from c' to c, respectively. From the instantaneous current-flow diagram shown in Figure 16.17a, we see that the situation is equivalent to having three coils placed on the x-y plane. The magnetic field is in the $\hat{z}$ direction going from the rotor to the stator through the air gaps and returning by way of the stator.

From Figure 16.18, we see that, at $\omega t = 60°$, I_a and I_b are positive and I_c is negative. Figure 16.17b shows the corresponding instantaneous current-flow diagram in the coils. This situation is equivalent to having three coils oriented in a direction that makes a 60° angle with the z axis. Thus, the magnetic flux is rotated in space by a 60° angle. Note that the coils do not move. Because of changes in the current, the coils in the stator produce a magnetic field as if they had been rotated by a 60° angle.

Figure 16.18 A set of three-phase currents. Current I_b lags current I_a by 120°, and current I_c lags current I_b by 120°. These currents, when fed to the stator coils shown in Figure 16.17, would produce a rotating magnetic field in the air gap of the motor.

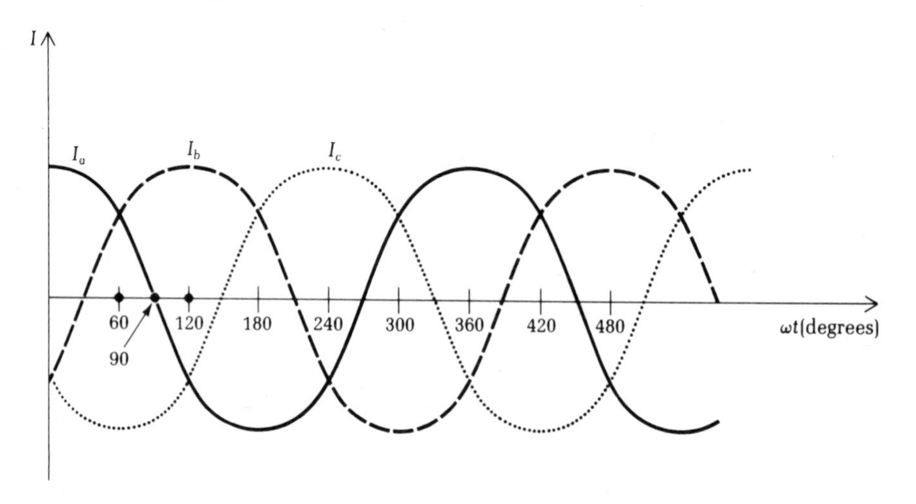

Figure 16.17c and Figure 16.17d show the magnetic field and the current-flow diagram at $\omega t = 90°$ and at $\omega t = 150°$, respectively. We conclude that by properly placing coils in a stator and by feeding them with three-phase currents, a rotating magnetic field can be excited in the rotor-stator air gap. The angular velocity of the rotation is synchronous with the angular frequency of the alternating three-phase current. If the frequency of the three-phase current is 60 Hz, then for the configuration shown in Figure 16.17, the speed of the rotating magnetic field is 3600 r/min. The speed can be reduced if two or more sets of coils are embedded in the stator so that they are periodic—for example, every 180° instead of every 360°. Then, the rotation speed of the magnetic flux may be reduced to 1800 r/min.

Induction Motors

An induction motor consists of a stator and a rotor, both made of ferromagnetic material of high permeability. Figure 16.19 shows a cross-sectional view of the motor. A set of coils embedded in the stator is fed by three-phase currents to produce a rotating magnetic field. The configuration of the stator coil is similar to that shown in Figure 16.17. The windings in the rotor are usually short-circuited. Therefore, we can model the rotor as a ferromagnetic cylinder with finite electric conductivity. Because the magnetic field is rotating, the magnetic field linking the short-circuited coils on the rotor changes with time. Because of

Figure 16.19 An induction motor. Stator coils generate a rotating magnetic field in the air gap. Currents are induced on the rotor, and these currents interact with the magnetic field to produce a torque.

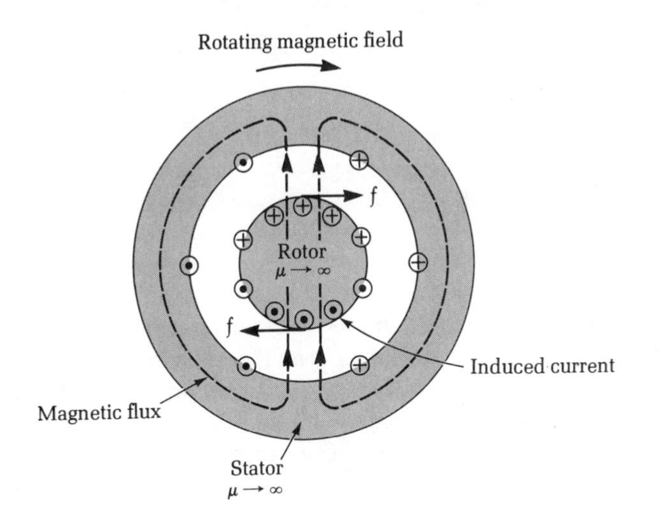

Faraday's law, an EMF is induced on the rotor and generates current in the rotor. This current interacts with the magnetic field and produces a force according to (13.26), and this force rotates the rotor.

As long as the rotor is rotating at a speed lower than the rotation speed of the magnetic field, the induced EMF is always there to produce current and torque. However, if the rotor catches up with the rotating magnetic field, then there is no relative motion between them, and the induced EMF is zero: no force is produced, and the rotor slows down. From the foregoing qualitative discussion, we see that the induction motor is an **asynchronous** machine. The rotation speed of an induction motor is always less than that of the rotating magnetic field, which, for the configuration shown in Figure 16.17 connected to a 60-Hz three-phase source, is 3600 r/min.

To analyze the induction motor quantitatively, let us make a few assumptions that will simplify the mathematics. First, we assume that the gap between the rotor and the stator is small compared with the radius of the rotor. Consequently, we can develop the motor into a linear motor, such as that shown in Figure 16.20. Second, we assume that the permeability of the stator and the rotor is infinity. Furthermore, we approximate the windings of the rotor by a thin sheet of conductor of conductivity σ and thickness δ.

The *rotating* magnetic field in the air gap can now be represented by a *traveling* magnetic field:

$$\mathbf{H} = \hat{x} H_x e^{-jkz} \tag{16.56}$$

The wave number k is determined by the periodicity of the winding Λ, $k = 2\pi/\Lambda$. For the winding shown in Figure 16.19, $\Lambda = 2\pi a$, where a is the radius of the

Figure 16.20 Developed diagram of the induction motor shown in Figure 16.19.

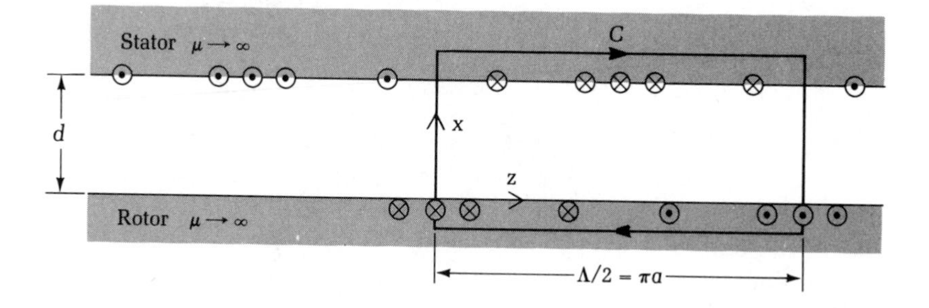

stator and $k = 1/a$. Let ω be the angular frequency of the electric power source; then $v_s = \omega/k = a\omega$ is the linear velocity of the magnetic field in the gap.

When the rotor is stationary, the traveling magnetic field induces an electric field in the rotor by virtue of Faraday's law. Using (2.28a) and (16.56), with $\partial/\partial x = \partial/\partial y = 0$ and $\partial/\partial z = -jk$, we obtain

$$-jk\mathrm{E}_{y1} = j\omega\mu_0 \mathrm{H}_x e^{-jkz}$$

or

$$\mathrm{E}_{y1} = -\frac{\omega\mu_0}{k}\mathrm{H}_x e^{-jkz} \tag{16.57}$$

Let the rotor be moving at a speed v_r in the $\hat{z}$ direction. Then the motion of the rotor in the magnetic field produces an additional induced electric field. Because

$$\boldsymbol{v}_r \times \mathbf{B} = \hat{y} v_r \mathrm{H}_x \mu_0 e^{-jkz}$$

we have

$$\mathrm{E}_{y2} = v_r \mathrm{H}_x \mu_0 e^{-jkz}$$

The total electric field induced on the rotor is the sum of E_{y1} and E_{y2}:

$$\mathbf{E} = \hat{y}\left(-\frac{\omega}{k} + v_r\right)\mu_0 \mathrm{H}_x e^{-jkz} = -\hat{y}(v_s - v_r)\mu_0 \mathrm{H}_x e^{-jkz}$$

This electric field induces a surface current $\mathbf{J}_{sr}$ in the rotor.

$$\mathbf{J}_{sr} = \sigma\delta\mathbf{E} = -\hat{y}\sigma\delta\mu_0(v_s - v_r)\mathrm{H}_x e^{-jkz}$$

Let us define

$$\mathbf{J}_{sr} = -\hat{y}\mathrm{J}_{sr0}\, e^{-jkz}$$

where

$$\mathrm{J}_{sr0} = \sigma\delta\mu_0(v_s - v_r)\mathrm{H}_x \tag{16.58}$$

To calculate the total $\hat{x}$-directed magnitude field H_x in the air gap, we integrate

Ampère's circuit law over the path C shown in Figure 16.20.

$$\oint_C d\boldsymbol{\ell} \cdot \mathbf{H} = \int_0^{\Lambda/2} dz(-\hat{y}) \cdot (\mathbf{J}_{sr} + \mathbf{J}_{ss}) \tag{16.59}$$

where $\mathbf{J}_{ss}$ is the surface current on the stator provided by the external power source. Let $\mathbf{J}_{ss}$ be expressed as follows:

$$\mathbf{J}_{ss} = -\hat{y}\mathrm{J}_{ss0}\, e^{-jkz} \tag{16.60}$$

where J_{ss0} is a constant determined by the strength of the external source. Because both the rotor and the stator have infinite permeability, there is no contribution to the line integral on the left-hand side except in the air gap. We obtain

$$\left[\mathrm{H}_x(z=0) - \mathrm{H}_x\left(z = \frac{\Lambda}{2}\right)\right] d = \frac{-1}{jk}(\mathrm{J}_{sr0} + \mathrm{J}_{ss0})(e^{-j\pi} - 1)$$

Because $\mathrm{H}_x(z=0) = \mathrm{H}_x$ and $\mathrm{H}_x(z = \Lambda/2) = \mathrm{H}_x e^{-j\pi}$, we find

$$\mathrm{H}_x = \frac{-j}{kd}(\mathrm{J}_{sr0} + \mathrm{J}_{ss0}) \tag{16.61}$$

From (16.58) and (16.61), we eliminate H_x to get

$$\mathrm{J}_{sr0} = -\frac{jR_m}{1 + jR_m} J_{ss0} \tag{16.62}$$

where the dimensionless parameter R_m is given by

$$R_m = \frac{\sigma\delta\mu_0}{kd}(v_s - v_r) \tag{16.63}$$

The total magnetic field follows from (16.61) and (16.62).

$$\mathrm{H}_x = \frac{-j/kd}{1 + jR_m} J_{ss0} \tag{16.64}$$

The magnetic field acts on the induced current sheet $\mathbf{J}_{sr}$ and gives rise to the time-average force on the rotor per period

$$\langle F \rangle = \frac{\ell}{2}\mathrm{Re}\left\{\int_0^{\Lambda} dz \mathbf{J}_{sr} \times \mu_0 \mathbf{H}^*\right\} = \hat{z}\frac{\ell\Lambda}{2kd}\frac{\mu_0 R_m}{1 + R_m^2}|\mathrm{J}_{ss0}|^2 \tag{16.65}$$

where ℓ is the length of the rotor in the axial direction.

The time-average torque produced by the force is

$$\langle T \rangle = 2a\langle F \rangle = \frac{2\pi a\ell}{k^2 d}\frac{\mu_0 R_m}{1 + R_m^2}|\mathrm{J}_{ss0}|^2 \tag{16.66}$$

Note that if the rotor's speed is the same as that of the rotating magnetic field, then $v_s = v_r$ and $R_m = 0$. This result would give $\langle F \rangle = 0$ and $\langle T \rangle = 0$ and would confirm the conclusion we drew in our qualitative discussion of the induction motor. Also note the existence of an optimum v_r that would give the maximum torque. The value of v_r that gives the maximum torque is the one that makes R_m equal to unity—that is,

$$\frac{\sigma\delta\mu_0(v_s - v_r)}{kd} = 1 \tag{16.67}$$

because $\langle T \rangle$ is maximum when $R_m = 1$. The maximum torque $\langle T \rangle_{\max}$ is given as follows:

$$\langle T \rangle_{\max} = \frac{\pi\mu_0 a\ell}{k^2 d} |\mathbf{J}_{ss0}|^2$$

It is independent of σ or δ, which determines the resistance of the rotor. However, the rotor speed v_r at which the maximum torque occurs is dependent on the rotor resistivity, as we see from (16.67). The torque-speed curves of the induction motor shown in Figure 16.21 clearly display these characteristics, which are typical of induction motors.*

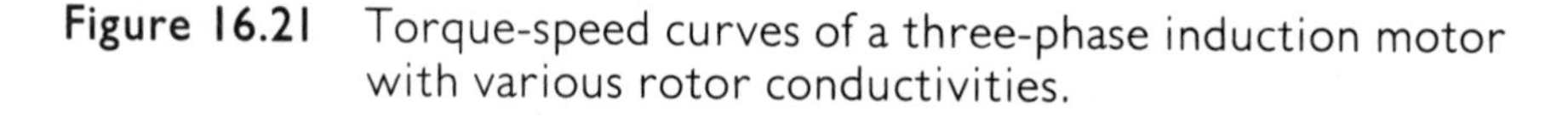

Figure 16.21 Torque-speed curves of a three-phase induction motor with various rotor conductivities.

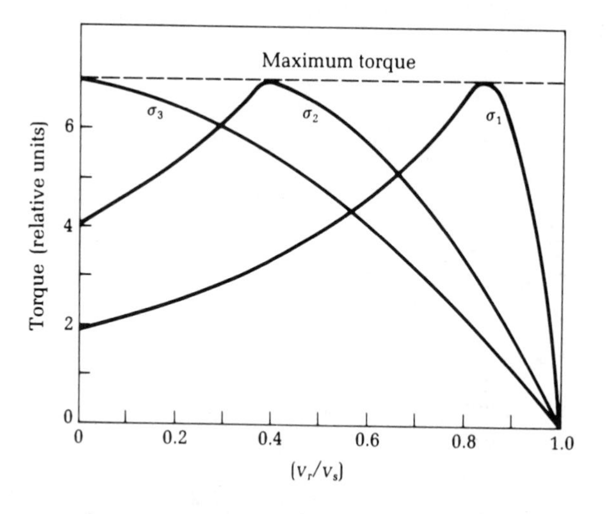

Synchronous Motors

Figure 16.22 shows a three-phase synchronous motor. The stator coils draw three-phase currents and produce a rotating magnetic field. The rotor consists of fixed magnetic poles produced by a dc source. The rotor shown in Figure 16.22 has one pair of magnetic poles. It has a magnetic moment $\boldsymbol{m}$ in the radial direc-

* A. E. Fitzgerald and C. Kingsley, Jr., *Electric Machinery*, New York: McGraw-Hill, 1952, p. 129.

Figure 16.22 A synchronous motor. The stator coils are fed with three-phase currents similar to the ones shown in Figure 16.17. The directions of the currents in the stator coils and the direction of the stator flux are shown for $t = 0$.

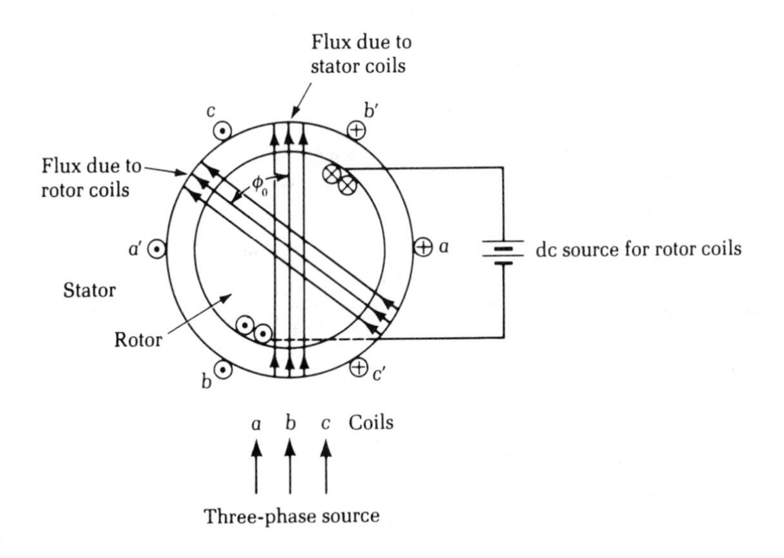

tion. The rotating magnetic field is also in the radial direction but makes a ϕ angle with respect to the magnetic moment of the rotor. According to (13.31), a torque is produced between $\boldsymbol{m}$ and the rotating magnetic field $\boldsymbol{B}$:

$$\boldsymbol{T} = \boldsymbol{m} \times \boldsymbol{B} = -\hat{z} mB \sin \phi \tag{16.68}$$

In the situation shown in Figure 16.22, the torque is pointing into the paper. Denote the angular speed of the rotating magnetic field as ω_m and that of the rotor as ω_r. The angle ϕ is then given by

$$\phi = \omega_m t - \omega_r t + \phi_0 \tag{16.69}$$

Thus, a constant torque will be produced only when the rotor is rotating at the same speed as the speed of the rotating magnetic field. Otherwise, according to (16.68) and (16.69), the time-average torque will be zero. This fact explains why the machine is called a synchronous motor.

A synchronous motor cannot be started by itself. A net torque is produced only if the motor is already rotating at the synchronous speed. To solve the start-up problem, the rotor of a synchronous motor is usually wound with an auxiliary coil called damper winding. This damper winding is short-circuited to allow induced currents to flow when the motor is being started. In other words, the motor is initially an induction motor. The rotor then accelerates almost to the synchronous speed during which time the dc source is disconnected from the rotor coil. The dc source is switched on when the rotor speed is almost to the synchronous speed. The torque is then given by

$$T(t) = mB \sin [(\omega_m - \omega_r)t + \phi_0] \tag{16.70}$$

If the difference between ω_m and ω_r is small (usually 5 to 10 r/min) and if the load and the inertia are not too great, the sine function will remain positive long enough for the rotor speed to catch up with the synchronous speed. This action is called "pulling the rotor into synchronism." At synchronism, $\omega_m = \omega_r$, and (16.70) becomes

$$T = mB \sin \phi_0$$

The angle ϕ_0 is called the **torque angle**. The torque angle is the angle between the magnetic moment of the rotor and the magnetic field of the stator at synchronism. This torque angle varies with the load. As the load is increased, the rotor momentarily slows to increase the torque angle and then resumes its synchronous speed. The maximum torque the motor can deliver is obtained when ϕ_0 is equal to 90°. This maximum torque is called the **pullout torque** of the motor. If the load requires a torque greater than the pullout torque, the rotor slows down steadily, and synchronism is lost.

The Magneplane

The magneplane is a high-speed transportation vehicle that flies approximately 0.3 m above a conducting track. It has no wheels and actually flies in the air. The lift is provided by **magnetic levitation**, and the propulsion is supplied by magnetic force based either on the principle of the induction motor or on that of the synchronous motor. We shall discuss magnetic levitation and propulsion separately.

The magneplane carries superconducting coils in which a large direct electric current I_v is maintained. We can understand the operating principle of the levitation force with the method of images. For conceptual simplification, assume that the vehicle coil is moving above a perfectly conducting plane at a height of $d/2$ (Figure 16.23a). The coil current can be considered due to positive charges moving in the direction of the current. The image charges are negative and move in the same direction, so that they constitute a current I_t opposite to the vehicle current. The distance between the vehicle-coil current and its image current is d (Figure 16.23b). By the image method, we have $I_t = I_v$. In practice, we do not have a perfectly conducting flat plane, and the image current will in general have a smaller amplitude.

To estimate the force between the two coils carrying opposite currents, we consider the case shown in Figure 16.24. The separation of the coils is small compared with their sizes. We model each pair and each side of the coils as two infinitely long wires carrying opposite currents. The magnetic field that is on the upper wire and due to the lower wire carrying I_t is given by (13.6):

$$\boldsymbol{H}_t = -\hat{z}\frac{I_t}{2\pi d}$$

Figure 16.23 **(a)** The magnetic levitation is provided for a magneplane by the interaction force between the coil on the magneplane and the track, which is modeled as a perfectly conducting plane. **(b)** The configuration in (a) is equivalent to this two-coil system.

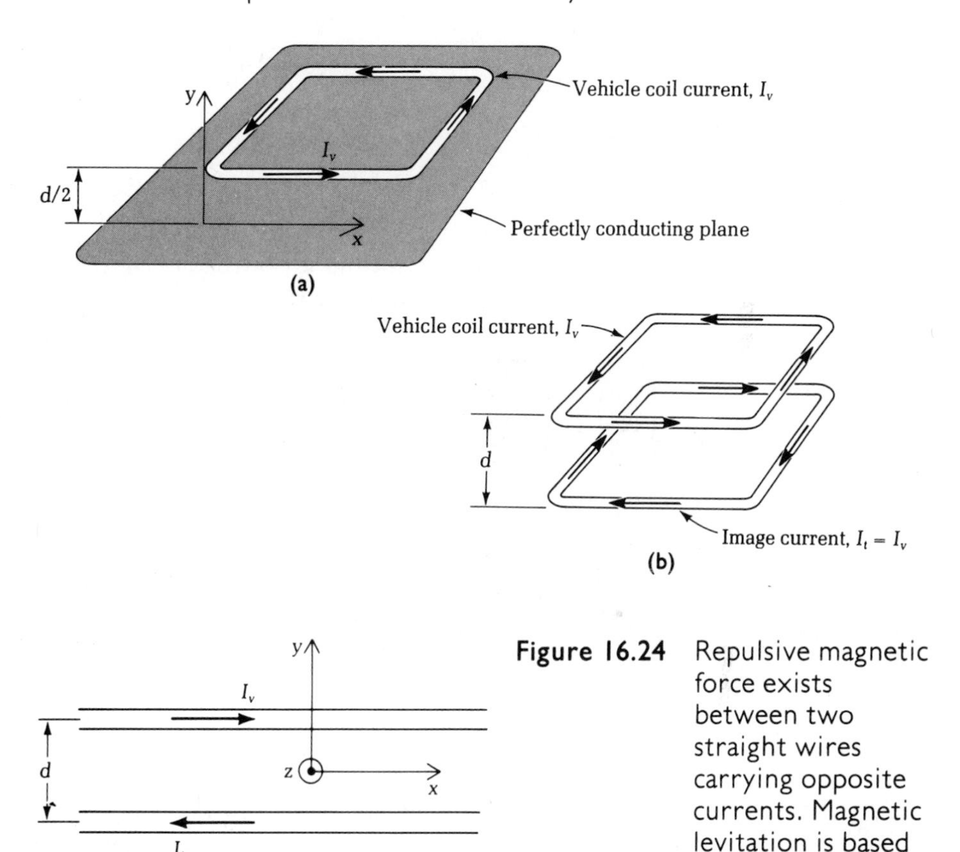

Figure 16.24 Repulsive magnetic force exists between two straight wires carrying opposite currents. Magnetic levitation is based on this principle.

The $-\hat{z}$ direction in the present case corresponds to $\hat{\phi}$ shown in Figure 13.2. The force that is on the upper wire and due to the field can be found from (13.26):

$$F = I_v \ell(\hat{x}) \times \frac{\mu_0 I_t}{2\pi d}(-\hat{z}) = \frac{\mu_0 I_t I_v \ell}{2\pi d}\hat{y} \tag{16.71}$$

The total force on a square coil is 4 times the above value:

$$F_{\text{coil}} = \hat{y}\frac{2\mu_0 I_t I_v \ell}{\pi d} \tag{16.72}$$

Note that the force is in the positive $\hat{y}$ direction. This force produces the magneplane's lift.

A magneplane's propulsion is based on the principle of synchronous motors. The track is the stator, and the vehicle is the rotor. Coils can be laid on the track in the same manner as those in a rotary synchronous motor. Figure 16.25 shows an example that corresponds to the diagram shown in Figure 16.22. The *a-a′* coil repeats itself periodically, as do the *b-b′* and *c-c′* coils. If these coils are fed with three-phase currents such as those shown in Figure 16.18, a traveling magnetic field is produced. This traveling magnetic field interacts with a dc magnetic moment on the vehicle and produces the propulsion force.

Figure 16.25 The magneplane is propelled by a linear synchronous motor. This diagram is developed from Figure 16.22. Note that the magnetic force tends to align the vehicle's south pole with the track's north pole. The distance ℓ_t and the frequency of the currents determine the synchronous speed of the magneplane.

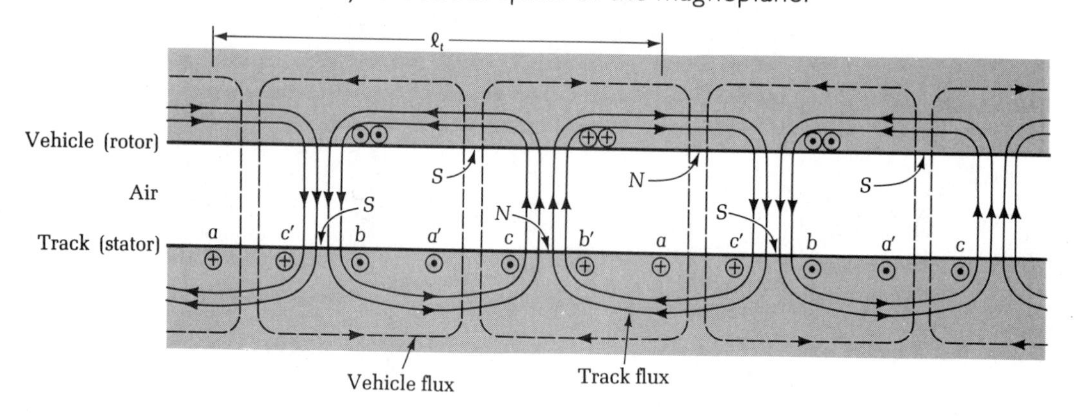

A magneplane being developed has a size similar to that of the Boeing 707 airliner's fuselage. It requires 5 MW of electric power to propel a 140-passenger vehicle once or twice a minute, in contrast with a conventional electrified railroad train that requires 50 MW of wayside power to propel an 800-passenger train once or twice an hour. The levitation and propulsion coils must carry very large currents, as is possible if these coils are made of hollow superconductors. Liquid helium is circulated inside the superconductors to maintain superconductivity.* Figure 16.26 shows a scaled model of a magneplane. A full scale 44-seat magneplane (also called a mag-lev train) has been tested on a 5.5 km track in Japan and attained a top speed of 500 km/hour (310 mph).**

* References on this subject: (1) H. H. Kolm and R. D. Thornton, "Electromagnetic flight," *Scientific American*, October 1973, pp. 17–25; (2) E. Ohno, M. Iwamoto, and T. Yamada, "Characteristics of superconductive magnetic suspension and propulsion for high-speed trains," *Proc. IEEE*, 6, no. 5 (1973): 579–786; (3) R. D. Thornton, Y. Iwasa, and H. H. Kolm. "The magneplane system," *Proc. 5th International Cryogenic Engineering Conference*, IPC Science and Technology Press, England.

** J. Hillkirk, "Japan's high-speed trains to give rail travel a lift," *USA Today*, January 31, 1989.

Figure 16.26 A 1/15 scale model of a magneplane being tested at the Massachusetts Institute of Technology. (Courtesy of Massachusetts Institute of Technology)

Example 16.12 *This example gives the magnitude of the magnetic levitation force.*

Assume that the vehicle coil of a magneplane is 1.0 m × 1.0 m and carries a current of 2×10^5 A. The separation between the vehicle and the track is 0.25 m. Find the levitation force.

Solution From (16.72), we find

$$F_{\text{coil}} = \frac{2 \times 4\pi \times 10^{-7} \times 2 \times 10^5 \times 2 \times 10^5 \times 1.0}{\pi \times 0.5} = 64 \times 10^3 \text{ N}$$

$$= 6.5 \text{ metric tons}$$

SUMMARY

1. When a time-varying magnetic field links through a circuit, an induced electromotive force (EMF) is generated in the circuit. The magnitude and polarity of the EMF are determined by Faraday's law.
2. When a circuit is placed in a time-varying magnetic field, a voltmeter may read different values even when it is connected to the same two nodes. The voltmeter reading depends on the deployment of the lead wires.
3. An EMF is induced on a wire that "cuts through" a magnetic field.

4. Magnetic energy is stored in a solenoid carrying a dc current. The coils in the solenoid are subject to an expansion force. The force can be obtained by calculating the change in the stored energy in the system with respect to the change in the radii of the coils.

5. An ideal transformer is the one for which the magnetic permeability of its core is infinity. As a result, the H field required to generate a certain amount of magnetic flux is negligible and there is no flux leakage.

6. The hysteresis loss is caused by the time-varying magnetic fields in a ferromagnetic material. It is determined by the hysteresis loop that characterizes the material.

7. An ac generator is a device that converts mechanical power to electric power. A dc motor is a device to convert electric power to mechanical power. From Faraday's law, we prove that the conversion is instantaneous; that is, energy is conserved at any given time.

8. We can generate rotating magnetic fields using three-phase currents to excite coils wound on a stationary "stator." This configuration is used in induction and synchronous motors.

Problems

16.1 A small circular loop of 5 mm radius is placed 1 m away from a 60-Hz power line. The voltage induced on this loop is measured at 0.6 microvolt. What is the current on the power line?

16.2 Assume that the current on the infinitely long line shown in Figure 16.1 is the triangular pulse shown in Figure P16.2. Find the induced voltage on the rectangular loop. Use the following data: $a = 2$ cm, $b = 4$ cm, and $d = 1$ cm.

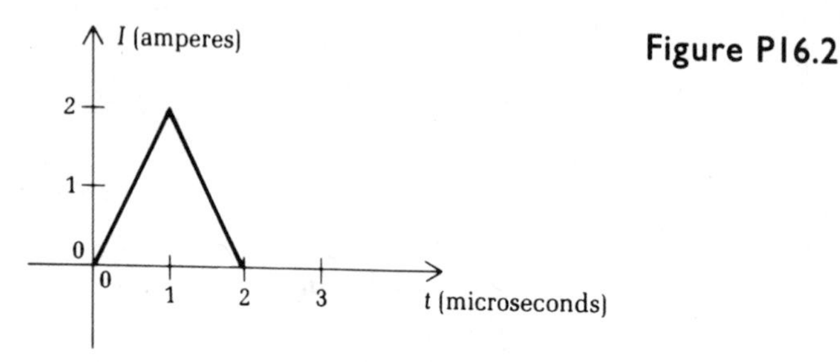

Figure P16.2

16.3 Consider the network shown in Figure P16.3. The magnetic flux is increasing at a rate of 0.5 Wb/s in the direction pointing into the paper. Find the readings of the voltmeters shown.

Figure P16.3

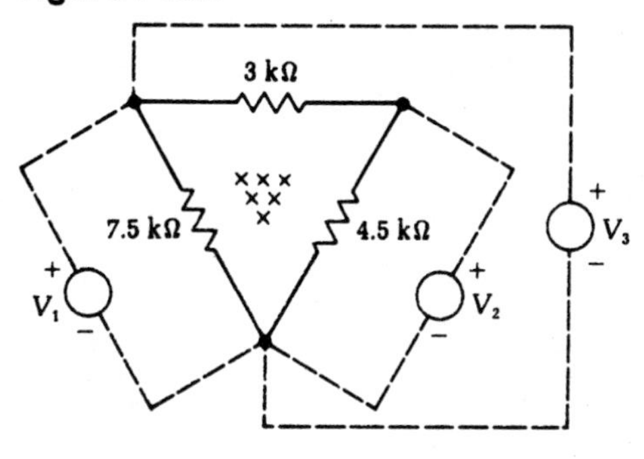

16.4 Find the readings of the voltmeters shown in Figure P16.4. The magnetic flux is increasing at a rate of 0.5 Wb/s in the direction pointing into the paper.

Figure P16.4

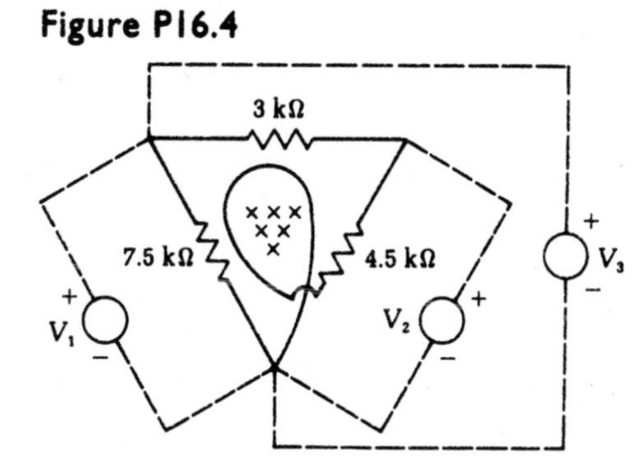

16.5 Four resistors form a circuit as shown in Figure P16.5. The toal magnetic flux linking the circuit is increasing at a rate of 0.5 Wb/s, in the direction pointing out of the paper.

(a) Find the direction and magnitude of the induced current in the circuit.

(b) Find the readings of the voltmeters V_1 and V_2.

Figure P16.5

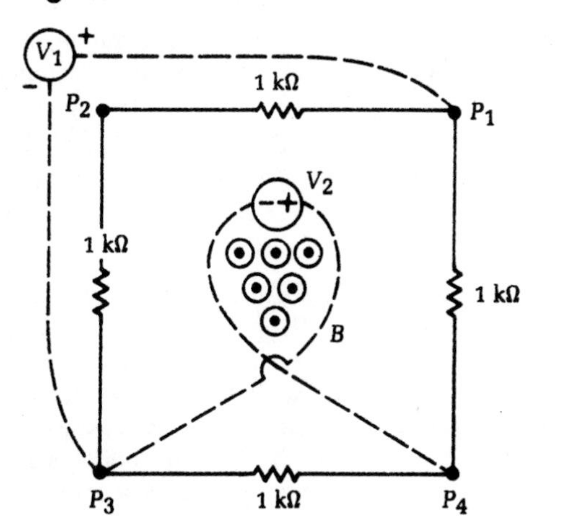

16.6 The circuit shown in Figure P16.6a is placed in a magnetic field with the flux pointing out of the paper. The time variation of the flux Ψ is shown in Figure P16.6b. The magnetic flux is limited to the central region bounded by points *ABCD*. Obtain and plot the voltage reading of the voltmeter as a function of time.

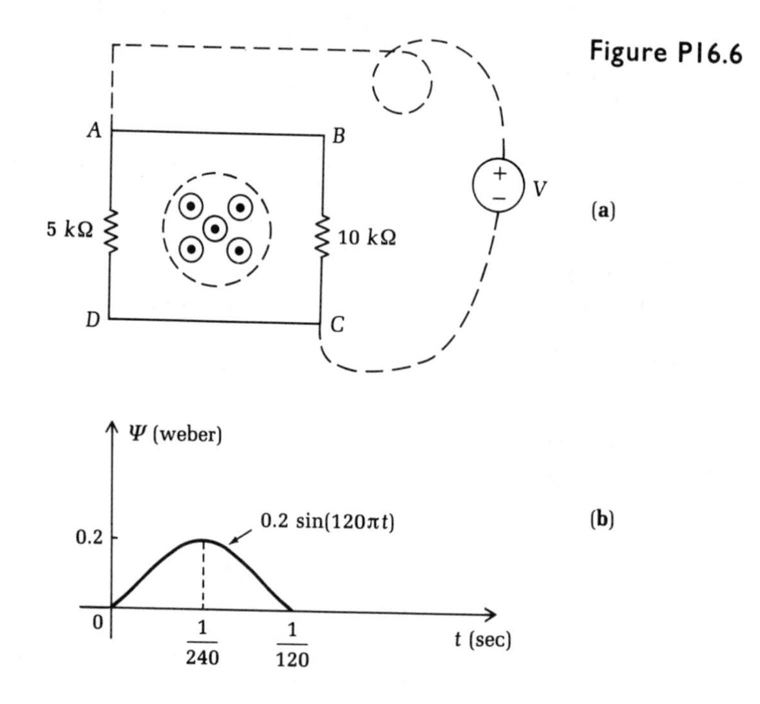

Figure P16.6

16.7 What is the EMF induced on a propeller blade that is 1.5 m long and is rotating at 10,000 r/min in the earth's magnetic field (0.5×10^{-4} Wb/m²)?

16.8 Find the total expansion force acting on the surface of an air-core solenoid that has 100 turns of coil and radius $a = 1$ cm, length $\ell = 10$ cm, and current $I = 10$ A.

16.9 Repeat Problem 16.8 for the case in which 100 turns of coil are wound over a ferromagnetic core with $\mu = 1000\,\mu_0$. The current is 10 mA, with $a = 1$ cm and $\ell = 10$ cm.

16.10 A magnetic core is made of a material whose hysteresis loop is shown in Figure P16.10. Note that this hysteresis curve is not a "square loop." To read the content of the core, two pulses are applied to the wires. The currents generate an H equal to 200 A/m. The core has an area of 3×10^{-7} m².

(a) What is the voltage induced in the sensing wire if the core is originally at the "zero" state (at point C)? Assume that switching from C to A is linear with time and that it is completed in a microsecond.

(b) What is the voltage induced in the sensing wire if the core is originally at the "one" state (at point A)? Assume that switching from A to A' is linear with time and that it is completed in 0.5 μs. This voltage is the "noise" voltage because it would ideally be zero if the hysteresis loop were a perfect square.

Figure P16.10 Ferrite core memory and its hysteresis loop.

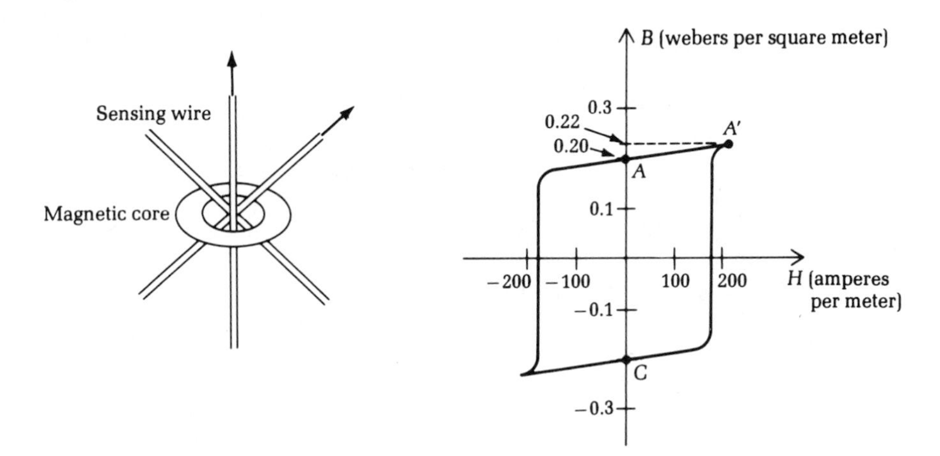

16.11 A transformer similar to the one shown in Figure 16.11 is made of a steel with relative permeability equal to 1100. The effective length of the core is 40 cm, and the flux density is $B = 0.3$ Wb/m^2. $N_1 = 100$, $N_2 = 1000$, $I_1 = 60$ A.

(a) Find I_2, assuming that the transformer is an ideal transformer.
(b) Find I_2, using (16.37).
(c) Compare the two answers.

16.12 Consider a magnetic circuit similar to the one shown in Figure 16.11. The effective length of the core is 0.4 m and its permeability is 2000 μ_0. The cross-sectional area of the core is 4×10^{-4} m^2. Let $I_1 = 10$ A, $I_2 = 24$ A, $N_1 = 50$, and $N_2 = 20$.

(a) Calculate the B field in the core. Give both the direction and the magnitude.
(b) If I_1 is ac with $f = 60$ Hz, what are $|V_1|$ and $|V_2|$? Assume that the magnetic flux always stays in the core without any leakage.

16.13 The primary coil of a transformer has 150 turns and the secondary coil has 450 turns. The effective length of the core is 0.5 m and the flux density in the core is 0.25 Wb/m^2. The transformer is similar to the one shown in Figure 16.11. Assume that $I_1 = 60$ A and there is no flux leakage.

(a) Find I_2, assuming ideal transformer condition.
(b) Find V_2, assuming ideal transformer condition and $V_1 = 110\cos(120\pi t)$.
(c) Find I_2, taking into consideration that the core material has a finite permeability equal to 1000 μ_0.
(d) The hysteresis loop of the core material has an area equal to 90 Wb-A/m^3. What is the power loss due to the hysteresis in the transformer? Assume that the core has a cross-sectional area equal to 4 cm^2.

16.14 Figure P16.14 shows a magnetization curve of a core used in a transformer. Notice that the hysteresis is negligible in this case and that the curve is linear in the range $0 \leq |H| \leq 150$ A/m but saturated when H is increased beyond this range. Let us now review Example 16.10. Because $|V_1| = \omega N_1 \Psi$, we want to use maximum Ψ in order to minimize the number of coils in the transformer. Using Figure P16.14, explain what will happen to the shape of the $\Psi(t)$, and consequently to the shape of $V_1(t)$, if Ψ is too high—for example, if Ψ is so high as to correspond to $B = 1.2\ \text{Wb/m}^2$.

Figure P16.14

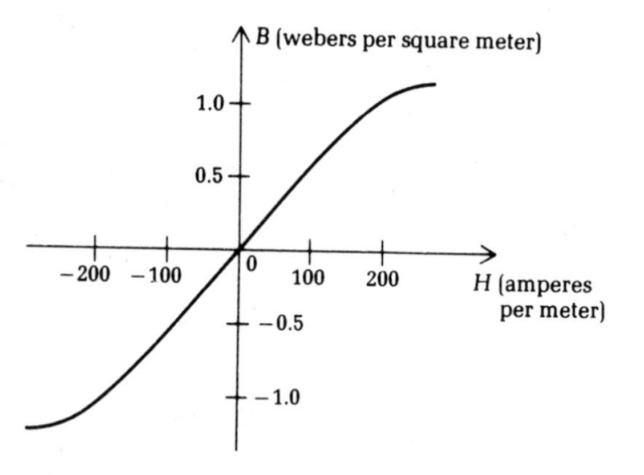

16.15 Estimate the approximate power loss attributed to hysteresis in the ferrite core shown in Figure P16.10 if the core is switched back and forth between "zero" and "one" states 1000 times in a second. Assume that the core has an average radius of 6×10^{-4} m and that its cross-sectional area is $3 \times 10^{-7}\ \text{m}^2$.

16.16 Three hysteresis loops for three different ferromagnetic materials are shown in Figure P16.16.

(a) If you had to use one of them in a power transformer, which one would you select? Why?

(b) If you had to use one of them for magnetic recording, which one would you select for best protection against accidental erasure of data by magnetic fields? Why?

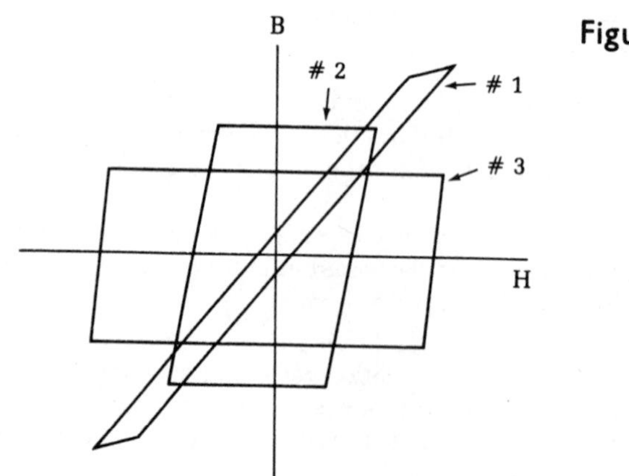

Figure P16.16

16.17 A uniform magnetic field of magnitude 0.3 T is rotating around the y axis. The rotation speed is 40π radians/s. A rectangular loop is placed on the x-y plane, as shown in Figure P16.17. At time t, the direction of the magnetic field makes an angle α with the x axis.

(a) What is the torque on the loop due to the magnetic field? Let $\alpha = 30°$.
(b) What is the mechanical power delivered to the outside load by this loop?
(c) What is the EMF induced in this loop? Is it in the same direction as the current or in the opposite direction?
(d) What is the electric power delivered by the external source to the loop? Use the result obtained in (d) and compare with the result obtained in (b).

Figure P16.17

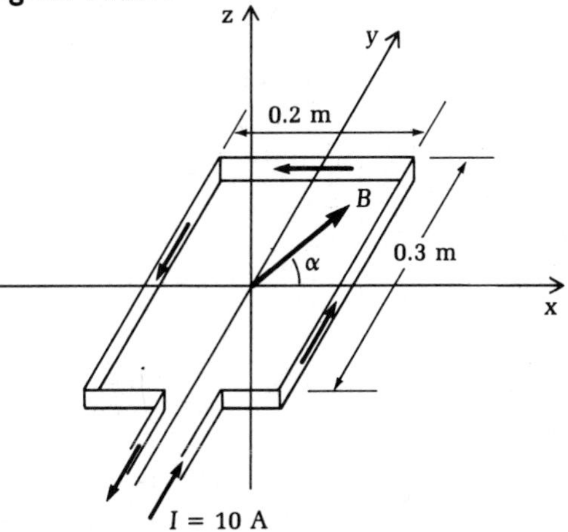

16.18 A coil carrying a dc current I_0 is rotating at an angular velocity ω in a constant magnetic field B, as shown in Figure P16.18. The area of the loop is A.

Figure P16.18

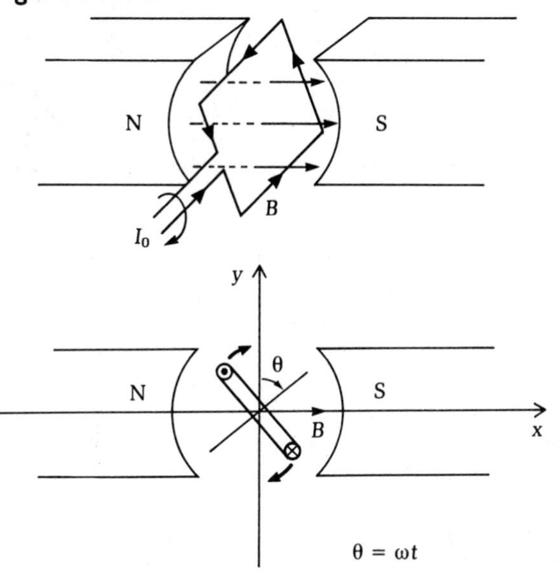

(a) Find the torque on the coil as a function of time.
(b) Find the mechanical power given out by the coil as a function of time.
(c) Find the magnetic flux Ψ going through the coil as a function of time.
(d) Find the EMF induced in the coil. Assume that there is only one turn of the coil.
(e) Find the total electric power delivered to the coil as a function of time. Show that energy is conserved at all times.

16.19 The coils *a*-*a*′, *b*-*b*′, and *c*-*c*′ shown in Figure P16.19 are fed by a three-phase source as shown in Figure 16.18. When I_a is positive, it goes into the paper at point "*a*" and comes out at "*a*′." Other currents are directed in the same way. Sketch the magnetic-flux lines at $\omega t = 0$ and $2\pi/3$ and find the direction and the speed of the rotating magnetic fields.

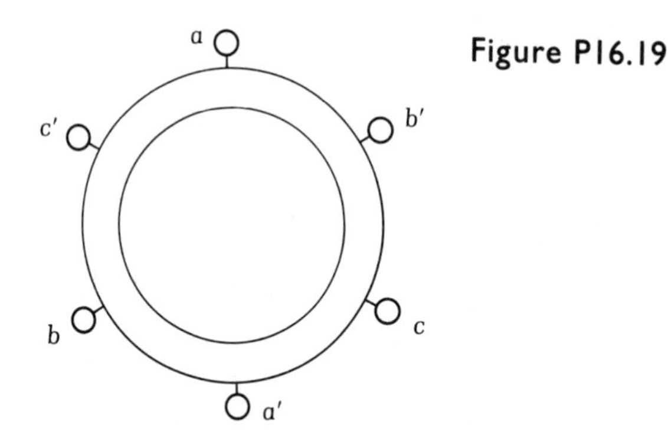

Figure P16.19

16.20 Show that the mechanical torque required to drive an ac generator is not constant with time or, to be exact, that it consists of a constant term and a term that varies sinusoidally with time with an angular frequency 2ω. What is the time average of the torque? Express the torque in terms of the area of the winding A, the current I, the magnetic-flux density B, and the phase angle α between the voltage and the current. Plot T as a function of t for $\alpha = 0$.

16.21 Figure 16.15 depicts an ac generator with a single coil being rotated in a constant magnetic field. It illustrates the operating principle of a single-phase ac generator. Let us now consider a three-phase ac generator. How would you physically arrange three sets of coils in order to generate three-phase electricity? To illustrate your design, sketch a diagram similar to Figure 16.15.

16.22 What is the total mechanical torque needed to drive the three-phase generator that you have designed for Problem 16.21? Express this torque as a function of time in terms of the appropriate parameters. Plot T as a function of time, and compare it with that obtained in Problem 16.20. Is the instantaneous mechanical torque "smoother" (does its time-average value fluctuate less) compared with that for a single-phase generator?

16.23 Design a coil configuration similar to the one shown in Figure 16.17. Design it in such a way that it will produce a rotating magnetic field in the armature-stator air gap and that the field will have an angular speed equal to $\omega/2$ when fed with the three-phase current shown in Figure 16.18. Prove that your design is correct by drawing instantaneous-current diagrams similar to those shown in Figure 16.17.

16.24 Show qualitatively that the torque generated by an induction motor may be varied by changing the resistance of the windings of the rotor. Figure 16.21 shows torque curves versus v_r/v_s with three different values of rotor conductivities. What are the relative magnitudes of σ_1, σ_2, and σ_3?

16.25 Refer to the synchronous motor shown in Figure 16.22. What happens when the torque angle is negative —that is, when the position of the rotor magnetic moment is ahead of the magnetic field?

16.26 Consider the coils of the magneplane's track shown in Figure 16.25. For the magneplane to travel at a speed of 250 km/h, what should be the distance ℓ_t in meters? Assume that the power is provided by a three-phase 60-Hz power line.

16.27 Find the voltage induced in the rectangular loop shown in Figure 16.1 if it is rotating about the axis parallel to the z axis located at $x = d + a/2$. Assume that the angular frequency of the rotation is ω and that the infinitely long wire carries a direct current of I amperes. Show that the induced EMF is not a pure sinusoidal voltage. It is approximately sinusoidal when $d \gg a$.

APPENDICES

Appendix A Frequently used symbols

Symbol	Quantity	Unit	Abbreviated unit
$\boldsymbol{A}$	vector potential	weber/meter	Wb/m
Å	atomic distance unit	1×10^{-10} meter	
$\boldsymbol{B}$	magnetic flux density	tesla or	T
		weber/meter^2	Wb/m^2
b_0	susceptance (normalized)	dimensionless	
C	capacitance	farad	F
	capacitance/length	farad/meter	F/m
$\boldsymbol{D}$	electric flux density	coulomb/meter^2	C/m^2
$D(\theta, \phi)$	directive gain of an antenna	dimensionless	
d_p	penetration depth of electromagnetic wave in dissipative medium	meter	m
$\boldsymbol{E}$	electric-field strength	volt/meter	V/m
e	magnitude of charge carried by an electron	1.602×10^{-19} coulombs	
$\boldsymbol{F}$	force	newton	N
$F(\theta, \phi)$	array factor	dimensionless	
f	frequency	hertz	Hz
f_c	cutoff frequency of a waveguide	hertz	Hz
	cyclotron frequency	hertz	Hz
G	conductance	mho or siemens	mho or S
$\boldsymbol{H}$	magnetic-field strength	ampere/meter	A/m
I	current	ampere	A
$\boldsymbol{J}$	electric-current density	ampere/meter^2	A/m^2
$\boldsymbol{J}_s$	surface electric-current density	ampere/meter	A/m
j	imaginary number defined as $j^2 = -1$	dimensionless	
k	wavenumber	1/meter	1/m
$\boldsymbol{k}$	vector wavenumber	1/meter	1/m
k_R	real part of k	1/meter	1/m
k_I	imaginary part of k	1/meter	1/m
L	inductance	henry	H
	inductance/length	henry/meter	H/m

Symbol	Quantity	Unit	Abbreviated unit
ℓ	distance, length	meter	m
$\boldsymbol{M}$	magnetic-moment density	ampere/meter	A/m
M_{12}, M_{21}	mutual inductance	henry	H
m	mass	kilogram	kg
$\boldsymbol{m}$ or m	magnetic moment	ampere · meter^2	A · m^2
N	number of electrons per volume of plasma	1/meter^3	1/m^3
	number of coil windings	dimensionless	
$\hat{\boldsymbol{n}}$	unit vector perpendicular to a surface or boundary	dimensionless	
$\boldsymbol{P}$	dipole-moment density	coulomb/meter^2	C/m^2
p	dipole moment	coulomb-meter	C-m
P_d	dissipated power/volume	watt/meter^3	W/m^3
Q	total charge on a capacitor	coulomb	C
q	charge	coulomb	C
R	resistance	ohm	Ω
R_I	reflection coefficient (perpendicular polarization)	dimensionless	
R_{II}	reflection coefficient (parallel polarization)	dimensionless	
r	spherical coordinate	meter	m
$\boldsymbol{r}$	position vector from the origin to a point	meter	m
$\boldsymbol{r}_{12}$	position vector from point 1 to point 2	meter	m
$\boldsymbol{S}$	Poynting vector	watt/meter^2	W/m^2
$\langle \boldsymbol{S} \rangle$	time-average Poynting vector	watt/meter^2	W/m^2
T	period of a time-harmonic signal	second	s
	torque	newton-meter	N-m
t	time	second	s
T_I	transmission coefficient (perpendicular polarization)	dimensionless	
T_{II}	transmission coefficient (parallel polarization)	dimensionless	
U_E	stored electric-energy density	joule/meter^3	J/m^3
U_H	stored magnetic-energy density	joule/meter^3	J/m^3
V	voltage	volt	V
	volume	meter^3	m^3
$\boldsymbol{v}$	velocity	meter/second	m/s
V_{AB}	voltage difference ($V_A - V_B$)	volt	V
W	work	joule	J
x	rectangular coordinate	meter	m
$\hat{\boldsymbol{x}}$	unit vector in x direction	dimensionless	
y	rectangular coordinate	meter	m
$\hat{\boldsymbol{y}}$	unit vector in y direction	dimensionless	
$Y_n(z)$	normalized admittance function	dimensionless	
z	rectangular coordinate	meter	m
$\hat{\boldsymbol{z}}$	unit vector in z direction	dimensionless	
Z_0	characteristic impedance of a transmission line	ohm	Ω
$Z(z)$	impedance function of a transmission line	ohm	Ω
Z_L	load impedance	ohm	Ω
$Z_n(z)$	normalized impedance function	dimensionless	
Γ_L	reflection coefficient at the load	dimensionless	
$\Gamma(z)$	generalized reflection coefficient	dimensionless	

Symbol	Quantity	Unit	Abbreviated unit
δ	skin depth	meter	m
$\tan\delta$	loss tangent	dimensionless	
ϵ	permittivity	farad/meter	F/m
ϵ_0	free-space permittivity	8.854×10^{-12}	F/m
ϵ_x	permittivity of an anisotropic medium for x-directed E field	farad/m	F/m
$\bar{\bar{\epsilon}}$	permittivity matrix of an anisotropic medium	farad/m	F/m
ϵ'	real part of ϵ		
ϵ''	imaginary part of ϵ		
η	intrinsic impedance of a medium	ohm	Ω
θ	spherical coordinate	radian	rad
θ_b	Brewster angle	radian	rad
θ_c	critical angle	radian	rad
κ	coupling coefficient	dimensionless	
λ	wavelength	meter	m
λ_c	cutoff wavelength	meter	m
λ_g	guide wavelength	meter	m
Λ	total magnetic flux linkage through N loops	weber	Wb
	periodicity of winding in rotating machines	meter	m
μ	permeability	henry/meter	H/m
μ_0	permeability of free space	$4\pi \times 10^{-7}$	H/m
μ_e	mobility of electron	meter²/second	m²/s
π	constant	3.14159265...	
ρ	cylindrical coordinate	meter	m
ρ_s	surface electric-charge density	coulomb/meter²	C/m²
ρ_v	volume electric-charge density	coulomb/meter³	C/m³
ρ_ℓ	line-charge density	coulomb/meter	C/m
ρ_r	resistivity	ohm-meter	Ω-m
σ	conductivity	siemens/meter or mho/meter	S/m or mho/m
σ_{total}	total scattering cross section	meter²	m²
$\sigma(\theta)$	scattering cross section	meter²	m²
$\sigma(\pi/2)$	back-scattering cross section	meter²	m²
Φ	scalar potential	volt	V
ϕ	cylindrical or spherical coordinate	radian	rad
	phase angle of a complex number	radian	rad
χ_e	electric susceptibility	dimensionless	
χ_m	magnetic susceptibility	dimensionless	
ψ	incremental phase angle between adjacent elements in a uniform linear array	radian	rad
Ψ	magnetic flux linkage through one loop	weber	Wb
ω	angular frequency	radian/second	rad/s
ω_c	cutoff angular frequency of a waveguide	radian/second	rad/s
	angular cyclotron frequency	radian/second	rad/s
ω_p	plasma frequency	radian/second	rad/s
Ω	ohm		

Appendix B Mathematical symbols

∇	gradient operator	$\langle\ \rangle$	time average
$\nabla\cdot$	divergence operator	$\leftrightarrow$	equivalent to
$\nabla\times$	curl operator	$\equiv$	identically equal to
∇^2	Laplacian operator	$\approx$	approximately equal to
$\perp$	perpendicular to	$n!$	$n! = 1 \times 2 \times 3 \cdots \times n$
$\parallel$	parallel to	$\text{Re}\{\ \}$	real part of the quantity in $\{\ \}$
*	complex conjugate		

Appendix C Prefixes

Prefix	Abbreviation	Meaning
Exa	E	10^{18}
Peta	P	10^{15}
Tera	T	10^{12}
Giga	G	10^{9}
Mega	M	10^{6}
Kilo	k	10^{3}
Hecto	h	10^{2}
Deka	da	10
Deci	d	10^{-1}
Centi	c	10^{-2}
Milli	m	10^{-3}
Micro	μ	10^{-6}
Nano	n	10^{-9}
Pico	p	10^{-12}
Femto	f	10^{-15}
Atto	a	10^{-18}

NOTE: A prefix is written directly before a unit without a dot. For example, mm means millimeter, but m·m means meter·meter or m^2.

Appendix D Physical constants

ϵ_0	permittivity of free space	8.8542×10^{-12} F/m
		$\approx \frac{1}{36\pi} \times 10^{-9}$ F/m
μ_0	permeability of free space	$4\pi \times 10^{-7}$ H/m
c	speed of light in vacuum	2.9979×10^{8} m/s
		$\approx 3 \times 10^{8}$ m/s
$-e$	charge of an electron	-1.602×10^{-19} C
m_c	mass of an electron at rest	9.11×10^{-31} kg
π		≈ 3.14159265

ANSWERS TO ODD-NUMBERED PROBLEMS

Chapter 1 **1.1(a)** $5 + j3$ **(b)** $11 + j$ **(c)** $-26 + j2$ **(d)** $-2.2 - j1.4$ **1.3** $\cos \omega t; \sin \omega t; 1$ **1.5** $\pm 2^{1/4} e^{j(\pi/8)}$ **1.7** Proof **1.9(a)** $\sqrt{2} \cos\left(\omega t + \frac{\pi}{4}\right)$ **(b)** $4 \cos(\omega t + 0.8)$ **(c)** $3 \cos\left(\omega t + \frac{\pi}{2}\right) + 4 \cos(\omega t + 0.8)$ **1.11(a)** $-6\hat{x} + 5\hat{y} + 2\hat{z}$ **(b)** $-10\hat{x} + 13\hat{y} - 4\hat{z}$ **(c)** -55 **(d)** $23\hat{x} + 22\hat{y} + 14\hat{z}$ **1.13** Proof **1.15** $\frac{1}{\sqrt{93}}(5\hat{x} - 8\hat{y} + 2\hat{z})$ **1.17** Proof omitted **1.19** Proof omitted **1.21** Proof omitted **1.23** $j\omega[(3 - j4)\hat{x} + 8(1 + j)\hat{z}]$ **1.25(a)** $(-1 + j3)\hat{y} + (1 + j3)\hat{z}$ **(b)** $2\hat{x} + (1 - j)\hat{y} + (1 + j)\hat{z}$ **(c)** -5 **(d)** $4\hat{x} - (1 + j3)\hat{y} + (-1 + j3)\hat{z}$ **1.27** parallel **1.29** Sketch omitted

Chapter 2 **2.1** $-6y\hat{x} - 3x^2\hat{y}\, 6z$ **2.3** Proof **2.5** Proof **2.7(a)** No. Example: $\bar{a} = \hat{x} \sin z + \hat{y} \sin y + \hat{z} \sin y$ at $(0, 0, 0)$ **(b)** No. Example: $\bar{E} = \hat{x} \sin z + \hat{y} \sin x$ on y axis **(c)** No. Example: $\bar{E} = \hat{x} \sin z$ on $z = 0$ plane. $\nabla \times \bar{E} \sim \partial\bar{B}/\partial t \neq 0$. **2.9** Not necessary. Example: $\bar{A} = \hat{x}$, $\bar{B} = \hat{y}$ **2.11** -1 **2.13** $2\ \text{C/m}^3$ **2.15** $k = \omega\sqrt{\mu\epsilon}$ **2.17** Proof omitted **2.19** $\nabla \times \mathbf{E} = j\omega\mathbf{B}$, $\nabla \times \mathbf{H} = \mathbf{J} - i\omega\mathbf{D}$, $\nabla \cdot \mathbf{B} = 0$ and $\nabla \cdot \mathbf{D} = \rho_v$ **2.21** Proof omitted **2.23** $\hat{z}, \hat{z}$ **2.25** Proof omitted **2.27** $U_H/U_E \approx 1.13 \times 10^7$

Chapter 3 **3.1** $3.6 \times 10^{-16}\ \text{W/m}^2$ **3.3** 3.84×10^8 m **3.5** $f = 4.74 \times 10^{14}$ Hz; $T = 2.11 \times 10^{-15}$ s; $k = 9.93 \times 10^6$ (1/m) **3.7** $H_0 = -\sqrt{\epsilon_0/\mu_0}\, E_0$, $k = \omega\sqrt{\mu_0 \epsilon_0}$ **3.9** No. **3.11(a)** R.H.C.P. (right-hand circular polarization) **(b)** R.H.C.P. **(c)** L.H.E.P. **(d)** L.P. **3.13** Proof: $a' = (a - jb)/2$, $b' = (a + jb)/2$ **3.15(a)** 1 **(b)** 1 **(c)** 1.58 **(d)** 2.12 **3.17** $\epsilon_r = 5.54$ **3.19** $k_I = 0.0094$(1/m); $d_p = 106.1$ m **3.21** $|E(0.1)/E(0)| = 0.618$; $\phi(0.1) - \phi(0) = -46.9°$ **3.23(a)** $E_x = e^{-0.5z} e^{-j0.5z}$ **(b)** $\mathbf{H} = \hat{y}(0.5 - j0.5) e^{-0.5z} e^{-j0.5z}$ **(c)** Sketch omitted **(d)** Sketch omitted **3.25** $v = 5.8 \times 10^{-3}$ m/s

Chapter 4 **4.1** 61.9° **4.3(i)** c **(ii)** f **(iii)** b **(iv)** a **(v)** d **(vi)** e **4.5** yes, cover the top surface with a circular patch of 1.68 cm radius **4.7** bevelled angle = 35.3°; mirror making 70.6° with $\hat{z}$ axis; $\hat{x}$ polarized (see Figure 4.11) **4.9(a)** $9\epsilon_0$ **(b)** 75 MHz

4.11 $\cos^{-1}\sqrt{\dfrac{\mu_1(\mu_2\epsilon_2-\mu_1\epsilon_1)}{\epsilon_1(\mu_2^2-\mu_1^2)}}$, $\cos^{-1}\sqrt{\dfrac{\epsilon_1(\mu_2\epsilon_2-\mu_1\epsilon_1)}{\mu_1(\epsilon_2^2-\epsilon_1^2)}}$ **4.13(a)** 80 cm in front of the plate **(b)** 2 V/m **4.15** $1.9994|E_0|$ **4.17** $\mathbf{H}^i=\hat{y}H_0e^{-jkx\sin\theta-jkz\cos\theta}$, $\mathbf{E}^i=(\hat{x}\cos\theta-\hat{z}\sin\theta)H_0\eta e^{-jkx\sin\theta-jkz\cos\theta}$, $\mathbf{H}^r=\hat{y}H_0e^{-jkx\sin\theta+jkz\cos\theta}$, $\mathbf{E}^r=-(\hat{x}\cos\theta+\hat{z}\sin\theta)H_0\eta e^{-jkx\sin\theta+jkz\cos\theta}$, where $\eta=\sqrt{\dfrac{\mu}{\varepsilon}}$ **4.19** $x=0.87$ m, $y=1.5$ m, 12.04 dB **4.21** $p_r=64.9\%$

Chapter 5

5.1 Proof **5.3** 1.875 kHz **5.5** $\bar{\mathbf{H}}=\hat{y}H_0e^{jkz}$, $k=\omega\sqrt{\mu\epsilon}$, $\bar{\mathbf{E}}=-\hat{x}H_0\eta e^{jkz}$, $\bar{\mathbf{J}}_s=\hat{z}H_0e^{jkz}$ on lower plate; $\bar{\mathbf{J}}_s=-\hat{z}H_0e^{jkz}$ on upper plate **5.7** 89.33 kW **5.9** Proof **5.11** 5.26–10.52 GHz for 2.85 × 1.262 (cm) waveguide, 21.1–42.2 GHz for 0.711 × 0.355 (cm) waveguide **5.13** 1.318 MW **5.15(a)** true **(b)** false **(c)** false **(d)** false **5.17** $\mathrm{E}_y=E_1\sin(\pi x/a)e^{jk_zz}$, $\mathrm{H}_x=(E_1k_z/\omega\mu)\sin(\pi x/a)e^{jk_zz}$, $\mathrm{H}_z=(jE_1\pi/\omega\mu a)\cos(\pi x/a)e^{jk_zz}$ where $k_z=[\omega^2\mu\epsilon-(\pi/a)^2]^{1/2}$ **5.19** $\theta=\tan^{-1}(na/mb)$ **5.21** 5.83 GHz **5.23** Proof omitted **5.25** (0.866, 0.5, 2) **5.27** Proof omitted **5.29** $A=4.93\hat{x}+7.46\hat{y}-3\hat{z}$ **5.31(a)** $\mathbf{E}^t=\hat{\rho}\dfrac{V_1}{\rho}e^{-jk_1z}$, $\mathbf{H}^t=\hat{\phi}\dfrac{V_1}{\eta_1\rho}e^{-jk_1z}$ **(b)** $V_0'=V_0(\eta_1-\eta_0)/(\eta_1+\eta_0)$, $V_1=2V_0\eta_1/(\eta_1+\eta_0)$

Chapter 6

6.1 2000 V **6.3** $(a_1/a_2)=(b_1/b_2)=\sqrt{1000}$ **6.5** $\mathbf{J}_s=\hat{z}\dfrac{V_0}{\eta a}e^{-jkz}$, $\mathbf{I}=\dfrac{2\pi V_0}{\eta}e^{-jkz}$

6.7(a) $|\mathrm{V}(z)|=2|V_+||\cos kz|$, $|\mathrm{I}(z)|=2\dfrac{|V_+|}{Z_0}|\sin kz|$ **(b)** Sketch omitted **(c)** ∞

6.9(a) 2.96 **(b)** $z=-0.35\lambda$ **(c)** 24.5% **6.11** 0.342λ **6.13** $d=0.172$ cm **6.15(a)** $1.25+j1.66$ **(b)** 0.54 **(c)** 1 **6.17(a)** $0.67+j1.33$ **(b)** $0.15-j0.09$ **6.19** $0.47-j0.3$ **6.21** Proof omitted **6.23** No, it cannot absorb any power **6.25** 48.6% **6.27** Sketch omitted **6.29** Sketch omitted **6.31** Sketch omitted **6.33** Proof omitted

Chapter 7

7.1 (0.75, 0.433, 0.5) **7.3** Proof omitted **7.5** $\hat{x}\cdot\hat{\theta}=\cos\theta\cos\phi$, $\hat{x}\cdot\hat{\phi}=-\sin\phi$, $\hat{y}\cdot\hat{r}=\sin\theta\sin\phi$, $\hat{y}\cdot\hat{\theta}=\cos\theta\sin\phi$, $\hat{y}\cdot\hat{\phi}=\cos\phi$, $\hat{z}\cdot\hat{r}=\cos\theta$, $\hat{z}\cdot\hat{\theta}=-\sin\theta$, $\hat{z}\cdot\hat{\phi}=0$ **7.7** Proof omitted **7.9(a)** $\mathrm{H}_\phi=\dfrac{10^3e^{-jkr}}{\eta r}\sin\theta$ (A/m) **(b)** $|\mathrm{E}_\theta|=10$ **(c)** $|\mathrm{E}_\theta|=10|\sin\theta|$ **(d)** $\langle S_r\rangle=0.13\sin^2\theta$ **7.11** Yes, improved to 18% **7.13** $\mathbf{E}=(-\hat{y})\dfrac{jkI\Delta z\eta e^{-jkx}}{8\pi x}$, linear

7.15(a) 0.314 V/m **(b)** 0.628 V/m **(c)** 6 V/m **7.17(a)** 1 **(b)** 1.5 **(c)** 1.64 **7.19** Six lobes; beam width = 19.2° along $\phi=0$; beam width = 26.4° along $\phi=41.8°$ **7.21** $D=4$ **7.23** Sketch omitted **7.25(a)** −90° **(b)** 6; 1.414; 0; 1.414 V/m **(c)** Sketch omitted **(d)** Sketch omitted

Chapter 8

8.1 $U_H/U_E\sim(kr)^2\ll 1$ **8.3** When background is dark, one sees light scattered by the smoke particles. Blue light is scattered more strongly than red light. Against a bright background, one sees light passing through the smoke. The blue light gets scattered, and red and yellow lights suffer less scattering. **8.5** Proof omitted **8.7** 360 km in radius **8.9** Actual speed is 126.5 km/h **8.11** Bandwidth = 59 kHz, 6.7 μs for 1 km resolution **8.13** Circular but opposite hand **8.15** No

Chapter 9

9.1 Exact: **(a)** 5.5302×10^{-10} V **(b)** 5.54244×10^{-12} V; Approximate: **(a)** 5.5426×10^{-10} V **(b)** 5.54256×10^{-12} V **9.3** $\Phi(0, 0, 0) = 0$

9.5 $(-\hat{r})\dfrac{1.44 \times 10^{-5}}{r^2}$ V/m **9.7** $E_y(0, 0, 0) = -2$ **9.9** Sketch omitted

9.11(a) $\hat{\rho}\dfrac{\rho_\ell}{40\pi\epsilon h}(0.0499)$ V/m **(b)** $\hat{\rho}\dfrac{\rho_\ell}{40\pi\epsilon h}(0.05)$ V/m **(c)** 0.2% **9.13(a)** $\bar{E} = 0$

(b) $\bar{E} = \hat{x}\rho_s/\epsilon$ **(c)** $\bar{\mathrm{E}} = 0$ **9.15** $\dfrac{10^{-6}}{\epsilon r^2}[2 - e^{-r}(r^2 + 2r + 2)]$

9.17 $\dfrac{10^{-6}}{\epsilon}(3e^{-1} - 1)$ Hint: $\displaystyle\int dre^{-r}(1 + r)/r^2 = -e^{-r}/r$

9.19 3 V, independent of path **9.21(a)** $\dfrac{q}{4\pi\epsilon c}$ **(b)** $\dfrac{q}{4\pi\epsilon c}$ **(c)** $\dfrac{q}{4\pi\varepsilon}\left(\dfrac{1}{r} + \dfrac{1}{c} - \dfrac{1}{b}\right)$

(d) $\dfrac{q}{4\pi\epsilon}\left(\dfrac{1}{a} + \dfrac{1}{c} - \dfrac{1}{b}\right)$ **9.23(a)** 4.03×10^{-6} C **(b)** 330 kV **(c)** 0.4 mA

Chapter 10

10.1 $-\frac{4}{9}q$ **10.3** 1.8×10^{-5} N (attractive) **10.5** It would land on the positive plate at $x = 4.77$ cm **10.7** $a < 0.36$ mm or $a > 26.3$ mm **10.9** $z = \pm 3.14$ cm
10.11(a) $v_0 = 1.874 \times 10^7$ m/s, $v_{0z} = 1.867 \times 10^7$ m/s, $v_{0x} = -0.163 \times 10^7$ m/s
(b) $x(t) = (8.78 \times 10^{14})t^2 - (0.163 \times 10^7)t$ m, $z(t) = (1.867 \times 10^7)t$ m
(c) $x = -3.52 \times 10^{-4}$ m, $z = 3 \times 10^{-2}$ m **10.13** Proof omitted

10.15 5.3×10^{-10} farad **10.17(a)** $\dfrac{A}{d}\left(\dfrac{\epsilon_1 + \epsilon_2}{2}\right)$ **(b)** $Q_1 = \dfrac{\epsilon_1 Q}{\epsilon_1 + \epsilon_2}$, $Q_2 = \dfrac{\epsilon_2 Q}{\epsilon_1 + \epsilon_2}$

10.19 $2\pi\Big/\left[\dfrac{1}{\epsilon_1}\ln(c/a) + \dfrac{1}{\epsilon_2}\ln(b/c)\right]$ **10.21** 4.97×10^{-6} J
10.23(a) $Q^2S/2A\epsilon_0$ **(b)** $-Q^2/2A\epsilon_0$ (attractive)

Chapter 11

11.1 $\Phi_3 = \Phi_1 + \Phi_2$ **11.3** $\Phi = [V_a \ln(b/\rho) + V_b \ln(\rho/a)]/\ln(b/a)$

11.5 $\Phi(\theta) = V_0 \ln\left(\tan\dfrac{\theta}{2}\right)\Big/\ln\left(\tan\dfrac{\theta_1}{2}\right)$, $\rho_s = -V_0\epsilon_0\Big/\left[r\ln\left(\tan\dfrac{\theta_1}{2}\right)\sin\theta_1\right]$

11.7 12.3 V **11.9** Sketch omitted **11.11** 17.5$\mu\mu$F/m **11.13** $a \approx 2.9$ cm

11.15 $\dfrac{q}{4\pi\epsilon}\left(\dfrac{1}{R_1} - \dfrac{a}{dR_2}\right) + \dfrac{q(a/d) + q_0}{4\pi\epsilon r}$, where $R_1 = (r^2 + d^2 - 2rd\cos\theta)^{1/2}$,

$R_2 = \left(r^2 + \dfrac{a^4}{d^2} - 2\dfrac{ra^2}{d}\cos\theta\right)^{1/2}$ **11.17** $(\rho_\ell)^2/[2\pi\epsilon(d - b)]$ where $d = a^2/b$, attractive **11.19** $(d/a)q^2/[4\pi\epsilon(d - b)^2]$, where $d = a^2/b$, attractive

11.21 $200\sin(2\pi x/a)\sinh(2\pi y/a)/\sinh(2\pi b/a)$

11.23 $(400/\pi)\displaystyle\sum_{n=\text{odd}}^{\infty}\frac{1}{n}\left\{\frac{\sin(n\pi x/a)\sinh(n\pi y/a)}{\sinh(n\pi b/a)} + \frac{\sin(n\pi y/b)\sinh(n\pi x/b)}{\sinh(n\pi a/b)}\right\}$

11.25

(a) $\dfrac{V_0}{\left(\dfrac{1}{a} - \dfrac{1}{b}\right)}\left(\dfrac{1}{r} - \dfrac{1}{b}\right)$ **(b)** $\dfrac{V_0}{\left(\dfrac{1}{a} - \dfrac{1}{b}\right)r^2}$ **(c)** $D_{r1} = \dfrac{V_0\epsilon_1}{\left(\dfrac{1}{a} - \dfrac{1}{b}\right)r^2}$ **(d)** $Q = \dfrac{2\pi V_0}{\left(\dfrac{1}{a} - \dfrac{1}{b}\right)}(\epsilon_1 + \epsilon_2)$

$D_{r2} = \dfrac{V_0\epsilon_2}{\left(\dfrac{1}{a} - \dfrac{1}{b}\right)r^2}$ $C = \dfrac{2\pi(\epsilon_1 + \epsilon_2)}{\left(\dfrac{1}{a} - \dfrac{1}{b}\right)}$

Chapter 12

12.1 $R = 2d/[A(\sigma_1 + \sigma_2)]$ **12.3(a)** $I \ln(a/\rho)/(2\pi\sigma_1\ell)$ **(b)** $I \ln(a/b)/(2\pi\sigma_1\ell) + I \ln(b/\rho)/(2\pi\sigma_2\ell)$
(c) $\frac{I}{2\pi\ell}\left[\frac{\ln(b/a)}{\sigma_1} + \frac{\ln(c/b)}{\sigma_2}\right]$ **12.5** 0.92×10^{-3} mho/m **12.7** 106 V
12.9 $\left(\frac{\sigma_2}{\sigma_1 + \sigma_2}\right) \times 100\%$ **12.11** 12.3 Ω-m **12.13** sketch omitted
12.15 $\rho_a = 600\left\{\frac{1}{6} - \frac{50}{d[(y + 6 - r_0 y/d)^2 + (70 - 70r_0/d)^2]^{1/2}} + \frac{50}{d[(y+6)^2 + (70)^2]^{1/2}}\right\}$, where $d = [(70)^2 + y^2]^{1/2}$, $r_0 = (50)^2/[(70)^2 + y^2]^{1/2}$

Chapter 13

13.1 $\hat{z}2\sqrt{2}I/(\pi b)$ **13.3** $\boldsymbol{H} = (-\hat{x})Jy$ for $|y| < \frac{d}{2}$, $\boldsymbol{H} = (-\hat{x})J(d/2)$ for $y > d/2$, $\boldsymbol{H} = \hat{x}J(d/2)$ for $y < (-d/2)$ **13.5** $H_\phi = \rho$ for $0 < \rho < a$; $(2\rho^2 - a^2)/\rho$ for $a < \rho < b$; $(2b^2 - a^2)/\rho$ for $\rho > b$. **13.7(a)** $(-\hat{z})Ia(dy)/[4\pi(a^2 + y^2)^{3/2}]$ **(b)** $(-\hat{z})I/(4\pi a)$ **(c)** 0 **(d)** $(-\hat{z})I/(4\pi a)$
13.9 $z_0 = \pm 0.27a$ **13.11** H_z at center is $9I/4\pi a$ **13.13** 2.8 MHz
13.15(a) $x = 0.04$ m, $z = -0.00725$ m **(b)** $-20.5°$ **13.17** 1.33×10^{-3} N/m (repulsive)
13.19(a) Loop should be placed horizontally or vertically in the east-west direction **(b)** 7.85×10^{-3} N-m **(c)** Vertically in north-south direction
13.21(a) $H_\rho = \frac{I\rho}{2\pi a^2}$ **(b)** $U_H = \frac{\mu I^2}{16\pi}$ **(c)** 5×10^{-8} H/m **13.23** 2.74×10^{-6} Joule/meter
13.25 $\frac{\mu_0}{2\pi}\left\{\frac{1}{4} + \ln(b/a) + \frac{c^4 \ln(c/b)}{(c^2 - b^2)^2} - \frac{c^2}{c^2 - b^2} + \frac{c^2 + b^2}{4(c^2 - b^2)}\right\}$

Chapter 14

14.1 Sketch **14.3** 0.21 weber/m², 100 A/m **14.5** 0.71 weber/m² **14.7** $x_3 y_1 = 1$, $x_3 y_2 = 0$, $x_3 y_3 = 1$, $x_3 y_4 = 0$, $x_1 y_2 = 1$, $x_2 y_2 = 1$, $x_4 y_2 = 1$ **14.9(a)** 1.10 mA **(b)** 1.16 mA **14.11** 8.32 A **14.13** 0.29 weber/m²

Chapter 15

15.1 $\mathbf{E}^{(0)} = 0$, $\mathbf{H}^{(0)} = -\hat{y}I_0 \cos(\omega t)/w$; $\mathbf{E}^{(1)} = -\hat{x}I_0 \omega\mu z \sin(\omega t)/w$, $\mathbf{H}^{(1)} = 0$; $\mathbf{E}^{(2)} = 0$, $\mathbf{H}^{(2)} = \hat{y}I_0 \omega^2 \epsilon\mu(\frac{1}{2})z^2 \cos(\omega t)/w$; $\mathbf{E}^{(3)} = \hat{x}I_0 \omega^3 \mu^2 \epsilon(\frac{1}{2})(\frac{1}{3})z^3 \sin(\omega t)/w$; $\mathbf{H}^{(3)} = 0$ **15.3** $U_H^{(1)} = \frac{\mu}{2}\omega^2\epsilon^2 V_0^2\left(\frac{1}{a^2}\right)\sin^2(\omega t)\left(\frac{1}{3}\right)aw\ell^3$, $U_E^{(1)} = 0$ **15.5** $U_E^{(0)} = 0$, $U_H^{(0)} = \frac{\mu}{2}\frac{I_0^2}{w^2}\cos^2(\omega t)(wa\ell)$
15.7 $E_\rho^{(0)} = V_0 \cos(\omega t)/[\rho \ln(b/a)]$, $I^{(0)} = 2\pi V_0 \sigma\ell \cos(\omega t)/\ln(b/a)$, $Q^{(0)} = 2\pi V_0 \epsilon\ell \cos(\omega t)/\ln(b/a)$, $I^{(1)} = -2\pi V_0 \epsilon\ell\omega \sin(\omega t)/\ln(b/a)$ **15.9** 11.1 s
15.11 $G_1 = \sigma_1 A/d_1$; $G_2 = \sigma_2 A/d_2$; $C_1 = \epsilon_1 A/d_1$; $C_2 = \epsilon_2 A/d_2$ G_1 and C_1 in parallel, then in series with G_2 and C_2 in parallel $Z = 1/(G_1 + j\omega C_1) + 1/(G_2 + j\omega C_2)$

Chapter 16

16.1 101 A **16.3** $V_1 = 0.25$ V, $V_2 = -0.15$ V, $V_3 = -0.25$ V **16.5(a)** 0.125 mA (clockwise) **(b)** $V_1 = -0.25$ V; $V_2 = -0.625$ V **16.7** 58.9 mV **16.9** 0.395×10^{-3} N
16.11(a) 6 A **(b)** 5.91 A **16.13(a)** 20 A **(b)** $330\cos(120\pi t)$ **(c)** 19.78 A **(d)** 1.08 W
16.15 0.16 mW **16.17(a)** $\bar{\mathrm{T}} = 0.156\hat{y}$ **(b)** $P_m = 19.59$ W **(c)** EMF $= -1.959$ V **(d)** $P_e = 19.59$ W **16.19** Speed $= \omega$ rad/s in counter-clockwise direction **16.21** Sketch omitted **16.23** Sketch omitted **16.25** It becomes a generator

16.27 $$\mathrm{V} = \frac{UIb}{2\pi}\frac{a\omega\left(d + \frac{a}{2}\right)}{\left(d + \frac{a}{2}\right)^2 - \left(\frac{a}{2}\cos\omega t\right)^2}\sin\omega t$$

BIBLIOGRAPHY

Crawford, Jr., F. S. *Waves.* New York: McGraw-Hill, 1965.

Feynman, R. P., Leighton, R. O., and Sands, M. *Lectures on Physics*, vol. 2. Reading, Mass.: Addison-Wesley, 1964.

King, R. W. P. *Fundamental Electromagnetic Theory.* New York: Dover, 1963.

Kong, J. A. *Theory of Electromagnetic Waves.* New York: Wiley, 1975.

Lorrain, P., and Corson, D. R. *Electromagnetism.* San Francisco: W. H. Freeman, 1978.

Paris, D. T., and Hurd, F. K. *Basic Electromagnetic Theory.* New York: McGraw-Hill, 1969.

Popovic, B. D. *Introductory Engineering Electromagnetics.* Reading, Mass.: Addison-Wesley, 1971.

Purcell, E. M. *Electricity and Magnetism.* New York: McGraw-Hill, 1963.

Ramo, S., Whinnery, J. R., and Van Duzer, T. *Fields and Waves in Communication Electronics.* New York: Wiley, 1965.

Seshadri, S. R. *Fundamentals of Transmission Lines and Electromagnetic Fields.* Reading, Mass.: Addison-Wesley, 1971.

Towne, D. H. *Wave Phenomena.* Reading, Mass.: Addison-Wesley, 1967.

INDEX

VECTOR IDENTITIES

$$\boldsymbol{A} \cdot (\boldsymbol{B} \times \boldsymbol{C}) = \boldsymbol{B} \cdot (\boldsymbol{C} \times \boldsymbol{A}) = \boldsymbol{C} \cdot (\boldsymbol{A} \times \boldsymbol{B})$$

$$\boldsymbol{A} \times (\boldsymbol{B} \times \boldsymbol{C}) = \boldsymbol{B}(\boldsymbol{A} \cdot \boldsymbol{C}) - \boldsymbol{C}(\boldsymbol{A} \cdot \boldsymbol{B})$$

$$\nabla(\Phi_1\Phi_2) = \Phi_1\nabla\Phi_2 + \Phi_2\nabla\Phi_1$$

$$\nabla \cdot (\Phi\boldsymbol{A}) = \boldsymbol{A} \cdot \nabla\Phi + \Phi\nabla \cdot \boldsymbol{A}$$

$$\nabla \cdot (\boldsymbol{A} \times \boldsymbol{B}) = \boldsymbol{B} \cdot (\nabla \times \boldsymbol{A}) - \boldsymbol{A} \cdot (\nabla \times \boldsymbol{B})$$

$$\nabla \times (\Phi\boldsymbol{A}) = \nabla\Phi \times \boldsymbol{A} + \Phi\nabla \times \boldsymbol{A}$$

$$\nabla \cdot (\nabla \times \boldsymbol{A}) = 0$$

$$\nabla \times \nabla\Phi = 0$$

$$\nabla \times \nabla \times \boldsymbol{A} = \nabla(\nabla \cdot \boldsymbol{A}) - \nabla^2\boldsymbol{A}$$

Divergence Theorem $\displaystyle\int_V (\nabla \cdot \boldsymbol{A})\, dV = \oint_s \boldsymbol{A} \cdot \hat{\boldsymbol{n}}\, ds$

Stokes' Theorem $\displaystyle\int_s (\nabla \times \boldsymbol{A}) \cdot \hat{\boldsymbol{n}}\, ds = \oint_c \boldsymbol{A} \cdot d\ell$